ARCTIC OCEAN

EURASIA

AFRICA

PACIFIC OCEAN

INDIAN OCEAN

AUSTRALIA

SOUTHERN OCEAN

ANTARCTICA

Highest dew point temperature
Dhahran, Saudi Arabia
July 8, 2003
35°C (95°F)

Highest barometric pressure
Agata, Siberia
December 31, 1968
1083.8 mb (32.03 Hg)

Largest freshwater lake by volume
Lake Baikal
23,600 km³ (5,700 mi³)

Lowest barometric pressure
Typhoon Tip
October 12, 1979
870 mb (25.63 Hg)

Highest land elevation
Mount Everest
8,848 m (29,029 ft)

Greatest rainfall in one year
Cherrapunji, India
August 8, 1860
26.5 m (80.8 ft)

Lowest land elevation
Dead Sea
−424 m (−1,391 ft)

Longest river
Nile River
6,671 km (4,145 mi)

Greatest ocean depth
Challenger Deep
−10,971 m (−35,994 ft)

Fastest wind
Barrow Isalnd, Australia
April 10, 1996
408 km/h (253 mph)

Lowest temperature
Vostok, Antarctica
July 21, 1983
−89.2°C (−128.6°F)

This newly released map, called a digital elevation model (DEM), is the most precise global-scale elevation map ever created. Using satellite data, every square inch of Earth's land surface elevation has been mapped to within 1 m (3.3 ft) accuracy. The highest elevations are dark brown. The lowest land elevations are near sea level and are light blue. The data in this map are used for climate and environmental research, surveying and mapping activities, urban development, and many other applications.

Living Physical Geography

SECOND EDITION

Bruce Gervais

California State University, Sacramento

w.h.freeman

Macmillan Learning

New York

Vice President, STEM: Daryl Fox
Program Director: Andrew Dunaway
Senior Program Manager: Jennifer Edwards
Director of Development: Lisa Samols
Development Editor: Erin Mulligan
Marketing Manager: Leah Christians
Marketing Assistant: Madeleine Inskeep
Senior Media Editor: Amy Thorne
Lead Content Developer: Emily Marino
Senior Media Project Manager: Chris Efstratiou
Director, Content Management Enhancement: Tracey Kuehn
Senior Managing Editor: Lisa Kinne
Senior Workflow Project Supervisor: Susan Wein
Senior Content Project Manager: Edward J. Dionne
Senior Project Manager: Aravinda Kulasekar Doss, Lumina Datamatics, Inc.
Director of Design, Content Management: Diana Blume
Design Services Manager: Natasha A. S. Wolfe
Interior Designer: Dirk Kaufman
Cover Design Manager: John Callahan
Art Manager: Matthew McAdams
Illustrations: Lachina Creative, Inc.
Senior Photo Editor: Sheena Goldstein
Photo Researcher: Jennifer Atkins
Copyeditor: Kitty Wilson
Production Supervisor: Lawrence Guerra
Composition: Lumina Datamatics, Inc.
Printing and Binding: LSC Communications
Cover Art: Erin Hanson

ISBN-13: 978-1-319-05688-9
ISBN-10: 1-319-05688-1

Library of Congress Control Number: 2018962188

W. H. Freeman and Company
One New York Plaza
Suite 4500
New York, NY 10004-1562
www.macmillanlearning.com

For my students

Brief Contents

Contents

(Courtesy Michael Davias)

(ESA/NASA)

(Dean Fikar/Getty Images)

(Mike Hollingshead/Alamy)

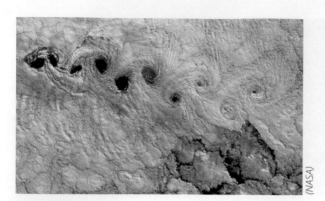

(NASA)

(Willoughby Owen/Getty Images)

(NASA/USGS)

(Nature Picture Library/Alamy)

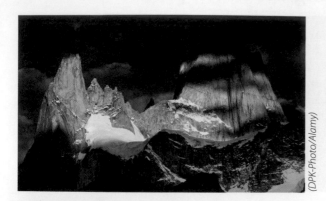

(DPK-Photo/Alamy)

(Carles Zamorano Cabello/Alamy)

(Peter Horree/Alamy)

(MediaProduction/Getty Images)

(travellight/Shutterstock)

(Jon Arnold/Danita Delimont Stock Photography)

(Bruce Gervais)

(Trevor Cole/Getty Images)

Preface

Living Physical Geography: The Big Picture

We are all living physical geography. Weather and climate strongly influence where we live and the types of crops farmers can grow and ultimately the foods we consume every day. Almost half the world's population lives within 150 km (93 mi) of the coast, mostly in large cities situated in bays and estuaries at the mouths of major rivers. Natural deposits of fossil fuels, running water in streams, sunlight, and wind provide the energy we use to produce the electricity that powers our electronics, cars, homes, and cities. Floods and drought, cold snaps and heat waves, volcanic eruptions and earthquakes, soil development and landslides all influence human beings in profound ways. Physical geography is now more relevant to society than ever before. Changes in air quality and climate, losses of habitat and species, shifting soil and water resource demands, and burgeoning renewable energy technologies are in the news daily and are all central to the science of physical geography.

The idea for this book originated with my desire to highlight the relevance of physical geography to students' daily lives and to address the most pressing environmental and resource issues that people face today. *Living Physical Geography* is unique in that it emphasizes how people change, and are changed by, Earth's physical systems. This approach creates a student-friendly context in which to understand Earth systems science and reveals the connections between Earth and people.

Three major themes are woven throughout this book:

1. **Earth is composed of interacting physical systems.** The atmosphere, the biosphere, water, and Earth's crust are major physical systems that interact with and affect one another. Energy from the Sun and energy from Earth's interior fuel changes in and among these systems.

2. **Earth is always changing.** The physical Earth is in a constant state of change on many different time scales. Within minutes the weather changes, over hours tides ebb and flow, across centuries rivers shift their channels, and over millions of years species evolve, mountains grow and are worn down, and whole continents move.

3. **The influence of people is important.** Earth's land surface, atmosphere, life, and oceans are extensively changed by people. It is not possible to study modern physical geography without considering the influences of human activity.

New to the Second Edition

This second edition of *Living Physical Geography* stands as a significant reworking of the first edition. While the topical sequencing and the ordering of the chapters remain largely the same, every chapter has been substantially updated with the latest information available about recent discoveries and significant events. For example, Chapter 7, "The Changing Climate," has been revised to provide the most recent analysis and data on the fast-moving topic of climate change. Throughout the second edition of *Living Physical Geography*, there are hundreds of new photographs, figures, and maps. In addition, familiar figures from the first edition have been reworked in this new edition to further strengthen their pedagogical effectiveness and visual appeal. The written content has been revised and updated to reflect current events and new understandings. Some text portions experienced minor revision, while others were extensively rewritten. Some text material in this second edition is completely new. A few examples include the new Chapter 4 Geographic Perspectives, "A Cloudy Science," about cloud seeding technology, the new "Weather Forecasting and Analysis" section in Chapter 6, and the new "Origins of Biodiversity" section in Chapter 8.

The Structure within Each Chapter

Each chapter has been designed carefully to facilitate student learning.

Learning Goals

At the start of each chapter, we list learning goals. Each numbered section of the chapter begins with a reiteration of the relevant learning goal. These learning goals break each chapter into manageable units for students while helping instructors focus on the learning outcomes that are important to them.

Chapter-Opening Photo

The opening spread of each chapter features a showpiece photo that vividly illustrates an important topic in the chapter; well-crafted captions pique student curiosity and draw students into the chapter.

The Human Sphere

Each chapter opens with a section titled "The Human Sphere." Each of these stories briefly explores the relationship between people and a physical phenomenon or process. The key goals of this feature are to illustrate the importance of people to physical geography and to demonstrate the relevance of physical geography to students' daily lives. Some examples of Human Sphere topics include the South Asian monsoon, China's "Airpocalypse," nuisance flooding, the collapse of the Maya, tsunamis, weathering on Mount Rushmore, and mammoth remains exposed by thawing permafrost.

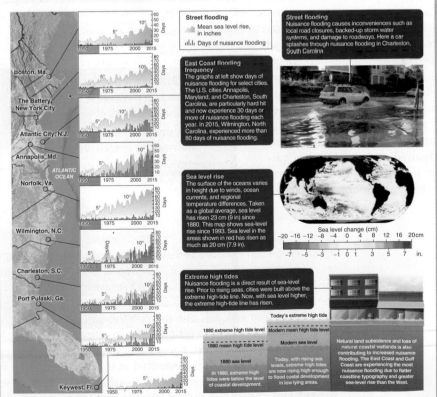

Figure 7.1 Nuisance flooding. *(Top right and right center. NOAA)*

Picture This **The Great Barrier Reef**

The Great Barrier Reef in northeastern Australia is the largest coral reef system in the world. It is a labyrinth of nearly 3,000 reef formations. The reef stretches more than 2,600 km (1,600 mi), is up to 60 km (40 mi) wide in some places, and covers some 350,000 km² (135,135 mi²) in area (red area of the inset map). This satellite image shows only the southern portions of the reef, off the central Queensland coast.

The Great Barrier Reef possesses some of Earth's richest biodiversity. It supports 350 to 400 species of corals, 1,500 fish species, 4,000 species of mollusks, some 240 species of birds, 30 whale species, and 6 sea turtle species.

In response to widespread declines in fish populations and poor reef health, the Great Barrier Reef Marine Park was established in 2004. Fishing was banned in 32% of the reef's area. Only 2 years after the protections were put in place, the reef's fish biomass had doubled, and the reefs had recovered in the protected areas. About 2 million tourists visit the reef each year, and the park generates over $3.4 billion from ecotourism annually. Australia has now enacted the Reef 2050 Plan, which will guide protection efforts and long-term sustainable uses of the Great Barrier Reef into the middle of this century.

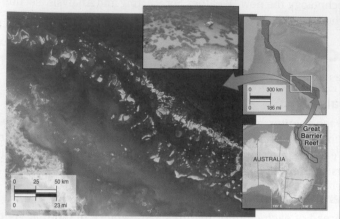

(Image courtesy NASA/GSFC/LaRC/JPL, MISR Team; Inset: Regien Paassen/Alamy)

Consider This

1. Weigh the pros and cons of restricting or banning fishing (or hunting or collecting) from a natural area to allow wildlife populations to recover.
2. Have you ever been an ecotourist? If so, where?

Picture This

In each chapter, the Picture This feature delivers pertinent and intriguing content that supplements the main text and illustrates a relevant principle. The wettest place on Earth, extreme climate events, mountaintop removal coal mining, and collapse sinkholes are examples of topics visited in this feature. Each Picture This includes two or three Consider This questions that students can answer by reading supporting text within the feature or the text just preceding it.

Scientific Inquiry

Each chapter has a figure titled "Scientific Inquiry" that reveals why scientists do what they do, how they assess what they know, and how they collect and interpret scientific data. The goal of this feature is to dispel the perception of science as being disconnected from students' daily lives or career options. Topics range from how stream gauges work and why they are important, to how contaminated groundwater aquifers are cleaned, to how data are collected from ice cores and other natural archives to reconstruct ancient climates.

Figure 2.5 SCIENTIFIC INQUIRY: All eyes on the sky. Understanding how the atmosphere is structured, how it functions, and how it is changing, as well as how weather can be accurately forecast, requires careful and precise measurements. Scientists use the most sophisticated instrumentation and computer analysis available today to better understand the atmosphere and forecast its behavior.

(NASA/Tim Marvel; inset balloon. NASA)

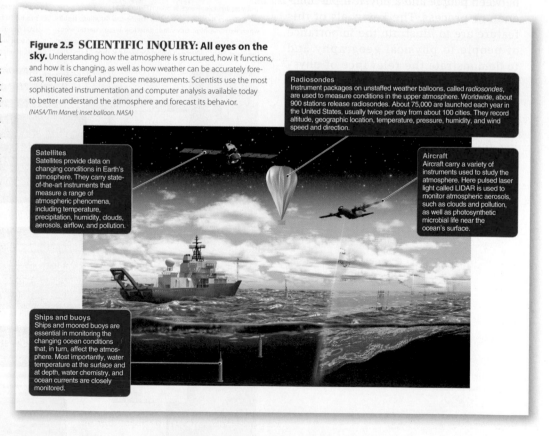

Satellites
Satellites provide data on changing conditions in Earth's atmosphere. They carry state-of-the-art instruments that measure a range of atmospheric phenomena, including temperature, precipitation, humidity, clouds, aerosols, airflow, and pollution.

Radiosondes
Instrument packages on unstaffed weather balloons, called *radiosondes*, are used to measure conditions in the upper atmosphere. Worldwide, about 900 stations release radiosondes. About 75,000 are launched each year in the United States, usually twice per day from about 100 cities. They record altitude, geographic location, temperature, pressure, humidity, and wind speed and direction.

Aircraft
Aircraft carry a variety of instruments used to study the atmosphere. Here pulsed laser light called LIDAR is used to monitor atmospheric aerosols, such as clouds and pollution, as well as photosynthetic microbial life near the ocean's surface.

Ships and buoys
Ships and moored buoys are essential in monitoring the changing ocean conditions that, in turn, affect the atmosphere. Most importantly, water temperature at the surface and at depth, water chemistry, and ocean currents are closely monitored.

Locator Maps

All photos taken in real geographic space are accompanied by a locator map. The purpose of these locator maps is to emphasize and familiarize students with the locations and spatial relationships of places visited in the book and to promote geographic literacy.

Geographic Perspectives

Each chapter concludes with a section titled "Geographic Perspectives." These sections are brief case studies that emphasize geographic thinking, spatial perspectives, and environmental issues. Topics explored in the Geographic Perspectives sections include wind and solar energy, fracking, dams on rivers, cloud seeding, rising sea level, and hurricane frequency and intensity. Geographic Perspectives sections encourage critical thought and assessment in four ways:

1. By providing context for a broader understanding of the material presented in the chapter

2. By illustrating the connections among seemingly disparate topics within a chapter and across chapters

3. By providing instructors with self-contained, manageable case studies that they can use to facilitate teaching and stimulate classroom discussion

4. By presenting a balanced view of contemporary environmental issues to encourage critical discussion, reflection, and independent conclusions

temperature, or precipitation amounts, or the likelihood of rain as a percentage. (A *qualitative* forecast, in contrast, does not specify numerically what the weather is expected to do.) *Short-term forecasts* cover 72 hours or less. *Medium-range forecast* models run out to 7 days. Any forecast longer than 7 days is a *long-range forecast*. Long-range forecasts are the least accurate because errors in the model are magnified by time. No weather model forecasts the weather with 100% accuracy because there will always be errors in the model, errors in the data entered, insufficient data input, and inherent chaos in atmospheric physics. That said, weather forecast accuracy has improved by about 1 day per decade, so that a 4-day forecast 10 years ago was as accurate as today's 5-day forecast.

range period of time, such as months to years or even decades in advance.

GEOGRAPHIC PERSPECTIVES

6.6 Are Atlantic Hurricanes a Growing Threat?

◎ Assess the current and potential vulnerability of the United States to major hurricanes.

The Atlantic hurricane season in the late summer and fall of 2017 stands out like no other hurricane season on record. Three major storms pummeled the United States. First, on August 25, Hurricane Harvey made landfall over Texas, dropping 164 cm (64.6 in)

Figure 6.31 2017 Hurricanes Harvey, Irma, and Maria. The 2017 Atlantic hurricane season was unusually active. Hurricanes Harvey, Irma, and Maria were the strongest of the season. *(Bottom left (background photo), Hillery Gatcher/EyeEm/Getty Images; top center, NASA/NOAA/UWM-CIMSS; William Stoika bottom center; Gerben Van Es/AFP/Getty Images; top right, NOAA; bottom right, STR/AFP/Getty Images)*

tion of being located where hurricanes frequently occur and having large populations living at or

Global warming has caused a sea-level rise of 23 cm (9 in) in the past century. This fact alone makes all

Figure 6.32 Hurricane activity in the United States. Each colored dot gives the hurricane return period—the number of years, on average, between major hurricanes (category 3 or higher)—for specified stretches of coast since records have been kept. The dots are spaced about 100 km (60 mi) apart. The lower the return period, the more vulnerable a stretch of coastline is. On the southern tip of Florida, for example, category 3 hurricanes have struck about every 19 years. Five particularly vulnerable large cities and metropolitan areas are indicated on the map. Coastal counties are mapped in light gray.

Online Geographic Analysis Activities

Every chapter has a new feature titled "Online Geographic Analysis" in which students are guided through real-time data from NOAA, NASA, and the USGS.

🅜 Sapling Plus

Resources Target the Most Challenging Concepts and Skills in the Course

This second edition of *Living Physical Geography* is accompanied by a media and supplements package that facilitates student learning and enhances the teaching experience. Fully loaded with our interactive e-book and all student and instructor resources, SaplingPlus is organized around a set of pre-built units available for each chapter in *Living Physical Geography*.

Created and supported by the author and other educators, SaplingPlus's instructional online homework drives student success and saves educators time. Every homework problem contains hints, answer-specific feedback, and solutions to ensure that students find the help they need.

Use the topographic map to answer the questions.

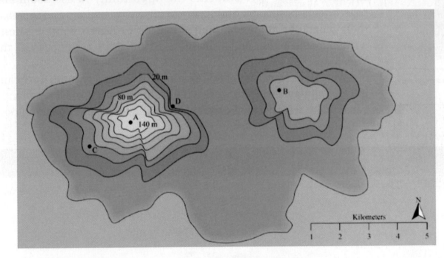

What is the contour interval?

○ 60 m

○ 60 ft

○ 20 ft

○ 20 m

What is the highest point on the map?

○ point D

○ point C

○ point A

○ point B

What is the lowest point on the map?

○ point A

○ point C

○ point D

○ point B

Between which two points does the steepest slope lie?

○ between point A and point D

○ between point B and point D

○ between point A and point C

⬛ LearningCurve

Put "testing to learn" into action. Based on educational research, LearningCurve really works. Game-like quizzing motivates students and adapts to their needs based on their performance. LearningCurve is the perfect tool to get students to engage before class and review after class. Additional reporting tools and metrics help teachers get a handle on what their classes know and don't know. LearningCurve is available in SaplingPlus and Achieve Read & Practice.

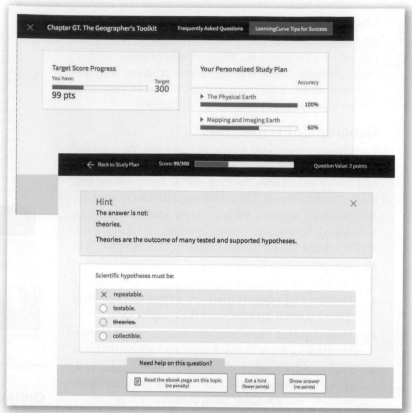

LearningCurve

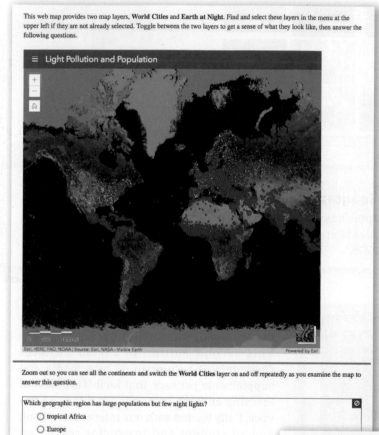

ArcGIS Web Map

Powered by ArcGIS Web Maps

Each chapter is accompanied by two to four ArcGIS Web Maps. The interactive maps give students the opportunity to use zoom and search tools, basemaps, and layers to answer questions and increase their geographic understanding of key topics. Assessment questions, written by the author and available in SaplingPlus, accompany all the Web Maps.

Web Maps topics include tornado paths on the ground, hurricane tracks, pollution levels for world cities, light pollution, sea level change, earthquake activity, cloud cover, biomes, and many additional topics.

Powered by ArcGIS Story Maps

Each chapter features one ArcGIS Story Map designed to compliment the key issues raised in the chapter. Story Map topics range from how to use maps and analyzing climate change evidence, to exploring the wolf ecology in Yellowstone National Park and dams and flood management.

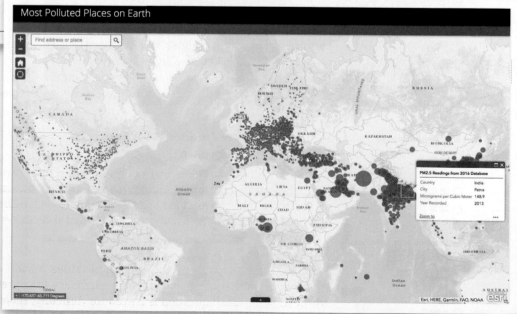

ArcGIS Story Map

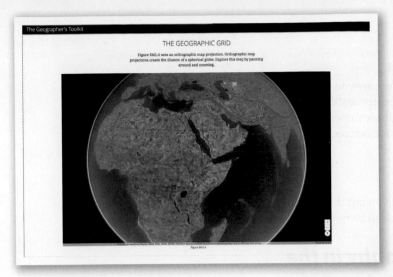

ArcGIS Story Map

The Story Maps are powerful pedagogical tools that allow students to delve deeper into selected topics. In each chapter, about 20 multiple-choice assessment questions have been written by the author and are provided for each Story Map. ArcGIS Assessment questions are available in SaplingPlus.

Animations and Video with Accompanying Assessment

More than 50 brief animations of key figures from the text step students through the most important and complex concepts and figures in each chapter. Many animations include a brief video clip to allow students to see the animation subject as it occurs in nature. Each animation is followed by assessment questions, available in SaplingPlus. Two to three videos related to cornerstone concepts from each chapter are also followed by assessment questions, available in SaplingPlus.

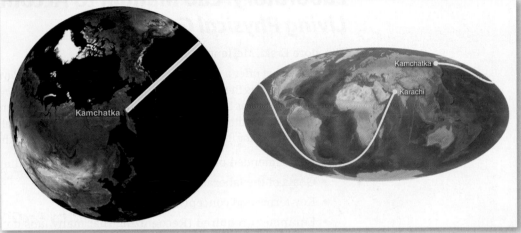

Animation

Exploring with Google Earth

Google Earth is an important pedagogical tool in *Living Physical Geography*. An "Exploring with Google Earth" activity is available in SaplingPlus for each chapter, complete with .kml files and assessment questions.

Pre-Lecture Reading Quizzes

Available in SaplingPlus for each chapter, these quizzes help ensure that students have read the material before attending class.

For instructors who want an affordable digital solution, Achieve Read & Practice is available with *Living Physical Geography*.

Achieve Read & Practice is the marriage of our LearningCurve adaptive quizzing and our mobile, accessible e-book, in one easy-to-use and affordable product.

Instructor Resources

The Hub for Active Learning

For this second edition, we've developed a suite of active learning resources to help instructors make their classes more engaging. Each chapter has a variety of activities based on the content in the text. The resources include handouts for students, instructor guides for each activity, and PowerPoint slide shows for each chapter.

Test Bank

The Test Bank, written by the author, provides a wide range of questions appropriate for assessing student comprehension, interpretation, analysis, and synthesis skills. Each question is tagged for difficulty; Bloom's Taxonomy level; book sections and page numbers; as well as learning goals.

Lecture Outlines for PowerPoint

Adjunct professors and instructors who are new to the discipline will appreciate these detailed companion lectures, perfect for walking students through the key ideas in each chapter. These rich, prebuilt lectures are written by the author and include all figure images, making it easy for instructors to transition to using the book in their classrooms.

Clicker Questions

Clicker questions allow instructors to jump-start discussions, illuminate important points, and promote better conceptual understanding during lectures.

Living Physical Geography in the Laboratory: Lab Manual to Accompany *Living Physical Geography*

Theodore Erski, McHenry County College

For schools that offer a physical geography laboratory, *Living Physical Geography in the Laboratory* is the ideal lab manual to accompany *Living Physical Geography*. The manual contains 30 lab activities, each broken down into four problem-solving modules, thus permitting lab instructors to customize the manual to fit the amount of time they have for their lab period. Each lab activity contains the following:

- Recommended pre-laboratory activity textbook reading
- Goals of the laboratory activities
- Key terms and concepts (from the textbook)
- Equipment required (Recognizing that many labs do not have access to expensive equipment, the manual focuses on activities that require only the most basic tools or equipment. Some problem-solving activities require more sophisticated equipment. Those activities are clearly separated into discrete modules so that instructors can skip them if the necessary equipment is not available.)
- Four problem-solving modules
- Summary of key terms and concepts for each lab

The activities in *Living Physical Geography in the Laboratory* require critical thinking, map and image analysis, data analysis, and occasionally math.

Acknowledgments by the Author

I could never have undertaken this journey were it not for the support from my family. Thank you, Nancy, for being an inspiration while I was working through and forming many ideas. Thank you, Katherine and Natalie, for your patience while dad was out in the "shed" writing, and thanks for helping me with photo selections. I am immensely grateful for my family's love of nature and enjoyment of visiting wild places. And thank you, Max, for our nightly walks, which cleared the head and refreshed the spirit.

It has been an honor to work with the staff at Macmillan. I am particularly grateful to my editor, Jennifer Edwards. Thank you, Jen, for bringing your leadership, knowledge, passion, and wonderful sense of humor to the project. I am deeply grateful for your whole-hearted support of *Living Physical Geography* each step of the way.

Thanks to senior project manager Aravinda Kulasekar Doss at Lumina Datamatics, Inc., senior content project manager Edward Dionne, and senior managing editor Lisa Kinne for their meticulous attention to all details great and small. I am thankful to design services manager Natasha Wolfe, who oversaw the design process as well as to senior design manager Heather Marshall at Lumina Datamatics, Inc., who did page makeup and helped assemble the pages in their final form, as well as to Matt McAdams, art manager, who oversaw the creation of the figures and illustrations. Thank you, Amy Thorne, for your patient "herding of the cats" and keeping us all on track with the job of producing media content. Thank you, Emily Marino, for help in developing the assessment program. Thanks to Sheena Goldstein for guiding the manuscript through photo research and permissions. Thank you to the Macmillan marketing team and to their colleagues in the field who have advocated for the value of this student-centered approach to *Living Physical Geography*.

It has been an honor and a great privilege to work with development editor Erin Mulligan and director of development Lisa Samols. Thank you, Erin, for bringing your talents, insight, expertise, and humor to the project and Lisa, for all your efforts overseeing the process. Thank you to artist Erin Hanson for allowing me to use her beautiful work, *Arizona Dusk*, for the book's cover. Photo researcher Jennifer Atkins worked tirelessly to find and acquire the photographs used throughout the book. The visual presentation in the book would not be what it is without Jennifer's hard work. I owe Dr. Chris Bone a debt of gratitude for his hard work in making the Web Maps and Story Maps. Thank you, Chris.

I appreciate the support provided by the Department of Geography and the College of Natural Sciences and Mathematics at California State University, Sacramento. I am also deeply appreciative of the input from my professional colleagues in geography for helping me hone the material. Finally, without my students, this book would not have been written. I wrote it for you. Thank you!

I am sincerely grateful to the following reviewers who contributed their knowledge, expertise, and time to *Living Physical Geography*:

Alexis Aguilar, Pasadena City College
Victoria Alapo, Metropolitan Community College, Fort Omaha
Faran Ali, University of North British Columbia
Jason Allard, Valdosta State University
Jake Armour, University of North Carolina, Charlotte
Geordie Armstrong, Santa Barbara City College
Augustine Avwunudiogba, California State University, Stanislaus
Randy Bertolas, Wayne State College
Owen Bettis, California State University, Chico
Tamara Biegas, University of Texas at San Antonio
Darcy Boellstorff, Bridgewater State University
Wayne Brew, Montgomery County Community College
Sara Bright, Cuyahoga Community College
Robin Buckallew, Central Community College
Michaela Buenemann, New Mexico State University
Ke Chen, East Tennessee State University
John Conley, Orange Coast College
Sam Copeland, State University of New York, Buffalo North Campus
William Courter, Santa Ana College
Richard Crooker, Kutztown University
Daryl Dagesse, Brock University
Carolyn Damato, Salem State University
Lincoln DeBunce, Blue Mountain Community College
Jeremy Dillon, University of Nebraska at Kearney
Taly Drezner, York University
Josh Durkee, Western Kentucky University
Robert Edsall, Carthage College
Tracy Edwards, Frostburg State University
Jean Ellis, University of South Carolina
Salvatore Engel-Di Mauro, State University of New York at New Paltz
Anilkumar Gangadharan, Kennesaw State University
William Garcia, University of North Carolina, Charlotte
Colleen Garrity, State University of New York at Geneseo
Greg Gaston, University of North Alabama
Christopher Gentry, Austin Peay State University
Dusty Girard, Brookhaven College
Brett Goforth, California State University, San Bernardino
J. Scott Greene, University of Oklahoma
Michael Grossman, Southern Illinois University, Edwardsville
Hillary Hamann, University of Denver
William Hansen, Worcester University
Jesse Hanson, Southwestern Connecticut State University

Paul Hanson, University of Nebraska, Lincoln
Megan Harlow, Orange Coast College
Mark Hecht, Mount Royal University
Delia Heck, Ferrum College
Patricia Heiser, Carroll College
April Hiscox, The University of South Carolina
Nancy Hoalst Pullen, Kennesaw State University
Chasidy Hobbs, University of West Florida
Margaret Holzer, Rutgers University
Chris Houser, Texas A&M University
Solomon Isiorho, Indiana University–Purdue University, Fort Wayne
Renee Jacobsen, California State University, San Bernardino
Brian Jones, University of Alberta
Don Jonsson, Austin Community College
Karen Jordan, University of South Alabama
Ranbir S. Kang, Western Illinois University
Wilberg Karigomba, Northwest Arkansas Community College
David Keellings, University of Alabama
Ryan Kelly, Bluegrass Community and Technical College
William Kelvey, Carroll Community College
John Keyantash, California State University, Dominguez Hills
Lawrence Kiage, Georgia State University
Michelle Kinzel Southwestern College
Kory Konsoer, Louisiana State University
Thomas Krabacher, California State University, Sacramento
Jack Kranz, Los Angeles Valley College
Barry Kronenfeld, Eastern Illinois University
John Lacozza, University of Manitoba
Matthew Letts, University of Lethbridge
Michael Lewis, University of North Carolina at Greensboro
Fuyuan Liang, Western Illinois University
Karl Lillquist, Central Washington University
Bruce Lindquist, Leeward Community College
Jennifer Lipton, Central Washington University
Thomas Loder, Texas A&M University
Julie Loisel, Texas A&M University
Kerry Lyste, Everett Community College
Taylor Mack, Louisiana Tech University
Laszio Mariahazy, Golden West College
Steven Marsh, University of the Fraser Valley
Yvonne Martin, University of Calgary
Blake Mayberry, Red Rocks Community College
Benjamin McDaniel, North Hennepin Community College
Larry McGlinn, State University of New York at New Paltz
Charles McGlynn, Rowan University
Armando Mendoza, Cypress College
Andrew Mercer, Mississippi State University
Peter Meserve, Fresno City College
Robert Milligan, Northern Virginia Community College, Woodbridge
Peter Muller, University of Miami
Steve Namikas, Louisiana State University
Elsa Nickl, University of Delaware
Ricardo Nogueira, Georgia State University
Abby Norton-Krane, Cuyahoga Community College
Thomas Orf, Las Positas College
Thomas Patterson, The University of Southern Mississippi
Barry Perlmutter, College of Southern Nevada

Arliss S. Perry, Wright State University
Michael Pool, Austin Community College
Sarah Praskievicz, University of Alabama
Dayna Quick, College of Marin
Abdullah F. Rahman, Indiana University
Rhonda Reagan, Blinn College
Max W. Reams, Olivet Nazarene University
Ronald Reynolds, Bridgewater State University
Stephanie Rogers, Auburn University
Shouraseni Sen Roy, University of Miami
Ruth Ruud, Cleveland State University
Erin Saffell, Arizona State University
David Sallee, University of North Texas/Tarrant County College
Ginger Schmid, Minnesota State University
Jason Senkbeil, University of Alabama
Alireza Shahvari, Towson University
David Shankman, University of Alabama
Andrew Shears, Kent State University
Susan Slowey, Blinn College
Erica A. H. Smithwick, Pennsylvania State University
Nicholas Sokol, University of South Carolina
Kristin Sorensen, South Plains College
Jane Southworth, University of Florida

Brian Steinberg, Northern Virginia Community College
Robert Stewart, Lakehead University
Suzanne Stewart, Northeast Lakeview College
Jeremy Tasch, Towson University
Sabina Thomas, Santa Barbara City College
Tak Yung (Susanna) Tong, University of Cincinnati
Jill Trepanier, Louisiana State University
Jon Van de Grift, Metropolitan State University of Denver
Cornelis J. van der Veen, University of Kansas
Paul Vincent, Valdosta State University
Scott Walker, Northwest Vista College
Megan Walsh, Central Washington University
James Wanket, California State University, Sacramento
Brad Watkins, University of Central Oklahoma
Theresa Watson, Central New Mexico Community College
Kevin Weakley, Northwest Arkansas Community College
Michele Wiens, Simon Fraser University
Forrest Wilkerson, Minnesota State University
Angie Wood, Chattanooga State Community College
Ken Yanow, Southwestern College
Sonja Yow, Eastern Kentucky University
Danlin Yu, Montclair State University
Catherine Zajaczkowski, Oakland Community College

About the Cover Artist

Erin Hanson

Erin Hanson began painting early in life. A high school scholarship took her to Otis College of Art in Los Angeles, where she immersed herself in figure drawing. After graduating from UC Berkeley, where she studied bioengineering, Hanson began her professional career as a landscape painter. She finds inspiration in the brilliantly colored cliffs of Zion, Canyon de Chelly, Monument Valley, and other locales in the southwestern United States. Erin Hanson's abstract landscape paintings capture the ever-changing palettes of color, texture, and emotional vibrancy of earth, sky, water, and life in these unique natural places.

About the Author

Bruce Gervais is a professor of geography at California State University, Sacramento. He holds a doctorate in geography from the University of California at Los Angeles. Bruce's research focus is in paleoclimatology. For his doctoral research at UCLA, he studied ancient climates by using tree rings and fossil pollen preserved in lake sediments on the Kola Peninsula in northwestern Russia. He has published 12 peer-reviewed research papers detailing his work in Russia and California. Bruce enjoys spending his free time backpacking, in the yoga studio, and with his family. He welcomes your comments and can be reached at gervais@csus.edu.

Bruce Gervais

Living
Physical
Geography

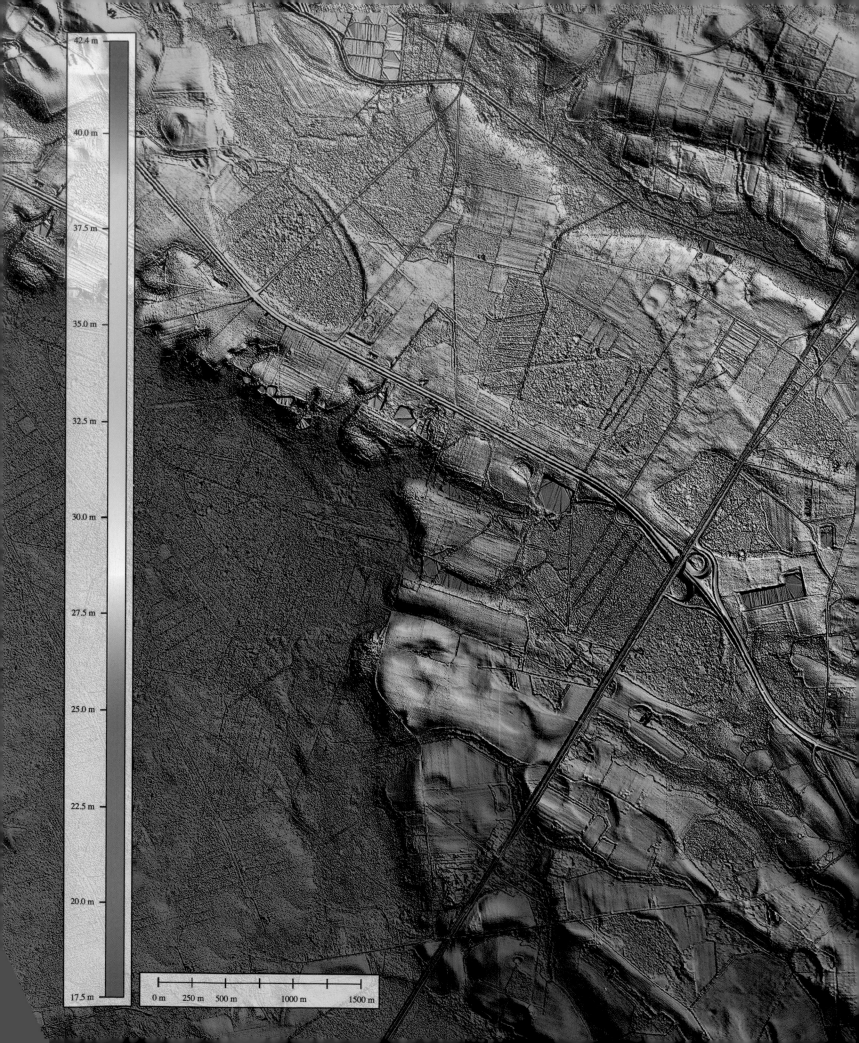

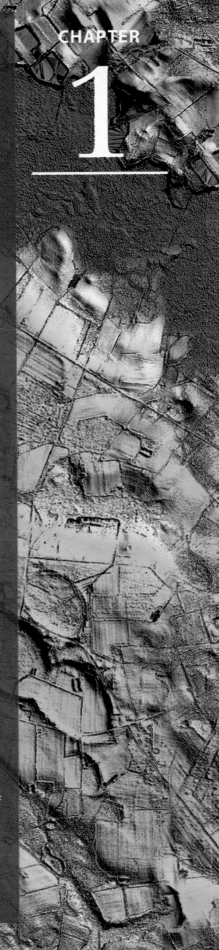

1

The Geographer's Toolkit

Chapter Outline *and Learning Goals* ◎

1.1 Welcome to Physical Geography!

◎ Define physical geography.

1.2 Scales of Inquiry

◎ Explain the various scales of geographic inquiry.

1.3 Energy and Matter in Earth's Physical Systems

◎ Describe the flow of energy and matter through Earth's physical systems.

1.4 Mapping Earth: The Science of Spatial Awareness

◎ Use the geographic grid coordinate system to identify locations on Earth's surface and distinguish among different types of map scales and map projections.

1.5 Imaging Earth

◎ Discuss how technologies such as satellite sensors and radar are used to study and portray Earth systems and processes.

1.6 Geographic Perspectives: The Scientific Method and Easter Island

◎ Apply the scientific method to study Easter Island's history of human settlement.

This image shows the land surface in Cordova, North Carolina, as it would appear without vegetation and in "false" color. Color is used to show elevation above sea level, and the image's *legend* (far left) specifies the relationship between color and elevation. For example, the bright blue areas at the top of the image are 40 m (131 ft) in elevation. The image's *bar scale*, or graphic map scale (bottom left), shows distance. A stream channel runs down the middle of the image, and the straight lines are roads and highways. The oval shapes are called "Carolina bays." There are hundreds of thousands of these natural features along the Atlantic seaboard, but most of them are now hidden beneath development. Scientists do not know what formed Carolina bays. The data used to make this *digital elevation model* (DEM) were acquired using LIDAR (light detection and ranging) technology.

(Courtesy Michael Davias)

To learn more about these geographic technologies, see Section 1.4.

1.1 Welcome to Physical Geography!

◎ **Define physical geography.**

Have you ever wondered why deserts are barren and dry and tropical rainforests are lush and wet? Why Hawai`i has such delightfully pleasant winters but Alaska's are brutally cold? Why the planet experiences winter and summer? How millions of tons of water can be held aloft in a thunderstorm and then fall to the ground as rain? Why tornadoes form? Whether humans are changing the climate? Why there are no polar bears in the Southern Hemisphere or penguins in the Northern Hemisphere? Why mountains form and how they are worn down? What causes volcanic eruptions and earthquakes? Such questions stem from a fundamental curiosity about the natural world around us. They are all questions about Earth's physical geography, and they are all questions explored in *Living Physical Geography*.

Physical geography is the study of Earth's living and nonliving physical systems and how these systems change naturally through space and time or are changed by human activity. A **system** is a set of interacting parts or processes that function as a unit. (A *process* is a stepwise progression of events.) The human body, for example, is composed of multiple systems, including the skeletal system, circulatory system, and nervous system. Similarly, a leaf, a tree, and even a whole forest are each examples of systems. One of the major themes that threads through the pages of *Living Physical Geography* is how Earth's natural physical systems have changed in the past, how they are changing today, and how they may change in the future.

Physical geography is nested within the larger discipline of **geography**: the study of the spatial relationships among Earth's physical and cultural features and how they develop and change through time. Geography emphasizes the role of spatial relationships between people and the physical world to gain insight into cultural and physical phenomena. Physical geography specializations include *biogeography* (study of plants and animals), *meteorology* (study of weather), *climatology* (study of climate), and *geomorphology* (study of landscapes). The counterpart to physical geography is *human geography*, which focuses on human landscapes and often goes by different names, such as *political geography*, *economic geography*, *urban geography*, and *transportation geography*. Often, physical geography and human geography overlap, and some scientific endeavors, such as the study of *natural hazards* and *remote sensing*, incorporate physical and human geography equally.

Regardless of the area of specialization, geographers universally tend to ask "where," "why," and "how" questions: Where is it? Why is it there? How did it get there? How long has it been there? Geographers also tend to be *holistic* in their thinking—taking into account all elements of a problem and paying particular attention to connections and interrelationships. Recall the question Why are polar bears only in the Northern Hemisphere? A geographer considering this question would mention relevant factors such as climate, water, other organisms, evolutionary history, biology, and ecology.

In this book, the role of people is considered within the discussion of nearly every topic, and this book is titled *Living Physical Geography* to reflect this important theme. We are surrounded by weather, climate, agriculture, streams, coastlines, elevation, plants, and animals. All of the material goods we rely on are connected to natural resources derived from Earth's physical systems. The materials that meet our needs, such as food, homes, cars, phones, computers, and clothing, were all once raw natural resources found in Earth's physical systems. People modify Earth's natural environments to grow or manufacture these materials.

The world shapes people, and we shape the world. It is difficult to find environments that are not at least in part **anthropogenic**: created or influenced by people. People are an active force of change. Earth's surface, its atmosphere, its oceans, and its organisms have been profoundly transformed in many ways by people in just the past few hundred years **(Figure 1.1)**. Every page of *Living*

1600　　2018

Figure 1.1　Anthropogenic landscapes. Have you ever wondered what New York City looked like before it was a city? In this image, Manhattan, New York, is shown as it appears today on the right and digitally reconstructed as it looked prior to European settlement 400 years ago on the left. Manhattan was once temperate deciduous forest and coastal estuaries. Now it is one of the most densely developed landscapes in the world. To learn more about how this image was created, see Figure 1.29 on page 27. *(Markley Boyer / WCS / Yann-Arthus Bertrand / Corbis)*

Physical Geography was written with this in mind. The health and well-being of the human species are intertwined with Earth's natural and anthropogenic environments. *Living Physical Geography* studies the physical Earth, the places humans occupy on Earth, and our active role in shaping Earth and in driving its rapidly changing environments.

Living Physical Geography explores the human transformation of Earth's physical landscapes through the lens of science. Science is fundamental to the discipline of physical geography and to all aspects of this book. Later in this chapter, Section 1.6 illustrates the process of science by exploring the history of Easter Island.

1.2 Scales of Inquiry

◎ **Explain the various scales of geographic inquiry.**

Geographers employ two types of scale: spatial scale and temporal scale. Different spatial and temporal scales provide varied perspectives on physical phenomena. **Spatial scale** refers to the physical size, length, distance, or area of an object such as a cloud or a rainforest. Spatial scale also pertains to the physical space occupied by a process such as migration of a species or movement of sand along a coastline. **Temporal scale** refers to the window of time used to examine phenomena and processes as well as the length of time over which they develop or change. Together, spatial

and temporal scales reveal important information about Earth's physical systems. Employing the two scales together provides a unique perspective that underpins the study of physical geography.

Spatial Scale: Perspective in Space

Imagine your campus or your neighborhood. When you do, you are thinking on a local spatial scale. Now zoom out mentally and imagine the city where you live, or the state, or even the entire country or continent. These are all examples of different spatial scales. Thinking on a local spatial scale involves more detail, such as what building a classroom is in or where a house in a neighborhood is found. On broader spatial scales, there is less local detail, and more geographic space is covered. Broader spatial scales allow for a clearer view of the bigger picture and of regional context. Geography is at its very heart a spatial science. Geographers think spatially, and they often use maps to represent spatial ideas.

A **map** is a flat two-dimensional representation of Earth's surface. A map can be drawn at any spatial scale. **Large-scale perspective** makes geographic features large to show more detail. **Small-scale perspective** makes geographic features small to cover broad regions. A map at a local scale, such as a college campus map that shows individual buildings, is a large-scale map. A small-scale map, in contrast, includes a large area of Earth's surface, such as a continent or a hemisphere. **Figure 1.2** illustrates how

Figure 1.2 Spatial scale. Our perspective on the Namib Desert in southwestern Africa changes as spatial scale changes. Different spatial scales reveal different geographic patterns and processes and stimulate different kinds of questions. Images 3 and 4 are both developed from satellites. *(1. Johnny Haglund/Getty Images; 2. imageBROKER/Superstock; 3. Jacques Descloitres, MODIS Rapid Response Team, NASA/GSFC; 4. NASA Goddard Space Flight Center Image by Reto Stöckli [land surface, shallow water, clouds]. Enhancements by Robert Simmon [ocean color, compositing, 3D globes, animation].)*

Large scale   **SPATIAL SCALE**   Small scale

  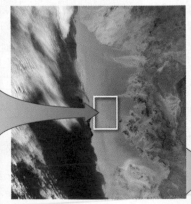 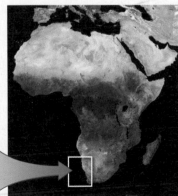

1 Do sand dunes move? How deep Is this sand dune? Why is there no vegetation on the dunes?

2 How does the wind form these dune patterns? Where did the sand come from? Are sand dunes found only in deserts?

3 Why does this desert occur near the ocean? Why is southwestern Africa arid?

4 Why is desert found north and south of central Africa but not in central Africa, which is green and fertile? Does this pattern occur on other continents?

Figure 1.3 Green Sahara. This 7,000- to 9,000-year-old giraffe petroglyph (rock engraving) is in the Sahara Desert in Niger, Africa. Giraffes require woodlands and standing pools of water to drink from. The petroglyph portrays a climate that was once much wetter than is found at this location today. It also illustrates how environments can change over time. *(© Frans Lemmens/Lithium/age fotostock)*

pattern of aridity around the world caused by patterns of atmospheric circulation. Such a shift in spatial thinking provides a way for us to see how phenomena or processes relate to one another. Physical geography focuses on phenomena that range in size from meters to the entire planet.

Temporal Scale: Time as a Perspective

There is an old adage that time changes everything. This is a particularly important idea in physical geography. It is difficult to see clouds moving in the sky unless you keep your eyes fixed on them. Time-lapse video, however, shows clouds slipping quickly across the sky in layers or billowing upward into huge heaps. On human time scales, most landscapes appear to be static (unchanging). On longer time scales, such as hundreds to thousands or millions of years, landscapes change and evolve, just like clouds.

Earth's physical landscapes are merely one frame in a continuing landscape of change. Mountains lift up and then erode away; continents split apart as new ocean basins form. Natural climate cooling creates massive ice sheets that cover whole continents, and once-vegetated regions turn to barren desert. Most of the Sahara, for example, is barren today, without surface water and vegetation. Some 7,000 years ago, however, that desert was a *savanna* woodland (see Section 9.2) with many large lakes and was home to many animals, including crocodiles, hippos, giraffes, lions, and humans **(Figure 1.3)**.

The temporal scale is particularly relevant to anthropogenic changes in Earth's environments.

different spatial scales lead to different perspectives and different levels of inquiry in the context of the Namib Desert.

It is easy to know what a thing is at a large spatial scale (consider a sand dune), but seeing the small-scale patterns and processes that produced it can be difficult from the large-scale perspective. At a small spatial scale, a geographic pattern begins to emerge: The Namib Desert is part of a broader

> **Video**
>
> *Annual Arctic Sea Ice Area*
>
> Available at www.saplinglearning.com
>
> ▶

Figure 1.4 Arctic sea ice loss.

(A) This image pair shows the North Pole's shrinking sea ice coverage between September 1979 and September 2017. In total, 2.4 million km² (920,000 mi²) of sea ice was lost during this period, an area nearly a quarter the size of the United States. (Measurements in this book are listed in both metric and U.S. customary units. Appendix I, "Unit Conversions," provides conversion information and a discussion of measurement.) (B) Yearly sea ice coverage data for the same time period are graphed here. Although there are year-to-year fluctuations, the overall trend is a sea ice decline of 13% per decade. At this rate of loss, the Arctic Ocean will be ice-free in summer by 2050. The Arctic Ocean hasn't been ice-free in the summer for over 3 million years. *(@ NASA's Scientific Visualization Studio)*

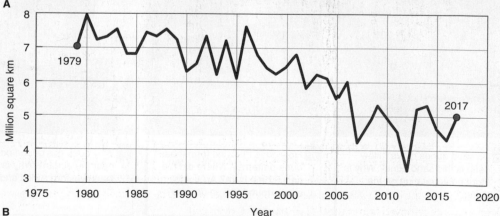

Perhaps nowhere is this clearer than in the Arctic. The North Pole consists of open ocean with no land. Because it is so cold there, the ocean's surface freezes to a thickness of about 2 m (6 ft). The extent of the Arctic sea ice is greatest in winter. It is at its geographic minimum in September. Today Arctic sea ice is on the decline as the climate changes and the planet warms **(Figure 1.4)**.

Physical geography explores phenomena and processes across temporal scales that range from minutes to millions of years. Some, such as earthquakes, occur in minutes. Others, such as Arctic sea ice changes, take months. Still other phenomena, such as the development of mountain ranges and new ocean basins, take millions of years. Some phenomena occur over many time scales. Climate change, for example, occurs over decades to millions of years.

1.3 Energy and Matter in Earth's Physical Systems

◎ Describe the flow of energy and matter through Earth's physical systems.

The movement of energy and matter is involved in every aspect of the science of physical geography. The interactions between energy and matter determine the characteristics of Earth's physical systems, including weather and climate, the kinds of organisms that occupy an area, the types of activities in which humans engage, the growth and erosion of mountain ranges, the formation of glaciers and streams, and the occurrence of earthquakes and volcanic eruptions. This section explores energy and matter in more detail.

Energy and Matter

Energy is the capacity to do work on or to change the state of matter. **Matter** is any material that possesses mass and occupies space. Energy works on matter by moving it, by changing its temperature, or by changing its state, such as when ice melts to liquid. Matter commonly exists in three states: solid, liquid, or gas. To change the state of matter, energy must be added to it or removed from it. Water is a particularly important form of matter in physical geography. **Figure 1.5** illustrates water in its different states. As the caption explains, energy is crucial to the transformation between states.

Several forms of energy affect Earth systems. **Radiant energy** is the energy of electromagnetic waves, such as light or X-rays. The Sun emits radiant energy that passes through Earth's atmosphere. A portion of that energy is absorbed by Earth's atmosphere and surface. When that

1. Solid (ice)
In solid ice, water molecules are bonded together to form a rigid structure.

2. Liquid
In liquid water, the molecular bonds continually form and break.

3. Gas
In gaseous water vapor, water molecules are not bonded together. Instead, they dart freely about.

Mount Robson

Berg Glacier

Berg Lake

Canada
Mount Robson, British Columbia
45° N
U.S.
30° N
120° W   90° W   75

Figure 1.5 States of water. Mount Robson (elevation 3,954 m [12,972 ft]), in British Columbia, is the highest peak in the Canadian Rockies. Berg Glacier flows down the mountain and into Berg Lake. Here, water exists in its three states: solid ice in Berg Glacier, liquid water in Berg Lake, and water vapor (a gas) that is invisible in the atmosphere. To change the state of water from solid ice to liquid water, energy must be added to the water. To change the liquid water to water vapor, more energy must be added. To change its state from gas to liquid and then from liquid to solid, energy must be removed from water. *(Jason Puddifoot/Getty Images)*

radiant energy is absorbed, it is converted to *heat energy*. (Section 3.2 explores solar energy in this context further.) **Photosynthesis** is a process by which plants, algae, and some bacteria convert the Sun's radiant energy to stored chemical energy. **Chemical energy** is energy in a substance that can be released through a chemical reaction. All living organisms use chemical energy to move and to carry out metabolic functions. Food is a form of chemical energy. Gasoline is also a form of chemical energy. When burned, that energy works to move a car. **Geothermal energy** (heat from Earth's interior) moves entire continents and heaves and buckles Earth's crust into mountain ranges. Lightning produced within a thunderstorm is *electrical energy*.

Two other important categories of energy in physical geography are *potential energy* and *kinetic energy*, which are both types of *mechanical energy*. Potential energy is stored in an object or a material. A boulder perched over a cliff about to fall is an example of potential energy. Kinetic energy is the energy of movement. A boulder that is falling down a cliff and smashing into other rocks is an example of kinetic energy.

As energy flows through Earth's physical systems, it moves matter, and it changes its form in the process. **Figure 1.6** examines how the Sun's radiant energy changes from one form to another in Earth's physical systems.

Earth's Physical Systems

As stated earlier, geographers tend to study the world holistically. They see a phenomenon as a whole—interconnected, with many moving parts, and never in isolation. This approach lends itself remarkably well to studying the physical Earth. Nothing in Earth is ever truly isolated, and this is particularly true with Earth's systems. The air, clouds, trees, people, soil, water, and rocks—all of these are connected in complex, interacting systems. Earth's four main physical Earth systems—the *atmosphere,* the *biosphere,* the *lithosphere,* and the *hydrosphere*—and some of their major characteristics are explored next.

The Atmosphere

The **atmosphere** is the layer of gases surrounding Earth that extends outward from the surface of Earth for hundreds of kilometers. The atmosphere contains more than a dozen different gases but consists mainly of molecules of nitrogen (N_2) and oxygen (O_2). The atmosphere performs many functions crucial to life, such as providing life-sustaining oxygen as well as gases that block the harmful rays of the Sun. The atmosphere also moderates temperatures on Earth's surface, making life on land possible.

The Biosphere

The **biosphere** is all of life on Earth. Humans are part of the biosphere. The biosphere extends from

1 The Sun's radiant energy excites water molecules, causing them to *evaporate* into the atmosphere. When the evaporated water *condenses* to liquid, clouds form.	**2** Rain from the clouds falls to Earth. The rain collects in a stream channel and forms a flowing river with kinetic energy that can erode canyons over time.	**3** A dam creates a *reservoir* of water with potential energy. The reservoir's potential energy is converted to electrical energy when water is released from the base of the dam, causing turbines there to spin and generate electricity.	**4** Electricity from the dam is used to charge the battery in this electric vehicle. A portion of the electrical energy is converted to the kinetic energy of motion and heat by car movement. The car's headlights convert the electrical energy to radiant energy.

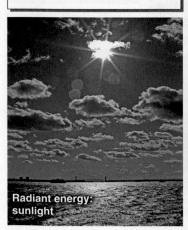
Radiant energy: sunlight

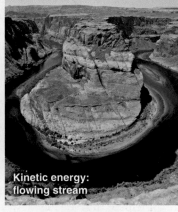
Kinetic energy: flowing stream

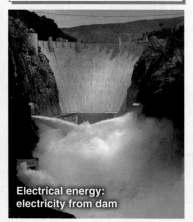
Electrical energy: electricity from dam

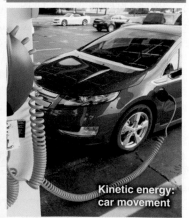
Kinetic energy: car movement

Energy flow →

Figure 1.6 A day in the life of solar energy. This graphic follows the path of solar radiant energy as it changes form and works on matter. *(1. Evan Kafka/Getty Images; 2. Marco Brivio/Getty Images; 3. U.S. Dept of Interior - Bureau of Reclamation; 4. Rebecca Cook/Reuters/Newscom)*

deep within the **crust** (the rigid outermost portion of Earth's surface) to high in the atmosphere. Far below ground in Earth's crust live bacteria that are adapted to extreme pressure and heat. Likewise, at 16 km (10 mi) above Earth's surface, tiny fern spores and microscopic bacteria and pollen float on air currents above the continents and oceans. Most of the biosphere, however, can be found on the land's surface and within the first few hundred meters of the surface of the oceans.

The Lithosphere

Humans walk on, build homes on, and grow food on the lithosphere. The **lithosphere** is Earth's rigid crust plus the heated layer beneath it. The lithosphere extends down to about 100 km (62 mi). It is fractured into 14 large *plates* (pieces) that slowly move, creating volcanoes, mountains, and earthquakes. The lithosphere regulates and determines the atmosphere's chemistry over time scales of millions of years through volcanic eruptions and the weathering of rocks.

The Hydrosphere

One unique characteristic of Earth compared with other planets is the quantity of its liquid water. Earth has immense liquid oceans, which cover 71% of the planet's surface and are a little over 4 km (2.5 mi) deep on average. The **hydrosphere** encompasses all of Earth's water in its three states: solid, liquid, and gas. It includes water in the oceans, in the ground, and in organisms. The hydrosphere also includes the **water vapor** (water in its gaseous state) in the atmosphere and water present in the atmosphere as clouds, which are composed mostly of liquid water. Water also occurs in its solid form as ice in snow, sea ice, glaciers, and ice sheets. This frozen portion of the hydrosphere is called the **cryosphere**. Water is present in, and plays roles in, the other three systems, so henceforth we will not treat the hydrosphere as a separate system.

Energy and matter flow and cycle within and between Earth's systems. Energy from the Sun and Earth's internal geothermal energy do work on Earth's systems and change matter within the systems. The atmosphere, biosphere, lithosphere, and hydrosphere are called *open systems* because they exchange energy and matter and affect one another. For example, water cycles from the hydrosphere where it is a liquid to the atmosphere as a gas after it has evaporated. In the atmosphere, water may condense to form clouds and rain or snow. Once on the ground, water may enter the biosphere and sustain plants and animals. Water also enters the lithosphere as it is absorbed into the ground as *groundwater*.

Earth System Feedbacks

Earth's spheres are dynamic systems that change over a wide range of spatial and temporal scales. Changes within one system always produce changes in other parts of the same system. Similarly, Earth's natural systems never operate in isolation: Changes in one system always cause changes in other systems. The interactions within a single system and between different systems are called *system feedbacks*.

There are two types of system feedbacks: positive feedbacks and negative feedbacks. A **positive feedback** is a process by which interacting parts in a system destabilize the system. A **negative feedback** is a process by which interacting parts in a system stabilize the system. For example, due to human activity, the atmosphere at Earth's surface has warmed over 1°C (1.5°F) during the past century. The warmer atmosphere is able to retain more water vapor, and water vapor is a *greenhouse gas* that warms the atmosphere (see Section 2.1). Therefore, the warming trend caused by human activity leads to more water vapor in the air, which leads to yet more warming:

Global warming → More water vapor → More warming (go back to start)

Notice that this positive feedback loops back on itself. The initial warming of the atmosphere triggers further warming. In other words, there is an upward trend in the atmosphere's temperature due to the feedback process. Positive feedbacks always result in a trend—but not necessarily an upward trend. Positive feedbacks can also create cooling trends for Earth (see Figure 7.3).

Trends caused by positive feedbacks can be slowed or stopped by negative feedbacks. For example, as the atmosphere's water vapor increases, the clouds in the atmosphere also increase. If the increased cloudiness causes cooling by reflecting sunlight and shading Earth's surface, then the warming trend can be slowed or stopped by the presence of more clouds:

Global warming → More water vapor → More clouds → Slowed warming

Thus, the warming triggered a negative feedback as increased cloudiness reflected sunlight and slowed the warming trend.

All feedback systems are complex interacting processes, and the two presented here in isolation are overly simplified for the sake of clarification. Scientists do not anticipate that the global warming trend caused by human activity will be stopped by an increase of cloudiness. One reason is that clouds also act to absorb heat from Earth's surface and cause further warming. Another reason is that there are many other more powerful positive feedbacks that are acting to further enhance the warming trend. Section 7.1 discusses these factors in further detail.

The Structure of *Living Physical Geography*

Each part of this book, *Living Physical Geography*, is about a different aspect of the flow of energy and matter through Earth's physical systems. In Part I and Part II, we discuss how solar radiant energy flows through the atmosphere and biosphere, powering the dynamic atmosphere and fueling the metabolism of living organisms. In Part III, we look closely at geothermal energy and how it builds up the lithosphere and creates variations in the shape and physical character of Earth's surface. In Part IV, we witness how solar energy and gravity tear down the lithosphere through the work of moving water (such as streams and coastal waves), flowing ice (glaciers), and wind. **Figure 1.7** provides a visual diagram of the structure of *Living Physical Geography*.

Chapter 1 Geographer's Toolkit

PART I
Atmospheric Systems: Weather and Climate

Chapter 2 Portrait of the Atmosphere
Chapter 3 Solar Energy and Seasons
Chapter 4 Water in the Atmosphere
Chapter 5 Atmospheric Circulation and Wind Systems
Chapter 6 The Restless Sky: Severe Weather and Storm Systems
Chapter 7 The Changing Climate

PART II
The Biosphere and the Geography of Life

Chapter 8 Patterns of Life: Biogeography
Chapter 9 Climate and Life: Biomes
Chapter 10 Ocean Ecosystems
Chapter 11 Soil and Water Resources

PART IV
Erosion and Deposition: Sculpting Earth's Surface

Chapter 16 Weathering and Mass Movement
Chapter 17 Flowing Water: Fluvial Systems
Chapter 18 The Work of Ice: The Cryosphere and Glacial Landforms
Chapter 19 Water, Wind, and Time: Desert Landforms
Chapter 20 The Work of Waves: Coastal Landforms

PART III
Tectonic Systems: Building the Lithosphere

Chapter 12 Earth History, Earth Interior
Chapter 13 Drifting Continents: Plate Tectonics
Chapter 14 Building the Crust with Rocks
Chapter 15 Geohazards: Volcanoes and Earthquakes

Figure 1.7 *Living Physical Geography's* **structure.** This book emphasizes the work of energy in Earth's physical systems. The four parts of *Living Physical Geography* focus on solar energy in the atmosphere (Part I), solar energy in the biosphere (Part II), geothermal energy building the lithosphere (Part III), and solar energy wearing down the lithosphere through erosion (Part IV). The chapters in each part are arranged in a sequence that shows the development of topics within each system. Water is everywhere on Earth, and water plays an essential role within each of these systems. The hydrosphere is therefore not treated as a separate system but is instead included in each topic.

Figure 1.8 Solar energy in the atmosphere. Differences in solar heating across Earth's surface set the atmosphere in motion, and solar energy evaporates water vapor into the atmosphere. Chapter 4, "Water in the Atmosphere," discusses this process in detail. In this way, the Sun's radiant energy is transformed into the deadly kinetic energy of thunderstorms (such as this one in east-central Oklahoma) and tornadoes (such as this one in Manchester, South Dakota). *(Top. David McGlynn/Getty Images; left. Dave Reede/Robert Harding Picture Library; right. Gene Rhoden/Weatherpix/Getty Images)*

Atmospheric Systems: Weather and Climate

The Sun's radiant energy flows through and does work on Earth's physical systems. This is the focus of Part I of *Living Physical Geography*. The Sun generates immense amounts of radiant energy. A tiny fraction of this energy is intercepted by Earth. This intercepted energy warms Earth's atmosphere, land surface, and oceans, and sets the atmosphere in motion, creating weather and climate systems. **Weather** is the state of the atmosphere at any given moment and is made up of ever-changing events on time scales ranging from minutes to weeks. **Climate**, on the other hand, is the long-term average of weather and the average frequency of extreme weather events. All atmospheric phenomena, ranging from gentle breezes to deadly tornadoes, are driven by solar energy. **Figure 1.8** illustrates the link between the Sun's energy and Earth's weather.

The Biosphere and the Geography of Life

Almost all life on Earth is a physical manifestation of solar energy. Solar radiant energy streams through the atmosphere and enters the biosphere as it is converted to chemical energy by plants through photosynthesis. This chemical energy is then transferred to other living organisms through their food, as illustrated in **Figure 1.9**. With only a few exceptions, all life on Earth, the subject of Part II of this book, is dependent on solar energy.

Tectonic Systems: Building the Lithosphere

In Part III of this book, we consider geothermal energy and topographic relief. The highest

Web Map

Earthquakes over Time

Available at
www.saplinglearning.com

Figure 1.9 Solar energy in the biosphere. Plants convert radiant energy from the Sun to chemical energy, which they use for growth. The keel-billed toucan (*Ramphastos sulfuratus*), whose range is from southern Mexico to Venezuela (inset map), acquires chemical energy directly from plants in the form of berries or indirectly as it eats insects that have eaten plants. In Chapter 8, "Patterns of Life: Biogeography," we discuss the movement of energy and matter through the biosphere. *(Left. David McGlynn/Getty Images; center. Bruce Gervais; right. Sue Flood/Getty Images)*

Figure 1.10 Geothermal energy and the lithosphere. Earth's internal heat energy moves and breaks Earth's crust through the process of plate tectonics, creating surface features such as the Alps of Switzerland (left) and volcanoes in Iceland (right). This topic is introduced and discussed in Chapter 13, "Drifting Continents: Plate Tectonics." *(Left. Svetoslava Slavova/Getty Images; right. © Michel Detay/Getty Images)*

mountains and the deepest ocean basins are created by Earth's geothermal energy. Through the process of *plate tectonics* (discussed in Chapter 12, "Earth History, Earth Interior"), geothermal energy causes Earth's heated interior to slowly move. This movement grinds the plates of the lithosphere against one another, lifting and buckling mountains in the process and creating Earth's topographic relief. Plate movement also creates deep-sea trenches, triggers deadly earthquakes, and powers violent volcanic eruptions **(Figure 1.10)**.

Erosion and Deposition: Sculpting Earth's Surface

As geothermal energy builds Earth's relief, solar energy works to reduce relief and flatten Earth's surface through erosion, a major focus of Part IV of

Figure 1.11 Solar energy and erosion. The erosive forces of flowing water and flowing ice, both parts of the hydrosphere, are manifestations of solar energy and gravity. Streams and ice in glaciers flow downslope and erode the land surface. Erosion is discussed in Chapter 17, "Flowing Water: Fluvial Systems," and in Chapter 18, "The Work of Ice: The Cryosphere and Glacial Landforms." Left: Shown here, water cuts into and flows over a plateau, creating Angel Falls in Venezuela (the highest waterfall in the world). Right: One of the many glaciers in Glacier Bay, Alaska, has created a valley through erosion. *(Left. FabioFilzi/Getty Images; center. David McGlynn/Getty Images; right. Jim Wark/airphotona.com)*

Picture This Living Physical Geography: Granite Lake

Atmosphere
The atmosphere envelops Earth with a layer of gases, mostly nitrogen and oxygen. Solar energy streams through the atmosphere, warming the atmosphere and evaporating water from the oceans. Clouds are composed of tiny drops of liquid water and ice crystals that form from water vapor. If the liquid cloud droplets grow large enough, rain will fall from the clouds.

Biosphere
Plants in the biosphere use water that falls from the atmosphere and, through photosynthesis, convert the Sun's radiant energy to chemical energy. The chemical energy of the plants streams through and fuels the chemical processes required by living organisms. These hikers will eat the converted solar energy in the form of sandwiches and potato chips for lunch.

Lithosphere
The hikers are sitting on a large granite boulder. The granite rocks in the Sierra Nevada formed from molten magma several million years ago. The magma cooled and the rocks formed miles deep within the crust. Over time, geologic forces pushed the rocks upward, and the forces of erosion eventually exposed them on the surface.

Hydrosphere
Glaciers formed many times over the Sierra Nevada mountains during colder periods in Earth's history. By about 10,000 years ago, most of the glaciers melted away due to natural climate change. The ice created depressions in the granite bedrock, and water from streams filled the depressions, forming lakes such as Granite Lake, pictured here.

This photo shows two young hikers overlooking Granite Lake in the Sierra Nevada mountains in California. In outdoor settings, even ones that are not as dramatic as this one, Earth's physical systems are always visible.

Consider This
1. Cite an example of how Earth's physical systems overlap and interact with one another.
2. Explain the following statement: "Animals (including people) are solar powered."

(Bruce Gervais)

---

the book. **Erosion** is the scouring and stripping away of rock fragments loosened by weathering through flowing water, ice, or wind **(Figure 1.11)**. **Deposition** is the laying down (depositing) of sediments that are being transported by flowing water or ice.

Over geologic time, rivers may carve deep canyons into plateaus that were uplifted by geothermal energy. Glaciers of flowing ice gouge and grind deep valleys (see Section 18.2). These erosive forces work to flatten out what geothermal energy works to build as we discuss in the **Picture This** feature above.

1.4 Mapping Earth: The Science of Spatial Awareness

◎ Use the geographic grid coordinate system to identify locations on Earth's surface and distinguish among different types of map scales and map projections.

Geography is a spatial science, and no tool is better at portraying spatial information than a map. To be effective and accurate, maps must be *georeferenced*

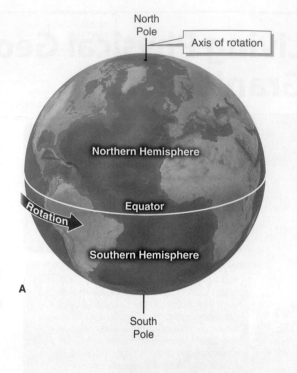

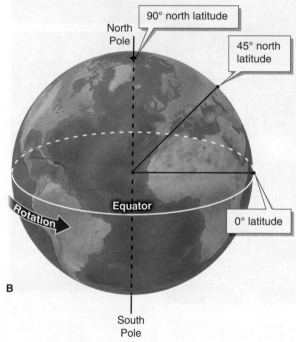

Figure 1.12 Earth rotation and latitude.

(A) The equator is halfway between the poles and divides Earth into the Northern Hemisphere and the Southern Hemisphere. (B) Latitude is measured in angles away from the equator. The latitude 45 degrees north marks the halfway point between the equator and the North Pole. The equator is at 0 degrees latitude, and the poles are at 90 degrees latitude north and south. Latitudes do not exceed 90 degrees.

Animation

Earth Rotation and Latitude

Available at www.saplinglearning.com

(▶)

(precisely located) in real geographic space. The geographic grid coordinate system is used to precisely identify—to georeference—the part of Earth's surface that is being mapped.

The Geographic Grid

Try telling someone how to get to your home without providing street names and addresses. It isn't easy to do. It is similarly difficult to identify where things are on Earth's surface without the geographic grid. The **geographic grid** is a coordinate system that uses latitude and longitude to identify locations on Earth's surface. The geographic grid allows us to pinpoint any location on Earth's surface.

Latitude

Earth rotates on an imaginary axis that runs through both poles. With the North Pole up, Earth rotates eastward (or left to right), parallel to lines of latitude. **Latitude** is the angular distance measured from Earth's center to a point north or south of the equator. The **equator** is the line of latitude that divides Earth into two equal halves. The equator is exactly perpendicular to Earth's axis of rotation, and it creates the Northern Hemisphere and the Southern Hemisphere **(Figure 1.12)**.

Points of the same latitude connected together form a line called a **parallel**. *Latitude* is the name of the angle; *parallel* is the name of the line. Parallels are imaginary circles that run parallel to the equator and are named for their latitude; for example, the parallel at 40 degrees north is the 40th north parallel **(Figure 1.13)**.

We refer to latitude in degrees (°), minutes ('), and seconds ("). Each degree is divided into 60 minutes. Each minute is divided into 60 seconds. Degrees of latitude are approximately 111 km (69 mi) apart, minutes are 1.9 km (1.2 mi) apart, and seconds are 31 m (102 ft) apart.

There are three major zones of latitude: the *tropics, midlatitudes,* and *high latitudes*. The **tropics** are the geographic region located between 23.5 degrees north and south latitude. Midlatitudes and high latitudes are less well defined but generally are recognized as meeting at the 55th parallel. In addition to these three major zones of latitude, geographers often refer to two subzones—*subtropical* and *polar*—as well **(Figure 1.14)**.

Longitude

The **prime meridian** is the counterpart to the equator: It is the 0 degree starting point from which all other lines of longitude are determined. **Longitude** is angular distance as measured from Earth's center to points east or west of the prime meridian.

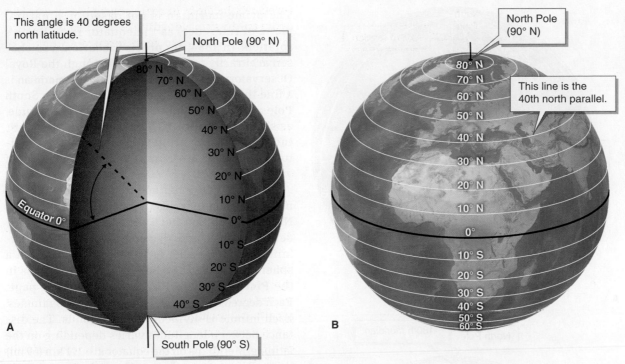

This angle is 40 degrees north latitude.

North Pole (90° N)

80° N
70° N
60° N
50° N
40° N
30° N
20° N
10° N
0°
10° S
20° S
30° S
40° S

Equator 0°

A

South Pole (90° S)

North Pole (90° N)

This line is the 40th north parallel.

80° N
70° N
60° N
50° N
40° N
30° N
20° N
10° N
0°
10° S
20° S
30° S
40° S
50° S
60° S

B

Figure 1.13 Latitude and parallels.
(A) Latitude is the measured angle in relationship to the equator. (B) Parallels are points of latitude connected together to form a line and are named by their latitude.

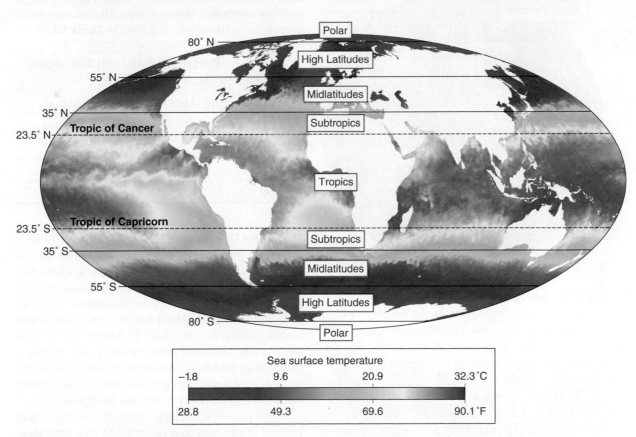

80° N — Polar
High Latitudes
55° N — Midlatitudes
35° N — Subtropics
23.5° N — **Tropic of Cancer**
Tropics
Tropic of Capricorn
23.5° S — Subtropics
35° S — Midlatitudes
55° S — High Latitudes
80° S — Polar

Sea surface temperature

−1.8	9.6	20.9	32.3 °C
28.8	49.3	69.6	90.1 °F

Figure 1.14 Zones of latitude. This world map includes satellite measurements of the ocean surface temperature for July 2, 2013. Surface ocean temperature roughly corresponds with the major zones of latitude. Orange indicates areas with surface water temperatures up to 32°C (90°F). Violet indicates surface water near freezing (0°C [32°F]). Surface seawater temperature is high in the tropics. Around the 55th parallel, at the boundary of the midlatitudes and high latitudes, the water transitions to near-freezing temperatures. At high latitudes, seawater is always near freezing.

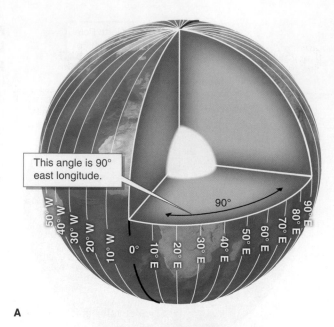

A

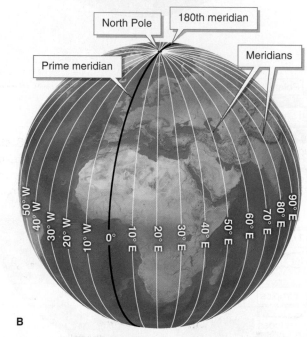

B

Figure 1.15 Longitude and meridians. (A) Longitude is determined by the angular distance from the prime meridian, which runs through Greenwich, England. Traveling east from the prime meridian, we pass through the Eastern Hemisphere, which ends at 180 degrees. Traveling west from the prime meridian, we pass through the Western Hemisphere, which also ends at 180 degrees. (B) Meridians are lines created by connecting points of longitude. The prime meridian (0 degrees longitude) and a portion of the 180th meridian (180 degrees longitude) are shown. All meridians terminate at the poles.

The prime meridian is not based on a natural plane of reference, as the equator is on Earth's rotation. In 1884, its precise location was chosen arbitrarily to pass directly through the Royal Observatory in Greenwich, England. A **meridian** is a line that runs from the North Pole to the South Pole and connects points of the same longitude, as shown in **Figure 1.15**. Like parallels and latitudes, meridians are the counterparts to longitude. Meridians are central to the development of world time zones.

Like latitude, longitude is given in degrees (°), minutes ('), and seconds ("). Just as latitude defines Northern and Southern hemispheres, longitude defines Eastern and Western hemispheres. Determining how far east or west something is on a sphere can sometimes be tricky, as we discuss in the **Picture This** feature on the following page. Each degree of longitude is divided into 60 minutes. Each minute is divided into 60 seconds. The distance between longitudes varies depending on the latitude. Meridians on the equator are 111 km (69 mi) apart. Because meridians all converge at a single point at the poles, there is zero distance between meridians at the poles, as shown in **Table 1.1**.

Table 1.1 **Distances between Meridians from the Equator to the Poles**		
LATITUDE	**DISTANCE (MI)**	**DISTANCE (MI)**
0°	111	69
30°	96	60
60°	56	35
90°	0	0

Using the Geographic Grid

Parallels and meridians together make up the geographic grid system. Geographic grid coordinates give latitude first, followed by longitude. As an example, the coordinates for Mount Whitney in California, the highest point in the continental United States, are 36°34'42" N, 118°17'32" W. This set of coordinates reads: "Thirty-six degrees, thirty-four minutes, and forty-two seconds north latitude; one hundred eighteen degrees, seventeen minutes, and thirty-two seconds west longitude."

Alternatively, *decimal degrees* are often used instead of minutes and seconds. In this approach, minutes and seconds are converted to decimals. Using decimal degrees, Mount Whitney is located at 36.57857°, –118.29225°. The coordinates for the Northern Hemisphere and Eastern Hemisphere are positive numbers; for the Southern Hemisphere and Western Hemisphere, coordinates are negative numbers. A decimal degree value of five decimal

Picture This Geography Quiz

What is the northernmost U.S. state? Most people would say that the northernmost state is Alaska, and that's correct. However, many might guess that the southernmost state is Florida. That is not true. Hawai`i, which is in the tropics, is farther south than Florida. That the westernmost state is Alaska is probably not a surprise to anyone familiar with U.S. geography. But which state lies farthest east? This is a surprisingly difficult question. In terms of the geographic grid and lines of longitude, the answer is not Maine. It surprises many people to find out that Alaska is much farther east than Maine. Alaska's Aleutian Islands arc westward and cross 180 degrees longitude into the Eastern Hemisphere. The easternmost scrap of U.S. land is Semisopochnoi Island, located at 179 degrees east longitude. Maine's farthest point is 66 degrees west longitude—not even close to the Eastern Hemisphere. Therefore, Alaska is the northernmost, westernmost, and easternmost U.S. state, all thanks to the geographic grid!

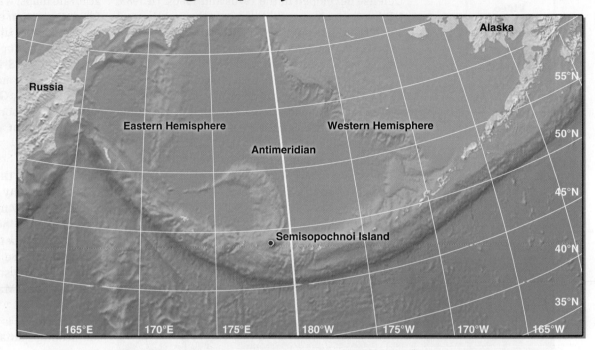

Consider This

1. Why is Alaska the easternmost U.S. state?
2. If only the *contiguous* U.S. states (*lower 48*)—that is, the states that share touching borders—are considered, which are the westernmost and easternmost states?

---

places, as given in the example on the previous page, is accurate to about 1 m (3.3 ft).

The latitudes and longitudes together create an *x–y* coordinate system that can be used to locate any point on the planet precisely. When finding your way on the geographic grid, start on the *origin* (at 0°, 0°). From the origin, first find the latitude and then find the longitude **(Figure 1.16)**.

The Global Positioning System (GPS)

Not too many years ago, when people got lost while driving, their options were to use a paper map or ask somebody for directions to get their bearings. Now there are navigation systems that provide directions by pinpointing our positions using the geographic grid and displaying our locations on a map. To track our positions, these systems use a global navigation satellite system (GNSS). There are several major GNSSs in use or currently in development, including China's *BeiDou* (or Compass) system, Europe's *Galileo* system, and Russia's *GLONASS* (Global Navigation Satellite System).

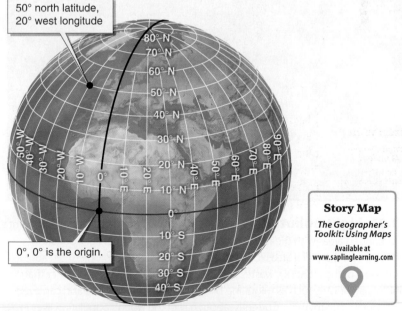

Story Map

The Geographer's Toolkit: Using Maps

Available at
www.saplinglearning.com

Figure 1.16 Using the geographic grid.
This globe includes both latitude and longitude. To find 50° N by 20° W, from the origin, first go north 50 degrees and then go west 20 degrees.

The system in use in the United States is the **Global Positioning System (GPS)**. The GPS is a global navigation system that the U.S. Department of Defense developed in the 1970s and 1980s. In 1983, GPS was made available free to the public, and since then, its commercial application has grown exponentially. Virtually any moving vehicle or aircraft is now equipped with a GPS receiver.

The system uses more than 30 satellites operated by the U.S. government to determine the geographic coordinates of any location. Each satellite uses radio waves to transmit its presence to GPS ground receivers such as those in our cars, tablets, cameras, and smartphones. When signals from four satellites are received, a ground receiver can calculate its location on Earth, often to within several feet.

In physical geography, common applications of GPS include precisely georeferencing satellite images on Earth's surface to create extremely accurate maps, tracking the movements of animals for conservation efforts, and measuring the movement of Earth's lithospheric plates in order to monitor volcanoes for ground movement and potential eruptions. GPS is also commonly employed in meteorology to monitor and forecast weather. And now, of course, GPS technology allows everyone with a smartphone to pinpoint the nearest coffee shop, restaurant, or gas station.

Maps

Cartography is the science and art of map making. Cartography and maps are as old as civilization itself. They may have their beginning in lines scratched into the sand long ago to communicate the spatial dimensions of trails, food, or danger. Today maps are created based on sophisticated computer analysis of immense data sets referred to as *big data*. Maps are central to all geographic inquiry.

Using maps is the most efficient means of communicating spatial information. All maps have one feature in common: They reduce the size of geographic space onto the map's surface. While maps may be printed on a flat surface such as paper, now more often spatial information is displayed on computer monitors or portable device screens, where it can be easily changed and updated. Maps can portray statistical and quantitative information, such as population numbers, temperature averages, atmospheric pressure differences across a region, or tornado frequency in the midwestern United States. They show patterns of physical phenomena such as mountain ranges, seafloor features, weather, earthquake activity, animal migration, and deforestation. Cultural phenomena such as regions under artificial lighting can also be mapped **(Figure 1.17)**. The term *map* is also applied to concepts such as the "map of the human genome" and a "computer network map." These are not maps in a geographic sense, however, because they lack a relationship to geographic space.

Spatial data and their analysis through *geovisualization* tools, such as interactive online mapping services, are now widely available to the public. Such tools allow users with little or no mapping experience to develop customized maps or explore various aspects of Earth's surface. Examples of geovisualization tools range from Google Maps, where one can get directions to a select destination, to Google Earth, where one can explore a wide range of Earth features, to sophisticated spatial data and analysis service companies such as Esri (Environmental Systems Research Institute) **(Figure 1.18)**.

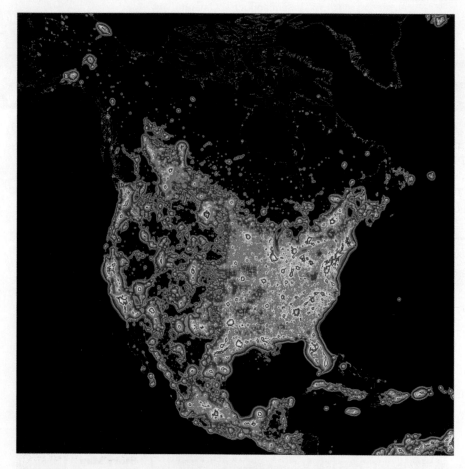

Figure 1.17 Light pollution. This map of North America shows nighttime light pollution. Urbanized areas with greater populations have much higher light pollution (white and red areas) than rural areas (blue areas). Light pollution is a problem because it obscures our view of the night sky, wastes energy by directing light upward rather than down, confuses organisms such as migrating birds and nesting sea turtles, and can even disrupt our natural sleep rhythms. Currently, about half of the world's population lives in urbanized areas with strong light pollution. By the year 2100, about 80% of the world's population will live in urbanized areas. *(From Fabio Falchi et al. Sci Adv 2016;2:e1600377, American Association for the Advancement of Science. © Falchi et al, some rights reserved; exclusive licensee American Association for the Advancement of Science. This work is distributed under CC BY-NC (http://creativecommons. org/licenses/by-nc/4.0/))*

Figure 1.18 Spatial awareness. Sheng Tan, a spatial analyst living in Sacramento, California, wears a T-shirt representing the data and mapping company Esri. As the shirt implies, spatial awareness is central to the art of cartography and to geographic inquiry. Esri provides geovisualization tools, such as Web Maps and Story Maps, that allow users to build customized interactive maps. You can build your own customized maps with Esri on the digital platform that accompanies this book. See the preface for more information. *(Courtesy Bruce Gervais)*

Map Projections

Maps are always abstract representations of the real world. They therefore sometimes run into problems. One such problem is the unavoidable distortion of space on small-scale maps, such as a continent or the world. Maps are flat, two-dimensional representations of Earth's surface.

A globe is a three-dimensional replica of the planet. Large-scale (local) maps do not suffer from significant distortion. World maps, on the other hand, cannot portray the globe's curved surface without significantly distorting the shapes and sizes of continents and ocean basins. To illustrate this distortion, imagine that you have a soccer ball with Earth printed on it. A traditional soccer ball has 32 patches that are stitched together to make a sphere. Each patch represents a large-scale map with little distortion. However, when all the patches are placed together to map the world, it is impossible to lay out these patches on a flat piece of paper without causing gaps between the patches in the oceans or continents **(Figure 1.19)**.

Only a globe can keep the sizes and shapes of Earth's features accurate. But a globe has many limitations:

- Only half a globe can be viewed at one time.
- The edges of the visible half are distorted.
- Globes are always at a small scale, so it is impossible to show details clearly.
- Globes are more expensive to make than maps and less versatile in terms of representing different themes, such as vegetation cover or earthquake patterns.
- Globes do not fit well into pockets or drawers. They cannot be printed in books.

Because of the limitations of globes, flat world maps will always be needed. So the problem of how to handle the gaps that result when the globe is transferred to a flat surface remains. Cartographers address that problem by using *map projections*.

Types of Map Projections

Cartographers use map projections to transfer the global surface onto a flat map surface. This process is called "projecting" because, before computers, cartographers used a light to project the outline of a wire globe onto a flat piece of paper and then traced the shadows onto the paper. The type of projection that resulted depended on how

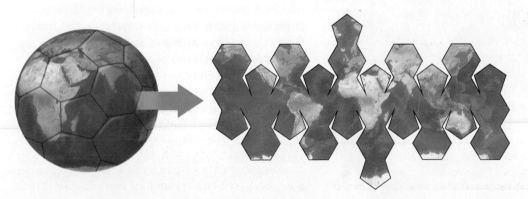

Figure 1.19 A sphere converted to a flat map. When the surface of a sphere is represented on a piece of flat paper, gaps result.

Figure 1.20 Map projections. Each projection type has a different application and use. The red line or dot represents the *tangent* where the map touches the globe. Map distortion increases with distance from the tangent.

Cylindrical   Conic   Azimuthal

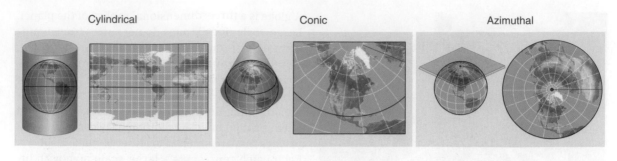

A

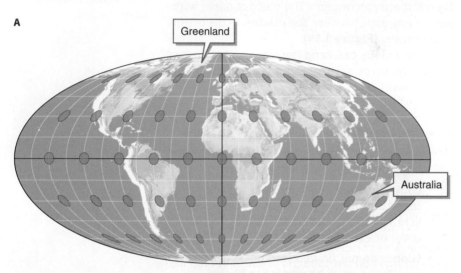

Greenland

Australia

B

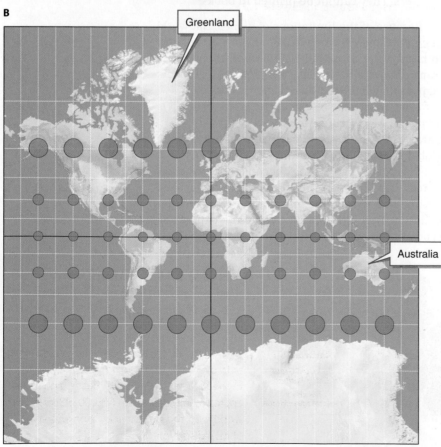

Greenland

Australia

the paper was positioned in relationship to the light. There are three basic kinds of projections: *cylindrical, conic,* and *azimuthal* **(Figure 1.20)**.

Equal-Area and Conformal Maps
There are two ways to address the gaps that result from stretching the global surface onto a flat piece of paper: by preserving areas and sacrificing shapes or by preserving shapes and sacrificing areas. An **equal-area map projection** preserves the true areas of continents at the expense of their true shapes. A **conformal map projection** preserves the true shapes of continents at the expense of their true areas **(Figure 1.21)**. Only globes are both conformal and equal-area representations of Earth. Unlike a globe, no single world map projection can show both shapes and areas accurately.

Figure 1.21 Map distortion. (A) Mollweide equal-area projection. The orange circular areas are all approximately the same size and cover roughly the same geographic area, no matter where on the map they are located. This equalization of area is accomplished by compressing and distorting regions near the poles. Note that the orange areas at high latitudes are distorted ellipses compared with those at low latitudes. This Mollweide equal-area map shows the true areas of Earth's surface: Note, for example, that Australia is over three times larger than Greenland. The Mollweide projection is used throughout *Living Physical Geography*. (B) Mercator conformal projection. Even though they are of different sizes, all the orange areas on this map are perfect circles and cover the same amount of Earth's surface. Compared with those in the tropics, the circles (and continents) at high latitudes are larger because those areas have been stretched. Notice that Greenland appears far larger than Australia in this projection. Antarctica's size is also greatly exaggerated. *(© 2012 Google, © 2012 Cnes/Spot image, U.S. Department of State Geographer, Image © 2012 GeoEye)*

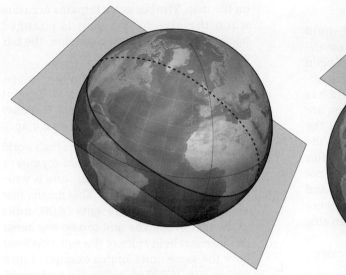

Figure 1.22 Great and small circles.
When Earth is divided into two equal halves with an imaginary flat sheet, the point of contact on Earth's surface traces a great circle route (shown in red). When the flat sheet does not divide Earth into two equal halves, a small circle route results (shown in blue).

Great Circle Routes

Cartographic distortion on a world map is apparent in many aspects, including when we consider global travel routes. For efficiency, airplanes and ships follow straight lines called great circle routes whenever possible. A **great circle** is a continuous line that bisects the globe into two equal halves, such as the equator. A great circle represents the shortest distance between two points on Earth. In contrast, *small circles* do not bisect the globe into two equal halves. A small circle is a continuous line that forms a circle smaller than the equator. All parallels other than the equator form small circles on the globe. The equator is the only parallel that is a great circle. All meridians are half of a great circle, and when two opposing meridians are combined, they create a great circle **(Figure 1.22)**.

Cartographic distortion on a world map becomes obvious when routes of long-distance travel, such as that of an airplane or a ship, are plotted. Great circle routes often do not appear as straight lines on world maps because of map distortion, as illustrated in **Figure 1.23**.

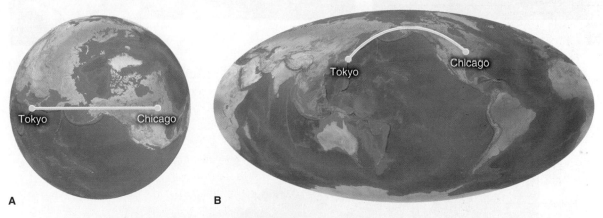

A   B

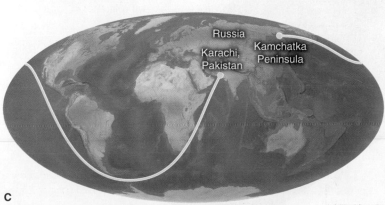

C

Figure 1.23 Distorted great circle routes. (A) A great circle route between Tokyo, Japan, and Chicago, Illinois, is plotted on the globe-like map. (Note that this is not a world map because we can see only one half of the globe.) (B) The same route is plotted on a world map. Great circle routes on many world maps appear curved and inefficient because of map distortion.

(C) This great circle route is the longest straight path over water. It runs from Karachi, Pakistan, to the Kamchatka Peninsula, Russia, and is approximately 32,000 km (20,000 mi) long. On a world map, the straight line is curved because of map distortion.

Animation
Distorted Great Circle Routes
Available at
www.saplinglearning.com

Map Scale: How Far Is It?

Of course, maps are smaller than the real-world distances they represent. So it is usually essential to know how much the real world has been reduced on a map. A **map scale** performs this function by specifying how much the real world has been reduced. For example, imagine that you are new to a college campus and you notice on the campus map that a river runs nearby. You decide to walk on your lunch break to see the river. After about 20 minutes of walking, you still have not reached the river. Frustrated, you turn around and race back to campus for your next class. Had there been a scale on the map, you might have been able to more accurately judge the distance.

There are three different types of map scales:

1. A **bar scale** (or **graphic map scale**) uses a simple line segment to depict real-world distances. Bar scales are intuitive to use. The length of the line indicates distances on the map. The bar scale remains accurate when the size of the map is changed because as the map's size changes, the bar scale line changes along with it.

2. **Verbal map scales** are also intuitive to use. A verbal scale uses a written statement such as "1 inch represents 1 mile." It is easy to measure or estimate an inch on a map.

3. A **representative fraction scale** uses real-world ground distances in fractional form. For example, a common fractional scale is written as 1/24,000 or 1:24,000. This means that 1 unit on the map represents 24,000 units in the real world. The unit can be any linear distance, but both sides of the equation must have the same units. In this example, 1 inch represents 24,000 inches in the real world. Note that it would be incorrect to say that 1 inch represents 24,000 centimeters because centimeters are not the same unit as inches.

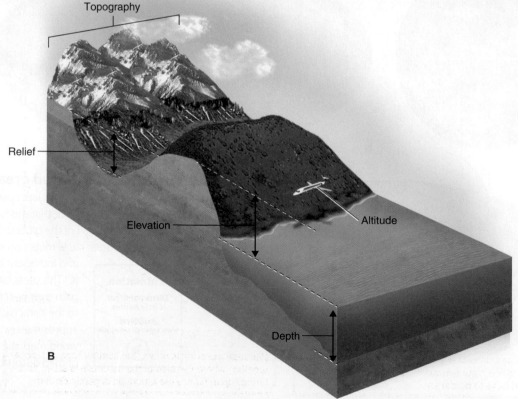

Figure 1.24
Topographic relief.

(A) Mount Everest (on the left) is 8,848 m (29,029 ft) above sea level in elevation, and the Dead Sea (on the right) is 427 m (1,401 ft) below sea level in elevation. They are the two most extreme elevations on Earth's land surface.
(B) Topographic terms used throughout *Living Physical Geography* are illustrated in this diagram. *(Left. STR/ Getty Images; right. SEUX Paule/ Getty Images)*

In addition to a map scale, a good map has several other elements that increase the map's effectiveness and reliability. A map should have a title that reflects the map's theme, a legend that clarifies any symbols or colors used, a direction arrow pointing north, and a date. The date allows the user to judge if the map is current.

Mapping Earth's Surface

Geothermal energy within Earth has in some places sunken the crust into valleys and in other places lifted it into mountains **(Figure 1.24A)**. These surface features are collectively called topography. **Topography** is the shape and physical character of Earth's surface in a region. **Elevation** is the vertical distance of a land surface above or below mean sea level. Similarly, **relief** is the relative difference in elevation between two or more points on Earth's surface. Areas with low relief are flat, and areas with high relief are mountainous. *Altitude*, in contrast, is height of an object above sea level, such as an airplane or a balloon, which is not in contact with the ground. *Depth* is the vertical distance below a water body's surface, such as a lake or the ocean **(Figure 1.24B)**.

There are many ways to represent Earth's surface topography on a map. Geographers commonly use *topographic maps* with **contour lines**, which are lines of equal elevation in relation to sea level. The increment between contour lines is the *contour interval*. **Figure 1.25** provides an example of

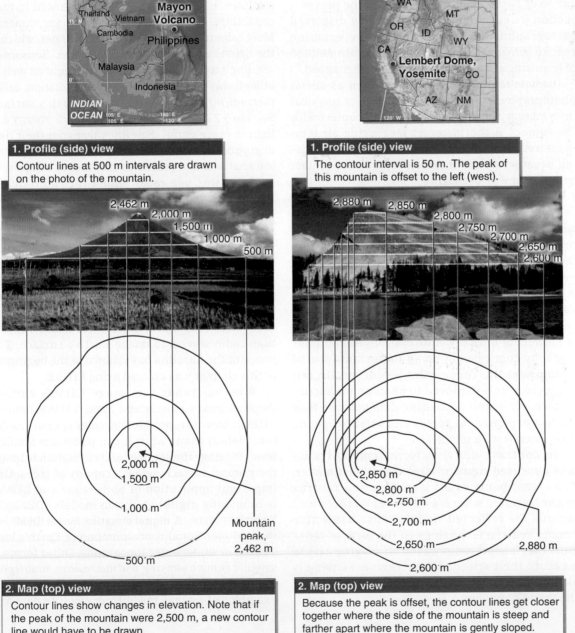

A

1. Profile (side) view
Contour lines at 500 m intervals are drawn on the photo of the mountain.

2,462 m
2,000 m
1,500 m
1,000 m
500 m

2,000 m
1,500 m
1,000 m
500 m

Mountain peak, 2,462 m

2. Map (top) view
Contour lines show changes in elevation. Note that if the peak of the mountain were 2,500 m, a new contour line would have to be drawn.

B

1. Profile (side) view
The contour interval is 50 m. The peak of this mountain is offset to the left (west).

2,880 m
2,850 m
2,800 m
2,750 m
2,700 m
2,650 m
2,600 m

2,850 m
2,800 m
2,750 m
2,700 m
2,650 m
2,600 m
2,880 m

2. Map (top) view
Because the peak is offset, the contour lines get closer together where the side of the mountain is steep and farther apart where the mountain is gently sloped.

Figure 1.25 Contour lines.

(A) The base of Mayon volcano in the Philippines is approximately 500 m (1,640 ft) above sea level. The peak is 2,462 m (8,677 ft) above sea level in elevation. In this example, the contour interval is 500 m. (B) The base of Lembert Dome, in Yosemite National Park, is 2,600 m (8,528 ft) above sea level. The contour lines are closest together on its steepest slope. If you were to use a topographic map to plan a hike to the top of this mountain, you would probably decide that the side where the contour lines are farthest apart is the easiest approach to the peak. *(Left. © Rolly Pedrina/Oriental Touch/Robert Harding; right. Nicholas Pavloff/Getty Images)*

Animation
Contour Lines
Available at
www.saplinglearning.com

the use of contour lines. (For more on topographic maps, see Appendix II, "Topographic Maps.")

Contour lines always connect back to themselves, forming a closed loop. Topographic maps often do not show this, however, because the lines are cut off by the edges of the map. Anywhere on Earth, if you walked on the ground along a contour line, you would eventually wind up where you started. Contour lines converge only on vertical cliffs, and they never cross each other.

1.5 Imaging Earth

◉ Discuss how technologies such as satellite sensors and radar are used to study and portray Earth systems and processes.

Without data, there could be no scientific investigation of Earth's natural systems. In the previous section we examined how data can be displayed cartographically. In this section we examine remote sensing, a particularly important method of acquiring geospatial data that can be mapped.

Remote sensing technologies, such as aerial photography and satellite data, collect physical information and data without direct contact with the subject. A thermometer measuring air temperature, in contrast, is not a form of remote sensing because the thermometer is in direct contact with the air it is measuring.

There are two modes of remote sensing: passive and active. *Passive remote sensing* uses sensors to receive energy that is radiated or reflected from an object, such as Earth's surface. Taking a photograph with a camera is an example of passive remote sensing of light that is reflected from the surface. Remote sensing of Earth's surface goes back to World War I, when aerial photography became an important means of gathering military information. Aerial photography continues to be an important form of remote sensing. Today, pilot-less drone aircraft are increasingly deployed to photograph Earth's surface and acquire scientific data because they are less expensive to deploy and safer to operate than aircraft with pilots.

In contrast, *active remote sensing* transmits a focused signal toward the subject matter. The signal bounces off a surface and returns to the sender, where a receiver detects and records the reflected signal. Police use active remote sensing technology in the form of radar guns that "shoot" radio waves off moving cars to measure their speed. Active remote sensing is also used in sports such as baseball to measure the speed of pitches.

Satellite Remote Sensing

Satellite remote sensing is an essential tool in acquiring data about Earth and its systems. The satellite era began in 1957 with Russia's *Sputnik* satellite, and the first weather satellite, *TIROS-1*, entered orbit in 1960. Now there are more than 1,200 operating satellites in orbit. Many of these satellites have revolutionized our ability to monitor Earth's rapidly changing surface, and no other generation has had so much data so freely available. Satellite remote sensing, GPS, and the Internet have provided access to important data about Earth.

Just as a digital camera—like the ones we all have on our smartphones—creates images using electronic sensors (rather than photographic film), satellites use electronic sensors to generate precisely georeferenced data that are used to create digital images of Earth's surface. The sensors on satellites, however, are far more advanced in their capabilities than the ones on our phone cameras. Most cameras can sense only *visible light*, which is the light that we can see with our eyes. Sensors on orbiting satellites can sense visible light as well as other forms of energy, such as heat radiation, called *thermal infrared*, emitted from Earth's surface. Section 3.2 discusses the Sun's radiant energy and light in more detail. Satellites transmit their data to ground-based receiving stations, where the data are analyzed and processed into digital imagery.

Satellites use **radar** (short for *radio detection and ranging*) active remote sensing to study Earth's surface. Radar transmitters mounted on satellites (and airplanes) send rapid pulses of radio waves and receive the return signals reflected off Earth's surface to create detailed and precisely georeferenced three-dimensional reconstructions of surface topography. *LIDAR* (*light detection and ranging*) is a type of remote sensing similar to radar that uses light pulsed from a laser, rather than radio waves, to image Earth's surface. The image of the Carolina bays shown at the beginning of this chapter was created using LIDAR.

Whereas radar can image Earth's surface through heavy clouds and haze, LIDAR cannot. LIDAR, however, produces greater *spatial resolution* (detail) than radar; it can penetrate shallow water to image the seafloor and vegetation to image the ground surface below a canopy of trees. One important application of both radar and LIDAR is in making digital elevation models (DEMs) of Earth's surface. A **digital elevation model (DEM)** is a three-dimensional representation of Earth's land surface or underwater topography. Other forms of satellite remote sensing use *microwave, multispectral*, and *infrared* energy. **Figure 1.26** shows some examples of images made from satellite data and

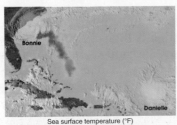

Sea surface temperature (°F)
75 80 85 90

Microwave

Satellites transmit microwave energy to Earth, and sea surface temperatures are calculated based on the energy reflected off the surface. Microwaves penetrate clouds, haze, and night, allowing for data collection at any time. This microwave image uses "false" colors to reveal a cool streak of ocean temperatures in blue left in the wake of Hurricane Bonnie near Florida.

Multispectral

Multispectral imagery allows satellites to detect light reflected off Earth's surface that is both visible and invisible to the human eye. This image uses natural colors to show algae blooms (in light green areas) in Lake St. Clair, which straddles the border between Ontario and Michigan.

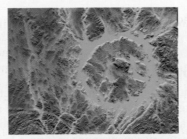

Radar and LIDAR

Satellites pulse radio waves using radar and light using LIDAR to Earth. Scientists construct digital elevation models (DEMs) of the surface from the reflected energy. This image shows Manicouagan Crater in Canada, which was formed 214 million years ago from an asteroid impact. The crater is 60 km (40 mi) across.

Thermal infrared

Satellites can sense thermal infrared (heat radiation) emitted by Earth's surface. Thermal infrared can "see" Earth's surface through haze, clouds, and night. Here, Hawai'i's Mauna Loa volcano is shown in thermal infrared. Yellow areas of lava are warmer than purple, blue, and green areas.

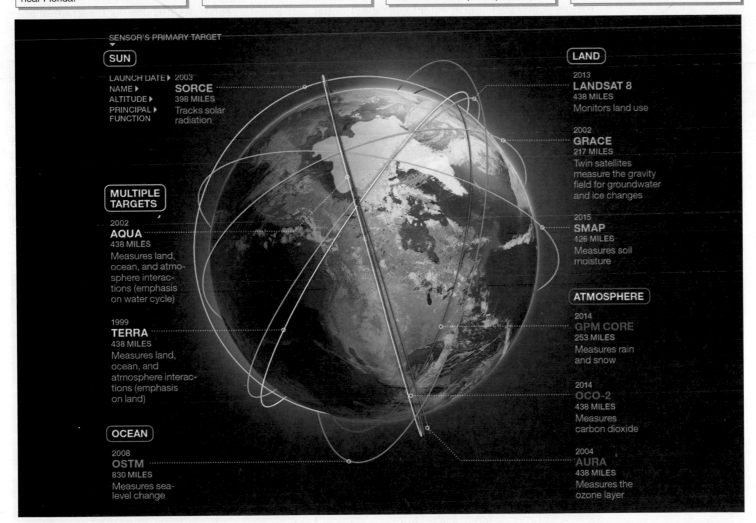

SENSOR'S PRIMARY TARGET

SUN

LAUNCH DATE ▸ 2003
NAME ▸ **SORCE**
ALTITUDE ▸ 398 MILES
PRINCIPAL ▸ Tracks solar
FUNCTION   radiation

MULTIPLE TARGETS

2002
AQUA
438 MILES
Measures land, ocean, and atmosphere interactions (emphasis on water cycle)

1999
TERRA
438 MILES
Measures land, ocean, and atmosphere interactions (emphasis on land)

OCEAN

2008
OSTM
830 MILES
Measures sea-level change

LAND

2013
LANDSAT 8
438 MILES
Monitors land use

2002
GRACE
217 MILES
Twin satellites measure the gravity field for groundwater and ice changes

2015
SMAP
426 MILES
Measures soil moisture

ATMOSPHERE

2014
GPM CORE
253 MILES
Measures rain and snow

2014
OCO-2
438 MILES
Measures carbon dioxide

2004
AURA
438 MILES
Measures the ozone layer

Figure 1.26 Satellite remote sensing. The current network of Earth-monitoring satellites is one of the greatest scientific achievements in modern history. (A) Satellites collect many different types of data, ranging from light that the human eye can see, to invisible heat energy, to reflected microwave and radio energy. Each technique offers unique information for different applications. (B) A constellation of Earth-observing satellites continually monitors the planet. Shown here are the names, dates of launch, and primary functions of a number of particularly critical Earth-observing satellites that are active today. Each one circles the globe up to 16 times per day, collecting a wide variety of data. The satellites are grouped by their primary targets: the Sun, multiple targets, the oceans, the land surface, and the atmosphere. (*Top far left. Image courtesy TRMM Project, Remote Sensing Systems, and Scientific Visualization Studio, NASA Goddard Space Flight Center; top center left. NASA Earth Observatory images by Joshua Stevens, using Landsat data from the U.S. Geological Survey; top center right. Courtesy SRTM Team NASA/JPL/NIMA; top far right. ASTER Instrument Team, ASTER Project, NASA; bottom. Monica Serrano/National Geographic Creative*)

Web Map

Remote Sensing

Available at
www.saplinglearning.com

a number of important Earth-monitoring satellites in use today.

Doppler Radar and Sonar

Ground-based **Doppler radar** is an active remote sensing technology that is widely used in the study of the atmosphere and meteorology. Doppler radar transmits *microwave* energy from a tower to measure the velocity and direction of movement of particles of precipitation within a cloud. Ground-based Doppler radar has a higher spatial resolution than Doppler radar used from satellites but has lower spatial coverage. **Figure 1.27** provides a ground-based Doppler radar image of Hurricane Harvey (2017).

Another active remote sensing technique is *sonar* (*sound navigation and ranging*). Sonar has been invaluable in creating high-spatial-resolution seafloor *bathymetry* (underwater topography). Sonar works by sending a pulse of sound from a transmitter aboard a ship or an autonomous underwater vehicle. The sound pulse reflects off the seafloor and back to a receiver on the ship or vehicle. The reflected data are used to create a georeferenced digital elevation model of the seafloor **(Figure 1.28)**.

The oceans are vast, and only about 10% of the seafloor has been mapped using sonar. The remainder has been mapped from satellites using LIDAR and radar. Using radar, scientists can indirectly infer seafloor features. The radar can precisely determine the sea level height. Because water follows gravity, and massive mountain ranges increase the gravitational field around them, the presence of mountains can be inferred where the sea level is domed slightly higher. Although these differences of sea level are only a matter of a few centimeters, satellite radar can detect them. Compared to sonar, satellite-based mapping of the seafloor is at very low spatial resolution.

Geographic Information Systems

The big data generated from remote sensing platforms such as satellites, airplanes, ships, and ground-based Doppler radar would be unmanageable without computers. Large volumes of spatial data are processed and analyzed using a **geographic information system (GIS)**. A GIS uses computers to capture, store, analyze, and display spatial data. Like maps, a GIS provides visual tools to help users gain a better understanding of the spatial relationships among geographic phenomena. But unlike traditional maps, which display all their information at once and are static (unchanging) once they are made, GIS maps are interactive. A GIS integrates the stacked layers of spatial data into a single dynamic unit, and the layers of data can be chosen for display. Each layer has a specific purpose and makes a specific contribution in helping an analyst achieve his or her goals. GIS map users can interact with the spatial data within the database and can continually update the data and refine the map to explore different questions and generate new ones. **Figure 1.29** illustrates an application of GIS mapping of Manhattan.

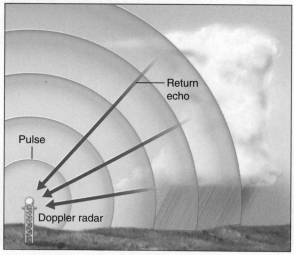

A

B

Figure 1.27 Doppler radar. (A) Ground-based Doppler radar works by beaming microwaves, which reflect off falling precipitation (rain, snow, hail, and sleet) in the sky. An image is developed based on the reflected energy. (B) Hurricane Harvey's spiral structure is clearly visible in this August 26, 2017, Doppler radar image of Texas near Houston. Reds and yellows indicate heavy rainfall; greens and blues show regions of relatively less rainfall. Hurricane Harvey brought historic rainfall amounts and catastrophic flooding to Houston and surrounding regions. The distance across this map is 400 km (250 mi). *(NWS/ NOAA)*

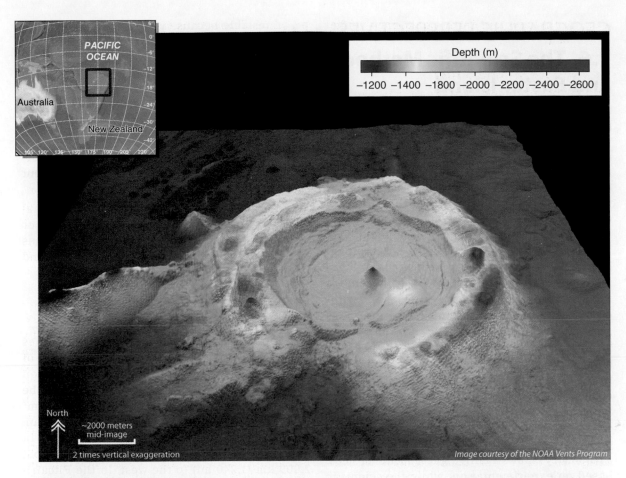

Figure 1.28 Sonar mapping. Sonar mapping of the ocean basins reveals the world's longest mountain ranges and deepest valleys hidden beneath the sea. This image shows "Volcano O," a collapsed volcano, known as a *caldera*, on the seafloor in the Lau Basin, east of Australia. Depths of the feature are shown in "false" color. The seafloor at the center of the volcano is about 2,000 m (6,500 ft) in depth, and the caldera is roughly 10 km (6 mi) in diameter. *(NOAA)*

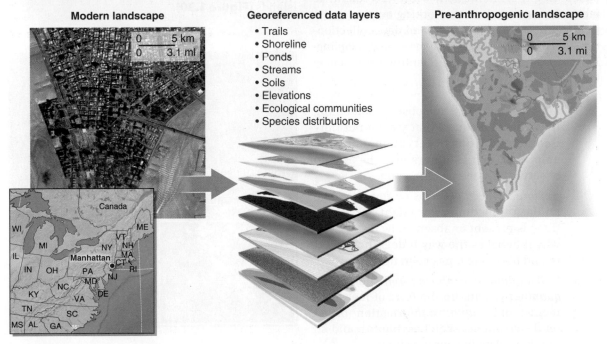

Figure 1.29 Reconstructing Manhattan. This map pair, made from a GIS, shows Manhattan as it appears today (left) and as it appeared 400 years ago. Several layers of data, including shoreline, soil, elevation, and vegetation data, were combined to create this reconstruction. To learn more about this GIS project, go to https://welikia.org. Both maps are zoomable and interactive, and the web page provides downloadable data and resources for researchers, teachers, and students. *(WCS / Mannahatta Project)*

GIS is one of the fastest-growing fields in geography and in many public and private sectors of society as well. Landscape management, species conservation, habitat monitoring and restoration, water quality monitoring, fire management, natural hazard management, pollution mapping, urban zoning, transportation planning, aviation routing, and traffic analysis are just a few of the many applications of GIS. Many students who study geography and gain GIS skills are well prepared to enter a career in a GIS-related field and are often in great demand among employers.

GEOGRAPHIC PERSPECTIVES
1.6 The Scientific Method and Easter Island

◎ Apply the scientific method to study Easter Island's history of human settlement.

Imagine someone telling you that intelligent alien life from a distant planet is on Earth and actively communicating with him or her. Your first responses might be skepticism of the claim and concern for the person's mental health. If the person persisted, your second response might be to ask for evidence. Anyone can make extraordinary claims, but not everyone can back up those claims with *empirical* (observable) evidence. Evidence is the foundation of all scientific inquiry, and without it there is no science but only competing claims and arguments won by those who talk the loudest.

The *scientific method* is an approach to developing a rational and logical understanding of the world by evaluating evidence and establishing evidence-based facts. The scientific approach is the foundation of all sciences. The application of the scientific method depends on the type of research undertaken. For example, the scientific method in a laboratory follows a fairly rigid sequence of steps, based on experimentation, while the scientific method used by scientists working outside in the field might involve sampling and data collection, question exploration, and then more field sampling. For physical geography, the scientific method may loosely follow this sequence:

1. *Make observations:* You observe something that interests you or that you do not fully understand. Your observations must be empirical; that is, others must be able to observe the same phenomenon.

2. *Explore intellectual curiosity:* You articulate and ask questions. You might ask why a thing is present or absent, how it got there, why it behaves the way it does, how old it is, and how long it has been there.

3. *Collect data and information:* You collect *quantitative* data (in the form of measurements) and *qualitative* information (based on descriptions, sketches, photos, and so on) that relate to your questions.

4. *Develop a hypothesis:* The information you collect may stimulate a *hypothesis*, which is a well-thought-out, testable idea that poses an answer to a question. Scientific hypotheses must be testable. For example, a hypothesis about a tree's age is

testable because you can count the number of annual rings in the tree's trunk, but a hypothesis about whether a tree feels happiness is not testable. Hypotheses can be developed before or after data collection and information gathering.

5. *Test the hypothesis:* In order to be accepted by others, your hypothesis must be supported by the data and information. If the information you gathered does not support your hypothesis, you must develop a new one.

6. *Further inquiry:* Whether your original hypothesis was supported by the data or not, new information and new questions often lead to new ideas and more data and information collection.

Easter Island's Mysterious Moai: Observations and Curiosity

The unusual history of Easter Island provides a good case study to illustrate the sequence of observations, curiosity, data and information collection, and hypothesis testing integral to the scientific method. Easter Island is a small, 166 km^2 (64 mi^2) volcanic island in the South Pacific Ocean. A person can walk around the entire island in half a day. The closest human neighbors are about 2,000 km (1,240 mi) to the west, in the Pitcairn Islands **(Figure 1.30)**.

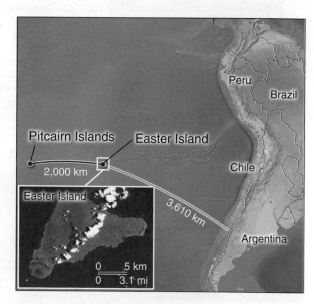

Figure 1.30 Easter Island. Easter Island is about 2,000 km (1,240 mi) from the nearest inhabited islands, the Pitcairn Islands, and 3,610 km (2,250 mi) from Chile. Easter Island is called Rapa Nui in the local Polynesian language. *(NASA image created by Jesse Allen, Earth Observatory, using data obtained from the University of Maryland's Global Land Cover Facility.)*

On Easter Day in 1722, the Dutch explorer Jacob Roggeveen was the first European to visit Easter Island. Roggeveen found a population of about 3,000 people who had no boats and no wood and yet had somehow built the hundreds of large stone statues, called *moai*, scattered across the island **(Figure 1.31)**.

When asked who had made the statues, the islanders said the statues "walked" to their present locations. They did not retain a cultural memory of how the statues were moved to their current positions from the distant stone quarries where they were obviously carved. Roggeveen visited for only one day, and he was unable to solve the mystery of the statues' creators.

A person could hypothesize that the statues were once alive and walked to their present positions. This hypothesis is invalid, however, because it is easy to declare false: There is no empirical evidence of statues ever having walked. Clearly, people had carved the statues in stone quarries on the island and somehow moved them to all parts of the island. But how? The islanders had no modern machinery, rope, or even wood. After Roggeveen's visit, several unanswered questions arose that awaited scientific examination:

- Where did the people on Easter Island come from?
- How long had the islanders lived on Easter Island?
- How did island inhabitants make and transport the huge stone statues without the obvious resources and technology?

The first question was answered as anthropologists in the twentieth century began piecing together the story of Polynesian expansion across the tropical Pacific Ocean over the course of several thousand years. Linguistic characteristics suggest that Easter Island was colonized by people living on the nearest group of islands, the Pitcairn Islands, which are some 2,000 km (1,240 mi) distant. But when did they come to Easter Island, and how did they make and move the statues?

Key Evidence from Pollen: Data, Information, and a Hypothesis

In 1977, British geographer J. R. Flenley was one of the first scientists to study the natural history of Easter Island. His research team took cores of sediments from the marshes on the island and analyzed the pollen in the cores. Pollen from plants had been trapped in the marshes and preserved. The upper layers of the marshes reflect modern vegetation. The deeper sediments were deposited further back in time and contain ancient pollen

Figure 1.31 Easter Island moai. When Europeans first reached Easter Island in 1722, islanders told them that they did not know who had made the hundreds of giant statues. The stone carvings weigh about 13 metric tons (over 28,500 lbs) on average and stand about 4 m (13 ft) tall. Some are much larger. *(Art Wolfe/Getty Images)*

from long ago. Changes in pollen types through time reveal the island's dynamic ecological history.

As they analyzed the pollen data, Flenley's team found that, much to their amazement, the data indicated the following two important points:

- The island had once been thickly forested with some 23 species of trees. However, by the late 1600s, open grasslands dominated the island, and the forests were almost entirely gone.
- The loss of forest began in the year 1250, some 50 years after the island was first settled.

The pollen data prompted many new questions, including these: Did the islanders destroy the island's forests? If so, how? This also led to new ideas about the statues. Scientists suspected that trees were somehow used to transport the statues across the island. This idea led to a testable hypothesis. Could logs be used to move the statues? To test the hypothesis, trees were cut and extensive track and roller systems were made out of tree logs.

Log Rollers and "Walking" Statues: Hypothesis Testing

Is it possible to move the massive moai statues using logs as rollers? To test this idea, researchers returned to Easter Island and attempted to move the statues using only materials that would have been available to the original inhabitants. Although their efforts using this technique yielded mixed results, after much labor, the researchers were eventually able to move and erect a statue.

But just because the statues can be moved using rollers doesn't mean that is how they were moved. An alternate technique of "walking" the statues by lassoing them around the head with long ropes and carefully balancing and rocking

Figure 1.32 Statue walking. All of the statues on the island have a convex (rounded) base with sharp edges that would facilitate the method of "walking" the statues forward using ropes. The bases of many statues are also deeply pitted and gouged, suggesting that they were possibly transported this way. *(Courtesy Terry L. Hunt)*

Figure 1.33 Rat-gnawed palm seeds. These centuries-old palm seeds from Easter Island were all gnawed by rats. Rat predation of tree seeds was the primary force driving the loss of the island's forests. *(Courtesy Terry L. Hunt)*

them back and forth to move them has also been tried with success on a 4.5-metric-ton replica statue. The rocking and walking method requires considerably less effort, fewer people, and no log rollers **(Figure 1.32)**. If this method was used, it might also explain why the islanders refer to the statues as having "walked" to their present locations around the island.

Key Evidence from Garbage: Further Inquiry

Garbage preserved from ancient and even modern societies offers a treasure trove of data and information about the people who generated the trash. Subsequent scientific research on Easter Island focused on analyzing the material found in preserved garbage mounds, called *middens*, discarded by its early inhabitants to study what they ate and how they lived. Scientists learned from the garbage evidence that the earliest inhabitants ate dolphins (porpoises), sea turtles, seals, shellfish, and fish. This diet depended on the use of boats made from trees. The islanders no longer ate seafood after the forests were gone; in the absence of trees, the island's inhabitants could no longer build boats and so lost their access to the sea.

Rat bones were also common in the garbage middens throughout the island. The Polynesian settlers brought the Polynesian rat (*Rattus exulans*) with them to Easter Island. A surprising related discovery was that every palm seed that the researchers found had been gnawed and destroyed by rats **(Figure 1.33)**. This indicates that the rats played a key role in decimating the forests. And the rats may have served a dual destructive purpose, destroying both the trees' seeds and their bird pollinators. Although there is no direct evidence, the rats certainly preyed on the eggs and chicks of nesting birds as well, as they are observed to do on remote oceanic islands today. Birds were essential pollinators of the island's trees, and without them, many tree species couldn't pollinate and reproduce.

Putting Together the Evidence: Easter Island's Story

One of the hallmarks of the scientific process is that scientists follow the data to an outcome rather than having an outcome in mind first and accepting only those data that support it. Ideally there is no place for personal bias or political desires in the scientific process. A researcher follows the data wherever they lead—as long as the data are valid and are analyzed logically and rationally and without bias. Similarly, whenever there are two competing hypotheses with equal scientific support, the simplest hypothesis with the fewest assumptions that fits the evidence should be chosen. This is called the *principle of parsimony*. Considered as a whole, the scientific evidence led to the following statements about Easter Island:

- Easter Islanders may have transported the statues using trees as rollers, or they may have "walked" the statues with ropes. Because walking with ropes would have been easier and would have required fewer people, it is the favored explanation.

- The loss of forests on the island certainly relates to humans because the timing of the first arrival of Polynesians coincides with the decline of the island's forests. People cut trees for various uses, such as building boats and shelters. However, rats were also a major factor in tree mortality; the rats ate tree seeds and preyed on bird pollinators the humans brought with them.

Until recently, scientists overlooked the devastating ecological harm that the rats appear to have inflicted on the island. As a result, the Easter Islanders were characterized as foolishly cutting their forests to build and transport statues. The purpose of the statues was assumed to be a form of political competition among rival chiefdoms on the island. But this interpretation does not fit the new evidence. Rather than being seen as reckless stewards of their environment, the Easter Islanders are now seen as victims of the ecological destruction unleashed on the island as the rat population increased uncontrollably and ravaged the birds and forests.

Although we may never know why the large statues were built, many scientists now think their purpose was to honor deceased people of significance in the island's history. This conclusion is based on examining the cultural history of similar but smaller statues on other Polynesian islands in the Pacific Ocean.

Further evidence shows that the islanders adapted to the loss of forests by using a farming technique called *lithic mulching*, in which rocks are piled to create screens against the incessant wind that can damage vulnerable crop seedlings in a treeless environment. Despite the loss of trees, the Easter Islanders managed to live sustainably on the island and maintain a population of perhaps as many as 3,000 until the time of Roggeveen's arrival.

After contact with European explorers, the Easter Island society collapsed to a few hundred due to most of its inhabitants being exported to Peru as slaves and the remainder being exposed to alcohol and Old World diseases such as smallpox and tuberculosis. A summary of the chronology of Easter Island prehistory is presented in **Figure 1.34**.

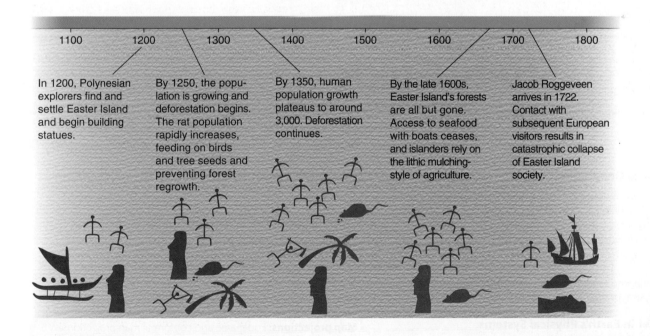

Figure 1.34 Easter Island chronology. Easter Island was colonized approximately in the year 1200, and the island's population grew to some 3,000 people. The culture lived sustainably on the island until European contact in the year 1722.

How Is a Hypothesis Different from a Theory?

A scientific hypothesis is a narrowly defined testable idea, such as "Can a 13-ton statue be moved by people using only log rollers and ropes?" or "Did rats eat the palm seeds?" These statements can be proved or disproved. The word *theory*, on the other hand, is often misused to describe an idea or a belief. In a formal scientific sense, a theory is a broad narrative constructed from many tested and supported hypotheses. Importantly, in science, theories are not beliefs: They are logical statements based on a preponderance of carefully evaluated evidence. Climate change—a long-term change in the weather—for example, is sometimes referred to as "just a theory." This statement is misleading because the theory of climate change is supported by a preponderance of evidence worldwide. Climate scientists nearly universally interpret the evidence the same way: The data show that Earth's climate is rapidly changing (see Chapter 7, "The Changing Climate").

Sometimes, different scientists evaluate the same evidence and come to different conclusions. If this happens, it is usually because the evidence is poor or there isn't enough of it. In time, such conflicts are typically resolved with new evidence.

Successful theories hold up as new evidence becomes available through time, but no theory can ever be "proven." Any theory can fail if new evidence contradicts it. When this occurs, either the theory must be adjusted or it must be discarded wholesale. Scientists are always revising and refining theories as new evidence becomes available. Two scientific theories about Easter Island's human history are "Easter Island society lived sustainably up until contact with the Europeans" and "Deforestation was caused mostly by rats, not people." These theories could change if new evidence contradicts them.

Some theories are called *unifying theories* because they explain a wide range of seemingly unrelated phenomena. The theory of plate tectonics (see Chapter 13, "Drifting Continents: Plate Tectonics"), for example, is a unifying theory that explains a wide range of physical phenomena, such as volcanoes, earthquakes, and mountain building. Likewise, the theory of evolution (see Chapter 8, "Patterns of Life: Biogeography") underpins the medical, biological, and ecological sciences.

In some cases, theories are so strong that no one reasonably thinks that new facts will ever refute them. In such a case, the theory is elevated to a *scientific law*. The law of gravity is an example. If apples or anything else were ever to fall upward, the law of gravity would have to be modified or even abandoned.

Chapter 1 Study Guide

Focus Points

1.1 Welcome to Physical Geography!

- **Geography:** Geography emphasizes spatial relationships to gain insight into cultural and physical phenomena.
- **Physical geography:** Physical geography explores Earth's physical systems and human influences on those systems.

1.2 Scales of Inquiry

- **Scale:** Different spatial and temporal scales provide varied perspectives on physical phenomena.

1.3 Energy and Matter in Earth's Physical Systems

- **Earth systems:** Earth's four major physical systems are the atmosphere, biosphere, lithosphere, and hydrosphere.
- **System feedbacks:** Systems experience positive and negative feedbacks. Positive feedbacks create instability within a system, and negative feedbacks stabilize systems.

- **Solar energy:** Solar energy flows through the atmosphere and biosphere. The Sun evaporates water, which precipitates on land and erodes the land surface as it flows.
- **Geothermal energy:** Geothermal energy builds Earth's surface relief through the process of plate tectonics.

1.4 Mapping Earth: The Science of Spatial Awareness

- **The geographic grid:** The geographic grid, which features both latitude and longitude, is a coordinate system that can be used to locate any point on the planet and define zones of latitude.
- **GPS:** The Global Positioning System provides coordinates for any location on Earth.
- **Map projections:** Equal-area and conformal map projections address the problem of map distortion.
- **Maps and map scale:** Maps portray spatial information efficiently. The degree to which Earth's surface has been reduced on a map is indicated by the map scale.

1.5 Imaging Earth

- **Remote sensing:** Remote sensing provides crucial information about Earth's changing physical systems. Satellite sensors, aerial drones, ground-based Doppler radar, and ship-based sonar are important remote sensing technologies.
- **GIS:** Geographic information systems allow geographers to process and analyze large volumes of data for spatial analysis and cartography.

1.6 Geographic Perspectives: The Scientific Method and Easter Island

- **Scientific method:** The scientific method is an evidence-based procedural framework that improves our understanding of the natural world.
- **Theory:** A theory is constructed from many hypotheses that have been tested and supported by data.

Key Terms

anthropogenic, 4
atmosphere, 8
bar scale (graphic map scale), 22
biosphere, 8
cartography, 18
chemical energy, 8
climate, 11
conformal map projection, 20
contour line, 23
crust, 9
cryosphere, 9
deposition, 13
digital elevation model (DEM), 24
Doppler radar, 26
elevation, 23
energy, 7
equal-area map projection, 20
equator, 14
erosion, 13
geographic grid, 14
geographic information system (GIS), 26
geography, 4
geothermal energy, 8
Global Positioning System (GPS), 18
great circle, 21
hydrosphere, 9
large-scale perspective, 5

latitude, 14
lithosphere, 9
longitude, 14
map, 5
map scale, 22
matter, 7
meridian, 16
negative feedback, 9
parallel, 14
photosynthesis, 8
physical geography, 4
positive feedback, 9
prime meridian, 14
radar, 24
radiant energy, 7
relief, 23
remote sensing, 24
representative fraction scale, 22
small-scale perspective, 5
spatial scale, 5
system, 4
temporal scale, 5
topography, 23
tropics, 14
verbal map scale, 22
water vapor, 9
weather, 11

Concept Review

1.1 Welcome to Physical Geography!

1. Define physical geography.

2. What does *anthropogenic* mean?

1.2 Scales of Inquiry

3. Compare spatial scale and temporal scale. How is each used in geography?

4. Compare a large scale to a small scale. Which makes surface features appear larger on a map?

5. How does a change in scale lead to a change in perspective and in the types of questions that can be asked and answered?

1.3 Energy and Matter in Earth's Physical Systems

6. What is matter? In what three states does matter occur? Give examples.

7. How is the term *energy* defined?

8. What type of energy gives motion to the atmosphere?

9. Give examples of positive and negative system feedbacks. Which type of feedback produces instability in a system?

10. Define the terms *atmosphere, biosphere, lithosphere, hydrosphere,* and *cryosphere.*

12. What types of energy are emphasized in physical geography?

13. Explain this statement: "Life is a physical manifestation of solar energy."

14. Where does the energy that builds up Earth's landscapes come from?

15. Where does the energy that erodes Earth's surface come from?

1.4 Mapping Earth: The Science of Spatial Awareness

16. What is the geographic grid? How are angular measurements used to form it?

17. What are parallels and meridians? Compare them to latitude and longitude. Give examples.

18. What and where are the three major global zones that are based on latitude? Briefly describe each of these zones.

19. What are the two subzones based on latitude?

20. What does GPS stand for? Describe what GPS is used for.

21. What does it mean to *georeference* a map?

22. What is a map? What kinds of phenomena can be mapped? Are all maps printed on paper?

23. What are two major types of map projections? How are they different from each other?

24. Compare a great circle route with a small circle route. Which is used in long-distance flights by aircraft, and why?

25. Compare a large-scale map with a small-scale map. Which would be used to map a single mountain?

26. What are five important elements for maps?

27. What are contour lines? Provide examples of their use.

1.5 Imaging Earth

28. What is remote sensing? Compare passive and active remote sensing.

29. List three examples of remote sensing technologies in wide use today.

30. What is a digital elevation model (DEM)?

31. What is GIS, and what is it used for?

32. What does it mean when we say that a map is *interactive*? Why is interactivity a desirable trait in maps?

1.6 Geographic Perspectives: The Scientific Method and Easter Island

33. What is the scientific method? What is it based on? Do all forms of scientific inquiry follow the scientific method in the same way?

34. What is a scientific hypothesis? Why must a hypothesis be testable, and how is a hypothesis tested?

35. What is a scientific theory? Can one ever be "proved"? Explain.

36. How did pollen data shed light on how Easter Island's statues were moved?

37. What evidence was found in the ancient garbage middens on Easter Island, and what did it indicate?

Critical-Thinking Questions

1. Look around you. If you are indoors, look out a window. How many of the three states of water are visible to you right now? Where is each found?

2. Which of the three types of map scales will be accurate when you view and zoom around on a street map on your smartphone? Why?

3. Many cell-phone and credit-card companies allow authorities to monitor the owner's movement. Do you think this an acceptable use of GPS technology? Explain.

4. Is the following statement a testable scientific hypothesis? "There is no connection between heat waves and the frequency of earthquakes." Explain.

5. Can you think of any parallels between Easter Island and today's global society with regard to anthropogenic changes to the natural environment?

Test Yourself

Take this quiz to test your chapter knowledge.

1. True or false? A world map is an example of a large-scale map.

2. True or false? Water enters the atmosphere through evaporation.

3. True or false? Matter and energy flow through Earth's systems.

4. True or false? Earth's geothermal energy tears down mountains, while solar energy builds them up.

5. True or false? The highest angle of longitude is 90 degrees.

6. Multiple choice: A storm derives its energy from

   a. geothermal energy.   c. solar energy.
   b. chemical energy.   d. electrical energy.

7. Multiple choice: The equator is an example of

   a. latitude.   c. a meridian.
   b. longitude.   d. a small circle.

8. Multiple choice: Which of the following would be useful in determining the rate of rainfall in a storm?

   a. LIDAR   c. Doppler radar
   b. sonar   d. satellite visible image

9. Fill in the blank: _____ specifies the relationship between distance in the real world and distance on a map.

10. Fill in the blank: The term _____ refers to something created for or influenced by people.

Online Geographic Analysis

Evaluating Landscape Change with Satellite Remote Sensing

Satellites have revolutionized our ability to detect and monitor physical changes to Earth's systems. Here we compare and analyze satellite image pairs developed by NASA at different times to reveal stark land use contrasts.

Activity

Go to NASA's Earth Observatory page, at https://earthobservatory.nasa.gov. In the "Search" field, type "Lake Mead." From the results list select "Visualizing the Highs and Lows of Lake Mead." Scroll down and click the "View Image Comparison" button (if this view is not already chosen). Use the slider to compare the satellite image pair and answer the following questions.

   1. What are the dates of the two satellite images?

   2. How did Lake Mead's water level change from the time the first image was taken to the time the second was taken?

   3. What caused this change?

Stay on NASA's Earth Observatory page. In the "Search" field left, type "San Francisco Bay. From Salt Production to Salt Marsh." Scroll down and click the "View Image Comparison" button (if this view is not already chosen). Use the slider to compare the satellite image pair and answer the following questions.

4. **What are the dates of the two satellite images?**

5. **How would you characterize the changes in the urban built environment between the two images? Was there heavy change or little change?**

6. **How would you characterize the changes in the San Francisco Bay shoreline? What human activity is causing these changes?**

Stay on NASA's Earth Observatory page. In the "Search" field, type "China's Pearl River." From the results list select "The World's Largest Urban Area." Scroll down and click the "View Image Comparison" button. Use the slider to compare the satellite image pair and answer the following questions.

7. **What are the dates of the two satellite images?**

8. **What do the white and gray areas represent? What do the green areas represent?**

9. **Compare the geographic extent of gray and white areas between the two images. How have they changed, and what caused the changes?**

10. **According to the information on the web page, what is the claim to fame of China's Pearl River Delta region?**

Picture This. *Your Turn*

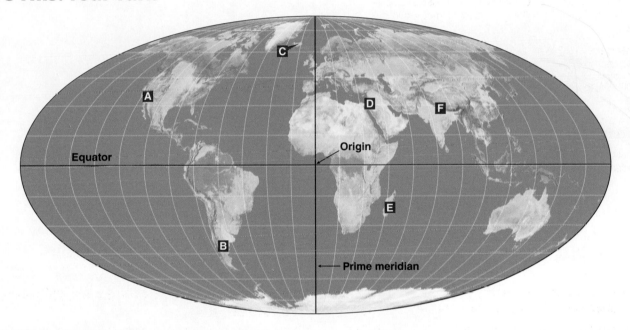

The Geographic Grid

The geographic grid is shown on this world map. The equator and the prime meridian are labeled. Latitudes and longitudes are 15 degrees apart. Using the origin (at 0°, 0°) as a starting point, identify the following locations on the map. (Each of these locations corresponds to a letter on the map.) Only degrees (not minutes and seconds) are given for each location:

The Dead Sea, Israel	31° N, 35° E
The Himalayas, Nepal and Tibet	30° N, 79° E
Iceland	64° N, 18° W
Madagascar	19° S, 46° E
Mount Whitney, California	36° N, 118° W
Patagonia, Argentina	41° S, 69° W

For animations, interactive maps, videos, and more, visit www.saplinglearning.com.    SaplingPlus

The Sun's heating of Earth's surface and water vapor in the atmosphere give rise to gentle breezes and gale-force winds, sunny afternoons and torrential downpours. Part I discusses the role of solar energy and water in atmospheric systems and explores Earth's changing climate system.

CHAPTER 2
Portrait of the Atmosphere

An envelope of gases surrounds Earth, making weather possible and Earth habitable.

CHAPTER 3
Solar Energy and Seasons

Solar energy and seasons create temperature differences and atmospheric motion.

CHAPTER 4
Water in the Atmosphere

Solar energy evaporates water into the atmosphere, forming water vapor and precipitation.

CHAPTER 5
Atmospheric Circulation and Wind Systems

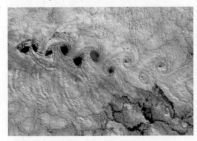

Air pressure differences arising from the Sun's heating drive global wind and precipitation patterns.

CHAPTER 6
The Restless Sky: Severe Weather and Storm Systems

Atmospheric motion organizes into different systems, from local sea breezes to deadly storms.

CHAPTER 7
The Changing Climate

Earth's climate system changes naturally and from human activity.

(Chapter 2: ESA/NASA; Chapter 3: Dean Fikar/Getty Images; Chapter 4: Mike Hollingshead/Alamy; Chapter 5: NASA Earth Observatory images by Joshua Stevens and Jesse Allen, using Landsat data from the U.S. Geological Survey and VIIRS data from the Suomi National Polar-orbiting Partnership; Chapter 6: Willoughby Owen/Getty Images; Chapter 7: Victor Liu, www.victorliuphotography.com.)

Portrait of the Atmosphere

Chapter Outline *and Learning Goals* ◎

2.1 Composition of the Atmosphere

◎ Describe the gases and other materials that make up the atmosphere.

2.2 The Weight of Air: Atmospheric Pressure

◎ Explain what causes air pressure and how air pressure changes vertically within the atmosphere.

2.3 The Layered Atmosphere

◎ Name and describe the atmosphere's layers.

2.4 Air Pollution

◎ Identify major atmospheric pollutant types and discuss their effects on human health.

2.5 Geographic Perspectives: Refrigerators and Life on Earth

◎ Assess the effects of anthropogenic pollutants in the ozonosphere and the anticipated condition of the ozonosphere in the coming decades.

This remarkable photograph of a sunset over the western Pacific Ocean was taken by Alexander Gerst from an altitude of 400 km (249 mi) while aboard the International Space Station. From this height, the "top" of the atmosphere can be seen where the thin envelope of blue air fades to the black of space about 100 km (62 mi) above sea level. In the foreground, towering *cumulonimbus* clouds, remnants of Typhoon Halong, cast long shadows in the setting Sun.

(ESA/NASA)

To learn more about the characteristics of the atmosphere, go to Section 2.1.

THE HUMAN SPHERE China's "Airpocalypse"

In recent decades, China has had the world's fastest-growing economy. A large part of the economic growth is the result of manufacturing and exporting goods to other countries. But growth has come at a steep a price: Much of eastern China now experiences some of the world's most polluted air and water. Industrial factories, vehicular emissions, coal-burning power plants, and loose environmental regulations all contribute to fouling the air and water. Today, urban centers in this area routinely have some of the world's most hazardous air.

In Beijing in January 2013, in an event the media termed "airpocalypse," the air quality shot off the charts on the international *Air Quality Index* (*AQI*) scale (see page 54)—to a level 30 times higher than the maximum level considered safe by the World Health Organization (WHO). On this scale, which ranges from 0 to 500, anything above 300 is considered too hazardous to safely breathe; in January 2013, Beijing's rating was 755. In November 2015, officials in the nearby city of Shenyang recorded an AQI reading of 1,400. In households in many large Chinese cities, indoor air purifiers are as common as washing machines and refrigerators. As heavy haze smothers cities during the almost routine orange and red air quality alerts, schools are closed, flights are grounded, and most people stay indoors **(Figure 2.1)**.

Fortunately, China's air quality is gradually improving. China has declared a "war on pollution," and officials have begun restricting driving and cracking down on polluting factories, forcing them to clean up or face closure. At the same time, China is becoming a world leader in clean energy usage. Its transition from dirty fossil fuels, such as coal, to clean, pollution-free renewable energies, such as wind and solar, is happening faster and on a larger scale than anywhere else in the world.

A

B

Figure 2.1 Beijing air pollution. (A) The photo on the left was taken from an apartment window on a clear day. The photo on the right is the same view on a polluted day. (B) To raise awareness of Beijing's unhealthy air, a Chinese artist, nicknamed "Brother Nut," vacuumed the air for 100 consecutive days. During that time he collected enough *particulate matter* pollution to make a brick. (*A. Imaginechina via AP Images; B. VCG/Getty Images*)

2.1 Composition of the Atmosphere

◎ **Describe the gases and other materials that make up the atmosphere.**

The **atmosphere** is the envelope of gases that surrounds Earth. The atmosphere (which we also refer to simply as *air*) is an important part of Earth's life-support systems. It provides us with oxygen. It shields us from cancer-causing ultraviolet (UV) radiation. Without the atmosphere, Earth would be as sterile as the Moon. All of Earth's water would have boiled away to space long ago. It warms the surface by about 15°C (59°F) and also prevents extremes of day and night temperature swings.

Compared to the size of the planet, the atmosphere is quite thin (see the chapter-opening photo).

If Earth were the size of a basketball, most of the atmosphere would be only about the thickness of a single piece of paper. If you were to travel straight up from Earth's surface, how far would you have to travel to get to the end of the atmosphere and the beginning of space? This question is not easy to answer because the atmosphere does not end abruptly but instead gradually fades. Half the mass of the atmosphere is located between Earth's surface and 5.6 km (3.5 mi) altitude, and by about 190,000 km (120,000 mi) above Earth's surface—about halfway to the Moon—the atmosphere is all but gone. One widely accepted definition of the "top" of the atmosphere is 100 km (62 mi), an altitude known as the **Karman line (Figure 2.2)**.

Gases in the Atmosphere

The different gases that make up the atmosphere are not evenly mixed throughout. The atmosphere can be divided into two major "layers" based on its chemistry: the *heterosphere* and the *homosphere*. At altitudes above 80 km (50 mi), in a region of the atmosphere called the heterosphere, there are a variety of atomically lightweight gases that are not well mixed. The lightest gases, hydrogen and helium, are concentrated in the uppermost reaches of the heterosphere, while the heaviest gases, nitrogen and oxygen, are concentrated at lower altitudes within the heterosphere. In contrast, the homosphere occurs at altitudes below 80 km (50 mi). Gases are well mixed in the homosphere, and, regardless of the altitude, the proportions of different gases are roughly equal.

There are, however, some important exceptions where certain gases occur in uneven concentrations within the homosphere. Water vapor, for example, is concentrated near Earth's surface, particularly over the warm, tropical oceans from which water readily evaporates. Likewise, some pollutants are concentrated at the surface, such as nitrogen dioxide (NO_2) in urban areas and carbon monoxide (CO) and carbon dioxide (CO_2) where fires are burning. Later in this chapter we explore another way to divide the atmosphere into layers, based on temperature changes.

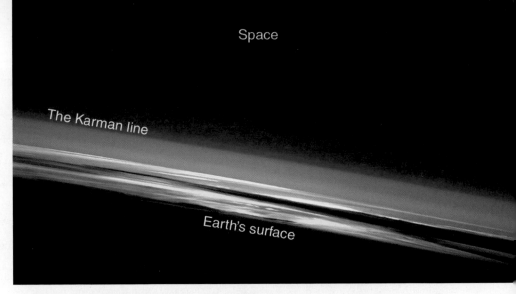

Figure 2.2 The "top" of the atmosphere. In this photo from the International Space Station, the edge of the atmosphere, or *limb*, can be seen. Over 99.99% of the atmosphere's mass lies below the Karman line at 100 km (62 mi) altitude. Above the Karman line, the atmosphere is extremely thin, and there are no longer enough air molecules to provide lift to aircraft; instead, vehicles must orbit the planet to remain aloft. *(NASA Earth Observatory)*

Permanent and Variable Gases

The gases in the atmosphere reflect and are a result of the dynamic interactions among Earth's physical systems. Some gases change in their proportions, and some of them don't. There are two groups of gases in the atmosphere: *permanent gases* and *variable gases* (also referred to as *trace gases*). Permanent gases are called "permanent" because their proportions change only a little. In contrast, the proportions of variable gases do change. Nearly all of the atmosphere is composed of only three permanent gases: nitrogen (N_2), oxygen (O_2), and argon (Ar). **Table 2.1** shows their proportions in the atmosphere as they occur in dry and unpolluted air and lists other permanent gases that occur in minute amounts.

Variable gases, in contrast, exist in extremely small quantities and do change in their proportions. There are many variable gases, but only those, called **greenhouse gases**, that absorb and emit heat (or *thermal infrared*) energy are included in this discussion. **Table 2.2** lists six

Table 2.1	Permanent Gases
PERMANENT GAS	**PERCENTAGE BY VOLUME**
Nitrogen (N)	78.08
Oxygen (O)	20.95
Argon (Ar)	0.93
Neon (Ne)	0.0018
Helium (He)	0.0005
Hydrogen (H)	0.00005
Xenon (Xe)	0.000009

Table 2.2	Variable Greenhouse Gases	
VARIABLE GAS	**PERCENTAGE BY VOLUME**	**PPM***
Water vapor (H_2O)	0 to 4	
Carbon dioxide (CO_2)	0.0408	408
Methane (CH_4)	0.00018	1.8
Nitrous oxide (N_2O)	variable	variable
Ozone (O_3)	variable	variable
CFCs and HFCs**	variable	variable

*ppm = parts per million. For example, 408 ppm carbon dioxide means that for every 1 million molecules of air, 408 are carbon dioxide molecules.

**Chlorofluorocarbons (CFCs) and hydrofluorocarbons (HFCs). CFCs and HFCs are manufactured by people and do not form naturally.

Table 2.3	Sources and Sinks for Atmospheric Gases	
GAS	**SOURCE(S)**	**SINK(S)**
Nitrogen (N_2)	Volcanic eruptions, decaying and burning organic matter, weathering of rocks	Rain, biological activity
Oxygen (O_2)	Photosynthesis	Decay of organic matter, weathering of rocks
Water vapor (H_2O)	Evaporation from oceans, photosynthesis, volcanic eruptions	*Condensation* and *deposition*, which convert H_2O to liquid and solid (ice) states
Carbon dioxide (CO_2)	Volcanic eruptions, decay of organic matter, respiration, burning of fossil fuels (coal, oil, and natural gas)	Photosynthesis, oceans, chemical reactions
Methane (CH_4)	Anaerobic (oxygen-free) bacterial decomposition	Absorption by ultraviolet radiation
Nitrous oxide (N_2O)	Soil bacterial processes, agricultural and industrial activities, burning fossil fuels	Absorption by ultraviolet radiation
Ozone (O_3)	Ultraviolet radiation, burning fossil fuels	Absorption and breakdown by ultraviolet radiation
CFCs and HFCs	Anthropogenic	Absorption and breakdown by ultraviolet radiation

prominent greenhouse gases and their current concentrations in the atmosphere. As we will see in Section 3.3, greenhouse gases are responsible for Earth's natural *greenhouse effect*. Without naturally occurring greenhouse gases, Earth's average atmospheric temperature would be approximately –18°C (0°F) compared to the much warmer 15°C (59°F) that it is now. Human activities, such as burning fossil fuels and deforestation, are increasing greenhouse gas concentrations in the atmosphere, creating the *anthropogenic greenhouse effect* and climate change. Chapter 7, "The Changing Climate," is devoted to this important topic.

Gas Sources and Sinks

The types of gases found in the atmosphere are extremely important. When we look at other planets in the solar system, we see worlds with radically different atmospheric chemistries and, as a result, radically different physical conditions. The proportions of gases in Earth's atmosphere are just right for supporting life on Earth.

Human activity is changing Earth's chemistry by modifying the sources and sinks of gases in the atmosphere. The terms *source* and *sink* describe how gases enter and exit the atmosphere or how they are changed into different gases. Like water that enters a bathtub from the faucet (the *source*) and exits down the drain (the *sink*), gases enter the atmosphere through a source, and they leave the atmosphere through a sink. Sources and sinks can be either physical environments (such as a lake or wetland) or processes (such as a chemical reaction) or both. Many gas sources are **anthropogenic**—created or influenced by people **(Table 2.3)**.

Aerosols in the Atmosphere

Aerosols, which are also called *particulate matter*, are another important ingredient of the atmosphere. **Aerosols** are microscopic solid or liquid particles suspended in the atmosphere. They are not gases because they are made up of matter in a solid or liquid, rather than a gaseous state. Clouds, for example, are composed of aerosols; they are made up of cloud droplets and tiny ice crystals. **Cloud droplets** are microscopic drops of liquid water found in clouds.

Video
Atmospheric Aerosols
Available at
www.saplinglearning.com

Other common aerosols in the atmosphere include windblown dust, pollen, spores, bacteria, emissions produced by human activities such as farming and industrial pollutants, salt particles from the oceans, volcanic ash, and smoke. Aerosols are concentrated near the ground in the areas where they are produced, and when concentrations are high, as often happens in urban areas, they can be a health hazard. Aerosols are discussed further in Section 2.4, in the context of air pollution. Section 4.6 discusses the crucial role aerosols play in cloud and precipitation processes.

2.2 The Weight of Air: Atmospheric Pressure

◎ **Explain what causes air pressure and how air pressure changes vertically within the atmosphere.**

Although it might not seem to be the case at first glance, the atmosphere and the oceans have much in common. The gaseous atmosphere and the liquid

Figure 2.3 Air pressure. (A) The weight of a column of water is greater above a person at the bottom of the deep end of a swimming pool than at the bottom of the shallow end. The water pressure is greater in the deep end than in the shallow end. (B) Similarly, the depth of the atmosphere and, therefore, the weight of a column of air are greater for a person at sea level than for a person in the mountains.

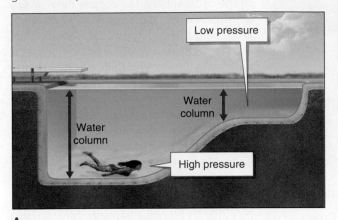

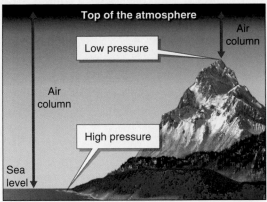

A

B

ocean are both *fluids*, which means they flow. The molecules in the oceans are mostly water and are more closely packed together in a relatively dense liquid form. In contrast, the molecules in the atmosphere, which are mostly nitrogen and oxygen, are farther apart from one another in their less dense, gaseous state. Furthermore, just as water flows in streams, air flows in currents in the atmosphere, creating wind. (See Section 5.2 for a more detailed discussion of air pressure and wind.)

Air Pressure

Wind is the result of differences in air pressure across a geographic region. **Air pressure**, which is the force exerted by molecules of air against a surface, is created by the weight of the atmosphere. *Weight* is defined as the mass of an object multiplied by the acceleration (strength) of gravity. Because it has mass and, consequently, weight, gravity keeps the molecules in the atmosphere pinned to the planet. The total weight of the atmosphere is estimated to be 5×10^{15}, or 5 quadrillion metric tons! It is remarkable that we live beneath the crushing weight of the atmosphere, but it is undetectable to us under most circumstances.

Molecules in a gaseous state move quickly. They constantly collide with and bounce off one another. At sea level, a single air molecule collides with some 10 billion other air molecules every second. It also collides with any other object it comes in contact with, such as trees, the ground, birds, water, and people. Each time a molecule in the air hits any surface, it pushes ever so slightly against it. The collective force (push) of air molecules creates pressure. Except for a few locations below sea level (such as the Dead Sea, where the elevation is −427 m [−1,401 ft]), air pressure is greatest at sea level, and air pressure always decreases as altitude increases.

To illustrate this point, **Figure 2.3** makes the comparison between pressure in a swimming pool and pressure in the atmosphere.

Air Density and Pressure

Because molecules in a gas are far apart, their density can increase as they are compressed. Air is compressed when it is forced into a bicycle or car tire inner tube, for example. In a similar way, the molecular density of the atmosphere increases as it is compressed by its own weight. Increased air density contributes to air pressure because in denser air, there are more molecules to exert force against objects, creating air pressure. This increased air density is the result of the weight of the overlying mass of air. This idea is illustrated in **Figure 2.4** on the following page.

Measuring Air Pressure

One of the essential elements of science is measurement. Measurement provides *quantitative* (numeric) information about phenomena. One way to express air pressure quantitatively is in kilograms per square centimeter (kg/cm^2) or in pounds per square inch (lb/in^2, or *PSI*). At sea level, there is approximately 1 kg pressing on every square centimeter of all objects (or 14.7 lb pressing on every square inch). In other words, a column of air that is 1 cm^2 in area and extends from sea level all the way up to the top of the atmosphere would, on average, weigh 1 kg (or 14.7 lb for a 1 in^2 column of air). (See Section 5.1 for a detailed discussion of how air pressure is measured and the relationship between air pressure and weather systems.)

The atmosphere weighs down on our bodies, so why are we not crushed by its weight, as we would be crushed beneath the weight of a large boulder? The answer is that a large boulder is a solid and

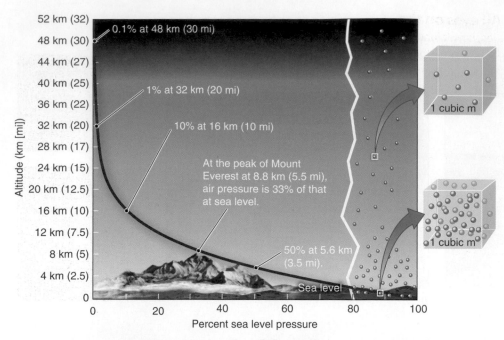

Figure 2.4 Air density, pressure, and height. Air pressure decreases exponentially with altitude. Air density and pressure are greatest near sea level, where air molecules are most tightly squeezed together. At higher altitudes, air molecules are less compressed, so air has lower density and lower pressure. If a 1-cubic-meter (1 m³) box (inset diagrams) were filled with air at sea level and another one were filled with air higher in the atmosphere, the air in the sea-level box would have a far higher molecular density and therefore greater air pressure. At 5.6 km (3.5 mi), air pressure is 50% of that at sea level, and atop Mount Everest, the highest land elevation, it is only 33% of that at sea level.

Animation

Air Density, Pressure, and Height

Available at
www.saplinglearning.com

▶

would press against us only from above. Air and water, on the other hand, are fluids. These fluids flow around (and inside) our bodies. As a result, the atmosphere (and water, when we are swimming) pushes down, sideways, and up against our bodies. We experience the same pressure from all angles simultaneously. The cells in our bodies exert outward pressure in balance with the inward pressure exerted by the atmosphere, and as a result, we normally cannot feel the pressure of the atmosphere.

The pressure between our inner ears and the atmosphere is normally in equilibrium. When the outside pressure changes as we drive up into mountainous areas or fly, this equilibrium is temporarily disrupted until our ears adjust to the change of pressure. These pressure differences push on the eardrum and can be quite uncomfortable.

2.3 The Layered Atmosphere

◎ **Name and describe the atmosphere's layers.**

To understand how the atmosphere works and why it changes, it is important to first understand the chemical composition of the atmosphere and what creates air pressure. Another major element of the atmosphere is its layered physical structure.

As we will see throughout this book, many of Earth's physical systems are composed of layers. Earth's internal structure, from the core to the outer crust, is made up of layers. In the hydrosphere, the oceans have layers distinguished by light penetration, temperature, and life. In tropical rainforests, vegetation forms many different canopy and understory layers. We learned earlier in this chapter that two layers of the atmosphere are defined by chemistry: the heterosphere and the homosphere. In the context of meteorology, a more important and useful way to define layers of the atmosphere is by differences of temperature.

Scientists use a variety of tools to study the structure of the atmosphere and to understand the physical systems that drive Earth's dynamic atmosphere. The oceans cover 71% of Earth's surface and interact with the atmosphere in important ways. Therefore, studies of the atmosphere require that the oceans be studied as well. Satellites, airplanes, **radiosondes** (unstaffed weather balloons), ships, and ocean *buoys* are all examples of tools scientists use to monitor Earth's atmosphere and oceans, as shown in **Figure 2.5**.

Atmospheric Layers Based on Temperature

If you have ever been high in the mountains, you probably noticed that the air at higher elevations is much cooler than air at lower elevations. But if you were able to rise high above Earth's surface, you might be surprised to find that around 80 km (50 mi) altitude, the temperature begins rising again and eventually reaches over 2,000°C (3,630°F) around 600 km (370 mi) above Earth's surface!

Four main *thermal divisions* of the atmosphere are based on temperature changes and altitude: The **troposphere** extends from Earth's surface up to about 12 km (7.5 mi), on average. It is the lowest layer of the atmosphere, and it is the layer in which all weather occurs. The topmost boundary of the troposphere is the **tropopause**. The next layer up, the **stratosphere**, is between 12 and 50 km (7.5 and 30 mi) above Earth's surface. The stratosphere contains a permanent **temperature inversion** because it is heated as it absorbs radiation from the Sun. A temperature inversion is a layer of the atmosphere in which air temperature increases with increased height (see also Section 2.4). Above the stratosphere, the **mesosphere** lies between 50 and 80 km (30 and 50 mi) above Earth's surface. The **thermosphere** is located from 80 to about 600 km (50 to 370 mi) above Earth's surface. **Figure 2.6** on page 46 illustrates these divisions and provides more detail about each.

Figure 2.5 SCIENTIFIC INQUIRY: All eyes on the sky. Understanding how the atmosphere is structured, how it functions, and how it is changing, as well as how weather can be accurately forecast, requires careful and precise measurements. Scientists use the most sophisticated instrumentation and computer analysis available today to better understand the atmosphere and forecast its behavior.

(NASA/Tim Marvel; inset balloon. NASA)

Radiosondes
Instrument packages on unstaffed weather balloons, called *radiosondes*, are used to measure conditions in the upper atmosphere. Worldwide, about 900 stations release radiosondes. About 75,000 are launched each year in the United States, usually twice per day from about 100 cities. They record altitude, geographic location, temperature, pressure, humidity, and wind speed and direction.

Satellites
Satellites provide data on changing conditions in Earth's atmosphere. They carry state-of-the-art instruments that measure a range of atmospheric phenomena, including temperature, precipitation, humidity, clouds, aerosols, airflow, and pollution.

Aircraft
Aircraft carry a variety of instruments used to study the atmosphere. Here pulsed laser light called LIDAR is used to monitor atmospheric aerosols, such as clouds and pollution, as well as photosynthetic microbial life near the ocean's surface.

Ships and buoys
Ships and moored buoys are essential in monitoring the changing ocean conditions that, in turn, affect the atmosphere. Most importantly, water temperature at the surface and at depth, water chemistry, and ocean currents are closely monitored.

The Weather Layer: The Troposphere

The troposphere is warmed from the bottom up by Earth's surface, which is warmed by sunlight. Because of this, higher in the troposphere, the air's temperature almost always decreases. Another reason it gets colder with altitude is that higher up, the air is less dense (thinner), and therefore there are fewer molecules to absorb and radiate heat energy, either from Earth's surface or directly from the Sun. The **environmental lapse rate** is the rate of cooling with increasing altitude in the troposphere. The average environmental lapse rate is 6.5°C per 1,000 m or 3.6°F per 1,000 ft (see Figure 2.6). This means that if you rise 1,000 m, you experience a temperature drop of 6.5°C. Keep in mind that this is an average lapse rate, and the atmosphere may cool more quickly or slowly, depending on the time and the location.

Although it normally gets colder higher up in the troposphere, this is not always the case. Under the right conditions in the lower troposphere, temperature inversions may develop. They often form in valleys, when a cold, dense mass of air forms close to the ground. Above this cold air mass, the temperature increases. Temperature inversions in the troposphere are usually no more than a few hundred meters thick, and a little higher up, the normal cooling trend resumes.

Temperature inversions in the atmosphere are significant in several ways. The temperature inversion in the stratosphere limits almost all clouds to the troposphere, as we'll see later in this section. Temperature inversions in the troposphere can trap pollutants near the ground surface (see Section 2.4). They also can create sleet and freezing rain during winter storms.

Roughly 80% of the mass (or the number of molecules) of the atmosphere is in the troposphere, and it is where Earth's weather occurs. All storms and precipitation, as well as almost all clouds, are limited to the troposphere. The troposphere experiences strong vertical mixing (or *convection*), which is upward movement of heated air. The troposphere ends, and the stratosphere begins, at the tropopause. Almost all clouds end at the tropopause,

Layers based on
temperature trends

Layers that absorb
harmful solar emissions

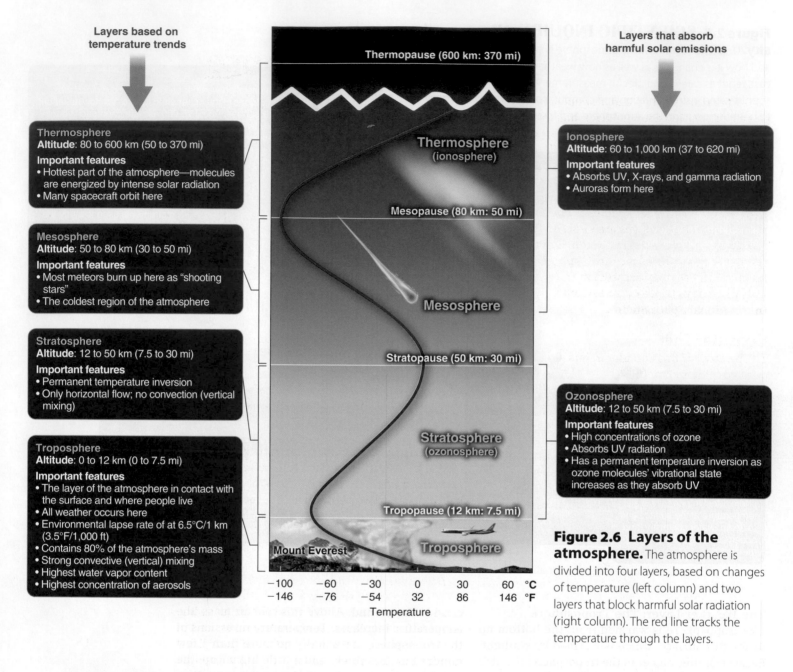

**Layers based on
temperature trends**

Thermosphere
Altitude: 80 to 600 km (50 to 370 mi)
Important features
• Hottest part of the atmosphere—molecules
  are energized by intense solar radiation
• Many spacecraft orbit here

Mesosphere
Altitude: 50 to 80 km (30 to 50 mi)
Important features
• Most meteors burn up here as "shooting
  stars"
• The coldest region of the atmosphere

Stratosphere
Altitude: 12 to 50 km (7.5 to 30 mi)
Important features
• Permanent temperature inversion
• Only horizontal flow; no convection (vertical
  mixing)

Troposphere
Altitude: 0 to 12 km (0 to 7.5 mi)
Important features
• The layer of the atmosphere in contact with
  the surface and where people live
• All weather occurs here
• Environmental lapse rate of at 6.5°C/1 km
  (3.5°F/1,000 ft)
• Contains 80% of the atmosphere's mass
• Strong convective (vertical) mixing
• Highest water vapor content
• Highest concentration of aerosols

**Layers that absorb
harmful solar emissions**

Ionosphere
Altitude: 60 to 1,000 km (37 to 620 mi)
Important features
• Absorbs UV, X-rays, and gamma radiation
• Auroras form here

Ozonosphere
Altitude: 12 to 50 km (7.5 to 30 mi)
Important features
• High concentrations of ozone
• Absorbs UV radiation
• Has a permanent temperature inversion as
  ozone molecules' vibrational state
  increases as they absorb UV

Thermopause (600 km: 370 mi)

Thermosphere
(ionosphere)

Mesopause (80 km: 50 mi)

Mesosphere

Stratopause (50 km: 30 mi)

Stratosphere
(ozonosphere)

Tropopause (12 km: 7.5 mi)

Mount Everest   **Troposphere**

| −100 | −60 | −30 | 0 | 30 | 60 | °C |
| −146 | −76 | −54 | 32 | 86 | 146 | °F |

Temperature

**Figure 2.6 Layers of the
atmosphere.** The atmosphere is
divided into four layers, based on changes
of temperature (left column) and two
layers that block harmful solar radiation
(right column). The red line tracks the
temperature through the layers.

where the air begins to warm due to the temperature inversion in the stratosphere. The temperature
inversion prevents clouds from rising, as discussed
in the **Picture This** feature on the following page.

As we have noted, on average the height of
the tropopause is 12 km (7.5 mi). But the height
of the tropopause varies depending on latitude. At
the equator, the tropopause is 18 km (11 mi) above
Earth's surface. At midlatitudes, it is about 12 km
(7.5 mi) above Earth's surface, and in the polar regions
it occurs at about 8 km (5 mi) in altitude **(Figure 2.7)**.

Why is the tropopause roughly twice as high
over the equator as over the poles? There are two
reasons. First, solar heating of the atmosphere
expands air and causes it to occupy more space.
Heating is greatest near the equator, in the tropics. Furthermore, the height of the tropopause
increases in summer and decreases in winter due

to changing temperatures in the atmosphere. In
addition, Earth rotates faster at the equator than
at higher latitudes because the circumference of
Earth is greatest near the equator (see Section 1.4).
Therefore, Earth spins faster near the equator compared to higher latitudes and produces a greater
centrifugal force near the equator. As a result, the
atmosphere near the equator bulges outward more
than that at higher latitudes, increasing its depth.

A Protective Shield: The Stratosphere

Above the tropopause is the stratosphere (see
Figure 2.6). As a result of its permanent temperature
inversion, there is little vertical mixing in the stratosphere—air flows in horizontal sheets instead—and
storms and clouds usually do not enter the stratosphere from below. The top of the stratosphere, the
stratopause, occurs at about 50 km (30 mi) in altitude.

Picture This The Tropopause

This towering cloud (called a *cumulonimbus*) near Lake Victoria in eastern Africa has grown from Earth's surface up to the base of the stratosphere and protrudes into it, producing a phenomenon called an *overshooting top*. Most of the cloud has not entered the stratosphere. Instead, it is splaying out horizontally at the tropopause, creating a flattened *anvil head* structure. The cloud cannot rise into the stratosphere because the cloud's relatively cold air is denser and heavier than the relatively warm air of the stratosphere. The land surface is masked by aerosols in the atmosphere. This photo was taken by an astronaut aboard the International Space Station.

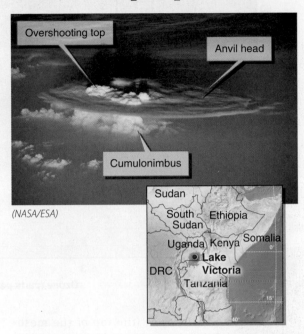

(NASA/ESA)

Consider This

1. The anvil of this cumulonimbus cloud occurs at what boundary line?
2. Given the latitude of this cloud, what is the altitude of the anvil structure?

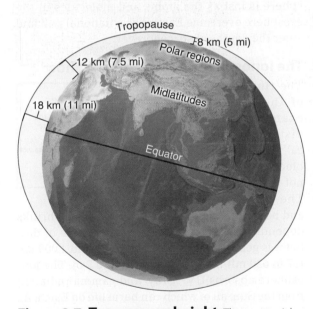

Figure 2.7 Tropopause height. The equatorial tropopause is located 10 km (6 mi) higher than the polar tropopause. Therefore, cloud tops in the tropics are potentially twice as high as cloud tops in polar latitudes. *(Image courtesy of the Earth Science and Remote Sensing Unit, NASA Johnson Space Center)*

There are very few clouds in the stratosphere, and the air is mostly free of atmospheric aerosols.

The Sun emits many *wavelengths* of light, including *gamma rays, X-rays, ultraviolet (UV), visible,* and *infrared.* (Each of these is discussed in more detail in Section 3.2.) The temperature inversion in the stratosphere is caused by the absorption of ultraviolet radiation by stratospheric ozone in the ozonosphere. **Ultraviolet (UV) radiation** is solar radiation that has shorter wavelengths than visible light and can cause skin cancer. The **ozonosphere** is a region of the stratosphere with high concentrations of ozone molecules, which absorb UV radiation and protect people on the ground **(Figure 2.8** on the following page**)**.

There are three wavelengths of UV radiation. UV-C occurs in the shortest wavelengths, and UV-A has the longest wavelengths; UV-B lies in the middle (see Section 3.2). The ozonosphere absorbs most of the UV-B and all of the UV-C. About 95% of all UV radiation at Earth's surface is UV-A. The strength of UV-A depends on elevation, latitude, air quality, and other factors. UV radiation causes sunburns, premature skin aging, and skin cancer, among other things. The effects of UV radiation and the critical role of the ozonosphere are explored further in the Geographic Perspectives: Refrigerators and Life on Earth at the end of this chapter.

Thin Air: The Mesosphere and Thermosphere
The mesosphere lies above the stratosphere. Although air density (as well as pressure) in the mesosphere is extremely low, *meteors* (space debris that enters the atmosphere; also known as *shooting stars*) heat up and vaporize as they encounter friction with the air molecules in the mesosphere. As in the troposphere, temperature decreases with height in the mesosphere, until it reaches about −90°C (−130°F) approximately 80 km (50 mi) above Earth's surface (see Figure 2.6).

**Figure 2.8
Stratospheric
ozone.** About 90% of all
ozone in the atmosphere
resides in the strato-
sphere. Ozone concen-
trations peak at about
35 km (22 mi) altitude.
Stratospheric ozone is
naturally produced, but
the slight increase of
ozone near the ground is
due to human activity.

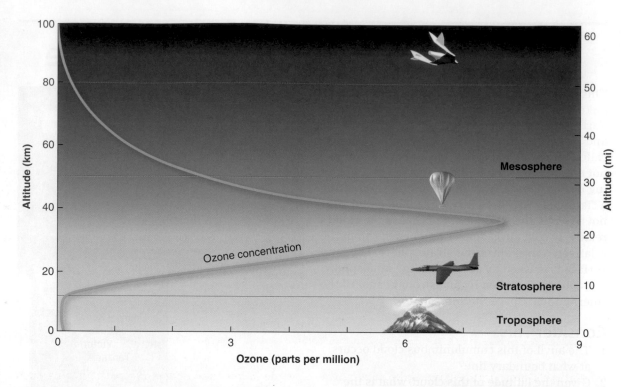

Above the *mesopause* (the top of the mesosphere) lies the thermosphere. As in the stratosphere, there is a permanent temperature inversion
in the thermosphere, caused by absorption of the
Sun's radiation by atmospheric gases. Temperatures
can reach 2,000°C (3,630°F) or even higher, depending on time of day and the intensity of solar activity.
What would the thermosphere feel like to a person?
The environment would feel extremely cold because
there is almost nothing there; the air is extremely
thin and the air molecules are exceedingly far
apart. As a result, even though the temperature of
the molecules is very high, there are so few of them
that very little heat would be transferred to a person's body.

The top of the thermosphere (the *thermopause*) ranges between 600 and 1,000 km (370 and
620 mi) above Earth's surface. Its altitude above
Earth's surface varies depending on the strength
of solar energy. When the Sun is in an active
sunspot cycle, high-energy X-rays and UV rays
heat the thermosphere, causing it to expand and
increase in thickness. Here, the air is so thin that
atoms and molecules travel some 10 km (6 mi), on
average, before colliding with one another. In the
troposphere, where molecular density is much
greater, molecules move only a millionth of a centimeter, on average, before colliding.

Above the thermopause lies a tenuous veil of
mainly hydrogen, helium, and oxygen atoms in
a region called the *exosphere*, which gradually
transitions to space some 190,000 km (120,000
mi) above Earth's surface. Most satellites orbit in
the thermosphere, well below the exosphere. For
this reason, not all scientists even agree that the
exosphere should be considered part of Earth's

atmosphere; some think it should instead be considered space. A small portion of Earth's atmosphere is lost as the atoms and molecules in the
exosphere overcome Earth's gravitational pull and
enter the depths of space forever.

The Ionosphere: Nature's Light Show

The final atmospheric region
of the atmosphere we will
examine, the ionosphere, is
distinguished by its electrical properties and its ability to
block harmful solar radiation,
not its temperature properties.
The **ionosphere** extends above
and below the thermosphere. It grows and shrinks
depending on solar activity and the time of day,
but it is generally located between 60 and 1,000 km
(37 to 620 mi) in altitude (see Figure 2.6). The ionosphere absorbs UV, X-ray, and gamma radiation
from the Sun, all of which can harm life on Earth. As
it absorbs this energy, nitrogen and oxygen become
ionized. Ionization is the process by which atoms and
molecules gain or lose electrons and become positively or negatively charged.

The ionosphere reflects certain radio waves
transmitted from Earth's surface, such as AM
radio and shortwave radio waves. Although radio
waves travel in straight lines only, they can travel
very long distances following the curved surface
of the planet because they reflect between the
base of the ionosphere and Earth's surface. Before
satellite communications, intercontinental radio
transmissions bouncing off the base of the ionosphere were the most important means of rapid
long-distance communication.

Video

*Exploring the
Ionosphere*

Available at
www.saplinglearning.com

Figure 2.9 Auroras. (A) The aurora borealis is best viewed from northern locations, such as near Sudbury, Ontario, Canada, where this display was photographed. (B) This photo, from the orbiting International Space Station, shows the aurora australis over the southern Indian Ocean. The solar panels and the body of the space station are visible in the upper portion of the photo. *(A. Mike Grandmaison/Getty Images; B. NASA Earth Observatory)*

Nature's most impressive light shows, the **aurora borealis** (northern lights) and the **aurora australis** (southern lights), occur in the ionosphere. Auroras form as fast-moving electrons discharged from the Sun (called the *solar wind*) collide with oxygen and nitrogen in the ionosphere. The electrons transfer their energy to the nitrogen and oxygen. When the energized molecules return to their normal state, they emit photons of light, creating beautiful displays. The colors of the aurora depend on the types of gases that are energized by the solar wind. Oxygen emits greenish-yellow or red, and nitrogen emits blue light.

Aurora displays coincide with solar flare activity on the Sun. Although auroras are occasionally seen at latitudes as low as the southern United States, the best place to see them is at high latitudes during the dark winter months **(Figure 2.9)**. Auroras are common near the poles because that is where Earth's magnetic field concentrates the solar wind's particles.

2.4 Air Pollution

◎ **Identify major atmospheric pollutant types and discuss their effects on human health.**

Each of us takes on average about 20,000 breaths each day, and Earth's atmosphere and the air we breathe are therefore incredibly important. The World Health Organization (WHO) estimates that 92% of the world's population breathes potentially unsafe air, and roughly 7 million people each year die prematurely because of air pollution. The majority of the fatalities are children and elderly people in *low-* and *middle-income* countries, such as China and India.

Air pollution is any harmful concentration of gases or aerosols in the atmosphere. Air pollution is often called *smog*. Although this term is a combination of *smoke* and *fog*, fog has nothing to do with air pollution.

Atmospheric Pollutants

An **air pollutant** is any airborne substance that occurs in concentrations high enough to endanger human health. Air pollutants, such as carbon monoxide, ozone, sulfur compounds, and carbon dioxide, originate from many different sources—natural as well as anthropogenic. Wildfires, for example, produce carbon monoxide, a toxic gas, and volcanoes release sulfur compounds. **Carbon dioxide (CO_2)** is a colorless, odorless gas also produced by the burning of fossil fuels and animal respiration. The U.S. Environmental Protection Agency (EPA) classifies carbon dioxide as a dangerous pollutant because it indirectly threatens human health. Although it is harmless when inhaled in low concentrations, it is a greenhouse gas that retains heat in the atmosphere and causes harmful climate change. Because CO_2 is such an important pollutant in the context of climate change, it is discussed in greater detail in Chapter 7, "The Changing Climate."

Most pollutants form as a byproduct of the burning of fossil fuels. **Fossil fuels** are the ancient remains of plants preserved in the lithosphere in the form of coal, oil, and natural gas. This section of the chapter focuses on the major anthropogenic pollutants that pose the greatest threat to human health: carbon monoxide (CO), nitrogen dioxide (NO_2), ground-level ozone (O_3), sulfur dioxide (SO_2), particulate matter (PM), and volatile organic

Table 2.4 Anthropogenic Pollutants and Sources

POLLUTANT	MAJOR SOURCES	HEALTH EFFECTS
Carbon monoxide (CO)	Motor vehicles	Headaches, slowed reflexes, drowsiness, death
Nitrogen dioxide (NO_2)	Motor vehicles, coal burning	Lung irritation, pulmonary disease (e.g., asthma, bronchitis)
Ground-level ozone (O_3)	Chemical reactions of NO_2 and VOCs in sunlight	Eye and throat irritation, respiratory disease (e.g., bronchitis and asthma), decreased crop and other plant growth
Sulfur dioxide (SO_2)	Coal burning, oil refineries	Respiratory symptoms (e.g., wheezing, shortness of breath), acid rain ecosystem damage
Particulate matter (PM)	Coal combustion, industrial processes, motor vehicles	Respiratory symptoms (e.g., cough, chest pain, difficulty breathing), pulmonary and cardiovascular disease
Volatile organic compounds (VOCs)	Motor vehicles, industry	Eye and skin irritation, nausea, headaches, damage to the liver, kidneys, and central nervous system

compounds (VOCs). **Table 2.4** provides a summary of each of the pollutants covered in this section.

There are two main categories of anthropogenic air pollutants: primary pollutants and secondary pollutants. A **primary pollutant**, such as carbon monoxide, enters the air or water directly from its source, like a car's tailpipe or a factory smokestack. A **secondary pollutant**, such as ground-level ozone, is not emitted directly from a source but instead forms through chemical reactions among primary pollutants.

Carbon Monoxide

Carbon monoxide (CO) is a toxic, odorless, and invisible gas. There are many natural sources of CO, including volcanic eruptions, forest fires, and bacterial processes as well as natural chemical reactions in the troposphere. These natural emissions far exceed anthropogenic emissions, but, in contrast to anthropogenic emissions, natural emissions are not concentrated locally. Instead, natural emissions form low *background* concentrations that are mixed and dispersed in the troposphere. As a result, natural emissions present little threat to people.

Anthropogenic carbon monoxide is formed mostly by the incomplete burning of carbon fuels (such as wood, natural gas, gasoline, or coal). In urban areas in the United States, automobiles are the main source of CO emissions, producing up to 95% of the CO in the air.

When breathed in high concentrations, CO reduces oxygen levels in the blood. Symptoms of CO poisoning include headaches, slowed reflexes, and drowsiness. In poorly ventilated areas, such as indoors or in tunnels, death by CO poisoning can happen within 1 to 3 minutes of exposure to very high concentrations of CO.

Nitrogen Dioxide

Nitrogen oxides are compounds of nitrogen and oxygen that form when nitrogen in the air reacts with oxygen during the high-temperature combustion of fossil fuels or during wildfires. Two important nitrogen oxides are *nitric oxide (NO)* and **nitrogen dioxide (NO_2)**, which together are generally referred to as *NOx*. Nitric oxide is nontoxic and harmless. But when it oxidizes (that is, when another oxygen atom is added to it), it forms NO_2. Nitrogen dioxide is significant because it produces a brown haze in polluted areas and can cause significant health problems for people. At high concentrations, NO_2 irritates lung tissue, and chronic exposure can lead to pulmonary diseases such as asthma and bronchitis. In the United States, metropolitan regions with traffic congestion account for almost 60% of NO_2 emissions. Coal burning accounts for 32% of U.S. NO_2 emissions.

Ground-Level Ozone

Ozone (O_3) is a secondary pollutant near Earth's surface. About 90% of ozone in Earth's atmosphere is found in the stratosphere. Without *stratospheric ozone*, harmful UV rays would be deadly to organisms living on land.

Ozone that forms near the land surface is called **ground-level ozone**. While *stratospheric ozone* is essential for life on land, ground-level ozone is not. Ground-level ozone also forms naturally when substances released by plants and soils are converted to ozone. The highest concentrations of ground-level ozone, however, are produced by human activity. Most ground-level ozone is **photochemical smog**, a pollutant that forms by the action of sunlight on tailpipe emissions. Ozone forms as sunlight breaks down NO_2 into NO + O. The oxygen (O) made available when NO_2 breaks apart combines with oxygen in the atmosphere (O_2), resulting in ozone (O_3).

Ultraviolet radiation breaks ozone apart within hours after it has formed. Therefore, ground-level ozone is typically highest in urban areas during the afternoon rush-hour traffic in summer, when sunlight is most direct and there is sufficient NO_2 available from tailpipe emissions.

Ozone irritates the eyes and nose, scars lung tissue, and can intensify respiratory ailments such as bronchitis, emphysema, and asthma. Ground-level ozone also decreases plant and crop growth by up to 10% and weakens plants so that they are more susceptible to drought and insect attacks. Ozone corrodes rubber and plastics unless they are specially treated. During the summertime in many urban areas, local authorities declare "Ozone Action Days" or "Clean Air Alerts" to warn residents to restrict their outdoor activity during high-ozone periods.

Sulfur Dioxide

Sulfur compounds (often denoted SO_x) are emitted from both natural and anthropogenic sources. The most significant sulfur compounds are *sulfur trioxide* (SO_3) and sulfur dioxide (SO_2). **Sulfur dioxide (SO_2)** is a primary pollutant gas that is produced by volcanoes as well as by burning of fossil fuels, particularly coal for generating electricity and petroleum for winter heating. Over 80% of anthropogenic

SO_2 emissions originate from the burning of coal to generate electricity. In high concentrations, SO_2 irritates lung tissue, causes difficulty breathing and chest pain, and makes respiratory illnesses such as bronchitis and asthma more acute.

When sulfur trioxide combines with nitrogen compounds, such as nitrogen dioxide, it forms droplets of *sulfuric acid* (H_2SO_4), which is a secondary pollutant. These droplets can then precipitate and fall to Earth as **acid rain** (also called *acid deposition*). Acid rain is any form of precipitation (including snow) that has a pH typically lower than 5.0, or any dry acidic particles that precipitate. Naturally acidic rain occurs downwind of volcanic eruptions. Anthropogenic acid rain, however, occurs mostly near industrial areas, particularly where coal is being burned. Over time, acid rain dissolves certain types of rock, such as limestone and marble, that are used in building construction and statues **(Figure 2.10)**. It also can change soil and water chemistry, causing extensive ecosystem damage. Most fish, amphibians, and *invertebrates*

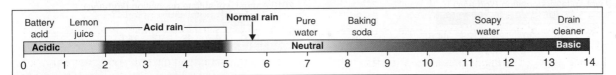

A

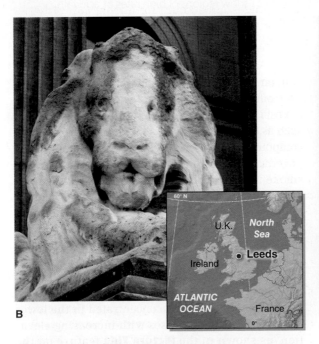

B

C

Figure 2.10 Acid rain damage. (A) The pH scale ranges from 0 to 14. A pH value of 7 is neutral; pH values above 7 are basic, or alkaline, and values below 7 are acidic. Normal rainfall is naturally acidic, with a pH of about 5.6 because it absorbs carbon dioxide from the air. Acid rain has a pH value of less than 5.0. Extremely acidic rain can have a pH value as low as 2.0, close to that of lemon juice. (B) This statue in Leeds has been dissolved by acid rain caused by the United Kingdom's long history of burning coal. (C) Anthropogenic acid rain weakened these Fraser fir (*Abies fraseri*) and red spruce (*Picea rubens*) trees in Mount. Mitchell State Park, North Carolina. Once trees are weakened by acid rain, insects can more easily attack them, causing additional forest damage. *(B. Ryan McGinnis/Alamy; C. Will & Deni McIntyre/Getty Images)*

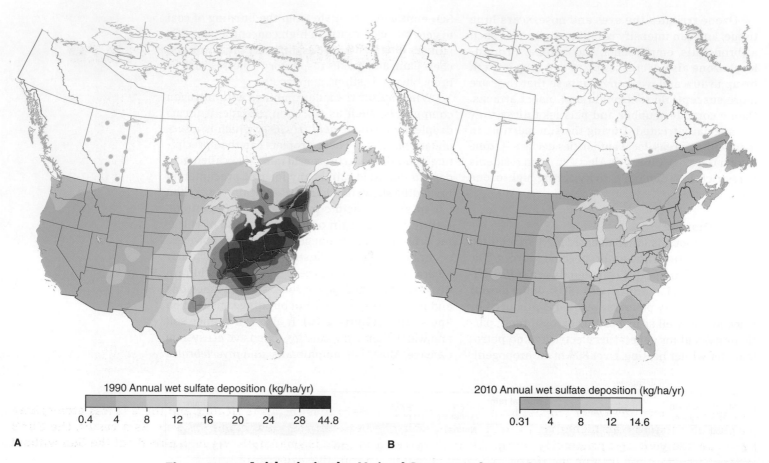

1990 Annual wet sulfate deposition (kg/ha/yr)

0.4   4   8   12   16   20   24   28   44.8

A

2010 Annual wet sulfate deposition (kg/ha/yr)

0.31   4   8   12   14.6

B

Figure 2.11 Acid rain in the United States and Canada. These maps compare acid rain amounts (wet sulfate deposition) in North America for 1990 and 2010. Red areas show regions of high rates of acid rain. Between the two time periods, emissions of sulfur compounds and subsequent acid rain decreased 71%. *(National Atmospheric Chemistry [NAtChem] Database and the National Atmospheric Deposition Program [NADP], 2014)*

(such as snails and clams) are unable to live in water with a pH of less than 5.

The eastern United States and Canada have a long history of acid rain and ecosystem damage due to industrial emissions from coal burning. There has, however, been much improvement as a result of the passage of the U.S.–Canada Air Quality Agreement in 1991. As a result of this agreement, sulfur compound emissions in the United States and Canada have been greatly reduced. In recent years, life has begun returning to lakes that were made nearly sterile from acid rain pollution due to significant improvements in air quality and the resulting decrease in acid rain **(Figure 2.11)**.

Particulate Matter

Particulate matter (PM) consists of microscopic aerosols (liquid and solid particles) suspended in the atmosphere. The term *particulate matter* is often used in the context of air pollution, while the term *aerosol* is usually used in a broader sense to include natural atmospheric particles such as liquid cloud droplets as well as human-made pollution. Particulate matter can be either a primary pollutant or a secondary

pollutant, depending on whether it was directly emitted from a source as a solid particle, such as soot from a diesel truck, or resulted from a chemical reaction, such as droplets of sulfuric acid forming from sulfur compounds. Particulates can be natural or anthropogenic; they include fine volcanic ash, pollen, fire smoke, salt particles from ocean spray, and wind-blown dust. Black carbon soot is a type of particulate matter that results from the burning of fossil fuels, vegetation, and animal dung (an important source of cooking fuel in many developing nations). Particulate matter emissions in the United States come mainly from coal burning (about 40%), industrial processes (about 23%), and transportation (about 20%).

Particulate matter is concentrated in the lower troposphere and decreases with increasing elevation, as shown in the **Picture This** feature on the following page.

The smaller the particulate matter size, the more harmful it is to human health. Particle size is typically given in micrometers (μm; 1 μm = 1 millionth of a meter). PM_{10} is particulate matter of 10 μm diameter. $PM_{2.5}$ is particulate matter of 2.5 μm diameter or less. For a sense of scale, we compare these particles

Picture This Particulate Matter Haze

@ Sea level   @ 4,000 ft   @ 10,000 ft

(Rob Ratkowski)

For most regions, the lower troposphere is dirtier with particulates than the air higher up. Even in Hawai`i, which many people view as a pristine paradise, this is the case. These three photos, taken on the Hawaiian island of Maui, illustrate this point. When a finger is held at arm's length at sea level, the solar *aureole* (halo) remains visible, but at higher elevations (photo on the far right), the same finger held at the same distance completely obscures the Sun and its glowing halo. The size of the Sun's aureole is greatest at sea level because we are looking through more particulate matter, water vapor, and heat from the surface, which scatter sunlight, creating a glow around the Sun. At higher elevations the air is cleaner, and there is less atmosphere between the viewer and the Sun. As a result, the Sun's aureole is smaller. (Never look directly at the Sun without protective eyewear.)

Consider This

1. Assume that the sea-level photo was taken in an urban area on the small island of Maui. What types of particulate would you expect to find there? Explain.
2. Do astronauts in orbit see an aureole around the Sun? Why?

---

to a human hair in **Figure 2.12** on the following page. Any inhaled particles smaller than 10 μm may move deep into the lungs and then into the bloodstream, where they can cause or worsen physical ailments, including bronchitis, asthma, various cancers, and heart disease. Particulate matter is removed from the atmosphere as it settles out by gravity. It is also removed as falling rain and snow carry it to Earth's surface.

Volatile Organic Compounds

Volatile organic compounds (VOCs) (also called *hydrocarbons*) are toxic compounds of hydrogen and carbon. Examples of VOCs include methane, butane, propane, and octane. These primary pollutants may irritate the eyes, nose, throat, and skin. They can also lead to liver, kidney, and central nervous system damage. Anthropogenic VOC emissions come mostly from industrial processes and from automobiles, when gasoline is burned incompletely and when gasoline evaporates at gas stations. Natural decomposition of organic material such as vegetation and animal waste also contributes dispersed low concentrations of methane to the atmosphere. VOCs react with other compounds in the air, creating toxic secondary pollutants and contributing to ozone formation.

Concentrating Pollutants: Topography and Temperature Inversions

Once pollutants enter or form in the atmosphere, topography and temperature inversions can dramatically increase their concentrations. Many large cities are situated in valleys surrounded by mountains. In such a natural basin, air pollutants can concentrate by preventing or diminishing wind that would otherwise transport the pollutants away. Temperature inversions often form

Figure 2.12 Particulate matter. (A) PM₂.₅ and smaller particulates are not filtered by the nose and mouth and pass into the lungs, where they can cause respiratory ailments. Here particulate matter is compared to the size of a human hair and grains of sand. (B) Particulate matter concentrations are expressed as micrograms per cubic meter of air, or µg/m³. Clean air generally has PM₂.₅ concentrations of 5 µg/m³ or less.

Air is considered unhealthy when the PM₂.₅ concentration rises above 30µg/m³ for more than 24 hours. This world map shows PM₂.₅ averaged from 2010 to 2012. It excludes particulate matter from windblown dust over the deserts and coastal salt spray. Gray areas have no data. Note that the data scale is logarithmic: Concentrations in China are nearly 100 times greater than those in the United States and Europe.

(A. U.S. Environmental Protection Agency; B. Data from van Donkelaar et al. / NASA Earth Observatory)

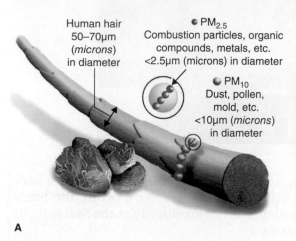

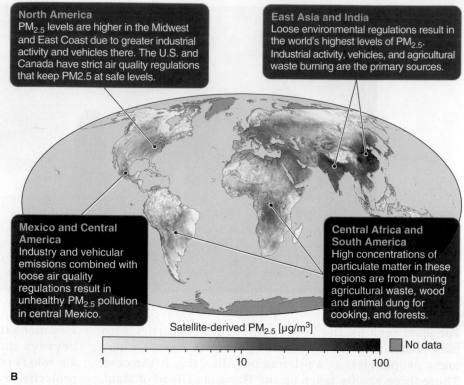

> **North America**
> PM₂.₅ levels are higher in the Midwest and East Coast due to greater industrial activity and vehicles there. The U.S. and Canada have strict air quality regulations that keep PM2.5 at safe levels.

> **East Asia and India**
> Loose environmental regulations result in the world's highest levels of PM₂.₅. Industrial activity, vehicles, and agricultural waste burning are the primary sources.

> **Mexico and Central America**
> Industry and vehicular emissions combined with loose air quality regulations result in unhealthy PM₂.₅ pollution in central Mexico.

> **Central Africa and South America**
> High concentrations of particulate matter in these regions are from burning agricultural waste, wood and animal dung for cooking, and forests.

Satellite-derived PM₂.₅ [µg/m³]

1   10   100   No data

A   **B**

Web Map

Particulate Matter 2.5

Available at
www.saplinglearning.com

in valleys as cold, dense air sits near the ground. These temperature inversions can prevent air near the ground from rising higher. When pollutants are emitted into such an environment, they become concentrated near the ground. **Figure 2.13** illustrates this process happening in Salt Lake City, Utah, where temperature inversions resulting from local topography are particularly common.

Warning the Public

In chronically polluted areas, it is important to be aware of air quality during seasons or times of the day when pollution is worst. The U.S. EPA has created the Air Quality Index (AQI). Canada uses a similar system, called the Air Quality Health Index. These systems provide hourly updates on the concentration of major pollutants in all major metropolitan regions. These rankings are based on any given pollutant that affects the air quality the most on any given day. In winter, for example, air quality is typically based on particulate matter concentrations, while in summer it is based on ground-level ozone. The AQI scales air quality from 0 to 500. An AQI of 151 or greater is considered unhealthy for humans **(Table 2.5)**.

Regulating Clean Air

Imagine that you are given a choice between fixing a minor toothache today or ignoring the problem

Table 2.5	The Air Quality Index
AIR QUALITY INDEX (AQI) VALUES	**LEVELS OF HEALTH CONCERN**
0 to 50	Good
51 to 100	Moderate
101 to 150	Unhealthy for sensitive groups
151 to 200	Unhealthy
201 to 300	Very unhealthy

indefinitely. You could save money in the short term by ignoring the problem and living with a little dull pain. But over time, the pain would worsen, and a major dental issue would develop. Either way, the problem would have to be addressed because it would not go away, and ignoring the pain allows it to worsen. This is what air pollution is like. Ignoring air pollution allows it to worsen as populations grow, as people drive more cars, and as more fossil fuels are burned to meet growing energy demands. Faced with this choice, the U.S. Congress passed the Clean Air Act in 1963. It was amended in 1970, in 1977, and again in 1990. In Canada, the Clean Air Regulatory Agenda (CARA) was established in 2006. Virtually all other developed countries,

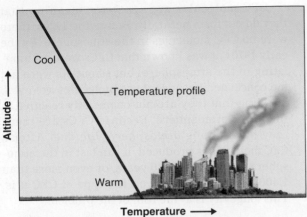

A. Normal temperature profile

Figure 2.13 Temperature inversion and smog.

(A) This normal temperature profile of the atmosphere shows decreasing temperature with altitude. This temperature profile allows pollutants to rise and be dispersed by winds. In a temperature inversion, air warms as altitude increases. This temperature profile prevents air from rising and allowing winds to carry pollutants away. (B) The air quality in Salt Lake City in winter is often poor due to persistent temperature inversions caused by cold, dense air settling in the valley in which the city is located. During peak winter inversion times, temperatures can be some 20°C warmer at 2,000 m altitude (40°F at 6,500 ft) than at ground level. Pollution levels in Salt Lake City often exceed the federal standards. *(Roland Li)*

B. Temperature inversion

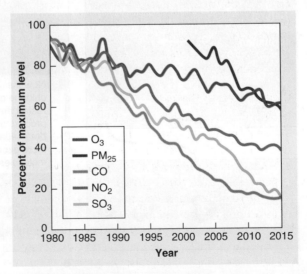

including countries of the European Union, China, and India, have environmental regulations to clean the air. These regulations impose limits on emissions of air pollutants and have been effective at reducing air pollution, saving lives, and reducing health care–related costs. **Figure 2.14** shows the declining trends in emissions of five major pollutants in the United States between 1980 and 2015.

Critics of clean air regulations argue that the technology necessary to reduce emissions is expensive and that compliance hurts the economy. They argue that enforcing clean air regulations moves jobs overseas, where it is cheaper to manufacture goods because pollution regulations are less stringent.

Proponents of clean air regulations argue that, in the long run, clean air is cheaper than dirty air. The U.S. EPA expects total U.S. spending of $65 billion to enforce the Clean Air Act for the period 1970–2020. These measures, however, provide an estimated savings of $2 trillion in health care–related costs in the United States for the same period. Furthermore, the lives saved and the improvements in the quality of life resulting from the regulations cannot easily be assigned monetary values.

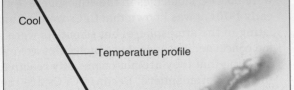

Figure 2.14 U.S. air pollution trends, 1980–2015.

Emissions of five major air pollutants have decreased in the United States as a result of the Clean Air Act. Values are given as percentages of their maximum values for the time period. There are no reliable $PM_{2.5}$ data before the year 2000. Pollution levels were reduced by over 80% for CO and SO_2 between 1980 and 2015, even though there were 95 million more people in 2015. *(Data from U.S. EPA)*

Web Map

Pollution Levels in the U.S.

Available at
www.saplinglearning.com

GEOGRAPHIC PERSPECTIVES
2.5 Refrigerators and Life on Earth

◎ **Assess the effects of anthropogenic pollutants in the ozonosphere and the anticipated condition of the ozonosphere in the coming decades.**

Before the 1930s, it was a wise practice to keep refrigerators outside and far away from the house because they occasionally blew up. Refrigerator coolants consisted of a mix of volatile and explosive chemicals—such as ammonia, methyl chloride, and sulfur dioxide—which often leaked and sometimes poisoned people, or even detonated. The safety of refrigeration increased with an invention that changed the world: **CFCs (chlorofluorocarbons)**, a class of ozone-degrading compounds used mainly as refrigerants, aerosol propellants, and fire retardants.

In 1930, General Motors and DuPont held the patent on a new chlorofluorocarbon molecule, developed with research driven primarily by the need for a better refrigerator. They gave the trademark name Freon to the new molecule. Freon and other related CFCs were nontoxic, cheap to make, and not explosive. Within 5 years of their introduction, kitchen refrigerators using Freon were the standard.

CFC molecules are *synthetic*, meaning that they do not form naturally. Before the 1920s, there were no CFC molecules in the atmosphere. By the early 1970s, it was known that CFCs were accumulating in the atmosphere, but scientists were not too concerned because these molecules are *inert*, meaning that they are not chemically reactive in the lower atmosphere. Because CFCs are inert, they have a long *atmospheric lifetime*. After a CFC molecule is produced, it can last in the atmosphere for 60 years, 200 years, or even more than 1,000 years, depending on the type of CFC it is. This longevity is at the root of the problem.

A Landmark Paper and an International Protocol

In 1974, Mario Molina and Sherwood Rowland, two professors at the University of California, Irvine, published a landmark paper that radically changed our understanding of CFCs and the ozonosphere. (Molina and Rowland would later receive the Nobel Prize for this work.) In the paper, they announced that their research showed that CFCs linger in the troposphere for decades. Furthermore, as CFCs mix in the atmosphere, they slowly make their way to the stratosphere. Once in the stratosphere, CFCs are broken down by ultraviolet (UV) radiation, and one of their components, chlorine, reacts with ozone, breaking the ozone into oxygen atoms and molecules. **Figure 2.15** illustrates the process of natural ozone formation

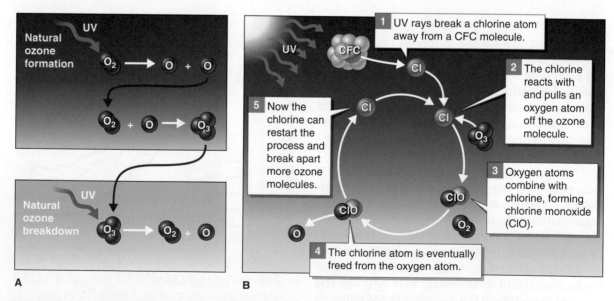

Figure 2.15 Stratospheric ozone formation and destruction. (A) Stratospheric ozone forms naturally when UV radiation from the Sun breaks apart an oxygen molecule (O_2) into two atoms of oxygen (O). One of these atoms combines with a different oxygen molecule, forming ozone (O_3). The ozone breaks down naturally when UV radiation breaks ozone molecules back down to oxygen. (B) Ultraviolet radiation breaks apart CFCs, releasing chlorine that reacts with and breaks apart ozone. A single chlorine atom can convert several hundred thousand ozone molecules to oxygen.

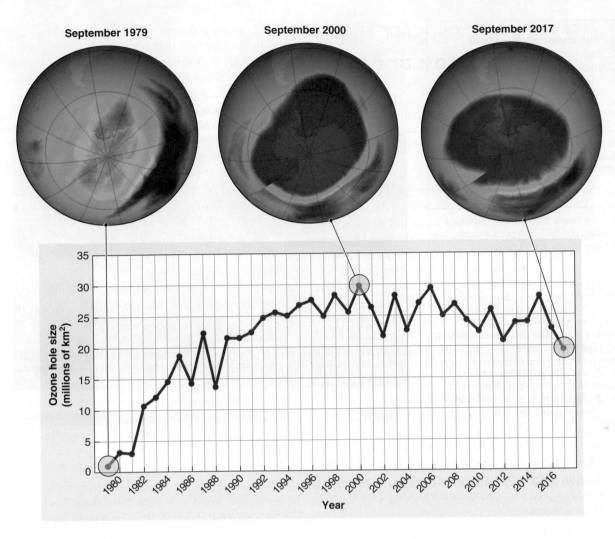

September 1979   September 2000   September 2017

Figure 2.16 Ozonosphere hole history, 1979 to 2017. Data for the ozone "hole" size are graphed here, starting in 1979. There was no hole in September 1979 (left map). The largest ozone hole ever recorded for a single day was 29.9 million km² (11.5 million mi²) in September 2000 (middle map). Since its maximum extent in 2000, the hole's size has been gradually decreasing. The map on the far right shows the extent of the hole for September 2017— 19.6 million km² (7.6 million mi²), the smallest hole since 1988. The maps are scaled by Dobson units (DU), which indicate the concentration of ozone in the stratosphere. Blues and purples indicate the thinnest ozonosphere, and greens, yellows, and reds indicate a healthy ozonosphere. The boundaries of the ozone hole are defined to be where Dobson units are 220 or lower. *(NASA Ozone Watch)*

and the CFC-related breakdown of ozone in the stratosphere.

Chlorine from CFCs destroys ozone in the ozonosphere. Without the protective layer of ozone, life on land would be burned by ultraviolet rays from the Sun. Molina and Rowland sounded the alarm that, in a span of only decades, CFCs had compromised Earth's UV shield that had protected the biosphere for over 2 billion years. Millions of tons of CFCs in the lower troposphere were working their way up to the stratosphere, breaking apart to form chlorine, and destroying ozone molecules.

Satellite and ground-based measurements have provided key data about the response of the ozonosphere to these chemicals. These data show that a "hole" in the ozonosphere—a region in the stratosphere where the concentration of ozone is reduced to below 220 *Dobson units* **(Figure 2.16)**— formed over Antarctica. The thinning is greatest over Antarctica because polar stratospheric clouds called *nacreous clouds*, named for their shimmering mother-of-pearl quality, form in the cold air during the Antarctic spring (September) and assist in the process of chlorine reactions with ozone.

Thinning is less severe over the North Pole because the stratosphere there is not as cold, so polar stratospheric clouds do not form there as easily.

In response to this problem, 189 countries initially crafted and signed on with *The Montreal Protocol on Substances that Deplete the Ozone Layer*. The international agreement was ratified in 1987, and it went into effect in 1989. The protocol banned CFC production outright in some countries and mandated phaseouts in other countries. The protocol also banned dozens of other *ozone-depleting substances* (*ODSs*) that are composed in part of chlorine atoms. As a result, concentrations of chlorine (from CFCs) in the stratosphere over Antarctica peaked in the early 1990s and have been on the decline ever since. By about 2050, they are expected to be reduced to 1980 levels. As the CFCs slowly diminish, the ozonosphere will repair itself.

Human Health Effects of a Thinning Ozonosphere

Although CFC production has been mostly phased out, the long residence time of CFCs in the troposphere means that their effects will be with

Exposure Category	Index number	Sun protection messages
Low	<2	Low damage. Safe to be outdoors.
Moderate	3–5	Moderate risk. Take precautions when outdoors. Use sunscreen SPF 30+. Seek shade in midday Sun.
High	6–7	Protection needed. Avoid Sun exposure. Use sunscreen SPF 30+ on exposed skin. Seek shade during midday.
Very high	8–10	Burn risk very high. Avoid Sun exposure between 10 a.m. and 4 p.m. If outdoors, wear hat, long sleeves, and sunscreen. Snow and white sand can reflect sunlight and double UV exposure.
Extreme	11+	Burn risk extreme. Avoid Sun exposure between 10 a.m. and 4 p.m. If outdoors, wear hat, long sleeves, and sunscreen. Snow and white sand can reflect sunlight and double UV exposure.

A

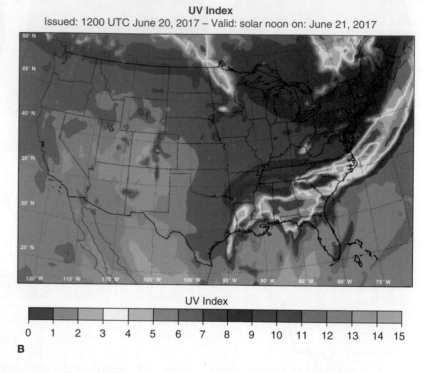

B

Figure 2.17 UV Index and map. (A) The *UV Index* ranks the UV level risk from 1 to 11+ and by color. A higher number indicates greater risk. (B) Each day the EPA publishes a color-coded map of the UV radiation forecast for the United States and southern Canada. Dark blue and green areas have the lowest UV levels, and violet and bright blue have the greatest UV levels. UV levels are calculated based on latitude, elevation, cloud cover, and stratospheric ozone amounts. *(U.S. Environmental Protection Agency)*

us for several more human generations. As long as there are CFCs in the atmosphere, the ozonosphere will remain thinner than its natural level, and we should be cautious of how much UV radiation we expose ourselves to. To assist in evaluating the risk of exposure for any given day and location, the U.S. EPA publishes a *UV Index* and forecast map for the United States and Canada **(Figure 2.17)**.

The cause of most cases of skin cancer is exposure to UV-A radiation. The incidence of all types of skin cancer has been on the rise since the 1970s, although the increase in cases is partly a result of better awareness and diagnosis. In the United States, *melanoma* (the most serious form of skin cancer) increased 800% among women and 400% among men between 1970 and 2009. The risk related to burning and skin damage by UV radiation varies depending on skin complexion. People with lighter complexions are more vulnerable to damage from the Sun than people with darker complexions. All complexions are vulnerable to skin cancer from overexposure to UV radiation.

UV radiation suppresses normal cellular immune responses by damaging or killing immune-response cells in the skin, which can, in turn, affect the growth of skin cancers and tumors. Medical research also indicates that UV radiation decreases the proportion of immune-response white blood cells in circulation in the bloodstream and may trigger the activity of dormant viruses. Cumulative exposure to UV radiation can also cause a condition called cataracts, in which the lenses of the eyes become an opaque milky white and severely impair vision.

Ecosystem Health Effects of Ozonosphere Thinning

Plants are stressed by and grow more slowly when they are exposed to increased levels of UV radiation. The result is reduced agricultural production. Scientists are working to increase tolerance for UV radiation in staple crops such as rice, wheat, corn, and soybeans. But it isn't just crops that are affected by ozone depletion. Almost all life on Earth depends on plants to convert radiant energy from the Sun to the chemical energy that fuels the biosphere. When plants grow more slowly, this chain of connectivity is affected.

A Crisis Averted

Antarctic ozonosphere thinning peaked in 2000 and is showing a gradual trend of improvement. If the Montreal Protocol had not been enacted, and we had continued producing CFCs, computer modeling studies indicate that the world's ozonosphere could have been completely destroyed a little after the middle of the twenty-first century.

The Montreal Protocol provides a good example of the international community working together to solve a problem that affects everyone. Although some ODSs—such *nitrous oxide* or the pest-control chemical *methyl bromide* used in agriculture, particularly on strawberries—are still produced, over 98% of ODS production worldwide has been phased out since the Montreal Protocol's enactment.

CFCs have largely been replaced with HFCs (hydrofluorocarbons), which do not degrade ozone. Unfortunately, however, both CFCs and HFCs are potent greenhouse gases. Molecule for molecule, they absorb thousands of times more heat energy than carbon dioxide and have contributed to about 17% of anthropogenic global warming. (In Section 3.3, we discuss greenhouse gases and Earth's greenhouse effect in more detail.) In this sense, ozonosphere thinning and global warming are interlinked. To address the growing problem of climate change, the Montreal Protocol was amended in 2015 with the aim of phasing out HFCs by midcentury.

Chapter 2 Study Guide

Focus Points

2.1 Composition of the Atmosphere

- **Types of gases:** Gases in the atmosphere are permanent gases or variable gases. Nitrogen and oxygen comprise about 99% of the atmosphere.
- **Greenhouse gases:** Greenhouse gases absorb and radiate heat in the atmosphere and play a crucial role in Earth's climate system.
- **Sources and sinks:** Gases enter and exit the atmosphere through natural and anthropogenic sources and sinks.

2.2 The Weight of Air: Atmospheric Pressure

- **Air pressure:** Because it has mass, the atmosphere is held down against the planet's surface by gravity, which gives it weight and creates air pressure.
- **Air density:** Air density and pressure are greatest near sea level and decrease as altitude increases.

2.3 The Layered Atmosphere

- **Thermal divisions:** The atmosphere consists of four layers, based on changes in temperature with altitude: the troposphere, stratosphere, mesosphere, and thermosphere.
- **Troposphere:** The troposphere gets colder with increasing altitude, and there is strong vertical mixing. Earth's weather occurs in the troposphere.
- **Stratosphere:** Storms and most clouds do not enter the stratosphere, and there is little vertical mixing. Ozone molecules in the stratosphere absorb ultraviolet radiation.

2.4 Air Pollution

- **Types of air pollution:** Anthropogenic air pollutants are the result of fossil fuel burning and include CO, NO_2, ground-level ozone, SO_2, particulate matter, and VOCs.
- **Concentration of air pollution:** Air pollution is concentrated in valleys when there are temperature inversions.

- **The Clean Air Act:** Emissions of air pollutants in the United States and Canada have been reduced since the passage of regulations to clean the air.

2.5 Geographic Perspectives: Refrigerators and Life on Earth

- **The ozonosphere:** Without the ozonosphere, life on land would be burned by ultraviolet radiation.
- **CFCs:** Anthropogenic chlorine atoms from CFCs break down ozone molecules in the stratosphere and threaten the ozonosphere.
- **Montreal Protocol:** After the enactment of the Montreal Protocol, the global manufacturing of CFCs was mostly phased out, and the ozonosphere is repairing itself.

Key Terms

acid rain, 51
aerosols, 42
air pollutant, 49
air pressure, 43
anthropogenic, 42
atmosphere, 40
aurora australis, 49
aurora borealis, 49
carbon dioxide (CO_2), 49
carbon monoxide (CO), 50
CFCs (chlorofluorocarbons), 56
cloud droplets, 42
environmental lapse rate, 45
fossil fuels, 49
greenhouse gas, 41
ground-level ozone, 50
ionosphere, 48
Karman line, 41

mesosphere, 44
nitrogen dioxide (NO_2), 50
ozone (O_3), 50
ozonosphere, 47
particulate matter (PM), 52
photochemical smog, 50
primary pollutant, 50
radiosonde, 44
secondary pollutant, 50
stratosphere, 44
sulfur dioxide (SO_2), 51
temperature inversion, 44
thermosphere, 44
tropopause, 44
troposphere, 44
ultraviolet (UV) radiation, 47
volatile organic compound (VOC), 53

Concept Review

The Human Sphere: China's "Airpocalypse"

1. Why do urban areas in eastern China have such poor air quality?

2.1 Composition of the Atmosphere

2. What are the three major permanent gases in the atmosphere? What are some of the variable gases?

3. What are sources and sinks for gases? Give examples.

4. What are aerosols? Give examples.

2.2 The Weight of Air: Atmospheric Pressure

5. What is air pressure, and how much is there at sea level, in kilograms per square centimeter? In PSI?

6. Why is air pressure greater near Earth's surface than at higher altitudes?

2.3 The Layered Atmosphere

7. What is the Karman line, and why is it significant?

8. List the four layers of the atmosphere, based on changes in temperature with altitude, and give the major characteristics of each.

9. What are the two layers of the atmosphere that block harmful solar radiation? At what altitudes in the atmosphere are they found?

10. What are the major characteristics of the troposphere? How do the temperature and air pressure of the troposphere change with altitude?

11. What is the tropopause? What is the altitude of the tropopause at midlatitudes? What is its altitude at the equator and at the poles?

12. What is the range of altitudes for the stratosphere? How does temperature change with altitude in the stratosphere?

13. Compare and contrast airflow in the stratosphere with that in the troposphere.

14. What are auroras? Where do they occur in the atmosphere? Over what regions? What causes them?

15. Why don't clouds from the troposphere extend up into the stratosphere?

2.4 Air Pollution

16. What is the source of most anthropogenic atmospheric pollutants?

17. Cite examples of primary pollutants, their sources, and their effects on human health.

18. What are secondary pollutants? Which secondary pollutants are discussed in this chapter?

19. How is the ozone molecule both beneficial and harmful to people? Explain.

20. How is particulate matter hazardous to human health?

21. What atmospheric and topographic factors make air pollution worse?

22. What is the U.S. Clean Air Act? When was it enacted? Has it been successful? How?

2.5 Geographic Perspectives: Refrigerators and Life on Earth

23. What are CFCs, and what were they used for?

24. How long do CFCs last in the troposphere? What happens when they reach the stratosphere?

25. What is the ozonosphere, and how do CFCs affect it?

26. What are the projected trends in stratospheric chlorine concentrations for this century?

27. What are nacreous clouds, and how are they related to ozonosphere thinning over Antarctica? Why has there been less ozonosphere loss over the Arctic than over Antarctica?

28. What are the connections between CFC production and human and ecological health?

29. When was the Montreal Protocol enacted, and what was its initial aim?

30. What recent amendment has been added to the Montreal Protocol, and why?

Critical-Thinking Questions

1. Why is air pollution in the United States and Canada decreasing even though the population, the numbers of cars, and energy consumption are all increasing?

2. What is the general air quality where you live? If you do not know the answer, how can you find it?

3. Acid rain is naturally produced by volcanoes. Given that acid rain is natural, why is natural acid rain not a hazard but anthropogenic acid rain is?

4. Construct an argument between critics and supporters of clean air regulations, with one side stating that these regulations are too costly to enact and the other side stating that it is too costly not to enact them. With which side do you identify more?

5. Can you think of other pressing global environmental problems besides ozonosphere thinning that the world currently faces? Explain.

Test Yourself

Take this quiz to test your chapter knowledge.

1. True or false? The atmosphere is composed mostly of nitrogen.

2. True or false? Aerosols are gases in the atmosphere.

3. True or false? Air density and air pressure decrease as altitude increases.

4. True or false? In the stratosphere, temperature decreases with an increase in altitude.

5. True or false? The size of the hole in the ozonosphere peaked in the 1970s and has since been declining.

6. Multiple choice: Which pollutant is most deadly in high concentrations?

a. Ozone
b. Carbon monoxide
c. Sulfur dioxide
d. Particulate matter

7. **Multiple choice:** In which atmospheric layer is most ultraviolet radiation from the Sun absorbed?

   a. The troposphere
   b. The stratosphere
   c. The mesosphere
   d. The thermosphere

8. **Multiple choice:** In which layer of the atmosphere do auroras form?

   a. Troposphere
   b. Mesosphere
   c. Thermosphere
   d. Ionosphere

9. **Fill in the blank:** The _____ mandated the phaseout of CFCs by all industrialized countries.

10. **Fill in the blank:** The vast majority of anthropogenic air pollutants originate from usage of _____.

Online Geographic Analysis

Gathering Real-Time Air Quality Data

Several online tools allow anyone with an Internet connection to monitor air quality in real time nearly anywhere in the world. This Geographic Analysis explores a few such tools.

Activity

Go to https://www.airnow.gov. Examine the national map that is provided on the home page and answer the following questions.

1. **Which regions have the highest AQI readings?**

2. **What cities are in the "top five" list provided on the home page (the Highest 5)? Among them, what is the highest AQI reading?**

3. **Click on the city with the highest AQI score. Scroll down and examine the measured levels for different pollutants. Which pollutant is highest and mostly accounts for the polluted air?**

4. **Now enter your zip code in the home page. What is the AQI score for your city?**

5. **Scrolling down, determine which pollutant is at the highest level.**

Now go to http://www.aqicn.org/map/. This a world map of real-time AQI scores.

6. **Zoom out until you can see all the continents. Which country or region appears to be the most polluted?**

7. **Zoom in to that country or region using the zoom function. What are the highest AQI scores you can find? In which country are they found?**

8. **Compare the AQI scores you found in Question 7 to the highest scores you found in the United States in Question 2 above. Which is higher? Can you explain why one country would have higher AQI values than another?**

9. **Click the highest AQI value you found on the world map in Question 7 and then click on the "full report" in the pop-up window. Which type of pollutant is highest and is causing the high AQI score?**

Next, go to https://www.epa.gov/sunsafety/uv-index-1. This is the EPA's UV Index page.

10. **Enter your city in the search box. What is your UV Index today? What factors do you think are influencing it most (making it low or high)?**

11. **On the same search results page, there are two tabs: "Daily UV Index" and "Hourly UV Index." Click the Hourly UV Index page. During which hours will the UV Index be highest?**

Picture This. *Your Turn*

The Atmosphere's Thermal Divisions

Fill in the blanks on the diagram using each term only once. In addition, indicate whether the temperature is increasing or decreasing with altitude.

1. Mesopause

2. Mesosphere

3. Stratopause

4. Stratosphere

5. Thermopause

6. Thermosphere

7. Tropopause

8. Troposphere

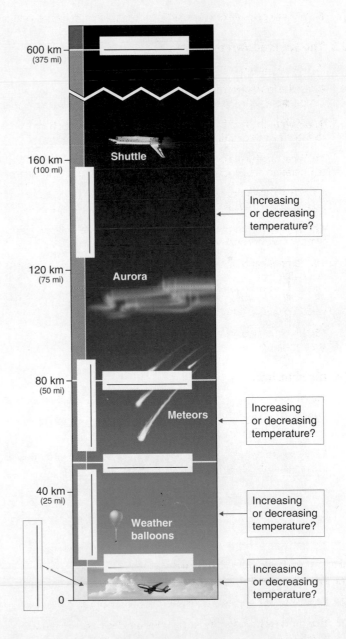

Solar Energy and Seasons

Chapter Outline *and Learning Goals* ◎

Purple clouds and lupine wildflowers (*Lupinus* spp.) and a double rainbow add splashes of color in the Palouse region in eastern Washington. Rainbows form only where drops of water are present in the atmosphere. Each water drop separates white sunlight into different wavelengths of light that our eyes perceive as color.

(Dean Fikar/Getty Images)

Go to Section 3.3 to learn more about how rainbows form, why objects such as wildflowers are colored the way they are, and other optical phenomena in the atmosphere.

THE HUMAN SPHERE | Hot Cities

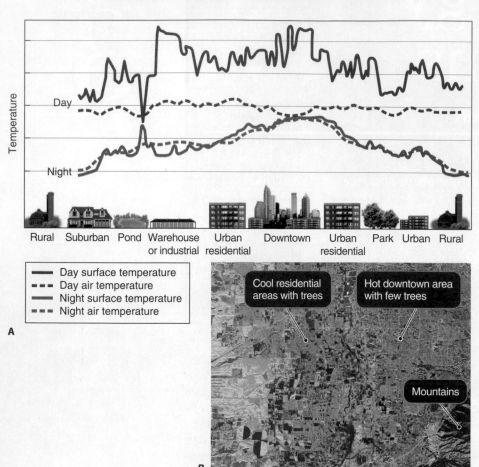

A

B

Figure 3.1 Urban heat island. (A) This graph compares typical temperature contrasts between different surfaces common to cities and the areas surrounding those surfaces. Daytime ground surface temperatures (solid red line) are higher than daytime air temperatures (dotted red line). At night, both ground and air temperatures (blue lines) are roughly the same, but the temperature contrast between urban and rural is much greater than during the day because city asphalt and concrete radiate heat after the Sun has set. (B) This satellite thermal infrared image of Salt Lake City, Utah, shows greater detail of the opening image for this chapter. Red areas are hotter streets and buildings in the downtown center, and green and blue areas are cooler surrounding residential areas with more trees. *(NASA)*

Cities cover only about 1% of Earth's surface, but about half of the world's 7.6 billion people live in cities. Urban areas consume 75% of the world's electricity and account for 80% of the world's economic productivity. By the year 2100, some 9 billion people, 85% of the global population, will live in cities. This is significant for many reasons, but one particularly important reason is that these urban dwellers are going to be living increasingly heat-stressed lives because cities run significantly warmer than surrounding rural areas. This phenomenon is referred to as the **urban heat island**. The urban heat island occurs when natural surfaces are replaced by materials that absorb sunlight and retain heat, such as concrete and asphalt. Cities of 1 million people or more can experience summer daytime temperatures up to 3°C (5.4°F) warmer than the surrounding areas. On calm nights, the temperature difference can be as much as 12°C (22°F) **(Figure 3.1)**!

As Earth's atmosphere warms and the world becomes increasingly urbanized, more people will be exposed to heat-related problems. Hot cities require more electricity to cool, experience greater air pollution due to temperature inversions (see Section 2.3), and suffer higher rates of human mortality. Cities around the world are already taking active measures to address the urban heat island problem. These measures include creating more green spaces, planting more trees along sidewalks, installing *green rooftops* (vegetated building rooftops), and using construction materials that reflect sunlight and absorb less heat.

3.1 Temperature and Heat

◎ **Explain the difference between temperature and heat and describe the four ways heat moves.**

Heat energy is fundamentally important in physical geography. In the process of *plate tectonics* (see Chapter 13, "Drifting Continents: Plate Tectonics"),

heat from Earth's interior drives the movement of the lithosphere, creating volcanoes, mountains, and earthquakes. The Sun's energy drives the *hydrologic cycle* (see Chapter 4, "Water in the Atmosphere"). Heat from the Sun keeps the oceans liquid rather than ice, and because the oceans are liquid, water is able to evaporate from them. The evaporated water condenses into clouds that precipitate the water back to Earth's surface. As the precipitated water flows back downslope to the ocean, it erodes Earth's surface. The Sun's energy also powers the biosphere (see Chapter 8, "Patterns of Life: Biogeography") as photosynthetic organisms convert sunlight to food. Similarly, all life on Earth requires the presence of liquid water and, therefore, heat energy to survive. Heat is sometimes mistaken for temperature, but they are not the same.

What Is Temperature?

What exactly is temperature, and how do we measure it objectively? **Temperature** is defined as the average speed of molecular movement (or level of excitation) within a substance or an object, and it is measured with a *thermometer.* Molecular movement is a form of kinetic energy. (Matter is made of molecules and atoms, but the term *molecules* will be used in this discussion as a generalization.) Molecules move energetically in objects (whether they are solid, liquid, or gas) with high temperatures and relatively slowly in objects with low temperatures.

In the early 1700s, the German scientist Daniel Gabriel Fahrenheit developed the Fahrenheit scale (abbreviated F), which is now part of the *U.S. customary system.* A few years afterward, the Swedish astronomer Anders Celsius introduced the Celsius scale (abbreviated C), which is part of the *metric system* of weights and measures (see Appendix I: Unit Conversion). Water freezes at 0°C (32°F) and boils at 100°C (212°F). Most scientists worldwide use the Celsius scale. The public in most countries also uses the Celsius scale. In the United States, however, the public uses the Fahrenheit temperature scale.

Students in the United States will therefore find it helpful to be able to convert the temperature scales back and forth. Each Celsius degree is 1.8 times larger than a Fahrenheit degree. The formula for converting °C to °F is:

$$°F = (1.8 \times °C) + 32$$

The formula for converting °F to °C is:

$$°C = \frac{(°F - 32)}{1.8}$$

To quickly estimate the conversion from Celsius to Fahrenheit, double and add 30. To quickly estimate the conversion from Fahrenheit to Celsius, subtract 30 and divide by 2. For example:

$$10°C \text{ is about } 50°F: (10 \times 2) + 30 = 50$$

$$50°F \text{ is about } 10°C: \frac{(50 - 30)}{2} = 10$$

If we were to somehow lower the temperature of an object to the point at which the molecules no longer moved, we would reach 0 *kelvins,* or *absolute zero.* The Kelvin scale, developed by the British scientist Lord Kelvin (1824–1907), is abbreviated K. The

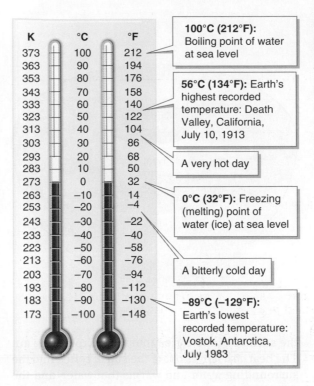

100°C (212°F): Boiling point of water at sea level

56°C (134°F): Earth's highest recorded temperature: Death Valley, California, July 10, 1913

A very hot day

0°C (32°F): Freezing (melting) point of water (ice) at sea level

A bitterly cold day

–89°C (–129°F): Earth's lowest recorded temperature: Vostok, Antarctica, July 1983

Figure 3.2 The Kelvin, Celsius, and Fahrenheit scales.
The thermometer at the right reads 0 degrees Celsius, or 32 degrees Fahrenheit, which is equivalent to 273 kelvins, as shown on the thermometer on the left.

coldest natural place known is the Boomerang Nebula in space. Its measured temperature is 1.15 kelvins, or –272°C (–457°F). Researchers in the space sciences, where subjects of inquiry can be very cold, prefer the Kelvin scale because it has no negative numbers, which facilitates calculations. Earth scientists, physical geographers, and the public use the *Fahrenheit* and *Celsius* scales **(Figure 3.2)**.

What Is Heat?

Temperature, as we have seen, is the average speed of molecular movement in a substance. **Heat** is the energy transferred between materials or systems due to their temperature differences. Molecules are too small to see, but the kinetic energy of their movement can be felt as heat. With a temperature increase, objects feel hotter because heat flows from objects of high temperature to objects of low temperature.

To illustrate this, imagine two identical ceramic mugs half full of hot coffee of the same temperature. Because each has the same temperature and the same amount of coffee, each has the same amount of internal energy.

Now imagine pouring all the coffee from one mug into the other so that one mug is full and the other is empty. The temperature of the full mug of coffee did not change, but its amount of energy doubled because it now has twice as much mass (or coffee).

Now say that you carefully set the mug of hot coffee in a bowl of ice water, making sure no water spills into the coffee. Heat flows out directly from

Figure 3.3 Heat transfer. A cup of hot coffee immersed in ice water quickly loses its energy out the sides of the cup to the cooler water. The greater the difference of temperature between two objects, the more quickly heat flows between them. This is why ice water cools the coffee faster than room-temperature tap water. When the coffee and the water are the same temperature, they exchange heat evenly.

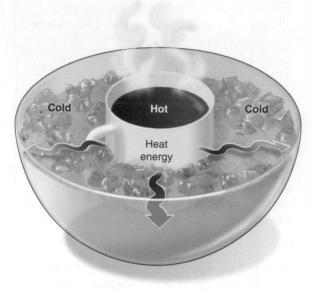

the surface of the coffee into the surrounding air. The heat *also* moves out of the cup of coffee into the surrounding water and warms that water, and the coffee's temperature rapidly decreases **(Figure 3.3)**.

How Does Heat Move?

Heat is transferred through space and Earth's physical systems in four ways: by conduction, convection, advection, and radiation.

Conduction is the process by which energy is transferred through a substance or between objects that are in direct contact with one another. Conduction occurs in solids, liquids, and gases. Remember

that when the coffee was poured into the ceramic mug, the temperature of the mug increased as the coffee's heat was transferred to the molecules in the mug. In conduction, heat always flows from objects of higher temperature to objects of lower temperature. Heat transfers faster when the temperature difference is greater and when the material through which it is traveling is a good *conductor* of heat. Metals are good heat conductors, and they usually feel cool even at room temperature because they readily conduct heat away from your hand. Still air is a poor conductor but a good insulator. An *insulator* is any material that inhibits energy from transferring through it. That is why warm jackets and the fur of animals in cold climates are thickly padded with air.

You may have noticed that when you open the door of a hot oven, a blast of hot air hits your face. This is an example of convection. **Convection** is the transfer of heat through the vertical movement of mass within a fluid (a gas or liquid). Gases and liquids move and flow easily in response to differences in temperature. For example, when the ground is warmed by sunlight, it heats the air above it. This heated warm air expands and becomes less dense, and then it rises. **Advection** is the horizontal movement of some property of the atmosphere, such as heat, humidity, or pollution. **Figure 3.4** illustrates both convection and advection.

Radiation is the process by which wave energy travels through the vacuum of space or through a physical medium such as air or water. The atmosphere is set in motion by the radiant energy Earth absorbs from the Sun, the topic of the next section.

Figure 3.4 Convection and advection. This figure shows the advection of moist air (air with water vapor) as wind moves horizontally across a cool ocean surface without changing temperature. As the advecting air moves over warmer land, it may be heated by the warm surface. As a result, it can rise, creating convection.

3.2 The Sun's Radiant Energy

◉ **Describe solar energy and its different wavelengths.**

Warm sunlight and heat from a fire are examples of radiant energy. **Radiant energy**, or *electromagnetic radiation*, is energy that is propagated in the form of electromagnetic waves, including visible light and heat. All forms of radiant energy have both electrical and magnetic properties; they are therefore referred to as *electromagnetic energy*. Visible light is only a small portion of the radiant energy emitted by the Sun.

Like all other radiant energy, sunlight travels at *light speed*, 300,000 km (186,000 mi) per second. The Sun is, on average, 149.6 million km (92.96 million mi) away from Earth, so sunlight takes 8.5 minutes to reach Earth. Other objects in the night sky, such as stars and galaxies, are much farther away, and the light they emit takes tens to thousands of years even or longer to reach Earth. The light we see shining from the stars in the sky was made long ago and reflects not what the stars look like today but what they looked like in the past, when their light was first emitted.

Photons and Wavelengths

All matter in the universe emits *photons*, which are packets of radiant energy. You emit photons, as do light bulbs, computer screens, microwave ovens, trees, rocks, plants, and water—everything.

Photons travel in waves; the distance between the peaks of two waves is called the *wavelength*. Longer wavelengths have less energy than shorter wavelengths. The **electromagnetic spectrum (EMS)** is the full range of wavelengths of radiant energy, as shown in **Figure 3.5**. The Sun emits radiation in wavelengths across most of the EMS.

Objects with higher temperatures emit more photons at shorter wavelengths than objects with lower temperatures (a phenomenon called *Wien's law*). In addition, objects with higher temperatures emit photons at higher rates than objects with lower temperatures (known as the *Stefan-Boltzmann law*). The Sun, for example, has a surface temperature of 5,800°C (10,500°F), and at that temperature, it emits energy with wavelengths centered on 0.5 μm (the visible portion of the EMS). Earth's lower atmosphere and surface, with an average temperature of 15°C (59°F), emit electromagnetic radiation with wavelengths centered on 10 μm (the thermal infrared portion of the EMS). Being much cooler,

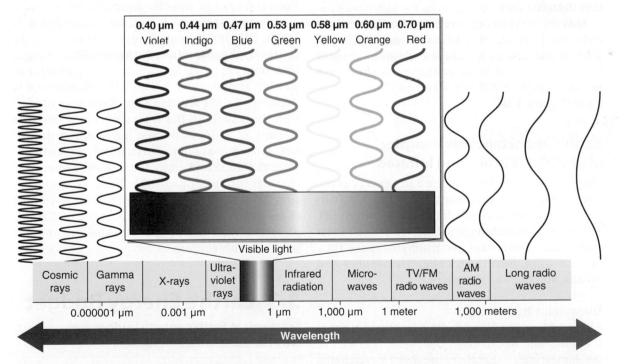

Figure 3.5 The electromagnetic spectrum. Radiant energy travels in waves. The human eye can detect only *visible* wavelengths. In the visible portion of the spectrum (shown in the inset), violet has the shortest wavelengths, at 0.40 μm, and red has the longest wavelengths, at 0.70 μm. Shorter-than-visible wavelengths, shown on the left, include cosmic rays, gamma rays, X-rays, and ultraviolet rays. (Cosmic rays originate in distant exploded stars in space, not from the Sun.) Longer-than-visible wavelengths, shown on the right, include infrared radiation, microwaves, and TV and radio waves.

Figure 3.6 Solar and terrestrial radiation. Most of the Sun's electromagnetic radiation ranges from ultraviolet waves to short infrared waves. The amount of energy the Sun emits is greatest in the visible portions of the spectrum, centered on 0.5 μm wavelengths, the wavelengths we see as green. The radiation emitted from Earth peaks at 10 μm wavelengths. (Earth's radiation intensity is greatly exaggerated here so that it can be compared to the Sun's energy. If it were scaled proportionally to the Sun's emissions, Earth's emissions would look flat.)

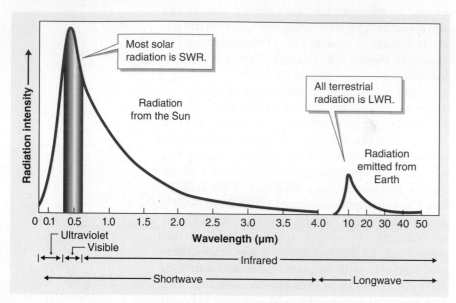

Earth emits energy at a rate about 200,000 times less than the Sun.

All radiation emitted by Earth (called *terrestrial radiation*) is **longwave radiation (LWR)**. LWR is defined as having wavelengths greater than 4 μm. Most solar radiation is **shortwave radiation (SWR)**. SWR has wavelengths less than 4 μm **(Figure 3.6)**.

Earth's Important Wavelengths: Ultraviolet, Visible, and Infrared

The vast majority—99%—of the Sun's energy emissions are in three wavelength categories: ultraviolet (UV) radiation, visible light, and infrared (IR). These three wavelength categories are the most important to physical geography because they provide light, heat the planet's atmosphere and oceans, and fuel the biosphere.

Ultraviolet Radiation
About 10% of the Sun's radiant energy output is UV radiation. Recall from Chapter 2, "Portrait of the Atmosphere," that there are three wavelengths of UV radiation: UV-A, UV-B, and UV-C. As we learned in Section 2.3, all UV-C and most UV-B radiation is absorbed by ozone molecules high in the stratosphere. Roughly 95% of the UV radiation reaching the surface is UV-A. Clouds and aerosols

also play important roles in determining how much UV radiation reaches Earth's surface. Thick clouds can absorb almost all UV radiation, and other atmospheric aerosols also scatter and absorb it. Generally, higher elevations receive more UV radiation because there are few air molecules to scatter and absorb UV rays, resulting in more intense surface UV exposure. In mountains, UV radiation exposure is further increased when a cover of bright snow blankets the ground. Clean snow can reflect up to 95% of UV radiation striking it.

Visible Light
Visible light, the portion of the electromagnetic spectrum that we can see, has wavelengths between 0.40 and 0.70 μm (see Figure 3.5). Less than half— 44%—of the Sun's radiation is concentrated in the visible wavelengths. When all visible light colors are combined, they blend into white.

Infrared Radiation
Infrared radiation (IR) has wavelengths longer than visible light. Infrared wavelengths range between 0.75 and 1,000 μm and about half of the Sun's energy output is in infrared wavelengths. There are different types of infrared energy. *Near infrared* is slightly longer than the visible portion of the EMS, and we cannot see it or feel it. TV remote controls use near infrared. In contrast, **thermal infrared (thermal-IR)** is longer than near infrared, and we can feel (sense) it as warmth and measure it with a thermometer; it is therefore called **sensible heat**. Examples of sensible heat include the warmth from a fire, the warmth of a sidewalk on a hot summer day, or heat from a heat lamp. Earth absorbs shortwave solar radiation and emits the absorbed energy as thermal infrared or sensible heat. This conversion from shortwave radiation to longwave heat radiation is exceedingly important in Earth's physical systems, as we will see in the next section of this chapter.

Video
Infrared: More Than Your Eyes Can See
Available at
www.saplinglearning.com

3.3 Earth's Energy Budget
◎ **Explain Earth's energy budget and why the atmosphere circulates.**

The fraction of the Sun's radiant energy that Earth intercepts is **insolation**, or *incoming solar radiation*. According to satellite measurements, about 1,367 W/m² of solar energy reaches the top of the atmosphere. This amount of intercepted energy is the *total solar irradiance*. The total solar irradiance is the total amount of measured solar radiation across all wavelengths per square meter. Not all of this solar energy striking the top of the atmosphere makes it through to Earth's surface, however. Insolation may be transmitted, reflected, scattered, and absorbed by the air, rocks, plants, and water. In the following paragraphs we describe each of these processes in more detail.

Transmission is the unimpeded movement of electromagnetic energy through a medium such as air, water, or glass. Ocean water transmits sunlight to limited depths. Some materials transmit only certain wavelengths of electromagnetic radiation. The atmosphere, for example, absorbs most ultraviolet and infrared wavelengths but transmits visible wavelengths. Untreated glass, on the other hand, transmits visible light and UV-A and absorbs UV-B and UV-C wavelengths.

Reflection may occur when solar radiation strikes physical matter, such as aerosols, gases, or the planet's surface. If it is reflected, the radiation returns back in the general direction from which it came, and its wavelength remains the same. If, for example, a photon in the UV-A wavelength is reflected off snow, it returns back into the atmosphere and possibly back out to space as a photon in the UV-A wavelength, unchanged.

Scattering is similar to reflection, but with scattering, solar radiation is redirected in random directions. When light is scattered, its electromagnetic wavelength does not change; only its direction of travel changes. This process creates *diffuse light*. Diffuse light seems to wrap around objects and come from every direction and is particularly pronounced at dawn and dusk **(Figure 3.7)**.

Albedo is the reflectivity of a surface, given as the percentage of incoming radiation that it reflects. Lighter-colored surfaces, such as snow, have a higher albedo than darker surfaces, such as vegetation. A perfectly reflective mirror would have an albedo of 100%. The albedo of Earth, taken as a whole, is 31%. The 69% of insolation not reflected back to space is absorbed. Air, water, plants, and the ground all absorb electromagnetic energy. **Absorption** is the ability of an object or material to assimilate electromagnetic energy and convert it to another form of energy, usually thermal infrared energy.

The planet's surface albedo varies considerably from region to region. Surfaces with a low albedo absorb more insolation than do objects with a high albedo. For example, clean snow absorbs about 10% of insolation and reflects the remaining 90%. Older and dirtier snow is less reflective. **Figure 3.8** illustrates the different albedos of various surfaces.

Figure 3.7 Diffuse light. This photo shows Mont Blanc massif and Lac Blanc in France before sunrise. The pink and orange diffuse glow often seen in mountainous regions at sunrise and sunset is called *alpenglow*. It results from scattering of light by the atmosphere. *(K Irlmeier/AGE Fotostock)*

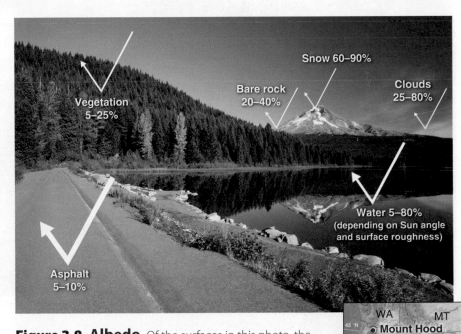

Figure 3.8 Albedo. Of the surfaces in this photo, the clean snow has the highest albedo, and the road surface and forest vegetation have the lowest. Water's albedo has a wide range because it depends on the choppiness of the water and the angle of the Sun. The albedo of clouds varies as well, depending on the type of clouds present. Oregon's Mount Hood is visible in the background. *(© Craig Tuttle/Design Pics/Corbis)*

The colors our eyes perceive are the result of reflection, refraction, scattering, and absorption of visible wavelengths of light. This concept is explored in the **Picture This** feature below.

Except in the case of incandescent lava, bioluminescent organisms, and artificial lighting by people, Earth does not emit visible light; it only reflects it. Earth is visible from space because it reflects visible sunlight that strikes it. **Figure 3.9** explains how scientists measure Earth's reflectivity.

Once a photon is absorbed, it ceases to exist in its original state. It is instead converted to another form of energy. The solar energy absorbed by plants is converted to chemical energy and thermal infrared. Solar panels absorb solar energy and convert it to electrical energy (see the Geographic Perspectives at the end of this chapter) and thermal infrared as they are warmed by sunlight. Earth's surface and atmosphere absorb solar radiation and convert it to thermal infrared.

In general, the more shortwave energy an object can absorb (or the lower its albedo), the more longwave thermal infrared energy it can radiate. How much shortwave radiation an object can absorb and retain as heat depends on its albedo and its heat capacity. Urbanized regions are often significantly

Picture This **The Colors of the Rainbow**

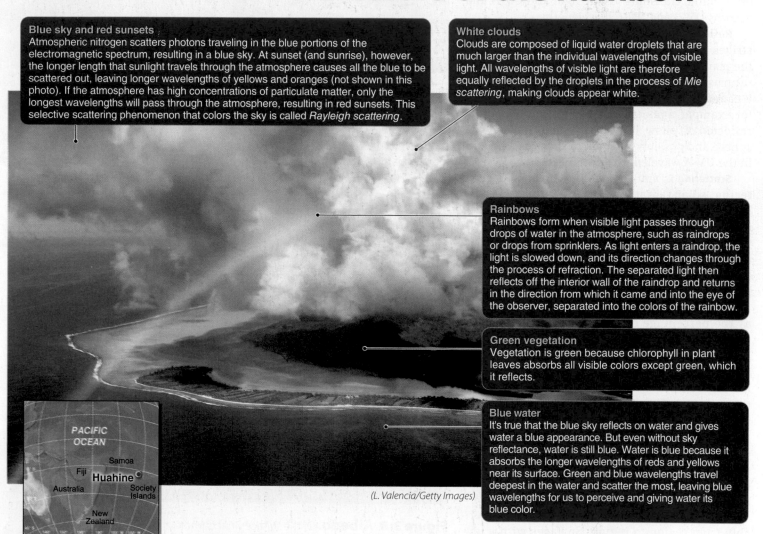

Blue sky and red sunsets
Atmospheric nitrogen scatters photons traveling in the blue portions of the electromagnetic spectrum, resulting in a blue sky. At sunset (and sunrise), however, the longer length that sunlight travels through the atmosphere causes all the blue to be scattered out, leaving longer wavelengths of yellows and oranges (not shown in this photo). If the atmosphere has high concentrations of particulate matter, only the longest wavelengths will pass through the atmosphere, resulting in red sunsets. This selective scattering phenomenon that colors the sky is called *Rayleigh scattering*.

White clouds
Clouds are composed of liquid water droplets that are much larger than the individual wavelengths of visible light. All wavelengths of visible light are therefore equally reflected by the droplets in the process of *Mie scattering*, making clouds appear white.

Rainbows
Rainbows form when visible light passes through drops of water in the atmosphere, such as raindrops or drops from sprinklers. As light enters a raindrop, the light is slowed down, and its direction changes through the process of refraction. The separated light then reflects off the interior wall of the raindrop and returns in the direction from which it came and into the eye of the observer, separated into the colors of the rainbow.

Green vegetation
Vegetation is green because chlorophyll in plant leaves absorbs all visible colors except green, which it reflects.

Blue water
It's true that the blue sky reflects on water and gives water a blue appearance. But even without sky reflectance, water is still blue. Water is blue because it absorbs the longer wavelengths of reds and yellows near its surface. Green and blue wavelengths travel deepest in the water and scatter the most, leaving blue wavelengths for us to perceive and giving water its blue color.

PACIFIC OCEAN

Samoa
Fiji   **Huahine**
Australia   Society
                 Islands
New
Zealand

(L. Valencia/Getty Images)

This aerial photograph of Huahine, part of the Society Islands in French Polynesia, shows a world awash in tropical sunlight and color. Colors are produced when visible light is reflected, refracted, scattered, and absorbed.

Consider This

1. Clouds are white because they scatter all wavelengths equally. Why are objects black?
2. What would change in this photo if Rayleigh scattering scattered mostly green wavelengths of light?

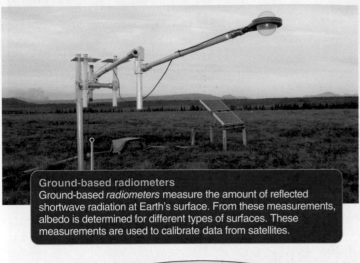

Ground-based radiometers
Ground-based *radiometers* measure the amount of reflected shortwave radiation at Earth's surface. From these measurements, albedo is determined for different types of surfaces. These measurements are used to calibrate data from satellites.

Satellite radiometers
Using scanning radiometers, satellites provide a comprehensive portrait of Earth's albedo and energy budget. They measure albedo over the entire surface of Earth, but at a lower spatial resolution than ground-based radiometers. An artist's rendering of the *Terra satellite* is shown here

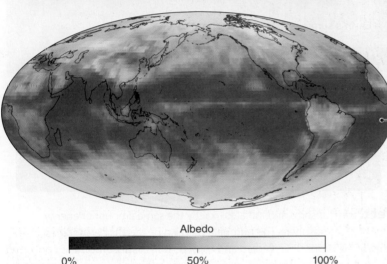

Earth albedo map
This image shows reflection of shortwave radiation by Earth's surface and atmosphere. It was compiled by the Terra satellite for March 2005 as measured by Clouds and Earth's Radiant Energy System (CERES) instrumentation. The reflectivity of light areas is high and for dark areas it is low. High latitudes with extensive snow, ice, and cloud cover are the most reflective areas of the planet. The equatorial band of reflectivity is caused by persistent cloudiness. The dark oceans absorb a large proportion of incoming solar energy.

Albedo

0%   50%   100%

Figure 3.9 Measuring Earth's reflectivity. Monitoring Earth's albedo is essential to climate studies. For example, decreasing snow cover due to global warming is decreasing Earth's albedo, allowing Earth's surface to absorb and be warmed by sunlight rather than reflecting it and remaining cool. Scientists monitor planetary albedo from the ground and from orbiting *satellites.   (Top left. Martin Shields/Getty Images; top right and bottom. NASA Earth Observatory)*

Web Map
Global Albedo
Available at
www.saplinglearning.com

warmer than surrounding rural areas in summer due to the urban heat island effect. The Human Sphere section at the beginning of this chapter discusses this phenomenon in more detail.

The Great Balancing Act

Earth's surface temperature increases as Earth absorbs energy from the Sun, and it decreases as Earth radiates heat to the atmosphere and to space. Internal geothermal energy from the planet's core warms Earth's surface only an insignificant amount. If the Sun were to stop shining, the temperature of Earth's surface would plunge as the planet quickly radiated away its heat. Each night, the Sun effectively does stop shining, and the surface temperature drops because Earth loses more energy than it absorbs. Each morning, the Sun rises again, and Earth's surface—and the atmosphere above it—warms again because it absorbs more energy than it radiates.

Just as a bank account's balance is the result of deposits and withdrawals, the temperature of Earth's surface and atmosphere is the result of a balance between incoming and outgoing energy. As long as the amount of energy absorbed is equal to the amount of energy radiated, Earth's temperature is stable. That stable temperature is the **radiative equilibrium temperature**. Taken as a whole and including the warming effects of the atmosphere caused by the greenhouse effect, Earth's radiative equilibrium temperature is 15°C (59°F). The two closest planets to Earth have very different radiative equilibrium temperatures. For Venus the radiative equilibrium temperature is 462°C (864°F), and for Mars it is –55°C (–67°F).

Although the amount of insolation absorbed from the Sun and the amount of longwave infrared radiated back to space are exactly balanced, several complex and important interactions are involved within Earth's systems that affect this

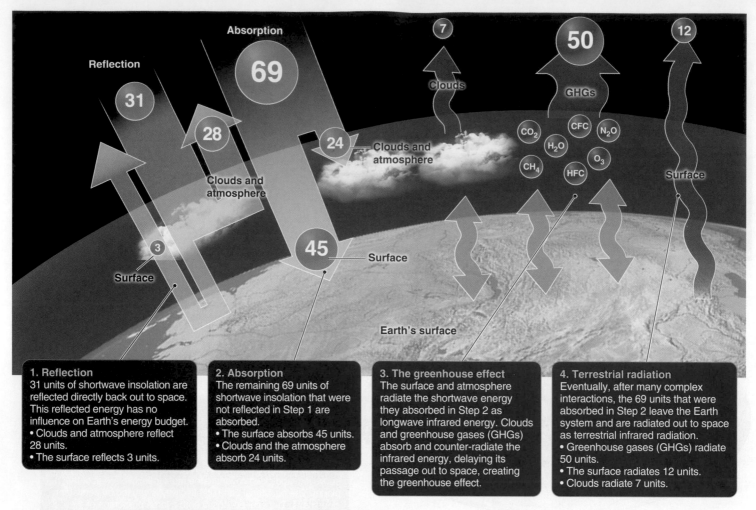

1. Reflection
31 units of shortwave insolation are reflected directly back out to space. This reflected energy has no influence on Earth's energy budget.
• Clouds and atmosphere reflect 28 units.
• The surface reflects 3 units.

2. Absorption
The remaining 69 units of shortwave insolation that were not reflected in Step 1 are absorbed.
• The surface absorbs 45 units.
• Clouds and the atmosphere absorb 24 units.

3. The greenhouse effect
The surface and atmosphere radiate the shortwave energy they absorbed in Step 2 as longwave infrared energy. Clouds and greenhouse gases (GHGs) absorb and counter-radiate the infrared energy, delaying its passage out to space, creating the greenhouse effect.

4. Terrestrial radiation
Eventually, after many complex interactions, the 69 units that were absorbed in Step 2 leave the Earth system and are radiated out to space as terrestrial infrared radiation.
• Greenhouse gases (GHGs) radiate 50 units.
• The surface radiates 12 units.
• Clouds radiate 7 units.

Figure 3.10 Energy budget and the greenhouse effect. Earth absorbs and radiates exactly the same amount of energy. This figure illustrates what happens when 100 units of solar energy enter Earth's atmosphere. The numbers can also be read as percentages. For example, of the 100% (or 100 units) of energy entering Earth's systems, 31% (31 units) is reflected and 69% (69 units) is absorbed and converted to infrared energy. These numbers are a planetary average, and any given region will have different values. For example, the average reflectance of Earth's surface is only 3%, but areas covered with fresh snow can reflect as much as 90% of insolation.

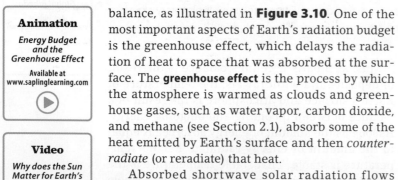

Animation
Energy Budget and the Greenhouse Effect
Available at
www.saplinglearning.com
▶

Video
Why does the Sun Matter for Earth's Energy Budget?
Available at
www.saplinglearning.com
▶

balance, as illustrated in **Figure 3.10**. One of the most important aspects of Earth's radiation budget is the greenhouse effect, which delays the radiation of heat to space that was absorbed at the surface. The **greenhouse effect** is the process by which the atmosphere is warmed as clouds and greenhouse gases, such as water vapor, carbon dioxide, and methane (see Section 2.1), absorb some of the heat emitted by Earth's surface and then *counter-radiate* (or reradiate) that heat.

Absorbed shortwave solar radiation flows through and does work on Earth's systems in many ways. It warms the circulating ocean currents, it evaporates water, and it is essential to photosynthesizing organisms, such as plants, which use it to grow. **Figure 3.11** illustrates how energy absorbed by the surface is eventually transferred back to the atmosphere.

Heat is not "trapped" through the greenhouse effect. If it were, no heat would radiate out to space, and it would instead remain within Earth's systems. If heat were trapped in this manner, Earth's atmosphere would have long ago become extremely

hot and inhospitable for life. Conversely, if it were not for the presence of greenhouse gases in the atmosphere, the greenhouse effect would not occur. Without the effects of greenhouse gases, Earth's radiative equilibrium temperature would be about −18°C (0°F), cold enough to freeze the oceans solid. Humans have modified the greenhouse effect by adding greenhouse gases to the atmosphere, creating the *anthropogenic greenhouse effect*; this important topic is discussed in Section 7.3.

The Global Heat Engine

For the planet as a whole, incoming SWR and outgoing terrestrial LWR radiation are equal. But the amount of incoming and outgoing energy are not equal for most geographic regions. Because Earth is a sphere, insolation is most intense within the tropical and subtropical latitudes. Between 37 degrees north and south, there is a net surplus of energy. Insolation becomes more diffuse (weakened) at high latitudes, and, consequently, at latitudes higher than 37 degrees, there is a net deficit of energy (**Figure 3.12**).

Figure 3.11 Surface-to-atmosphere energy transfer. Earth's surface absorbs 45 units of shortwave solar energy (see Figure 3.10). This energy eventually is transferred back to the atmosphere or directly out to space: The land surface radiates 22 of the 45 units of energy. Of those 22 units, 12 go directly out to space, while the other 10 are absorbed by gases and clouds in the atmosphere. Of the remaining 23 units of energy absorbed by Earth's land surface, 19 units work to melt ice and evaporate water to form water vapor, and 4 units work to conduct and convect heat to the atmosphere.

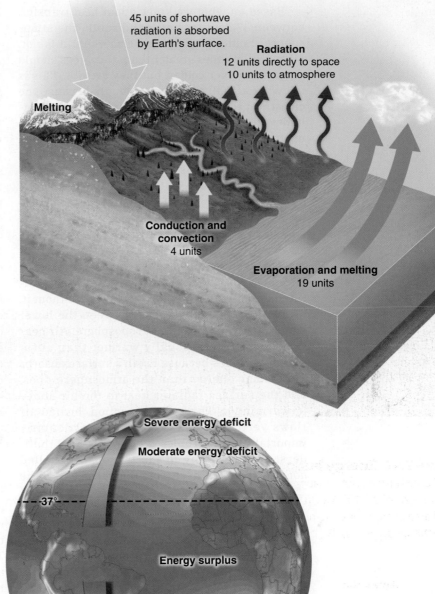

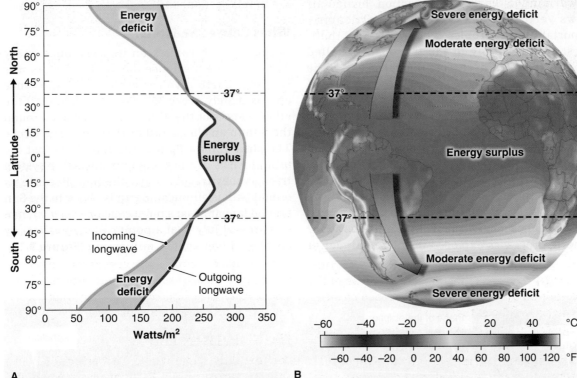

A

B

Figure 3.12 Energy balance and the global heat engine. (A) On average, only near 37 degrees latitude does the amount of insolation at the top of the atmosphere equal the amount of heat radiated by Earth. Equatorward of 37 degrees, there is a surplus of energy because Earth absorbs more solar energy than it radiates as longwave infrared energy. Poleward of 37 degrees, there is a net deficit of energy because Earth radiates more longwave infrared energy than it absorbs from the Sun. (B) This map shows air temperature (in degrees Celsius) near the surface averaged for the period 1979 to 2015. The tropics have energy surpluses, the midlatitudes have moderate energy deficits, and the polar regions have severe energy deficits. The arrows represent the direction of heat flow out of the tropics. *(Original data source: ECMWF ERA-Interim reanalysis. Image made from the website ClimateReanalyzer.org, courtesy of the Climate Change Institute, University of Maine, USA.)*

Based on what we know about heat transfer, between 37 degrees north and south, the atmosphere should be growing warmer, but it is not. Likewise, latitudes greater than 37 degrees should be growing ever colder, but they are not. This is not what actually happens, though, because the **global heat engine** moves heat from low to high latitudes and low to high altitudes as a result of heating inequalities across Earth's surface.

Heat energy at low latitudes is converted to kinetic energy as it is advected (transported horizontally) poleward by the atmosphere and the oceans. Roughly 75% of heat advected from the tropics to higher latitudes occurs by atmospheric flow (or wind). The remaining 25% of heat is advected from low to high latitudes in seawater by ocean currents. The global heat engine circulates heat worldwide and keeps Earth's temperature far more uniform than it would be without it.

Heat is also imbalanced between the lower troposphere and the upper troposphere. Air near Earth's surface is usually warmer than air at higher altitudes because Earth's surface absorbs more solar energy than the atmosphere does, and the surface radiates heat to the air above it, warming it. Through convection, heated air flows vertically upward. This is particularly important in the tropics, where powerful thunderstorms reaching to the tropopause, called *hot towers,* inject heat from the surface into the stratosphere.

The global heat engine transports immense amounts of heat. Together, movement of heat toward the poles and to higher altitudes creates the global heat engine that puts the atmosphere and oceans in motion. Advective movement of water vapor, as we will see in Chapter 4, "Water in the Atmosphere," also plays a vitally important role in the global heat engine and generating winds. The global heat engine plays a role in all the atmospheric processes explored in the remainder of Part I (Chapters 4, "Water in the Atmosphere" to 7, "The Changing Climate").

3.4 The Four Seasons

◎ **Explain what causes seasons.**

We learned in the previous section that solar radiation intensity varies according to latitude and over the course of a day as the Sun rises and sets. Solar radiation intensity at any given location also varies over the course of a year due to the changing of the seasons. There are two types of seasons. The *meteorological seasons* are the cyclical changes in weather over a year. Some regions, such as the interior of North America, experience extreme meteorological seasons, or *seasonality,* while other regions, such as the tropics, experience the same weather all year and have low seasonality. In contrast to meteorological seasonality, everywhere on the planet experiences *astronomical seasons.* These are the cyclical changes in the position of the Sun and stars in the sky through the year. The astronomical seasons cause the meteorological seasons, as we will see in this section.

What Causes Seasons?

Because Earth *rotates* on its axis, once every 24 hours we experience day and night. As it is rotating, Earth *revolves* around the Sun, once approximately every 365 days. Earth and all the other planets of the solar system revolve around the Sun in an *ellipse* rather than in a true circle. The **plane of the ecliptic** is the flat plane that the orbital paths of the planets in the solar system trace as they travel around the Sun. Because its orbital path is elliptical, Earth is closer to the Sun around January 3 (at a point called *perihelion*) than it is around July 4 (at a point called *aphelion*) by about 5 million km (3.1 million mi) **(Figure 3.13)**.

**Figure 3.13
Aphelion and perihelion.** Earth's orbit around the Sun follows an ellipse rather than a circle. Because Earth is closest to the Sun (at the perihelion) about January 3, sunlight at that time is about 7% stronger than when Earth is farthest from the Sun on July 4 (at the aphelion).

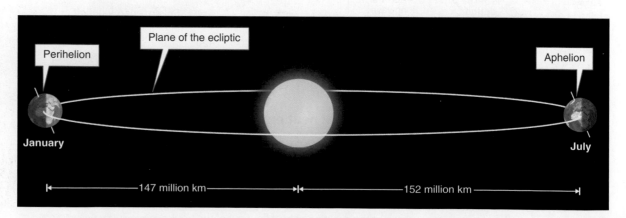

The average distance between Earth and the Sun is called an *astronomical unit,* or *AU* (1 AU = 149.6 million km [92.96 million mi]). The changing distance between Earth and the Sun does not cause the seasons, however. If it did, it would be warmest in the Northern Hemisphere when Earth is closest to the Sun in January.

The seasons are caused by the changing length of daylight hours over the course of a year and by the changing position of the Sun's most vertical position as it crosses the sky each day. Both of these factors, in turn, are controlled by the tilt of Earth's axis. The following four components of Earth's movement and position are necessary to understand how tilt causes seasons:

1. *Earth's revolution and the plane of the ecliptic:* Earth's revolution around the Sun follows the plane of the ecliptic. Earth does not stray above or below this plane. Each revolution takes 365 days, 5 hours, 48 minutes, and 46 seconds (or 365.25 days) to complete.

2. *Axial tilt:* If the North Pole were pointed straight "up," the rotational axis of Earth would be perpendicular to the plane of the ecliptic, and the angle of its tilt would be zero. Instead, Earth's North Pole is tilted 23.5 degrees from the vertical **(Figure 3.14)**.

3. *Parallel axis:* As Earth orbits the Sun, the angle of Earth's axis remains constant. It is always tilted 23.5 degrees in the same direction and is parallel to itself in all locations along its orbital path (see Figure 3.13).

4. *Subsolar point:* Because Earth's surface is curved, the Sun's rays are exactly perpendicular (directly overhead) to Earth's horizontal surface at only one point at or near noon. This point is called the **subsolar point (Figure 3.15)**.

As Earth rotates on its axis, the subsolar point moves along a constant line of latitude. Earth rotates 1,670 km/h (1,037 mph) at the equator. This means that when the subsolar point is at the equator, it is moving 1,670 km/h. In other words, if you wanted to keep up with the subsolar point, keeping the Sun straight overhead, you would have to travel 1,670 km/h westward along the equator!

The subsolar point moves gradually north and south over the year as a result of the tilt of Earth's axis. It never moves farther north than 23.5 degrees north latitude (a parallel called the **Tropic of Cancer**) or farther south than 23.5 degrees south latitude (a parallel called the **Tropic of Capricorn**). The latitude of the subsolar point is the *solar declination.* For example, when the subsolar

point is at the equator, the solar declination is 0 degrees. When the subsolar point is at 10 degrees north latitude, the solar declination is 10 degrees north, and so on. As Figure 3.15 shows, the latitude of the subsolar point is at a right angle from the **circle of illumination**, the line separating night from day, where sunrise and sunset are occurring.

Story Map

What Causes Seasons?

Available at
www.saplinglearning.com

Figure 3.14 Axial tilt. Earth's axis is always tilted in the same direction, 23.5 degrees away from vertical in relation to the plane of the ecliptic.

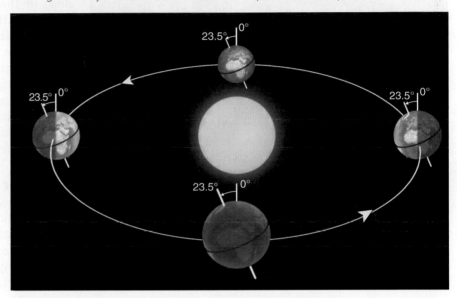

Figure 3.15 The subsolar point. The Sun's rays are parallel to one another when they reach Earth. Because Earth is spherical, they are exactly perpendicular to Earth's horizontal surface at only one point, the subsolar point. Person C, who is standing on the equator at noon, is standing on the subsolar point.

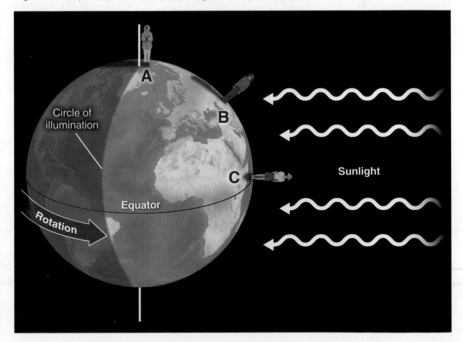

Picture This **The Circle of Illumination**

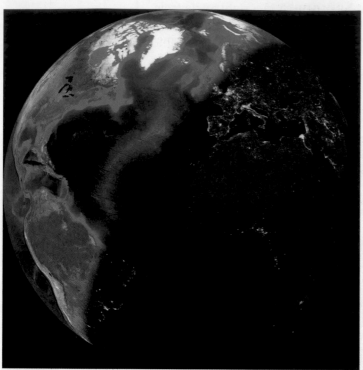

(Earth image was produced by Earth and Moon Viewer (https://www.fourmilab.ch/earthview/) derived from NASA Blue Marble, Terra/MODIS cloudless Earth and Black Marble night lights images.)

The beauty of sunrises and sunsets has inspired humans for as long as we have walked Earth. For most of our history, human activity was controlled by the passage of the circle of illumination as well. It is the dividing line between night and day. People were active by day and inactive at night. But that is no longer the case. Now we have artificial night lighting that allows us to be active any time. Earth is the only planet that glows on the night side. Night lighting is so prevalent that the outlines of the continents in populated areas, such as in Italy and Spain in this image, are clearly discernible. Large cities, like London, Paris, and Cairo, stand out. Can you find them in the image at right? In order to create a globe without clouds, numerous cloud-free satellite images were stitched together to create this *composite image*. It shows the position of the circle of illumination for June 28, 2017, at 6:19 p.m. Eastern Standard Time. The shadow night will sweep across the Atlantic Ocean and into North America.

Consider This

1. How many times in a 24-hour period will the circle of illumination pass over you? Is there anywhere on Earth where this answer may change? (Hint: Continue reading this section.)

2. How many times have you passed through the circle of illumination in your lifetime?

Figure 3.16 Latitude and sunlight intensity. Sunlight enters the atmosphere perpendicular to Earth's surface only in the tropics. Outside the tropics, the same amount of solar energy is distributed across a greater surface area, creating more diffuse sunlight. In addition, at higher latitudes sunlight passes through a greater distance of atmosphere than at lower latitudes. This longer path length reduces the intensity of sunlight because aerosols (such as clouds and particulate matter) and gases attenuate (weaken) sunlight. At location A, sunlight is diffuse because of the low solar altitude (60°) and the long distance of travel sunlight takes through the atmosphere. Location B, in contrast, has more intense sunlight because of the high solar altitude (90°) and the short path length of sunlight through the atmosphere.

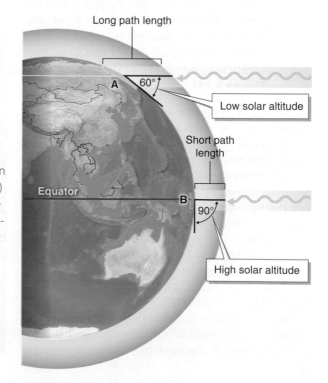

The **Picture This** feature on the facing page discusses the circle of illumination further.

The subsolar point determines the **solar altitude** (or *angle of incidence*), which is the noontime angle of the Sun above the horizon, in degrees. Sunlight on the horizon is at 0 degrees, and sunlight that is straight overhead is at 90 degrees, the highest solar altitude possible. The *path length,* which is the amount of atmosphere that sunlight must pass through to reach the surface, determines the intensity of the Sun **(Figure 3.16)**. The solar altitude decreases (the noontime Sun gets closer to the horizon) at higher latitudes, causing sunlight to become more diffuse (spread out). Because of this diffusion, high latitudes receive less solar energy than low latitudes, which also means they are colder.

How Does the Migration of the Subsolar Point Affect Seasonality?

The subsolar point, where sunlight is most intense, moves from the Tropic of Cancer to the Tropic of Capricorn, then back again, over a 12-month period. Because the latitude of the subsolar point changes, from our perspective on the ground, the Sun's path across the sky changes with the time of year **(Figure 3.17)**.

The **December solstice** occurs around December 21 each year, when the subsolar point arrives at the Tropic of Capricorn. This seasonal marker is also

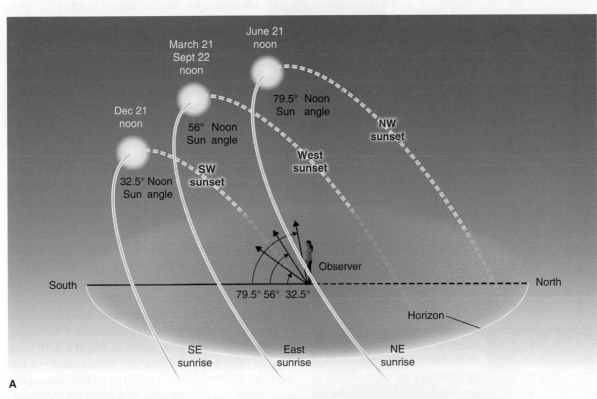

Figure 3.17 Topanga sunsets.

(A) The position of sunrise and sunset and the solar altitude change as the seasons change. At 34 degrees north latitude (where the accompanying photos were taken), the solar altitude ranges from 32.5 degrees on the December solstice to 79.5 degrees on the June solstice. (B) Three different photos looking west from Topanga Canyon in southern California (located at 34° N latitude) were taken on December 21, March 21, and June 21 and stitched together. They show the changing location of the setting Sun for this location. The yellow dotted lines show the approximate path of the Sun across the sky on each day. *(David Lynch)*

Figure 3.18 Seasonal markers. Because Earth's axial tilt is 23.5 degrees from vertical, the subsolar point (shown as a dot and not visible at point 3 in the diagram) migrates between the Tropic of Cancer and the Tropic of Capricorn, across 47 degrees of latitude, in a 6-month period.

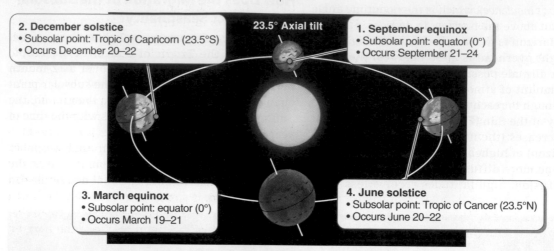

2. December solstice
• Subsolar point: Tropic of Capricorn (23.5°S)
• Occurs December 20–22

23.5° Axial tilt

1. September equinox
• Subsolar point: equator (0°)
• Occurs September 21–24

3. March equinox
• Subsolar point: equator (0°)
• Occurs March 19–21

4. June solstice
• Subsolar point: Tropic of Cancer (23.5°N)
• Occurs June 20–22

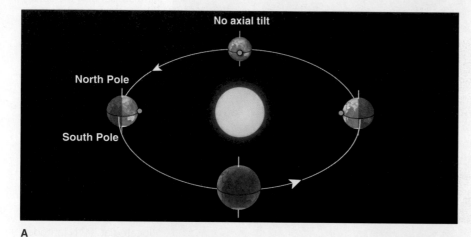

No axial tilt

North Pole

South Pole

A

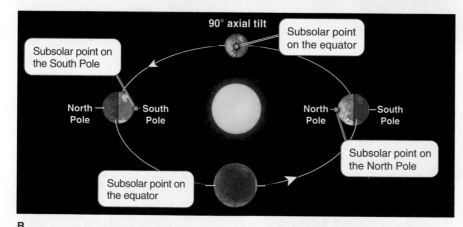

90° axial tilt

Subsolar point on the equator

Subsolar point on the South Pole

North Pole — South Pole

North Pole — South Pole

Subsolar point on the North Pole

Subsolar point on the equator

B

Figure 3.19 Tilt and seasonality. (A) If Earth's axis had no tilt, the subsolar point (shown as an orange dot and not visible on the nearest globe) would always be on the equator, no matter the time of year. In this scenario, there would be no seasonality. (B) If Earth's axis were tilted 90 degrees, the subsolar point (the orange dot) would migrate from the South Pole to the North Pole and then back again within the period of 1 year. This would create extreme seasonal change.

called the *winter solstice* in the Northern Hemisphere and the *summer solstice* in the Southern Hemisphere. After the December solstice, the subsolar point migrates northward. The **March equinox** (also called the *spring equinox* or *vernal equinox* in the Northern Hemisphere and the *fall equinox* or *autumnal equinox* in the Southern Hemisphere) occurs when the subsolar point crosses the equator on or around March 20. Around June 21, the subsolar point arrives at the Tropic of Cancer, marking the **June solstice** (or *summer solstice* in the Northern Hemisphere). From there the subsolar point migrates back south. Three months later, around September 22, the subsolar point crosses the equator again, marking the **September equinox** (or *fall equinox* in the Northern Hemisphere). These seasonal markers are illustrated in **Figure 3.18**. You can find the solar declination (the latitude of the subsolar point) for any day of the year by using the *analemma* chart provided in Appendix IV, "Analemma Chart."

The importance of the tilt of Earth's axis to seasons is revealed when we imagine what would happen if that tilt were changed significantly. As axial tilt increases, so does seasonality **(Figure 3.19)**.

How Does Earth's Tilt Affect Day Length?

Just as Earth's axial tilt causes the subsolar point to move across latitudes, it also causes the number of daylight hours to change. To illustrate, we will first look at what happens to day length on the equinoxes.

Day Length on an Equinox

On an equinox, all locations on Earth (except the poles) receive 12 hours of daylight and 12 hours of darkness. At these times, all locations on Earth rotate through equal day and night sides of the planet, as shown in **Figure 3.20**.

Day Length on the December Solstice

The situation is quite different on the December solstice, when the North Pole is pointed away from the Sun, and the subsolar point is directly over the Tropic of Capricorn, at 23.5 degrees south latitude. The equator always receives 12 hours of day and night, but the rest of the planet's day length varies.

The way day length varies depends on the hemisphere. Day length is shortest in the Northern Hemisphere and longest in the Southern Hemisphere on the December solstice. At the **Arctic Circle** (at 66.5° N latitude) and northward, there are 24 hours of darkness. Similarly, at the **Antarctic Circle** (at 66.5° S latitude) and southward, there are 24 hours of daylight **(Figure 3.21)**.

Day Length on the June Solstice

The June solstice marks the longest day length in the Northern Hemisphere and the shortest day length the Southern Hemisphere. On June 21 from the Arctic Circle to the North Pole, the Sun never sets. Because the Sun sets only briefly if at all in the high Arctic (or in the Antarctic in December), the high latitudes are often referred to as the "land of the midnight Sun" **(Figure 3.22** on the following page**)**.

If you were to stand on the equator for 1 year, you would see the solar altitude at noon range between 66.5 degrees north (in June) and 66.5 degrees south (in December). In other words, in the tropics, the noon Sun is always high in the sky near 90 degrees, and seasonality is negligible. On the other hand, if you were to stand on the North Pole for 1 year, you would see the noon solar altitude range from 23.5 degrees above the horizon

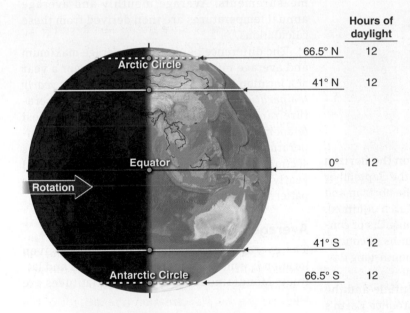

	Hours of daylight
66.5° N Arctic Circle	12
41° N	12
Equator 0°	12
41° S	12
Antarctic Circle 66.5° S	12

Rotation

Figure 3.20 Equinox day length. On the equinoxes in March and September, the subsolar point is at the equator, so the circle of illumination (which is always 90 degrees from the latitude of the subsolar point) passes through the poles. This position makes the length of day and night equal. Therefore, any point on Earth except the poles takes 12 hours to rotate through the daylight side and 12 hours to rotate through the night side. At the poles on the equinoxes, the half-disk of the Sun traces the line of the horizon where the horizon is flat, such as over the North Pole's sea ice.

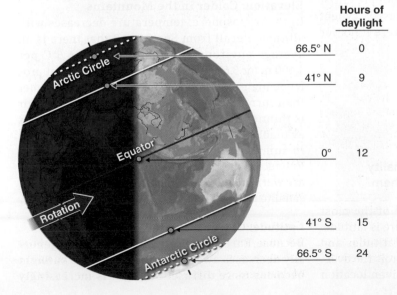

	Hours of daylight
66.5° N Arctic Circle	0
41° N	9
Equator 0°	12
41° S	15
Antarctic Circle 66.5° S	24

Rotation

Figure 3.21 December solstice day length. Imagine that you are standing at 66.5 degrees south, on the Antarctic Circle. At that location, you never cross the circle of illumination as Earth rotates, and the Sun does not set at your location. Now imagine standing at 41 degrees south latitude. Here you rotate across the night side of the planet for 9 hours, and you are on the daylight side for 15 hours. On the equator, day and night are equal. At 41 degrees north latitude, day length is 9 hours, and night length is 15 hours. At 66.5 degrees north latitude, on the Arctic Circle, you receive no direct sunlight because that location does not cross the circle of illumination into the lighted side of the globe as Earth rotates. Locations just north of the Arctic Circle experience twilight for several hours as the Sun approaches, but the Sun does not rise above the horizon.

Figure 3.22 June solstice day length. On the June solstice, the situation is reversed relative to the December solstice (see Figure 3.21). As you travel north, daylight increases until you reach the Arctic Circle at 66.5 degrees north. Locations within the Arctic Circle do not cross the circle of illumination into the night side of the globe, and as a result, day length is 24 hours. Traveling southward, day length decreases. Anywhere within the Antarctic Circle, 66.5 degrees south, the night is 24 hours long.

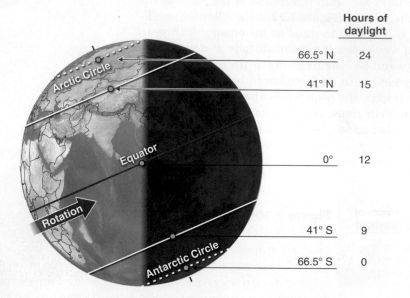

	Hours of daylight
66.5° N	24
41° N	15
0°	12
41° S	9
66.5° S	0

to the south (in June) to 0 degrees on the horizon (on the September equinox). After the September equinox, the Sun would dip below the horizon and not return for 6 months (until the March equinox). In other words, the poles receive 6 months of continuous darkness, followed by 6 months of continuous daylight, as if there were a 6-month-long day, followed by a 6-month-long night.

Together, the changing solar altitude and the changing length of daylight hours produce Earth's seasonality. Warm summer days are the result of both more intense sunlight and increased daylight hours. And this would not happen were it not for the tilt of Earth's axis.

3.5 Patterns of Temperature and Seasonality

◎ Describe Earth's surface seasonality patterns and explain what causes them.

We learned in Section 3.4 that one of the most important determinants of temperature is latitude: The tropics are warm, and the midlatitudes and high latitudes are cold. But what controls the overall air temperature pattern of any given location

on Earth? Death Valley, California, is always hot in summer; in fact, it holds the record for the highest air temperature ever officially recorded on Earth: On July 10, 1913, it reached 56°C (134°F)—in the shade! Earth's lowest official air temperature, −89°C (−129°F), was recorded in Vostok Station, Antarctica, at a scientific research base in July 1983. Why is Death Valley so hot? Why is Antarctica so cold?

Recording air temperatures gives us critical data for answering these questions as well as for monitoring climate change. Temperature data are compiled by the Global Historical Climatology Network (GHCN) and are recorded daily at more than 7,000 meteorological stations in 180 countries and territories **(Figure 3.23)**. Each station has at least 10 years of data. For each station, an average daily temperature is calculated from maximum and minimum daily temperature measurements. Average monthly and average annual temperatures are then derived from these calculations.

The difference between the average maximum and average minimum temperatures over a year at a location is referred to as that location's *yearly temperature range*. Earth's maximum temperature range (the difference between the highest temperature, in Death Valley, and the lowest temperature, in Antarctica) is 145°C (265°F). In this section, we use average annual temperatures and yearly temperature ranges to explore geographic patterns of temperature and seasonality.

Average Annual Temperature Patterns

The average annual temperature at any given location is controlled mainly by elevation and latitude. Mountainous areas and high latitudes are cold.

Elevation: Colder in the Mountains

In the troposphere, temperature decreases with altitude. Recall from Section 2.3 that there is an average environmental lapse rate of 6.5°C per 1,000 m (or 3.6°F per 1,000 ft). Mountains that protrude high into the troposphere are always cooler than surrounding lowland regions because the air is thinner and, consequently, less heat energy is available. High mountains are often snowcapped in summer, even in tropical locations, such as Ecuador, where adjacent areas of low elevation are warm. **Figure 3.24** illustrates this point with a satellite thermal image of South Asia.

Latitude: Colder near the Poles

Because Earth is spherical, temperature generally decreases away from the equator as sunlight becomes more diffuse. Sunlight is increasingly

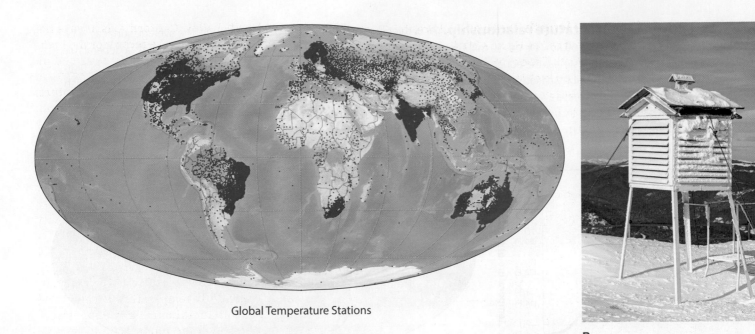

Global Temperature Stations

A   B

Figure 3.23 Map of global temperature stations. (A) This world map shows the locations of official GHCN meteorological stations that record air temperatures. Station density decreases in remote areas and in less economically developed countries. (B) According to the World Meteorological Organization, official instrument shelters (shown here) must be 1.2 to 3 m (3.9 to 9.8 ft) off the ground and shielded from direct sunlight. They also cannot be in the shade of a building, tree, or mountain. *Thermistors* (digital thermometers) within the shelter record and transmit their data automatically. *(Cosmin-Constantin Sava/Alamy)*

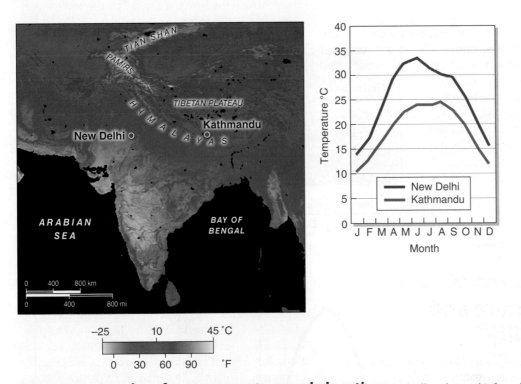

Figure 3.24 Land surface temperature and elevation. A satellite thermal infrared image shows average January land surface temperatures for South Asia during the period 2001–2010. Reds and oranges, which depict warm temperatures, are found mainly at low elevations, and blues, which show cold temperatures, are found mainly in mountainous regions. The monthly air temperature for Kathmandu, Nepal, is cooler than that for New Delhi, India, because it is at a higher elevation (climate diagram at right). Kathmandu is located at 1,400 m (4,600 ft) elevation, and New Delhi is at 216 m (709 ft). *(NASA)*

Figure 3.25 Latitude–temperature relationship. (A) Between 20 degrees and 80 degrees latitude, the average surface temperature changes roughly 1°C per 160 km (1°F per 50 mi). The effects of elevation have been removed from this graph. (B) As a general rule, average annual temperatures are highest near the equator and decrease toward higher latitudes. But there are many exceptions. Areas of high elevations, such as the Andes and the Tibetan Plateau, are colder than their surrounding regions, regardless of the latitude. In some areas, such as the Sahara Desert, temperatures increase away from the equator because of the persistently cloudless skies and intense sunlight. Antarctica is Earth's coldest region because the cold *circumpolar current* circulates around the continent and inhibits warm water from reaching it. Antarctica also has a high average elevation: 2,300 m (8,200 ft). *(NASA)*

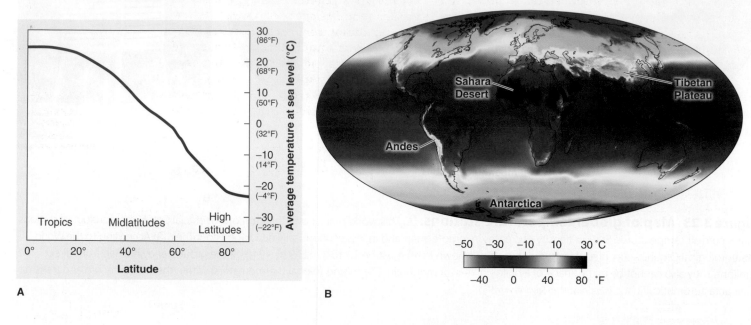

A   B

Figure 3.26 Seasonality graphs. (A) Verkhoyansk, Russia, holds the Guinness world record for the greatest temperature range on Earth. The lowest temperature recorded there is −67.8°C (−90°F) in February and the highest is 37.3°C (99.1°F) in July; thus, the extreme temperature range for the city is 105.1°C (189.1°F). In contrast, the average temperature ranges from −45°C (−50°F) in January to 16°C (62°F) in July; thus, the average yearly temperature range is 61°C (112°F). (B) Singapore's average monthly temperatures vary by only a few degrees. Every month brings warm tropical temperatures.

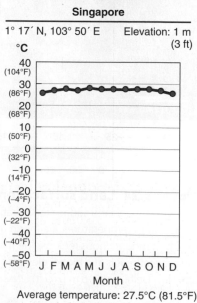

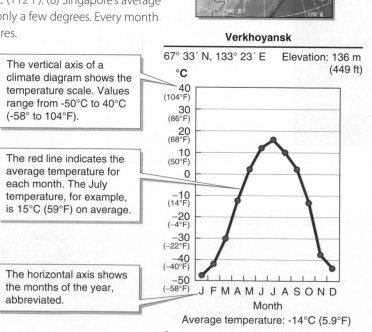

The vertical axis of a climate diagram shows the temperature scale. Values range from -50°C to 40°C (-58° to 104°F).

The red line indicates the average temperature for each month. The July temperature, for example, is 15°C (59°F) on average.

The horizontal axis shows the months of the year, abbreviated.

Verkhoyansk
67° 33′ N, 133° 23′ E   Elevation: 136 m (449 ft)
Average temperature: -14°C (5.9°F)
A

Singapore
1° 17′ N, 103° 50′ E   Elevation: 1 m (3 ft)
Average temperature: 27.5°C (81.5°F)
B

diffuse outside the tropics, resulting in lower surface temperatures at higher latitudes **(Figure 3.25)**.

Seasonality Patterns

Imagine that you live in a subarctic climate where the average January temperature is −47°C (−53°F). It is so cold that you must leave your car running all day and all night during the winter months to avoid having the engine freeze and seize up. On a cold day, boiling water thrown into the air comes down as ice crystals. If your house were not heated, there would be no liquid water in it, and any food containing water would be frozen solid. Come summer, the average temperature at the same location is roughly 15°C (59°F), and in the past, heat waves have reached the upper 30s in Celsius (90s in Fahrenheit). In contrast, the temperature in coastal locations in the tropics changes very little in temperature **(Figure 3.26)**.

Verkhoyansk's extreme seasonality illustrates two important factors that determine seasonality for most places: latitude and proximity (nearness) to the oceans. As a general rule, high latitudes have a greater annual temperature range than low latitudes. In addition, *continental* (inland) locations have a greater annual temperature range than *maritime* (coastal) locations. This increase in seasonality with distance from the oceans is called the **continental effect**. **Figure 3.27** compares climate diagrams for several different cities to

Web Map

Land Surface Temperature

Available at www.saplinglearning.com

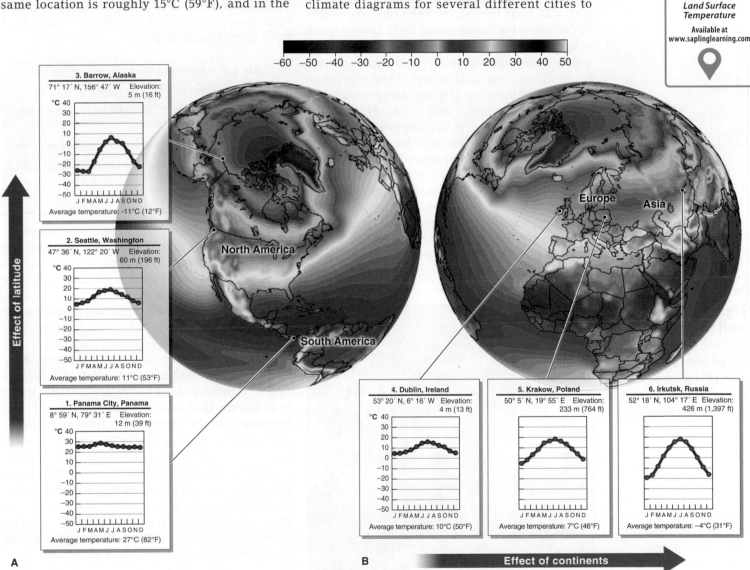

Figure 3.27 Seasonality by latitude and ocean influence. These maps show average yearly air temperatures near the surface between 1979 and 2015 for the Northern Hemisphere. (A) Seasonality increases as latitude increases. These three cities are all at sea level on the coast but at different latitudes. Barrow has much greater seasonality than Seattle, and Panama City has almost no seasonality. The map shows the average annual temperature, which dramatically decreases northward for each location. (B) Seasonality increases farther from the coast due to the continental effect. These three cities are all at about the same latitude and are at a similar elevation but have very different annual temperature ranges. Dublin is coastal and has mild seasonality, and Irkutsk is far inland with severe seasonality. Krakow is in between geographically and climatically. These three cities have the same average yearly temperature of about 10°C (50°F). *(Original data source: ECMWF ERA-Interim reanalysis. Image made from the website ClimateReanalyzer.org, courtesy of the Climate Change Institute, University of Maine, USA.)*

illustrate the effects of latitude and continentality on the yearly temperature range.

Why Do Inland Regions Have Greater Seasonality than Coastal Regions?

There are four main factors that cause the continental effect: (1) the specific heat of water, (2) the evaporation of water, (3) the transparency of water, and (4) the mixing of water.

You may have noticed at the beach that the sand becomes warm in the afternoon sunlight, but the water remains cool and refreshing. If you were to go to that same beach early in the morning, before sunrise, you might find that the sand is cooler than the water. The water hardly changes temperature over a 24-hour period, but the sand varies from warm during the day to cool at night. Land is quick to heat up and quick to cool down, but seawater's temperature changes much less. In addition, compared with the seasonal swings of temperature on land, the temperature of water changes relatively little between summer and winter. **Figure 3.28** shows the differences in daytime heating between land and water using satellite thermal imagery.

The cause of these temperature differences between land and water is the difference in their capacities to retain and store heat energy. The capacity of an object to retain and store heat energy depends on what materials it is composed of, such as water or rock, and how much of the material there is. Water, for example, retains more heat than the same volume of rock. Likewise, a larger rock composed of the same material as a smaller rock will retain more heat energy.

Specific heat (or *specific heat capacity*) is a term that is used to compare thermal capacities of different materials. **Specific heat** is the heat required to raise the temperature of 1 gram of any material by 1°C. For example, it takes 1 calorie of energy to heat 1 gram of water 1°C. Therefore, water has a specific heat of 1. However, it takes only about one-fifth of a calorie to raise 1 gram of dry sand 1°C. Dry sand, therefore, has a specific heat of 0.2, one-fifth that of water **(Table 3.1)**. If we add 1 calorie to 1 gram of sand, its temperature increases by about 5°C. Thus, when the same amount of sunlight shines down on both land and water equally, land heats up much more than water because land has a lower specific heat.

Table 3.1   Specific Heat Properties of Different Materials

SUBSTANCE	SPECIFIC HEAT (Cal/g × °C)
Pure water	1.0
Wet mud	0.6
Dry sandy clay	0.33
Dry quartz sand	0.2

Evaporation, water transparency, and water mixing are also important in moderating the temperature of water. Evaporation cools water or any moist surface and prevents it from becoming warmer than it would otherwise become. This cooling happens because liquid water absorbs the heat energy of the sunlight striking it, causing the water to evaporate. Evaporation then carries the water molecules away (and the heat energy they absorbed), cooling the water's surface (see Section 4.2). Because land has relatively little water to evaporate, land heats up more in sunlight than the oceans do. Because water is transparent, sunlight can pass through several hundred meters of water and warm it. Land is opaque, and sunlight does not penetrate it. Similarly, convection mixes the surface water warmed by sunlight with cooler water at greater depths. Land is rigid and cannot mix, so the land surface heats up faster and to a higher temperature. **Figure 3.29** summarizes and illustrates these factors and their influence.

Ocean Currents and Seasonality

As part of the global heat engine, warm ocean currents originating in the tropics carry immense amounts of heat toward the poles. When warm ocean currents reach high latitudes, some of their heat is transferred to the atmosphere. A good example of such an *ocean–atmosphere heat transfer* in the Northern Hemisphere is the Gulf Stream, which flows from the southern tip of Florida along the

Figure 3.28 Land and water heating.

Land and surface water temperatures stand in sharp contrast in this thermal infrared image, made during a heat wave in the western United States on May 2, 2004. The land has become hot during the daytime, but the water remains cool. Notice that the inland mountains in California and in eastern Nevada (far right) have remained cooler because they are at higher elevations. *(NASA)*

NEVADA

CALIFORNIA

PACIFIC OCEAN

| 0 | 100 | 200 km |
| 0 | 100 | 200 mi |

| 15 | 35 | 65 °C |
| 60 | 100 | 150 °F |

Web Map

Sea Surface Temperature

Available at www.saplinglearning.com

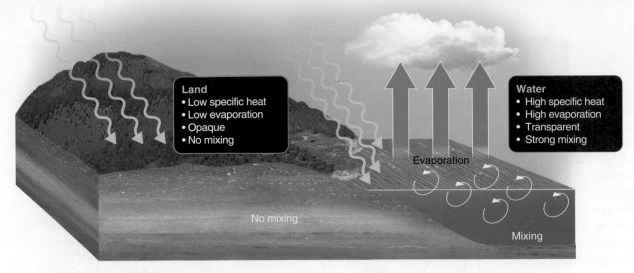

Figure 3.29 Factors that influence land and water temperatures. Land and water have different physical characteristics that result in their differences of temperature.

coast of eastern North America and into the North Atlantic Ocean. The warm water transfers its heat to the atmosphere when it reaches the North Atlantic, warming regions in northern Europe downwind.

The Gulf Stream transports more water than all the rivers of the world combined. Some 30 million m³ (1 billion ft³) of water pass by Florida each second. This amount increases to some 150 million m³ (5.3 billion ft³) per second by the time the current reaches Nova Scotia, Canada. As a comparison, the Amazon River, the largest river in

the world, has an average flow of about 200,000 m³ (7.1 million ft³) per second.

Warm ocean currents raise the average annual temperature and reduce the yearly temperature range. Cold currents lower the annual average temperature and also reduce the yearly temperature range. The Gulf Stream's moderating effect on yearly temperature ranges can be seen when coastal stations at the same latitude are graphed and compared **(Figure 3.30)**. Because it transports so much heat, the Gulf Stream has many effects

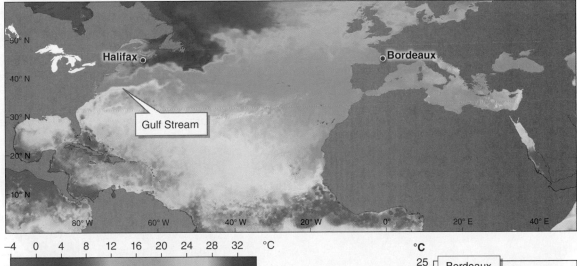

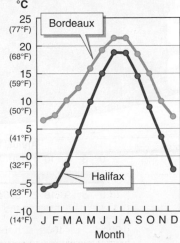

Figure 3.30 Ocean currents and seasonality. This satellite thermal infrared image was made June 7, 2017. It shows surface water temperatures in degrees Celsius. Orange areas are warmest. Light gray areas are continents. The Gulf Stream warms northern Europe as it delivers its tropical heat to the North Atlantic Ocean and then to the atmosphere. Halifax, Canada, and Bordeaux, France, are at the same latitude (44° N) and both are coastal towns at sea level, but they have very different seasonality. Bordeaux is influenced by the heat delivered to the atmosphere by the Gulf Stream more than Halifax. Bordeaux therefore experiences less seasonality. *(NASA/JPL)*

Figure 3.31 Effect of prevailing winds on seasonality. New York City is dominated by prevailing winds that originate in the interior of the North American continent. Although New York City is coastal, the air masses it receives are continental and, consequently, its yearly temperature range is considerable. Crescent City, California, in contrast, is dominated by prevailing winds that originate over the Pacific Ocean and, consequently, it has a maritime climate. Its yearly temperature range is small. The map shows average January temperatures during the period 2001–2010. *(NASA)*

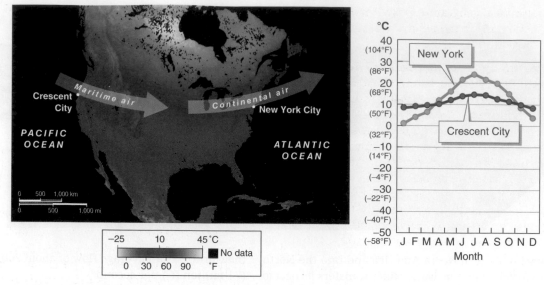

Figure 3.32 Global annual temperature range map. When we sum up the factors that control seasonality, we can see that the geographic pattern that emerges is affected by latitude, the continental effect, ocean currents, and prevailing winds. *(NASA)*

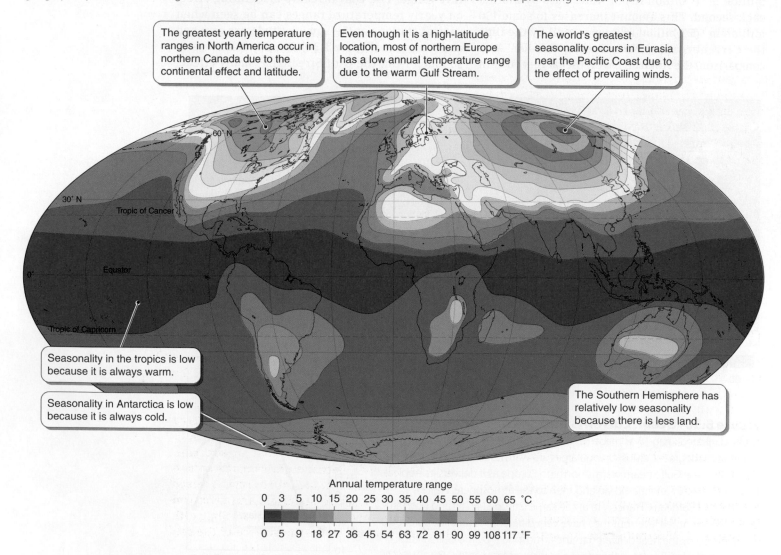

The greatest yearly temperature ranges in North America occur in northern Canada due to the continental effect and latitude.

Even though it is a high-latitude location, most of northern Europe has a low annual temperature range due to the warm Gulf Stream.

The world's greatest seasonality occurs in Eurasia near the Pacific Coast due to the effect of prevailing winds.

Seasonality in the tropics is low because it is always warm.

Seasonality in Antarctica is low because it is always cold.

The Southern Hemisphere has relatively low seasonality because there is less land.

Annual temperature range

on physical geography, from hurricane formation (Section 6.3) to the behavior of the global climate system (Section 7.2).

Prevailing Wind and Seasonality

New York, New York, and Crescent City, California, are both coastal cities located at 40 degrees north latitude. We might expect, therefore, that both would have maritime climates (influenced by the ocean), but they do not. Although the average annual temperature for both cities is about 12°C (53°F), these two cities have very different yearly temperature ranges. New York, on the East Coast, has a climate that is strongly influenced by the continent (called a *continental* climate). Crescent City, in California, has a maritime climate. As shown in **Figure 3.31**, the direction of the *prevailing winds* controls this pattern.

Generally, because the prevailing wind direction is from the west, west coasts have maritime climates, and east coasts have continental climates. This pattern is strongest at midlatitudes. It is weakened in the Southern Hemisphere (because landmasses there are small), in polar regions (because the water is very cold), and in the subtropics and tropics (because both land and sea are warm).

As we have seen in this section, many factors combine to control temperatures at any given location on the planet, but some play larger roles than others, depending on the geographic location. **Figure 3.32** shows the global patterns of Earth's annual temperature ranges and summarizes the main factors determining those patterns.

GEOGRAPHIC PERSPECTIVES
3.6 The Rising Solar Economy

◎ **Assess the role of sunlight as a carbon-free energy source.**

The world has entered an era of rapid growth in harnessing sunlight for energy. The job market in the solar industry is among the fastest growing in the country. In 2017, roughly 250,271 people were employed in the solar industry in the United States. Most of these jobs are in the manufacture and maintenance of solar panel equipment.

Solar energy is a renewable energy technology. **Renewable energy** is energy that comes from sources that—unlike the fossil fuels we discussed at length in Chapter 2, "Portrait of the Atmosphere"—are not depleted when used. Solar and wind energy are driving the current renewable energy revolution. Examples of other renewable energies are hydroelectricity (from streams), wave (from ocean waves), geothermal (from ground heat), combustible *biomass* and *biofuels* (such as agricultural waste and ethanol from corn), and waste materials (such as landfill gas). Also unlike fossil fuels, renewable energy sources do not transfer ancient carbon from the lithosphere into the atmosphere, where it forms carbon dioxide. These technologies are, instead, *carbon-free* or *carbon-neutral*, depending on their type.

Fossil fuels currently provide about 80% of the world's energy needs, and 60% of this comes from coal. Fossil fuels are a *finite* energy source: They will eventually run out. The push for renewable energy is driven mostly by concern about climate change. When burned, fossil fuels emit the greenhouse gas carbon dioxide (among other pollutants; see Section 2.4), which warms the atmosphere. In addition, other forces are driving the surge in solar energy on a global scale, including economic development and the improvement of peoples' lives in regions where *conventional energy* (from fossil fuels) is prohibitively expensive, intermittent, harmful to human health, or unavailable.

Solar Energy

There are many ways to convert the Sun's radiant energy to an energy form people can use. *Solar thermal power* involves heating water by running it through black tubing heated in sunlight and feeding the heated water directly into a home's shower or faucet or holding tank. The *concentrated solar power* (*CSP*) method uses mirrors to focus the Sun's energy on a material, usually sodium. The heated material is used to boil water, and the resulting steam turns a turbine to generate electricity. *Solar architecture* uses construction materials and building orientation to maximize or minimize heating from the Sun.

Photovoltaics (*photo* means "light" and *voltaic* means "relating to electricity") are by far used the most to harness the Sun's energy. Over 97% of the electricity generated from sunlight involves photovoltaic cells (PV cells). A **photovoltaic cell (PV cell)** is a *semiconductor* (a solid material with lower conductivity than most metals) that converts sunlight directly into electricity. Silicon, the second most abundant element in Earth's crust, is mixed with other conductive materials and arranged into two thin sheets to make a photovoltaic cell. The Sun's photons strike the surface of the cell and force electrons to flow through the layered materials, creating an electrical current. A single photovoltaic cell generates only a small amount of electricity. About 40 to 60 photovoltaic cells are combined into a weatherproof case to form a *solar panel*, and a working group of solar panels make up a *solar array* **(Figure 3.33A)**. A solar array for a home consists of about 10 to 20 solar panels. Large industrial-scale solar energy facilities can have several thousand solar panels or even millions of them **(Figure 3.33B)**.

The Global Context

At the beginning of 2018, global cumulative solar electricity generating capacity (peak output) was approximately 370 gigawatts (GW), about 2% of the world's electricity capacity. China far

surpassed all other countries **(Figure 3.34A)**. In some regions photovoltaics are achieving *grid parity*. This means that, without government subsidies, solar energy costs the same as or less than conventional energy. In countries including Chile, Australia, and Mexico, unsubsidized photovoltaic energy is now less expensive than energy from coal. Photovoltaic energy is on track to become the least expensive energy source by 2025 **(Figure 3.34B)**—followed closely by wind energy.

In theory, photovoltaics and concentrated solar power could meet the world's energy needs. Nearly 89,000,000 GW of solar energy reaches Earth's surface—far more energy than is possible with any other renewable source **(Table 3.2)**. The world's demand for energy is 15,000 GW, and by 2050 that demand is expected to have doubled to 30,000 GW.

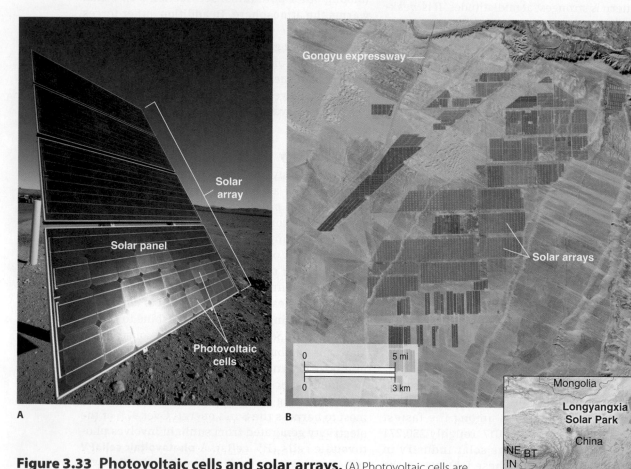

Figure 3.33 Photovoltaic cells and solar arrays. (A) Photovoltaic cells are mounted in weatherproof cases to make solar panels. They maintain their effectiveness for decades and do not require maintenance. (B) This Landsat 8 image shows one of the largest solar arrays in the world, the Longyangxia Solar Park in western China. There are nearly 4 million solar panels in this image. The facility covers 27 km² (10 mi²), and it is still growing.
(A. GIPhotoStock/Science Source; B. NASA Earth Observatory images by Jesse Allen, using Landsat data from the U.S. Geological Survey. Caption by Adam Voiland.)

Figure 3.34 Photovoltaic capacity by country and photovoltaic cost. (A) This graph shows the top 10 countries producing electricity from photovoltaics in gigawatts (GW) for 2016. China's electricity generation from photovoltaic panels is nearly double that of any other country. (B) As solar energy capacity increases, the cost to produce this type of renewable energy drops. Better technology, increased demand for renewables, and increased efficiency in production drive down the costs of photovoltaic cells. Solar is expected to be the least expensive form of energy by about 2030.

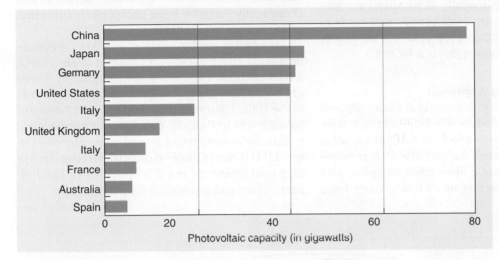

A

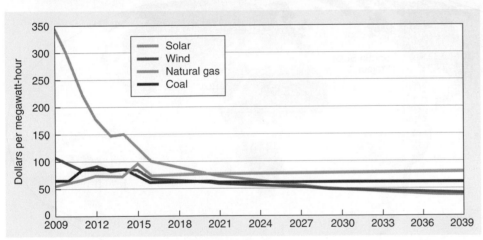

B

Table 3.2	**Renewable Energy Capacity**
RENEWABLE ENERGY TYPE*	**TOTAL THEORETICAL MAXIMUM CAPACITY (IN GIGAWATTS)**
Solar	89,000,000
Wind	190,000
Biomass	92,000
Geothermal	42,000
Hydroelectric	4,700

*__Biomass__ energy is generated by the burning of organic material, such as agricultural wastes or plant oils. *Geothermal* energy comes from Earth's internal heat. *Hydroelectric* energy is generated by the flow of rivers by means of turbines on dams.

The problem, however, is that generating even 15,000 GW of electricity from photovoltaics would require 860,000 km² (332,000 mi²) of solar panels—an area roughly equivalent to British Columbia or two Californias. Finding this much available land for conversion to photovoltaic panels would be a major challenge.

Even the most efficient photovoltaic cells convert only 44% of sunlight striking them to electricity. Unless we dramatically increase the efficiency of photovoltaic cells, meeting all of the world's energy needs with photovoltaics isn't likely because they simply would take up too much land surface. This fact underscores the need for other renewable energy sources in addition to the Sun.

Decentralized versus Centralized Solar Power

There are two broad approaches to capturing and converting sunlight to electricity: *decentralized solar* and *centralized solar*. Decentralized solar disperses solar power production in areas where people live and use it locally. Centralized production (also called *utility-scale* production) concentrates power production geographically into large industrial facilities and exports it for use.

The Decentralized Approach

Although rooftops could provide at most only 20% of the electricity needs of the United States, they are an excellent space resource for placing solar panels. In an idealized decentralized solar economy, every household, apartment complex, and business would generate its own electricity from its own rooftop. If any one rooftop generated too much power, it could be traded or sold.

In addition to solar panels, there are also photovoltaic paints and flexible plastic films. As the technology continues to improve, we may someday be able to cover our houses, buildings, and cars with these materials. Even our backpacks and clothes may someday have photovoltaic films that power our electronic devices.

Roadways and parking lots also provide important space to convert sunlight to electricity. Roads are often exposed to strong sunlight. There are 4.3 million km (2.7 million mi) of paved roads in the United States and 415,600 km (258,200 mi) of paved roads in Canada. Corridors of photovoltaic panels are already being placed along highways in several U.S. states. Solar panels in parking lots are also now providing the dual purpose of shading parked cars and generating clean electricity.

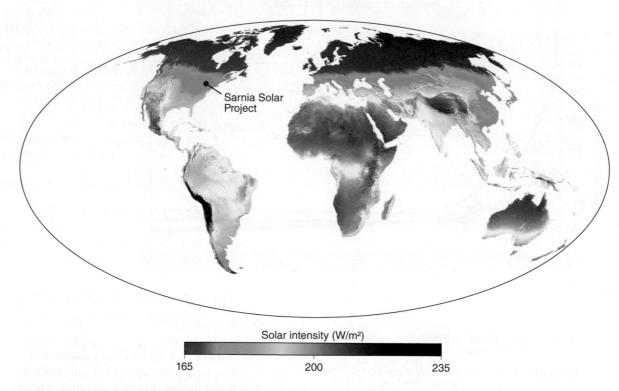

Solar intensity (W/m²)

165   200   235

Figure 3.35 Solar intensity map. This map shows the intensity of solar radiation (in W/m²) striking the ground. The Sun's intensity on the ground depends on latitude, elevation, and the persistence of cloud cover. Low latitudes that are cloud-free most of the time receive the most intense sunlight. Examples of such areas are the Sahara and the Kalahari deserts in Africa, the U.S. Southwest, and northern Australia. Midlatitude locations at high elevations that are cloud-free, particularly the Andes and the Tibetan Plateau, also receive abundant solar energy. *(2014 3TIER by Vaisala)*

The Centralized Approach

Many centralized solar facilities in the United States are located in the southwestern deserts, where the skies are often cloud-free. Solar energy is more intense at lower latitudes worldwide, particularly in arid regions, where cloud cover is scarce much of the year **(Figure 3.35)**.

Although Figure 3.35 may suggest that solar energy production is effective only in deserts, this is not the case. The Sarnia Solar Project in Ontario, Canada, for example, generates 97 megawatts (MW), enough electricity to meet the needs of 12,000 homes.

Many proposed centralized solar facilities have met stiff opposition in the United States. Why?

Because there is little "empty" land to use. What appears to be a barren desert can be prime habitat for organisms such as the endangered California desert tortoise (*Gopherus agassizii*), which is protected by federal law. Many solar farm projects have been bogged down in litigation because they take habitat from rare or endangered plants and animals. Proposals for solar projects that degrade the natural habitat of threatened species rarely win these court battles.

Given the urgent need for carbon-free energy and the breakneck pace of solar development, both centralized and decentralized solar as well as other renewable energy technologies will play increasingly larger roles in our energy future.

Chapter 3 **Study Guide**

Focus Points

3.1 Temperature and Heat

- **Temperature and heat:** Temperature is the average kinetic state of molecules; heat is the transfer of energy between substances with different temperatures.
- **Heat transfer:** Heat is transferred through conduction, convection, and radiation.
- **Heat:** In conduction, heat always travels from objects of higher temperature to objects of lower temperature.

3.2 The Sun's Radiant Energy

- **Electromagnetic radiation:** Electromagnetic radiation is emitted by all objects, and it travels in waves.
- **The Sun's emissions:** Because of its high temperature, the Sun emits mostly shortwave radiation, most of it in ultraviolet, visible, and infrared wavelengths.
- **Earth's emissions:** Earth mainly emits longwave thermal infrared radiation because of the planet's relatively low temperature.

3.3 Earth's Energy Budget

- **Insolation:** Incoming solar radiation can be transmitted, scattered, reflected, or absorbed.
- **Albedo:** The albedo of Earth as a whole is 31%. This means that 69% of insolation is absorbed.
- **Absorption:** Most absorbed solar radiation is converted to heat and warms Earth's surface and atmosphere.
- **Radiative equilibrium temperature:** Earth's radiative equilibrium temperature, 15°C (59°F), is a result of the balance between incoming solar and outgoing terrestrial radiation.

- **Greenhouse effect:** The greenhouse effect warms the atmosphere and keeps Earth's temperature suitable for life.
- **Global heat engine:** There is a net surplus of heat between 37 degrees latitude north and south, and there is a net deficit of heat at higher latitudes. These heating inequalities drive movement in Earth's atmosphere and oceans.

3.4 The Four Seasons

- **Cause of seasons:** The tilt of Earth's axis causes the subsolar point to migrate between the Tropic of Cancer and the Tropic of Capricorn, causing seasons.
- **Solar altitude:** The solar altitude determines the intensity of sunlight.
- **Equinox day length:** On the equinoxes, all locations on Earth experience 12 hours of day and 12 hours of night.
- **Solstice day length:** On the June solstice, day length increases northward and decreases southward. The reverse occurs for the December solstice.

3.5 Patterns of Temperature and Seasonality

- **Average annual temperature:** The average annual temperature decreases at higher elevations and at higher latitudes.
- **Yearly temperature range:** The yearly temperature range increases at high latitudes and interior continental locations.
- **Ocean currents:** Warm ocean currents from low latitudes raise the average annual temperature and reduce the annual temperature range.
- **Midlatitude coastal patterns:** At midlatitude locations, the west coasts of continents have maritime climates, and the east coasts of continents have continental climates because of the direction of the prevailing wind, which blows from west to east.

3.6 Geographic Perspectives: The Rising Solar Economy

- **Renewable energy:** Solar energy is a renewable resource that is rapidly being developed in response to climate change and economic development in poor rural regions.
- **Solar energy:** There are many different ways to convert sunlight to usable energy. Photovoltaic cells that convert sunlight to electricity are the most important solar energy technology.
- **Decentralized and centralized:** Decentralized solar energy is locally produced, and centralized solar energy is exported from a central facility. Both approaches are important to growing the solar and renewable energy economy.

Key Terms

absorption, 69
advection, 66
albedo, 69
Antarctic Circle, 79
Arctic Circle, 79
circle of illumination, 75
conduction, 66
continental effect, 83
convection, 66
December solstice, 77
electromagnetic spectrum (EMS), 67
global heat engine, 74
greenhouse effect, 72
heat, 65
infrared radiation (IR), 68
insolation, 68
June solstice, 78
longwave radiation (LWR), 68
March equinox, 78
photovoltaic cell (PV cell), 87
plane of the ecliptic, 74

radiant energy, 67
radiation, 66
radiative equilibrium temperature, 71
reflection, 69
renewable energy, 87
scattering, 69
sensible heat, 68
September equinox, 78
shortwave radiation (SWR), 68
solar altitude, 77
specific heat, 84
subsolar point, 75
temperature, 65
thermal infrared (thermal-IR), 68
transmission, 69
Tropic of Cancer, 75
Tropic of Capricorn, 75
urban heat island, 64
visible light, 68

Concept Review

The Human Sphere: Hot Cities

1. What is the urban heat island? What causes it?

2. What measures can be taken to address the urban heat island problem?

3.1 Temperature and Heat

3. What is the difference between temperature and heat?

4. In what three ways is heat energy transferred? Give real-world examples of each.

3.2 The Sun's Radiant Energy

5. What is electromagnetic radiation? Does a rock or tree emit electromagnetic energy? Explain.

6. Which radiates longer wavelengths of electromagnetic energy, the Sun or Earth? Explain why.

7. Which type of radiation has shorter wavelengths, infrared radiation or visible light? Green light or yellow light? Ultraviolet or visible light?

8. When the visible portion of the EMS is blended together, what color do we perceive?

9. In the context of reflection, refraction, scattering, and absorption, explain the coloration of each of the following: blue sky, red sunsets, green grass, white clouds, blue water, and rainbows.

3.3 Earth's Energy Budget

10. Describe what happens to insolation (incoming solar radiation) when it is transmitted, scattered, reflected, and absorbed.

11. What is albedo? Give examples of surfaces and objects with high and low albedos. What is Earth's overall albedo?

12. What does the statement "Earth's energy budget is balanced" mean?

13. What is a radiative equilibrium temperature? What would happen to Earth's radiative equilibrium temperature if more energy came into the atmosphere than left the atmosphere?

14. What is Earth's radiative equilibrium temperature?

15. What is the greenhouse effect? Describe how it works in relationship to Earth's energy budget.

16. Where on the planet (in the context of latitude and altitude) is there a surplus of heat, and where is there a deficit of heat?

17. Why is the lower atmosphere, near Earth's surface, warmer than the upper atmosphere?

18. What is the global heat engine? What causes it?

3.4 The Four Seasons

19. What are aphelion and perihelion, and when do they occur?

20. What is axial tilt, and why is it the most important factor causing seasons?

21. What is the subsolar point? What would happen to it if there were no axial tilt? What would happen to it if axial tilt were 90 degrees?

22. What happens to seasonality when axial tilt is increased?

23. In relation to the subsolar point, how are the tropics defined?

24. Describe day length and the position of the subsolar point for each of the four seasonal markers:

    June solstice
    December solstice
    September equinox
    March equinox

25. Why is 1 "day" 6 months long at the poles?

3.5 Patterns of Temperature and Seasonality

26. Describe the relationship between average annual temperature and changes in elevation and latitude.

27. How do latitude and the continental effect influence the yearly temperature range?

28. What factors cause seasonality to increase farther inland?

29. Why does the Southern Hemisphere have smaller average seasonal fluctuations (on average) than the Northern Hemisphere?

30. How is seasonality affected in regions under the influence of warm ocean currents?

31. Compare the general pattern of annual temperature ranges at midlatitudes on west coasts and on east coasts. Which has a higher temperature range, and why?

3.6 Geographic Perspectives: The Rising Solar Economy

32. Compare and contrast renewable energy and nonrenewable energy. Give examples of each.

33. Why are fossil fuels a problematic source of energy?

34. What different methods are used to harness sunlight into usable energy? Which is used the most?

35. Compare centralized and decentralized solar energy production. What are some pros and cons of each approach?

Critical-Thinking Questions

1. With regard to melting snow and ice, how could climate warming change Earth's albedo? What would happen to the planetary radiative equilibrium temperature as a result?

2. At the latitude where you live, is there a net heat surplus or a net heat deficit?

3. What aspects of your life depend on the seasons or somehow relate to them?

4. What is the yearly temperature range where you live? Is the climate influenced more by oceans or by the continental effect? How does latitude factor into your location's seasonality?

5. Assuming that you own your home, would you put solar panels on your rooftop? Why?

Test Yourself

Take this quiz to test your chapter knowledge.

1. True or false? Seasonality increases as axial tilt decreases.

2. True or false? The highest latitudes the subsolar point can reach is 23.5 degrees north and south.

3. True or false? Temperature and heat are the same phenomenon.

4. True or false? The greenhouse effect keeps Earth's atmosphere warm and habitable.

5. Multiple choice: On June 21, the subsolar point is

a. on the equator.
b. at 23.5 degrees north.
c. over the Tropic of Capricorn.
d. at 23.5 degrees south.

6. Multiple choice: Which of the following absorbs most SWR from the Sun?

a. the continents
b. clouds
c. the atmosphere
d. the oceans

7. Multiple choice: Which of the following locations would have the lowest average annual temperature?

a. at sea level in the tropics
b. in the mountains in the tropics
c. at sea level at high latitudes
d. in the mountains at high latitudes

8. Multiple choice: Which of the following locations would have the highest annual temperature range?

a. inland low latitudes
b. coastal low latitudes
c. inland high latitudes
d. coastal high latitudes

9. **Fill in the blank:** _____ causes the sky to be blue.

10. **Fill in the blank:** Energy transfer by _____ occurs when a fluid such as the atmosphere or oceans circulate and mix.

Online Geographic Analysis

NASA Temperature Maps

In this exercise, we analyze air temperature maps.

Activity

Go to https://www.strategies.org/interactive/. Hover over the map or click around on it until you reach the "Land Surface Temperature" map. Click this map to zoom in.

1. **Read the map's legend. What do dark red areas indicate? What do white areas indicate? What units of temperature are used?**

2. **In general, how would you describe the pattern of temperature across latitude? How do temperatures at high latitudes compare to those at low latitudes?**

3. **In general, how would you describe the pattern of temperature across elevation? How do temperatures at high elevations compare to those at low elevations?**

4. **What is the hottest continent on Earth?**

Close this map and, hovering and clicking around, open the "Mid-Tropospheric Temperature" map.

5. **Read the map's legend. What do red areas indicate? What do blue and purple areas indicate? What units of temperature are used in this map?**

6. **Read the description of this map. How high above Earth's surface are the mapped temperatures?**

7. **What are the similarities between this map and the "Land Surface Temperature" map?**

Close this map and, hovering and clicking around, open the "Sea Surface Temperature" map.

8. **What do the colors on the map represent? What units of temperature are used in this map? What period of time does the map cover?**

9. **Click North America to zoom in. Compare the sea surface temperatures for the East Coast of the United States, the West Coast of the United States, and southern Canada. How are the coastal water temperatures in these areas different, and why? (Hint: For the "why" part of this question, refer to Section 3.5.)**

10. **What effect do these water temperature differences have on air temperatures on both coasts during the time period mapped?**

Picture This. *Your Turn*

Seasonality

Using what you have learned from reading this chapter, match the climate diagrams with the locations given on the map. Each diagram shows temperatures for every month of the year. Hint: If temperature drops during Northern Hemisphere summer months (May through August), it is a Southern Hemisphere location. Use each climate diagram only once.

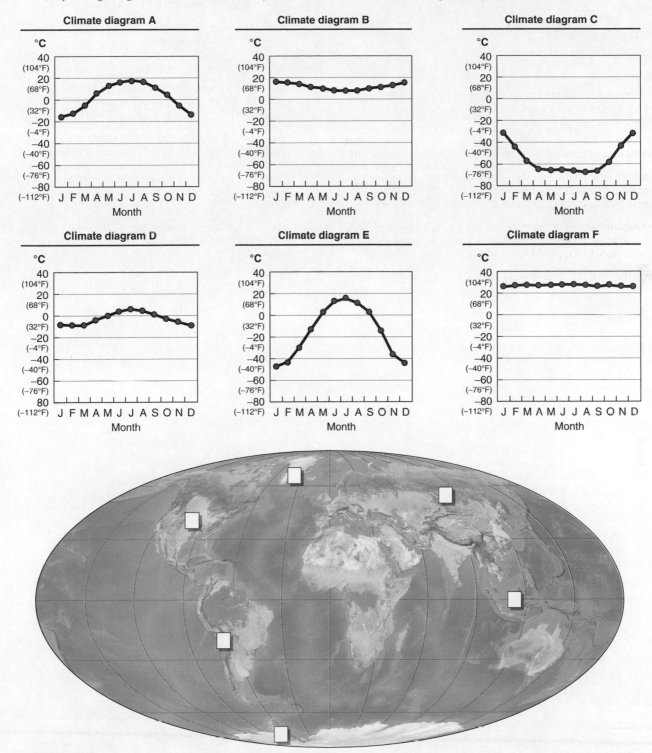

Water in the Atmosphere

Chapter Outline *and Learning Goals* ◎

A deluge of rain pours out of a severe thunderstorm in central Nebraska. Most clouds are made of microscopic liquid *cloud droplets*. Cloud droplets are so small and so light that updrafts of air keep them suspended in the air. A single raindrop is composed of millions of cloud droplets that have merged together. The heavy rain coming from this thunderstorm is the result of the merging of countless cloud droplets.

(Mike Hollingshead/Alamy)

Section 4.6 explores this interesting topic further.

THE HUMAN SPHERE Killer Heat

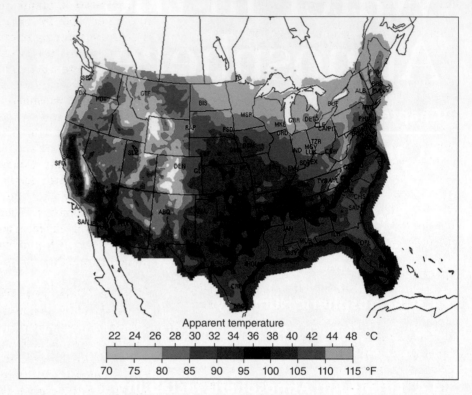

Figure 4.1 Apparent temperature map.
This map shows apparent temperatures for the United States and southern Canada for June 29, 2017. Apparent temperatures exceeded 38°C (100°F) in the Gulf States and climbed as high as 43°C (110°F) from Florida to North Carolina. Heat waves with apparent temperatures exceeding 38°C (100°F) for the Gulf States are routine each summer. *(NOAA)*

"It's not the heat, it's the humidity." When most people think of heat, they think of high temperatures. But the most intense heat occurs when both the air temperature and the humidity are high. The *apparent temperature* is the temperature it feels like, not the actual air temperature. When sweat evaporates from our bodies, it carries body heat away and provides cooling relief. But when the humidity is high during hot weather, the rate of evaporation slows, making it feel even hotter than it really is.

The *dew point*, which is further explained on page 108, is a good indicator of atmospheric humidity. Although there is no official record of apparent temperatures, official dew point temperatures are recorded. Dew points above 21°C (70°F) are uncomfortably hot. The highest official dew point ever recorded was 35°C (95°F) on July 8, 2003, in the city of Dhahran, Saudi Arabia. The air temperature reached 42°C (108°F) that day. Together, the dew point and air temperature resulted in an apparent temperature of 81°C (178°F)! More recently, in July 2015 in the Iranian city of Bandar Mahshahr,

the apparent temperature reached 74°C (165°F). Although such extreme apparent temperatures are rare, apparent temperatures do routinely enter the low triple digits, particularly in the eastern United States **(Figure 4.1)**.

High heat coupled with high humidity is the number-one meteorological (weather-related) killer in the United States: Each year, about 400 people lose their lives as a result of high apparent temperatures. In the Chicago and Milwaukee areas in 1995, more than 750 people died from heat-related causes. Major heat waves gripped Europe in 2003, 2006, and again in 2018, with death tolls totaling more than 73,500. In 2010, Russia saw more than 56,000 heat-related fatalities. In India and Pakistan, more than 5,000 people died in a heat wave that same year. In all these cases, high dew points amplified the effects of high air temperatures, creating deadly conditions. According to the Intergovernmental Panel on Climate Change, heat waves are becoming more frequent as carbon dioxide levels in the atmosphere increase due to human activity and the world warms.

4.1 The Hydrologic Cycle

◎ **Explain what the hydrologic cycle is and why it is important.**

Earth is a water planet. Over 71% of the planet's surface is covered by liquid oceans. Another 10% of Earth's land surface is covered by glacial ice, mostly in Greenland and Antarctica. Water is everywhere. It circulates within the planet's physical systems, and it has a profound effect everywhere it moves. Water is lifted from the oceans, transported through the atmosphere, and precipitated on the continents. It flows through the atmosphere above us in invisible plumes as water vapor and as visible clouds. Water flows through streams and lakes, and it moves through the ground, the pore spaces in rocks, and all living organisms. We examine water in this chapter because it is so fundamental to nearly all atmospheric processes.

Liquid water is heavy: It weighs 1 kg/L (8.3 lb/gal). A single small cumulus cloud, made of microscopic liquid droplets, can contain millions of liters of water and weigh several hundred tons. Hurricanes are much larger and weigh trillions of tons. How does such weight defy gravity and remain suspended in the atmosphere? The answer to this question lies in the ability of water to shift between the three states of matter—solid, liquid, and gas—at temperatures found on Earth's surface.

States of Water: Solid, Liquid, and Gas

Water shifts between solid, liquid, and gaseous states through melting, evaporation, condensation, and freezing **(Figure 4.2)**. **Evaporation** is the change in the state of water from liquid to water vapor (a gas) from a body of water. In the process of **transpiration**, plants also change liquid water into water vapor by "breathing" it out the pores on their leaves during photosynthesis. Conversely, **condensation** is the change in state from water vapor to liquid water. Through evaporation, the Sun's energy heats and lifts trillions of tons of water into the atmosphere, water molecule by water molecule. That water vapor condenses into microscopic liquid cloud droplets or is deposited onto ice crystals, which are suspended in the atmosphere by updrafts of air. **Clouds** are composed of these microscopic water droplets and ice crystals that are visible only when they are grouped together in very large numbers. If conditions are right, snow, rain, and hail may form within clouds, resulting in precipitation. **Precipitation** is solid or liquid water that falls from the atmosphere to the ground.

The Hydrologic Cycle: Water on the Move

There would be no liquid water (and therefore no life) on land were it not for the hydrologic cycle. Every glass of freshwater we drink came from water that was once salt water in the oceans before it was desalinated in the hydrologic cycle. Water also flows through Earth's crust, sustaining plants, providing *groundwater* for people, and facilitating the movement of the crust in the important process of plate tectonics (see Chapter 13, "Drifting Continents: Plate Tectonics"). Clouds, made of water, strongly modify Earth's climate system and help to make the temperature suitable for life. These elements of water are illustrated in the **hydrologic cycle**—the circulation

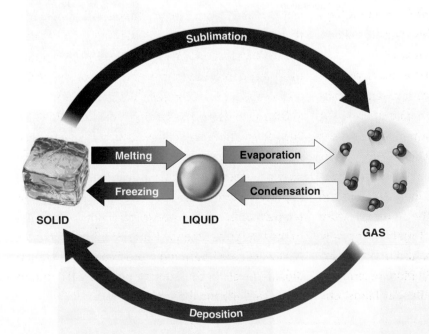

Figure 4.2 Physical changes in water. Water readily shifts between the solid, liquid, and gaseous states. Two additional physical changes, which skip the liquid state of water entirely, are possible: Through *sublimation*, ice shifts directly to water vapor; through *deposition*, water vapor forms ice directly.

of water within the atmosphere, biosphere, lithosphere, and hydrosphere **(Figure 4.3)**.

The hydrologic cycle is entirely solar powered. The process of evaporating water from Earth's surface (see Figure 3.11) consumes approximately 20% of all solar energy absorbed by Earth. Evaporation from the oceans provides about 85% of the water vapor in the atmosphere. Transpiration from plants and evaporation from surface streams and lakes together account for the remaining 15% of atmospheric water vapor. Together, evaporation and transpiration create a single process of moving water vapor to the atmosphere called **evapotranspiration**.

4.2 Properties of Water

◉ **Describe water's ability to absorb, transport, and release** *latent heat* **energy and the role of latent heat in cloud formation and weather events.**

The movement of water through the hydrologic cycle is one of the most important processes governing and modifying Earth's physical systems. A crucially important aspect of water is its ability to absorb, transport, and release immense quantities of heat energy in the atmosphere. The powerful winds in hurricanes and tornadoes, for example, come from the energy released from water vapor in the atmosphere as it condenses to liquid. What properties of water allow it to generate such strong winds?

Water Molecules and Hydrogen Bonds

Water molecules form when hydrogen and oxygen atoms bond to each other. Water molecules attach to one another because of their electrical *polarity*: The hydrogen end of a water molecule has a weak positive charge, and its oxygen end has a weak negative charge. One water molecule's positive end attaches to another water molecule's negative end, forming a **hydrogen bond,** as illustrated in **Figure 4.4**. The bonds holding the hydrogen and oxygen atoms together to form a water molecule are far stronger than the hydrogen bonds that join water molecules.

The physical state of water depends on the number of hydrogen bonds and the strength of the hydrogen bonds. Ice has more hydrogen bonds than liquid water. The hydrogen bonds in ice create a lattice, in which each molecule is locked firmly to its neighbor. The hydrogen bonds in liquid water are weaker, and they continually stretch and break and then re-form. Water vapor has few

Figure 4.3 The hydrologic cycle. In the hydrologic cycle water changes among the three states of matter and flows through all of Earth's physical systems. Water resides in four great reservoirs (storage areas): the oceans (the hydrosphere), ice and snow (the cryosphere), the atmosphere, and the ground (or lithosphere).

Animation

The Hydrologic Cycle

Available at
www.saplinglearning.com

Video

Earth's Water Cycle

Available at
www.saplinglearning.com

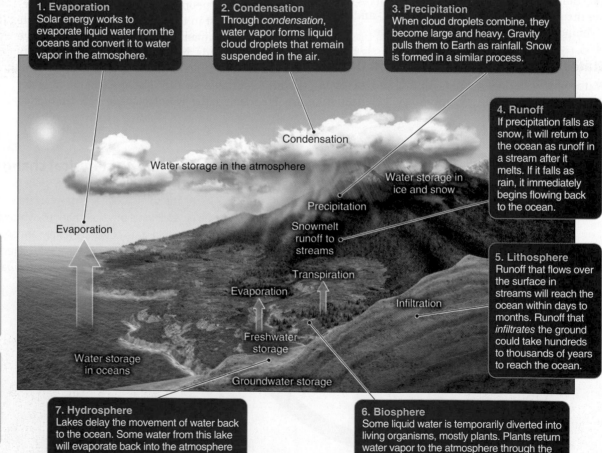

1. Evaporation
Solar energy works to evaporate liquid water from the oceans and convert it to water vapor in the atmosphere.

2. Condensation
Through *condensation*, water vapor forms liquid cloud droplets that remain suspended in the air.

3. Precipitation
When cloud droplets combine, they become large and heavy. Gravity pulls them to Earth as rainfall. Snow is formed in a similar process.

4. Runoff
If precipitation falls as snow, it will return to the ocean as runoff in a stream after it melts. If it falls as rain, it immediately begins flowing back to the ocean.

5. Lithosphere
Runoff that flows over the surface in streams will reach the ocean within days to months. Runoff that *infiltrates* the ground could take hundreds to thousands of years to reach the ocean.

6. Biosphere
Some liquid water is temporarily diverted into living organisms, mostly plants. Plants return water vapor to the atmosphere through the process of *transpiration*.

7. Hydrosphere
Lakes delay the movement of water back to the ocean. Some water from this lake will evaporate back into the atmosphere rather than flow to the ocean.

Condensation

Water storage in the atmosphere

Water storage in ice and snow

Evaporation

Precipitation

Snowmelt runoff to streams

Transpiration

Evaporation

Infiltration

Freshwater storage

Water storage in oceans

Groundwater storage

or no hydrogen bonds, and the water molecules move freely about (see Figure 1.5).

Water molecules join to other water molecules through *cohesion* (by means of hydrogen bonds), and water molecules join to other objects through *adhesion*. Cohesion and other properties of water contribute to the central role of water in many physical processes on Earth **(Figure 4.5)**.

Latent Heat of Water: Portable Solar Energy

As shown in Figure 4.5, water has many properties that make it a critical compound in Earth's physical systems. Perhaps the most important property of water is its ability to absorb radiant solar energy, transport that energy, and release it as *latent heat*

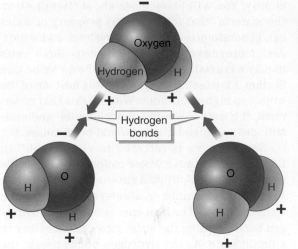

Figure 4.4 Hydrogen bonds join water molecules to each other. A water molecule consists of two hydrogen atoms bonded to one oxygen atom. The oxygen side of the molecule has a weak negative charge, while the hydrogen side has a weak positive charge. These electrical charges create polarity and allow weak hydrogen bonds to form between water molecules.

Figure 4.5 Properties of water. Water plays a central role in many physical processes on Earth. *(Top left. Kajo Merkert/Getty Images; top right. NASA/JPL; bottom left. © Jim Wark/airphotona.com; bottom right. Don Johnston/Getty Images)*

Density
As the temperature of water approaches the freezing point, the density of the water decreases. Because the near-freezing water is less dense than liquid water, it rises to the surface, where it freezes. Ice is 9% less dense than liquid water and also floats. If ice were more dense than liquid water, it would sink and melt more slowly. If that happened, even large bodies of water in cold regions would become frozen solid.

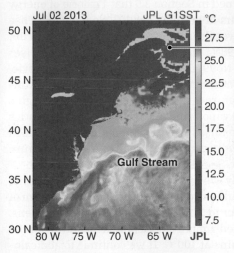

Specific heat
Water has a high specific heat; this means that it can absorb a large amount of heat without a significant change in its temperature (see Section 3.5). In other words, water heats up and cools down slowly. So the oceans are immense reservoirs of thermal energy, and they strongly moderate coastal temperatures and the global climate system. The warm Gulf Stream ocean current, shown here, transports a great quantity of heat northward along the East Coast.

The universal solvent
Water is known as the universal solvent. Almost everything dissolves in it. This property of water is essential for weathering and for carrying nutrients to plants. Shown here is the *confluence* (junction) of the Milk (left) and Missouri Rivers in Montana. Both streams carry high concentrations of dissolved materials. Dissolved material is invisible in water. The Milk River's color comes from suspended particles, not dissolved material.

Cohesion
Water is attracted to this leaf through *adhesion*. It is attracted to itself through *cohesion* because of the electrical polarity of water molecules. The wax on this leaf has no electrical charge, so water forms a bead as it "sticks" to itself more than to the leaf. Water's cohesive and *capillary* (wicking) properties are necessary for the formation of precipitation. Cohesion allows plants on land to take up water through their tissues against the pull of gravity.

energy. You will learn more about latent heat in the material that follows. This property of water can generate powerful storm systems and winds. And, more importantly, this property allows water to play a crucial role in the global heat engine (see Section 3.3; page 73) by transporting heat out of the tropics to higher latitudes. Without this heat movement, the tropics would be much hotter, and mid-latitudes and high latitudes would be far colder.

Solar energy is referred to as "portable" in this section because water molecules essentially carry it from one physical environment to another. When the Sun shines on a water body, it heats the water's surface. The heat energy breaks the hydrogen bonds holding the water molecules together in a liquid state. As the hydrogen bonds break, the water evaporates. The evaporated water, in turn, is advected (transported) away on the winds, where it will condense and release that heat energy that it "carried" when it was in vapor form.

We learned in Section 3.5 that 1 calorie of energy raises 1 gram of liquid water 1°C. So it makes sense that when we add 1 calorie to 1 gram of water (1 cm³, or 1 ml) at 98°C, the temperature of the water rises from 98°C to 99°C. When we add 1 more calorie, the temperature rises to 100°C (212°F, water's boiling temperature at sea level). The water then begins boiling and vaporizing (evaporating). These calories we added to raise the water's temperature are called *sensible heat* because they change the temperature of the water. Sensible heat is heat that can be felt.

Logically, we might think that our next calorie of energy should raise the temperature of our 1 gram of boiling water to 101°C. But that is not what happens. When we add 1 more calorie, the temperature of the water remains at 100°C. If we continue to add calories, the water temperature still remains at 100°C. But even though its temperature does not rise, it also continues to change from liquid water into vapor (or steam). These calories absorbed by the water that allow it to change state without producing a change in temperature are referred to as latent heat. **Latent heat** is the energy that is absorbed or released during a *state change*, such as evaporation to vapor or condensation to liquid. Latent heat energy changes the state of the water—not the temperature of the water. Unlike sensible heat, latent heat cannot be felt.

Only after 540 calories have been added to that 1 gram of water will all the water have changed state and transformed from a liquid into a gas. At this point, if we could somehow capture that gas and add another calorie to it, its temperature would finally rise from 100°C to 101°C. This added last calorie is sensible heat because it changes the temperature of the water. The first 540 calories are latent heat calories because they work to break the hydrogen bonds in liquid water—to change its physical state—rather than to raise its temperature **(Figure 4.6)**.

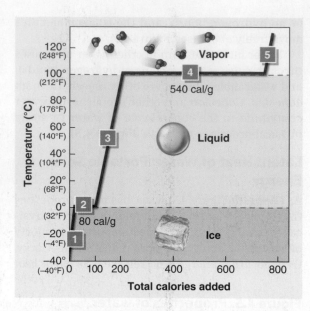

Figure 4.6 The changing states of water.

The red line shows the temperature of 1 gram of water and the number of calories that are added to or removed from the water as it changes state. (1) As energy is added to ice, it raises the temperature of the ice. Each 0.5 calorie added raises the ice's temperature 1°C. (2) When the ice's temperature reaches 0°C, it begins melting. It remains at 0°C until 80 calories of latent heat have been added and all the ice has melted. (3) Once all the ice has melted, each calorie added raises the temperature of the water 1°C. (4) When the temperature reaches 100°C, the water begins boiling. It remains at 100°C until 540 calories of latent heat have been added and all the water has vaporized. (5) Once all the water has vaporized, each additional calorie raises the vapor's temperature 1°C.

Only 0.5 calorie of sensible heat is required to raise the temperature of 1 gram of ice 1°C because ice has a lower specific heat than liquid water. As the ice is warmed, once it reaches 0°C (32°F), its temperature remains at 0°C until all the ice has melted and turned to liquid. The energy added to ice to melt it but not change its temperature is the *latent heat of melting*. The energy used to vaporize liquid is the *latent heat of vaporization*. Similarly, as water vapor condenses and turns back into a liquid, it releases its stored latent heat as *latent heat of condensation*. The energy removed from liquid water to freeze it back into solid ice is the *latent heat of freezing* **(Figure 4.7)**.

Water's temperature does not have to be raised to 100°C for it to evaporate. We can prove this by spraying a warm summer sidewalk with a garden hose. The water does not boil, but it does quickly evaporate. Even after winter rainstorms when it is cold outside, puddles on the streets and sidewalks evaporate.

When energy is absorbed by an individual water molecule at the surface of a body of water, the molecule may obtain enough energy to break the hydrogen bonds with its neighbors and evaporate. The energy involved in this process is the *latent heat of evaporation* (see Figure 4.7). This evaporation process occurs in natural outdoor settings at temperatures well under 100°C. Boiling, in contrast, occurs when the water temperature is 100°C, as in a pot of water on the stove.

It takes more calories to evaporate water at temperatures below 100°C. Depending on the temperature of the water, the amount of energy required to evaporate it varies from about 540 calories to 600 calories per gram of water (see Figure 4.7). In most outdoor settings, roughly 580 calories are absorbed from the environment for each gram of water evaporated.

Evaporation is a cooling process. The temperature of the surrounding environment drops as heat is absorbed from it and carried away in the air by the evaporating water molecules. Your body feels cooler when your sweat evaporates because it takes some 580 calories of heat to evaporate 1 g of sweat. Many, but not all, of those calories come from your body. (Note that the calories we are using for these measurements are not the same as the calories used to measure the energy content of food, which are given in *kilocalories,* or *kcal*; 1 kilocalorie is equal to 1,000 calories.)

In contrast to evaporation, condensation is a warming process. Roughly 580 calories of sensible heat per gram of water are released to the surrounding environment as water vapor evaporates and becomes liquid water (see Figure 4.7). This

Figure 4.7 Latent heat and water's physical states. Changes in the physical state of water require the addition or removal of latent heat. For example, for each gram, the transition from ice to water absorbs 80 calories, and the transition from water to vapor absorbs up to 600 calories.

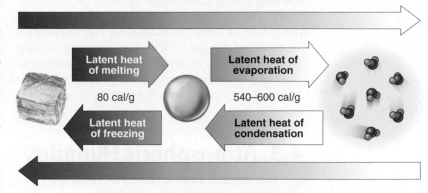

Heat is absorbed from the environment by water molecules as latent heat.

Latent heat of melting   80 cal/g

Latent heat of freezing

Latent heat of evaporation   540–600 cal/g

Latent heat of condensation

Heat is released back to the environment by water molecules as sensible heat.

released heat is sensible heat because it warms the environment into which it is released.

The most important setting where latent heat is released is in clouds. For each gram of water vapor that is condensed to liquid, about 580 calories are released to the surrounding air, warming the interior of the cloud. The warmed cloud interior has the tendency to rise because it is less dense and more buoyant than the relatively cooler air outside the cloud. This is why many clouds billow and rise higher into the air **(Figure 4.8)**.

3. Cloud development
As the cloud's interior warms, it becomes less dense than the air outside the cloud and rises upward. The base of the cloud remains at the same altitude, but the cloud grows taller.

2. Condensation
As the water vapor condenses it forms liquid cloud droplets. For every gram of water that condenses about 580 calories of latent heat energy are released as sensible heat that warms the cloud's interior.

1. Evaporation
For every gram of water that evaporates about 580 calories are absorbed as latent heat. but the cloud grows taller.

Figure 4.8 Latent heat in clouds. Condensation releases immense amounts of heat energy. Cumulus-type clouds billow upward and grow in vertical height because their interiors become warmer than the surrounding environment when latent heat is released through condensation. *(imagenavi/Getty Images)*

Latent heat released within clouds is important because the latent heat creates rising columns of air called *updrafts* within clouds. As these updrafts pull in humid air near Earth's surface, water vapor condenses and releases yet more latent heat within the cloud, further strengthening the updrafts. Without these updrafts generated by latent heat, precipitation could not form, nor could the powerful winds of tornadoes and hurricanes occur (see Chapter 6, "The Restless Sky: Severe Weather and Storm Systems"). In other words, the "fuel" for most storm systems is evaporated water and its latent heat, which ultimately is derived from the Sun.

4.3 Atmospheric Humidity

◎ **Distinguish among the different ways of expressing atmospheric humidity and explain each method.**

Humidity, the water vapor content of the air, plays a crucial role, contributing to the natural greenhouse effect in Earth's climate system; without water vapor in the atmosphere, the greenhouse effect would be much diminished and Earth would be unbearably cold (see Section 3.3). Furthermore, humidity is central to almost all meteorological processes. Without humidity, there would be no clouds, no precipitation, and almost no weather systems.

Our bodies are very good sensors of humidity. Why does air sometimes feel sticky and wet and at other times thin and dry? When it is humid, a hot day feels hotter because the humidity reduces our bodies' ability to cool off by evaporating sweat. (Also, when it is windy, a cold day feels colder because the *wind carries away our body heat*.) What the temperature feels like to a person (rather than what it reads on a thermometer), called the **apparent temperature,** can be ascertained by using *wind chill* and *heat-index* charts (Appendix III, "Wind Chill and Heat-Index Charts"). It is not colder or hotter in either of these situations; it only feels that way to people.

All air at Earth's surface is at least in part composed of water molecules and has humidity. Evaporation increases humidity. (Transpiration from plants also increases humidity, but in this discussion only the term *evaporation* will be used.) Condensation reduces humidity because it changes water vapor to liquid. (In this discussion, the term *condensation* will also include the process of deposition.) Evaporation and condensation are always occurring at the same time in Earth's lower atmosphere and surface waters. When evaporation exceeds condensation, *net evaporation* is occurring, and the air is cloud-free. When condensation exceeds evaporation, *net condensation* is occurring, and clouds form. All clouds form in environments where there is net condensation. Net condensation only occurs where air has reached **saturation**: that is, the point at which the air's *water vapor content* is equal to or greater than the air's *water vapor capacity*.

Recall from Section 4.1 that evaporation from the oceans accounts for 85% of the water vapor in the atmosphere, and transpiration from plants accounts for 15%. Accordingly, the most humid air is found in the tropics, near warm oceans, where evaporation rates are high **(Figure 4.9)**. In tropical rainforests, however, most of the humidity comes from plant transpiration, not evaporation.

Meteorologists measure atmospheric humidity to better understand how meteorological processes work and to forecast the weather. **Hygrometers** of several types measure water vapor in the atmosphere. *Sling psychrometers*, for example, are large, handheld instruments that measure humidity using thermometers. Human or animal hair is also used to measure humidity. Humidity changes the length of hair, and the length can be calibrated to give a precise humidity value. Electronic sensors measure humidity by measuring electrical conductivity across a metal plate. These sensors are used in *radiosondes* (weather balloons) because they are small, light, and can track rapid changes in humidity. Radiosondes transmit

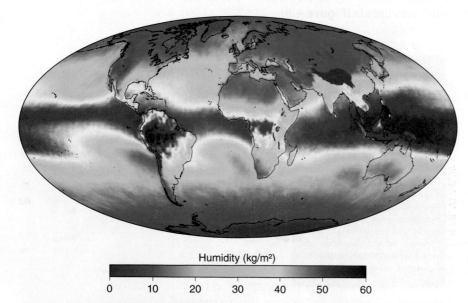

Humidity (kg/m²)

| 0 | 10 | 20 | 30 | 40 | 50 | 60 |

Figure 4.9 Global humidity map. This map illustrates the average water vapor content of the atmosphere for 2009 (in kg) for a 1 m² column of air extending from Earth's surface to the top of the atmosphere. The tropics are most humid. Polar regions and deserts, such as the Sahara, have the lowest humidity. The Tibetan Plateau in Eurasia stands out with low humidity because of its high elevation and resulting cold, dry air. *(NASA)*

data in real time to a ground-based receiving station. Satellites also remotely sense humidity in the atmosphere from high above Earth. They have the advantage of providing continuous and regional coverage.

We are most familiar the term *relative humidity* to indicate how humid the air is. But relative humidity can be misleading: It does not necessarily indicate how much actual water vapor is in the air. The driest air on Earth, found in Antarctica, has almost no water vapor but still has a very high relative humidity. To sort this out, we consider four measures of humidity: vapor pressure, specific humidity, relative humidity, and dew point.

Vapor Pressure

Vapor pressure is the portion of air pressure exerted exclusively by molecules of water vapor.

Recall from Section 2.2 that the weight of the atmosphere exerts 1 kg of pressure on every square centimeter (1 kg/cm^2[14.7 lb/in^2]) of all objects at sea level. A **millibar (mb)** is a measure of atmospheric pressure that is conceptually identical to kilograms per square centimeter (or pounds per square inch). Both units refer to the force exerted by air molecules as they push against a surface. Millibars are commonly used by meteorologists and other scientists because they are finer units that can express small changes in atmospheric pressure. (A *hectopascal,* or *hPa,* is the metric equivalent of a millibar.)

The average weight of the atmosphere at sea level exerts 1013.25 mb of pressure. Almost all of this pressure is caused by nitrogen and oxygen molecules, but a small portion of this pressure is caused by molecules of water vapor. The portion of air pressure exerted exclusively by molecules of water vapor is vapor pressure. Imagine stuffing a full water bottle into a backpack already filled with books. The extra "pressure" exerted by the water bottle in addition to the pressure exerted by the books and other items in the backpack is equivalent to vapor pressure.

Recall that evaporation and condensation occur continuously throughout the atmosphere, but net condensation occurs only when air is saturated. The vapor pressure at which saturation occurs, called the **saturation vapor pressure**, varies with air temperature. Saturation vapor pressure is important because it indicates how close the air is to saturation at any given temperature. As shown in **Figure 4.10**, for every 10°C increase in air temperature, saturation vapor pressure approximately doubles. In other words, as air temperature increases, more water vapor must be added for saturation to occur.

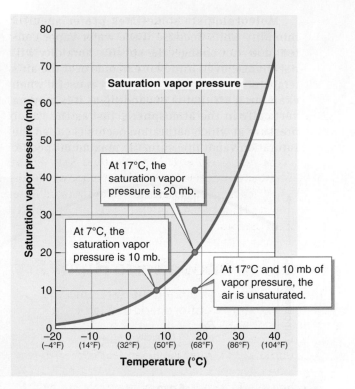

Figure 4.10 Saturation vapor pressure graph.

The blue line on this graph represents saturation vapor pressure, which is the amount of water vapor that must be in the air at a given temperature for saturation to occur. For example, if the air temperature is 7°C, there must be 10 mb of vapor pressure for the air to be saturated and clouds to form. The yellow area, in contrast, represents air that is unsaturated and cloud-free. If the air temperature is 17°C and the vapor pressure is 10 mb, the air is unsaturated. For the air to be saturated and form clouds, either the temperature must drop to 7°C or the vapor pressure (humidity) must rise to 20 mb, or a combination of the two must occur.

Specific Humidity

Another measure of humidity evaluates the water vapor content of an air sample by mass. **Specific humidity** is the water vapor content of the atmosphere, expressed in grams of water per kilogram of air (g/kg):

$$\text{specific humidity} = \frac{\text{mass of water vapor}}{\text{mass of an air sample}}$$

If, for example, there are 12 g of water vapor in 1 kg of air, then the specific humidity is 12 g/kg. Specific humidity is like vapor pressure in that both indicate the water vapor content of the air. The only difference is how each is expressed. One is expressed in grams per kilogram (specific humidity), and the other is expressed in millibars (vapor pressure). Specific humidity and vapor pressure change only when water vapor is added to or removed from the atmosphere. Condensation lowers them, and evaporation raises them.

Meteorologists sometimes prefer specific humidity units because if the water vapor content does not change, the specific humidity will not change, even if the air expands or if the air's temperature changes. This quality is useful when examining air that is expanding as it is moving vertically in the atmosphere. Just as the vapor pressure at which saturation occurs is called the saturation vapor pressure, the maximum specific humidity possible is called the *saturation specific humidity*. As with saturation vapor pressure, as air temperature increases, so too does the saturation specific humidity.

Relative Humidity

Most daily weather reports to the public use relative humidity when discussing humidity. The relative humidity is useful in that in warm weather, it

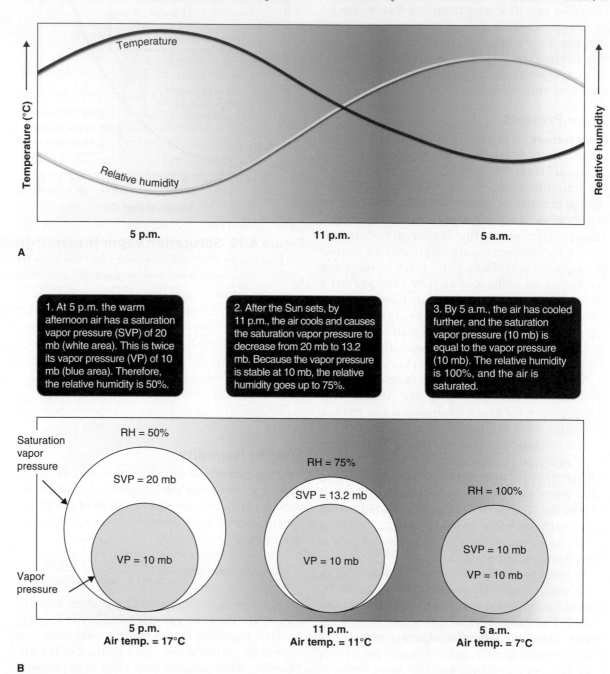

1. At 5 p.m. the warm afternoon air has a saturation vapor pressure (SVP) of 20 mb (white area). This is twice its vapor pressure (VP) of 10 mb (blue area). Therefore, the relative humidity is 50%.

2. After the Sun sets, by 11 p.m., the air cools and causes the saturation vapor pressure to decrease from 20 mb to 13.2 mb. Because the vapor pressure is stable at 10 mb, the relative humidity goes up to 75%.

3. By 5 a.m., the air has cooled further, and the saturation vapor pressure (10 mb) is equal to the vapor pressure (10 mb). The relative humidity is 100%, and the air is saturated.

Figure 4.11 Relative humidity and temperature. (A) Air temperature and relative humidity rise and fall (inversely) over a 24-hour period. (B) This example uses saturation vapor pressure (SVP) and vapor pressure (VP) to illustrate how relative humidity changes throughout the day. Note that in this example, the vapor pressure (water vapor content of the air; represented with blue circles) is not changing. Only the saturation vapor pressure (the air's capacity for water vapor; represented with white circles) is changing because the air temperature is changing.

gives a sense of how muggy the air feels. **Relative humidity (RH)** is defined as the ratio of water vapor content to water vapor capacity, expressed as a percentage. Relative humidity indicates how close to saturation the air is. The closer to saturation the air is, the higher the relative humidity and the more humid the air feels.

Temperature is the most important variable in determining relative humidity. Temperature and relative humidity are inversely related (that is, when one goes up the other goes down). Warm air has a greater water vapor capacity than cold air (see Figure 4.10). If the air temperature rises, the water vapor capacity of the air also rises. In terms of its water vapor capacity, the air also has proportionally less water vapor as the temperature increases (as long as no water vapor is evaporated into the air and the level of water vapor remains constant).

We can express relative humidity using this formula:

$$\frac{\text{relative}}{\text{humidity}} = \frac{\text{vapor pressure}}{\text{saturation vapor pressure}} \times 100$$

Figure 4.11 illustrates how relative humidity may change over a 24-hour period. If the temperature of air is lowered, the saturation vapor pressure (water vapor capacity) decreases. As cooling continues, the air will eventually become saturated. When air is saturated, the relative humidity is 100%. When the relative humidity is 100%, net condensation occurs, and clouds can form.

We can apply the relative humidity formula to an **air parcel**, defined as a body of air with uniform humidity and temperature. Refer to Figures 4.10 and 4.11 as you work through this section.

Imagine a parcel of air at 17°C with a vapor pressure of 10 mb. At this temperature, the saturation vapor pressure is 20 mb. With these numbers, we can calculate the RH of the air parcel:

At 17°C: RH = 10 mb/20 mb × 100 = 50%

If the temperature of the air parcel decreases to 11°C, the relative humidity will increase because there is a new saturation vapor pressure of 13.2 mb:

At 11°C: RH = 10 mb/13.2 mb × 100 = 75%

If the temperature drops even further, the air becomes saturated:

At 7°C: RH = 10 mb/10 mb × 100 = 100%

Notice that the 10 mb vapor pressure in the air parcel held stable because water vapor was not added to or removed from the atmosphere. Only the air temperature changed. Relative humidity does not indicate the absolute water vapor content of the air. Vapor pressure and specific humidity do. Instead, relatively humidity only tells us how close to saturation the air is. To illustrate this, polar air often has high relative humidity but almost no water vapor (low vapor pressure) because it is very cold. Warm tropical air, in contrast, has high relative humidity because it is extremely humid (high vapor pressure) **(Figure 4.12)**. In the tropics immense amounts of water are evaporated from the warm tropical oceans and give the tropics a muggy feel.

Changes in vapor pressure also affect relative humidity. Increases in evaporation resulting in increases in relative humidity can sometimes saturate the air. The two examples in the **Picture This** feature on page 108 show how air can become saturated by raising the vapor pressure.

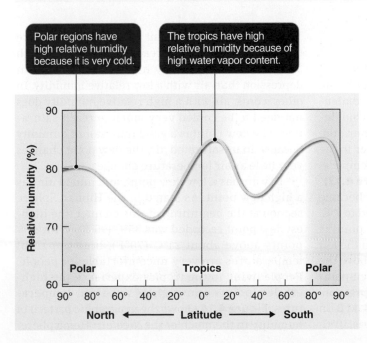

Polar regions have high relative humidity because it is very cold.

The tropics have high relative humidity because of high water vapor content.

Figure 4.12 Relative humidity and latitude. This graph shows roughly equivalent relative humidity values at the equator and 80 degrees latitude north and south. High relative humidity does not necessarily indicate high water vapor content in air. It indicates, instead, how close to saturation the air is.

Picture This **Saturation by Raising Vapor Pressure**

A *Evaporation fog* is rising from this stream in the Sierra Nevada in California. Because the stream is warmer than the chilly early morning air, water evaporates from the stream's surface and raises the vapor pressure to the saturation vapor pressure. Condensation and visible liquid cloud droplets result. *(Bruce Gervais)*

B The same phenomenon happens on cold days when people exert themselves physically and evaporate sweat into cold air. Evaporated water (sweat) raises the vapor pressure to the saturation vapor pressure, and a cloud forms. Those are liquid cloud droplets over the athlete's head, not water vapor. *(AP Photo/Alex Brandon, File)*

Relative humidity can be increased to 100% if vapor pressure rises enough to equal the saturation vapor pressure. This happens more readily when the air is cold and has a low saturation vapor pressure. Remember that water vapor is invisible. The "steam" (condensation) seen in the photos above is a cloud composed of liquid cloud droplets.

Consider This

1. Would evaporation fog continue to form if the water temperature of the stream decreased? Explain.
2. Could the cloud over the athlete's head have formed if the air were warm? Explain.

Dew Point

Why does a can of cold soda or a glass of an ice-cold beverage become wet on the outside? The water on the outside of the container is *condensate* (*dew*). Condensate forms because the cold liquid lowers the temperature of the container and, therefore, the air surrounding the container to the **dew point** (or *dew point temperature*), the temperature at which air becomes saturated **(Figure 4.13)**.

In Figure 4.11B, we considered air that became saturated when its temperature was lowered to 7°C. For that parcel of air, therefore, the dew point is 7°C. In the same figure, in the first panel on the left the air temperature was 17°C, 10 degrees above the dew point. The difference between the air temperature and the dew point is the **dew-point depression**. In the example illustrated in Figure 4.11B in the first panel, the dew-point depression is 10°C.

The greater the dew-point depression, the more the air must be cooled to reach saturation. Air with a high relative humidity has a lower dew-point depression than air with a low relative humidity. In other words, air with a high relative humidity does not need to be cooled very much to reach saturation. The dew point is a good indicator of humidity because in unsaturated air, the dew point changes very little as air temperature changes.

Dry air has a low dew point, and humid air has a high dew point. As stated in the Human Sphere section at the beginning of this chapter, the highest dew point recorded was 35°C (95°F). With dew points above about 21°C (70°F), high apparent temperatures are very uncomfortable for people. People living in the tropics experience the highest dew points and highest apparent temperatures. **Figure 4.14** shows the geographic pattern of humidity in the much of the Western Hemisphere.

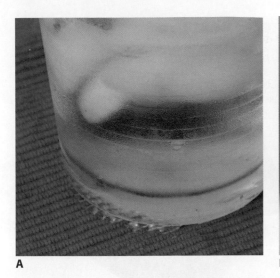

A

B

Figure 4.13 Condensation and frost. (A) The cold water in this glass chilled the glass and the air temperature around it to the dew point. Consequently, condensate formed in the saturated air around the glass. (B) When the dew point is at or below freezing, deposition (the state change from gaseous water vapor to solid ice) forms crystals of ice called *hoarfrost* or *frost*. These needles of frost accumulated on hawthorn berries in Bearsden, Scotland. They are about 1 cm (0.5 in) in length and formed in temperatures of about –5ºC (23ºF).   *(A. Bruce Gervais; B. By Ian Miles-Flashpoint Pictures/Alamy)*

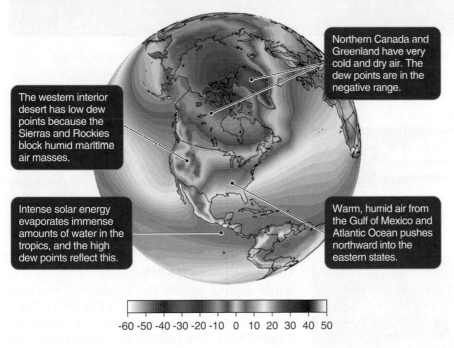

Northern Canada and Greenland have very cold and dry air. The dew points are in the negative range.

The western interior desert has low dew points because the Sierras and Rockies block humid maritime air masses.

Intense solar energy evaporates immense amounts of water in the tropics, and the high dew points reflect this.

Warm, humid air from the Gulf of Mexico and Atlantic Ocean pushes northward into the eastern states.

-60 -50 -40 -30 -20 -10 0 10 20 30 40 50

Figure 4.14 North American dew points.
This map shows lines of equal dew point in degrees Celsius for much of the Western Hemisphere averaged between 1979 and 2015. Green areas, for example, have dew points near 10ºC (50ºF). *(Original data source: ECMWF ERA-Interim reanalysis. Image made from the website ClimateReanalyzer.org, courtesy of the Climate Change Institute, University of Maine, USA.)*

4.4 Lifting Air: Atmospheric Instability

◎ **Describe the role of atmospheric instability in cloud formation.**

As we saw in Section 2.3, the troposphere experiences strong vertical mixing. That mixing is a result of *unstable air* moving higher in the atmosphere. *Atmospheric instability* is important because it transports heat from Earth's surface, particularly in the warm tropics, to higher altitudes. This *convectional heat transport* is a key component of the global heat engine (discussed in Section 3.3) and is part of the process of moving heat energy. Atmospheric instability is also the underlying force behind many meteorological processes. We will return to discuss instability in more detail later in this section.

Uneven heating of Earth's surface by the Sun creates warm and cool parcels of air. Like hot-air

Figure 4.15 A hot-air balloon festival in Albuquerque, New Mexico. Heating of the air inside these balloons makes it less dense than the air outside. The balloons become buoyant and rise. *(Bobby Bank/Getty Images)*

Figure 4.16 Air pressure on a sealed bag. This sealed bag of potato chips expanded because the atmospheric pressure decreased around it as the bag was taken from sea level to 2,130 m (7,000 ft) in elevation. *(Bruce Gervais)*

balloons **(Figure 4.15)**, warm air parcels tend to rise higher; in contrast, cool air parcels remain near the surface. Warm air rises because it is less dense and more buoyant than the air surrounding it. Cool air sinks because it is more dense and less buoyant than the surrounding air.

Rising Air Is Cooling Air: The Adiabatic Process

While driving up into the mountains, you may have noticed that sealed bags of chips or flexible plastic bottles expand at higher elevations **(Figure 4.16)**. The same process that causes the air in these containers to expand acts on rising air parcels in the atmosphere.

As air parcels rise, they expand, and as they expand, they cool. In the atmosphere, air pressure decreases with height (see Section 2.2). As an air parcel rises, the air pressure outside it decreases, and the air molecules inside it push the parcel outward, causing it to expand. This expansion uses energy, and as a result, the temperature inside the parcel decreases. Conversely, when an air parcel descends in the atmosphere, it is compressed into a smaller volume. When an air parcel is compressed, the air temperature within it increases.

Figure 4.17 Adiabatic temperature change. As air parcels move vertically in the atmosphere, they cool or warm because of changes in their volume. Unsaturated parcels cool and warm at the same dry adiabatic rate of 10°C per 1,000 meters.

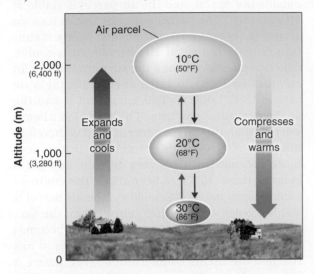

Temperature changes within air parcels resulting from changes in their volume are *adiabatic* temperature changes. **Adiabatic cooling** is the cooling of an air parcel through expansion. **Adiabatic warming** is the warming of an air parcel through compression.

At what rate does the temperature of an air parcel change? The **dry adiabatic rate** is the rate of temperature change in an unsaturated parcel of air. The air temperature changes 10°C per 1,000 m (or 5.5°F per 1,000 ft) of change in altitude. That is, it cools 10°C for every 1,000 meters it rises, and it warms 10°C for every 1,000 meters it descends. Temperature changes at the dry adiabatic rate are illustrated in **Figure 4.17**.

Recall that the *environmental lapse rate*, described in Section 2.3, refers to the decrease in air temperature *outside* the air parcel as altitude increases. The average environmental lapse rate is 6.5°C per 1,000 meters (or 3.6°F per 1,000 feet), but this rate varies considerably for any given location at any given time. The environmental lapse rate is important here because for an air parcel to continue rising higher on its own accord, it must be warmer than the air surrounding it. The temperature of the air outside the air parcel is determined by the environmental lapse rate. The temperature of the air inside the air parcel is determined by adiabatic temperature changes, which are independent of the temperature outside the air parcel.

Forming Clouds: The Lifting Condensation Level

Saturated air and clouds form in two ways: (1) when the humidity increases (through evapotranspiration) until the air is saturated or (2) when the air temperature decreases until the air is saturated. Almost all clouds form the second way—as a result of the air cooling to the dew point. In other words, net condensation in the atmosphere is almost always caused by air being lifted and cooled to its dew point. In Section 4.2, we learned that as air temperature drops, the relative humidity increases. It follows that if an air parcel rises and cools, its relative humidity will increase. If the parcel's temperature decreases to the dew point, its relative humidity will become 100%. The air inside the parcel will become saturated, and cloud formation will follow. The altitude at which the temperature inside the air parcel reaches the dew point is the **lifting condensation level (LCL)**. Most clouds have flat bases that correspond to the LCL because the LCL occurs at a constant altitude above Earth's surface, as shown in **Figure 4.18**.

3. Above the LCL, the parcel continues cooling as it rises. The air remains saturated because the air temperature is equal to the dew point.

2. At the LCL, the air temperature within the parcel has dropped to the dew point, and condensation begins.

1. Below the LCL, a warm rising air parcel is unsaturated because its air temperature is higher than the dew point.

Lifting condensation level (LCL)

Figure 4.18 The lifting condensation level. As air parcels rise, they cool adiabatically. As the temperature within them decreases, the relative humidity rises. At the lifting condensation level, the relative humidity becomes 100%, and the air within the parcel becomes saturated and condensation forms cloud droplets. *(Jason Edwards/Getty Images)*

Figure 4.19 Dry and moist adiabatic rates.

Starting on the ground, an unstable air parcel at 30°C rises, expands, and cools at the dry adiabatic rate of 10°C per 1,000 meters. The parcel's dew point of 10°C is reached at 2,000 meters (the LCL), and a cloud can form. Above 2,000 meters, the air parcel is saturated and cools at a moist adiabatic rate of 6°C per 1,000 meters.

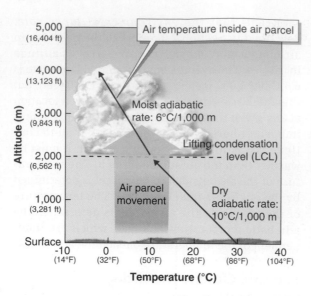

Above the LCL, water vapor condenses into liquid cloud droplets, and clouds form. In Section 4.2, we learned that whenever water undergoes a change in its physical state, heat is either absorbed from the surrounding environment (during evaporation) or released to the surrounding environment (during condensation). Above the LCL, condensation releases latent heat, which slows the rate of adiabatic cooling to the **moist adiabatic rate**—the rate of cooling in a saturated air parcel. The moist adiabatic rate is usually about 6°C per 1,000 meters (3.3°F per 1,000 feet). The air parcel continues to cool as it lifts and expands above the LCL, but it cools less quickly due to the release of latent heat. The moist adiabatic rate is always less than the dry adiabatic rate, which is 10°C per 1,000 meters. How much less it is depends on how much condensation is occurring within the cloud and, subsequently, how much latent heat is being released. This is important because more condensation releases more latent heat, and more latent heat warms the cloud, making it more unstable. **Figure 4.19** illustrates this shift in adiabatic rates.

Three Scenarios for Atmospheric Stability

In a **stable atmosphere**, the interior of air parcels is cooler and denser than the surrounding air. As a result, the air parcels do not rise of their own accord. They must be swept over a mountain range or over a cold mass of heavy, dense air in order to rise. In an **unstable atmosphere,** air parcels are warmer and less dense than the surrounding air and rise on their own. Assuming that there is sufficient water vapor in an air parcel, clouds and thunderstorms form in unstable air parcels (see Figure 4.19). A *conditionally unstable* atmosphere occurs where air parcels are stable while

unsaturated near the ground and unstable when saturated higher up. These three stability scenarios are illustrated in **Figure 4.20**.

In Figure 4.20, in Scenario 1, the environmental lapse rate is 5°C/1,000 m. The air is warmer outside the parcel, and the air parcel is stable at all altitudes. The parcel does not rise of its own accord (although it rises if pushed over a mountain range) because at all altitudes, it is cooler, denser, and heavier than the surrounding air. In contrast, in Scenario 2, the environmental lapse rate is 12°C/1,000 m. The air parcel is cooling at the dry adiabatic rate. The parcel is always warmer than the surrounding air and is therefore unstable. A cloud does not develop in this scenario because the air is very dry (the dew point is lower than –10°C). In Scenario 3, the environmental lapse rate is 9°C/1,000 m. The air parcel is stable (cooler, denser, and heavier than the surrounding air) below 1,000 meters, and it becomes unstable at 1,500 meters. At 1,000 meters, it has reached the lifting condensation level, where it has cooled to its dew point of 20°C. Now that it has cooled to its dew point, condensation is occurring. The latent heat that is released through condensation causes it to cool less quickly as it rises: at a moist adiabatic rate of 5°C per 1,000 meters. Relative to the outside air temperature, the parcel is warmer, and it is unstable at 1,500 m because of the latent heat released within the cloud. This air parcel is conditionally unstable. By definition, conditional instability occurs when the environmental lapse rate is between the dry rate of cooling (10°C/1,000 m) and the moist rate of cooling (in this example 5°C/1,000 m).

Just as adiabatic cooling results as an air parcel expands, adiabatic warming follows when an air parcel is compressed. As moving air meets the *windward* side of a mountain (the side facing the prevailing wind), air is forced upward and cools adiabatically. As air flows back down the *leeward* side of the mountain (the side sheltered from the prevailing wind), it warms adiabatically. **Figure 4.21** on page 114 illustrates this process.

Adiabatic warming on the leeward side of a mountain forms a dry environment called a **rain shadow**. Arid regions are found wherever tall mountains create a rain shadow effect. Hawai'i, for example, is seldom associated with aridity, but much of the island chain experiences rain shadow desert-like conditions, as shown in **Figure 4.22** on page 114.

Four Ways Air Lifts and Clouds Form

As we have just seen, the lifting of air is fundamental to air saturation and cloud formation because rising air cools. There are four ways that air can rise higher in the atmosphere and cool:

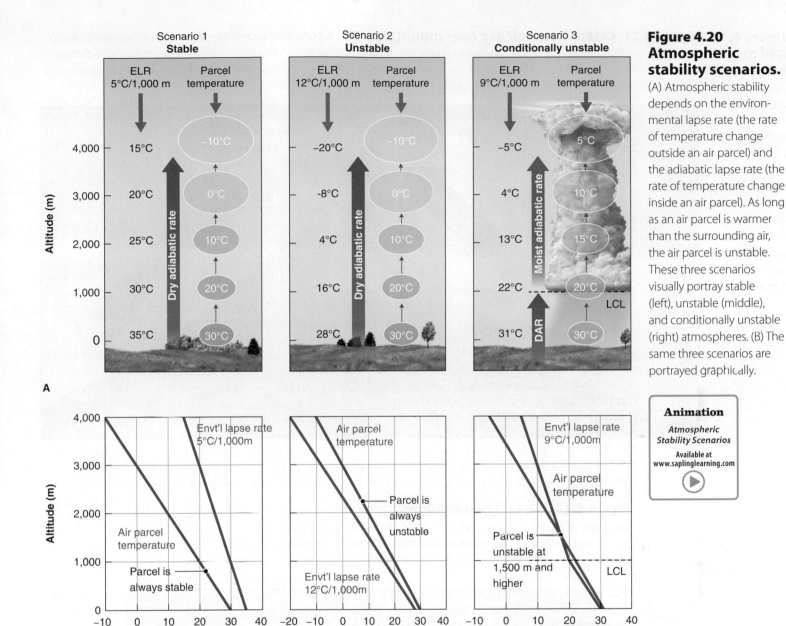

Figure 4.20 Atmospheric stability scenarios.

(A) Atmospheric stability depends on the environmental lapse rate (the rate of temperature change outside an air parcel) and the adiabatic lapse rate (the rate of temperature change inside an air parcel). As long as an air parcel is warmer than the surrounding air, the air parcel is unstable. These three scenarios visually portray stable (left), unstable (middle), and conditionally unstable (right) atmospheres. (B) The same three scenarios are portrayed graphically.

Animation
Atmospheric Stability Scenarios
Available at
www.saplinglearning.com

convective uplift, orographic uplift, frontal uplift, and convergent uplift (**Figure 4.23** on page 115).

Convective Uplift
The Sun heats Earth's surface unevenly. Earth's heated surface, in turn, warms the atmosphere unevenly. Warmed air forms unstable air parcels, which rise. Figure 4.18 shows a cumulonimbus cloud that has formed through this process, called **convective uplift**. Cumulus clouds typically form from convective uplift and result in showery *convective precipitation*.

Orographic Uplift
Orographic uplift is the lifting of air over mountains (see Figure 4.21). Clouds formed by air moving up the windward side of mountains are *orographic clouds*, and the resulting precipitation is *orographic precipitation*. A rain shadow forms on the downwind, leeward side of a mountain. Orographic precipitation occurs wherever there are mountains, even in desert regions.

Frontal Uplift
In Earth's atmosphere, geographically extensive and homogeneous regions of air called *air masses* (much larger than air parcels) form (see Section 6.1). Warm and cold air masses come into contact at midlatitudes, forming a transition zone (a *front*) between cold polar air and warm subtropical air. When warm and cold air masses meet, they do not mix. Instead, warm air flows over cold air

Figure 4.21 Latent heat release over mountains. Clouds often form on the windward side of mountains due to adiabatic cooling. The leeward side of mountains is typically dry and relatively warm because of adiabatic warming. Note that on the windward side at sea level in this example the air temperature is 22°C, but on the leeward side it is 8°C warmer. The release of latent heat caused this difference in temperature.

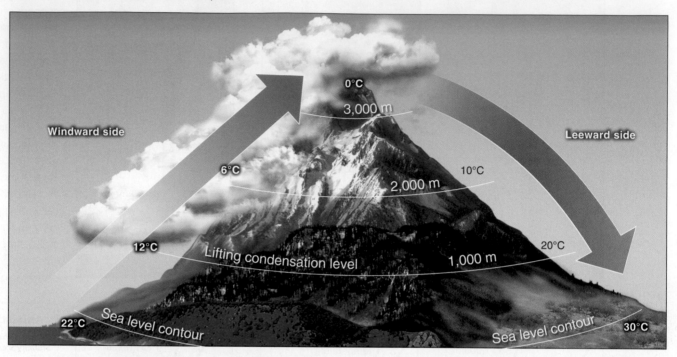

Figure 4.22 Hawai'i rain shadows. Most of the Hawaiian Islands experience a rain shadow effect, as can be seen in this map of 2011 annual rainfall totals. Annual precipitation of 20 to 50 cm (8 to 20 in), shown in the brightest red areas, is considered semiarid. *(Giambelluca, T.W., Q. Chen, A.G. Frazier, J.P. Price, Y.-L. Chen, P.-S. Chu, J.K. Eischeid, and D.M. Delparte, 2013: Online Rainfall Atlas of Hawai'i. Bull. Amer. Meteor. Soc. 94, 313–316, doi: 10.1175/BAMS-D-11-00228.1)*

Web Map

Rain Shadow on Hawai'i

Available at
www.saplinglearning.com

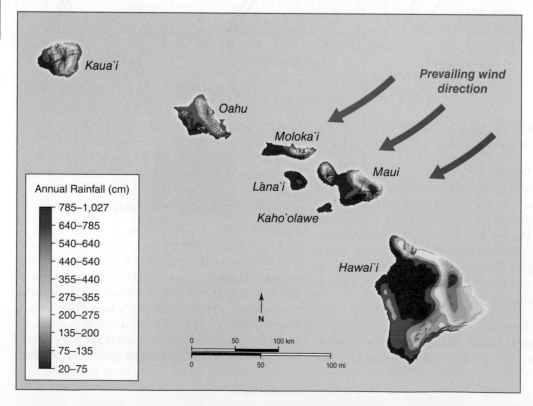

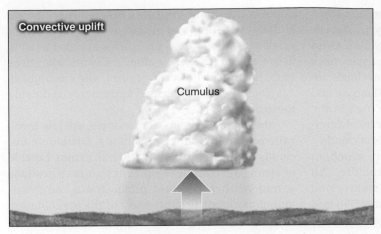

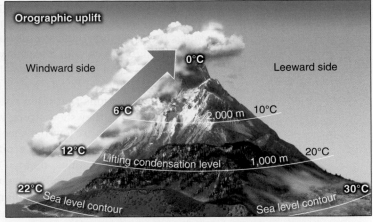

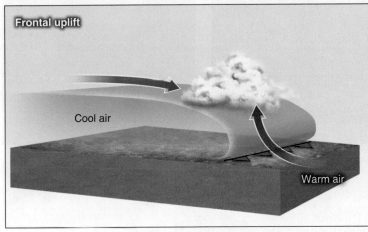

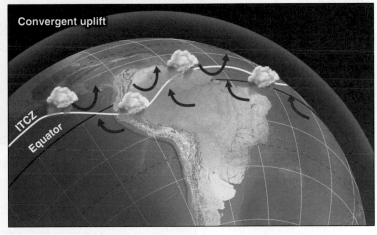

Figure 4.23 Four types of uplift. Air is lifted in the atmosphere through (A) convective uplift, (B) orographic uplift, (C) frontal uplift, and (D) convergent uplift.

because it is less dense and more buoyant. This process is **frontal uplift**, which often results in *frontal precipitation* at midlatitudes and high latitudes. Frontal weather systems are discussed in detail in Section 6.4.

Convergent Uplift

The fourth mechanism that causes air to rise occurs where surface winds converge in a geographic region of low atmospheric pressure. In **convergent uplift**, air rises as a result of converging airflow.

Convergent uplift is common in cyclonic storm systems that form at midlatitudes (see Section 6.4). It is also common in the tropics, where prevailing winds converge from the north and south in a region called the *intertropical convergence zone (ITCZ)*. The ITCZ forms a band of low pressure that encircles Earth (illustrated in Figure 4.23D). Along the ITCZ, converging and rising air strengthens convective uplift and thunderstorms. Because of this, in the tropics, convective and convergent uplift blend together into a single process.

4.5 Cloud Types

◉ **Identify and describe the major cloud types.**

Clouds are a wonderful addition to the beauty of nature. They delight the eye in ever-changing images of differing colors, textures, and shapes. We do not have to travel to exotic places to see clouds; they are overhead everywhere. From downtown urban areas to the most remote places on the globe—even in the most arid locations—clouds are present and visible.

Cloud Classification

In 1803, Luke Howard, an English naturalist, developed the cloud classification scheme still used today. The World Meteorological Organization published the *International Cloud Atlas* in 1956, based on Howard's classification scheme.

Although there are an infinite variety of cloud shapes, all clouds can be organized into four major categories. Following the system that is used to name living organisms, Howard employed Latin

Web Map

Global Cloud Cover

Available at
www.saplinglearning.com

words to create three different groups of clouds, based on their appearance: *cirrus* (wispy, feathered), *cumulus* (heaped, puffy), and *stratus* (layered). **Cirrus** clouds are high clouds with a feathery appearance that are composed of ice crystals. **Cumulus** clouds are dome-shaped, bunched clouds, with a flat base and billowy upper portions that often rise high in the troposphere; that is, they show strong *vertical development*. **Stratus** clouds are low, flat sheets of clouds.

A later addition was the *alto* group, which, in the context of cloud names, means "middle." An altocumulus cloud, for example, is a cumulus cloud that forms between 2,000 and 7,000 m (6,500 and 23,000 ft). *Nimbus* (rain) is included in a cloud type

name if the cloud produces precipitation. Later, Howard's system was modified to include height categories. *Cloud form* categories are also used to describe and categorize clouds. Clouds take three basic forms: *cirriform*, *stratiform*, and *cumuliform*. All clouds can be described and categorized using these terms.

Nimbostratus clouds, for example, are low-level, rain-producing sheets of clouds. **Cumulonimbus** clouds are heaped-up rain and hail clouds capable of strong vertical development that may produce severe weather. Together, nimbostratus and cumulonimbus clouds produce most of the precipitation that reaches the ground. **Figure 4.24** illustrates

Figure 4.24 Cloud categories. Eleven common cloud types are found within the four major cloud categories. *(Top left. Philip and Karen Smith/ Getty Images; top center. No Photo, No Life!/Getty Images; top right. Mark A. Schneider/Science Source; bottom. Dave Reede/Getty Images)*

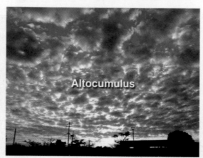

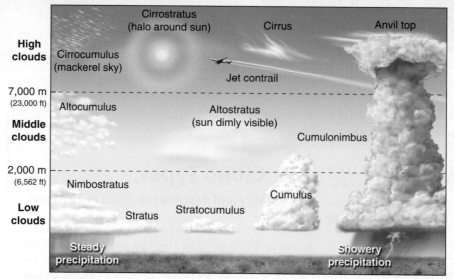

CLOUD GROUP	ALTITUDE	CLOUD TYPE	CLOUD FORM	DESCRIPTION
Cirrus	Above 7,000 m (23,000 ft)	**Cirrus**	Cirriform	Wispy, thin
		Cirrocumulus	Cirriform	Thin, bunched layer
		Cirrostratus	Cirriform	Thin sheet, produces Sun or Moon halo
		Jet contrail	Cirriform	Linear clouds produced by jet engine exhaust
Alto	2,000 to 7,000 m (6,500 to 23,000 ft)	**Altocumulus**	Cumuliform	Middle-level, bunched
		Altostratus	Stratiform	Sheeted, Sun slightly visible
Stratus	Up to 2,000 m (6,500 ft)	**Stratus**	Stratiform	Whole sky gray, Sun obscured
		Stratocumulus	Stratiform	Low, bunched
		Nimbostratus	Stratiform	Whole sky gray, precipitating
Cumulus	2,000 to 23,000 m (6,500 to 75,000 ft)	**Cumulus**	Cumuliform	Fair weather, puffy
		Cumulonimbus	Cumuliform	Severe weather, heavy rain

A

B

Figure 4.25 Radiation fog. (A) This satellite image shows the Central Valley of California under an extensive blanket of radiation fog that developed as the cold valley air became saturated. Radiation fog in this region creates dangerous driving conditions due to poor visibility. (B) This radiation fog developed in Alhandra, Portugal, along the Tagus River valley. Alhandra's central church protrudes above the fog. The photo was taken at 8 a.m., before the warm sunlight began evaporating the fog. *(A. NASA/GSFC; B. © António Ferreira Pinto)*

the 11 common cloud types found within the four major cloud categories.

Fog

Have you ever wondered what it would be like to walk inside a cloud? Perhaps you already have. **Fog** is a stratus cloud on or near the ground that restricts visibility to less than 1 km (0.62 mi).

All types of fog form through non-adiabatic cooling, where the surface of Earth cools the air above it to the dew point. There are many kinds of fog, including *orographic fog*, *freezing fog*, and *evaporation fog* (see the Picture This feature on page 108). The two most common types are radiation fog and advection fog.

Radiation fog (or *valley fog*) is fog that results when the ground radiates its heat away at night. The cooled ground cools the air above it. If the air temperature near the ground decreases to the dew point, net condensation (and fog) occur **(Figure 4.25)**.

Unlike radiation fog, **advection fog** forms as moist air moves over a cold surface, such as a lake or a cold ocean current, and the air temperature drops to the dew point. Fog over the ocean is always advection fog. There is perhaps no more famous example of advection fog than that in San Francisco **(Figure 4.26)**.

Figure 4.26 Advection fog. Advection fog often forms off the California coast near San Francisco as humid air moves over the cold California Current. As air moves over cold water, it is cooled to its dew point and becomes saturated, forming fog. *(Rob Kroenert/Getty Images)*

4.6 Precipitation: What Goes Up . . .

◎ **List the different kinds of precipitation and explain how each forms.**

One of the most important aspects of clouds is their ability to bring freshwater to Earth's continents. Why do some clouds make precipitation and other do not? And why do some clouds make rain, but others make snow or hail? Tiny bits of dust, such as bacteria, salt, smoke, pollen, volcanic ash, industrial pollution, as well as other aerosols can act as condensation nuclei. Clouds make precipitation if there is enough water vapor in the atmosphere. The type of precipitation a cloud produces depends on the temperature of the cloud and the temperature of the air beneath the cloud through which the precipitation falls.

Making Rain: Collision and Coalescence

Rain starts with condensation nuclei. **Condensation nuclei** are small particles, roughly 0.2 μm in diameter, that provide a surface on which water vapor can condense. Tiny bits of dust, bacteria, salt, smoke, pollen, volcanic ash, industrial

pollution—these and other aerosols can act as condensation nuclei. A single cubic meter of air may contain 5 to 10 billion condensation nuclei. In saturated air, water vapor condenses onto condensation nuclei to form cloud droplets. Sometimes there are too few condensation nuclei available, and water molecules have no surface on which to condense. In these situations, water remains in a vapor state, and the relative humidity can rise 1% or 2% above 100%, creating a *supersaturated* atmosphere.

Cloud droplets can never fall to Earth as they are far too light. Instead, they merge and grow heavier through collision and coalescence. It takes over a million cloud droplets to create a raindrop heavy enough to fall to Earth. **Collision and coalescence**, the process by which cloud droplets merge to make raindrops **(Figure 4.27)**, occurs in clouds at air temperatures of about −15°C (5°F) and warmer. The relatively warm temperatures needed for collision and coalescence occur at all latitudes, but they are particularly widespread in the tropics and subtropics. Without water's cohesive property, illustrated in Figure 4.5, cloud droplets would not merge; there would be clouds but no rainfall.

Normally, water freezes at 0°C (32°F), but the microscopic size of a cloud droplet allows it to enter a *supercooled* state below 0°C. Water in a supercooled state freezes quickly as it comes into contact with a solid object on the ground. Below about −10°C (14°F), a mixture of ice crystals and supercooled droplets occurs. Below about −40°C (or −40°F), all cloud droplets freeze. Supercooled cloud droplets are essential in the process of forming snow.

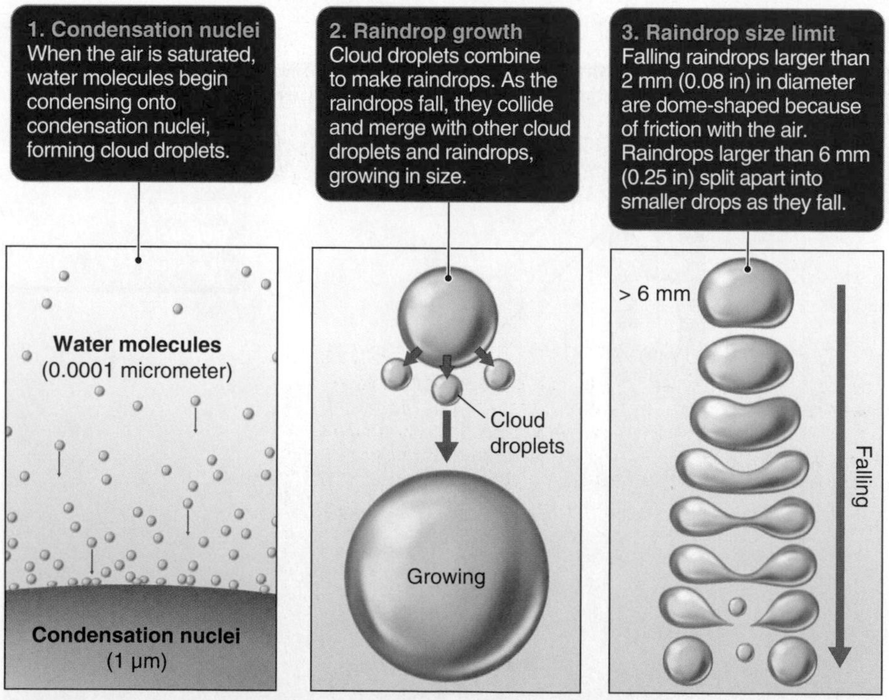

Making Snow: The Ice-Crystal Process

The **ice-crystal process** (or *Bergeron process*) forms snow in clouds in which the temperature is 0°F or below. If the snow formed in the ice-crystal process melts to water before it reaches the ground, rainfall results. Snow can melt into rain even at high latitudes. Most winter precipitation at middle and high latitudes, where clouds are typically well below freezing, results from the ice-crystal process. The ice-crystal process is somewhat more complex and less well understood than the collision-and-coalescence process that forms raindrops.

Through the ice-crystal process, snow forms from water molecules in water vapor as they are deposited directly on *ice nuclei*, as illustrated in **Figure 4.28**. Ice nuclei differ from condensation nuclei in that they are platelike or angular in shape. Particles such as salt crystals, some bacteria, and clay minerals can act as ice nuclei. Humans even

1. Condensation nuclei
When the air is saturated, water molecules begin condensing onto condensation nuclei, forming cloud droplets.

2. Raindrop growth
Cloud droplets combine to make raindrops. As the raindrops fall, they collide and merge with other cloud droplets and raindrops, growing in size.

3. Raindrop size limit
Falling raindrops larger than 2 mm (0.08 in) in diameter are dome-shaped because of friction with the air. Raindrops larger than 6 mm (0.25 in) split apart into smaller drops as they fall.

Water molecules
(0.0001 micrometer)

Condensation nuclei
(1 μm)

Cloud droplets

Growing

> 6 mm

Falling

Figure 4.27 Making raindrops. Water molecules condense onto condensation nuclei to make cloud droplets. Cloud droplets collide (crash into each other) and coalesce (join together) to make raindrops. Contrary to popular belief, raindrops are not tear-shaped but instead are spherical when they are small and become domed and elongated as they grow larger. The size of raindrops is limited to less than 6 mm (0.25 in) because as larger raindrops fall through the atmosphere, they split apart.

make artificial ice nuclei in order to "seed" clouds to increase their precipitation. The Geographic Perspectives at the end of this chapter discusses the science of *cloud seeding* further. In cold clouds, the saturation vapor pressure is slightly lower around ice crystals than it is around cloud droplets. Under these conditions, water vapor is deposited on an ice crystal more readily than it condenses on a cloud droplet. Once it is snowing, *contact freezing*,

illustrated in **Figure 4.29**, also plays a significant role in snow formation.

In *warm clouds* (or *water clouds*), the temperature is above 0°C, there is only liquid water in the cloud, and only the collision-and-coalescence process occurs. In *cold clouds* (or *ice-crystal clouds*), the temperature is below 0°C throughout, and the ice-crystal process dominates. Warm clouds always produce rain. Cold clouds may produce snow or

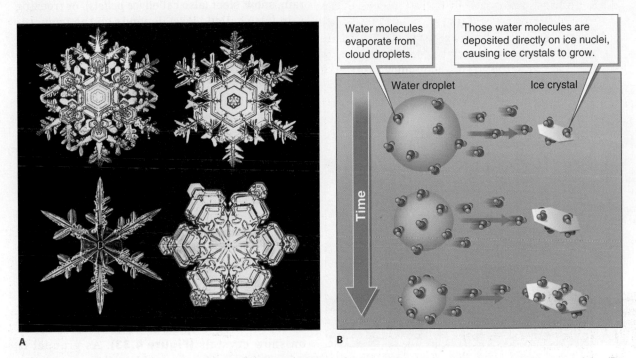

Figure 4.28 Making snow through deposition. (A) Snowflakes grow when water vapor, evaporated from supercooled cloud droplets, is deposited directly on an ice crystal, causing it to grow. (B) All snowflakes are hexagons, reflecting the six-sided structure of the hydrogen bonds between water molecules (see Figure 1.5). *(Robin Treadwell/Science Source)*

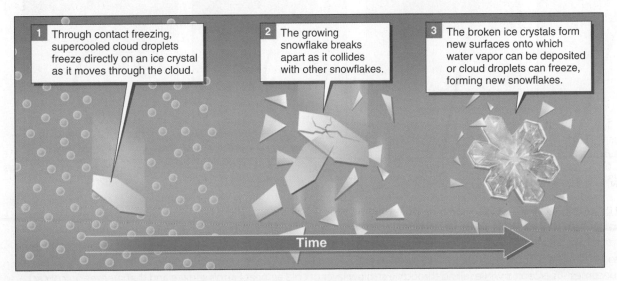

Figure 4.29 Making snow through contact freezing.

Figure 4.30 Warm, cold, and mixed clouds. In the tropics cumulus and stratus clouds are warm clouds, but at middle and high latitudes where it is colder these clouds are often cold clouds, particularly in winter. Cumulonimbus clouds are always mixed clouds at all latitudes. Cirrus clouds are always cold clouds at all latitudes and in all seasons.

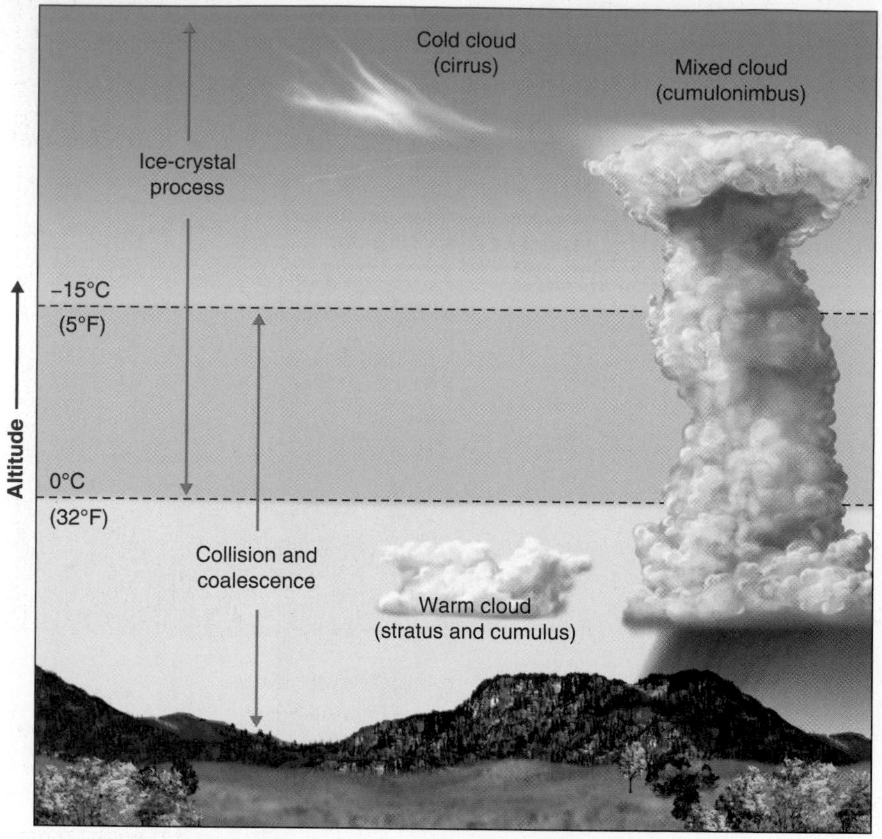

rain, depending on the temperature of the air between the cloud and the ground. Clouds with a range of temperatures, and in which both processes occur, are called *mixed clouds* (**Figure 4.30**).

Four Types of Precipitation

With the exception of hail, all types of precipitation are produced either through the collision-and-coalescence process or the ice-crystal process. Whether precipitation takes the form of rain, snow, sleet (also called *ice pellets*), or freezing rain (also called *glaze*) depends on the *temperature profile* of the atmosphere between the cloud and the ground, as illustrated in **Figure 4.31**.

The world record for most snowfall in a single season was set on Mount Baker, in Washington State, which received 28.96 m (95 ft) in the 1998–1999 season. How was that snow depth measured? There is an established protocol for measuring snow and rain. Some of the techniques are described in **Figure 4.32**.

Making Hail

Hail—hard, rounded pellets of ice—precipitates out of cumulonimbus clouds with strong vertical airflow. Any pellet of ice that is 5 mm (0.20 in) or more in diameter is considered a *hailstone*. Anything smaller is *sleet*. Hail formation starts in the cold upper reaches of cumulonimbus clouds as small pellets of ice called *graupel*, which form as supercooled droplets of water freeze directly on snow crystals (**Figure 4.33**). As graupel is

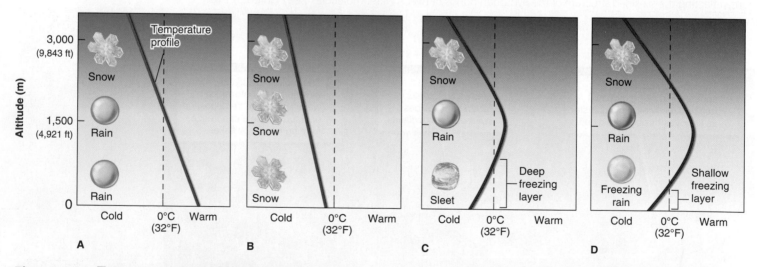

Figure 4.31 Temperature profiles and precipitation. These four different temperature profiles at different geographic locations result in four different types of precipitation. (A) Rain forms as snow melts before reaching the surface as it falls into warm air. (B) Snow results where the air is freezing all the way to the surface. (C) Sleet (or ice pellets) forms as rain is frozen in a deep layer of cold air near Earth's surface. (D) Freezing rain (or glaze) forms as rain becomes supercooled while falling through a shallow layer of cold air near Earth's surface and then freezes on contact with surface objects.

How it works
Rain is measured with a *rain gauge* that is marked in centimeters and inches. If 1 cm of rain falls it would cover the ground to a depth of 1 cm if it did not flow into the soil or downslope. Automatic rain gauges use a similar principle but weigh the amount of water rather than measure its height. They can transmit data to receivers automatically.

Potential problems
Rain gauges can be inaccurate in heavy wind or if they are too close to obstructions, such as buildings or trees.

How it works
Like a rain gauge, a snow gauge captures snowfall from the air. After the snow has fallen, the chamber is heated, and the melted water is measured. Typically, the snow-to-water ratio is 10:1; this means that 10 cm of snow will melt into 1 cm of liquid water. But the ratio varies depending on the water content of the snow.

Potential problems
Snow gauges suffer from inaccuracies as snow blows around and into and out of the gauge.

How it works
A simple hollow coring tube can be used to measure snow depth and take samples for analysis of water content.

Potential problems
This technique is labor intensive. The threat of avalanches can pose a danger to field workers.

How it works
Remote sensing techniques such as ground-based Doppler radar and satellite-based radar create precipitation estimates by examining return echoes and cloud-top temperatures. This image shows Hurricane Harvey on August 25, 2017, as it made landfall on the Texas coast. It is a composite image made with data from surface radar and infrared satellite sensors. Harvey brought record amounts of rain. The red regions show precipitation rates of up to 8.2 cm (3.2 in) per hour.

Potential problems
Remote sensing precipitation estimates are less accurate than direct measurements.

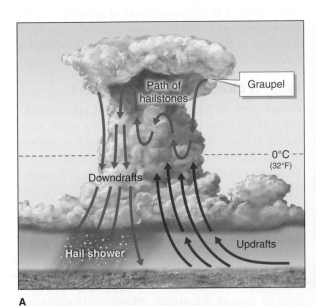
Rain gauge   Snow gauge   Snow stick   Radar remote sensing

Figure 4.32 SCIENTIFIC INQUIRY: How is precipitation measured? Water managers must measure precipitation amounts to prepare for potential droughts or floods. Precipitation measurements are also central to understanding how climate change may be affecting precipitation. *(Far left. Photo by Ray Martin; left. National Weather Service Portland; right. AP Photo/Rich Pedroncelli; far right. NASA/JAXA, Hal Pierce)*

Video
Gravity Recovery and Climate Experiment (GRACE) Freshwater Movements
Available at
www.saplinglearning.com

A

Graupel
Path of hailstones
0°C
(32°F)
Downdrafts
Hail shower
Updrafts

B

WA   MT
OR   ID   WY
Black Rock Desert
NV   UT   CO
CA
AZ   NM

Figure 4.33 Hail formation. (A) Strong updrafts keep hailstones suspended in the cloud, allowing them to accumulate more ice. (B) In this photo, a thunderstorm in the Black Rock Desert in Nevada is producing a deluge of heavy rain and hail. The hail is visible as the white curtain descending from the base of the cumulonimbus cloud. *(George Post/Science Source)*

Figure 4.34 The world's largest hailstone.

This is the world's largest officially recorded hailstone, which fell in Vivian, South Dakota, on July 23, 2010. It weighed nearly 1 kg (2 lb). *(NOAA)*

transported through the cloud, supercooled raindrops and cloud droplets freeze onto it and add to its mass. Updrafts keep hailstones suspended in the cloud. As long as the force of the updraft is greater than the gravitational force pulling at a hailstone, the hailstone remains suspended in the cloud. The longer the growing hailstone remains aloft, the larger and heavier it becomes.

Most hailstones are less than 2 cm (0.75 in) in diameter. Sometimes, however, they are much larger. In the late afternoon of July 23, 2010, in Vivian, South Dakota, the world's largest officially recorded hailstone fell to Earth in a farm field **(Figure 4.34)**. As recorded by a National Oceanic and Atmospheric Administration (NOAA) official, it measured 21.9 cm (8.63 in) in diameter—about the size of a bowling ball—and weighed 0.88 kg (1.94 lb).

Hail destroys crops and property, costing taxpayers more than a billion dollars per year. Most of this damage comes as hail pits fruit and disfigures it, making it undesirable to consumers in grocery stores, or damages young plants. And large hailstones can be deadly. Hailstones the size of tennis balls or larger fall to Earth at about 160 km/h (100 mph). The process of cloud seeding, discussed in the next section, has been tried with some success to reduce the size of hailstones in an effort to minimize crop damage.

GEOGRAPHIC PERSPECTIVES
4.7 A Cloudy Science

◎ **Assess the techniques and effectiveness of cloud seeding.**

Imagine if people could control the weather. What would that be like? What if we could stop or weaken deadly storms like Hurricane Harvey or Hurricane Irma that struck Texas and Florida in 2017 and caused billions of dollars in damage and tragic loss of human life? Would we relieve droughts and floods by making more rainfall where it is needed and less where it isn't? Would we even schedule the weather for outdoor events like national parades and important sporting events? Who would control it? Would wealthy and powerful governments divert rainfall away from poorer countries? Would the military *weaponize* the weather, creating droughts, flooding rains, and fog to weaken adversaries?

The technology to engineer the weather at this level doesn't yet exist, and perhaps it never will. Nonetheless, whether you're for it or against it, weather modification is a serious ongoing scientific effort. Scientists have been cloud seeding in the United States since the 1940s, with varying degrees of success. **Cloud seeding** is the introduction of artificial substances, such as silver iodide, to modify or enhance precipitation from clouds. The United States, Canada, China, Israel, India, Australia, and many other countries routinely cloud seed to modify their weather. Cloud seeding is a form of *climate engineering* (see Section 7.5) and is part of the broader topic of geoengineering. **Geoengineering** is the deliberate, global-scale modification of Earth's environments to improve the living conditions for people.

What Is Cloud Seeding Used For?

Cloud seeding has a wide range of applications. First and foremost, it is used to wring more water out of storms. But it also used to reduce the size of hail. Large hailstones pit crops and make produce visually unappealing and therefore economically worthless. By reducing the size of hailstones, crop damage can also be reduced. Cloud seeding has also been used at airports to reduce fog and improve visibility. It has even been tested in diminishing the strength of hurricanes. Seeding the outside edges of a hurricane may prevent moisture and latent heat energy from reaching the central area of the storm, where that energy can cause more damage. Ski resorts frequently cloud seed to enhance snowpack and draw more skiers to their resorts. Beijing, China, officials used cloud seeding to keep the city rain-free during the opening ceremonies of the 2008 Summer Olympic Games **(Figure 4.35)**. This Geographic Perspectives focuses on the use of cloud seeding to produce more precipitation from storms.

How Does Cloud Seeding Work?

There are two methods of cloud seeding. In the first method, scientists cool the temperature of the clouds by introducing *dry ice*. Dry ice has a temperature of –78°C (–109°F). Extremely fine shavings of dry ice are dropped from a plane into the clouds. The dry ice lowers the saturation vapor pressure within the clouds by lowering the temperature of the air in the clouds. This process facilitates deposition in the ice-crystal process. As snow falls through warmer air near the ground, it may melt into rain. In theory, more snow and rain result from this effort.

A second, more common way to seed clouds is by introducing *silver iodide* into the clouds. Silver iodide molecules have a six-sided, hexagonal structure, and these six-sided molecules function as freezing nuclei. Water vapor is deposited on these nuclei and forms precipitation through the ice-crystal process described in Section 4.6. Silver iodide can be introduced into clouds from aircraft **(Figure 4.36)**. More often it is emitted into the air from ground-based burners and carried up into the clouds by updrafts.

Figure 4.35 Cloud seeding in the 2008 Olympic Games.
A member of Beijing's Xiangshan Weather Modification Practice Base loads a cannon with rockets armed with silver iodide. To ensure that the 2008 Olympic Games opening ceremony would not be rained on, Beijing spent several million dollars, employed some 40,000 people, and fired more than 1,000 rockets from 20 stations located on the city outskirts. They claim this enhanced the rainfall outside the city and diminished it over the capital, where the opening ceremony was held. *(China Photos/Getty Images)*

Figure 4.36 Cloud seeding from airplanes. (A) These planes, which carry silver iodide in special flares, are cloud seeding over North Dakota. As the planes fly through the clouds, the flares are burned, releasing the silver iodide. (B) This diagram illustrates the process of cloud seeding from planes. Silver iodide works best in temperatures between –7°C and –20°C (19° and –4°F). Ground-based flares also release silver iodide, and updrafts carry it into the clouds. *(inga Spence/Alamy)*

1. Silver iodide molecules have a six-sided structure that acts as a freezing nucleus.

2. Water vapor is deposited onto the silver iodide through the ice-crystal process.

Silver iodide

–7°C to –20°C

Water droplets

3. Deposition of water vapor and resulting snow releases latent heat. This enhances updrafts, which pull more moisture into the cloud.

Moist air

Ground-based station

A

B

What Are the Dangers of Cloud Seeding?

There are many potential pitfalls in cloud seeding. Cloud seeding does not produce more overall precipitation from the atmosphere; rather, it changes the geography of precipitation. If a region is successfully cloud seeded and more precipitation falls in that area, less will fall in areas downwind. There is considerable regulatory uncertainty and potential litigation surrounding weather modification.

Also, there have been concerns about injecting silver iodide into the air because it returns to the surface in rainfall and enters the soil. Long-term studies, however, have not found levels of silver in the soil in areas that have been clouded seeded to be any greater than natural background levels.

There is also a risk to pilots. Aircraft flying into storms run the risk of having their wings ice over. Some 30% of cloud seeding missions have to turn back because dangerous ice accumulates on their wings. Nevada's Desert Research Institute is testing the use of drones to seed clouds. Drones are safer and less expensive than aircraft and more effective at dispersing the cloud-seeding agent into the clouds.

How Effective Is Cloud Seeding?

Statistically, cloud seeding does not work. In 2003, the National Academies of Sciences found that there was no convincing evidence that cloud seeding produced a significant increase in precipitation and called for more research to be done to learn more about its effectiveness. It is very difficult to know definitively whether cloud seeding works because it is impossible to do a controlled experiment in the atmosphere. Scientists simply don't know whether a given storm that was cloud seeded produced more precipitation due to cloud seeding or due to Mother Nature.

In response to the 2003 National Academies of Sciences statement, the Wyoming Weather Modification Pilot Project in the Medicine Bow Range initiated a cloud-seeding study in two adjacent areas. One area was seeded, and the other was *the control* (unseeded). Historically, both areas had received roughly the same amount of precipitation. After gathering data for 9 years, the study concluded that the seeded area received up to 15% more precipitation than the control area. A similar study in Tasmania came to roughly the same conclusion.

On the other hand, cloud-seeding studies that show no statistically significant results are too numerous to list. In sum, many researchers believe that only under the right atmospheric conditions and in the right geographic locations does cloud seeding produce meaningfully more precipitation.

Despite the uncertainty, considerable money is spent on cloud seeding. China spends about $100 million a year on cloud seeding. In the United States, government agencies and private utilities spend about $15 million each year. California's utility companies have been seeding every year since the 1950s to enhance the snowpack in the Sierra Nevada. These companies generate and sell hydroelectricity (electricity generated from streams), so the more water flowing downhill, the higher their production and revenue. The Los Angeles Department of Water and Power has laid out a half million dollars each year since the 1950s to seed clouds over the San Gabriel Mountains; it claims that the seeding boosts precipitation up to 20%.

Why spend money on cloud seeding if the results are uncertain? It all boils down to the increasingly pressing need for freshwater resources. By some estimates, 4 billion people, two-thirds of the world's population, face serious water shortages during several months of the year. Most of them are in India and the Middle East. The most acute water scarcity in the United States is in the Southwest **(Figure 4.37)**. Water resources will continue to shrink as population grows and climate change makes droughts more severe and prolonged. Given the seriousness of water resource issues, many think that even the limited returns on cloud-seeding investment make it worthwhile.

Figure 4.37 Water scarcity map. This map shows water scarcity. It indicates the number of months each area is certain to experience water scarcity. *Water scarcity* is defined as the withdrawal of water from streams and wells faster than it is replenished. Dark red areas represent the greatest scarcity. In the dark red areas each month of the year people use water faster than it is replenished. *(From Sci Adv. 2016 Feb; 2(2): e1500323.* © *Mekonnen MM, Hoekstra AY, some rights reserved; exclusive licensee American Association for the Advancement of Science. This work is distributed under CC BY-NC (http://creativecommons.org/licenses/by-nc/4.0/).)*

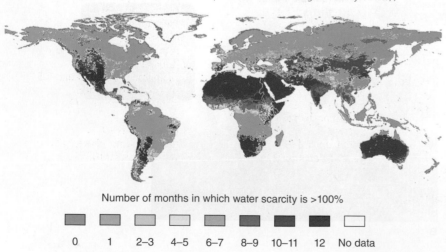

Number of months in which water scarcity is >100%

0   1   2–3   4–5   6–7   8–9   10–11   12   No data

Chapter 4 Study Guide

Focus Points

The Human Sphere: Killer Heat

- **Dangerous humidity:** High atmospheric humidity makes high air temperatures feel even hotter, creating dangerous heat waves that often take many lives.

4.1 The Hydrologic Cycle

- **States of water:** Water changes its physical state (solid, liquid, gas) when energy is added or removed.
- **The hydrologic cycle:** In the hydrologic cycle, water circulates through the atmosphere, hydrosphere, cryosphere, biosphere, and lithosphere.
- **Water in the atmosphere:** About 85% of water vapor in the atmosphere originates from evaporation from the oceans. The remaining 15% comes from plant transpiration.

4.2 Properties of Water

- **Characteristics of water:** When water changes its physical state, hydrogen bonds linking water molecules are either formed or broken.
- **Evaporation and condensation:** Evaporation is a cooling process because it absorbs heat from the environment. Condensation is a warming process because it releases latent heat to the environment.
- **Net condensation:** Clouds form in environments of net condensation.

4.3 Atmospheric Humidity

- **Relative humidity:** Changes in relative humidity happen primarily because of air temperature changes. As air cools, relative humidity rises. Saturated air has a relative humidity of 100%.
- **Humidity indicators:** High vapor pressure, specific humidity, and dew point all indicate a high atmospheric water vapor content. Relative humidity does not reliably indicate the water vapor content of the atmosphere.

4.4 Lifting Air: Atmospheric Instability

- **Atmospheric instability:** Unstable air parcels lift and cool. This is the common way air becomes saturated and clouds form.
- **Lifting condensation level:** Above the lifting condensation level, condensation releases latent heat and slows the rate of adiabatic cooling to the moist adiabatic rate.
- **Adiabatic warming and mountains:** The leeward side of a mountain is often warmer and drier than the windward side due to the release of latent heat through condensation.
- **Four forms of uplift:** Air is lifted in four ways: convective uplift, orographic uplift, frontal uplift, and convergent uplift.

4.5 Cloud Types

- **Cloud composition:** Clouds are composed of suspended microscopic liquid cloud droplets and ice crystals.
- **Cloud forms:** Clouds take one of three forms: cirriform, stratiform, or cumuliform.

- **Clouds that bring precipitation:** Nimbostratus and cumulonimbus clouds bring precipitation. Cumulonimbus clouds can produce severe weather.
- **Fog:** Fog forms where surface air has been cooled to the dew point. Most fog is either radiation fog or advection fog.

4.6 Precipitation: What Goes Up . . .

- **Condensation and ice nuclei:** Cloud droplets form around condensation nuclei (such as dust, pollen, or particulate pollution) suspended in the atmosphere. Ice crystals grow around ice nuclei (such as salt spray and clay particles).
- **Rain and snow:** Rain is formed through collision and coalescence of cloud droplets in warm clouds. Snow is formed through the ice-crystal process in cold clouds.
- **Types of precipitation:** The temperature profile of the atmosphere determines which form of precipitation occurs. Falling snow may become rain, sleet, or freezing rain as it falls from a cloud to the ground. Hail forms in cumulonimbus clouds.

4.7 Geographic Perspectives: A Cloudy Science

- **Cloud seeding:** Cloud seeding is used to increase precipitation for areas that need it.
- **Uncertain science:** The safety and effectiveness of cloud seeding are uncertain, but many countries and states practice cloud seeding because they feel it works.

Key Terms

adiabatic cooling, 111
adiabatic warming, 111
advection fog, 117
air parcel, 107
apparent temperature, 104
cirrus, 116
cloud, 99
cloud seeding, 122
collision and coalescence, 118
condensation, 99
condensation nucleus, 118
convective uplift, 113
convergent uplift, 115
cumulonimbus, 116
cumulus, 116
dew point, 108
dew-point depression, 108
dry adiabatic rate, 111
evaporation, 99
evapotranspiration, 100
fog, 117
frontal uplift, 115
geoengineering, 122
hail, 120

humidity, 104
hydrogen bond, 100
hydrologic cycle, 99
hygrometer, 104
ice-crystal process, 118
latent heat, 102
lifting condensation level (LCL), 111
millibar (mb), 105
moist adiabatic rate, 112
nimbostratus, 116
orographic uplift, 113
precipitation, 99
radiation fog, 117
rain shadow, 112
relative humidity (RH), 107
saturation, 104
saturation vapor pressure, 105
specific humidity, 105
stable atmosphere, 112
stratus, 116
transpiration, 99
unstable atmosphere, 112
vapor pressure, 105

Concept Review

The Human Sphere: Killer Heat

1. How is apparent temperature different from the actual temperature?

2. What does the dew point indicate?

3. What is the most important meteorological threat to human life in the United States?

4.1 The Hydrologic Cycle

4. What is the hydrologic cycle? Describe how water moves through it.

5. Give examples of how the hydrologic cycle is important.

4.2 Properties of Water

6. In what three physical states does water occur?

7. What are hydrogen bonds, and how do they differ in the three physical states of water?

8. What must be added to or removed from water to break or form hydrogen bonds?

9. What does the statement "water vapor has the highest internal energy state of water's states" mean? What is the source of that energy?

10. Why is evaporation a cooling process? What is being cooled?

11. Why is condensation a warming process? What is being warmed?

12. What is atmospheric saturation? What causes air to become saturated?

13. How many calories of latent heat must be added to or removed from 1 g of water for it to melt or freeze?

14. How many calories of latent heat must be added to or removed from 1 g of water for it to evaporate or condense in temperatures lower than 100ºC?

4.3 Atmospheric Humidity

15. What is saturation? What happens to water vapor when the air becomes saturated?

16. Compare specific humidity, vapor pressure, and dew point in terms of what units each measure uses and how each can be increased or decreased.

17. What is saturation vapor pressure? How can it be increased or decreased?

18. What is relative humidity? How can it be increased or decreased?

19. Does a high relative humidity always mean that the air has a high water vapor content? Explain.

20. What is the dew point? What happens when air temperature reaches the dew point?

4.4 Lifting Air: Atmospheric Instability

21. What are adiabatic temperature changes, and what causes them? Compare them with temperature changes caused by the environmental lapse rate.

22. Compare an unstable air parcel with a stable air parcel. What determines stability?

23. Compare the dry adiabatic rate of cooling with the moist adiabatic rate. Why does adiabatic cooling occur at two different rates? Explain why the moist rate of cooling is less than the dry rate.

24. What is the lifting condensation level? Where is it, and what forms there?

25. Compare the windward and leeward sides of a mountain in terms of adiabatic temperature changes and precipitation amounts. What is a rain shadow? Why does a rain shadow form?

26. What are the four major ways air is lifted and cooled adiabatically to the dew point? Describe how each works.

27. Under what physical conditions does convergent uplift take place?

4.5 Cloud Types

28. What are clouds composed of? List the four major cloud categories.

29. Which cloud type creates a thickly overcast sky with no precipitation?

30. Which clouds are composed of ice crystals? Why are they composed of ice?

31. What are jet contrails? What forms them, and where do they form? (Hint: See Figure 4.24.)

32. Which types of clouds form precipitation? Which types bring severe weather?

33. How do radiation fog and advection fog form?

4.6 Precipitation: What Goes Up . . .

34. What are condensation nuclei and ice nuclei? For what processes are each required?

35. Compare and contrast how rain and snow form through the collision-and-coalescence and ice-crystal processes.

36. Define the following terms: *warm clouds*, *cold clouds*, and *mixed clouds*. Which precipitation-forming processes occur in each?

37. What instruments are used to measure rain and snow?

38. Compare the temperature profiles in the atmosphere necessary to form rain, snow, sleet, and freezing rain.

39. Which type of cloud produces hail? Explain how hail forms in this cloud.

4.7 Geographic Perspectives: A Cloudy Science

40. What is cloud seeding? What two methods are used to do it?

41. What is geoengineering?

42. What are some of the applications of cloud seeding? What is the most common use of cloud seeding?

43. What are the risks of cloud seeding?

44. Is cloud seeding effective? Explain.

45. Why is cloud seeding so potentially important for society? And why is it controversial?

Critical-Thinking Questions

1. The hydrologic cycle is solar powered. Does it stop working at night, after the Sun has set?

2. Which of the four types of air lifting occur where you live? What kinds of information do you need to answer this question?

3. Explain why "steam" fills the air in the bathroom if you take a hot shower on a cold day. What is that steam, and why does it form?

4. How can warm, humid tropical air have a lower relative humidity than cold, dry polar air?

5. Explain how fog can form over a cold pond in warm air and over a warm pond in cold air. Describe the relative temperatures of the water and air. Use any of the humidity variables you have learned about in this chapter.

Test Yourself

Take this quiz to test your chapter knowledge.

1. **True or false?** When temperature goes up, relative humidity goes down.

2. **True or false?** Water cannot move through the hydrologic cycle as a gas.

3. **True or false?** Below the lifting condensation level, the air temperature is lower than the dew point temperature.

4. **True or false?** Evaporation is a warming process.

5. **True or false?** When an air parcel is cooler than the surrounding air, the parcel is stable.

6. **Multiple choice:** Which of the following is true when relative humidity is 100%?

    a. The air temperature and the dew point temperature are the same.
    b. Clouds and fog are likely.
    c. Vapor pressure = saturation vapor pressure.
    d. All of the above are true.

7. **Multiple choice:** Which of the following is likely to form when rain falls through a very shallow layer of freezing air near the ground?

    a. rain
    b. hail
    c. freezing rain
    d. sleet

8. **Multiple choice:** Which of the following will not change if the water vapor content increases? (Assume that temperature does not change.)

    a. specific humidity
    b. dew point
    c. relative humidity
    d. saturation vapor pressure

9. **Fill in the blank:** The _____ process occurs in cold clouds and results in snow.

10. **Fill in the blank:** A _____ occurs on the leeward side of a mountain and results in arid conditions there.

Picture This. *Your Turn*

Explain the Cloud in the Photo

1. Which of water's three states is represented by the cloud?

2. Why do we breathe out clouds on cold days?

3. What is the relative humidity inside this cloud?

4. What evidence is there that net evaporation is happening outside the cloud?

Online Geographic Analysis

Gathering Real-Time Air Quality Data

In this exercise we analyze water vapor and clouds in the atmosphere.

Activity

Go to https://svs.gsfc.nasa.gov/3732, NASA's Scientific Visualization Studio web page. This page displays two videos of relative humidity. The top video shows the summer of 1988, and the bottom video shows relative humidity in summer 1993. Click the play button for the top video.

1. **What do the moving orange fields indicate about relative humidity? What do the purple fields indicate?**

2. **Compare the overall relative humidity in Mexico to that in the United States and Canada. Which is higher?**

3. **How would you characterize the relative humidity along the equator?**

4. **Compare the global pattern of relative humidity shown in Figure 4.9 to what you see in this video. How are they similar?**

Go to https://weather.msfc.nasa.gov/GOES/, NASA's Interactive Global Geostationary Weather Satellite Images web page. Scroll down to the "GOES East—Full Disk" category and click the "Band 14 Longwave Infrared" map. This map shows current cloud-top heights in the Western Hemisphere.

5. **What is the difference between black and gray areas on the map and areas of color on the map?**

6. **Which colors are used to show the highest/coldest cloud tops?**

Keep the previous web page open. In a new tab, go to https://pmm.nasa.gov/gpm/imerg-global-image, NASA's Precipitation Measurement Missions web page. This map shows current precipitation rates. Click the map to make it larger.

7. **Which colors of the map indicate the greatest precipitation rates?**

8. **Compare this map to the cloud-top height map open in the other tab. What is the relationship between cloud-top height and precipitation rate?**

9. **Which kind of clouds are producing the highest rate of precipitation? (See Figure 4.24 for help.)**

(Enigma/Alamy)

Atmospheric Circulation and Wind Systems

Chapter Outline *and Learning Goals* ◎

5.1 Air Pressure and Wind

◎ Explain how surface air pressure changes and describe three controls on wind speed and direction.

5.2 Global Atmospheric Circulation Patterns

◎ Illustrate global patterns of atmospheric pressure, winds, and precipitation.

5.3 Wind Systems: Mountain Breezes to the Asian Monsoon

◎ Identify local and regional wind systems and explain how they form.

5.4 El Niño's Wide Reach

◎ Discuss how El Niño forms and describe its global effects.

5.5 Geographic Perspectives: Farming the Wind

◎ Assess the challenges and potential of wind as a clean energy source.

Cloud patterns reveal wind patterns. This June 2017 Landsat 8 satellite image captured spiraling cloud patterns, called "von Kármán vortices," on the lee side of Tristan da Cunha, a small volcanic island in the South Atlantic Ocean. Von Kármán vortices form where prevailing winds are diverted around high topographic features that block the wind. This image is shown in false color to better highlight the differences between clouds and Earth's surface.

(NASA Earth Observatory images by Joshua Stevens and Jesse Allen, using Landsat data from the U.S. Geological Survey and VIIRS data from the Suomi National Polar-orbiting Partnership).

To learn more about surface wind patterns around the world, go to Section 5.2.

THE HUMAN SPHERE Life under the South Asian Monsoon

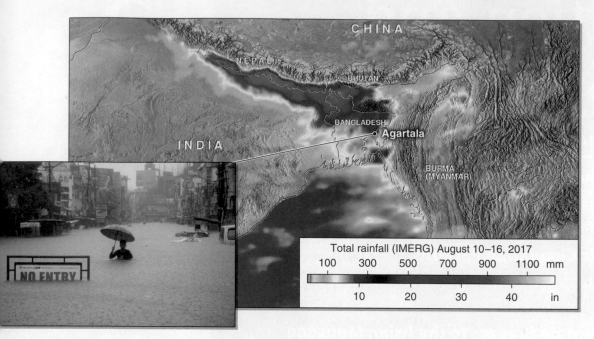

Figure 5.1 Drought and flooding in India. From August 10 to August 16, 2017, heavy monsoon rains flooded and temporarily displaced about 45 million people in India, Bangladesh, and Nepal, taking the lives of more than 1,200 people. The heaviest precipitation (in purple on this rainfall map) was created by orographic lifting and occurred along the windward slope of the Himalayas. The inset photo of a man wading through a street shows the extent of the 2017 flooding in Agartala, the capital of Tripura State in India. *(NASA/JAXA, Hal Pierce; inset. Arindam Dey/AFP/Getty Images)*

Total rainfall (IMERG) August 10–16, 2017
100   300   500   700   900   1100 mm
10   20   30   40   in

Humans have always been at the mercy of the weather. This is particularly true for South Asia, where deadly drought or flooding rainfall hinges on which way the wind blows. It is no exaggeration to say that "water is life," and the monsoon over South Asia brings winter drought followed by life-giving summer rainfall. A **monsoon** is a seasonal reversal of winds, characterized by moist summer *onshore wind* (blowing from sea to land) and dry winter *offshore wind* (blowing from land to sea).

Monsoons occur in many different parts of the world, including in the southwestern United States. Asia has two monsoon systems: the *South Asian monsoon*, which affects India, and the *East Asian monsoon*, which affects Indonesia, northern Australia, southern China, Korea, and Japan. The South Asian monsoon is the world's most dramatic monsoon. India lies at the geographic heart of the South Asian monsoon, but Pakistan, Bangladesh, and Nepal are also affected by it. In most of India, winters are long and dry, and May, June, and July can bring torrential rainfall. India relies on the summer rains to grow its crops, refill its wells and groundwater aquifers, and generate hydroelectricity from rain-swollen rivers.

Over half of India's population works in agriculture, and agriculture accounts for 15% of the country's gross domestic product (GDP). The hydroelectricity generated from streams provides 25% of the country's electricity. When less rain falls, crops wither, agricultural productivity slows, and food must be imported. As rivers run low, electricity becomes more expensive, and electrical blackouts occur. All of these drought-caused issues are detrimental to the economy.

But too much rain spells trouble as well. Farmland and city streets flood, mudslides bury villages in mountainous terrain, and outbreaks of water-borne disease such as dengue fever, malaria, and cholera strike. The 2017 monsoon brought a weeklong burst of unusually heavy rainfall and severe flooding **(Figure 5.1)**. We discuss the spatial variability of monsoon rainfall and the mechanisms responsible for the monsoon in more depth in Section 5.3.

5.1 Air Pressure and Wind

◎ **Explain how surface air pressure changes and describe three controls on wind speed and direction.**

The wind plays an important role in our lives. Many plants rely on the wind to disperse their pollen and seeds. The wind transports clouds and moisture from the oceans to the continents, delivering life-sustaining precipitation. Tornadoes and hurricanes are ultimately just wind, but they can destroy homes and even flood entire cities. In the effort to fight climate change, we are increasingly making carbon-free electricity generated from the wind to power our cars, light our homes, and run our cities. In the broader context of the global heat engine, the wind makes Earth more livable. The wind distributes heat energy around the globe. Friction from the wind against the ocean's surface generates ocean currents that transport tropical heat energy to colder high latitudes. Warm places would become far warmer and cold places would become far colder were it not for wind.

Wind is an expression of solar energy. If the Sun were to stop shining, most atmospheric motion would soon cease. Variations in the heating of Earth's surface create differences of air pressure, which in turn, create wind.

Measuring Air Pressure

Air pressure is crucially important to the science of meteorology and weather processes. Air pressure varies across Earth's surface, and variations in air pressure drive atmospheric motion and determine the type of weather that will occur. Generally, high pressure is associated with fair weather, and low pressure is associated with potentially stormy weather. Air pressure is measured using a **barometer**. For this reason, air pressure is also called *barometric pressure*. **Figure 5.2** shows two types of barometers.

Below about 8 km (5 mi) altitude, air pressure drops roughly 10 mb for each 100 m (328 ft) increase in elevation. Meteorologists must remove the effects of elevation to make meaningful comparisons of air pressure among different geographic regions. *Station pressure* is barometric pressure that has not been mathematically adjusted to sea level. In contrast, **sea-level pressure** is barometric pressure that has been adjusted to sea level. Weather stations always report sea-level pressure rather than station pressure. **Figure 5.3** on the following page illustrates how station pressure is adjusted to sea-level pressure.

Extremes of Air Pressure

Because air pressure increases at lower elevations, Earth's highest average air pressure is

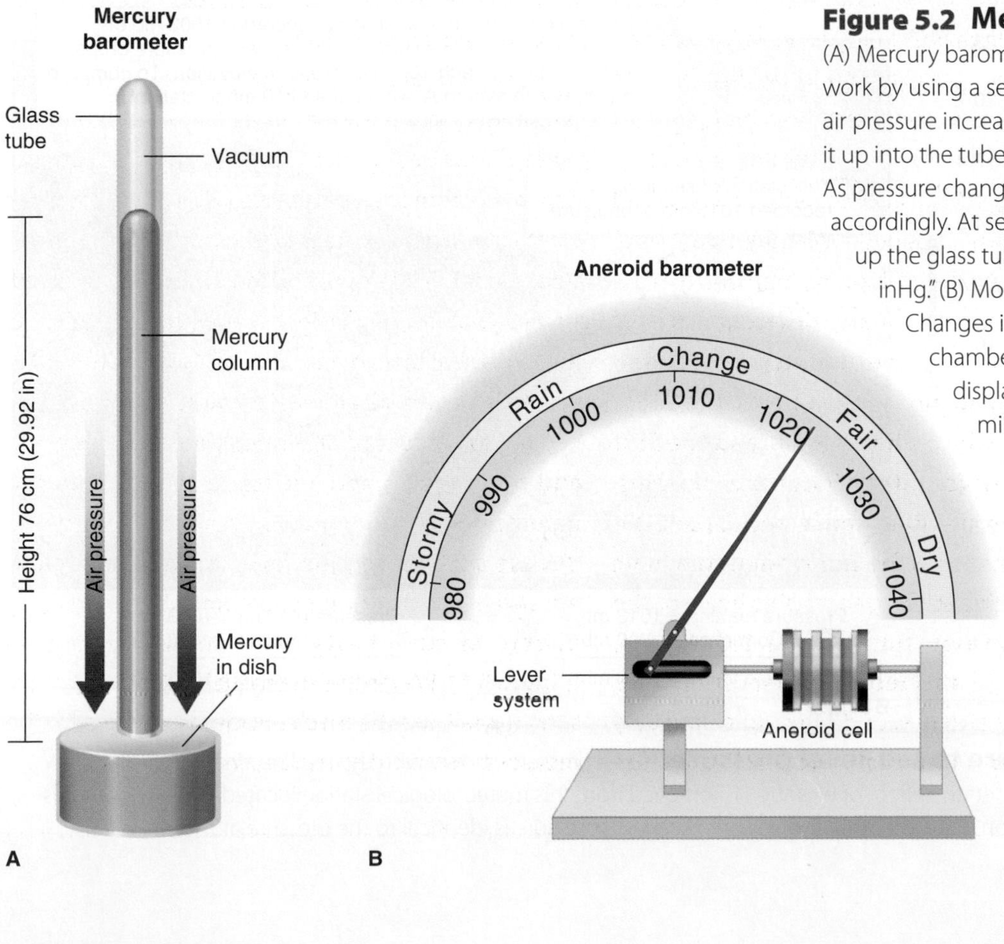

Figure 5.2 Mercury and aneroid barometers.
(A) Mercury barometers are the oldest type of barometer. They work by using a sealed glass tube partially filled with mercury. As air pressure increases, it exerts force on the mercury and pushes it up into the tube. The tube is marked in centimeters and inches. As pressure changes, the height of the mercury column changes accordingly. At sea level, the mercury is pushed 76 cm (29.92 in) up the glass tube, on average. This pressure is written as "29.92 inHg." (B) Most modern barometers are the aneroid type. Changes in air pressure change the volume of the sealed chamber, which moves a needle or changes a digital display. Aneroid barometers usually give pressure in millibars. Average pressure at sea level is 1013.25 mb (or 1 kg/cm^2 [14.7 lb/in^2]; see Section 2.2). Canada and several other countries use kilopascal units (kPa; 1 kPa = 10 mb). Average pressure at sea level is 101.33 kPa.

found at the lowest land elevation—the Dead Sea. The Dead Sea's elevation is 430 m (1,412 ft) below sea level, and the average (unadjusted to sea level) barometric pressure there is 1065 mb (31.44 inHg). This is much higher than Earth's average sea-level pressure of 1013 mb (29.92 inHg). The highest pressure ever recorded was in the Agata region of Russia (66°53′N, 93°28′E, elevation 261 m [856 ft]) on December 31, 1968. Official barometers there recorded a sea-level pressure of 1083.8 mb (32 inHg).

The lowest sea-level pressure recorded was in the eye of a *tropical cyclone* (tropical cyclones are called *typhoons* in Southeastern Asia and *hurricanes* in North America). On October 12, 1979, a barometer dropped out of an airplane into the eye of Typhoon Tip in the western Pacific Ocean measured sea-level air pressure of 870 mb (25.69 inHg). In 2013, the barometric pressure in Typhoon Haiyan was measured at 860 mb (25.40 inHg). Officials are still considering whether this measurement should be recognized as a new world record. Similarly, scientists know that air pressure is considerably lower in the vortex of tornadoes, but there are no official readings from tornadoes due to the dangers associated with getting a barometer into the center of a tornado.

Kinds of Air Pressure

In this chapter, we consider differences in sea-level air pressure across geographic space. Horizontal differences in pressure are much more subtle than vertical differences. They are important because even the slightest horizontal differences cause the wind to blow.

Why does sea-level air pressure change across geographic space? Meteorologists identify two types of air pressure and categorize them based on the cause of pressure change: thermal air pressure and dynamic air pressure.

Thermal Air Pressure

Thermal air pressure is air pressure that results when air heats up or cools down. Warm air is associated with low pressure, and cold air is associated with high pressure. When air in a sealed container heats up, its kinetic energy (rate of molecular movement) increases and, because the air molecules cannot escape the container, they push more vigorously against its walls. As a result, the air pressure within the chamber increases. In the atmosphere, the opposite occurs: When the temperature of air in the atmosphere increases, its kinetic energy increases, and because there are no walls to contain it, its volume (not its pressure) expands.

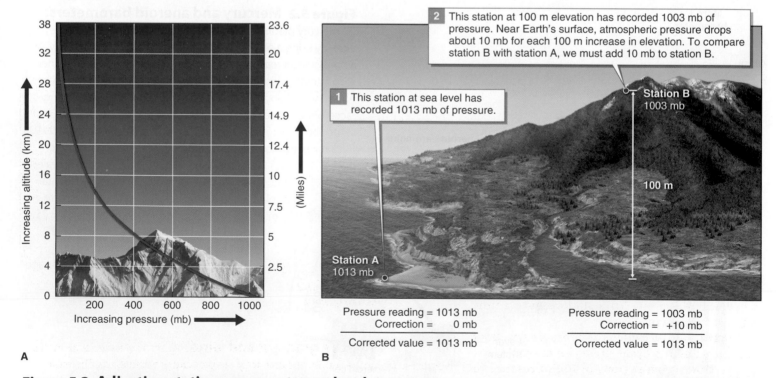

A

B

Figure 5.3 Adjusting station pressure to sea-level pressure. (A) Air pressure, shown with the red line, changes fastest within about 8 km (5 mi) of Earth's surface. (B) When the effect of elevation is removed from this meteorological station located in the mountains (by subtracting 10 mb for every 100 m of elevation gain), the resulting adjusted sea-level pressure is identical to the pressure at the coast.
(Akash9792/shutterstock)

This expansion reduces the molecular density of the air (because the molecules are more spread out) and lowers the air pressure. This results in buoyant rising air parcels. Conversely, when the temperature of air in the atmosphere decreases, molecular movement decreases. As a result, the density of air molecules increases, and so does the air pressure.

Dynamic Air Pressure

Dynamic air pressure is pressure that is caused by air movement. The relationship between air temperature and air pressure is not constant because dynamic air pressure can override thermal air pressure. As air flows and moves around in Earth's atmosphere, it creates dynamic air pressure that can override thermal air pressure.

In places where air compresses and becomes more dense, dynamic high pressure results. Dynamic high pressure, for example, forms in subtropical latitudes centered on 30 degrees north and south, creating the *subtropical high* zone (see Section 5.2). Likewise, in places where the air flows upward vertically, the air near Earth's surface "stretches," and its molecular density decreases, resulting in dynamic low pressure. Dynamic low pressure forms near the equator with the formation of the ITCZ, and near 60 degrees latitude north and south with the formation of the *subpolar low* pressure system (see Section 5.2). Dynamic low pressure can also form downwind of mountain ranges.

Wind Speed and Direction

Whether thermal or dynamic, air pressure differences across geographic space drive the wind. But why are some winds hurricane force, while others are gentle breezes? Why do some winds spiral, while other travel in straight lines? Three forces control wind speed and direction: pressure-gradient force, the Coriolis effect, and friction.

Pressure-Gradient Force

When there are horizontal differences in pressure across a region, a **pressure-gradient force** is created. The *pressure gradient* is the change in air pressure across Earth's surface. The pressure gradient is the most important factor that sets the atmosphere in motion and determines wind speed and direction. Without a pressure gradient, there is no wind.

Figure 5.4 illustrates the pressure-gradient concept using a bicycle tire as an example. Air always flows from regions of high pressure to regions of low pressure. The greater the pressure gradient between the bicycle tire inner tube and the outside air—in other words, the more the tube is inflated—the faster air flows out of the tube when

punctured. Air will stop flowing out of the tube when the air pressure inside the tube is equal to the air pressure outside it.

Although horizontal pressure changes on Earth's surface are far more subtle than the changes that occur when air rushes out of a punctured bicycle tube, we can similarly envision the atmosphere's molecular density as being higher in some regions and lower in others **(Figure 5.5)**. Air flows from regions of relatively high molecular

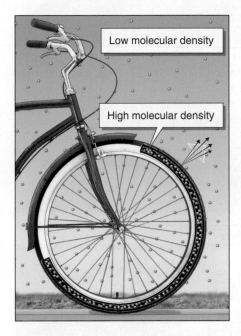

Figure 5.4 Bicycle inner tube air pressure. The dots here represent air molecules. In a tire's inner tube, air molecules are packed together at a high density, and the air pressure inside the tube is high. When the tube's valve is opened or the tube is punctured, air molecules rush out of this high-pressure environment into the low-pressure environment of the surrounding air. Air always flows down the pressure gradient from high to low pressure.

Low molecular density

High molecular density

Uniform pressure

High pressure

Low pressure

Pressure gradient

Wind direction

• Uniform air pressure
• No pressure gradient
• No wind

A

B

Figure 5.5 Pressure gradient and wind. (A) The molecules of air in this geographic region are distributed uniformly. No pressure gradient exists. Therefore, there is no wind. (B) The air molecules decrease in density across this geographic region from left to right. The horizontal pressure gradient is steep and runs from areas of high density to areas of low density. The air molecules move down this steep pressure gradient, creating wind.

density and high pressure into regions of relatively low molecular density and low pressure. The greater the pressure difference between regions (the steeper the pressure gradient), the faster the air flows between them. In regions where little pressure change occurs (where the pressure gradient is shallow), the air moves slowly.

Meteorologists map sea-level pressure gradients using isobars. An **isobar** is a line connecting points of equal pressure on a map. Isobars are the most important means of mapping horizontal pressure differences in Earth's atmosphere. Isobars are quantitative representations of the changing molecular density of the air over a geographic region **(Figure 5.6)**.

There are two key points to remember about isobars **(Figure 5.7)**: (1) The pressure gradient runs perpendicular (90 degrees) to isobars, and (2) where isobars are close together, the wind blows fast because the pressure gradient is steep. Where isobars are far apart, there is little or no wind because the pressure gradient is shallow.

The Coriolis Effect

The winds in hurricanes form a tight spiral rotation because Earth rotates on its axis. Earth's rotation creates the Coriolis effect, which causes objects that travel great distances to follow curved paths rather than straight lines. The **Coriolis effect** (or *Coriolis force*) is the perceived deflection of moving objects in relation to Earth's surface. This deflecting effect is named after the French scientist Gaspard-Gustave de Coriolis (1792–1843). The effect causes a perceived deflection in the direction of flowing fluids, such as wind and ocean currents, and of flying objects traveling long distances, such as airplanes, relative to Earth's surface **(Figure 5.8)**.

In the Northern Hemisphere, moving objects veer to the right because of the Coriolis effect. In the Southern Hemisphere, they veer to the left. The direction of travel of the object does not affect the direction of deflection. The Coriolis effect is greatest at high latitudes, less at middle latitudes, and absent within about 5 degrees latitude north and south of the equator. **Figure 5.9** on page 136 illustrates the influence of the Coriolis effect on wind direction in relationship to isobars.

The Effects of Friction

As air flows down a pressure gradient, it is slowed by frictional drag near Earth's uneven surface. As friction slows the wind, the Coriolis effect deflects the wind less. Mountains, forests, buildings, and even ocean waves all slow the wind.

The **friction layer**—the area of the atmosphere where wind is slowed by friction with Earth's surface—generally extends about 1,000 m (3,280 ft) above the surface. Above the friction layer, the wind moves more quickly. The height of the friction layer varies across Earth's surface. For example, the oceans have a relatively smooth surface, and the friction layer extends only a few hundred meters above the water.

Winds that blow high above the effects of surface friction are *upper-level winds* (also called *winds aloft*). Winds aloft always flow faster and straighter than surface winds and, because they

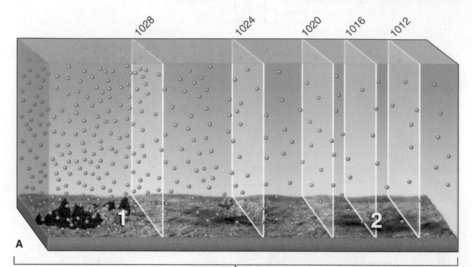

A

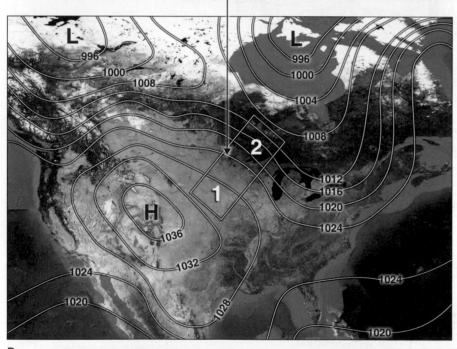

B

Figure 5.6 Visualizing isobars. (A) In this scenario, atmospheric pressure decreases from left to right. A partition is drawn for every 4 millibars of pressure change. Location 1 is between the 1028 mb and 1032 mb isobars. Location 2 is on the 1016 mb isobar. Therefore, the pressure gradient runs from Location 1 (where pressure is higher) to Location 2 (where pressure is lower). (B) When the three-dimensional image from A is transferred onto a flat two-dimensional map, the partitions become lines of equal pressure, or isobars, and the pressure gradient that runs from Location 1 to Location 2 is directed to the northeast. *(Reto Stöckli, NASA Earth Observatory)*

are fast moving, they are deflected more by the Coriolis effect. At roughly 5,000 m (16,400 ft), the Coriolis effect deflects winds so strongly that the winds do not cross isobars, resulting in geostrophic winds. **Geostrophic winds** are high-altitude winds that are strongly deflected by the Coriolis effect and move along a path parallel to the pressure gradient rather than across it. The important effects of geostrophic winds on midlatitude storms are discussed further in Section 5.2.

Measuring, Naming, and Mapping the Wind

Measuring and communicating information about the direction and strength of winds is essential to the science of meteorology. **Anemometers** measure wind speed based on the speed of propeller rotation when exposed to the wind. Faster winds cause faster rotation of the propeller, which is calibrated to kilometers per hour (km/h), miles per hour (mph), meters per second (mps), or knots (kt). (A *knot* is equal to 1 nautical mile per hour, 1.86 km/h, or 1.15 mph.)

A **wind vane** (or *weather vane*) measures wind direction with a fin mounted on a vertical rod. The wind pushes the fin such that it aligns with the wind. As shown in **Figure 5.10** on the following page, wind vanes and anemometers are often combined into a single instrument, an **aerovane**.

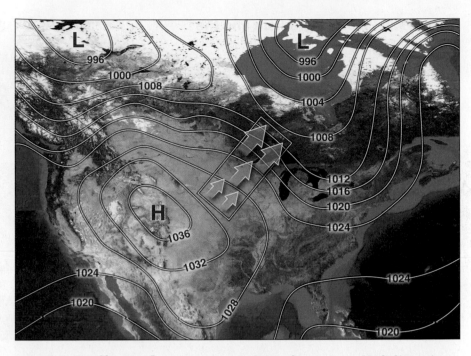

Figure 5.7 Effects of pressure-gradient force on wind direction and speed. The pressure gradient direction is always perpendicular to isobars and always runs from areas of high pressure to areas of low pressure. Widely spaced isobars show a weak pressure gradient and relatively slower wind (small blue arrows). Closely spaced isobars show a steep pressure gradient and relatively faster wind (large blue arrows). *(Reto Stöckli, NASA Earth Observatory)*

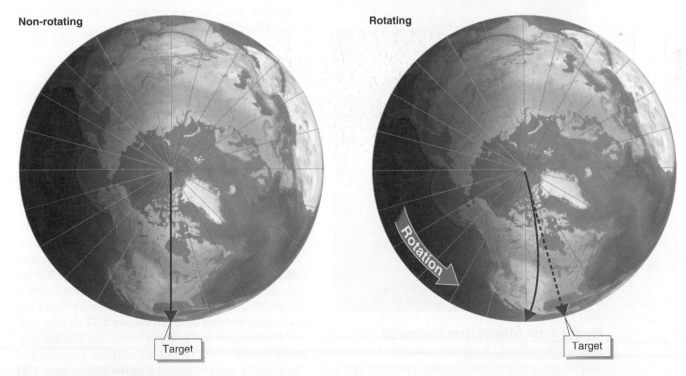

Figure 5.8 Coriolis effect. The red arrow traces the path of a flying object. On a non-rotating Earth (left), objects would travel in straight lines over Earth's surface, and the resulting path would be straight. On a rotating Earth (right), objects still travel in straight lines, but Earth rotates beneath them, resulting in curved paths (solid line).

Figure 5.9 Pressure gradient and the Coriolis effect. As air flows down the pressure gradient, the Coriolis effect deflects it to the right in the Northern Hemisphere. The blue arrows show the wind direction resulting from both the pressure-gradient force and the Coriolis effect. Where isobars are far apart, winds are slow, and Coriolis deflection is relatively weak. Where isobars are close together, wind speed increases, and so does Coriolis deflection. *(Reto Stöckli, NASA Earth Observatory)*

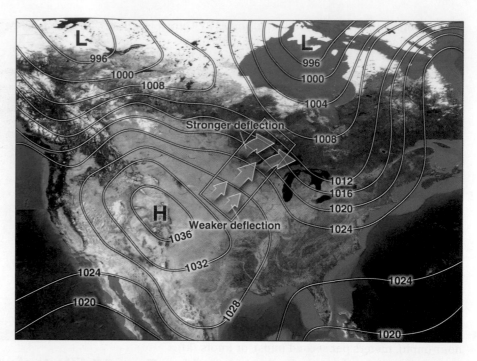

Animation

Pressure Gradient and the Coriolis Effect

Available at
www.saplinglearning.com

Figure 5.10 Measuring the wind. This aerovane is mounted on an ocean buoy. Aerovanes on ocean buoys collect wind data and feed the data automatically to satellite receivers. The satellites then relay the data to land-based computers for weather reports and analysis. (Ocean buoys collect many other types of data as well.) *(NOAA)*

Winds are named based on the direction of their flow. For example, a wind coming from the west is called a *west wind* or *westerly*. The **prevailing wind** is the direction the wind blows most frequently during a specified window of time. An **onshore wind** is a coastal wind that flows from sea to land. In contrast, an **offshore wind** is a coastal wind that flows from land to sea. Winds are portrayed on maps using *wind barbs, wind vectors,* and *streamlines* **(Figure 5.11)**.

Cyclones and Anticyclones

Together, the three controls on wind speed and direction—pressure-gradient force, the Coriolis effect, and friction—create different wind speeds and directions as altitude changes. The changes in wind speed at different altitudes result in rotating meteorological systems called cyclones and anticyclones. A **cyclone** is a system in which air flows toward a low-pressure region of the atmosphere, creating counterclockwise circulation in the Northern Hemisphere and clockwise circulation in the Southern Hemisphere. An **anticyclone** is a system in which air flows away from a high-pressure region, creating clockwise circulation in the Northern Hemisphere and counterclockwise circulation in the Southern Hemisphere. These systems are explained in more detail in **Figure 5.12** on page 138.

Cyclones are particularly important weather systems because most storms are cyclonic systems, and they can be quite powerful **(Figure 5.13** on page 138**)**. Cyclonic storms are covered in greater detail in Chapter 6, "The Restless Sky: Severe Weather and Storm Systems."

Figure 5.11 SCIENTIFIC INQUIRY: How is wind named and mapped?

Naming the wind

Compass and degree headings
Winds are named for the direction they blown from. A south wind is a 180° wind. Zero degrees represents calm conditions.

North
360°
NW
315°   45°   NE
West 270° 90° East
225°   135°
SW   180°   SE
South

Coastal winds
Onshore winds blow from sea inland. Offshore winds blow from the land out to sea.

Onshore wind

Offshore wind

Prevailing winds
The prevailing wind comes most frequently from a certain direction over a defined period of time. These pines are flagged (windswept) by the prevailing wind.

(Bruce Gervais)

Wind rose
A wind rose displays prevailing wind direction using compass headings. This wind rose shows that the prevailing wind direction is from the northwest.

NW
N
NE
W   E
SW   SE
S

Mapping the wind

Streamlines
Streamlines use continuous lines with arrows indicating wind direction. Wind speed is shown with background colors. Here, orange and pink areas have the highest wind speeds.

Wind vectors
Wind vectors are disconnected arrows that show the direction and speed of the wind. The wind vector length and background color indicate wind speed. Here, light purple areas indicate the fastest wind.

Wind barbs
The lengths and shapes of the flags on wind barbs specify wind speed. These barbs are also color-coded to portray wind speed. The fastest wind speeds are shown in red and orange.

(NASA/JPL)

(NASA/JPL)

(Remote Sensing Systems/NASA)

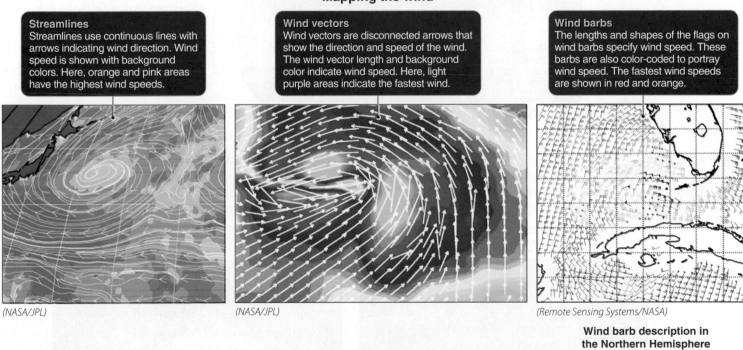

Wind barb description in the Northern Hemisphere

Full-barb
10 knots   Half-barb
5 knots
From the west
at ~15 knots

From the
southeast
at ~75 knots
Flag
50 knots
From the
south at
~50 knots

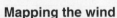

**Figure 5.12
Cyclones and
anticyclones.**
Together, the three
controls on wind speed
and direction create
three-dimensional,
rotating cyclones and
anticyclones.

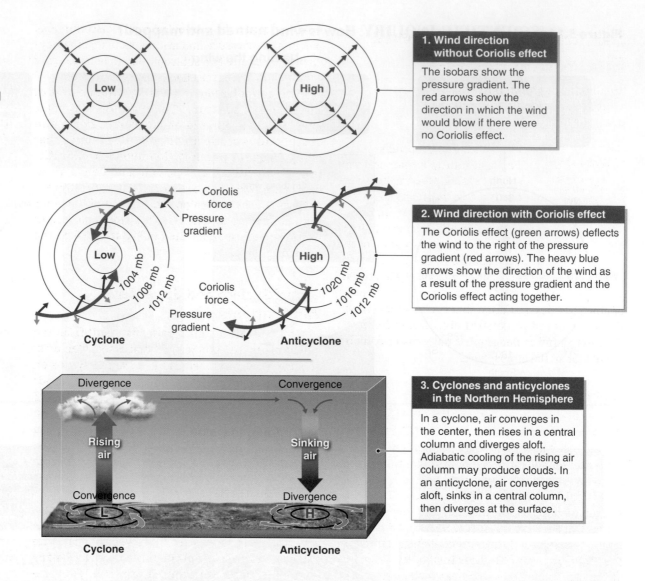

**1. Wind direction
without Coriolis effect**

The isobars show the
pressure gradient. The
red arrows show the
direction in which the wind
would blow if there were
no Coriolis effect.

2. Wind direction with Coriolis effect

The Coriolis effect (green arrows) deflects
the wind to the right of the pressure
gradient (red arrows). The heavy blue
arrows show the direction of the wind as
a result of the pressure gradient and the
Coriolis effect acting together.

**3. Cyclones and anticyclones
in the Northern Hemisphere**

In a cyclone, air converges in
the center, then rises in a central
column and diverges aloft.
Adiabatic cooling of the rising air
column may produce clouds. In
an anticyclone, air converges
aloft, sinks in a central column,
then diverges at the surface.

Figure 5.13 Cyclonic rotation. Tropical cyclones (or hurricanes) develop strong counterclockwise rotation in the Northern Hemisphere
as air flows toward an area of low pressure and is deflected by the Coriolis effect. Here two images of Hurricane Maria are shown just off the coast
of Florida after it devastated Puerto Rico in late September 2017. (A) This isobar map shows the storm's center of low pressure. Wind direction and
counterclockwise rotation are illustrated with arrows. (B) A satellite image shows the clouds of Hurricane Maria in the visible spectrum. The arrows
show the storm's rotation.   (A. Earth image was produced by Earth and Moon Viewer (https://www.fourmilab.ch/earthview/) using NASA Blue Marble, Terra/MODIS cloudless Earth
and Black Marble night lights images; B. NASA Goddard MODIS Rapid Response Team)

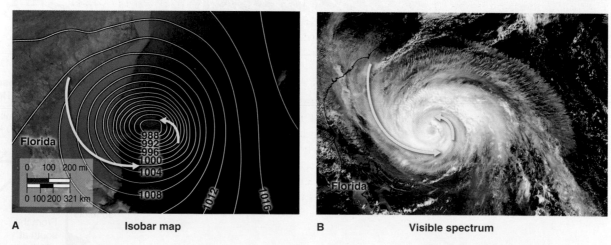

5.2 Global Atmospheric Circulation Patterns

◎ Illustrate global patterns of atmospheric pressure, winds, and precipitation.

Story Map

Atmospheric Circulation

Available at
www.saplinglearning.com

Physical geographers study Earth at multiple temporal and spatial scales. Geographers recognize two patterns of atmospheric circulation: primary circulation and secondary circulation. *Primary circulation* of the atmosphere refers to persistent airflow patterns that occur on a global scale; *secondary circulation* refers to regional and local patterns of atmospheric circulation that are relatively short-lived compared to general circulation features. We next examine geographic patterns of primary circulation of the atmosphere.

Global Pressure Belts

As we saw in Section 3.4, sunlight is strongest at the subsolar point. This strong sunlight heats the air. Each day as the subsolar point migrates along a parallel (or constant line of latitude) in the tropics, it creates areas of unstable heated air parcels and resulting thermal low pressure. Thunderstorms develop from this thermal low pressure through convective uplift (see Section 4.4), and the latent heat released through condensation within the thunderstorms enhances instability. This area of convective uplift located near the equator is the **ITCZ (intertropical convergence zone**; also called the *equatorial trough*). The ITCZ is a discontinuous band of thermal low pressure and thunderstorms that encircles the planet in the tropics.

The unstable air of the ITCZ rises as far as the tropopause (the base of the stratosphere). Because of the temperature inversion in the stratosphere, rising air parcels become stable at about 16,000 m (52,500 ft) as they encounter warm air at the tropopause (see Section 2.3). The ascending air then begins moving toward the poles because it cannot move into the stratosphere.

As the air moves poleward, it begins cooling, and the Coriolis effect deflects it eastward and downward to the ground. As a result, some (but not all) of the air is directed downward to Earth's surface at about 30 degrees latitude. As the air descends at 30 degrees latitude, it is compressed and warmed adiabatically (see Section 4.4). This compression creates an area of high dynamic air pressure called the **subtropical high**, a discontinuous belt of high pressure and aridity made up of anticyclones. There are two subtropical high belts roughly centered on 30 degrees north and south latitude.

The **subpolar low** is a discontinuous belt of dynamic low pressure. There are two subpolar lows. Each is centered on 60 degrees north and south, and is made up of cyclonic systems that bring frequent precipitation due to frontal lifting (see Section 4.4). The cold and dense air at each pole forms an area of thermal high pressure called the **polar high**. The dry air of the polar high creates polar deserts. This global pattern of pressure belts is illustrated in **Figure 5.14** on the following page.

Global Surface and Upper-Level Wind Patterns

Earth's prevailing wind patterns result from the global pressure belts we just discussed. The most geographically extensive and consistent winds on Earth's surface are the trade winds. **Trade winds** are easterly surface winds found from the ITCZ to the subtropical high, between 0 degrees and 30 degrees north and south latitude. They form as air descends at 30 degrees latitude from the tropopause and reaches Earth's surface. From the surface at 30 degrees latitude, the air either travels toward the equator or toward the poles. As the air moves along the surface toward the equator, the Coriolis effect deflects it, creating easterly trade winds **(Figure 5.15** on page 141**)**. The trade winds are most pronounced over the oceans where the effect of friction is minimal. They are weak or even absent over most inland regions. Early travel between Europe and Africa and the Americas was made possible by these winds, and their name is derived from this history. In the Northern Hemisphere, they are called the *northeasterly trade winds*. In the Southern Hemisphere, they are the *southeasterly trade winds*. As the trade winds travel to the equator from the subtropical high, they form a circulation loop called a *Hadley cell*.

The **doldrums** are a low-wind region near the equator associated with the ITCZ, where the trade winds meet. This region is generally not windy because air is rising upward rather than flowing horizontally. "The doldrums" is a reference to the lack of useful wind to move sailing ships to their destinations. The **horse latitudes** are a low-wind region centered on 30 degrees north and south, associated with the subtropical high, which is an area of gently subsiding air with little horizontal flow. As

Figure 5.14 Global pressure belts. This illustration shows the pattern of Earth's global belts of pressure. The same pattern occurs in the Southern Hemisphere, but only the subtropical high of the Southern Hemisphere is visible in this view.

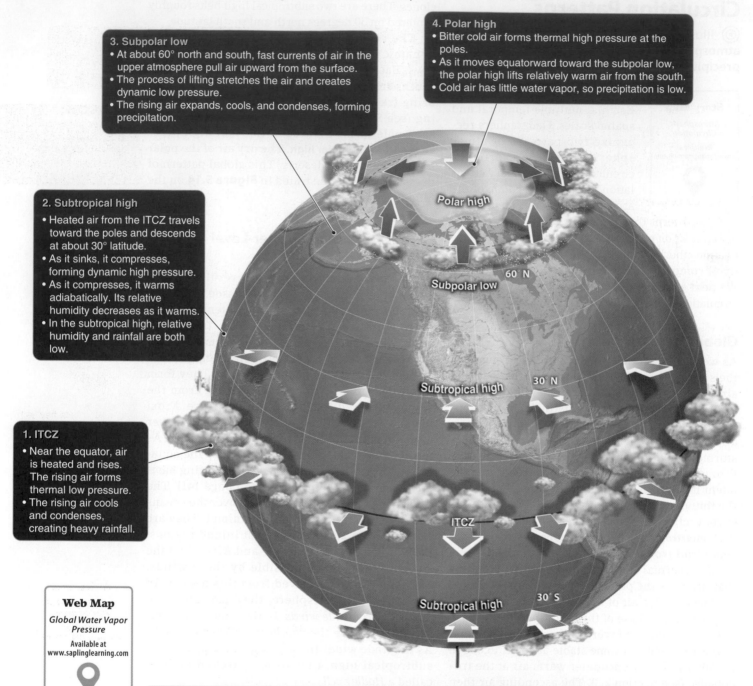

3. Subpolar low
• At about 60° north and south, fast currents of air in the upper atmosphere pull air upward from the surface.
• The process of lifting stretches the air and creates dynamic low pressure.
• The rising air expands, cools, and condenses, forming precipitation.

4. Polar high
• Bitter cold air forms thermal high pressure at the poles.
• As it moves equatorward toward the subpolar low, the polar high lifts relatively warm air from the south.
• Cold air has little water vapor, so precipitation is low.

2. Subtropical high
• Heated air from the ITCZ travels toward the poles and descends at about 30° latitude.
• As it sinks, it compresses, forming dynamic high pressure.
• As it compresses, it warms adiabatically. Its relative humidity decreases as it warms.
• In the subtropical high, relative humidity and rainfall are both low.

1. ITCZ
• Near the equator, air is heated and rises. The rising air forms thermal low pressure.
• The rising air cools and condenses, creating heavy rainfall.

Web Map

Global Water Vapor Pressure

Available at
www.saplinglearning.com

Animation

Global Pressure Belts

Available at
www.saplinglearning.com

Global Pressure Belts

PRESSURE ZONE	TYPE OF PRESSURE	APPROX. LOCATION	CLIMATE
1. ITCZ	Thermal low	The equator	Warm and rainy
2. Subtropical high	Dynamic high	30° north and south	Warm and dry
3. Subpolar low	Dynamic low	60° north and south	Cold and snowy
4. Polar high	Thermal high	90° north and south	Very cold and dry

Figure 5.15 Prevailing global wind patterns. The global pattern of prevailing winds is determined by the global pressure systems and the Coriolis effect. Winds in the Southern Hemisphere are the same as those in the Northern Hemisphere. Only the southeasterly trade winds are shown here for the Southern Hemisphere.

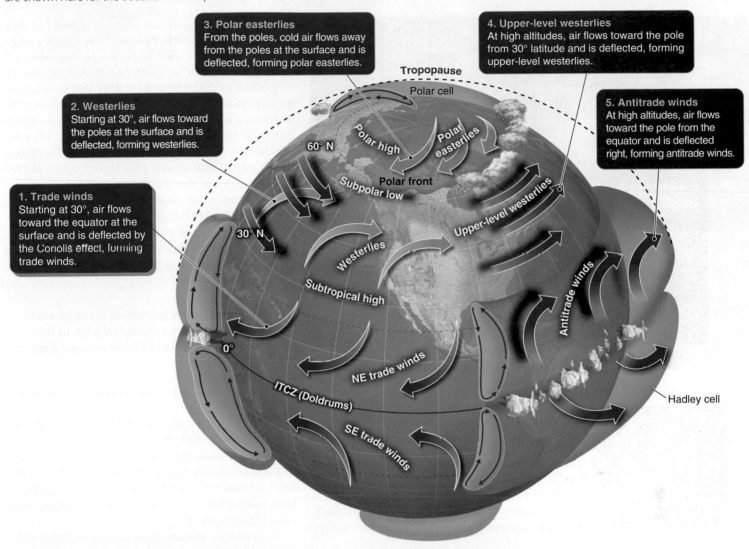

3. Polar easterlies
From the poles, cold air flows away from the poles at the surface and is deflected, forming polar easterlies.

4. Upper-level westerlies
At high altitudes, air flows toward the pole from 30° latitude and is deflected, forming upper-level westerlies.

2. Westerlies
Starting at 30°, air flows toward the poles at the surface and is deflected, forming westerlies.

5. Antitrade winds
At high altitudes, air flows toward the pole from the equator and is deflected right, forming antitrade winds.

1. Trade winds
Starting at 30°, air flows toward the equator at the surface and is deflected by the Coriolis effect, forming trade winds.

Tropopause
Polar cell
Polar high
Polar easterlies
60° N
Polar front
Subpolar low
30° N
Upper-level westerlies
Westerlies
Subtropical high
Antitrade winds
0°
NE trade winds
ITCZ (Doldrums)
Hadley cell
SE trade winds

Global Wind Patterns in Both Hemispheres		
LATITUDE RANGE	**SURFACE WINDS**	**UPPER-LEVEL WINDS**
0° to 30°	Trade winds	Antitrade winds
30° to 60°	Westerlies	Upper-level westerlies
60° to 90°	Polar easterlies	

Animation
Prevailing Global Wind Patterns.
Available at www.saplinglearning.com

the **Picture This** feature on page 142 illustrates, the horse latitudes can be extremely arid.

Blowing from west to east, the **westerlies** are the prevailing surface winds in both hemispheres between the subtropical high and the subpolar low. They form as the descending air of the subtropical high flows poleward along the surface and is deflected by the Coriolis effect. In a similar manner, the polar high exports cold, dense air equatorward. Coriolis deflection creates **polar easterlies**

that are cold and dry. Like the trade winds, the westerlies and polar easterlies are more pronounced over water than over landmasses.

The trade winds and westerlies will have an important role to play in the growing technology of wind power because they occur where there are large human populations in need of energy. The Geographic Perspectives in Section 5.5 explores the role of wind power production in addressing human energy needs.

Picture This **The Empty Quarter**

(Buena Vista Images/Getty Images)

The Empty Quarter, or *Rub' al Khali*, of the southeastern Arabian Peninsula, is Earth's largest sand "sea." Located between 16 degrees and 23 degrees north latitude and covering nearly 650,000 km² (250,000 mi²), it is the southernmost extension of the vast Arabian Desert wilderness on the Arabian Peninsula. Some sand dunes of the Empty Quarter rise 250 m (820 ft) above the surrounding desert floor. Between 9,000 and 6,000 years ago, the climate of the region was wetter due to increased summer monsoon rainfall. During that time, many permanent lakes formed, and woodland vegetation grew (also see Figure 19.3). Subsequently, increased aridity due to natural climate change transformed the landscape, and seas of sand reclaimed the area.

Consider This

1. Based on what you have learned in this chapter up to this point, explain why the Empty Quarter of the Arabian Peninsula is so arid.

2. What do you think happened to the strength and position of the subtropical high in this region about 9,000 to 6,000 years ago? Explain.

Upper-Level Wind Patterns and Jet Streams

Recall from Section 5.1 that upper-level winds (*winds aloft*) are above the friction layer and flow faster than surface winds due to the lack of friction. The Coriolis effect deflects upper-level winds more than it does the relatively slow surface winds,

resulting in geostrophic winds that flow parallel to the pressure gradient rather than across it. The upper-level westerlies and antitrade winds, shown in Figure 5.15, are geostrophic winds. The upper-level airflow between 60 degrees and the poles in both hemispheres creates another circulation loop, called the *Polar cell*.

Situated on the margins of each of these cells and embedded within the geostrophic flow are fast ribbons of air called *jet streams*. Jet streams are discontinuous bands of high-velocity geostrophic winds that flow west to east and encircle the planet at the tropopause. They are several hundred kilometers wide and just a few kilometers deep. Commercial aircraft may fight the headwinds of the jet streams when traveling westward against them or enjoy their fuel- and time-saving benefits when traveling eastward with them.

There are four jet streams, two in each hemisphere. The *subtropical jet stream* is found at the boundary of the Hadley cell and meanders between 20 and 50 degrees latitude in both hemispheres. The **polar jet stream** occurs on the boundary of the Polar cell and meanders between 45 and 70 degrees latitude in both hemispheres. The polar jet stream extends up to the tropopause, but it is usually strongest between 9,000 to 12,000 m (30,000 to 40,000 ft) altitude, where wind speeds can exceed 400 km/h (250 mph) **(Figure 5.16)**.

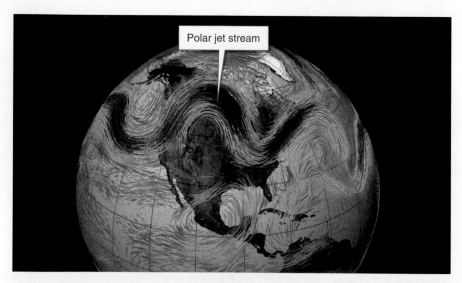

Figure 5.16 The polar jet stream. This image shows the polar jet stream at an altitude of 15,000 m (49,200 ft). The fastest winds—up to 360 km/h (224 mph)—are in orange and red areas. *(NASA Goddard Space Flight Center)*

Figure 5.17 Rossby waves. (A) A *trough* forms where Rossby waves bend equatorward, and a *ridge* forms where Rossby waves bend poleward. Rossby wave ridges often cause above-average temperatures on the ground and generally bring fair weather. Rossby wave troughs bring below-average temperatures and potentially stormy weather. (B) This map shows temperature anomalies (departures from the average) between December 26, 2017, and January 2, 2018. Red areas are warmer than average, and blue areas are colder than average. This unusually deep Rossby wave trough dipped as far south as the Gulf of Mexico and produced a severe storm that meteorologists nicknamed a "bomb cyclone," in reference to its intensity. The storm created blizzard conditions for much of the East Coast and wind gusts up to 96 km/h (60 mph). *(NASA Earth Observatory maps by Jesse Allen, based on MODIS land surface temperature data provided by the Land Processes Distributed Active Archive Center)*

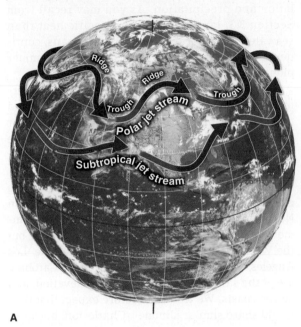

A

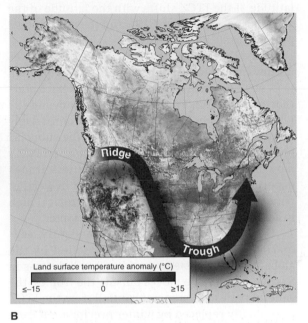

B

Jet streams are caused by the temperature contrasts on the edges of the Polar cell and the Hadley cell. The temperature contrasts (and, as a result, the pressure gradients) are greatest in winter and spring, and consequently the jet streams flow fastest during those seasons. Jet streams play a significant role in midlatitude meteorology because they steer storm systems and can strengthen storms by increasing instability within them (see Section 6.4). Occasionally, the subtropical jet stream merges with the polar jet stream, bringing potentially severe weather as a result.

Geostrophic winds and the jet streams within them trace wavy paths rather than straight lines. The curves or undulations in the geostrophic winds and jet streams are called **Rossby waves**. Rossby waves determine the nature of weather at the midlatitudes **(Figure 5.17)**.

There are always three to six Rossby waves in the Northern Hemisphere polar jet stream. They are seldom stationary, continually changing their geographic position and their depth. Rossby waves are also part of the global heat engine discussed in Section 3.3. As they bend north and south, they provide *meridional* (south–north along a meridian) heat transport from the equator to the poles. In so doing, they export surplus heat from the tropics and deliver it to higher latitudes. **Figure 5.18** illustrates how *polar outbreaks* (often informally called "the polar vortex") occasionally develop as a Rossby wave trough deepens, is pinched off, and moves southward.

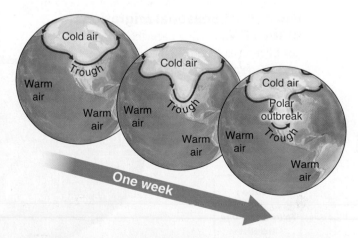

Figure 5.18 Polar outbreak. Rossby waves change their position and depth through time. The blue line represents the polar jet stream. A deep Rossby trough, and then a polar outbreak, develops over the Great Plains in this time sequence spanning a week.

Seasonal Shifts of Global Pressure

In this chapter, we have largely referred to global pressure systems as being centered and fixed on key lines of latitude. For instance, the ITCZ is portrayed in Figure 5.14 as lying on the equator. However, seasons influence the global pattern of pressure systems and surface-level and upper-level winds. The ITCZ shifts its latitude as it tracks the subsolar point (see Section 3.4). Generally, the latitude of the ITCZ shifts with the latitude of the subsolar point **(Figure 5.19)**, and the other global pressure systems shift with it.

In the Northern Hemisphere, as the ITCZ moves northward in summer, it brings heavy precipitation to Central America and into southern Mexico. Farther north in Mexico, where the influence of the ITCZ lessens, the climate becomes progressively more arid. The subtropical high and the subpolar low also move north and south with the ITCZ's movement. This is the reason the midlatitudes experience a marked seasonal contrast in weather patterns: They are affected by the subtropical high in summer and by the subpolar low in winter.

Global precipitation and, consequently, vegetation patterns are mainly controlled by the global pressure systems (which we will discuss in Section 9.1). As one moves outside the tropics, the general pattern of summer rainfall from the ITCZ is gradually replaced by winter precipitation from the subpolar low, as shown in **Figure 5.20**.

The latitudinal precipitation patterns controlled by the ITCZ, the subtropical high, the subpolar low, and the polar high generally hold true for much of the world, but there are many exceptions. Recall from Chapter 4, "Water in the Atmosphere," that the windward sides of mountain ranges, for instance, are typically much wetter than surrounding lowlands, regardless of their latitude. Orographic lifting on the windward side results in more precipitation than the leeward, rain shadow side (see Section 4.4). Another exception that breaks this tidy latitudinal pattern is the influence of landmasses.

The Influence of Landmasses

The continents play an important role in determining the patterns of global atmospheric circulation and precipitation. As you may recall from Section 3.3, land has a lower specific heat than seawater. Therefore, relative to the oceans, continental interiors become hotter during the summer months and colder during the winter months.

This seasonal heating and cooling creates thermal low-pressure regions over land in summer and thermal high-pressure regions over land in winter **(Figure 5.21** on page 146**)**. Because these pressure systems are seasonal and do not last all year, they are called *semipermanent pressure systems*.

At a larger spatial scale, we can examine the effects of semipermanent pressure systems on North American weather in more detail. In summer, the East Coast of the United States is humid, but the West Coast is relatively arid. For example, Los Angeles, California, and Charleston, South Carolina, are at the same latitude, at the same elevation, and on the coasts. We would therefore expect that they would share similar climates. Charleston, however, has a much wetter summer climate than Los Angeles due to the influence of the *Bermuda high* and the *Pacific high*. The Bermuda high is a semipermanent anticyclone in the western Atlantic, and the Pacific high is a semipermanent anticyclone in the eastern Pacific. Both systems are part of the subtropical high pressure belt that is centered on 30 degrees latitude **(Figure 5.22** on page 147**)**.

Figure 5.19 Seasonal migration of the ITCZ. The ITCZ tracks the subsolar point. It reaches its northernmost extent in July at 25 degrees north latitude, passing over Southeast Asia. It reaches its southernmost extent in January at about 20 degrees south latitude, passing over northern Australia.

Video

Seasonal Rainfall Shifts across Africa

Available at
www.saplinglearning.com

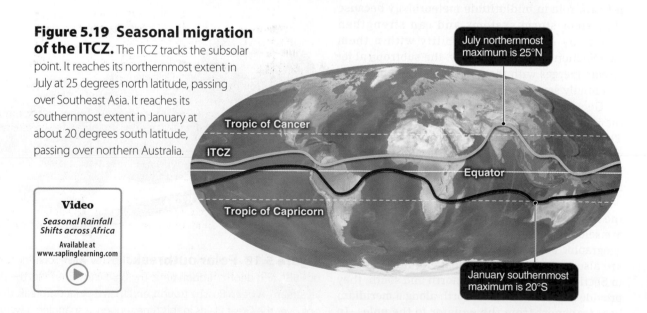

July northernmost maximum is 25°N

Tropic of Cancer

ITCZ

Equator

Tropic of Capricorn

January southernmost maximum is 20°S

Figure 5.20 Global precipitation patterns. In this image from NASA's Blue Marble Next Generation series, land colors are true color and show what the land surface would look like to a person in space if there were no clouds or atmosphere. Darker green areas are vegetated and have wet climates. Lighter tan areas are sparsely vegetated and arid. *(Reto Stöckli, NASA Earth Observatory)*

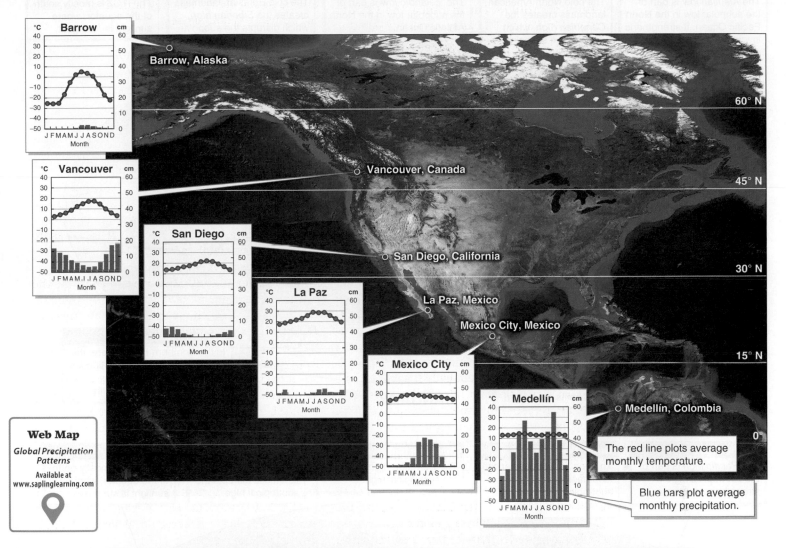

Web Map

Global Precipitation Patterns

Available at
www.saplinglearning.com

Climate Diagram Information

Barrow, Alaska: 71° N; sea level	Barrow is dominated by the polar high. During summer, a weakened subpolar low brings some precipitation.
Vancouver, Canada: 49° N; sea level	Vancouver is dominated by the subpolar low for most of the year. The subpolar-low rainfall diminishes during summer as the subtropical high moves north.
San Diego, California: 32° N; sea level	San Diego is mainly influenced by subtropical-high aridity, but it is far enough north that the polar jet stream can bring some rainfall during the winter months.
La Paz, Mexico: 24° N; 27 m (89 ft) elevation	La Paz is dominated year-round by the subtropical high. Some infrequent ITCZ-assisted thunderstorms reach this far north during late summer.
Mexico City, Mexico: 19° N; 2,421 m (7,943 ft) elevation	The ITCZ is near Mexico City from about May to October. In the winter, the ITCZ moves south of the equator, and the subtropical high brings aridity.
Medellín, Colombia: 6° N; 1,495 m (4,905 ft) elevation	Medellín receives heavy rainfall from the ITCZ in all months. The ITCZ passes over Medellín twice a year, in approximately May and October.

Figure 5.21 Semipermanent pressure systems. Isobars are used to show the average global sea-level pressure in (A) January and (B) July.

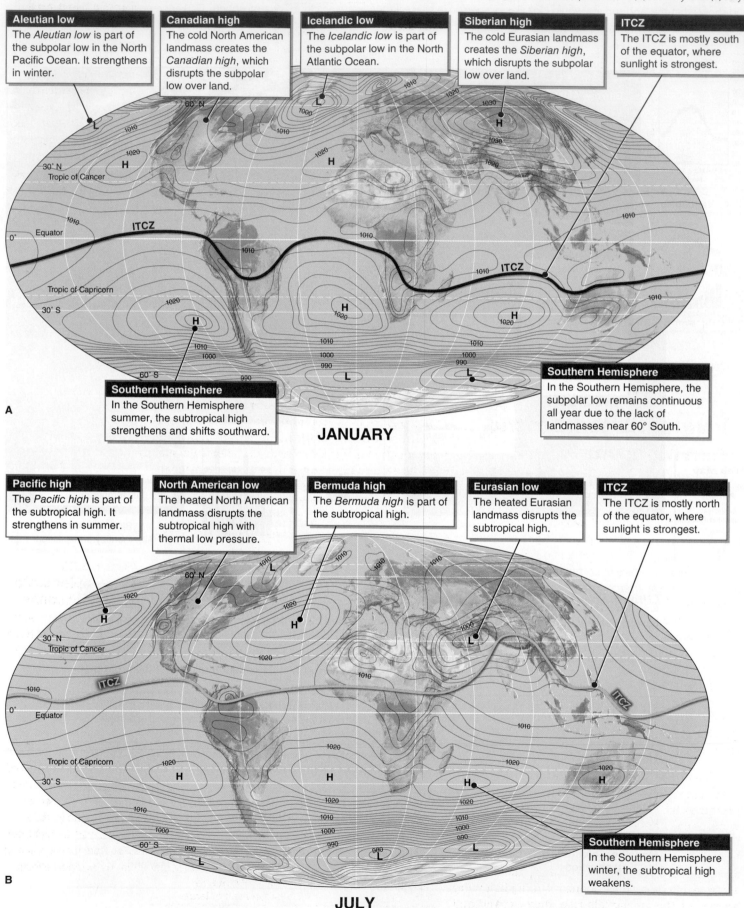

Aleutian low
The *Aleutian low* is part of the subpolar low in the North Pacific Ocean. It strengthens in winter.

Canadian high
The cold North American landmass creates the *Canadian high*, which disrupts the subpolar low over land.

Icelandic low
The *Icelandic low* is part of the subpolar low in the North Atlantic Ocean.

Siberian high
The cold Eurasian landmass creates the *Siberian high*, which disrupts the subpolar low over land.

ITCZ
The ITCZ is mostly south of the equator, where sunlight is strongest.

Southern Hemisphere
In the Southern Hemisphere summer, the subtropical high strengthens and shifts southward.

Southern Hemisphere
In the Southern Hemisphere, the subpolar low remains continuous all year due to the lack of landmasses near 60° South.

JANUARY

Pacific high
The *Pacific high* is part of the subtropical high. It strengthens in summer.

North American low
The heated North American landmass disrupts the subtropical high with thermal low pressure.

Bermuda high
The *Bermuda high* is part of the subtropical high.

Eurasian low
The heated Eurasian landmass disrupts the subtropical high.

ITCZ
The ITCZ is mostly north of the equator, where sunlight is strongest.

Southern Hemisphere
In the Southern Hemisphere winter, the subtropical high weakens.

JULY

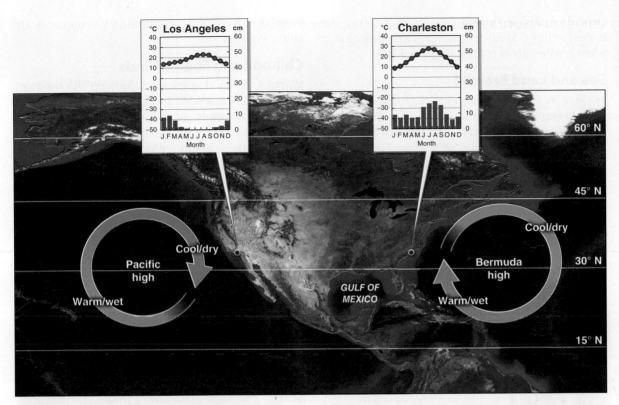

Figure 5.22
Pacific high and Bermuda high.

In summer, the Bermuda high brings humid air up from the Gulf of Mexico and into the eastern half of North America. Except for the mountains that receive orographic precipitation, most of western North America during summer, in contrast, is arid because it is under the influence of the Pacific high that brings relatively cold and dry air from the Gulf of Alaska and over California. *(Reto Stöckli, NASA Earth Observatory)*

5.3 Wind Systems: Mountain Breezes to the Asian Monsoon

◎ **Identify local and regional wind systems and explain how they form.**

In this chapter we have explored how the atmosphere is set in motion by surface pressure differences that result from uneven heating of Earth's surface by the Sun. In this section, we examine secondary circulation of the atmosphere: local *microscale* systems, larger *mesoscale* systems, and *synoptic-scale* systems that cover broad regions **(Figure 5.23)**. The spatial and temporal scales of wind systems are related: Phenomena that cover broad geographic scales tend to persist for a long time, while short-lived events tend to be localized and geographically restricted. Later in the chapter we explore El Niño, which begins as a reversal in wind direction in the tropical Pacific Ocean and often grows to have a global influence on Earth's weather.

Mountain and Valley Breezes

Mountain and valley breezes are microscale breezes produced by heating and cooling differences in mountainous areas. In summer, mountain slopes that face toward the afternoon Sun are heated and form warm, buoyant parcels of air. As this warmed air rises, it creates a pressure gradient and draws in air from adjacent valley floors, resulting in a **valley breeze**. As the air parcels rise, they expand and

cool adiabatically (see Section 4.4).If there is sufficient vapor pressure in the rising air parcels, their temperature may drop to the dew point, and condensation will follow. Clouds such as cumulus and cumulonimbus formed from this condensation may produce afternoon summer thunderstorms in mountainous regions as valley breezes develop.

This situation is reversed after the Sun sets and upper elevations cool faster than lower elevations. Cold, dense, and heavy **mountain breezes**

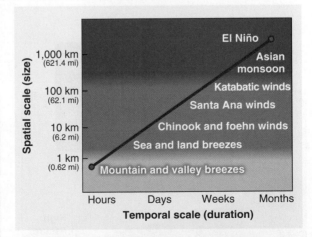

METEOROLOGICAL SCALE	APPROXIMATE SIZE
Microscale	Up to 2 km (1.2 mi)
Mesoscale	Up to several hundred kilometers
Synoptic scale	Several hundred kilometers or larger

Figure 5.23
Geographic scale of wind systems.

Microscale systems, such as local breezes, last hours and cover up to about 2 km. Mesoscale systems, such as Santa Ana winds, last for days and are up to several hundred kilometers in extent. Synoptic-scale systems, such as the South Asian monsoon, last months and span several hundred kilometers or more. El Niño (with its counterpart, La Niña), persists more than a year and often has global effects.

flow downslope through canyons, finding the lowest valleys. Mountain breezes are strongest in winter, when air is coldest **(Figure 5.24)**.

Sea and Land Breezes

Sea breezes and **land breezes** are microscale to meso-scale breezes created by heating and cooling differences between water and land **(Figure 5.25)**. Sea and land breezes are most pronounced in the tropics because of strong daytime heating of the land. They also form in the midlatitudes, where cold ocean water temperatures contrast with warm temperatures just inland. They do not form in high latitude regions due to insufficient daytime heating of the land. Given their localized nature, sea and land breezes can easily be disrupted by synoptic-scale storm systems.

Chinook and Foehn Winds

Warm and dry downslope winds on the leeward side of the Rocky Mountains are called **chinook winds**. **Foehn winds** (pronounced FEH-rn) are the name for the same phenomenon in the European Alps, and there are other local names for this type of wind. As we saw in Section 4.4, the leeward side of a mountain range is typically warmer and drier than the windward side. This difference is due to the release of latent heat and precipitation on the windward side and adiabatic heating as air flows down the leeward side **(Figure 5.26)**.

Figure 5.24 Mountain and valley breezes.

Valley breeze
Daytime heating warms the Sun-facing slopes, creating unstable air parcels and low pressure. As a result of the pressure gradient, air flows from valleys upslope. If the rising air parcels cool to the dew point, clouds or thunderstorms may develop.

Low pressure

High pressure

Mountain breeze
At night, cold heavy air sinks downslope from the upper slopes, forming cool mountain breezes.

High pressure

Low pressure

12 hours

Figure 5.25 Sea and land breezes.

Sea breeze
By day, the land becomes warmer than the ocean, creating thermal low pressure over land and drawing air inland. Air flows from sea to land, creating an onshore sea breeze.

Low pressure

High pressure

Sea breeze

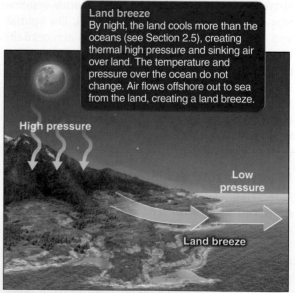

Land breeze
By night, the land cools more than the oceans (see Section 2.5), creating thermal high pressure and sinking air over land. The temperature and pressure over the ocean do not change. Air flows offshore out to sea from the land, creating a land breeze.

High pressure

Low pressure

Land breeze

12 hours

2 The air is cooled to the dew point, resulting in condensation and the release of latent heat.

3 The air descends the leeward side. It compresses and warms adiabatically.

1 Air flows up over the windward side of a mountain range. As it does so, it expands and cools adiabatically.

4 Warm leeward-side chinook and foehn winds form in the mountain rain shadow.

Figure 5.26 Chinook and foehn winds. Warm chinook and foehn winds form as air flows up, over, and back down the leeward side of a mountain range.

Chinook winds often come in sharp contrast to cold winter conditions. With the arrival of chinook winds, temperatures can rise 20°C (36°F) or more within a matter of minutes. These winds quickly melt and sublimate snow. They can bring rapid relief from winter cold and benefit livestock by uncovering prairie grass and humans by clearing railway tracks of snow.

Santa Ana and Diablo Winds

Santa Ana winds and **Diablo winds** sometimes create a major fire hazard for coastal regions in California and northern Baja California, Mexico. Santa Ana winds are offshore winds that form in southern California. In northern California, winds that result from the same process are called Diablo winds. Both Santa Ana and Diablo winds are offshore winds that occur from fall through spring and usually peak in fall. They form as high pressure develops in the Great Basin of Nevada **(Figure 5.27)**.

As the cool desert air, with a temperature just over 10°C (low 50s Fahrenheit), flows from the Great Basin plateau to sea level, it compresses and warms adiabatically. Because these winds originate in the desert, they have a very low relative humidity. As they flow from the high desert downslope to sea level, the air is compressed and heated adiabatically. By the time they reach sea level, their temperature may approach the low 30s Celsius (low 90s Fahrenheit), with relative humidity in the single digits.

The winds can fan wildfires, caused by people accidentally or intentionally, and carry the plumes of smoke hundreds of miles offshore. Occasionally catastrophe strikes, especially with rural residences, which are most vulnerable to the wildfires **(Figure 5.28** on the following page**)**.

Katabatic Winds

Katabatic winds, or *gravity winds* (*katabatikos* is Greek for "downhill movement"), form mainly over high mountain plateaus or ice sheets, where intensely cold, dense, and heavy air forms and drains downslope by the force of gravity. Katabatic wind speeds can exceed those of hurricanes. Gravity also plays a part in downslope mountain breezes, but the term *katabatic wind* is reserved for very cold and strong winds. For example, in January 1984 in Yosemite National Park, California, winds generated from cold mountain drainage exceeded 185 km/h (115 mph), killing one person with a falling tree. These were considered katabatic winds rather than a mountain breeze because of their high speed and low temperature. Greenland and Antarctica routinely experience powerful katabatic winds **(Figure 5.29** on the following page**)**.

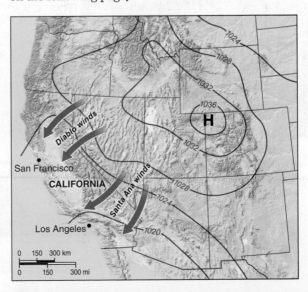

Figure 5.27 Santa Ana and Diablo winds. Santa Ana and Diablo winds develop when high pressure forms over the high plateau of Nevada's Great Basin and relatively low pressure forms off the California coast.

Figure 5.28 Santa Ana and Diablo fires. This pair of satellite images shows smoke from the Tubbs Fire burning during Diablo winds in northern California on October 8, 2017 (top), and the Thomas Fire burning during Santa Ana winds in southern California on December 5, 2017 (bottom). The strong offshore airflow is apparent as the smoke blows out over the Pacific Ocean. The Tubbs Fire was the most destructive wildfire in California history, totaling $1.2 billion in damage, destroying 5,300 structures, and killing 22 people. The Thomas Fire began on December 4, 2017, and burned for over a month. This fire became the largest wildfire in California history, burning 1,140 km^2 (440 mi^2). *(Top left. NASA Earth Observatory images by Joshua Stevens, using MODIS data from LANCE/EOSDIS Rapid; top right. Noah Berger; bottom left. NASA Earth Observatory images by Joshua Stevens; bottom right. © Ray Ford/Noozhawk)*

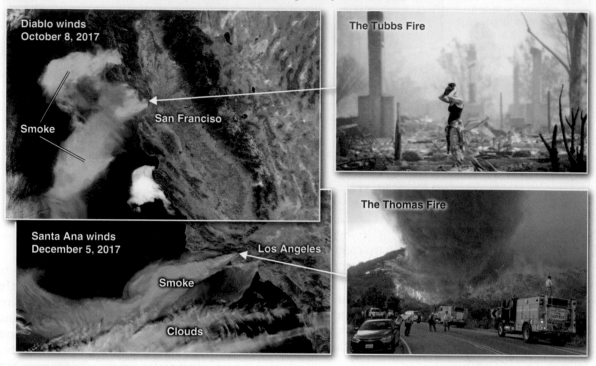

Figure 5.29 Katabatic winds. (A) The katabatic winds that flow into Terra Nova Bay, Antarctica, push sea ice into long lines. Arrows show the direction of the wind. (B) Katabatic winds form as cold heavy air spills downslope, as illustrated here. *(NASA Earth Observatory image created by Jesse Allen and Robert Simmon)*

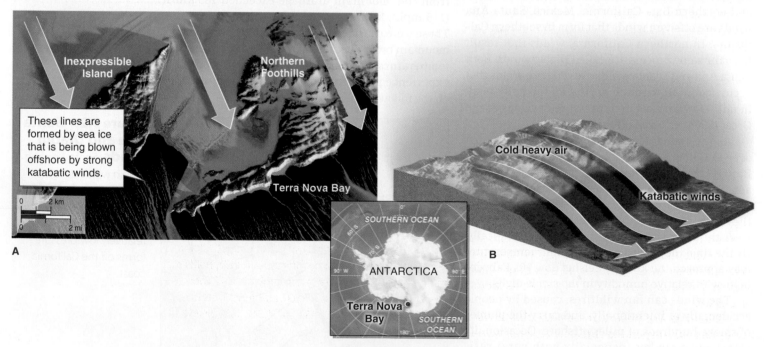

Monsoon Winds and the Asian Monsoons

In the Human Sphere section at the beginning of this chapter, we discussed monsoon winds. Monsoons are seasonal reversals of wind. All monsoons consist of a *summer monsoon*, a warm and moist onshore airflow, and a relatively cool *winter monsoon*, an offshore flow of dry air that originates in the continental interior. Monsoons occur throughout the world, including tropical Asia and Australia, Indonesia, India, eastern equatorial Africa and South America, and the southwestern United States and northern Mexico.

The Asian monsoons (both the South Asian monsoon and the East Asian monsoon—see the Human Sphere section on page 130) are the largest and strongest monsoon systems in the world. About half the world's population lives in tropical Asia and experiences only two seasons: a warm and rainy summer and a slightly cooler but dry winter. In summer, Asian monsoon rains do not necessarily fall every day. Dry *break periods* can last a few weeks to more than a month. When these break periods are prolonged, drought and crop failures can result. Too little rain can mean food and water shortages for tens of millions of people. When the break periods are short, excessive rain results, and flooding, soil erosion, and waterborne disease outbreaks strike. The place recognized as the wettest place on Earth is in India, subject to the influence of the South Asian monsoon **(Figure 5.30)**.

The timing and strength of the summer monsoon rains depend on the strength of onshore airflow and the position and strength of the ITCZ. **Figure 5.31** on the following page shows the summer and winter pressure differences that create the monsoon system, as well as the three synoptic-level controls of the monsoon over India.

The meteorology of the Asian monsoon is complex, and scientists cannot predict with accuracy how it will behave from one year to the next. Onshore airflow, orographic uplift, and the ITCZ affect one another and provide overlapping influence on the strength, duration, and timing of the South Asian monsoon. For example, when the ITCZ migrates over the high Tibetan Plateau, the plateau acts like a chimney and causes heated air to be injected high into the troposphere, further pulling in air from below and strengthening onshore airflow and precipitation. Extensive snow cover on the plateau can reduce the strength of this chimney effect by cooling the air and making it more stable. This diminishes the strength of the summer monsoon.

Scientists are uncertain how the South Asian monsoon will respond to climate change. A warmer

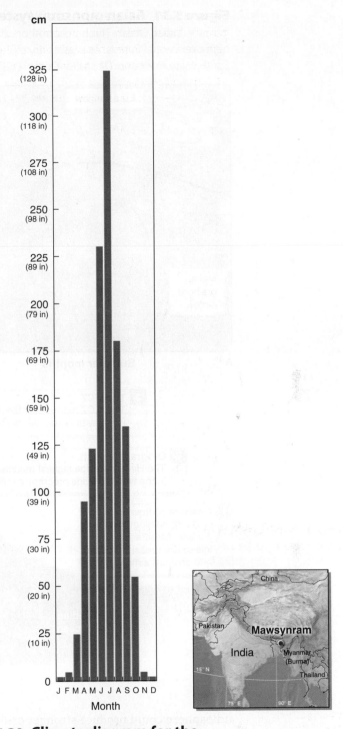

Figure 5.30 Climate diagram for the wettest place on Earth. Mawsynram, India, is recognized by Guinness World Records as the wettest place on Earth. This climate diagram graphs its average yearly rainfall of 1,187 cm (467.4 in or 38.9 ft). It is located in the Khasi Hills in Meghalaya State, at about 1,400 m (4,560 ft) elevation. Nearby Cherrapunji, India, holds the official record for the maximum rainfall in a 12-month period. Between August 1, 1860, and July 31, 1861, Cherrapunji recorded 2,646.1 cm (1,041.77 in, or 86.8 ft) of rainfall.

Figure 5.31 Asian monsoon system. (A) In summer, the ITCZ draws air north (shown with red arrows) from the Indian Ocean. This humid onshore airflow brings summer rains. (B) In winter, air flows away from the Siberian high over central Eurasia. As a result, dry offshore winds develop. (C) There are three synoptic-scale controls on the South Asian monsoon: (1) onshore airflow, (2) orographic uplift, and (3) the ITCZ.

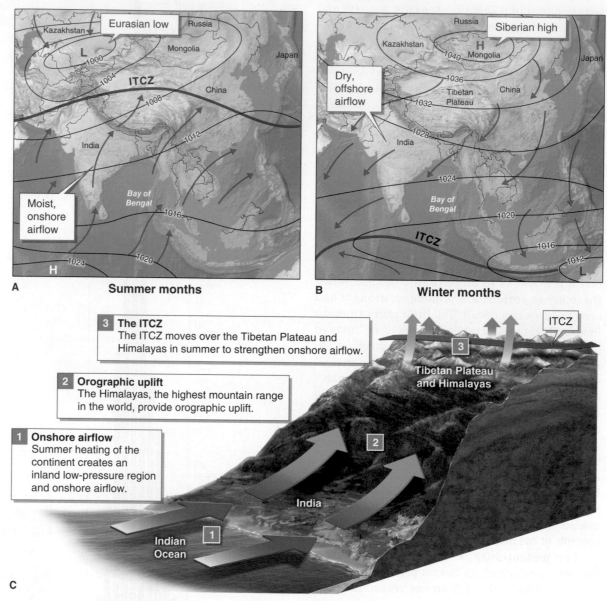

A   Summer months

B   Winter months

3 The ITCZ
The ITCZ moves over the Tibetan Plateau and Himalayas in summer to strengthen onshore airflow.

2 Orographic uplift
The Himalayas, the highest mountain range in the world, provide orographic uplift.

1 Onshore airflow
Summer heating of the continent creates an inland low-pressure region and onshore airflow.

C

atmosphere could produce stronger onshore flow as the land heats up more. Also, the onshore flow could become moister because warm air has a high water vapor capacity. These two factors should result in heavier rainfall in lowland areas and heavier snowfall in the Himalayas. Heavier snowfall in the Himalayan highlands, on the other hand, could diminish the monsoon's strength by weakening the ITCZ and its effects. But all things considered, scientists anticipate that the Asian monsoons will become stronger and bring heavier rainfall as the atmosphere continues to warm as a result of climate change in the coming decades.

5.4 El Niño's Wide Reach

◎ **Discuss how El Niño forms and describe its global effects.**

For generations, Peruvians have reaped the bounty of the sea. Commonly around Christmastime, their catches decrease as cold, nutrient-rich coastal surface water disappears and is replaced by nutrient-poor warm water from the equatorial regions. In other seasons, cold water *upwells* (moves from the seafloor to the surface), bringing nutrients from the seafloor to the sunlit surface of the ocean, supporting high marine productivity and one of the world's largest anchovy fisheries. This upwelling stops, however, when the warm water arrives again in late December. The warm water normally lingers a few weeks to a few months, and then the cold water returns, and the upwelling and fishing resume. Given the timing of this event, it was named *El Niño* after the Christ child.

El Niño is a periodic shift in the state of Earth's climate caused by the temporary slackening and reversing of the Pacific equatorial trade winds and increased surface temperatures in the seas off coastal Peru. Usually, El Niño lasts only a few weeks and does not reach beyond the coast of western South America. During some El Niños, however, the cold waters do not immediately return. Instead, the warm surface water persists and gets warmer. When this happens a global *El Niño event* develops that can have a profound influence on weather around the world.

El Niño events occur randomly, about every 3 to 7 years on average. They develop in March through June and reach peak intensity between December and April, when equatorial sea surface temperatures are highest. By July, they usually dissipate, although some weak El Niño events have lasted as long as 4 years.

The Global Influence of El Niño Events

The global influence of El Niño events illustrates the connections among the oceans, the atmosphere, and climate patterns around the world. For instance, during El Niño years, tropical cyclone activity usually decreases in the Atlantic Ocean and increases in the Pacific Ocean. Tornadoes in North America increase. The Asian monsoon often weakens, resulting in dangerous drought and potential food shortages for millions. El Niño's typical influence is shown in **Figure 5.32**.

El Niño Event Development

Figure 5.33 on the following page illustrates the development of an El Niño event. Slackening and

Figure 5.32 El Niño's global influence. The effects of strong El Niño events reach around the globe, from Australia, where they bring drought and wildfires, all the way to the southeastern United States, where more severe weather can be expected.

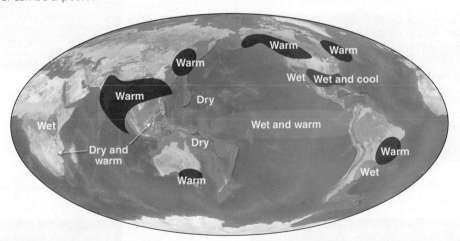

El Niño's Wide Reach	
IF YOU LIVE IN	**DURING AN EL NIÑO EVENT YOU MIGHT EXPECT**
Southwestern United States	Wet and stormy weather
Great Plains and southeastern United States	More thunderstorms and tornadoes
Northwestern United States, western and central Canada	Dry and warm weather
Northeastern United States, eastern Canada	Dry and warm weather
U.S. East Coast	Fewer tropical cyclones (hurricanes)
Southeastern Asia	More tropical cyclones (typhoons)
Australia, Indonesia	Drought and fires
India	Warmer temperatures
Peru and Ecuador	Flooding rains

reversal of the trade winds in the Pacific is one of the most prominent changes leading to an El Niño event.

Toward the end of an El Niño event, the pattern of atmospheric pressure returns to normal. After some El Niño events, the pressure pattern returns to an "enhanced normal," a phenomenon called **La Niña**. During La Niña, the low pressure over Indonesia is deeper than normal, and the high pressure near western South America is higher than normal. A La Niña event does not always follow an El Niño event, but when a La Niña event does occur, it typically lasts about 9 months to a year.

El Niño and La Niña events and the changes they cause in the climate system are collectively referred to as *ENSO* (the *El Niño–Southern Oscillation*). **Figure 5.34** depicts the history of El Niño and La Niña events since 1965.

Scientists do not know what triggers El Niño, and they cannot reliably predict El Niño events more than 6 months in advance. During the past 50 years, El Niño and La Niña events have occurred about half of the time. The other half of the time, ocean conditions have been near normal. Given how often these patterns develop, many scientists question whether "normal" conditions are simply transitional states and the ENSO system is normal. Furthermore, recent research on corals indicates that El Niño intensity has been greater in recent decades than at any other time in the past 7,000 years. These data are leading scientists to question the potential influence of anthropogenic climate change on El Niño.

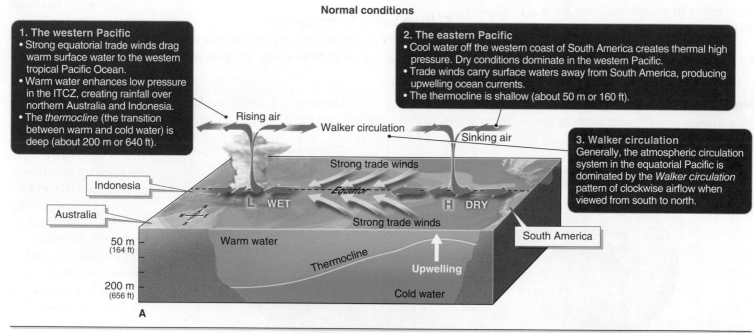

Normal conditions

1. The western Pacific
• Strong equatorial trade winds drag warm surface water to the western tropical Pacific Ocean.
• Warm water enhances low pressure in the ITCZ, creating rainfall over northern Australia and Indonesia.
• The *thermocline* (the transition between warm and cold water) is deep (about 200 m or 640 ft).

2. The eastern Pacific
• Cool water off the western coast of South America creates thermal high pressure. Dry conditions dominate in the western Pacific.
• Trade winds carry surface waters away from South America, producing upwelling ocean currents.
• The thermocline is shallow (about 50 m or 160 ft).

3. Walker circulation
Generally, the atmospheric circulation system in the equatorial Pacific is dominated by the *Walker circulation* pattern of clockwise airflow when viewed from south to north.

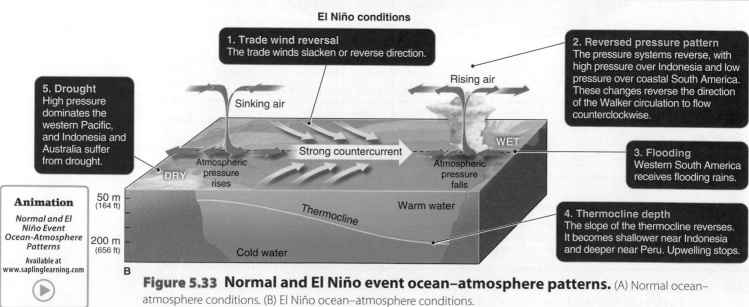

El Niño conditions

1. Trade wind reversal
The trade winds slacken or reverse direction.

2. Reversed pressure pattern
The pressure systems reverse, with high pressure over Indonesia and low pressure over coastal South America. These changes reverse the direction of the Walker circulation to flow counterclockwise.

5. Drought
High pressure dominates the western Pacific, and Indonesia and Australia suffer from drought.

3. Flooding
Western South America receives flooding rains.

4. Thermocline depth
The slope of the thermocline reverses. It becomes shallower near Indonesia and deeper near Peru. Upwelling stops.

Animation
Normal and El Niño Event Ocean-Atmosphere Patterns
Available at www.saplinglearning.com

Figure 5.33 Normal and El Niño event ocean–atmosphere patterns. (A) Normal ocean–atmosphere conditions. (B) El Niño ocean–atmosphere conditions.

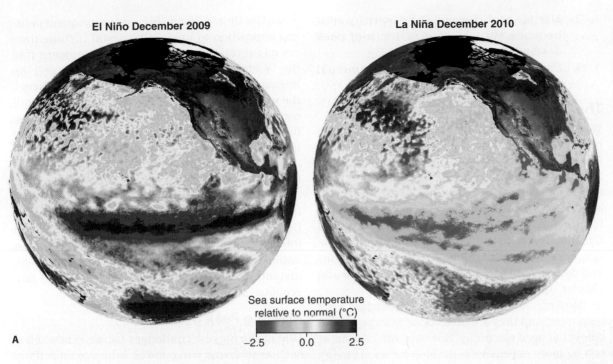

El Niño December 2009

La Niña December 2010

Sea surface temperature relative to normal (°C)

−2.5   0.0   2.5

A

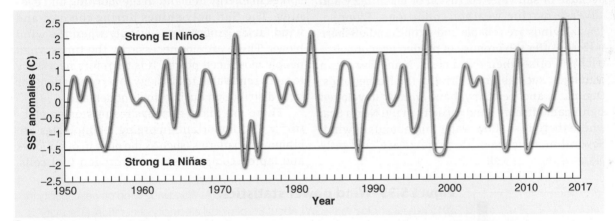

B

Figure 5.34 El Niño and La Niña sea surface temperatures. (A) Sea surface temperature anomalies in the equatorial Pacific Ocean during El Niño and La Niña events stand in stark contrast. El Niño's most prominent feature is the migration of warm water to the eastern end of the Pacific basin. During a La Niña event, this warm water is replaced by unusually cold water. (B) The green line in this graph shows sea surface temperature (SST) anomalies (departures from the long-term average) in the eastern tropical Pacific Ocean from 1950 to 2017. Degrees are in Celsius. Temperatures 1.5°C above the long-term average indicate strong El Niño events, marked with the red line. Temperatures 1.5°C below the long-term average, marked with the blue line, indicate strong La Niña events. Since 1950, there have been about seven strong El Niño and La Niña events. The most recent and strongest El Niño occurred in 2015. *(PO.DAAC/NASA/JPL)*

GEOGRAPHIC PERSPECTIVES
5.5 Farming the Wind

◎ **Assess the challenges and potential of wind as a clean energy source.**

Wind turbines convert the kinetic energy of the wind into usable electricity. A wind turbine is a spinning rotor that converts the kinetic energy of moving air to electrical energy. Electricity generated from the wind is clean renewable energy: It never runs out and it does not generate carbon dioxide that causes climate change. Furthermore, compared to fossil fuels, wind farms have a relatively small environmental impact (although they are not impact-free). Like solar power (see Section 3.6), wind power (electricity generated from wind turbines) is front and center in the renewable energy movement. Wind is one of the fastest-growing energy technologies, and

China leads the way in wind power development **(Figure 5.35** on the following page).

In 2016, there were more than 53,000 wind turbines in operation in the United States, with a capacity of 82.1 GW, meeting roughly 6% of the U.S. electricity supply. By 2050, wind capacity in the United States is expected to grow to more than 400 GW, which would be enough to power 100 million homes and supply 35% of the electricity consumed in the United States. There is certainly sufficient wind energy available for the task. Texas alone has enough wind energy to meet all of the current U.S. electrical demand. But is this a realistic goal? What growing pains will accompany this expansion? As wind energy continues to develop, three significant problems must be addressed:

1. *Geography*: Electricity must be moved from where it is generated to where it is in demand.

2. *Meeting peak demand*: The electricity must be made available during times of peak demand.

3. *Environmental impacts*: Environmental problems must be addressed.

The Geography of Wind

Wind turbine technology has greatly improved in recent years. The main challenge is not in building wind turbines but in finding places to install them. Situating wind farms can be challenging for several reasons, particularly because places that are windy are not always near large cities, where the demand for electricity is high. In the United States and Canada, the consistently fastest winds, the westerlies, occur in the Great Plains, where populations and demand for energy are low (**Figure 5.36** and chapter-opening image on pages 128–129).

Most offshore coastal areas have abundant wind resources, and they are often close to large coastal cities that need electricity. However, offshore areas can be more expensive to develop for wind energy because of saltwater corrosion of machinery and adverse marine weather conditions. Even so, coastal winds are reliable and strong, and offshore turbines offer the promise of producing more electricity to offset their extra costs. Many European countries, such as Great Britain, the Netherlands, Denmark, and Germany, have already developed significant offshore wind capacity in the North Sea, where there is shallow water and abundant wind. Several more mega-scale projects are currently planned there as well.

In the United States, in contrast, frequent political opposition to unsightly coastal turbine towers has stymied offshore wind development (see the "Wumps and NIMBYs" section on page 158). Because of the opposition to offshore wind farms, the windy Great Plains is already playing, and will continue to play, an important role in developing this renewable energy resource in the United States in the coming decades (**Figure 5.37** on page 158).

Given its geographic distance from population centers, Great Plains wind power must be transported over long distances, which can result in considerable power losses along transmission lines. This problem is being addressed through the modernization of the U.S. power grid. Once the nation's grid is improved, moving electricity long distances will become increasingly efficient and economically viable.

Meeting Peak Demand

One of the biggest challenges facing renewables, such as solar and wind power, is how to meet daily spikes in energy demand in the morning and evening. The Sun only shines during the day, and wind farms generate power only when the wind blows. These are not necessarily the times when people most need power. It is therefore necessary to store renewable technology "overgeneration" for later distribution during times of demand.

There are many approaches to storing electricity. Two important emerging technologies are lithium ion batteries (such as the ones in our phones and laptop computers) and hydrogen fuel cells.

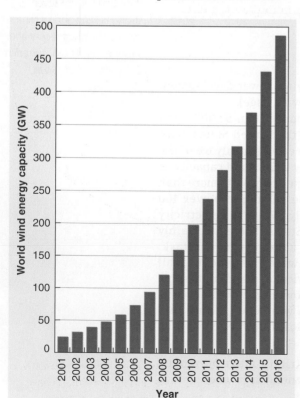

Figure 5.35 Wind power statistics. (A) Worldwide wind power capacity in 2016 was 486.7 gigawatts (GW), about 8% of global electricity generation. ("Capacity" is the maximum amount of power that can be generated.) (B) These are the top 10 countries by wind power capacity in 2016. China tops the list, installing more than one large wind turbine per hour. It installed nearly 10,000 large turbines in 2015. One of every three wind turbines in the world is in China.

Top-10 Countries in Wind Power (2016)		
COUNTRY	**TOTAL INSTALLED CAPACITY (GW)**	**% OF WORLD TOTAL**
China	168.7	34.7
United States	82.1	16.9
Germany	50.0	10.3
India	28.7	5.9
Spain	23.1	4.7
United Kingdom	14.5	3.0
France	12.1	2.5
Canada	11.9	2.4
Brazil	10.7	2.2
Italy	9.3	1.9
Rest of world	75.6	15.5
Total	**486.7**	**100**

A

B

Although these technologies have significant short-comings, their performance is improving and their costs are dropping rapidly. As a result, they are increasingly employed to store wind- and Sun-generated electricity. There are other approaches to storing wind energy, including pumping water uphill to a reservoir, then releasing the water to spin a turbine that produces electricity. Another method of storing wind energy is geologic compression. Using wind-generated electricity, air can be pumped under high pressure into geologic reservoirs, causing the air to become compressed. The compressed air is released and used to spin a turbine to make electricity when needed.

Environmental Impacts of Wind Power

Two significant environmental factors that have stood in the way of wind power development have been the potential of turbines to kill birds and bats and concerns about the effects of wind farms on people who live nearby.

Birds and Bats

Rotating wind turbines kill bats, migratory song-birds, and large birds of prey such as hawks. Most bird kills occur through direct collision with the spinning rotor blades. Supporters of wind power argue, however, that these bird deaths, while unwelcome, pale in comparison to the birds killed by domestic cats, windows (which birds accidentally fly into), and the other factors summarized in **Table 5.1**.

Table 5.1	Global Annual Bird Kill Estimates
CAUSE	**YEARLY FATALITIES**
Building window collisions	Up to 1 billion
Domestic cats	Up to 1 billion
Power line electrocutions and collisions	Up to 175 million
Agricultural pesticides	67 million to 90 million
Automobiles	60 million to 80 million
Communications towers	5 million to 6.8 million
Wind turbines	100,000 to 440,000

Bat kills by wind turbines are also a problem. According to studies by the National Wind Co-ordinating Committee, bird kills, on average, are 0.6 to 7.7 per turbine per season nationwide. For bats, the number is 3.4 to 47.5 per turbine. A single wind

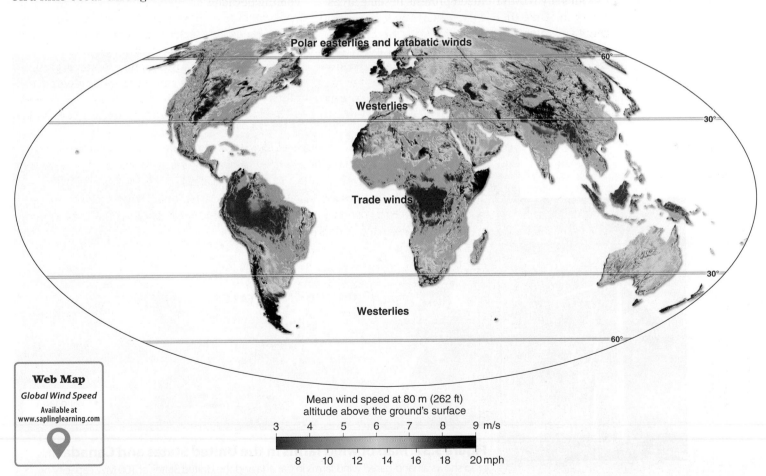

Figure 5.36 Wind speed world map. This map shows average surface wind speeds over land. Red areas have the fastest wind speeds. Global surface wind zones are drawn here to identify wind systems on each continent. North America, northern Europe, and eastern Eurasia have reliable and strong westerlies. Antarctica is not mapped. Refer to the chapter-opening image on pages 128–129 to see greater detail of North America. *(2014 3TIER by Vaisala)*

farm in West Virginia was killing an estimated 4,000 bats each year during the fall migration.

Given the importance of these animals in ecosystems, declines in the numbers of insect-eating songbirds, bats, and large raptors could be a serious concern. Slowing or shutting down the turbines during peak migratory periods or on low-wind nights when bats are most active could be a solution. Such efforts have documented an 80% reduction in bat kills.

Wumps and NIMBYs

People who live near large wind farms sometimes hear strange, deep rhythmic pulses, or "wumps." This problem has been somewhat successfully addressed with new variable-pitch (angle) rotor blades, which can be adjusted to minimize noises.

More significant is the NIMBY ("not in my back yard") problem. Many people want clean, renewable energy, but at the same time, they do not want tall wind turbines dominating the horizon. Wind farms can divide communities. Wind towers are large and conspicuous and, to many, diminish the natural beauty of a region. Offshore wind development in the United States, in particular, has been stymied by NIMBY citizen protests in some areas.

The $2.5 billion Cape Wind Project, for example, proposed to develop 130 wind turbines, each 137 m (440 ft) tall from the water's surface to blade tip. The project was located within sight of Martha's Vineyard in Nantucket Sound, off Cape Cod, Massachusetts. It was stalled out for more than 15 years in litigation, until the project was finally cancelled in 2017. Many wealthy landowners, including the Kennedy family, opposed the project. Although they favor renewable energy development, they do not want it marring their views of the ocean. The 30-MW Block Island Wind Farm in Rhode Island is the only functioning offshore wind-power facility in the United States.

The Energy Portfolio

Renewable energy sources such as wind and solar power are dependent on weather, and weather is undependable. All renewable energy sources face this problem of intermittency. The solution lies in a diverse array of storage technologies and carbon-free energy sources. Although not without its problems, energy from the wind will play an increasingly important role in addressing climate change and meeting people's energy needs in the coming decades.

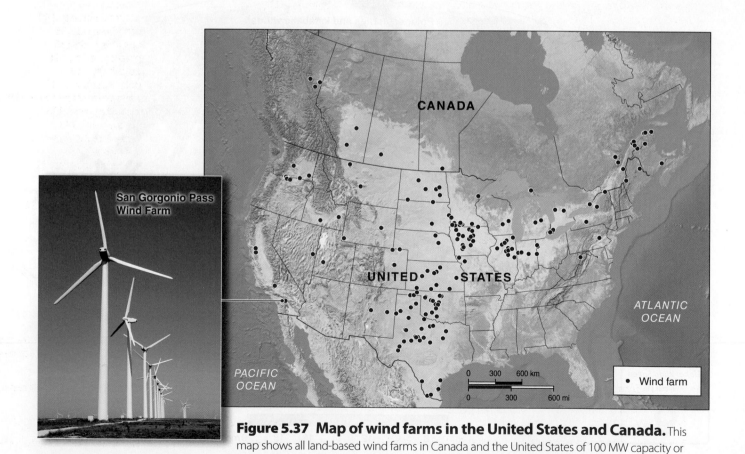

Figure 5.37 Map of wind farms in the United States and Canada. This map shows all land-based wind farms in Canada and the United States of 100 MW capacity or higher. The greatest concentration of wind turbines is in the southern Great Plains and the Midwest in the United States, while Canada has developed wind resources mostly in the east. The inset photo shows wind turbines in California's San Gorgonio Pass Wind Farm. *(Eye Ubiquitous/Getty Images)*

Chapter 5 Study Guide

Focus Points

5.1 Air Pressure and Wind

- **Causes of pressure:** Warm air creates low pressure, and cold air creates high pressure, but dynamic air pressure can override thermal air-pressure changes.
- **Cause of the wind:** Wind is caused by unequal heating of Earth's surface and the release of latent heat within clouds.
- **Wind controls:** Pressure-gradient force, the Coriolis effect, and friction control wind speed and direction both at the surface and aloft. Air flows from areas of high pressure to areas of low pressure.
- **Wind names:** Most winds are named for the direction from which they blow.
- **Cyclones and anticyclones:** Cyclones and anticyclones are important meteorological systems created by deflected flowing wind. Cyclones bring potentially stormy weather.

5.2 Global Atmospheric Circulation Patterns

- **Global pressure belts:** The four global pressure systems are the ITCZ (rainy and warm), the subtropical high (dry and warm), the subpolar low (rainy or snowy and cold), and the polar high (dry and very cold).
- **Surface winds:** The trade winds, the westerlies, and the polar easterlies are the prevailing global surface winds.
- **Effect of seasons:** The strength and latitude of the global pressure systems are strongly affected by the seasonal shift of the subsolar point and by landmasses. All the systems shift north in June and south in December.
- **Semipermanent pressure:** The Bermuda high is a semipermanent pressure system that brings summer precipitation to the East Coast of the United States. The Pacific high brings aridity to the West Coast.

5.3 Wind Systems: Mountain Breezes to the Asian Monsoon

- **Mountain and sea breezes:** Daytime heating of land creates upslope valley breezes and onshore sea breezes. Nighttime cooling of land reverses the wind, creating downslope mountain breezes and offshore land breezes.
- **Chinook and Santa Ana winds:** Chinook and foehn winds and Santa Ana winds flow downslope and heat adiabatically.
- **Katabatic winds:** Fast and frigid katabatic winds form in sloped terrain where cold air drains downslope by gravity.

- **The South Asian monsoon:** The ITCZ, orographic uplift by the Himalayas, and heating of the Indian landmass generate the South Asian monsoon, creating the world's heaviest rainfall.

5.4 El Niño's Wide Reach

- **El Niño events:** El Niño events rearrange moisture and weather patterns for many regions, causing drought and flooding for many parts of the world. El Niño events bring fewer hurricanes for the Atlantic Ocean and often weaken the Asian monsoon.
- **La Niña events:** A La Niña event often follows an El Niño events and creates "enhanced normal" meteorological conditions for affected areas.

5.5 Geographic Perspectives: Farming the Wind

- **Wind energy:** Wind is a promising clean and renewable energy source.
- **Wind energy hurdles:** There are many hurdles to wind energy development, including storage and transport of the generated energy, bird and bat kills, and the NIMBY problem.

Key Terms

aerovane, 135
anemometer, 135
anticyclone, 136
barometer, 131
chinook wind, 148
Coriolis effect, 134
cyclone, 136
Diablo winds, 149
doldrums, 139
dynamic air pressure, 133
El Niño, 153
foehn wind, 148
friction layer, 134
geostrophic wind, 135
horse latitudes, 139
isobar, 134
ITCZ (intertropical convergence zone), 139
katabatic wind, 149
La Niña, 154
land breeze, 148

monsoon, 130
mountain breeze, 147
offshore wind, 136
onshore wind, 136
polar easterlies, 141
polar high, 139
polar jet stream, 142
pressure-gradient force, 133
prevailing wind, 136
Rossby wave, 143
Santa Ana winds, 149
sea breeze, 148
sea-level pressure, 131
subpolar low, 139
subtropical high, 139
thermal air pressure, 132
trade winds, 139
valley breeze, 147
westerlies, 141
wind vane, 135

Concept Review

The Human Sphere: Life under the South Asian Monsoon

1. What is a monsoon? Why do monsoons bring wet summers and dry winters?

2. Why is the monsoon so important to India's 1.3 billion people?

5.1 Air Pressure and Wind

3. How is air pressure measured?

4. Compare station pressure with sea-level pressure. Why must station pressure always be adjusted to sea-level pressure on surface weather maps?

5. How do thermal air pressure and dynamic air pressure create differences in air pressure? Explain each.

6. What are isobars, and what do they represent?

7. What happens to wind speed when isobars are closer together or farther apart?

8. What are the three controls on wind speed and direction? How does each work?

9. What instruments measure wind speed and direction?

10. How are winds named? What is a prevailing wind?

11. What are three ways to portray wind on a map?

12. What are geostrophic winds? Where do they occur?

13. Describe the circulation systems of cyclones and anticyclones. In which direction does air rotate in each of these systems in the Northern Hemisphere?

14. Why do cyclones bring potentially stormy weather but anticyclones never do?

5.2 Global Atmospheric Circulation Patterns

15. What are the four global pressure systems? Describe the atmospheric circulation that results in each of these systems.

16. What is the ITCZ, and what forms it?

17. Describe the moisture and temperature characteristics of each global pressure belt and describe where each is located geographically.

18. Describe the prevailing surface winds: the trade winds, the westerlies, and the polar easterlies. Where does each occur? Explain why they occur where they do.

19. What and where are the doldrums? What and where are the horse latitudes? What causes them?

20. Where are the subtropical jet stream and the polar jet streams? What causes them?

21. What are Rossby waves? Of what significance are their ridges and troughs to the weather?

22. How does the latitude of the four global pressure systems change over the course of a year?

23. How do landmasses affect global pressure systems in winter? In summer? In this regard, why does the East Coast of the United States receive summer rainfall, while the West Coast does not?

5.3 Wind Systems: Mountain Breezes to the Asian Monsoon

24. What is the relationship between the geographic scale and the duration of wind systems? Give examples.

25. Review the major wind systems. Where is each located, and what causes each?

26. What hazards do Santa Ana winds and Diablo winds present?

27. What three factors create the South Asian monsoon? Describe the summer weather of the South Asian monsoon as well as the winter weather of the South Asian monsoon. Explain why they are different.

5.4 El Niño's Wide Reach

28. What influences does El Niño have on global climate and weather?

29. What is an El Niño event? Compare and contrast the ocean–atmosphere system in the equatorial Pacific in a normal year and in an El Niño event year.

30. What is the El Niño–Southern Oscillation (ENSO)?

5.5 Geographic Perspectives: Farming the Wind

31. How is the kinetic energy of the wind converted to electrical energy?

32. Where is the windiest area of the United States? What is the inherent problem with using this region as a source of wind power?

33. Why is transporting and storing wind energy so important to the growing wind energy market?

34. What methods are available to store wind energy?

35. How are birds and bats killed by wind turbines? What are some solutions to this problem?

36. What does NIMBY stand for? How does this term relate to wind power development?

37. Generating power from wind energy is a response to what larger problem? Does wind energy completely solve that problem?

Critical-Thinking Questions

1. Look back on this chapter's opening image of wind speed in North America. Find where you live on this map. How would you characterize the windiness of where you live? What causes the windiness where you live? (Refer to Figures 5.15 and 5.33 for help.)

2. How might an El Niño event influence the weather where you live?

3. Explain the precipitation and temperature pattern where you live in the context of Figures 5.20 and 5.21. Draw a climate diagram for your location.

4. Landmasses can get warmer as global temperatures rise. How could this change the strength of the Asian monsoon during summer? How might it affect precipitation from the monsoon?

5. If you owned a home overlooking the ocean, would you oppose or support a commercial-scale offshore wind turbine project that you could see from your house? Explain.

Test Yourself

Take this quiz to test your chapter knowledge.

1. True or false? An anemometer measures wind speed.

2. True or false? Cyclones rotate clockwise in the Northern Hemisphere.

3. True or false? A sea breeze is an onshore breeze.

4. True or false? Chinook winds experience adiabatic heating.

5. True or false? El Niño events occur randomly about every 3 to 7 years.

6. Multiple choice: Which of the following is not one of the three controls on wind speed and direction covered in this chapter?

a. friction   c. longitude
b. pressure-gradient force   d. Coriolis effect

7. Multiple choice: Which of the following surface winds is found in midlatitude locations, such as the United States?

a. trade winds   c. polar easterlies
b. westerlies   d. antitrade winds

8. **Multiple choice:** Excluding coastal areas, where is there the most wind energy in the United States?
    a. the Great Plains
    b. the West Coast
    c. the Northeast
    d. the Northwest

9. **Fill in the blank:** _____ is an instrument used to measure wind direction.

10. **Fill in the blank:** _____ are used to show changing pressure on a map.

Online Geographic Analysis

Wind Patterns

In this exercise, we examine surface wind patterns that are displayed in real time.

Activity

Go to https://earth.nullschool.net. Click on or touch the globe and pull it so North America is visible. Double-click or use two fingers on a touchscreen to zoom in until North America fills the screen.

1. **Click on or touch the map. A display box appears on the left, showing wind speeds where you touched. (You can click/touch the units to change them.) In what U.S. state or Canadian province are winds fastest?**

2. **Given what you learned in this chapter, do the surface wind patterns you see on this map correlate with the surface wind patterns shown in this chapter? Explain.**

3. **To pan around on the globe, grab and move the globe. Compare the extent of rotation in the wind between the tropics and the high latitudes. How are they different, and what accounts for this difference?**

4. **Examine the pattern of westerlies and trade winds. What happens to them when they encounter landmasses? Why do you think this happens?**

Click the "Earth" tab in the lower left to expand the controls menu. From the "Height" tab select "500" and click the "Earth" tab again to minimize the window. You are now viewing geostrophic winds where the air pressure is 500 mb, at about 5,000 m (16,000 ft) in altitude.

5. **How has the wind speed changed at this altitude? Why did it change?**

6. **What is the prevailing wind direction over California? What is the prevailing wind direction over New York?**

7. **Why did the prevailing wind direction change at a higher altitude?**

8. **Where on the map is the wind fastest? How fast is it?**

9. **Zoom out until you can see the entire Northern Hemisphere. You should see fast, narrow bands of wind colored in red and orange. What are these narrow bands called?**

10. **How many ridges and troughs can you find? Where are they?**

11. **Can you predict were on the surface weather is most likely to be stormy?**

Picture This. *Your Turn*

Global Pressure and Wind Systems

This satellite image shows the Western Hemisphere on July 31, 2017. Major latitudes are drawn. Green areas are vegetated, and tan areas are arid. Place the labels listed below in the appropriate boxes on the image.

Doldrums
Horse latitudes
ITCZ
Southeasterly trade winds
Subpolar low
Subtropical high
Westerlies

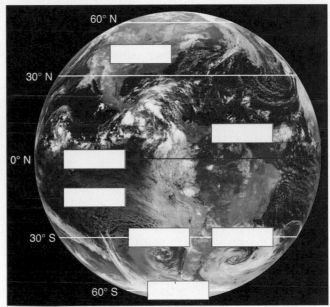

(NASA/NOAA GOES Project)

For animations, interactive maps, videos, and more, visit www.saplinglearning.com.

Sapling Plus

The Restless Sky: Severe Weather and Storm Systems

Chapter Outline *and Learning Goals* ◎

This tornado, photographed in rural Oklahoma, extends downward from the base of a severe thunderstorm. A cloud of debris has formed where the tornado touches the ground and tears into the landscape. About 1,200 tornadoes form in the United States each year.

(Willoughby Owen/Getty Images)

To learn more about tornadoes and how they form, turn to Section 6.2.

THE HUMAN SPHERE The EF5 Tornado

Figure 6.1 Tornado damage. This aerial photo shows the damage done by the May 20, 2013, Moore, Oklahoma tornado. A tornado warning system that alerted people to find safety before the tornado struck saved many lives. *(Jocelyn Augustino/FEMA)*

The most powerful and locally intense storm on the planet is the EF5 tornado (as rated on the enhanced Fujita scale; see Section 6.2). An EF5 tornado has wind speeds of more than 322 km/h (200 mph). There have been approximately 60 EF5 tornadoes in the United States over the past six decades, an average of one per year. In some years, there are none, and in other years, several develop during a single tornado outbreak over the course of a few days. Canada, the country with the second highest tornado activity after the United States, has seen only one EF5 tornado in its recorded history.

The EF5 tornado can destroy most human-built structures. One of the more recent EF5 tornadoes was the tornado that struck Moore, Oklahoma, on May 20, 2013, killing 23 people and injuring almost 400 others. It had a diameter of 2.1 km (1.3 mi), and its peak wind speeds were about 340 km/h (210 mph). **Figure 6.1** shows some of the damage done by this tornado.

6.1 Thunderstorms

◎ **Distinguish among three types of thunderstorms and describe the weather associated with each.**

Like all other atmospheric systems, storm systems are physical manifestations of the Sun's energy. Every weather system, without exception, is the Sun's handiwork. Storm systems derive their energy from solar heating of air near Earth's surface and by the condensation of water vapor that was evaporated into the atmosphere. Storm systems focus this energy, creating lightning in thunderstorms, powerful winds, flooding rains, and other forms of weather that can become significant threats to human lives and property. In this chapter, we delve deeper into the many types of storm systems that occur and learn more about how storm systems function and how they become violent and dangerous.

In Chapter 5, "Atmospheric Circulation and Wind Systems," we discussed the strong relationship between spatial and temporal magnitudes in wind systems. Recall that wind systems that cover broad geographic scales tend to persist for a long time, while short-lived phenomena tend to be localized and geographically restricted. Similarly, there is a direct relationship between the size and duration of a storm system **(Figure 6.2)**. In this section, we discuss the most localized microscale and mesoscale storm systems—thunderstorms. In subsequent sections, we examine more geographically extensive, synoptic-scale storm systems—tropical cyclones and midlatitude cyclones.

Thunderstorms are cumulonimbus clouds that produce lightning and thunder. We explore thunderstorms as isolated systems in this discussion to better understand how they function. It is important to keep in mind, however, that most thunderstorms do not occur in isolation. In reality, thunderstorms are embedded within larger synoptic-scale systems, such as tropical cyclones, midlatitude cyclones, and the ITCZ (intertropical convergence zone) in the tropics.

Each day, about 40,000 thunderstorms occur worldwide. At any given time, some 2,000 thunderstorms are in progress in the world. Most of them are the results of afternoon ground heating and subsequent convective uplift in the ITCZ. The southeastern United States, particularly Florida, has the highest frequency of thunderstorms in the

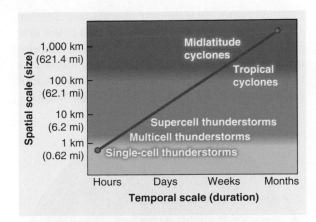

Figure 6.2 Spatial and temporal relationships of atmospheric systems.
At one end of the scale, single-cell thunderstorms are short-lived and occupy little geographic space. Most are meso-scale events. At the other end of the scale, synoptic-scale (geographically extensive) systems such as hurricanes and midlatitude cyclones potentially persist for weeks.

United States, as a result of convective uplift of warm and moist air masses that move north from the Gulf of Mexico. In North America, thunderstorms also form along frontal systems in the early spring (see Section 6.4) and due to orographic lifting along mountain ranges such as the southern Rocky Mountains **(Figure 6.3)**.

Air masses are essential components of thunderstorms and midlatitude cyclones. An **air mass** is a large region of air, extending over thousands

of kilometers, that is uniform in temperature and humidity. When air remains over a region for weeks or longer, it absorbs the characteristics of that region. For example, air over a warm desert becomes warm and dry. Air over a warm ocean becomes warm and humid.

Air mass types are referred to using two-letter abbreviations. Humidity characteristics are represented with the first (lowercase) letter, followed by temperature characteristics, which are represented by the second (uppercase) letter **(Figure 6.4** on the following page**)**. For example, mT refers to a maritime-tropical air mass. This particular type of air mass forms over an ocean in the tropics and is humid and warm. All thunderstorms form within or on the boundaries of mT air masses.

In the midlatitudes, the interactions among contrasting types of air masses form the basis of most weather systems. For example, weak thunderstorms form within mT air masses. More powerful severe thunderstorms form where mT air masses come into contact with cP air to the north. There are three types of thunderstorms: single-cell thunderstorms, multicell thunderstorms, and supercell thunderstorms. The two most important factors that determine thunderstorm type are atmospheric humidity and **wind shear** (changes in wind speed and direction with altitude).

Single-Cell Thunderstorms

Single-cell thunderstorms are almost always relatively mild, short-lived thunderstorms that last an

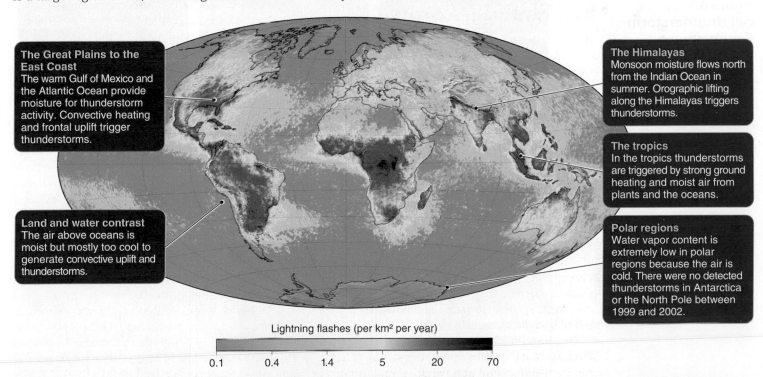

Figure 6.3 Thunderstorm frequency maps. Lightning indicates thunderstorm activity. This lightning-flash frequency map shows the density of lightning flashes per square kilometer per year worldwide. About 70% of all thunderstorms occur over land in the tropics. *(National Weather Service)*

hour or less. Only rarely do they generate weak tornadoes. Single-cell thunderstorms form within mT air masses where wind shear is weak, meaning winds near the ground and winds higher up are weak and flow in the same direction. These storms develop in the late afternoon when heating of the ground creates unstable air parcels (see Section 4.4). As these air parcels rise, they cool to the dew point, which results in condensation and cumulonimbus cloud formation. Single-cell thunderstorms typically experience a predictable sequence of growth, maturation, and dissipation (**Figure 6.5**).

Figure 6.4 Air mass types and source regions. This base map shows average temperatures for the winter months (December, January, February) between 2010 and 2015 in North America. During the winter months, cP air from the north descends southward over the continent. The cP air comes into contact with mP and mT air masses to the south. cT air masses form over the deserts of northern Mexico. *(Original data source: ECMWF ERA-Interim reanalysis. Image made from the website ClimateReanalyzer.org, courtesy of the Climate Change Institute, University of Maine, USA.)*

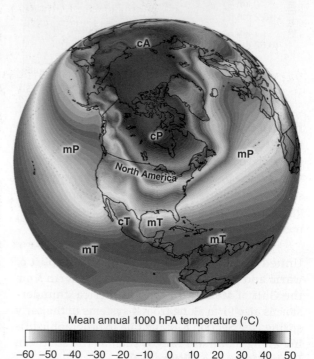

Mean annual 1000 hPA temperature (°C)

−60 −50 −40 −30 −20 −10 0 10 20 30 40 50

AIR MASS TYPE (ABBREVIATION)	SOURCE REGION/ DESCRIPTION
Continental arctic (cA)	The polar high/very cold, dry
Continental polar (cP)	The polar high/cold, dry
Maritime polar (mP)	The polar high and subpolar low/cold, humid
Continental tropical (cT)	The subtropical high/hot, dry
Maritime tropical (mT)	The ITCZ and the subtropical high/warm, humid

Figure 6.5 Single-cell thunderstorm development.

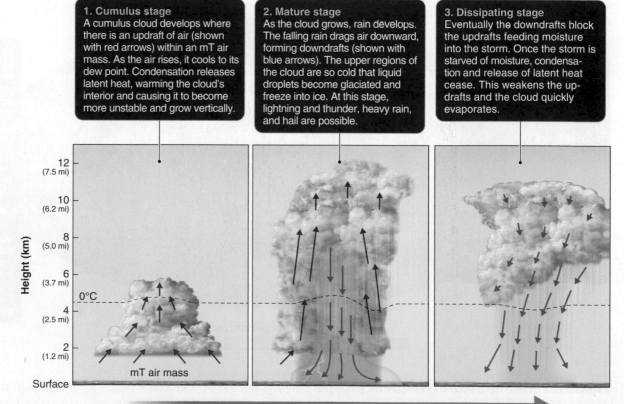

1. Cumulus stage
A cumulus cloud develops where there is an updraft of air (shown with red arrows) within an mT air mass. As the air rises, it cools to its dew point. Condensation releases latent heat, warming the cloud's interior and causing it to become more unstable and grow vertically.

2. Mature stage
As the cloud grows, rain develops. The falling rain drags air downward, forming downdrafts (shown with blue arrows). The upper regions of the cloud are so cold that liquid droplets become glaciated and freeze into ice. At this stage, lightning and thunder, heavy rain, and hail are possible.

3. Dissipating stage
Eventually the downdrafts block the updrafts feeding moisture into the storm. Once the storm is starved of moisture, condensation and release of latent heat cease. This weakens the updrafts and the cloud quickly evaporates.

Height (km)

12 (7.5 mi)
10 (6.2 mi)
8 (5.0 mi)
6 (3.7 mi)
4 (2.5 mi)
2 (1.2 mi)

0°C

mT air mass

Surface

1 hour

Multicell Thunderstorms

Multicell thunderstorms form under conditions of moderate wind shear; these conditions exist in areas with wind speeds of about 40 to 65 km/h (23 to 40 mph), where winds are flowing in different directions at different altitudes. They develop along Rossby wave troughs in the polar jet stream (see Section 5.2). These mesoscale systems consist of individual thunderstorm cells organized in long lines or clusters. Multicell thunderstorms often produce severe weather. They differ from single-cell thunderstorms in that they form along boundaries where mT and cP air masses meet (called *fronts*) rather than within the geographic boundaries of mT air masses, as single-cell thunderstorms do.

Multicell thunderstorms are sometimes severe. A **severe thunderstorm** produces either hail 2.54 cm (1 in) or greater in diameter, a tornado, or wind gusts of 93 km/h (58 mph) or greater. Only about 10% (10,000) of the 100,000 thunderstorms that form in the United States each year are classified as severe. Individual cells within a multicell thunderstorm persist for only about an hour (like single-cell thunderstorms), but as a whole, the system may persist for several days, depending on the strength of the polar jet stream above.

Multicell thunderstorm systems are arranged in clusters called *mesoscale convective systems* or linearly in squall lines. A **squall line** is a line of multicell thunderstorm cells that typically forms along a cold front in a midlatitude cyclone (see Section 6.4). Squall lines can extend for hundreds of kilometers and commonly produce severe weather **(Figure 6.6)**. Cold downdrafts spilling out of a squall line create a

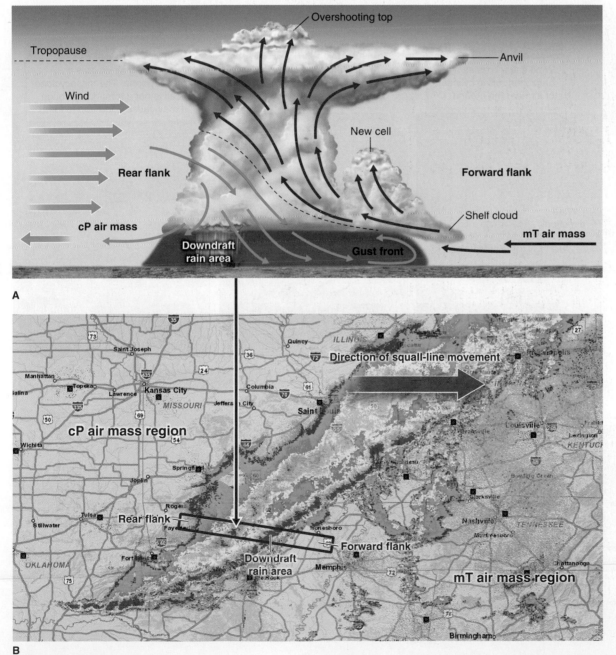

A

B

Figure 6.6 Anatomy of a squall line.

(A) This illustration shows a cross section of a multicell thunderstorm within a squall line. The squall line and multicell thunderstorms form along an advancing front of cold cP air, which is moving to the east (right). As the cP air advances into a warm mT air mass, it lifts the relatively less dense mT air upward on the forward flank (front of the storm). Heavy rain forms downdrafts on the rear flank of the storm. The downdrafts form a dry gust front and shelf cloud on the forward flank. As the gust front advances, it lifts warm and moist mT air, and a new thunderstorm cell arises. Over the course of an hour, this new cell will grow and replace the main thunderstorm. (B) A Doppler radar reflectivity image shows a squall line that formed on August 4, 2001, and stretched from Oklahoma to Ohio. The heaviest band of precipitation and severe weather is shown in red. The diagram in part A represents the red-boxed region of the squall line. *(NOAA)*

strong wind called a *gust front*. A *shelf cloud* often forms on the leading edge of the gust front. When a strong gust front produces wind speeds of 93 km/h (58 mph) or greater over 400 km (250 mi), it is called a *derecho* (from Spanish, meaning "straight"). Damage from derecho winds, often called "straight-line wind damage," can be as destructive as winds of tropical cyclones or tornadoes.

Supercell Thunderstorms

The least common but most powerful thunderstorm type is the **supercell thunderstorm**, sometimes called a rotating thunderstorm. The features that set a supercell apart from other thunderstorms are its duration and updraft. It can last several hours, and it contains a rotating cylindrical updraft, called a **mesocyclone**. The term *supercell* refers to the thunderstorm as a whole, and the term *mesocyclone* refers to the rotating cylindrical updraft within the supercell. Mesocyclones within supercells can be 2 to 10 km (1 to 6 mi) in diameter. Only about 25% of supercells produce tornadoes, but when they do, they are the most powerful and dangerous tornadoes.

Although supercell thunderstorms mostly form in isolation from other thunderstorms, they have been observed within squall lines. Supercells form over land where humid mT air masses come in contact with relatively dry cT air and relatively cold cP air. A well-developed trough in the polar jet stream is required to provide the vertical lifting and wind shear in these powerful storms. These conditions occur most often in spring in North America over the southern Great Plains in Kansas and Oklahoma, in the area known as *Tornado Alley*. **Figure 6.7** outlines how a mesocyclone develops within a thunderstorm.

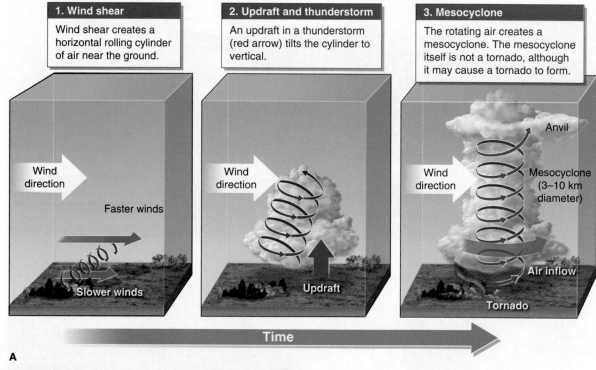

1. Wind shear
Wind shear creates a horizontal rolling cylinder of air near the ground.

2. Updraft and thunderstorm
An updraft in a thunderstorm (red arrow) tilts the cylinder to vertical.

3. Mesocyclone
The rotating air creates a mesocyclone. The mesocyclone itself is not a tornado, although it may cause a tornado to form.

Wind direction

Faster winds

Slower winds

Wind direction

Updraft

Wind direction

Anvil

Mesocyclone (3–10 km diameter)

Air inflow

Tornado

Time

A

MT
ND
MN
SD
Kadoka
WY
NE
IA
CO
ID

B

Figure 6.7 Supercell thunderstorm development. (A) A mesocyclone forms when strong updrafts vertically tilt a horizontal cylinder of rotating air. (B) The rounded striations (lines) seen here indicate mesocyclonic rotation in a supercell thunderstorm. This supercell was photographed near Kadoka, South Dakota. *(Mike Hollingshead/Getty Images)*

**Figure 6.8
Lightning.** This 20-second exposure of a lightning storm in Athens, Greece, shows both cloud-to-cloud and cloud-to-ground lightning bolts. Only about 25% of all lightning discharges are cloud-to-ground. *(© Christos Kotsiopoulos)*

Web Map

Lightning Strikes in U.S.

Available at www.saplinglearning.com

6.2 Thunderstorm Hazards: Lightning and Tornadoes

◎ **Discuss how lightning and tornadoes form and the hazards that each presents.**

All thunderstorms are potentially dangerous. Two particularly hazardous phenomena produced by thunderstorms are lightning and tornadoes.

Lightning

Lightning, the electrical discharge that all thunderstorms produce, is one of nature's most awe-inspiring and dangerous displays. Recent satellite data indicate that there are about 35 lightning flashes per second and more than 3 million flashes per day worldwide. Lightning may discharge between cumulonimbus clouds (*cloud-to-cloud* lightning), within a single cumulonimbus cloud, or between a cumulonimbus cloud and the ground (*cloud-to-ground* lightning) **(Figure 6.8)**.

In 1997, NASA, in partnership with the Japan Aerospace Exploration Agency (JAXA), launched the Tropical Rainfall Measuring Mission satellite. The satellite measured lightning-flash activity in the tropics (see Figure 6.3). From that data, scientists have concluded that the world's lightning hotspot is not in tropical Africa, as previously assumed. The **Picture This** feature on page 170 explores this further.

A bolt of lightning is about 2.5 cm (1 in) in diameter and about five times hotter than the surface of the Sun (30,000°C [54,000°F]). The air around a lightning

bolt becomes superheated and expands explosively, creating an acoustic shock wave called **thunder**.

When lightning strikes the ground, it follows channels of least resistance, particularly through objects with a high water content, such as wet soil or tree roots. Lightning often hits trees. In heavy rain, lightning may follow the paths of water running down the bark of trees, causing superficial damage to tree exteriors. When the bark is dry, however, lightning may travel instead through the tree's wet interior, superheating the interior and causing the tree to explode violently. Lightning's path can be seen when the heat of lightning fuses silica in sand into a glassy hollow tube called a **fulgurite**, shown in **Figure 6.9**.

Video

Spider Lightning

Available at www.saplinglearning.com

Figure 6.9 A fulgurite. Fulgurites are produced when heat melts sand into glass as lightning travels through the ground. This fulgurite is about 35 cm (14 in) long and was found near Queen Creek, Arizona. *(Stan Celestian)*

Picture This Earth's New Lightning Capital

After analyzing 16 years of data, in 2016 scientists concluded that the world's lightning capital is Lake Maracaibo in Venezuela. This area experiences 233 flashes of lightning per square kilometer per year. (In comparison, the second-place spot, in the Democratic Republic of the Congo in equatorial Africa, has 205 flashes per square kilometer each year.) Put a different way, Lake Maracaibo has an average of 260 days of thunderstorms each year, with each day of thunderstorm activity lasting roughly 9 hours, and producing up to 2,500 flashes of lightning.

The lightning forms most where the Catatumbo River empties into Lake Maracaibo. Locally, the lightning is called Catatumbo lighting. Catatumbo lightning forms as a result of the combination of warm water from Lake Maracaibo and cool air from the surrounding mountains. At night, cool air from the mountains spills downslope onto the lake, wedging beneath and lifting warm and humid air. This initiates uplift, condensation, and thunderstorm development. The surrounding mountains also trigger lighting through orographic uplift.

Consider This

1. If Lake Maracaibo's water were cooler, what effect would it have on the lightning activity in the area? Explain.
2. What would happen to lighting activity if the surrounding mountains were lower? Explain.

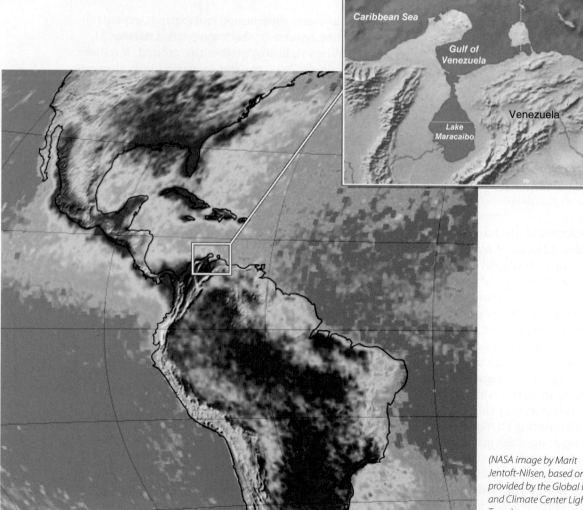

(NASA image by Marit Jentoft-Nilsen, based on data provided by the Global Hydrology and Climate Center Lightning Team)

What Causes Lightning?

We know that lightning forms when cumulonimbus clouds develop a separation of electrical charges, but the process is not yet fully understood. Scientists think that as ice crystals in a thunderstorm collide, friction creates electrical charges among these particles. Larger particles become negatively charged, and smaller particles become positively charged. With the assistance of air turbulence and gravity, lighter ice crystals with a positive charge migrate upward in the cloud, and heavier particles with a negative charge migrate to the lower parts of the cloud **(Figure 6.10)**.

Normally, Earth's surface has a negative charge, and the upper atmosphere has a positive charge. Because of the strong negative charge at the base of the thunderstorm, Earth's surface is relatively positive in its electrical charge.

Air is an insulator and inhibits the flow of electricity. As a result, the opposite electrical charges between the cloud and ground, or within the cloud, build up to tens of millions of volts. Eventually, two oppositely charged regions develop ionized channels of molecules through which electricity flows, and a bolt of lightning is discharged. This discharge equalizes the electrical charges for a short while. Soon, however, another pair of opposite charges builds up and results in another lightning discharge. Lightning usually "finds" objects that protrude highest from the ground, such as tall trees, mountaintops, towers, and buildings. But lightning can strike anywhere on Earth's surface.

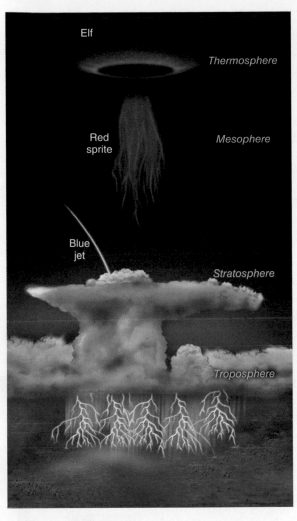

Figure 6.11 Elves, red sprites, and blue jets. Strong lightning storms sometimes produce a huge flat disk of red light called an *elf* that forms in the thermosphere, at an altitude of about 100 km (62 mi). *Red sprites* form just above a thunderstorm and tower some 80 km (50 mi) high. *Blue jets* sometimes burst out of a thunderstorm like fireworks, climbing high into the stratosphere.

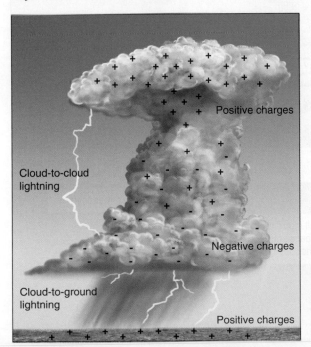

Figure 6.10 Electrical charges in a cloud.
Lightning results as opposite electrical charges develop in different regions of a cloud and on the ground.

Scientists have only recently discovered several new types of lightning-related phenomena: *elves, red sprites,* and *blue jets.* These brief flashes occur high above the thunderstorm and are typically visible only from high-flying planes and orbiting spacecraft. It isn't yet known how these electrical phenomena form **(Figure 6.11)**.

Staying Safe in Lightning

Being caught outside in lightning is an unforgettable and potentially deadly experience. About 500 people are struck by lightning each year in the United States, mostly in Florida. Only about 10% of lightning-strike victims are killed. The rest suffer injuries ranging from minor burns to permanent disabilities. The only safe place during thunderstorms is indoors. If you are caught outside during a thunderstorm, avoid open fields, the tops of hills, and boating on or swimming in water bodies. Remaining in a car is relatively safe, as lightning is likely to be conducted safely through the car's metal exterior and down into the ground.

Light travels at a speed of 300,000 km (186,000 mi) per second. Thus, the light from a flash of lightning reaches our eyes almost instantly. The speed of sound is much slower, however, and depends on characteristics of the atmosphere. In some cases, lightning is too distant for thunder to be heard, resulting in silent flashes of *heat lightning*. At sea level, sound travels at about 343 m (1,120 ft) per second. Because of the time lag between the lightning discharge and the sound of thunder, it is possible to calculate the distance of a bolt of lightning. After seeing the flash of lightning, count off in seconds until you hear the thunder. Divide the number of seconds you counted by three to get the lightning's distance in kilometers. Divide by five to get the lightning's distance in miles. A 3-second count means the storm is 1 km away, and a 5-second count means it is 1 mi away.

If lightning is within 10 km (6 mi) of your location, there will be approximately 30 seconds between lightning and thunder. The *30/30 rule* of lightning safety states that it is not safe to go outdoors if lightning is within the 30-second/10 km range and that it is best to wait until 30 minutes after the storm has passed to go back outside.

Tornadoes

A **tornado** is a violently rotating column of air that descends from a cumulonimbus cloud and touches the ground. A *funnel cloud*, in contrast, is a rotating column of air that descends from a cumulonimbus cloud and is not in contact with the ground. Tornadoes form in supercell thunderstorms, in squall lines, in cold fronts, and even in tropical cyclones. Meteorologists do not yet understand why only about 25% of supercell thunderstorms produce tornadoes or exactly how supercells make tornadoes. Scientists think that when the downdrafts flanking a mesocyclone constrict it and decrease its rotational diameter, tornadoes are more likely to form. If a tornado does form from a supercell, it usually descends from a **wall cloud**, a cylindrical cloud that protrudes from the base of the mesocyclone **(Figure 6.12)**.

Tornadoes are ranked using the **enhanced Fujita scale (EF scale) (Figure 6.13)**. Tornado strength is estimated based on the damage done to the

Figure 6.12 Tornadic supercell.

(A) A supercell thunderstorm contains a powerful rotating updraft, or mesocyclone. If a tornado forms, it will descend from the wall cloud at the base of the mesocyclone. (B) A tornado has formed from this supercell thunderstorm over Campo, Colorado. *(Mike Theiss/Getty Images)*

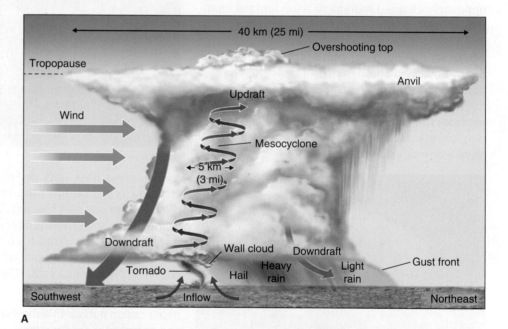

Figure 6.13 The enhanced Fujita scale. (A) The enhanced Fujita scale (EF scale) lists six tornado strength categories. (B) This photo shows some of the destruction from the May 4, 2007, Greensburg, Kansas, tornado. Over 90% of the town was destroyed. In addition to destroying structures, EF5 tornadoes generate lethal airborne debris, such as this fork that impaled a tree.   *(Mike Theiss/Getty Images)*

The Enhanced Fujita Scale

EF SCALE	ESTIMATED WIND SPEED, KM/H (MPH)	TYPICAL DAMAGE
EF0	105–137 (68–85)	Light damage. Some roof tiles peeled off. Tree branches broken.
EF1	138–177 (86–110)	Moderate damage. Roof tiles stripped. Windows broken.
EF2	178–217 (111–135)	Considerable damage. Frame houses shifted from foundations. Large trees snapped or uprooted.
EF3	218–266 (136–165)	Severe damage. Trains overturned. Cars lifted off the ground.
EF4	267–322 (166–200)	Devastating damage. Houses demolished. Cars thrown and small missiles generated.
EF5	Over 322 (200)	Incredible damage. Cars thrown through the air more than 100 m.

A

B

landscape. Directly measuring their wind speed is impossible to do safely. As with many other natural phenomena, the frequency and intensity of tornadoes are inversely related: Almost all (99%) tornadoes are EF3 and lower. Only 1% of all tornadoes in the United States are EF4 and EF5; however, those EF4 and EF5 tornadoes cause about two-thirds of tornado-related deaths.

Most tornadoes in the United States cause only light to moderate damage, and they stay on the ground no longer than 5 minutes. The months of April through July are the most active period for tornadoes **(Table 6.1)**.

Table 6.1 U.S. Tornado Characteristics

CHARACTERISTIC	TYPICAL	UNUSUAL
Time of year	April–July	Winter
Time of day	4 to 6 p.m.	Early morning
Diameter	50 m (160 ft)	1.5 km (1 mi)
Forward movement speed	48 km/h (30 mph)	113 km/h (70 mph)
Length of ground path	3 km (2 mi)	480 km (300 mi)
Duration	5 minutes	Up to 6 hours

Tornado Geography

Tornadoes do sometimes strike the same place twice. Despite very low odds, Codell, Kansas (current population about 2,700), was struck by tornadoes 3 years in a row—1916, 1917, and 1918—on May 20 each year!

About 1,200 tornadoes form in the United States each year, the highest of any country **(Figure 6.14A** on the following page**)**. Canada ranks second in tornado frequency, experiencing about 100 per year on average. Florida experiences the most tornadoes (4.7 per year for every 10,000 km² [3,860 mi²]) of any U.S. state. The majority of these are EF0 and EF1 tornadoes. Kansas, in the Great Plains, experiences the second highest frequency of tornadoes (4.5 per year for every 10,000 km²). Oklahoma is often mistaken as having the most tornadoes, but it is ranked seventh in tornado frequency.

Why does the United States experience so many tornadoes? Severe thunderstorms (and their tornadoes) form most where mT air masses collide with cP and cT air masses. The Great Plains region is relatively flat. As a result, mT air masses from the Gulf of Mexico interact with cT and cP air masses from the interior continent. In most other parts of the world, mountain ranges keep these air masses separate, but in the United States, these air masses converge on the Great Plains. Most EF4 and EF5 tornadoes in the United States occur between the Rocky Mountains and the Appalachian Mountains as far north as the Dakotas **(Figure 6.14B** on the following page**)**.

Warning the Public, Saving Lives

When a tornado touches ground in the United States, the National Weather Service issues an alert. Tornado sirens sound the alarm, and all media channels, including TV, radio, and the Internet, broadcast the alert. These alerts have saved many thousands of lives since they were first implemented in the mid-1950s **(Figure 6.15** on the following page**)**.

Figure 6.14
Tornado geography. (A) As this tornado risk map shows, tornadoes occur on every continent except Antarctica. The greatest frequency of tornadoes in the world occurs in the southwestern United States. (B) The region of the Great Plains that extends roughly from northern Texas to South Dakota is often called *Tornado Alley*. Warm maritime tropical (mT) air from the Gulf of Mexico, dry continental tropical (cT) air from the interior deserts, and cold continental polar (cP) air from the north frequently meet over the Great Plains. These air masses often stack on top of one another and together provide high humidity, strong instability, and wind shear. These conditions can give rise to supercell thunderstorms.

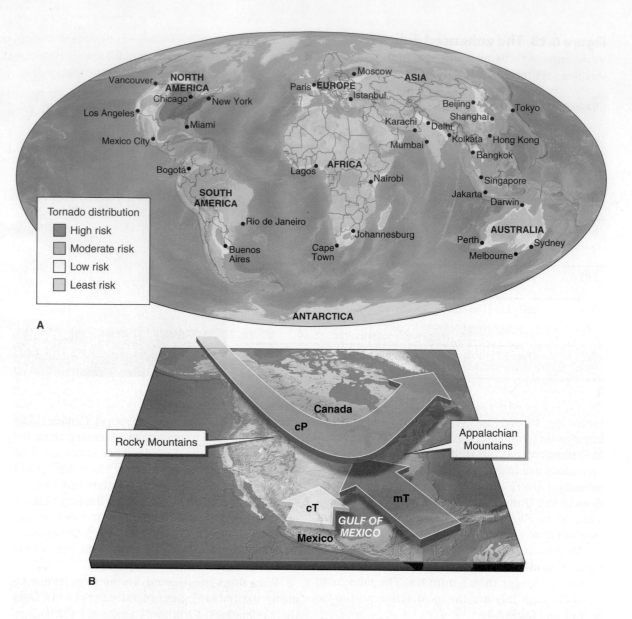

A

Web Map

Tornado Tracks

Available at
www.saplinglearning.com

Figure 6.15 Annual tornado-related deaths, 1890–2017. Since the 1930s, as a general trend, tornado-related deaths (blue bars) have decreased each year. (The red line shows the long-term trend.) Tornado fatalities have decreased even though populations have increased. This contradiction is explained by the development and application of Doppler radar to provide earlier warnings to people in harm's way. The year 2011 had an unusually high number of tornadoes and fatalities. The Joplin tornado in Missouri was particularly devastating, killing 158 people. The year 1925 stands out as the most deadly in U.S. history. That year had the single deadliest tornado in U.S. history, the "Tri-State Tornado," which killed nearly 700 people.

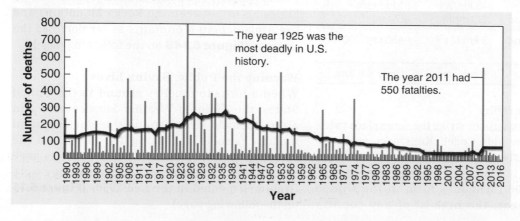

A **tornado watch** is issued when conditions are favorable for tornadic thunderstorms, usually by 10 a.m. on the day of potentially severe weather. A **tornado warning** is issued when a tornado or funnel cloud has been seen and called in to local authorities or when Doppler radar detects a strong near-surface rotation pattern (called a *hook echo signature*) **(Figure 6.16)**.

Not all storms that produce tornadoes develop a hook echo, and not all hook echoes indicate a tornado has formed. Nonetheless, hook echoes are a useful diagnostic of *tornadogenesis* (tornado formation), and tornado warnings triggered by hook echoes have saved thousands of lives and averted many more injuries. **Figure 6.17** discusses the development of Doppler radar in increasing tornado *lead time*.

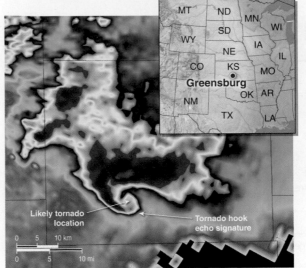

Likely tornado location

Tornado hook echo signature

0 5 10 km

0 5 10 mi

Figure 6.16 Hook echo signature. This Doppler radar reflectivity image shows the hook echo signature produced by the Greensburg, Kansas, tornado of May 4, 2007. The heaviest precipitation is orange and red. Pink indicates large hail. Areas with no data are in black. The hook echo pattern reflects rotation of the mesocyclone, not a tornado. *(National Weather Service, Dodge City, KS)*

1. Radar development
Before 1950, weather radar was not used, and there was no systematic way to detect tornadoes or to warn people of danger. Warnings were spread by word of mouth, radio, and telephone. In 1950, a network of 134 radar stations, developed as the "Tornado Project" by the Weather Bureau, provided radar coverage for areas of Kansas, Nebraska, and Oklahoma. As a result, the first-ever tornado watches were issued in 1953. This photo shows a typical Doppler radar unit. The radar equipment is inside the spherical housing shell.

2. The Union City tornado
During 1973 the "Union City" meteorologists first detected a *tornadic vortex signature* (TVS). Damage from that tornado is shown here. The TVS is an area of tight rotation several miles above the ground. It has tighter and smaller rotation than the mesocyclone's rotation and is a good indicator that a tornado is likely.

3. NEXRAD
Between 1990 and 1997 the U.S. government installed a network of 159 Doppler radar stations across the United States. The network is known as the NEXRAD (Next Generation Weather Radar) system. NEXRAD made high-resolution, real-time precipitation and wind data available to scientists for the first time, allowing them to spot early warning signs of tornadoes, such as the TVS pattern of rotation.

0 375 750 mi

Radar coverage
☐ 3,000 ft above ground level
☐ 6,000 ft above ground level
☐ 10,000 ft above ground level

4. Increasing lead-time
NEXRAD cannot predict where a tornado will form. However, it has made an important contribution in increasing tornado warning lead time --the amount of warning time before a tornado strikes a specific area --by spotting TVS rotation. Thanks to NEXRAD, the average lead time has increased from 5 minutes in the 1980s to 13 minutes today, and this additional warning has saved hundreds of lives each year. The goal is to eventually increase lead times to 30 minutes or more. Shown here is a radar hook echo of a tornadic thunderstorm on March 12, 2006, near Lincoln, Illinois. The distance across the image is about 12 miles.

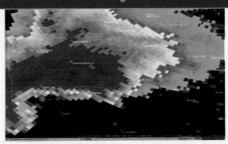

Figure 6.17 SCIENTIFIC INQUIRY: What is tornado lead time? Scientists cannot predict which thunderstorms will spawn tornadoes. They instead use Doppler radar to look for key indicators that occur before a tornado forms. *(Top left. Eric Kurth, NOAA/NWS/ER/WFO/Sacramento; top right. NOAA Photo Library, NOAA Central Library; OAR/ERL/National Severe Storms Laboratory (NSSL); bottom left. NWS/NOAA; bottom right. NWS)*

Table 6.2   Tornado Safety

WHAT TO DO	WHAT TO AVOID
Indoors	
Retreat to a basement or an interior room or hallway.	Windows, which are easily shattered into dangerous glass shards.
Retreat to the lowest floor.	Heavy objects such as refrigerators, which can fall on a person.
Crouch in a bathtub.	Exterior walls, which can fail in high winds.
Cover yourself with a mattress or sleeping bag.	Mobile homes, even those with anchored foundations, are unsafe in a tornado.
Outdoors	
Find low ground and lie face down with arms over head.	Cars. If caught in a car, park on the side of the road. Keep a seat belt on and put your head down below the windows.
	Bridges. Wind speed increases under bridges.

Those living where powerful tornadoes are possible should always have a 2-day (or more) stock of food and water stored away. **Table 6.2** provides basic guidelines for tornado safety.

6.3 Nature's Deadliest Storms: Tropical Cyclones

◎ **Explain how tropical cyclones develop, where they occur, and what makes them dangerous.**

It perhaps may come as no surprise that the planet's strongest storms are in the tropics, where solar energy is strongest. The fastest wind ever officially recorded was generated by the release of latent heat through condensation in Cyclone Olivia on April 10, 1996, on Barrow Island, Australia, located in the tropics at 20 degrees latitude. Olivia produced a record-setting wind gust of 408 km/h (253 mph).

A **tropical cyclone** is a cyclonic low-pressure system occurring in the tropics with sustained winds of 119 km/h (74 mph) or greater. Like all other cyclonic storm systems, tropical cyclones rotate counterclockwise in the Northern Hemisphere (clockwise in the Southern Hemisphere) around a region of low barometric pressure and are capable of producing meters of rainfall, heavy flooding, and damaging high winds. A single storm can devastate extensive coastal regions and cause hundreds of thousands of fatalities. To make matters worse, tornadoes often form in the severe thunderstorms that precede tropical cyclones.

Tropical cyclones go by many names, depending on the geographic region in which they occur. **Hurricane** is the North American and Central American name for a tropical cyclone. The word *hurricane* is derived from *Huracán/Juracán*, meaning "god of storms" in the Taino language of the Lesser Antilles islands in the Caribbean Sea. In Southeastern Asia tropical cyclones are called **typhoons**, and in countries bordering the Indian Ocean they are called *cyclones*, except in Australia, where they are called *tropical cyclones*.

Individual tropical cyclones are given identifying names, such as Hurricane Irma or Typhoon Jelawat (which means "carp"). Naming systems are different for each of the ocean basins where tropical cyclones occur. Atlantic Ocean hurricanes are named according to a 6-year alphabetical list of common first names in English, French, and Spanish. Every seventh year, the list starts over at the beginning. The names of destructive hurricanes, such as Hurricane Sandy and Hurricane Katrina, are taken off the list and replaced by other names starting with the same letter. The names of Pacific typhoons west of the 180th meridian describe a range of phenomena, including food, star constellations, plants, and wildlife.

Why Are Tropical Cyclone Winds So Fast?

Tropical cyclone winds are among the fastest on the planet. One key to their strength is their pinwheel-like structure. If air flowed directly into the central low pressure area without being deflected, the low pressure would "fill up," and the pressure gradient between the storm's center and farther out would lessen. Instead, the central low pressure and the steep pressure gradient are maintained as air flows in toward the low-pressure center and is deflected by the Coriolis effect (see Section 5.1). Tropical cyclones form a central area of lowest pressure called the **eye**. Air is gently sinking within the eye because cold and relatively dense air that forms near the tropopause sinks downward toward the surface within the eye. Surrounding the eye of the storm is the **eyewall**, the most severe part of the storm. Air is rising upward fastest in the eyewall, releasing immense amounts of latent heat as water vapor condenses. *Rain bands* are areas of heavy precipitation that extend outward from the eyewall **(Figure 6.18)**.

The strength of a tropical cyclone depends on how much water vapor condenses to liquid. Condensation is important because it releases latent heat into the storm (see Section 4.2). As the rotating airflow grows faster, it pulls in more moisture, which releases more latent heat and further strengthens the storm, creating positive feedback **(Figure 6.19)**.

The latent heat positive feedback in tropical cyclones results in the planet's strongest surface winds and, sometimes, extraordinary amounts of precipitation. Rainfall amounts, however, are more often a function of the tropical cyclone's forward speed than its strength. Cyclone Gamede moved over

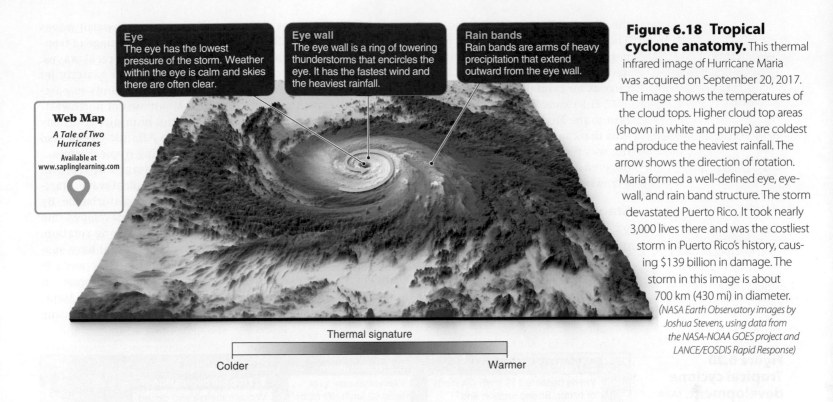

Eye
The eye has the lowest pressure of the storm. Weather within the eye is calm and skies there are often clear.

Eye wall
The eye wall is a ring of towering thunderstorms that encircles the eye. It has the fastest wind and the heaviest rainfall.

Rain bands
Rain bands are arms of heavy precipitation that extend outward from the eye wall.

Web Map

A Tale of Two Hurricanes
Available at www.saplinglearning.com

Thermal signature

Colder   Warmer

Figure 6.18 Tropical cyclone anatomy. This thermal infrared image of Hurricane Maria was acquired on September 20, 2017. The image shows the temperatures of the cloud tops. Higher cloud top areas (shown in white and purple) are coldest and produce the heaviest rainfall. The arrow shows the direction of rotation. Maria formed a well-defined eye, eye-wall, and rain band structure. The storm devastated Puerto Rico. It took nearly 3,000 lives there and was the costliest storm in Puerto Rico's history, causing $139 billion in damage. The storm in this image is about 700 km (430 mi) in diameter. *(NASA Earth Observatory images by Joshua Stevens, using data from the NASA-NOAA GOES project and LANCE/EOSDIS Rapid Response)*

Figure 6.19 Tropical cyclone latent heat positive feedback. Tropical cyclones derive their high wind speeds from latent heat positive feedback.

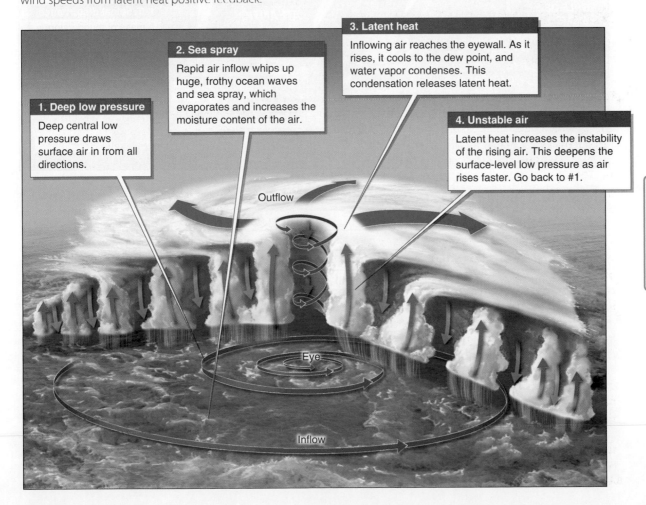

2. Sea spray
Rapid air inflow whips up huge, frothy ocean waves and sea spray, which evaporates and increases the moisture content of the air.

3. Latent heat
Inflowing air reaches the eyewall. As it rises, it cools to the dew point, and water vapor condenses. This condensation releases latent heat.

1. Deep low pressure
Deep central low pressure draws surface air in from all directions.

4. Unstable air
Latent heat increases the instability of the rising air. This deepens the surface-level low pressure as air rises faster. Go back to #1.

Outflow

Eye

Inflow

Animation

Tropical Cyclone Latent Heat Positive Feedback

Available at www.saplinglearning.com

the island of Réunion in the Indian Ocean in 2007. The storm brought 3.93 m (12.89 ft) of rain in a 3-day period. After 6 days of continuous rainfall, precipitation totals came to 5.51 m (18.07 ft), setting world records for rainfall totals during a single storm. In 2017, Hurricane Harvey brought 164 cm (64.6 in) of rain to the Houston area, a record amount for the area that caused catastrophic flooding there. Flooding was particularly excessive because the storm lingered over the area as its forward speed slowed to a crawl and the storm even reversed direction.

Stages of Tropical Cyclone Development

Recall from Section 5.2 that the trade winds of the tropics flow from east to west. As the trade winds flow across northern Africa, they may form a band of fast wind known as the *African easterly jet*. Rather than flowing straight from east to west, the airflow of the African easterly jet often

develops meridional (bending northward) waves of air called *tropical waves*. The first stage of tropical cyclones develops on these tropical waves. Some tropical waves in the African easterly jet are caused by the Ethiopian Highlands in eastern Africa and the Atlas Mountains of northwestern Africa. Temperature and humidity contrasts between the Sahara and the Atlantic Ocean also play an important role in forming tropical waves.

Figure 6.20A shows the stages of tropical cyclone development along a tropical wave, starting with the first stage, a tropical disturbance. By the time a tropical disturbance has developed into a tropical storm, the system has strong rotation. **Tropical storms** are cyclonic systems that have sustained winds between 62 and 118 km/h (between 39 and 73 mph). Once a tropical cyclone has formed, it may strengthen if conditions are favorable. Meteorologists use the **Saffir-Simpson scale**, a hurricane

Figure 6.20 Tropical cyclone development. (A) A tropical disturbance may develop into a tropical cyclone if conditions are right. This image shows the progression of a single tropical disturbance into a hurricane as it crosses the Atlantic Ocean. Although most Atlantic hurricanes form near the west coast of northern Africa, some form closer to the United States—for example, within the Caribbean Sea and the Gulf of Mexico. (B) The Saffir-Simpson scale of tropical cyclone intensity ranks storms based on measured wind speed.

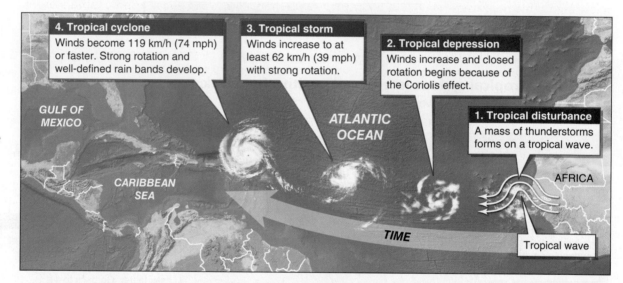

Stages of Tropical Cyclone Development

STAGE	WIND SPEEDS	OTHER CHARACTERISTICS
Tropical disturbance	Light	A mass of thunderstorms with no rotation
Tropical depression	Up to 61 km/h (38 mph)	Closed rotation begins
Tropical storm	62–118 km/h (39–73 mph)	Stronger rotation, heavy rainfall
Tropical cyclone	Over 118 km/h (73 mph)	Strong circulation, heavy rain. Identifiable eye, eyewall, and rain bands often evident.

A

The Saffir-Simpson Scale

CATEGORY	WIND SPEED (KM/H)	WIND SPEED (MPH)	PRESSURE (MB)	DAMAGE
1	119–153	74–95	980 or higher	Minimal
2	154–177	96–110	979–965	Moderate
3	178–208	111–129	964–945	Extensive
4	209–251	130–156	944–920	Extreme
5	252 or higher	157 or higher	919 or less	Catastrophic

B

ranking system based on measured wind speeds, to describe five categories of tropical cyclone intensity **(Figure 6.20B)**.

For a tropical cyclone to persist and strengthen, it must have an ample supply of warm seawater that readily evaporates into warm air. Generally, water temperature must be 26°C (80°F) or warmer. Cooler water does not evaporate as quickly and reduces the moisture supply to the storm. The high ocean waves whipped up by the storm will mix deeper and cooler seawater up to the ocean's surface. Because of this mixing from the waves, warm water must extend to about 60 m (200 ft) in depth. There must also be little to no wind shear, which tears tropical cyclones apart. In other words, winds from the sea's surface up to the tropopause and even into the stratosphere must flow in the same direction for tropical cyclones to form and persist.

A tropical cyclone normally occurs alone rather than in pairs or groups. The **Picture This** feature below highlights the rare formation of hurricane trios.

Monitoring Approaching Tropical Cyclones

As tropical cyclones approach populated areas, meteorologists monitor them closely to forecast where the cyclones will make *landfall* (move onshore) and how strong they will be when they do. Knowing where and when a tropical cyclone will strike, as well as its likely strength, is essential to efforts to save lives. In the United States, decisions to evacuate coastal areas are based on close monitoring of these storms by NASA, NOAA, and the National Weather Service. Satellites provide early detection and tracking of tropical cyclones. But they cannot provide enough detail to develop accurate forecasts and warnings. To compliment satellite observations, as hurricanes approach U.S. shores, aircraft drop small, 40 cm (16 in) long, 5.78 cm (2.75 in) diameter instrument platforms called **dropsondes** into the storms. These devices provide precise measurements of wind speed, pressure, humidity, and air temperature **(Figure 6.21** on the following page**)**.

Marine buoys are floating meteorological stations anchored to the seafloor that also provide important data about approaching storms. They record seawater temperature at the surface and, importantly, at depth. They also record meteorological conditions just above the water. Finally, when a tropical cyclone comes within 320 km (200 mi) of shore, Doppler radar can begin

Picture This Hurricane Trios

Every now and then a pair of tropical cyclones form and move together across the same ocean basin. But it is rare to see three tropical cyclones at one time in the same view. The top NASA image shows hurricanes Kilo, Ignacio, and Jimena simultaneously churning across the Pacific Ocean. When this image was taken on August 29, 2015, all three were major category 4 storms. Hurricanes that form in the eastern Pacific (near Mexico) are called "shy hurricanes" because they seldom strike land (and people). True to their name, these hurricanes dissipated in open water and caused no damage.

Then, amazingly, in 2017, another hurricane trio formed, this time in the Atlantic Ocean and the Gulf of Mexico. Hurricanes Katia (category 1), Irma (category 4), and Jose (category 3) are pictured here on September 8, 2017. Just after this image was acquired, Irma tore through the Florida peninsula. Throughout its course, the storm caused $63 billion in damage and took 134 lives.

Scientists do not know if these unusual events represent a new normal in hurricane activity and will soon be repeated or if they were extremely unusual events.

(NASA; NOAA/NASA GOES Project)

Consider This

1. What are the prevailing winds in the tropics? Which direction do they blow?

2. Given your answer to Question 1, which storm(s) in the top satellite image would have threatened the Big Island of Hawai'i most?

3. In the bottom satellite image, which hurricane has the most well-defined eye, and why?

imaging it, providing detailed information on wind strength, rainfall intensity, and direction of movement. Meteorologists use these data to develop wind speed and precipitation forecasts as the storm comes ashore.

Tropical Cyclone Geography

The subtropical highs play the important role of steering the direction tropical cyclones take as they move across the Atlantic Ocean. The Bermuda high, which is part of the subtropical high-pressure belt (see Section 5.2), determines whether hurricanes make landfall or hook harmlessly northward into the North Atlantic before they reach land. **Figure 6.22** summarizes the factors that control the paths of tropical cyclones and the resulting geographic extent of tropical cyclones worldwide.

Tropical Cyclone Hazards

Tropical cyclones present two key hazards: flooding and sustained high winds. Two types of flooding accompany tropical cyclones: inland flooding from rivers overflowing their banks and coastal flooding from storm surges.

Coastal flooding is the most dangerous aspect of these storms. The strong winds and low atmospheric pressure of a tropical cyclone cause a rise in sea level called a **storm surge (Figure 6.23)**. The storm surge is not a wave—it is a dome of

water that forms beneath tropical cyclones. As a storm surge comes ashore, it inundates low-lying coastal areas with seawater. The storm's high winds also create large waves that break on top of the storm surge.

High winds do not cause nearly as many human fatalities as flooding, but if people are caught outside in high winds, they are at risk of being struck by flying debris. High winds are most notable for causing property damage.

Inland flooding from tropical cyclones can sometimes result in considerable loss of life. For instance, when Hurricane Mitch struck Honduras and Nicaragua in 1998, it caused rivers there to overflow their banks and cut into steep cliff faces. This created a dangerous mix of mudslides and flooding. As a result of inland flooding, some 11,000 people died in that storm, and 1.5 million people were left homeless.

Worldwide, tropical cyclones are the main meteorological killer. The losses of life they inflict can be staggering **(Table 6.3** on page 182). Thanks to satellite technology, tropical cyclones no longer catch us by surprise, but this fact does not always prevent the loss of human lives. Some particularly deadly storms have occurred during the era of satellites. The most recent devastating tropical cyclone was in 2008 in Burma.

All but two of the disasters listed in Table 6.3 occurred in countries bordering the Indian Ocean. Burma, Bangladesh, and India have suffered the greatest losses of life. Presumably, such large death tolls could have been avoided by getting people out of harm's way. This is a challenge, however, as many people in these regions live at sea level near the coast, where there is no high ground. In some areas, dirt roads that turn to mud in the heavy rainfall also can make it impossible to evacuate everyone.

As **Table 6.4** on page 182 shows, the United States also experiences deadly tropical cyclones, but loss of life is far less there because of the ability to evacuate people in coastal areas to higher ground or inland.

Hurricane Katrina in 2005 caused the third largest number of fatalities of any U.S. hurricane. New Orleans lies in a topographic bowl that is below sea level and is easily flooded. When the hurricane struck the city, some of the protective walls surrounding it failed, and much of the bowl filled with water. Various obstacles to evacuation were also important factors that amplified that disaster.

Hurricane Watches and Warnings

The warning system for hurricanes in the United States is very similar to that used for tornadoes. A **hurricane watch** is issued 48 hours in advance for areas where tropical storm–force winds are possible. Once high winds develop, it can become

Figure 6.21 Dropsondes.

Dropsondes are released from aircraft flying into tropical cyclones. As a dropsonde descends through a storm, it records its position with a GPS unit, measures atmospheric data, and transmits those data back to the airplane.

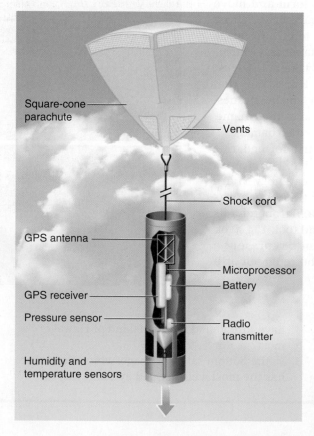

Square-cone parachute

Vents

Shock cord

GPS antenna

Microprocessor

Battery

GPS receiver

Pressure sensor

Radio transmitter

Humidity and temperature sensors

Figure 6.22 The geography of tropical cyclones. This map shows the paths of all tropical cyclones, tropical storms, and tropical depressions over the last six decades. TD = tropical depression. TS = tropical storm. *(Bill Rankin, www.radicalcartography.net)*

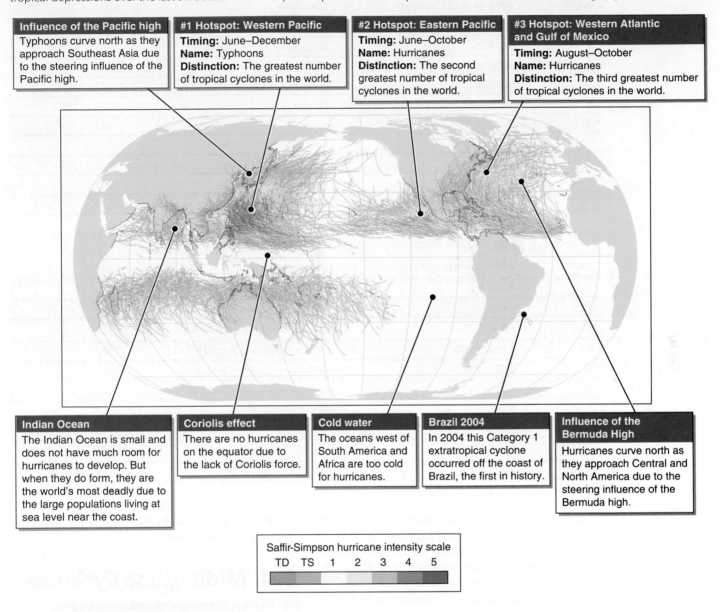

Influence of the Pacific high
Typhoons curve north as they approach Southeast Asia due to the steering influence of the Pacific high.

#1 Hotspot: Western Pacific
Timing: June–December
Name: Typhoons
Distinction: The greatest number of tropical cyclones in the world.

#2 Hotspot: Eastern Pacific
Timing: June–October
Name: Hurricanes
Distinction: The second greatest number of tropical cyclones in the world.

#3 Hotspot: Western Atlantic and Gulf of Mexico
Timing: August–October
Name: Hurricanes
Distinction: The third greatest number of tropical cyclones in the world.

Indian Ocean
The Indian Ocean is small and does not have much room for hurricanes to develop. But when they do form, they are the world's most deadly due to the large populations living at sea level near the coast.

Coriolis effect
There are no hurricanes on the equator due to the lack of Coriolis force.

Cold water
The oceans west of South America and Africa are too cold for hurricanes.

Brazil 2004
In 2004 this Category 1 extratropical cyclone occurred off the coast of Brazil, the first in history.

Influence of the Bermuda High
Hurricanes curve north as they approach Central and North America due to the steering influence of the Bermuda high.

Saffir-Simpson hurricane intensity scale
TD TS 1 2 3 4 5

Figure 6.23 Tropical cyclone storm surge. (A) About 95% of the storm surge (shown here in a cross section) is a result of strong winds that pile up the seawater. Decreased atmospheric pressure, which allows the sea to expand upward, contributes about 5% of the storm surge. Storm surges are typically 80 to 160 km (50 to 100 mi) wide. (B) This table lists the storm surge heights by tropical cyclone category. The storm surge height in a category 5 storm, for example, is 5.7 m (19 ft) or higher.

Storm movement

Wind-driven surge

Pressure-driven surge (5% of total)

Eye

Shore

A

B

SAFFIR-SIMPSON CATEGORY	STORM SURGE HEIGHT (METERS)	STORM SURGE HEIGHT (FEET)
1	1–1.7	3–5
2	1.8–2.6	6–8
3	2.7–3.8	9–12
4	3.9–5.6	13–18
5	5.7 or higher	19 or higher

Table 6.3	The 10 Deadliest Tropical Cyclones Worldwide		
RANK	APPROXIMATE DEATHS	LOCATION	DATE
1	500,000	West Bengal/Bangladesh	1970
2	300,000	Hooghly River, Bengal, India	1737
3	300,000	Haiphong, Vietnam	1881
4	300,000	Coringa, southern India	1839
5	200,000	Bengal	1584 or 1582
6	200,000	China	1975
7	200,000	Calcutta, Bengal, India	1876
8	140,000	Burma	2008
9	140,000	Bangladesh	1991
10	100,000	Arabian Sea, western India	1882

Table 6.4	The 10 Deadliest Hurricanes in the United States		
RANK	APPROXIMATE DEATHS	LOCATION	DATE
1	8,000	Galveston, Texas	1900
2	2,500	Lake Okeechobee, Florida	1928
3	1,836	Hurricane Katrina, New Orleans	2005
4	1,800–2,000	Louisiana and Mississippi	1893
5	1,000–2,500	Georgia and South Carolina	1893
6	700	Georgia and South Carolina	1881
7	638	New England	1938
8	600	Florida	1919
9	500	Georgia and South Carolina	1804
10	450	Corpus Christi, Texas	1919

impossible to prepare for the storm or to evacuate. Preparation includes obtaining plywood to cover windows, gathering items for a disaster-readiness kit, collecting valuables, stashing food and water, and reviewing the evacuation routes.

A **hurricane warning** is issued 36 hours in advance of the approaching storm for coastal regions where the hurricane is forecast to make landfall and tropical storm–force winds are imminent. Preparation at this stage includes fastening plywood over windows and evacuating. Local officials direct residents to immediately leave the area, and if that is not possible to take cover in the safest part of a home or storm shelter.

All countries that experience tropical cyclones now have a system for warning and evacuating coastal populations. India, historically one of the most at-risk countries in this regard, has greatly reduced its vulnerability with coordinated and mandatory evacuations. In December 2016, more than 16,000 people were safely evacuated from coastal areas of southern India when Cyclone Vardah struck. In total, India suffered 18 fatalities from that storm, a number that would have been in the thousands had the warning and evacuation system not been in place.

6.4 Midlatitude Cyclones

◎ **Review the major characteristics and stages of development of midlatitude cyclones.**

Recall from Section 5.1 that a *cyclone* is any meteorological system that rotates counterclockwise in the Northern Hemisphere (clockwise in the Southern Hemisphere) around a low-pressure center. When large cyclonic systems occur at midlatitudes, they are called **midlatitude cyclones** (or *extratropical cyclones*). They are also called *depressions, lows,* or *low-pressure systems,* names that indicate that they are geographically extensive regions of low barometric pressure. In contrast to tropical cyclones, which are fueled by the release of latent heat, midlatitude cyclones form as a result of temperature contrasts between air masses. In the Northern Hemisphere, these large storm systems, which move from west to east with the westerlies, affect the United States, most of Canada, Europe, and Asia.

Cyclogenesis (midlatitude cyclone formation) occurs from fall through spring between approximately 30 degrees and 70 degrees latitude in both hemispheres. There are two main settings in which cyclones form: downwind of mountain ranges and in areas where warm water is located downwind of cold water or land surfaces **(Figure 6.24)**. Cyclones cannot persist if they do not have *upper-level support*, meaning that the jet stream must form a Rossby wave trough over the region of cyclogenesis (see Section 5.2 and Figure 6.27). Troughs are key because as air flows through them from west to east, it accelerates and diverges. This creates a column of rising air that supports the surface level low.

Some 10 to 20 midlatitude cyclones are in progress at any given time worldwide. They are the largest storm systems on the planet, having a diameter of 1,600 km (1,000 mi) or more. In some cases, their wind speeds become as strong as tropical cyclones at sea level. In mountainous regions, they commonly produce hurricane-force winds.

Anatomy of a Midlatitude Cyclone

There are many different kinds of midlatitude cyclones, depending on the types of air masses that interact and the characteristics of geostrophic winds aloft. A midlatitude cyclone is typically composed of a warm front and a cold front. A **warm front** is produced when warm air advances on and flows over cooler, denser air. Warm fronts may bring precipitation, but they are rarely associated with severe weather. A **cold front** is a region where cold, dense air advances on relatively warm and less dense air. Cold fronts are sometimes associated with severe weather. The different densities of warm and cold air cause them to move over

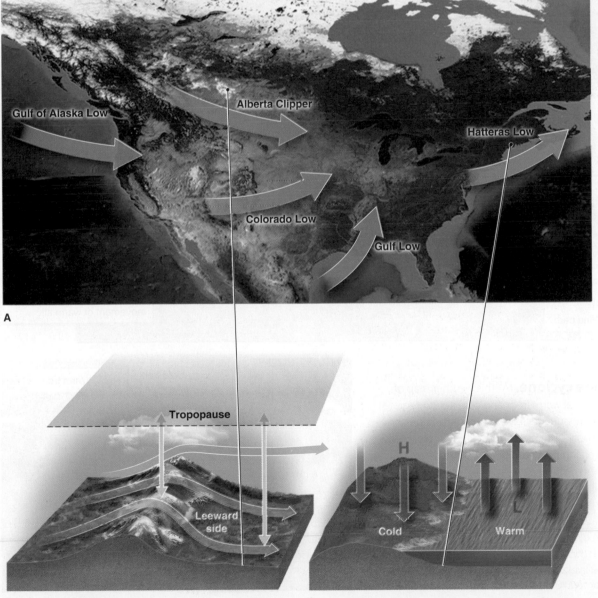

Figure 6.24 Cyclogenesis environments.

(A) Midlatitude cyclones form in five key areas: the Gulf of Alaska, the leeward side of the Rocky Mountains in Alberta and in Colorado, the Gulf of Mexico, and the northeastern United States. The purple arrows show the direction of movement of midlatitude cyclones after they have formed. (B) As air moves over and down mountain ranges, the distance between the surface and the tropopause (yellow arrows) increases on the leeward side, stretching the air and creating low pressure. Counterclockwise rotation from the Coriolis effect leads to cyclogenesis. Alberta clippers and Colorado lows form in this manner. (C) Cyclogenesis also occurs where the ocean is relatively warm compared to an adjacent landmasses or other regions of water. Gulf of Alaska low, Gulf low, and Hatteras low are names given to cyclones that form in these ocean environments. *(Reto Stöckli, NASA Earth Observatory)*

and under one another but not to mix together. Without this characteristic, frontal systems would not form. **Figure 6.25** illustrates how fronts are arranged in a midlatitude cyclone.

Effects of Midlatitude Cyclones on Weather

As a midlatitude cyclone moves over a region, the weather experienced in that region reflects the type of front moving through. In most cases, a warm front moves through the region first and is followed by a cold front. Warm fronts may produce nimbostratus clouds that bring steady precipitation.

Cold fronts are usually associated with cumulonimbus clouds that bring short bursts of rainfall and potentially severe weather. **Figure 6.26** diagrams the typical characteristics and weather patterns of frontal systems.

Life Cycle of a Midlatitude Cyclone

Like a single-cell thunderstorm or a hurricane, a midlatitude cyclone experiences predictable stages of growth, maturation, and dissipation. A midlatitude cyclone progresses through the stages of development and dissipation over a period of about 1 to

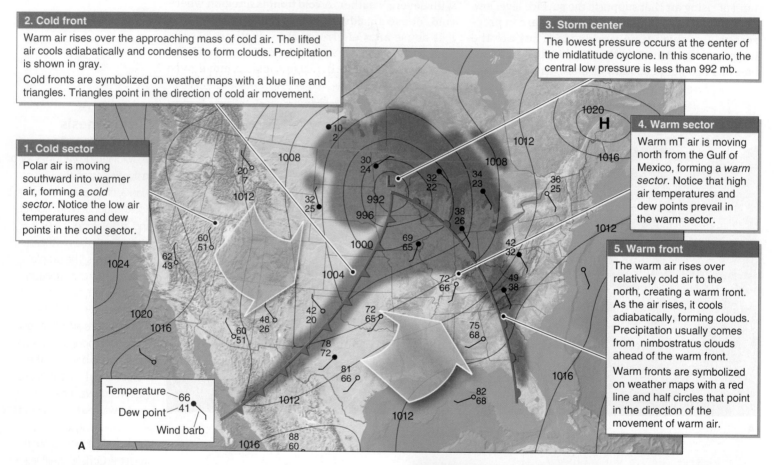

2. Cold front

Warm air rises over the approaching mass of cold air. The lifted air cools adiabatically and condenses to form clouds. Precipitation is shown in gray.

Cold fronts are symbolized on weather maps with a blue line and triangles. Triangles point in the direction of cold air movement.

1. Cold sector

Polar air is moving southward into warmer air, forming a *cold sector*. Notice the low air temperatures and dew points in the cold sector.

3. Storm center

The lowest pressure occurs at the center of the midlatitude cyclone. In this scenario, the central low pressure is less than 992 mb.

4. Warm sector

Warm mT air is moving north from the Gulf of Mexico, forming a *warm sector*. Notice that high air temperatures and dew points prevail in the warm sector.

5. Warm front

The warm air rises over relatively cold air to the north, creating a warm front. As the air rises, it cools adiabatically, forming clouds. Precipitation usually comes from nimbostratus clouds ahead of the warm front.

Warm fronts are symbolized on weather maps with a red line and half circles that point in the direction of the movement of warm air.

Temperature — 66
Dew point — 41
Wind barb

A

Figure 6.25 Midlatitude cyclone. (A) This weather map of North America shows how a typical warm front and cold front integrate to form a midlatitude cyclone. Pressure decreases toward the center of the system, as shown by the isobars; the lowest pressure, at the center, is labeled L for "low." Notice the changing air temperature and dew point ahead of and behind both fronts. After a warm front moves through, the air temperature and dew point rise. After a cold front moves through, the air temperature and dew point fall. The gray area shows cloudiness and precipitation. (B) This October 26, 2010, satellite image shows a midlatitude cyclone spanning much of eastern North America. It is located in the same position and is labeled with the same features illustrated in part A. *(NASA Earth Observatory imagery created by Jesse Allen, using imagery provided courtesy of the NASA GOES Project Science Office)*

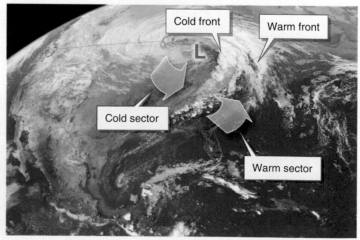

B

Figure 6.26 Cold and warm fronts and their weather patterns. (A) Warm front characteristics. (B) Cold front characteristics. (C) Warm and cold fronts compared.

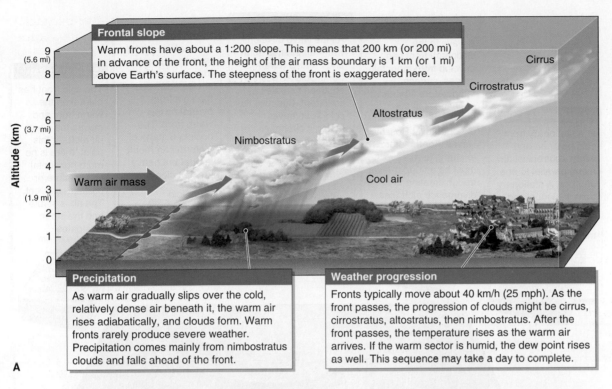

Frontal slope

Warm fronts have about a 1:200 slope. This means that 200 km (or 200 mi) in advance of the front, the height of the air mass boundary is 1 km (or 1 mi) above Earth's surface. The steepness of the front is exaggerated here.

Cirrus

Cirrostratus

Altostratus

Nimbostratus

Cool air

Warm air mass

Precipitation

As warm air gradually slips over the cold, relatively dense air beneath it, the warm air rises adiabatically, and clouds form. Warm fronts rarely produce severe weather. Precipitation comes mainly from nimbostratus clouds and falls ahead of the front.

Weather progression

Fronts typically move about 40 km/h (25 mph). As the front passes, the progression of clouds might be cirrus, cirrostratus, altostratus, then nimbostratus. After the front passes, the temperature rises as the warm air arrives. If the warm sector is humid, the dew point rises as well. This sequence may take a day to complete.

A

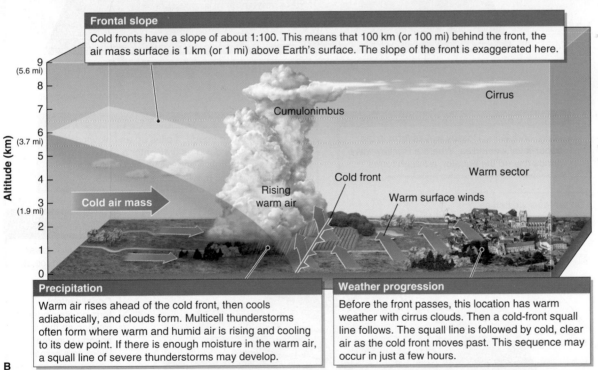

Frontal slope

Cold fronts have a slope of about 1:100. This means that 100 km (or 100 mi) behind the front, the air mass surface is 1 km (or 1 mi) above Earth's surface. The slope of the front is exaggerated here.

Cumulonimbus

Cirrus

Cold front

Warm sector

Rising warm air

Warm surface winds

Cold air mass

Precipitation

Warm air rises ahead of the cold front, then cools adiabatically, and clouds form. Multicell thunderstorms often form where warm and humid air is rising and cooling to its dew point. If there is enough moisture in the warm air, a squall line of severe thunderstorms may develop.

Weather progression

Before the front passes, this location has warm weather with cirrus clouds. Then a cold-front squall line follows. The squall line is followed by cold, clear air as the cold front moves past. This sequence may occur in just a few hours.

B

	WARM FRONT	COLD FRONT
Map symbol	●●●●	▲▲▲▲
Frontal slope	Shallow	Steep
Progression of clouds	Cirrus—alto—stratus	Cirrus cumulonimbus
Likely weather	Steady rain, snow, or ice	Intense showers, potentially severe weather
Duration of precipitation	A day or more	A few hours
Speed of movement	Relatively slow	Relatively fast

C

Figure 6.27 Midlatitude cyclone development. (A) A typical midlatitude cyclone undergoes several stages of development over the course of 1 to 2 weeks. (B) This diagram illustrates how Rossby waves at an altitude of about 5,000 m (16,400 ft) can maintain midlatitude cyclones and anticyclones. Rossby wave troughs provide essential upper-level support for midlatitude cyclones.

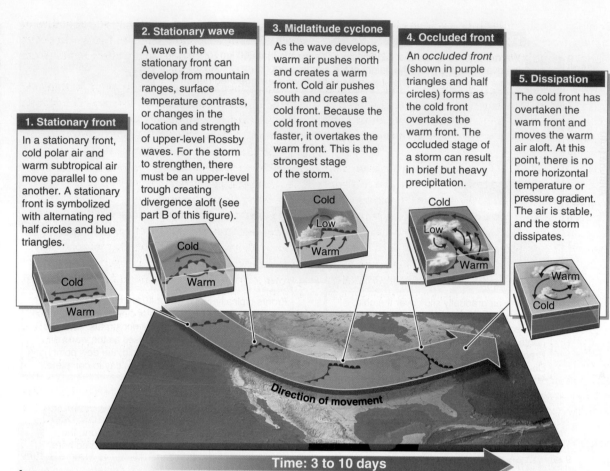

1. Stationary front

In a stationary front, cold polar air and warm subtropical air move parallel to one another. A stationary front is symbolized with alternating red half circles and blue triangles.

2. Stationary wave

A wave in the stationary front can develop from mountain ranges, surface temperature contrasts, or changes in the location and strength of upper-level Rossby waves. For the storm to strengthen, there must be an upper-level trough creating divergence aloft (see part B of this figure).

3. Midlatitude cyclone

As the wave develops, warm air pushes north and creates a warm front. Cold air pushes south and creates a cold front. Because the cold front moves faster, it overtakes the warm front. This is the strongest stage of the storm.

4. Occluded front

An *occluded front* (shown in purple triangles and half circles) forms as the cold front overtakes the warm front. The occluded stage of a storm can result in brief but heavy precipitation.

5. Dissipation

The cold front has overtaken the warm front and moves the warm air aloft. At this point, there is no more horizontal temperature or pressure gradient. The air is stable, and the storm dissipates.

Direction of movement

Time: 3 to 10 days

A

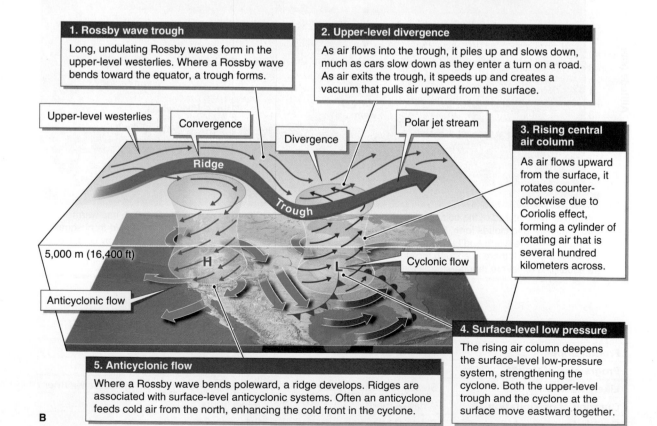

1. Rossby wave trough

Long, undulating Rossby waves form in the upper-level westerlies. Where a Rossby wave bends toward the equator, a trough forms.

2. Upper-level divergence

As air flows into the trough, it piles up and slows down, much as cars slow down as they enter a turn on a road. As air exits the trough, it speeds up and creates a vacuum that pulls air upward from the surface.

Upper-level westerlies

Convergence

Divergence

Polar jet stream

Ridge

Trough

3. Rising central air column

As air flows upward from the surface, it rotates counter-clockwise due to Coriolis effect, forming a cylinder of rotating air that is several hundred kilometers across.

5,000 m (16,400 ft)

H

L

Cyclonic flow

Anticyclonic flow

4. Surface-level low pressure

The rising air column deepens the surface-level low-pressure system, strengthening the cyclone. Both the upper-level trough and the cyclone at the surface move eastward together.

5. Anticyclonic flow

Where a Rossby wave bends poleward, a ridge develops. Ridges are associated with surface-level anticyclonic systems. Often an anticyclone feeds cold air from the north, enhancing the cold front in the cyclone.

B

2 weeks. Most midlatitude cyclones begin as *stationary waves* in the subpolar low **(Figure 6.27A)**. Although midlatitude cyclones do not all look and behave alike, temperature gradients (where cold air at high latitudes meets warm air at low latitudes) and, therefore, pressure gradients give rise to all of them. In addition, midlatitude cyclones must have upper-level support with a trough in the polar jet stream to persist.

If an upper-level Rossby wave trough is present (see Section 5.2), the low pressure at Earth's surface deepens (decreases), and the cyclonic system strengthens **(Figure 6.27B)**. Upper-level troughs maintain surface level low pressure by pulling air from the surface to higher altitudes.

After wintertime, midlatitude cyclones move over the Great Lakes, the cold air behind them often creates lake-effect snow. **Lake-effect snow** is heavy snowfall that results as cold air moves over large, relatively warm bodies of water, such as the Great Lakes **(Figure 6.28)**. Lake-effect snow happens where warm water readily evaporates and provides moisture to a cold atmosphere. Because cold air has a low water-vapor capacity, the moisture readily condenses and deposits to form snowflakes and heavy snowfall. Sometimes a single storm can produce meters of lake-effect snow. Although it is particularly pronounced in the Great Lakes region, lake-effect snow occurs wherever cold Arctic air masses move over large, relatively warm bodies of water. The highest snow totals in the world occur on the island of Hokkaido, Japan, after cold air from Asia moves over the Sea of Japan.

One particularly important type of midlatitude cyclone is called a **nor'easter**. These powerful storms bring blizzard-like conditions from the Mid-Atlantic states north to New England. A nor'easter originates as a Gulf low (see Figure 6.24) where mT air from the Gulf of Mexico meets cold air from the Great Plains. The name of these storms derives from the direction of the wind (from the northeast) that they bring to the regions where their precipitation falls.

Another important type of storm that forms at midlatitudes that has only recently received scientific attention is called an *atmospheric river*. Atmospheric rivers are narrow bands of atmospheric moisture that stretch thousands of kilometers and transport huge amounts of moisture from the tropics to the midlatitudes. A single atmospheric river may transport as much water as the world's largest rivers. As it passes over an area, it may bring torrential rains and heavy snowfall to higher elevations for several days or even weeks. There are three to seven atmospheric rivers present in the world at any given time. Californians call this phenomenon the "Pineapple Express," given that the storms are warm and originate near Hawai'i. The entire west coast of North America can be affected

Figure 6.28 Lake-effect snow. This map shows average annual snowfall totals in the Great Lakes region. Downwind (east) of the lakes, snowfall increases due to lake-effect snow. The Upper Peninsula of Michigan, bordering Lake Superior, receives up to 500 cm (200 in) of snow per year, on average.

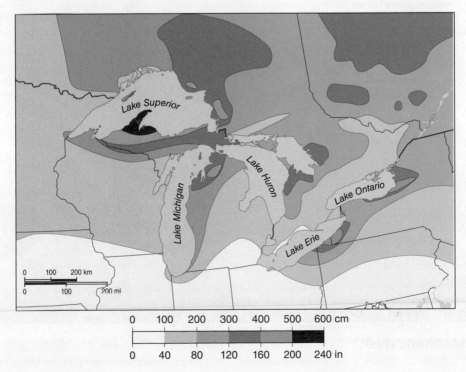

by atmospheric rivers, and they occasionally originate in the Gulf of Mexico and affect the East Coast of the United States as well. They also affect the British Isles and western Europe **(Figure 6.29)**.

6.5 Weather Forecasting and Analysis

◎ **Describe how weather forecast models are developed.**

Weather systems like the ones we have been discussing in this chapter seldom catch people by surprise. Satellites and Doppler radar are always scanning the atmosphere and continuously generating data for observation and analysis. Using these data, meteorologists have developed sophisticated computer models to predict or forecast the atmosphere's behavior. Most of us rely on weather forecasts every day in one way or another. On a weather app, we see a small graphic weather symbol, such as a sun or clouds, for each day of the week. This weather graphic is always accompanied by a high and low temperature for each day of the forecast, as well as a percent chance of precipitation if precipitation is possible. The weather forecast is usually fairly accurate for about 3 days into the future, and after that it becomes progressively less accurate. But how do meteorologists create these weather forecasts? There are two basic steps involved with weather forecasting: collecting weather data and running the data through computer models.

Collecting Weather Data

Data is the lifeblood of science. Weather forecasting is a particularly data-intensive effort. The meteorological community is very well organized in terms of integrating the big data from various sensing technologies and making it quickly available, often in real time. Forecasters rely on weather data to develop forecast models from a variety of sources:

- **Surface stations:** More than 10,000 land-based weather stations provide weather observations several times a day. In the United States, weather data comes from the Automated Surface Observing System (ASOS), which encompasses more than 800 automated weather stations that automatically and continuously feed weather data to the National Weather Service. Marine buoys also provide essential oceanic data.

- **Radiosondes:** More than 750 locations around the world launch radiosondes (weather balloons) to acquire weather data above Earth's surface. In the United States and Canada alone, more than 80 locations launch radiosondes twice daily.

- **Satellites:** *Geostationary satellites* (those that remain fixed over one location over the equator) and *polar-orbiting satellites* together provide 24-hour global coverage (see Figure 1.26).

Video

Worldwide Tour of Global Precipitation

Available at
www.saplinglearning.com

▶

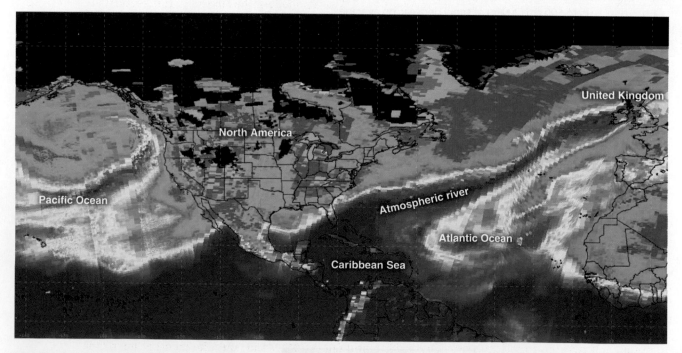

Figure 6.29 An atmospheric river. This atmospheric river formed during the first week of December 2015. It stretched across the Atlantic Ocean from the Caribbean Sea to the British Isles and brought considerable flooding to the United Kingdom, Ireland, and Norway. *(NOAA / CIRA)*

- **Doppler radar:** The 159 fixed Doppler radar units in the United States provide 24-hour precipitation and wind data. The National Weather Service also oversees a fleet of radar units mounted on trucks, known informally as "Doppler on wheels."

Each of these data-acquisition systems is seamlessly integrated by the National Weather Service in the United States and by the Meteorological Service of Canada in Canada to feed into computers that develop forecast models.

Developing Computer Forecast Models

The weather data that are collected are input into computer models, in a process called *numerical weather forecasting*. These models are three-dimensional grids that mimic the behavior of the atmosphere using a series of complex mathematical equations **(Figure 6.30)**.

Given the number of cells and meteorological variables within each cell, weather forecast models require several billion calculations in order to be advanced forward in time and forecast what the weather might do in the future. The work can only be done by supercomputers with extremely fast computational speeds. It takes about an hour to advance a weather forecast model 1 day.

A single forecast provides only one version of the future. There could be several different forecast outcomes, depending on the nature of the data that is fed into the model. To address this issue, meteorologists run several forecast models for the same area, in a process called *ensemble forecasting*. Ensemble forecasting is needed because small differences in the initial data input can create significant changes in the results of a forecast model. Individual human meteorologists assess the ensemble forecasts and, using professional experience and judgment, develop the forecast that they think is most accurate.

The final outcome of this work is a *quantitative forecast*. A quantitative forecast specifies the future temperature, or precipitation amounts, or the likelihood of rain as a percentage. (A *qualitative* forecast, in contrast, does not specify numerically what the weather is expected to do.) *Short-term forecasts* cover 72 hours or less. *Medium-range forecast* models run out to 7 days. Any forecast longer than 7 days is a *long-range forecast*. Long-range forecasts are the least accurate because errors in the model are magnified by time. No weather model forecasts the weather with 100% accuracy because there will always be errors in the model, errors in the data entered, insufficient data input, and inherent chaos in atmospheric physics. That said, weather forecast accuracy has improved by about 1 day per decade, so that a 4-day forecast 10 years ago was as accurate as today's 5-day forecast.

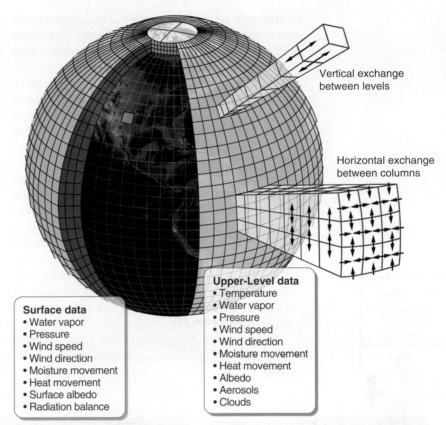

Vertical exchange between levels

Horizontal exchange between columns

Surface data
- Water vapor
- Pressure
- Wind speed
- Wind direction
- Moisture movement
- Heat movement
- Surface albedo
- Radiation balance

Upper-Level data
- Temperature
- Water vapor
- Pressure
- Wind speed
- Wind direction
- Moisture movement
- Heat movement
- Albedo
- Aerosols
- Clouds

Figure 6.30 Weather forecast model. Weather forecast models divide the study region (the whole globe, in this example) into gridded cells. If each cell is 10 km^2 (3.9 mi^2), 5 million cells will be needed to cover the whole planet's surface. In a typical forecast model, each cell has 100 levels (like floors in a building), and 10 meteorology variables (such as temperature, humidity, and pressure) are input into each cell. *(K. Cantner / AGI)*

Weather models, the subject of this section, are different from *climate models* (sometimes called *weather outlooks*). Weather and climate models use the same approach of data input and a gridded model, but they have different forecast goals. Weather models forecast conditions for a specific day and what weather conditions will be like at a certain time of day. Climate models, in contrast, develop a forecast for average conditions for a long-range period of time, such as months to years or even decades in advance.

GEOGRAPHIC PERSPECTIVES
6.6 Are Atlantic Hurricanes a Growing Threat?

◎ **Assess the current and potential vulnerability of the United States to major hurricanes.**

The Atlantic hurricane season in the late summer and fall of 2017 stands out like no other hurricane season on record. Three major storms pummeled the United States. First, on August 25, Hurricane Harvey made landfall over Texas, dropping 164 cm (64.6 in) of rain over a 4-day period; its torrential rains broke U.S. rainfall records for a single storm. The heavy rains put much of Houston and its suburbs under water, claimed some 90 lives, and caused $70 billion in damage. Next, Hurricane Irma made landfall over the Florida Keys on September 10, after tearing through several Caribbean islands. Irma ran up the length of Florida, leaving widespread destruction and inflicting $63 billion in damage. After Irma, hurricanes Jose, Katia, and Lee formed but caused little harm. Only Katia made landfall in Mexico as a weak category 1 storm; the other two dissipated far out in open water. Then on October 4 came Hurricane Maria. Maria, a powerful category 5 storm, packing 282 km/h (175 mph) winds, worked its way through the Caribbean. It was the strongest storm ever to strike Puerto Rico, where it left a trail of devastation, disabling most of the island's infrastructure—already weakened by Irma several weeks earlier—and its economy **(Figure 6.31)**.

There have been other notable tropical cyclones in recent years. Cyclone Winston in 2016 formed near Fiji in the Pacific Ocean. Winston's peak wind speeds were 285 km/h (180 mph), the fastest winds ever recorded in the Southern Hemisphere. In 2015 Hurricane Patricia formed in the Pacific Ocean off the southwestern coast of Mexico. Patricia achieved 345 km/h (215 mph) wind speeds, the second fastest wind speeds ever recorded in a tropical cyclone.

Increasing Damages and Costs

All countries where tropical cyclones occur, including the United States, are vulnerable to these storms. This vulnerability was made painfully obvious in 2005, when Hurricane Katrina struck and flooded nearly all of New Orleans. Because of natural settling and compaction of the sediments on which New Orleans was built, much of the city has been slowly sinking and is now below sea level. A fortress of seawalls and river levees surrounding the city are all that keep Lake Pontchartrain and the Mississippi River out of New Orleans.

Story Map

Devastating Hurricanes

Available at www.saplinglearning.com

Hurricane Harvey
- Maximum winds: 209 km/h (130 mph)
- Maximum category: 4
- Damage: $70 billion, extensive flooding throughout Houston area
- Fatalities: 83
- Broke rainfall record for a single storm in the United States.

Hurricane Irma
- Maximum winds: 298 km/h (185 mph)
- Maximum category: 5
- Damage: $63 billion, extensive wind damage to Florida
- Fatalities: 124
- Broke record for longest sustained high wind speed: 37 hours of winds of 298 kp/h.

Hurricane Maria
- Maximum winds: 282 km/h (175 mph)
- Maximum category: 5
- Damage: $139 billion; costliest in Puerto Rican history; extensive wind damage to Puerto Rico.
- Fatalities: nearly 3,000
- Fifth costliest hurricane on record.

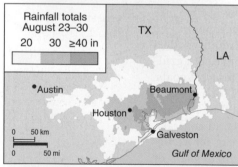

Puerto Rico before and after
Comparison of lights at night in Puerto Rico before (top) and after (bottom) Hurricane Maria. After Maria, nearly the entire island was without electricity.

About 9 trillion gallons of water fell across southeast Texas. If that water were a sphere it would be 2.5 miles (4.02 km) in diameter.

Figure 6.31 2017 Hurricanes Harvey, Irma, and Maria. The 2017 Atlantic hurricane season was unusually active. Hurricanes Harvey, Irma, and Maria were the strongest of the season. *(Bottom left (background photo). Hillary Gaschler/EyeEm/Getty Images; top center. NASA/NOAA/UWM-CIMSS, William Straka; bottom center. Gerben Van Es/AFP/Getty Images; top right. NOAA; bottom right. STR/AFP/Getty Images)*

The 6 m (20 ft) storm surge generated by Hurricane Katrina overtopped or collapsed many of the levees protecting New Orleans and flooded the city up to 8 m (25 ft) deep. The cost was more than 1,800 human lives and some $128 billion in damage (the second costliest in U.S. history). Then, in 2012, Superstorm Sandy, a hurricane combined with a nor'easter, inflicted $75 billion in damage in the New Jersey and New York coastal areas and flooded much of the New York City subway system. The 2017 hurricanes Harvey, Irma, and Maria left about $272 billion in damage and another major U.S. city under water.

The Most Vulnerable Cities in the United States

What U.S. cities are most vulnerable to hurricanes? Almost all coastal regions in the Gulf of Mexico, in the southeastern United States, and up the eastern coast of North America have been struck by hurricanes. Five of the most vulnerable large cities and metropolitan areas in the United States are New Orleans, Miami, St. Petersburg/ Tampa, Houston/Galveston, and Norfolk/Virginia Beach. Their vulnerability results from a combination of being located where hurricanes frequently occur and having large populations living at or just above (or below, in the case of New Orleans) sea level **(Figure 6.32)**.

Like New Orleans and the Houston area, Miami is at great risk of a hurricane catastrophe. The last time Miami was struck by a hurricane was in 1950, when Hurricane King made a direct hit on the city's downtown. In all, King caused 3 deaths and about 50 injuries. Although damage was considerable, there was little loss of life because fewer than half a million people lived in the Miami metropolitan region at the time. Miami has not suffered a direct strike by a hurricane since then, although Hurricane Andrew struck Homestead, a southern suburb, in 1992.

Since 1950, however, the population of south Florida has grown significantly, and it is projected to be one of the fastest-growing regions in the country over the next several decades. Much of the coastal area has felt the effects of urbanization. Some 5.5 million people live in Miami and the surrounding area. Scientists estimate that if a category 4 hurricane were to strike Miami today, it could cause some $90 billion in damage and bring significant flooding and loss of human life.

Climate Change and Hurricanes

Global warming has caused a sea-level rise of 23 cm (9 in) in the past century. This fact alone makes all

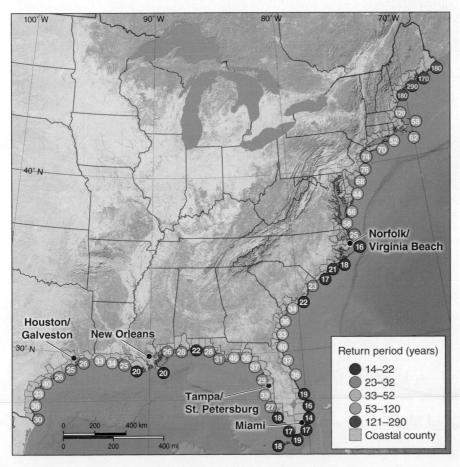

Figure 6.32 Hurricane activity in the United States. Each colored dot gives the hurricane *return period*—the number of years, on average, between major hurricanes (category 3 or higher)—for specified stretches of coast since records have been kept. The dots are spaced about 100 km (60 mi) apart. The lower the return period, the more vulnerable a stretch of coastline is. On the southern tip of Florida, for example, category 3 hurricanes have struck about every 19 years. Five particularly vulnerable large cities and metropolitan areas are indicated on the map. Coastal counties are mapped in light gray.

Return period (years)
- 14–22
- 23–32
- 33–52
- 53–120
- 121–290
- Coastal county

coastal cities more vulnerable to the storm surges of hurricanes. Because the air and oceans are warming, hurricanes can be expected to become more frequent and stronger. The recent unusually strong storms have led many to wonder whether these are the storms of a new era of anthropogenic climate change. Are tropical cyclones becoming more powerful as a result of a warming world?

So far there is little evidence to suggest that hurricanes are becoming more frequent and stronger. There is a natural cyclical pattern of Atlantic hurricane activity. **Figure 6.33A** shows the number of all hurricanes since 1851 and shows the number of storms that have been category 3 or greater. **Figure 6.33B** shows the Accumulated Cyclone Energy (ACE) index. The ACE index sums the total wind energy generated by hurricanes and creates an index: Years with multiple large hurricanes have a higher index number than years with just a few strong hurricanes. These data indicate that the 1950s and 1960s were active in terms of hurricanes. In the 1970s and 1980s, hurricane activity decreased. In the 1990s, it began to increase again.

There has been little to no change in the average number or the accumulated strength of Atlantic hurricanes, as measured in the ACE data. There are natural long-term cycles, and there are years with extreme events, but so far there is little

evidence of an upward trend. The data for other ocean basins in which tropical cyclones form also do not show any significant trend in increasing frequency or intensity of hurricanes.

Earth's climate is exceedingly complex. Many factors besides the temperature of the oceans and atmosphere influence tropical cyclone activity. El Niño events (discussed in Section 5.4), for example, decrease the number of Atlantic hurricanes by causing increased wind shear, which tears hurricanes apart. La Niña events, on the other hand, favor Atlantic hurricane development.

Taken as a whole, many of the computer forecast models for the coming decades indicate that there will be no overall increase in the number of tropical cyclones worldwide or hurricanes in the Atlantic Ocean, but the number of category 4 and 5 storms is expected to increase as a result of the increase in heat energy in the oceans and atmosphere. Compounding this, as sea levels continue to rise, the effects of coastal storm surges will worsen. Tropical cyclones are therefore expected to be more destructive and produce significantly more rainfall (up to 15% more on average). So although hurricanes Harvey, Irma, and Maria may not represent a trend of more hurricanes overall, they could represent an upward shift in the number of powerful Atlantic hurricanes reaching U.S. shores.

Figure 6.33 Atlantic hurricane activity, 1851–2017. (A) This chronology records all detected hurricanes (908 in total, 328 of them category 3 and above) that formed in the Atlantic Ocean from 1851 to 2017. Aircraft monitoring began in 1944, and satellite monitoring began in 1966. For the era before satellite monitoring, the record is far less reliable. Blue bars represent category 1 and 2 hurricanes. Gold bars indicate hurricanes that were category 3 and up. The red line illustrates long-term trends. (B) This graph shows the Accumulate Cyclone Energy (ACE) data for the same period. Data are from the National Atmospheric and Oceanic Administration (NOAA) Atlantic Hurricane Database.

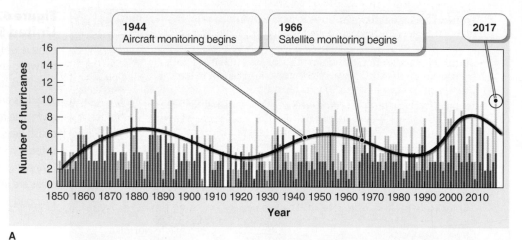

A

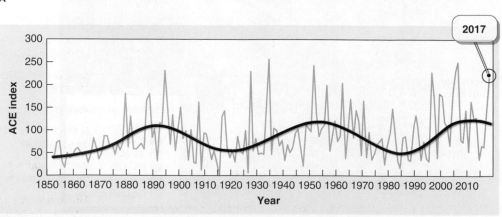

B

Chapter 6 Study Guide

Focus Points

6.1 Thunderstorms

- **Spatial and temporal scales:** Larger atmospheric systems (such as hurricanes) persist longer than smaller systems (such as single-cell thunderstorms).
- **Air masses:** Warm and humid mT air masses are essential for thunderstorm development.
- **Thunderstorm geography:** Thunderstorms are most frequent in the tropics over land because of land heating in the afternoon.
- **Thunderstorm types:** There are three types of thunderstorms: single-cell, multicell, and supercell thunderstorms. Multicell and supercell thunderstorms may bring severe weather.

6.2 Thunderstorm Hazards: Lightning and Tornadoes

- **Lightning and thunder:** Lightning is a discharge of electricity within a thunderstorm. Thunder is created as air is heated by lightning and rapidly expands.
- **Tornadoes:** Tornadoes form in thunderstorms, tropical cyclones, and cold fronts. The United States experiences the most frequent and strongest tornadoes in the world.

6.3 Nature's Deadliest Storms: Tropical Cyclones

- **Tropical cyclone structure:** A tropical cyclone consists of a calm eye, an eyewall of heavy wind and rain, and rain bands.
- **Tropical cyclone strength:** Tropical cyclones must have warm seawater to persist. They derive their strength from the latent heat positive feedback.
- **Stages of growth:** Tropical cyclones go through a series of stages of formation, from tropical wave, to tropical disturbance, to tropical depression, to tropical storm, to tropical cyclone.
- **Tropical cyclone geography:** Worldwide, hurricanes are restricted to tropical oceans. They do not occur within about 5 degrees latitude of the equator due to the lack of the Coriolis effect there.
- **Tropical cyclone hazards:** Coastal storm surge is the most dangerous aspect of tropical cyclones, particularly in the Indian Ocean.

6.4 Midlatitude Cyclones

- **Midlatitude cyclone effects:** Midlatitude cyclones bring storms to midlatitude regions, such as the United States and Canada, from fall through spring.

- **Warm and cold fronts:** Most midlatitude cyclones are composed of warm fronts that bring steady precipitation and cold fronts that bring bursts of short-lived showers.
- **Stages of development:** A midlatitude cyclone undergoes a sequence of growth, maturation, and dissipation over the span of days to weeks.

6.5 Weather Forecasting and Analysis

- **Weather forecast models:** Weather forecast models use computers to analyze weather data to create forecast models. Short-range forecasts are more accurate than long-range forecasts.

6.6 Geographic Perspectives: Are Atlantic Hurricanes a Growing Threat?

- **U.S. cities at risk:** New Orleans is the U.S. city most vulnerable to the effects of hurricanes. Miami is the second most at-risk largest metropolitan area.
- **Long-term trend:** Climate change could result in more and stronger hurricanes, but so far no such trend has been observed.

Key Terms

air mass, 165
cold front, 183
dropsonde, 179
enhanced Fujita scale (EF scale), 172
eye, 176
eyewall, 176
fulgurite, 169
hurricane, 176
hurricane warning, 182
hurricane watch, 180
lake-effect snow, 187
lightning, 169
mesocyclone, 168
midlatitude cyclone, 182
multicell thunderstorm, 167
nor'easter, 187
Saffir-Simpson scale, 178

severe thunderstorm, 167
single-cell thunderstorm, 165
squall line, 167
storm surge, 180
supercell thunderstorm, 168
thunder, 169
thunderstorm, 164
tornado, 172
tornado warning, 175
tornado watch, 175
tropical cyclone, 176
tropical storm, 178
typhoon, 176
wall cloud, 172
warm front, 183
wind shear, 165

Concept Review

6.1 Thunderstorms

1. Describe the relationship between the duration of an atmospheric system and its spatial scale.

2. What is an air mass? Describe the characteristics of an mT air mass. Where does it form?

3. What is a thunderstorm? What kind of cloud produces a thunderstorm? What are the three types of thunderstorms?

4. Describe the global geographic pattern of thunderstorms. Where are they most and least frequent, and why?

5. Describe the life cycle of a single-cell thunderstorm. How long do such storms last, and what causes them to dissipate?

6. How are multicell thunderstorms different from single-cell thunderstorms in terms of where they form and their duration?

7. What is a squall line, and how does it relate to multicell thunderstorms?

8. What is a supercell thunderstorm? How does it relate to a mesocyclone?

9. Are all supercell thunderstorms classified as severe? What makes a thunderstorm severe?

10. Explain how a supercell thunderstorm begins rotating and describe the general anatomy of this type of storm.

6.2 Thunderstorm Hazards: Lightning and Tornadoes

11. What is lightning? What kind of cloud produces it?

12. What is thunder? What causes it?

13. What is a fulgurite?

14. Is it safe to be in a car during a lightning storm? Is everywhere indoors safe from lightning?

15. How does lightning form?

16. What is a tornado? What system is used to rank tornado strength, and on what evidence is it based?

17. Describe the typical characteristics of tornadoes.

18. Where in the world are tornadoes most powerful and most frequent?

19. What factors explain why tornado activity is so high in North America?

20. Compare a tornado watch with a tornado warning. Under what conditions is each issued?

6.3 Nature's Deadliest Storms: Tropical Cyclones

21. What is the minimum wind speed required for a storm to be classified as a tropical cyclone?

22. What are the four major stages of tropical cyclone development?

23. What system is used to rank tropical cyclone strength? What are the rankings based on?

24. Draw and label the anatomy of a tropical cyclone. Where are winds fastest? Describe air movement direction in a tropical cyclone.

25. Explain the role of the tropical cyclone positive feedback in maintaining high wind speeds. Why is sea spray so important to this feedback (see Figure 6.19)?

26. Describe and explain the global geographic pattern of tropical cyclones.

27. What aspects of tropical cyclones are the most dangerous for communities?

28. Which ocean basin has the most dangerous tropical cyclones? What makes this area so dangerous?

29. What are hurricane watches and hurricane warnings? When is each issued?

6.4 Midlatitude Cyclones

30. What is a midlatitude cyclone? Where do midlatitude cyclones occur? How do cold fronts and warm fronts relate to them?

31. What are the two primary midlatitude cyclogenesis environments? Explain how midlatitude cyclones form in each of these environments.

32. Compare and contrast a cold front and a warm front. Discuss how each forms and the weather that is typically associated with each.

33. Describe the life cycle of a midlatitude cyclone, beginning with a stationary front and ending with an occluded front.

34. What is lake-effect snow? Where and how does it form?

35. What is a nor'easter? Where do these events form?

36. What is an atmospheric river? How are atmospheric rivers unusual, and what kind of weather do they bring?

6.5 Weather Forecasting and Analysis

37. How are the data used in forecasting the weather collected?

38. What is a weather model? How do these models work?

39. What is a quantitative weather forecast? Give an example of the information in such a forecast.

40. Which is more accurate: a 3-day weather forecast or a 5-day weather forecast? Why?

6.6 Geographic Perspectives: Are Atlantic Hurricanes a Growing Threat?

41. Which coastal states in the United States have the shortest hurricane return period?

42. In addition to New Orleans, which large U.S. metropolitan areas are most vulnerable to hurricanes?

43. In the context of climate change, why might a person reasonably expect that tropical cyclones should be getting more frequent and stronger?

44. How would you characterize Atlantic hurricane activity since 1944?

45. Are scientists certain how hurricane activity will change in the coming decades?

Critical-Thinking Questions

1. Based on your reading of this chapter, what, if any, atmospheric hazards occur where you live? Why do they occur there?

2. Do you think it would be possible to reduce the vulnerability of the Great Plains to tornadoes? What would need to be done to accomplish this?

3. What steps could be taken to reduce the vulnerability of the United States to hurricanes? How would these steps be different from those taken to address the threat from tornadoes?

4. What kinds of actions could reduce the vulnerability of countries bordering the Indian Ocean to tropical cyclones?

5. There is a saying in computer modeling: "garbage in, garbage out." Explain this statement in the context of weather forecasting accuracy and relate it to the importance of obtaining accurate weather data.

Test Yourself

Take this quiz to test your chapter knowledge.

1. **True or false?** Maritime tropical (mT) air masses are necessary for thunderstorm development.

2. **True or false?** Warm fronts usually bring steady rain but not severe weather.

3. **True or false?** Colorado has fewer thunderstorms than California.

4. **True or false?** Most lightning moves between cloud and ground.

5. **True or false?** Tropical cyclones usually do not catch people in populated areas by surprise.

6. **Multiple choice:** Which of the following can produce a fulgurite?

a. a valley or mountain breeze   c. a bolt of lightning
b. a tornado   d. a hurricane

7. **Multiple choice:** Which of the following locations is the safest in a lightning storm?

   a. at a park, under tall trees
   b. on a golf course
   c. indoors
   d. in a boat on a lake

8. **Multiple choice:** Which of the following is not a criterion used to categorize a thunderstorm as severe?

   a. tornado generation
   b. hailstone diameter
   c. wind gusts speed
   d. frequency of lightning

9. **Fill in the blank:** A _____ is a line of multicell thunderstorms.

10. **Fill in the blank:** The coastal sea-level rise caused by a hurricane is called _____.

Online Geographic Analysis

Tornado Analysis

In this exercise we analyze tornado activity in the United States.

Activity

Go to http://www.spc.noaa.gov/exper/envbrowser/. This a National Oceanic and Atmospheric Administration (NOAA) web page. This "U.S. Tornado Environment Browser" page provides tornado data for the United States. Click on the map to the left. Notice that data for the region you clicked are mapped and graphed to the right.

1. **Find and click in the center of Oklahoma. What time of year does tornado frequency peak?**

2. **For the area you clicked, how many EF5 tornadoes occurred during the period the data cover?**

3. **Now find and click on Minnesota. Compare the timing of peak tornado activity between Minnesota and Oklahoma.**

4. **What do you think causes the difference in timing of peak tornado activity between Minnesota and Oklahoma (refer to Question 3)?**

5. **Click around on the map until you think you've found the area with the most EF4 and EF5 tornadoes. Is it where you would expect it to be? Explain.**

Next go to https://coast.noaa.gov/hurricanes/. This is a NOAA web page that maps the tracks of hurricanes in the United States. Above the search bar, select the "Name/Year" tab and answer the following questions.

6. **Type in "Katrina." Why are there multiple hurricanes called Katrina in different years?**

7. **Select "Katrina 2005" and click the search icon. In the lower right on the map, click the "View Legend" icon to open it. Which Saffir-Simpson category ranking was Hurricane Katrina when it made landfall?**

8. **What was Katrina's maximum category? Where was it then?**

9. **What happened to Katrina's strength as it moved north and inland? Why did this happen?**

10. **In the lower left of the page is a graph showing wind speed and pressure in the storm. What is the relationship between these two variables?**

Picture This. *Your Turn*

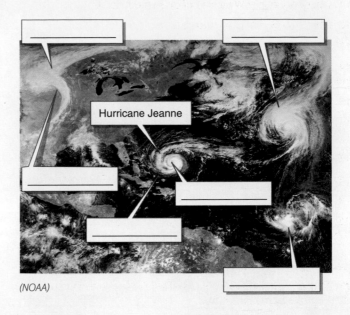

Hurricane Jeanne

(NOAA)

Name That Storm

In this satellite image from September 23, 2004, Hurricane Jeanne (labeled) and two well-organized midlatitude cyclones are visible. There is also one tropical disturbance visible in the image. Find and label these systems. Can you find Jeanne's eye and rain bands? Can you also find the cold sectors of the midlatitude cyclones? Fill in the boxes with the terms given here. Some terms will be used more than once.

Midlatitude cyclone
Rain bands
Cold sector
Tropical disturbance
Eye

For animations, interactive maps, videos, and more, visit www.saplinglearning.com.  Sapling Plus

The Changing Climate

Chapter Outline *and Learning Goals* ◎

7.1 The Climate System
◎ Differentiate between weather and climate and distinguish between climate forcings and climate feedbacks.

7.2 Climate Trends, Cycles, and Anomalies
◎ Explain the factors that cause climate to change and explain how scientists investigate ancient climates.

7.3 Carbon and Climate
◎ Compare the long-term and short-term carbon cycles and describe the role of human activity in changing those cycles.

7.4 Climate at the Crossroads
◎ Weigh the evidence of an anthropogenic greenhouse effect in the atmosphere and describe its consequences.

7.5 Geographic Perspectives: Fixing Climate: The 2-Degree Limit
◎ Assess the urgency of addressing climate change and different approaches to tackling the issue.

The methane bubbles in this photo are trapped in ice in Abraham Lake in the foothills of the Canadian Rockies in Alberta. Across the Northern Hemisphere, high-altitude lakes like Abraham are warming due to climate change. As a result, natural bacteria (called *methanogens*) that live in the sediments and produce methane are becoming more active. The methane the bacteria produce bubbles up through the water and freezes when it makes contact with colder water near the surface. The increased activity of methanogens is a major concern because methane is a very potent greenhouse gas that warms the atmosphere.

(Victor Liu, www.victorliuphotography.com)

To learn more about the structure of the atmosphere, see Section 1.4.

THE HUMAN SPHERE Nuisance Flooding

Some roads running through low-lying stretches of the East Coast are now almost routinely disappearing beneath 1 to 2 feet of seawater due to nuisance flooding **(Figure 7.1)**.

Nuisance flooding, also called "sunny-day flooding," results when extreme high tides inundate low-lying coastal areas. Nuisance flooding events are increasing as sea level rises around the globe. Since 1880 mean global sea level has risen 23 cm (9 in),

and when the tide is unusually high, nuisance flooding may result. By the end of this century, global sea level is expected to be 1 to 2 m (3.3 to 6.6 ft) higher than today. By that time coastal flooding will no longer be merely a nuisance; it will be a major threat to low-lying coastal development.

Web Map
Sea Level Rise
Available at
www.saplinglearning.com

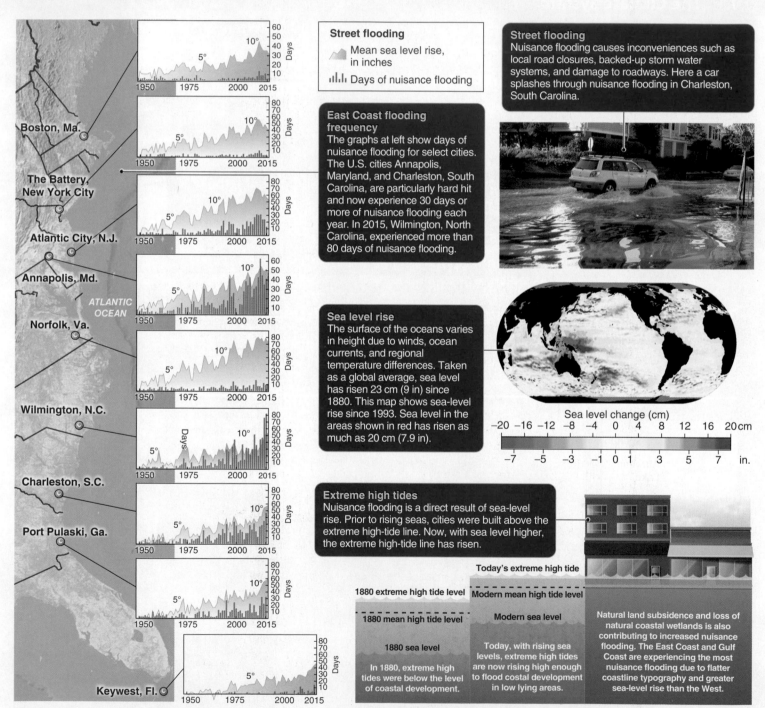

Street flooding
- Mean sea level rise, in inches
- Days of nuisance flooding

East Coast flooding frequency
The graphs at left show days of nuisance flooding for select cities. The U.S. cities Annapolis, Maryland, and Charleston, South Carolina, are particularly hard hit and now experience 30 days or more of nuisance flooding each year. In 2015, Wilmington, North Carolina, experienced more than 80 days of nuisance flooding.

Street flooding
Nuisance flooding causes inconveniences such as local road closures, backed-up storm water systems, and damage to roadways. Here a car splashes through nuisance flooding in Charleston, South Carolina.

Sea level rise
The surface of the oceans varies in height due to winds, ocean currents, and regional temperature differences. Taken as a global average, sea level has risen 23 cm (9 in) since 1880. This map shows sea-level rise since 1993. Sea level in the areas shown in red has risen as much as 20 cm (7.9 in).

Sea level change (cm)
−20 −16 −12 −8 −4 0 4 8 12 16 20 cm
−7 −5 −3 −1 0 1 3 5 7 in.

Extreme high tides
Nuisance flooding is a direct result of sea-level rise. Prior to rising seas, cities were built above the extreme high-tide line. Now, with sea level higher, the extreme high-tide line has risen.

Today's extreme high tide
1880 extreme high tide level
Modern mean high tide level
1880 mean high tide level
Modern sea level
1880 sea level
In 1880, extreme high tides were below the level of coastal development.

Today, with rising sea levels, extreme high tides are now rising high enough to flood coastal development in low lying areas.

Natural land subsidence and loss of natural coastal wetlands is also contributing to increased nuisance flooding. The East Coast and Gulf Coast are experiencing the most nuisance flooding due to flatter coastline typography and greater sea-level rise than the West.

Boston, Ma.
The Battery, New York City
Atlantic City, N.J.
Annapolis, Md.
ATLANTIC OCEAN
Norfolk, Va.
Wilmington, N.C.
Charleston, S.C.
Port Pulaski, Ga.
Keywest, Fl.

Figure 7.1 Nuisance flooding. *(Top right and right center. NOAA)*

7.1 The Climate System

◎ **Differentiate between weather and climate and distinguish between climate forcings and climate feedbacks.**

Earth does not have just one type of climate. It has dozens—equatorial rainforest, dry interior desert, frozen tundra, grassland—all of which are examples of different vegetation zones resulting from different climate types. Several systems are used to classify Earth's many types of climates. The system we use in *Living Physical Geography* is the Köppen climate classification system. The Köppen (pronounced KER-puhn) system is presented in Section 9.1, in the context of global vegetation patterns. The emphasis in this chapter is on the average state of Earth's climate as a whole and how it changes naturally and due to human activity.

What Is the Difference between Weather and Climate?

Weather and climate are not the same. One way to understand the difference is to think of the expression "Climate is what you expect, but weather is what you get." **Climate** is the long-term average of weather and the average frequency of extreme weather events. It is often calculated as an average of the previous 30 years of weather. For example, the average temperature or precipitation for a city can be calculated by averaging the past 30 years of temperature or precipitation data. **Weather** is the state of the atmosphere at any given moment and comprises ever-changing events on time scales ranging from minutes to weeks. Sunshine, rain showers, heat waves, thunderstorms, and clouds all are aspects of the weather.

Table 7.1 summarizes events that represent weather and climate. These events occur along a time continuum ranging from hours to tens of millions of years.

San Diego, CA
34° 42′ N, 117° 09′ W   Elevation: 26 m (85 ft)

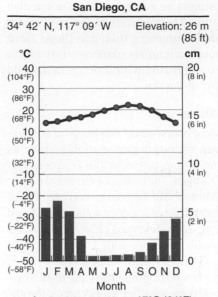

Average temperature: 17°C (64°F)
Average precipitation: 26 cm (10.3 in)

Tucson, AZ
32° 13′ N, 110° 55′ W   Elevation: 751 m (2,463 ft)

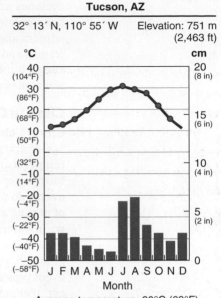

Average temperature: 20°C (69°F)
Average precipitation: 29 cm (11.6 in)

Figure 7.2 Climate diagrams for San Diego and Tucson. Although average annual temperatures and amounts of precipitation for these two cities are similar, the climate diagrams, which show average monthly temperature (red line) and precipitation (blue bars), reveal that their climates are quite different.

Weather and climate, while different, are related. Long-term weather observations such as temperature, precipitation, wind, and humidity are averaged to represent the climate of a given region. Simple annual averages of temperature and precipitation, however, do not fully describe the climate of a region. Take, for example, the average annual temperature and precipitation for San Diego, California, and Tucson, Arizona **(Figure 7.2)**. Judging by their annual averages, these two cities appear to have similar climates—but they do not. Remember that climate also includes the frequency of extreme events. For example, much of Tucson's rainfall comes from thunderstorms in July and August. In contrast, San Diego gets winter precipitation from midlatitude

Table 7.1	**Weather and Climate**	
PHENOMENA	**TEMPORAL SCALE**	**WEATHER OR CLIMATE?**
Cloudiness, rain shower, rainbow, sea breeze tornado	Minutes to hours	Weather
Night-and-day temperature difference	Days	Weather
Hurricane, midlatitude cyclone	Weeks	Weather
Winter, hurricane season, drought	Months	Climate
Asian monsoon	1 year	Climate
El Niño and La Niña	1 year or more	Climate
Younger Dryas*	1,000 to 10,000 years	Climate
Quaternary* glacial and interglacial cycles	10,000 to 1,000,000 years	Climate
Cenozoic* cooling	Millions of years	Climate

*These terms are defined in Section 7.2.

cyclones and almost no summer rain. Tucson also has a greater annual temperature range, with colder winters and hotter summers, than San Diego. Hard freezes and snow in winter are extremely rare in San Diego, but below-freezing winter temperatures do sometimes occur in Tucson.

Climate Change

Climate change occurs when the long-term average of any given meteorological variable—such as temperature or precipitation or the frequency of hurricanes—changes. Individual extreme weather events do not change the long-term average. Think of putting a single drop of water in a glass half-filled with water. One drop does not change the water level. But if enough drops are added, the water level gradually rises. Similarly, although a single weather event typically does not change the long-term average, if enough extreme weather events occur, the climate changes.

In recent years, weather stations, orbiting satellites, and ocean buoys have recorded innumerable extreme events, such as record-breaking heat waves and warm ocean water. Collectively, these extreme events are changing the long-term average. The official average global temperature record goes back to 1880. Since then the average temperature of the lower atmosphere has increased 0.99°C (1.78°F). Similarly, the surface of the oceans has warmed by about 0.56°C (1°F) in the past century. These gradual temperature trends in the atmosphere and oceans are climate change.

One question that frequently comes up is whether a single extreme event, such as Hurricane Irma, which caused widespread damage throughout most of Florida in 2017, was caused by climate change. Scientists know with certainty that the long-term average number of heat waves worldwide is increasing. The "extra" heat waves are a result of Earth's changing atmosphere. Yet separating the heat waves or storm events that would have occurred without global temperature change from those that were caused by increased atmospheric temperatures is scientifically challenging. Every now and then, however, natural events that occur are so far outside the normal climate system's behavior that they leave no doubt that Earth's climate is changing right before our eyes. The **Picture This** feature on the following page discusses one such event.

Climate Forcing and Feedbacks

Climate is a result of the movement of energy and matter between Earth's physical systems: the atmosphere, biosphere, lithosphere, hydrosphere, and cryosphere (introduced in Section 1.3). Recall that the cryosphere is the frozen portion of the hydrosphere, which includes glaciers and sea ice. Long-distance connections between Earth's different physical systems in different geographic areas are called *climate teleconnections*. El Niño's wide reach (see Section 5.4) is a good example of the role of such long-distance connections.

Earth's climate is controlled by two broad sets of factors. The factors that operate outside of and are independent of the climate system are **climate forcing factors**. The factors that arise within the climate system and are changed by the climate system are climate feedbacks (see Section 1.3).

The Sun is an example of a climate forcing factor. If the Sun it were to shine more intensely, it would force climate into a warmer state through *solar forcing*. Similarly, *volcanic forcing* may occur if a volcano erupts aerosols into the stratosphere. These aerosols may reflect sunlight and cool the planet's surface for a year or two. The Sun and volcanoes, like all other climate forcing factors, are not affected by changes in climate.

In contrast to climate forcing factors, climate feedbacks involve interacting parts of the climate system that affect one another and are strongly affected by climate. Recall from Section 1.3 that negative feedbacks maintain a system's stability and that positive feedbacks destabilize a system. There are many feedbacks in the climate system, some of which can support climate stability and others that can destabilize the climate system and cause climate change.

An example of a destabilizing positive feedback in the climate system is the **ice–albedo positive feedback**. Recall from Section 3.3 that Earth's surface albedo varies considerably from region to region; surfaces with a low albedo absorb more solar energy than surfaces with a high albedo. When the temperature of the atmosphere increases, more snow and ice are melted. Compared to snow and ice, bare ground and ice-free water have a lower albedo and absorb more solar energy. As more bare ground and ice-free water are exposed, more warming occurs, creating a positive feedback loop **(Figure 7.3)**.

Warming melts more snow and ice.

Exposed land and water absorb more **sunlight** and **convert** it to heat.

Figure 7.3 Positive feedback loop. *(NOAA)*

Picture This Greenland Burning

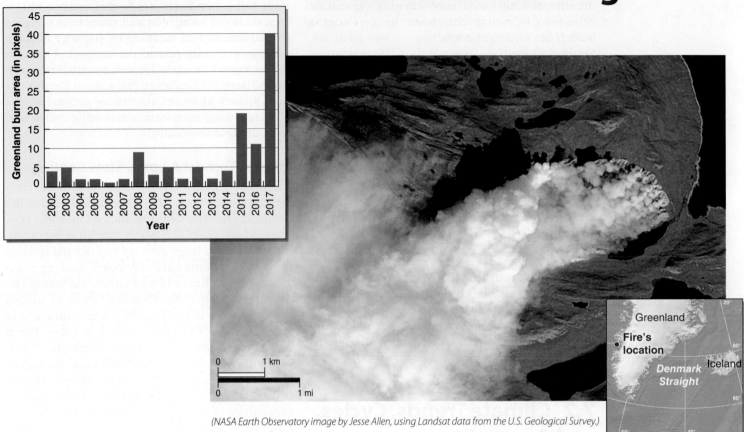

(NASA Earth Observatory image by Jesse Allen, using Landsat data from the U.S. Geological Survey.)

Scientists had never seen anything like it before. On July 31, 2017, satellites detected a large wildfire on the Greenland tundra that eventually burned 23 km² (9 mi²). This image, captured by Landsat 8, shows smoke from the fire on August 3, 2017. Over 80% of Greenland is covered by an ice sheet that is up to 3 km (2 mi) thick. Along the margins of the ice sheet, cold, wind-swept tundra vegetation is covered with snow for all but a few months of the year. Wildfires there are rare, and when they have burned in the past, they have always been small and short-lived. But the Arctic is warming at more than twice the rate of the middle and low latitudes, and an unusually warm and dry spell in 2017 in eastern Greenland created conditions that led to this surprising fire. Given that no thunderstorms were detected in the area, the Greenland fire was likely caused by a campfire.

NASA's Aqua satellite has tracked fires in Greenland since 2000. Its data (inset graph) show that fire activity there is on the rise. Because of climate change, high-latitude fire activity across the Northern Hemisphere is now greater than at any other time in the past 10,000 years.

Consider This

1. Why was the Greenland fire so significant, even though it was small compared to fires that occur at lower latitudes?
2. Given what you know about albedo (see Section 3.3), if fire activity increases in Greenland, what will happen to Greenland's albedo? How could this affect the ice sheet?

The ice–albedo positive feedback destabilizes the climate system and causes climate change by enhancing the warming trend that was already taking place. But the ice–albedo positive feedback can cause cooling as well. If, for whatever reason, there were a cooling trend in Earth's atmosphere, the ice–albedo positive feedback would enhance that cooling trend because more snow and ice would reflect more sunlight, resulting in further cooling. This happened 650 million years ago, creating a "Snowball Earth" state, with even the tropical oceans covered in ice.

In addition, feedbacks often trigger other feedbacks that accelerate the initial change. For example, warming temperatures increase the number of forest wildfires. Burning vegetation releases CO_2, which enhances the preexisting warming. This is the *wildfire–carbon feedback*. Similarly, extensive frozen soils called *permafrost soils* are typically common at high latitudes, such as in Siberia and Canada's Northern Territories. But in many places permafrost is now thawing and releasing CO_2 and methane (CH_4) because methane-producing bacteria become more active as the soils warm. Methane is a much more potent (25 times more potent) greenhouse gas than CO_2. The release of these greenhouse gases from thawing permafrost enhances the warming trend that is already under way. This is the *permafrost–carbon feedback*.

But positive climate feedbacks do not go on forever. They may be kept in check by other positive feedbacks pushing the system in the opposite direction. They can also be slowed by negative feedbacks that stabilize a changing system. We will return to the important role of climate forcing factors and feedbacks as we move through the remainder of this chapter.

7.2 Climate Trends, Cycles, and Anomalies

◎ Explain the factors that cause climate to change and explain how scientists investigate ancient climates.

Earth's climate continually changes. The climate that we are experiencing today only developed about 10,000 years ago. The time our current climate has been in existence has been a mere snapshot in Earth's 4.6-billion-year history. On long-term time scales of millions of years,

profound climate changes have stressed life on the planet and have even caused *mass extinction events,* in which over 75% of Earth's species went extinct. Smaller fluctuations of climate on short-term scales of hundreds and thousands of years are continually in motion. To see why climate changes and understand the context of what is happening to climate today, it is essential to examine the past. We can learn more about Earth's climate history by examining three natural modes of change: long-term trends, repeating cycles, and unpredictable anomalies.

Climate Trends: A Long, Slow Cooling

Earth's long history is divided into geologic time periods (see Section 12.2). One such period is the Cenozoic era. The **Cenozoic era** (or Cenozoic; pronounced see-no-ZO-ic) began 66 million years ago. It began when a large asteroid struck what is today the Yucatán Peninsula of Mexico, causing some 75% of life on Earth to go extinct—including the dinosaurs. Early in the Cenozoic, about 55 million years ago, the average global temperature was about 12°C (22°F) warmer than it is today. There was no ice at either of the poles. Atmospheric CO_2 concentrations were above 1,000 ppm, far higher than the present concentration of 408 ppm. (Recall from Section 2.1 that atmospheric carbon dioxide concentration is expressed in *parts per million*, or *ppm*.) After this warm period a long, slow cooling trend set in **(Figure 7.4)**.

The building and uplift of the Tibetan Plateau and the Himalayas are the leading explanation for the Cenozoic cooling trend. Gradual weathering and erosion of the uplifting mountain range caused CO_2 in the atmosphere to bond with other minerals that became dissolved in rivers. As the rivers flowed to the ocean the minerals that were dissolved in them were deposited and stored as sediments and rocks on the ocean floor. The Cenozoic cooling trend gradually progressed as this process removed CO_2

Figure 7.4 Cenozoic cooling trend. The atmospheric temperature has dropped steadily since the early Cenozoic. Temperatures are given as anomalies above or below today's average, defined as 0°C (32°C). The cooling trend started when India began colliding with Asia, forming the Himalayas and Tibetan Plateau. The Antarctic ice sheet formed by about 35 million years ago, and the Greenland ice sheet (and other ice sheets in the Northern Hemisphere that are now gone) formed by about 6 million years ago.

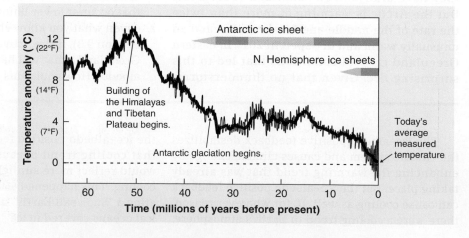

from the atmosphere. This process is still unfolding, and weathering of the Himalayas is still drawing CO_2 out of the atmosphere. In the short-term, however, CO_2 emissions from human activity are far surpassing the removal of atmospheric CO_2 through weathering.

Climate Cycles: A Climate Roller Coaster

By 2.6 million years ago, as the Cenozoic became progressively colder, ice sheets began growing in northern Europe and North America. Scientists refer to this period of ice growth over Europe and North America as the **Quaternary period** ice age (or Quaternary; pronounced kwa-TER-nery). We are still in the Quaternary today. During the Quaternary, the climate has experienced a series of up-and-down swings like a roller coaster, cycling back and forth between cold glacial periods and warm interglacial periods some 22 times.

Glacial and Interglacial Periods

A **glacial period** (or *stadial*) is an interval of cold climate within the Quaternary ice age. An **interglacial period** (or *interstadial*) is an interval of warm climate that occurs between glacial periods. Between 2.6 million and 1 million years ago, swings between warm and cool climates occurred on a 41,000-year cycle. Since about 1 million years ago, glacial periods have lasted roughly 90,000 years, followed by interglacial periods lasting about 10,000 years **(Figure 7.5)**.

Geologic time is separated into eras, which are subdivided into periods, which are in turn subdivided into epochs. We are currently in a warm interglacial called the **Holocene epoch** (pronounced HOL-o-seen). The Holocene, which began 11,700 years ago and continues today, is Earth's most recent interglacial. The Holocene is nested within the larger Quaternary period ice age, which in turn is nested within the larger Cenozoic era, which began 66 million years ago; we are currently in the Cenozoic Era, the Quaternary period, and the Holocene epoch **(Table 7.2)**. Within the Quaternary as a whole, the Holocene has been a period of unusually stable climate. The

Figure 7.5 Quaternary temperature cycles. Climate swings occurred on a 41,000-year cycle before about 1 million years ago and then transitioned to a 100,000-year cycle. Scientists are uncertain why the transition to a 100,000-year cycle occurred.

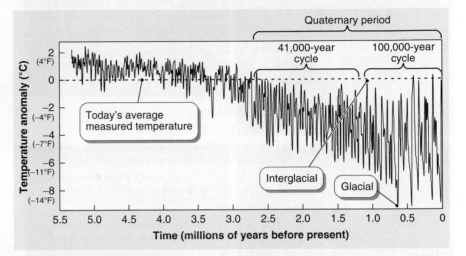

average length of warm periods over the past million years has been roughly 10,000 years, which means we are likely nearing the end of the Holocene interglacial. Following this same cyclical pattern, in the next few thousand years, Earth's climate should enter a glacial cooling trend that will last some 90,000 years.

Why has climate cycled back and forth between warm interglacial and cold glacial conditions about 22 times during the past 2.6 million years? The first person to contribute to our understanding of this phenomenon was the largely self-educated Scottish scientist James Croll (1821–1890). Interestingly, Croll worked as a janitor while he was developing his climate theory. He first conceived the idea that changes in Earth's orbit around the Sun could produce cold glacial periods. This idea was further developed and mathematically refined by the Serbian astrophysicist Milutin Milankovitch (1879–1958).

Milankovitch Cycles

Milankovitch identified three periodic changes in Earth's orbital relationship to the Sun that led

Table 7.2	**Current Geologic Time Periods**		
	WHEN IT BEGAN	**WHY IT BEGAN**	**CLIMATE CHARACTERISTICS**
Holocene epoch	11,700 years ago	Beginning of the current interglacial period	Warm and stable interglacial climate
Quaternary period	2.6 million years ago	Beginning of the growth of ice sheets in North America and Europe	Alternating between cold glacial and warm interglacial periods about 22 times
Cenozoic era	66 million years ago	Large asteroid impact in eastern Mexico, extinguished 75% of life	Gradual cooling trend beginning 55 million years ago

Figure 7.6 Milankovitch Cycles. Milankovitch cycles create glacial and interglacial periods.

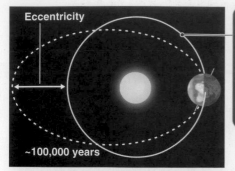

Eccentricity

~100,000 years

1. Orbital eccentricity
The shape of Earth's orbit around the Sun changes from circular to elliptical. The diameter of the ellipse changes by about 18 million km (11 million mi). It takes about 100,000 years for the orbital shape to go from circular, to elliptical, then back to circular. Note that the shape of the ellipse is greatly exaggerated here.

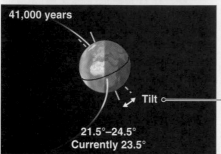

41,000 years

Tilt

21.5°–24.5°
Currently 23.5°

2. Tilt
The angle of Earth's axial tilt changes with respect to the plane of the ecliptic (see Section 2.1). The tilt shifts between 21.5 and 24.5 degrees from vertical then back to 21.5 degrees over a period of about 41,000 years. Greater axial tilt increases seasonality in the middle and high latitudes.

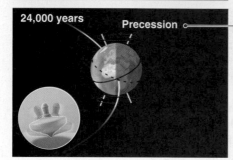

24,000 years   Precession

3. Precession
As Earth rotates on its axis, it wobbles like a spinning top. This wobble, called *precession*, operates on a cycle of roughly 24,000 years. Precession changes the timing of the seasons. For example, the Northern Hemisphere is currently pointed away from the Sun at perihelion, when Earth is closest to the Sun in January. At other times, the Northern Hemisphere points toward the Sun at perihelion.

Animation

Milankovitch Cycles

Available at
www.saplinglearning.com

to changes in the timing and distribution of solar heating across Earth's surface **(Figure 7.6)**. He asserted that these small changes in Earth–Sun orbital geometry, now called **Milankovitch cycles**, resulted in Quaternary glacial and interglacial cycles. Milankovitch developed a mathematical model that predicted climate cycles about every 100,000 years. His work went unrecognized for 50 years, until the mid-1970s, when new data from marine sediments led the scientific community to embrace his important theory.

Changing Earth–Sun orbital geometry is a process that operates outside, and is unaffected by, the climate system. Milankovitch cycles, therefore, constitute a climate forcing factor, referred to as *orbital forcing*.

Milankovitch cycles alter the intensity of sunlight over the seasons, but the Sun's energy output does not change significantly. The most important aspect of orbital forcing is that when it cools the Northern Hemisphere in summer (so that snow

and ice do not melt), snow accumulates and forms continental ice sheets. Ice sheets can only grow over land. Cooling the Southern Hemisphere summer cannot trigger ice sheet growth there because there is not enough land at mid and high latitudes where ice sheets develop. Internal feedbacks are important in amplifying the climate changes forced by Milankovitch cycles. As the ice sheets grow, the ice–albedo positive feedback causes further cooling. Glacial periods occur when the orbital distance from the Sun is greatest in the Northern Hemisphere summer and axial tilt is at a minimum.

Climate Anomalies: Unexpected Events

The pattern of climate change is like a layer cake. The bottom, foundation, layer is the long-term Cenozoic cooling trend. On top of that is a layer of the Milankovitch cycles, orbital forcing, and feedbacks. On top of that layer are unpredictable climate anomalies. The word *anomaly* refers to something that deviates from what is normal or expected. Anomalous climate events occur seemingly randomly through time and can be caused by both climate forcings and feedbacks. Changes in the Sun's output, volcanic eruptions, and changes in deep-ocean circulation are three examples of forcing and feedbacks that cause climate anomalies.

Changes in the Sun's Output

If **solar irradiance**—the Sun's total energy reaching Earth—were to increase, Earth would warm. If it decreased, Earth would cool. The amount of solar energy reaching Earth has been measured precisely since 1978 by satellites orbiting above the distorting effects of Earth's atmosphere. During this time, only small variations in the intensity of the Sun's output have been detected.

Sunspots are dark and relatively cool regions that migrate across the surface of the Sun. Approximately every 11 years, sunspot activity peaks. Dark sunspots are surrounded by regions of unusually high temperatures. As a result, increased sunspot numbers are correlated with a very slight overall increase in energy from the Sun reaching Earth. Ultraviolet radiation increases during high sunspot activity.

Although sunspots occur in cycles, not all of them have a discernible effect on Earth's climate. Instead of producing predictable temperature changes, these cycles produce unpredictable climate events. For example, low sunspot activity during a period 1645–1715, called the *Maunder Minimum,* coincided with a period of unusual cooling during the **Little Ice Age (LIA)**, a natural cooling period that extended from about 1350 to 1850 and was felt mostly in the Northern Hemisphere. High

Picture This The Little Ice Age and the Greenland Norse

The Scandinavian Norse (Vikings) settled coastal southern Greenland during the Medieval Warm Period in the 980s CE and developed a thriving society based on livestock rearing and trade. Perhaps as many as 5,000 Norse lived on Greenland at the peak of the settlement period. The Greenland colony was not self-sufficient. It was always too cold there to grow grain crops, and settlers relied on grain imports from northern Europe. The Greenland Norse traded polar bear hides, walrus ivory, butter, and wool for items they could not make or grow, such as grain, nails, wood, church bells, and wine.

Once the Little Ice Age set in, the growing summer sea ice made it increasingly difficult and dangerous to travel by boat from Greenland to trade with Europe. Worse, it became too cold on Greenland for settlers to grow food for themselves or for their livestock. The last written record of the Greenland Norse was a marriage document created in 1408. Sometime after that, their society collapsed, and they died or left Greenland. Those who could escape went to Iceland and mainland Europe, leaving behind gravesites and stone ruins such as the church of Hvalsey shown here.

(Danita Delimont/Getty Images)

Consider This

1. Compare the Greenland Norse to modern society in terms of their ability to anticipate upcoming changes in climate.
2. What are the similarities and differences between modern-day societies and climate change and the Greenland Norse and climate change?

---

sunspot activity may have triggered brief periods of warming, such as the **Medieval Warm Period (MWP)**, a naturally warm period that extended from about 950 to 1250 CE and was felt mostly in the Northern Hemisphere. The **Picture This** feature above discusses the LIA and MWP in the context of Norse settlers on Greenland.

Volcanic Eruptions

Large volcanic eruptions can cool Earth's surface temperature for 1 or 2 years. To cause cooling, ash and sulfur dioxide from the volcanic eruption must enter the stratosphere, where rainfall will not wash it out of the atmosphere. Only the largest volcanic eruptions are capable of spewing

material into the stratosphere. Sulfur dioxide from large volcanic eruptions combines with water to form reflective sulfuric acid droplets. These aerosols may remain suspended in the stratosphere for up to 5 years, where they reflect incoming sunlight, thereby cooling Earth's surface. In June 1991, Mount Pinatubo, in the Philippines, erupted, sending 10 km^3 (2.4 mi^3) of pulverized rock and ash into the upper atmosphere and causing climate cooling (**Figure 7.7** on the following page).

The eruption of Mount Tambora in Indonesia in 1815 was among the largest historical volcanic eruptions. It led to cold summers and failed crops in many regions of the world. It was so cold throughout northern Europe that

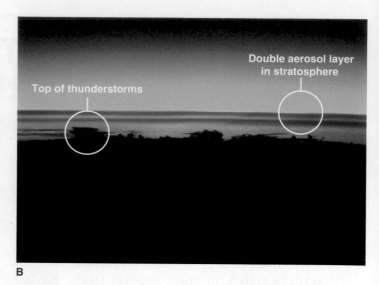

A **B**

Figure 7.7 1991 Mount Pinatubo eruption. (A) The eruption column of Mount Pinatubo reached 32 km (20 mi) into the atmosphere. The resulting veil of ash and sulfuric acid droplets spread around the world within weeks and remained suspended in the stratosphere for over 5 years, causing up to 1.3°C (2.3°F) of cooling in some regions. (B) This photograph of the atmosphere's limb was taken in 1991 by astronauts aboard the space shuttle. The photo shows a double layer of volcanic aerosols in the stratosphere, high above the tops of thunderstorms, which terminate at the tropopause. *(A. Arlan Naeg/AFP/Getty Images; B. NASA)*

1816 was called "the year without a summer." This event is explored further in Section 15.5. Volcanoes can also cause long-term trends of warming when they emit large quantities of CO_2 over time spans of hundreds of thousands to millions of years.

Changes in the Ocean Conveyor Belt
Before the 1980s, scientists thought that major shifts in climate happened smoothly and too slowly for humans to directly perceive. In the 1980s, however, new data taken from ice cores indicated that climate can change in a matter of decades or less. Like a precariously balanced bucket that, with a little nudging, tips over, the climate system reaches "tipping points." After these points, it may change quickly. This type of climate behavior is referred to as *nonlinear* because the initial changes in the system are slow at first and then accelerate as positive feedbacks take over and destabilize the system rapidly.

One example of nonlinear climate change is the **Younger Dryas**, a cold period that occurred between 12,900 and 11,600 years ago, just overlapping with the start of the Holocene, 11,700 years ago. (It was named after the cold-loving plant *Dryas octopetala,* also called mountain avens.) During this period, within a few decades or less, much of the Northern Hemisphere plunged into deep cold. Some places experienced colder temperatures than others. Ice cores retrieved from the summit of the Greenland ice sheet show that the average temperature there dropped about 15°C (27°F) and stayed low for about 1,300 years. **Figure 7.8** compares the Younger Dryas event with the relatively stable Holocene climate of the past 10,000 years.

What happened 12,900 years ago to cause temperatures to plunge so quickly in the Younger Dryas? There are several competing scientific

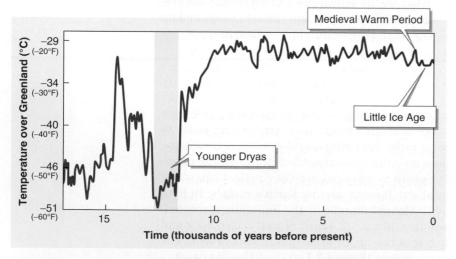

Figure 7.8 The Younger Dryas. The Younger Dryas cold period came and went abruptly. For comparison, the Medieval Warm Period and the Little Ice Age are also shown. Compared with the Younger Dryas, climate change in those events was insignificant.

Figure 7.9 The ocean conveyor belt system. The global system of ocean currents moves heat from the tropics to higher latitudes. *(Map by Robert Simmon, adapted from the IPCC 2001 and Rahmstorf 2002)*

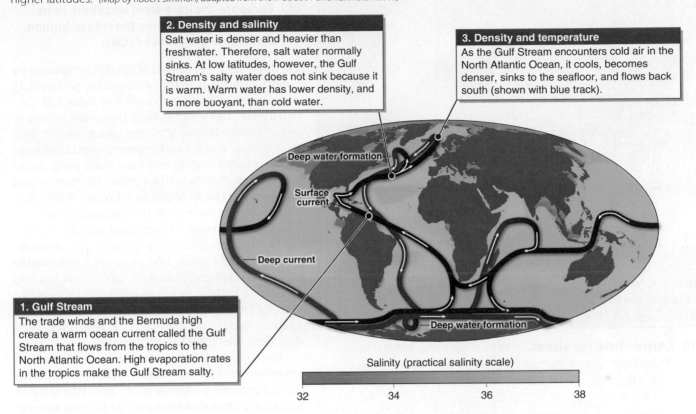

2. Density and salinity
Salt water is denser and heavier than freshwater. Therefore, salt water normally sinks. At low latitudes, however, the Gulf Stream's salty water does not sink because it is warm. Warm water has lower density, and is more buoyant, than cold water.

3. Density and temperature
As the Gulf Stream encounters cold air in the North Atlantic Ocean, it cools, becomes denser, sinks to the seafloor, and flows back south (shown with blue track).

Deep water formation

Surface current

Deep current

Deep water formation

1. Gulf Stream
The trade winds and the Bermuda high create a warm ocean current called the Gulf Stream that flows from the tropics to the North Atlantic Ocean. High evaporation rates in the tropics make the Gulf Stream salty.

Salinity (practical salinity scale)

32   34   36   38

ideas. One widely accepted explanation is that changes in the circulation of ocean currents caused it. Scientists think that an important part of Earth's climate system is the circulation of ocean water, which behaves something like a conveyor belt. At grocery stores, we often see conveyor belts that transport our groceries from near the cart to within easy reach of the cashier. The **ocean conveyor belt** (more technically referred to as the *Atlantic meridional overturning circulation* or *AMOC*) is the global system of surface and deep-ocean currents that transfers heat toward the poles. The ocean conveyor belt's flow depends on differences in the buoyancy of ocean water caused by differences in temperature and salinity. Colder, saltier water sinks to the depths and warmer, fresher water rises to the surface. **Figure 7.9** explains how this system works.

The Gulf Stream transports 25% of all global heat moving to higher latitudes. The faster the Gulf Stream flows, the more heat it delivers to the North Atlantic Ocean and to the atmosphere in the Northern Hemisphere. If the ocean conveyor belt system slows down or stops, the Northern Hemisphere gets colder. The ocean conveyor belt system greatly slowed or shut down entirely 12,900 years ago, plunging most of the Northern

Hemisphere into 1,300 years of frigid Younger Dryas climate. Once the Northern Hemisphere became cooler, the ice–albedo positive feedback maintained the Younger Dryas cooling that was initially triggered by changes in the ocean conveyer belt system.

Why would the ocean conveyer belt system slow or shut down? Evidence indicates that a massive influx of freshwater into either the North Atlantic Ocean or the Arctic Ocean forced the system to shut down. Why would freshwater shut down the system? Freshwater is more buoyant than salt water. If the Gulf Stream were freshened, it would become more buoyant and would no longer sink in the North Atlantic Ocean. This would slow down or stop the conveyor system altogether.

It would take a lot of freshwater to slow the ocean conveyor belt system. As **Figure 7.10** on the following page shows, that water probably came from the melting of the **Laurentide ice sheet**, the large ice sheet that covered much of North America during the most recent glacial period until about 12,000 years ago.

In April 2018, new research published in *Nature* found that during the last decade the conveyor system has been slowing down slightly.

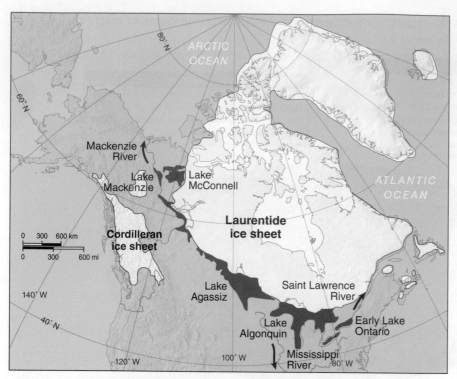

Figure 7.10 Laurentide ice sheet. At its maximum extent 20,000 years ago, the Laurentide ice sheet covered much of Canada. As it melted in response to Milankovitch forcing, the meltwater flowed out the Saint Lawrence River, the Mississippi River, and the Mackenzie River, adding freshwater to the oceans. At times during the melting process, ice dams created huge inland lakes (such as those shown here in dark blue). Scientists hypothesize that ice dams broke repeatedly over time and released massive pulses of freshwater into the North Atlantic Ocean. The freshwater slowed or stopped the ocean conveyor belt system and caused cooling events such as the Younger Dryas.

This has resulted in a persistent cold patch of water in the North Atlantic just south of Greenland (see Figure 17.18B). Scientists expect that the cooling produced by changes in the ocean conveyor belt system will be very slight compared to the warming caused by increased greenhouse gases in the atmosphere due to human activity.

Reading the Past: Paleoclimatology

How do scientists know what Earth's climate was like many thousands of years ago? Earth's climate history is recorded in various natural archives, including growth rings in trees, glaciers, and ocean sediments.

Scientists who study **paleoclimates,** which are Earth's ancient climates, are *paleoclimatologists*. They analyze natural Earth materials that record environmental changes in layers that have been left undisturbed. By analyzing these materials, they reconstruct ancient environments **(Figure 7.11)**.

7.3 Carbon and Climate

◎ Compare the long-term and short-term carbon cycles and describe the role of human activity in changing those cycles.

Earth's climate system is strongly influenced by greenhouse gases in the atmosphere, particularly water vapor and carbon dioxide (see Table 2.2). Carbon dioxide has a particularly important influence today because human activities emit more of it than ever before, and its anthropogenic emissions are growing faster than those of any other greenhouse gas. Carbon dioxide is like a global thermostat that controls the temperature in a house. When carbon dioxide is increased, the atmosphere warms; when carbon dioxide is decreased, the atmosphere cools. The Cenozoic cooling trend, for example, was caused by a gradual decrease of carbon dioxide in the atmosphere. Similarly, past warm periods in Earth's history were caused by an increase in atmospheric carbon dioxide. The remainder of this chapter focuses on the role of carbon and carbon dioxide in changing Earth's climate.

Carbon atoms move among Earth's physical systems through the **carbon cycle**. When bonds form between a carbon atom and two oxygen atoms, a carbon dioxide molecule (CO_2) is formed. When these bonds are broken, the carbon atom is freed from the oxygen atoms. In this section, we consider both carbon atoms and carbon dioxide molecules as they move through the carbon cycle.

To better understand the role of carbon in Earth's physical system, we divide the carbon cycle into a *long-term carbon cycle* and a *short-term carbon cycle*. The long-term carbon cycle involves the movement of carbon into and out of the lithosphere and takes millions of years to unfold. The short-term carbon cycle involves the movement of carbon among the oceans, the atmosphere, and the biosphere over spans of time from minutes to a few thousand years. (Section 7.4 includes a more detailed description of the short-term carbon cycle.)

The Long-Term Carbon Cycle

About 99.9% of Earth's carbon (65,500 billion metric tons) is stored in the lithosphere, where it is bonded with other elements to form different materials, including many types of rocks and fossil fuels (coal, oil, and natural gas). Coal is the preserved remains of ancient terrestrial (land) plants, and petroleum oil is formed from the remains of ancient marine (ocean) plankton. Natural gas is formed and found with both coal and oil. The other 0.1% of Earth's carbon is found in the oceans, atmosphere, and biosphere. Carbon moves from the atmosphere, oceans, and biosphere into the lithosphere through

Figure 7.11 SCIENTIFIC INQUIRY: How do paleoclimatologists reconstruct ancient climates? Natural archives of information about Earth's past can be found in a wide range of environments. In each of the settings shown here, layers record the changing environmental conditions. Layers can be formed through biological growth, as in tree rings and corals; through the settling of material by gravity, as in lake sediments, marine sediments, and snow on glaciers; or through nonliving growth, such as mineral deposits in caves. Paleoclimatologists extract cores of these materials and analyze the layers to gain an understanding of past environments and how they change through time.   *(From top left to bottom right. Bruce Gervais; Hickerson/FGBNMS/NOAA; Lonnie G. Thompson, Byrd Polar Research Center, Ohio State University; Auscape/UIG/ Getty Images; Rod Benson; William Crawford, IODP/TAMU; John Beck, IODP/TAMU)*

Tree rings

Corals

Glaciers

Cave Deposits

Temporal Resolution and Extent of Paleoclimate Records

MATERIAL	HOW FAR BACK IN YEARS?	TEMPORAL RESOLUTION
Tree rings	Thousands	Annual
Corals	Tens of thousands	Annual
Glaciers	Hundreds of thousands	Annual
Cave deposits	Hundreds of thousands	Decades
Lake sediments	Hundreds of thousands	Decades
Marine sediments	Millions	Centuries

Natural materials vary in how far back in time their records reach and in their *temporal resolution*: how focused in time the information is. Tree rings provide information for each year (annual resolution), while marine sediments provide information for increments of about 100 years or longer (centennial resolution).

Lake sediments

Marine sediments

weathering and erosion and through the burial and preservation of photosynthetic organisms on land and in the oceans. Carbon leaves the lithosphere and enters the atmosphere through volcanic eruptions and through the burning of fossil fuels.

Weathering and Erosion Remove Carbon from the Atmosphere

Carbon dioxide in the atmosphere combines with rainwater to form a weak acid called *carbonic acid* (H_2CO_3). Carbonic acid in rainwater slowly dissolves the rocks over which it flows, through the process of *chemical weathering*. This process releases calcium ions that rivers carry to the oceans. *Ions* are atoms or molecules with electrical charges that readily react with other particles. Calcium ions combine with bicarbonate ions (HCO_3^-) in seawater to create chalky white *calcium carbonate* ($CaCO_3$) sediments, which are similar to the deposits that accumulate on faucets in homes with hard, mineral-rich water. Organisms such as corals and clams also build their shells from calcium carbonate ions, pulling carbon from seawater in the process. When they die, their shells form layers of sediments as well. Over time, these sediments and remnants of shells, cemented together, form *carbonate rocks* such as limestone, locking away immense reserves of carbon in long-term storage reservoirs in the lithosphere **(Figure 7.12)**.

Photosynthesis Removes Carbon from the Atmosphere

The second way that carbon enters long-term storage reservoirs in the lithosphere is through photosynthetic organisms, which include plants, algae, and certain types of bacteria. These organisms convert the Sun's radiant energy to chemical energy through the process of photosynthesis.

During photosynthesis, they absorb carbon dioxide from the atmosphere, split the oxygen from it, and store the resulting carbon in their tissues.

Normally, after these photosynthetic organisms die, they soon decompose. The carbon in their tissues recombines with oxygen in the atmosphere to make carbon dioxide again. Under certain *anaerobic* (oxygen-free) conditions, however, these organisms and the carbon in their tissues is preserved rather than decomposed.

Several hundred million years ago, microscopic photosynthetic marine algae and bacteria (called *phytoplankton*) and terrestrial forests grew, died, and did not decompose. At that time, Earth was much warmer, and climate favored their preservation. Their carbon-rich remains accumulated and were preserved in marine sediments on the seafloor or in peat wetlands on land. Over millions of years, the preservation of these organisms gradually transferred carbon from the atmosphere and oceans into long-term storage in the lithosphere. This process of transferring carbon from the atmosphere to long-term storage in marine sediments is known as the *biological pump*. Today, the remains of these organisms are fossil fuels. As people burn them, the ancient carbon stored within them recombines with oxygen in the atmosphere and forms carbon dioxide.

The Short-Term Carbon Cycle

The short-term carbon cycle involves the movement of carbon among the oceans, the atmosphere, and the biosphere on time scales ranging from minutes to thousands of years. The oceans store the vast majority, 91%, of the carbon that moves through the short-term carbon cycle. The biosphere stores 7%, and the atmosphere stores only 2% of the carbon in the short-term carbon cycle.

The oceans readily absorb carbon dioxide from the atmosphere. About 50% of the carbon humans have put into the atmosphere by burning fossil fuels since the Industrial Revolution has been absorbed out of the atmosphere by the oceans.

Carbon does not take long to move between the biosphere, atmosphere, and oceans. For example, each breath we exhale moves carbon from the biosphere (our bodies) to the atmosphere. Whenever an organism dies and decomposes, bacteria recycle the carbon in its body back into the atmosphere and the soil, where that carbon becomes available to other organisms.

The Anthropogenic Carbon Cycle

During the 800,000 years before the Industrial Revolution, approximately the same amount of

Figure 7.12 Carbon transfer from the atmosphere to the lithosphere. This process of carbon removal from the atmosphere drove the long-term cooling trend of the Cenozoic era.

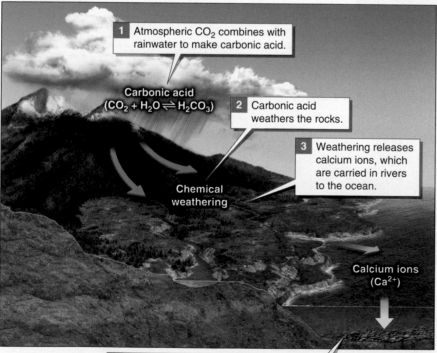

1 Atmospheric CO_2 combines with rainwater to make carbonic acid.

Carbonic acid
($CO_2 + H_2O \rightleftharpoons H_2CO_3$)

2 Carbonic acid weathers the rocks.

3 Weathering releases calcium ions, which are carried in rivers to the ocean.

Chemical weathering

Calcium ions
(Ca^{2+})

4 Calcium ions combine with bicarbonate ions in seawater to form calcium carbonate. Calcium carbonate is eventually converted to carbonate rocks, such as limestone, on the seafloor.

Figure 7.13 Natural transfer of carbon dioxide between the lithosphere and the atmosphere. Each year volcanic activity moves about 130 to 440 million metric tons of CO_2 into the atmosphere. Chemical weathering of rocks moves about the same amount of CO_2 from the atmosphere to the lithosphere.

Atmosphere

130–440 million metric tons CO_2 per year

130–440 million metric tons CO_2 per year

Ocean sediments

4 billion metric tons CO_2 per year

32 billion metric tons CO_2 per year

Deforestation

Fossil-fuel burning

A

Annual emissions from human activity

Annual emissions from volcanic activity

B

Figure 7.14 Anthropogenic transfers of carbon dioxide to the atmosphere. (A) Human activity now adds 35 to 40 billion metric tons of CO_2 to the atmosphere each year. Deforestation adds about 4 billion metric tons of CO_2 to the atmosphere each year. Fossil-fuel burning adds about 32 billion metric tons of CO_2 to the atmosphere each year. (B) Here natural volcanic CO_2 emissions (purple) are graphed next to anthropogenic CO_2 emissions (green). Natural volcanic emissions remain stable, while anthropogenic emissions have grown with each passing year.

carbon that entered the atmosphere through volcanoes left the atmosphere through weathering of rocks **(Figure 7.13)**. As a result, carbon dioxide levels in the atmosphere hovered around a steady state between 200 and 300 ppm.

During the *Industrial Revolution,* which began in roughly 1800, people first began burning large amounts of fossil fuels to run machines. Now human activity injects about 100 times more CO_2 into the atmosphere than all the world's volcanoes and other natural sources combined. In about 4 days, human activity emits as much CO_2 as an entire year's worth of natural emissions. By burning fossil fuels, people transfer carbon from long-term storage to the short-term carbon cycle. As **Figure 7.14** illustrates, fossil-fuel burning is the main human activity that adds CO_2 to the atmosphere, but deforestation and land-use changes also play a role.

7.4 Climate at the Crossroads

◎ **Weigh the evidence of an anthropogenic greenhouse effect in the atmosphere and describe its consequences.**

The transfer of carbon from long-term storage to the short-term carbon cycle has important implications for the climate system. Carbon dioxide in the atmosphere is a greenhouse gas and a climate forcing factor. It absorbs heat and increases the temperature of the atmosphere (see Section 3.3).

Human activity is increasing atmospheric CO_2 concentrations by over 2.5 ppm per year. Precise measurements of atmospheric CO_2 were begun by Charles Keeling in 1958 at Mauna Loa Observatory in Hawai'i. The observatory is far away from the effects of cities and pollution, and the measurements are taken upwind of any volcanic emissions.

Video

Following Carbon Dioxide Through the Atmosphere

Available at www.saplinglearning.com

Animation

Anthropogenic Carbon Emissions

Available at www.saplinglearning.com

Figure 7.15 Modern CO$_2$ concentrations. (A) The Keeling curve, shown here, graphs concentrations of CO$_2$ in the atmosphere since 1958. The red line is actual CO$_2$ measurements, which fluctuate with the seasons. In summer, values drop as plants grow and pull CO$_2$ from the atmosphere. In winter, values rise as plants lose their leaves, which decay and release stored carbon back into the atmosphere (inset graph). The black line is the annual average. (B) The rate of increase of atmospheric CO$_2$ concentrations is shown here. The black bars show the average annual rate of increase by decade. In the 1960s, CO$_2$ rose a little less than 1 ppm per year. By 2000–2010, the average annual rate of increase had doubled to 2 ppm per year. *(Data from NOAA)*

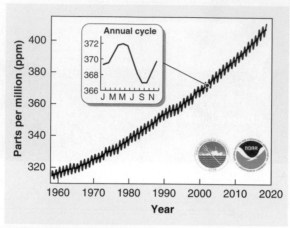

A. Atmospheric CO$_2$ at Mauna Loa Observatory

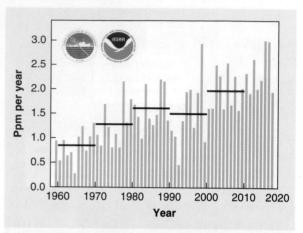

B. Annual mean growth rate of CO$_2$ at Mauna Loa

The **Keeling curve** is a graph showing the change in atmospheric CO$_2$ concentrations since 1958 **(Figure 7.15)**.

But what were atmospheric CO$_2$ concentrations before 1958, when Keeling and other scientists began measuring them? To find out, scientists have analyzed air bubbles in ancient ice from the Greenland and Antarctic ice sheets **(Figure 7.16)**. They have found that before 1800, CO$_2$ concentrations were much lower than they are today. Atmospheric CO$_2$ increased after societies began burning fossil fuels in large quantities.

Temperature and CO$_2$ always rise and fall together. Note that today's CO$_2$ level is far higher in comparison with any levels seen during the past 800,000 years, and the speed at which CO$_2$ is rising today is unprecedented **(Figure 7.17)**.

The causal relationship between atmospheric CO$_2$ and temperature is complex. Both influence each other. Rising temperature caused by orbital forcing increases atmospheric CO$_2$ because warmer oceans release stored CO$_2$. (Warm water holds less dissolved CO$_2$ than cold water.) Likewise, increased CO$_2$ in the atmosphere causes more warming through the greenhouse effect. Recent research of Antarctic ice cores suggests that, rather than one leading the other, CO$_2$ and temperature change together and affect one another synchronously.

The Warming Atmosphere

Today, human activities are inadvertently tinkering with Earth's greenhouse gas–controlled thermostat. Natural atmospheric CO$_2$ concentrations never exceeded 300 ppm during the last 3 million years. In the past 150 years, CO$_2$ concentrations have risen sharply to over 400 ppm due to human activities. Human emissions of CO$_2$ and other greenhouse gases into the atmosphere are creating an anthropogenic greenhouse effect (see Section 3.3).

Carbon dioxide gas is not the only greenhouse gas produced by people. Methane, nitrous oxide, and halogenated gases are far more efficient at causing warming than CO$_2$ and, like CO$_2$, all of them are currently far above their historic natural levels. Carbon dioxide is the most important anthropogenic greenhouse gas, however, because human activities emit more of it than any other greenhouse gas. **Table 7.3** provides the

Table 7.3	Greenhouse Gas Sources	
	ANTHROPOGENIC	**NATURAL**
Carbon dioxide (CO$_2$)	Burning fossil fuels (coal, oil, natural gas), deforestation, biomass fuel burning, fertilizers, cement production	Volcanic activity, vegetation decomposition
Methane (CH$_4$)	Livestock, fossil-fuel mining, anaerobic agricultural fields, burning vegetation	Bacterial processes in wetlands, mammals, and termite mounds
Nitrous oxide (N$_2$O)	Fertilizers, wastewater, fossil fuels, industrial activity, vehicles	Soil bacterial processes
Halogenated gases (CFCs, HFCs)	Industrially manufactured for various uses	None

anthropogenic and natural sources of important greenhouse gases.

As CO_2 and other greenhouse gas concentrations increase, so does the global temperature. Each passing decade is warmer than the last. In recent years, record temperatures have become routine. The years 2014, 2015, and 2016 each set records as the hottest year to date, and the year 2017 tied 2015 as the world's second hottest year. In July 2015, China set an all-time high of 50.8°C (122.5°F); in May 2016, India's highest temperature reached 51°C (123.8°F); in June 2017, Iran recorded 54°C (129°F); and in July 2017, Spain reached 47.3°C (117.1°F). As these extreme high temperatures events become more frequent, they are changing the long-term average. Although there is year-to-year fluctuation, this trend of planetary heating began in the early 1900s **(Figure 7.18A** on the following page**)**.

Earth's average surface air temperature in 2017 was 0.99°C (1.78°F) above the twentieth-century average, but the Arctic is warming at more than twice the rate of the rest of the world **(Figure 7.18B** on the following page**)**. The ice–albedo positive feedback (see Figure 7.3) is creating nonlinear rates of warming at high latitudes, a process known as **Arctic amplification**—the tendency of high-latitude regions to warm faster than the rest of the planet. Recent studies also indicate that dark *soot*, black dust from fossil-fuel combustion at lower latitudes, is driving a significant portion of the warming in the Arctic. As soot settles on ice, it darkens the white surface and lowers the albedo of the ice. As a result, the ice absorbs more solar radiation, which in turn causes more warming.

Given its accelerated rate of warming, the Arctic will continue to experience the greatest environmental shifts on the planet. Antarctica is also warming but not as quickly because it is relatively isolated by the *polar vortex* winds and the *Antarctic circumpolar current* that encircle the South Pole.

Global Temperature over the Past 800,000 Years

Temperature and CO_2 records from ice cores from Antarctica and Greenland extend back to 800,000 years and have given scientists a firm understanding of atmospheric chemistry and temperature during this time. Earth's average atmospheric temperature is higher now than at any other time in recorded history. But over the course of the past 800,000 years, there have been periods warmer than today. The most recent interglacial warm period was the *Eemian* (also called the *Sangamonian*), which occurred between 130,000 and 115,000 years ago. The average global

Figure 7.16 CO_2 concentrations since 1000 CE. (A) Scientists take ice cores from the Greenland and Antarctic ice sheets in segments. When the segments are placed end to end, the cores are up to 3 km (2 mi) long. Scientists then carefully analyze ancient gas bubbles preserved in the ice. (B) Ancient air from ice cores provides a basis for comparison with the chemistry of today's atmosphere. The light-green graph line shows data from ice cores; the dark-green graph line shows data from direct modern measurements. *(Left and center. © Reto Stoeckli; right. British Antarctic Survey/Science Source)*

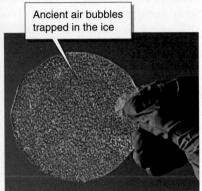

Greenland ice core

Ancient air bubbles trapped in the ice

A

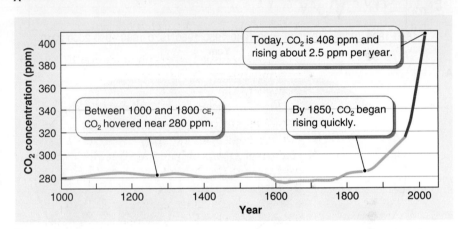

Today, CO_2 is 408 ppm and rising about 2.5 ppm per year.

Between 1000 and 1800 CE, CO_2 hovered near 280 ppm.

By 1850, CO_2 began rising quickly.

B

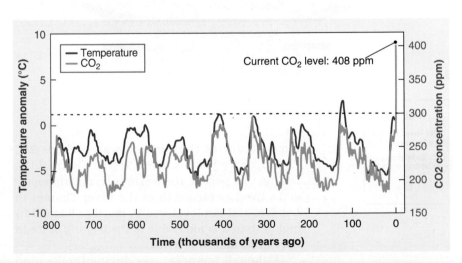

Figure 7.17 Long-term CO_2 and temperature fluctuations. This graph shows atmospheric CO_2 (green line) and temperature (red line), reconstructed from ice cores from Antarctica for the past 800,000 years. Natural CO_2 concentrations never surpassed 300 ppm (shown with the dotted line).

Figure 7.18 Modern temperature record and map. (A) This graph shows recorded global average yearly temperatures from 1880 to 2017. The temperature data are scaled as anomalies above or below the 1951–1980 average, which is defined as 0°C. The orange line shows the long-term average. (B) This world map shows average temperatures for 2017, given as anomalies above or below the 1951–1980 global average, which is defined as 0°C. With a few regional exceptions, everywhere on the map is orange or red, indicating that the world was warmer than average. Note, however, that the Arctic is warming fastest due to the ice–albedo positive feedback. Some Arctic areas were over 4°C (7°F) warmer than the average. *(GISTEMP Team, 2018: GISS Surface Temperature Analysis (GISTEMP). NASA Goddard Institute for Space Studies. Dataset accessed 20YY-MM-DD at https://data.giss.nasa.gov/gistemp/)*

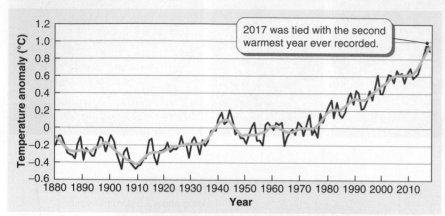

A

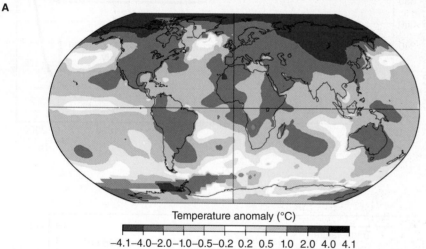

Temperature anomaly (°C)

−4.1 −4.0 −2.0 −1.0 −0.5 −0.2 0.2 0.5 1.0 2.0 4.0 4.1

B

temperature of the Eemian was perhaps 2°C (3.6°F) above today's average (see Figure 7.17). During that period, the warmth-loving hippopotamus lived as far north as the River Thames in London, and sea level was as much as 9 m (30 ft) higher than it is today due to the melting of ice in Greenland and Antarctica.

Although temperatures during the Eemian were warmer than today's, atmospheric CO_2 concentrations never rose above 300 ppm. Because CO_2 and temperature rise and fall together, it stands to reason that Earth's atmosphere could soon get warmer than it was in the Eemian because current atmospheric CO_2 concentrations are already far higher and are quickly rising. The **Picture This** feature on page 218 explores Earth's CO_2 and temperature history further.

Is Today's Warming Trend Natural?

Given the evidence, the current warming trend can be explained only by the current increase in atmospheric CO_2 concentrations caused by human activities. There is no known natural phenomenon that can account for this warming trend. The only data set that corresponds to and matches the warming trend since the early 1900s is carbon dioxide **(Figure 7.19)**.

A Strange New World

One question about climate change that often arises is this: "Climate change has happened before, so why should people be concerned now?" Any kind of climate change, whether natural or anthropogenic, can be destabilizing for human societies. There are 7.6 billion people living today, and the population may reach 9 billion by 2050. Earth is vulnerable to small changes in climate for a number of reasons. For example, scientists are concerned that climate change could disrupt agricultural output required to feed the world's population. Other changes could include major demographic, economic, political, and environmental shifts. Human societies have developed during 10,000 years of stable Holocene climate. Any change to the climate system, natural or anthropogenic, will challenge modern societies.

Positive Changes

A warming world could have limited positive aspects for some societies. Canada, for example, is already growing more grapes. It may increase its agricultural output of fruits and vegetables and could even begin growing citrus in the near term. England is at the northernmost limit of wine-grape growing, but that is quickly changing, and many growers there are also switching to grapes in anticipation of a viable and lucrative wine industry.

A new Arctic economy based on shipping, fishing, tourism, and petroleum and natural gas exploration is already opening up. According to the USGS (United States Geological Survey), the Arctic could provide some 30% of the world's natural gas in the coming years. Arctic shipping routes have been blocked by ice year-round until recently, but Arctic ice cover is rapidly diminishing, and these sea routes are now open for part of the year, and more and more cargo is moving through them every year **(Figure 7.20)**.

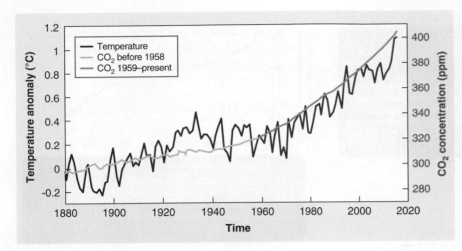

A

POTENTIAL CAUSE OF WARMING TREND	WHY IT DOES NOT EXPLAIN THE CURRENT WARMING TREND
El Niño	El Niño does not last decades.
Solar irradiance changes	The Sun's energy output is not increasing.
Volcanic eruptions	On short time scales, volcanic eruptions cause cooling.
Ocean conveyor belt	The speed of ocean conveyor belt circulation has not increased.
Milankovitch cycles	The warming is far too rapid, and current changes in orbital geometry should be cooling, not warming, the planet.
Mountain uplift and erosion	Uplift and erosion operate on time scales of millions of years.

B

Figure 7.19 CO$_2$ and temperature, 1880 to present. (A) Trends in CO$_2$ (green line) and temperature (red line) are shown here. Carbon dioxide directly measured (1959 to present) is shown with the dark green line. Earlier values of CO$_2$ (light green line) are derived from ice cores. The rise in CO$_2$ closely matches the rise in Earth's temperature. (B) This table lists other factors that cause climate change but do not account for the current warming trend.

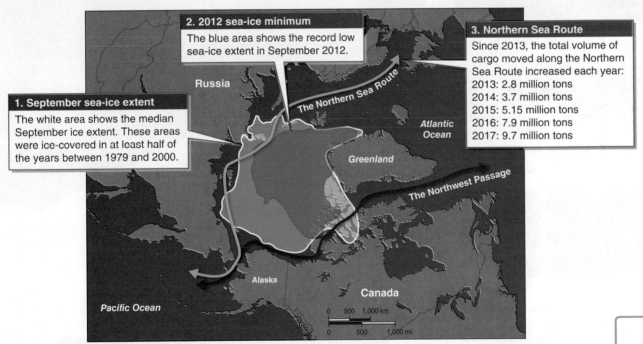

Figure 7.20 Arctic shipping routes. The Northwest Passage and the Northern Sea Route can offer a considerably shorter, faster, and less costly route for shipping traffic between the Atlantic and Pacific oceans. Before 2010, these routes were mostly covered with sea ice in summer. As the Arctic sea ice melts, however, these routes are opening up to shipping traffic. As the routes open up, political friction between northern countries bordering the Arctic Ocean (such as Russia, Norway, Canada, and the United States) is growing.

Video

Arctic Sea Ice Continues a Trend of Shrinking Maximum Extents

Available at www.saplinglearning.com

Negative Changes

The benefits of a warmer world are minor compared with the detrimental effects. With each passing year, evidence mounts that rapid shifts in Earth's physical systems are under way. These shifts raise serious concerns for human populations in the coming decades. **Figure 7.21** on the following pages presents some of the changes currently happening in Earth's physical systems. In the next 50 years, these changes are certain to continue creating profound challenges for people.

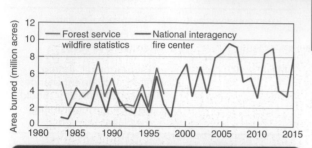

Percentage change

● < −80
● −60 to −80
● −40 to −60
● −20 to −40
○ 0 to −20
○ 0 to 20
● 20 to 40
● 40 to 60
● 60 to 80
● >80

Shrinking snowpack
As temperatures warm, snowpack melts earlier. This figure shows spring snowpack in the western United States between 1955 and 2017. Red circles indicate areas where average April snowpack has decreased. Blue circles indicate areas where snowpack has increased. Snowpack, which has declined in many regions by more than 80%, provides essential water resources for the arid western United States.

Shrinking glaciers
Mountain glaciers everywhere are melting in response to global warming. The graph above shows the decline of 40 reference glaciers located around the world since 1945. Negative values represent the decreased thickness of glacier ice, in meters, relative to the base year of 1945.

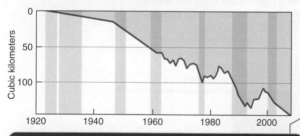

Increasing wildfires
Warmer temperatures and increased evaporation favor increased wildfire activity. This figure shows the area burned by wildfires in the United States between 1983 and 2015. The data in this graph above shows a trend of more area burning each year. Although these data do not show it, wildfires are also becoming more intense.

Groundwater loss
As temperatures rise, the frequency of drought has increased worldwide. Farmers rely on groundwater for their crops in times of drought when surface streams run dry. This graph shows the net trend in declining groundwater in California. During this 93-year period, over 143 cubic km (34 cubic mi) has been lost.

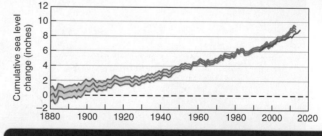

Rising sea level
Global sea level has risen about 23 cm (9 in) since 1880. The rate of sea-level rise is increasing; currently it is about 3.5 mm (0.14 in) per year. Sea-level rise is due to melting glaciers and thermal expansion of seawater as it warms. The shaded area shows the range of statistical uncertainty. Rising sea level threatens coastal populations.

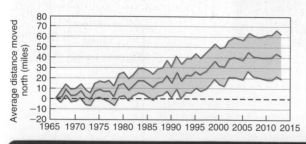

Shifting plant and animal ranges
As the atmosphere warms, plant and animal species are relocating to higher latitudes and elevations. The graph above shows the population movements of 305 bird species in North America from 1966 to 2013. The geographic center of these species shifted north by an average of about 40 mi (64 km). As species relocate, cities, highways, farms, and other human-built landscapes may impede their movement. Also, many species will be unable to relocate fast enough. For this reason, biologists anticipate that many species will become extinct as they attempt to move. The shaded area indicates the range of statistical uncertainty.

Story Map

The Changing Climate

Available at
www.saplinglearning.com

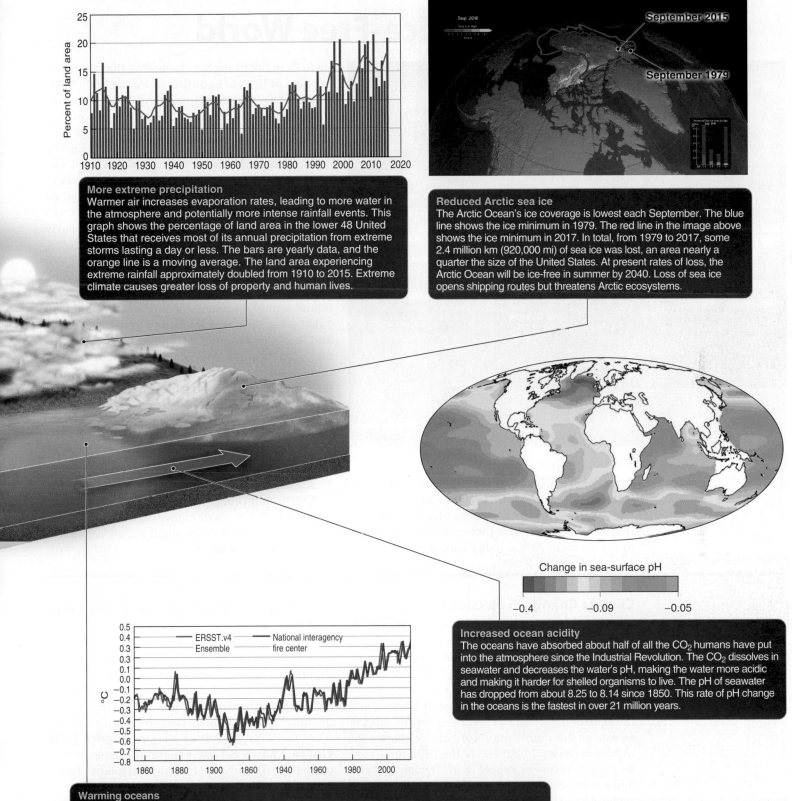

More extreme precipitation
Warmer air increases evaporation rates, leading to more water in the atmosphere and potentially more intense rainfall events. This graph shows the percentage of land area in the lower 48 United States that receives most of its annual precipitation from extreme storms lasting a day or less. The bars are yearly data, and the orange line is a moving average. The land area experiencing extreme rainfall approximately doubled from 1910 to 2015. Extreme climate causes greater loss of property and human lives.

Reduced Arctic sea ice
The Arctic Ocean's ice coverage is lowest each September. The blue line shows the ice minimum in 1979. The red line in the image above shows the ice minimum in 2017. In total, from 1979 to 2017, some 2.4 million km (920,000 mi) of sea ice was lost, an area nearly a quarter the size of the United States. At present rates of loss, the Arctic Ocean will be ice-free in summer by 2040. Loss of sea ice opens shipping routes but threatens Arctic ecosystems.

Increased ocean acidity
The oceans have absorbed about half of all the CO_2 humans have put into the atmosphere since the Industrial Revolution. The CO_2 dissolves in seawater and decreases the water's pH, making the water more acidic and making it harder for shelled organisms to live. The pH of seawater has dropped from about 8.25 to 8.14 since 1850. This rate of pH change in the oceans is the fastest in over 21 million years.

Warming oceans
As the atmosphere warms, so do the oceans. The global ocean has absorbed about 90% of the atmospheric heat energy caused by anthropogenic greenhouse gases. As a result, the oceans' surface temperature has risen 0.56°C (1°F) since 1854. With reduced sea ice at high latitudes, some high-latitude regions have even warmer water, some 5°C (9°F) above normal. The deep ocean is also warming. Warmer oceans threaten coral reefs and force marine species to shift poleward in search of cooler water. Warmer oceans also affect weather, such as by evaporating more water that provides energy to storm systems like hurricanes.

Figure 7.21 Signs of climate change. Shown here are just a few of the extensive physical changes that have been set in motion by global warming. *(NOAA)*

Picture This **An Ice-Free World**

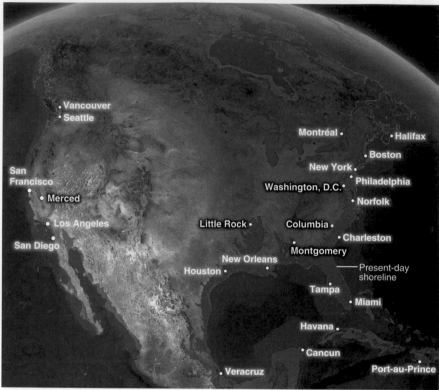

(National Geographic Creative)

up to 25 m (82 ft) higher. Assuming that the present rate of increase continues, by the end of this century, CO_2 will be approaching 1,000 ppm (see the Geographic Perspectives at the end of this chapter). The last time CO_2 was that high was about 55 million years ago, during the *Eocene epoch*. The world was ice-free, and sea level was on the order of 80 m (260 ft) higher than today.

This map illustrates what the current coastline of North America would look like if all the ice in Greenland and Antarctica were to melt. Nearly all the world's coastal cities would be submerged. Most of the Florida peninsula would be under 76 m (250 ft) of water. Cities labeled in blue would be under water. Other white-labeled cities, now far inland, would become coastal cities. It would take 1,000 years or longer for all the ice in Greenland and Antarctica to melt. Scientists know with certainty that sea-level rise of this magnitude will not happen anytime in the near future. Instead, this map illustrates how profoundly climate change can impact Earth's physical geography.

Consider This

1. Why did sea level rise so high during the Pliocene and Eocene? Where did all the water come from?

2. Scientists anticipate that at the end of the twenty-first century, sea level will have risen 1 to 2 m (3.3 to 6.6 ft) higher than today. What strategies and methods might society use to adapt to this change?

Atmospheric CO_2 today is roughly 400 ppm; the last time it reached this level was 3 million years ago—well before *Homo sapiens* had evolved. At that time, temperatures were 2°C to 3°C (3.6°F to 5.4°F) higher than today. Global sea level was

Climate Change Projections

The **Intergovernmental Panel on Climate Change (IPCC)** is the world's leading governing body on climate change. The IPCC, established in 1988, is a politically neutral body operating as part of the United Nations. The IPCC's mission is to provide objective, scientific, evidence-based statements about the state of the world's present and future climate, as well as the ramifications of climate change for human societies. In 2013–2014 the IPCC published its *Fifth Assessment Report* (with 16 chapters and more than 2,000 pages), summarizing the peer-reviewed published research of several thousand climate scientists from around the world.

Projecting what the state of the climate will be out to the year 2100 is one of the main goals of the IPCC. This work is accomplished by building general circulation models. A **general circulation model** is a mathematical simulation of the behavior of the atmosphere, oceans, and biosphere that can be used to create long-term climate projections.

General circulation models are developed largely in the same way as the models used in weather forecasting (see Section 6.5).

According to the IPCC's *Fifth Assessment Report,* by the end of this century, the carbon dioxide concentration will likely be somewhere between 550 and 900 ppm, and the global mean surface temperature is likely to have risen between 1.8°C and 4.2°C (3.2°F and 7.6°F) as a result **(Figure 7.22)**. **Figure 7.23** shows the geographic areas where temperature and precipitation are most likely to change. The range of uncertainty in the general circulation models results from unpredictable political and economic events. For example, the United States and a few other countries have been resistant to curbing their greenhouse gas emissions. Likewise, while a world leader in clean renewable energy development, China's emissions increased in the past decade due to its strong economic growth, the energy for which has come mostly from coal.

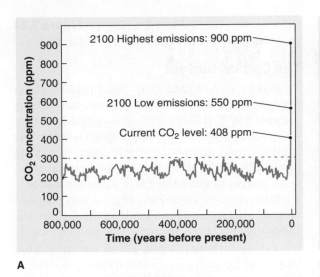

A

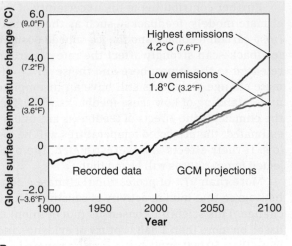

B

Animation

Carbon Dioxide Projections

Available at www.saplinglearning.com

Figure 7.22 Projections of CO₂ concentrations and temperature to 2100.

(A) Projections of atmospheric CO_2 concentrations to 2100 range from 550 ppm to 900 ppm. (B) The three estimates in this graph are based on emissions scenarios created by the IPCC: The "low emissions" scenario (called the RCP4.5 scenario by the IPCC) assumes that greenhouse gas emissions will grow slowly, peak by 2040, and then decline. In the "highest emissions" scenario (called the RCP8.5 scenario), annual anthropogenic greenhouse gas emissions double by 2050 and continue to rise throughout the twenty-first century, just as they did throughout the twentieth century. The orange line represents estimates with moderate emissions.

Web Map

Projected Temperature Change by 2050

Available at www.saplinglearning.com

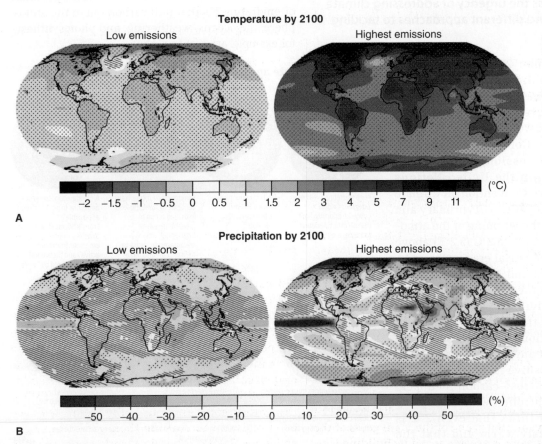

A

B

Figure 7.23 Temperature and precipitation change maps for 2100.

(A) The IPCC "low emissions" scenario (left) and "highest emissions" scenario (right) for temperature are mapped. Darker reds indicate higher temperatures. In the "highest emissions" model, temperatures everywhere are far above what they are today. The Arctic shows the greatest change. Stippling or striping indicates areas of high statistical confidence. (B) This map pair shows the IPCC "low emissions" scenario (left) and "highest emissions" scenario (right) for precipitation. Darker green and blue areas are projected to get wetter and tan areas drier. Much of Canada and the northeastern United States are projected to become wetter. The southwestern United States, the Amazon rainforest, and the Mediterranean are expected to become drier. Stippled and striped areas have the highest statistical confidence. *(Figure SPM.7 from Climate Change 2014: Synthesis Report. Contribution of Working Groups I, II and III to the Fifth Assessment Report of the Intergovernmental Panel on Climate Change [Core Writing Team, Pachauri, R.K. and Meyer, L. (eds.)]. IPCC, Geneva, Switzerland.)*

Further contributing to the uncertainty of the climate models, feedbacks—such as the permafrost–carbon feedback and the ice–albedo positive feedback—can strongly affect the rate of carbon released to the atmosphere and the severity climate change. Scientists still have an incomplete understanding of how these feedbacks will affect the climate. If the effects of feedbacks are underestimated, the projected temperatures will be too low. If their effects are overestimated, the projected temperatures will be too high.

More than 97% of professional climate scientists agree that human activity is causing climate change. This scientific consensus (not opinion) is based on more than a half century of evidence and more than 50,000 published scientific papers fact-checked by other experts. That people are causing climate change is not in dispute among scientists. Now the problem lies in how to best address the problem, the subject of the next section.

GEOGRAPHIC PERSPECTIVES
7.5 Fixing Climate: The 2-Degree Limit

◎ **Assess the urgency of addressing climate change and different approaches to tackling the issue.**

In December 2015, 197 countries agreed to the world's most ambitious globally coordinated plan to curb greenhouse emissions. The Paris Agreement (also called COP21, for Conference of the Parties, 21st session) is an agreement within the United Nations Framework Convention on Climate Change (UNFCCC). Its primary aim is to limit the warming of the atmosphere by 2100 to 2°C (3.2°F) above the pre-industrial era temperature. Only Syria and Nicaragua did not sign on to the agreement. (Syria is torn by civil war, and Nicaragua protested that the agreement was not ambitious enough.) Scientists think that anything above 2°C will trigger positive feedbacks, such as the permafrost–carbon feedback, and uncontrolled warming. Many scientists think that even 2°C is dangerously warm and that the aim should be to keep the warming under 1.5°C above pre-industrial levels. A secondary goal of the Paris Agreement, therefore, is to keep the warming limited to 1.5°C (2.7°F). **Figure 7.24** provides more details about the Paris Agreement.

The Carbon Budget

To make the 2-degree goal, the world must act very quickly. Because the atmosphere has already warmed 0.99°C (1.78°F) since pre-industrial times, we're already halfway to the 2-degree limit. The international scientific community, speaking through the IPCC, states that the total carbon emissions budget to meet the Paris Agreement goal is 1 trillion tons, or 1,000 PgC. (PgC = petagrams of carbon; 1 petagram [Pg] = 1 billion tons). Since the Industrial Revolution, humans have emitted roughly 550 PgC of carbon, causing 1°C of warming. That leaves 450 PgC left in the carbon emissions budget before we exceed the 1,000 PgC, 2-degree limit.

From 2001 to 2015, global carbon emissions rose from 7.8 PgC per year to 9.8 PgC per year. Currently, annual emissions of carbon are about 10 PgC. Assuming that this rate were to hold constant, in 45 years we would reach the limit of a cumulative total of 1,000 PgC emissions. Thereafter, in order to stay under 2°C warming, the amount of carbon entering the atmosphere would have to equal the amount naturally leaving the atmosphere through carbon sinks, a situation called *carbon neutral. Carbon sinks* are processes or environments that pull carbon out of the atmosphere; the oceans, weathering, and photosynthesis, for example, are carbon sinks.

The Paris climate agreement: key points

Temperatures 2100
· Keep warming "well below 2 degrees Celsius"
· Continue efforts to limit the rise in temperatures to 1.5 degrees Celsius"

Financing 2020-2025
· Rich countries must provide 100 billion dollars from 2020, as a "floor"
· Amount to be updated by 2025

Specialisation
· Developed countries must continue to "take the lead" in the reduction of greenhouse gases
· Developing nations are encouraged to "enhance their efforts" and move over time to cuts

Emissions goals 2050
· Aim for greenhouse gases emissions to peak "as soon as possible"
· From 2050: rapid reductions to achieve a balance between emissions from human activity and the amount that can be captured by "sinks"

Burden sharing
· Developed countries must provide financial resources to help developing countries
· Other countries are invited to provide support on a voluntary basis

Review mechanism 2025
· A review every five years. First mandatory world review: 2025
· Each review will show an improvement compared with the previous period

Climate-related losses
· Vulnerable countries have won recognition of the need for "averting, minimising and addressing" losses suffered due to climate change

© *AFP*

Figure 7.24 Paris Agreement key points. This graphic summarizes the major elements of the Paris Agreement. The historic agreement will go into effect in 2020. *(© AFP)*

Keep in mind that these numbers are not written in stone. For example, if the annual rate of emissions slowly decreases, then the time it takes to reach the 1,000 PgC limit will be longer; if the rate of emissions increases, then there is less time. Changes in the rate of global deforestation could also change the timetable. In addition, acceleration of positive feedbacks, such as the permafrost–carbon feedback and the wildfire–carbon feedback, could greatly reduce the time it takes to reach 2°C of warming.

Only the "low emissions" IPCC scenario will keep us under 2°C of warming (see Figure 7.22). Under this scenario, CO_2 emissions must slow and peak by 2040 and then decline by about 66% by the year 2100. All this must be done while the world population grows and economic development leads to greater demand for electricity (which is now mostly generated from coal) and cars (which now run mostly on petroleum) in developing countries with large populations, such as India and China.

In response to this challenge, the world is rapidly building a **green economy**—a sustainable economic system that has a small environmental impact and is based on carbon-free energy sources. For example, according to the U.S. Department of Energy, in 2016 solar energy producers employed twice as many workers as all the traditional fossil-fuel sectors combined. Electricity from the wind and from sunlight are two central pillars of the burgeoning green economy (see Sections 3.6 and 5.5). There will always be a use for fossil fuels (or *hydrocarbons*) in society, such as for machinery lubricants and manufacturing materials such as plastics. But electricity generated from fossil fuels, such as by burning coal, will have to be all but phased out by 2100 if we are to resolve our climate problem.

Unburned fossil fuels in the ground represent more than 2,800 PgC, an amount worth over $20 trillion. If the world is to avoid major climate change, these reserves of fossil fuels must remain in the ground. In other words, the fossil-fuel industries that own these reserves must forsake $20 trillion in revenue. At the same time, the up-and-coming green economy is becoming hugely profitable. This clash is at the heart of why the climate change issue has become so politically charged.

Curbing Carbon Emissions

How, specifically, do we limit carbon emissions? **Table 7.4** provides eight large-scale ways to reduce carbon emissions. Each of the rows in the table is an action that could reduce global carbon emissions by 1 PgC annually if it were fully implemented by 2050. Many other ways to reduce carbon are not shown on this table.

The large-scale carbon-reduction efforts like those in Table 7.4 simply won't happen unless they are legislated by government at the municipal, state, and national levels. Their implementation requires *top-down* regulations (as opposed to *bottom-up*, or *grassroots*, campaigns). Carbon taxes and cap-and-trade programs are other regulatory tools available to reduce carbon emissions.

Carbon Taxes

Carbon taxes are taxes levied on CO_2 emissions that exceed a predetermined level. They can be applied to a single factory, a city, or even a whole state or province. Much like water meters on houses reduce household water use, these taxes provide a strong incentive to reduce carbon emissions by switching to carbon-free energy. Carbon tax revenue can

Table 7.4 Carbon-Reduction Actions

ACTION	HOW TO DO IT
Cars: Double the fuel efficiency of all cars worldwide from 30 to 60 mpg.	Improve technology in vehicles. Phase out inefficient vehicles.
Cars: Halve the number of miles traveled by cars each year.	Increase telecommuting and improve city design and public transit.
Buildings: Reduce energy use in all buildings by 25%.	Improve insulation and building design. Improve efficiency of heating and cooling, lighting, and appliances.
Coal efficiency: Improve coal power plant efficiency from the current 40% to 60%.	Improve technology for burning coal and converting it to electricity.
Carbon capture and storage (CCS): Capture CO_2 emissions before they enter the atmosphere.	Retrofit (update) 800 large coal power plants with CCS technology.
Solar energy: Increase photovoltaic solar power capacity.	Increase today's photovoltaic capacity by 700 times. This would require 2 million hectares of land (see Section 3.6).
Wind energy: Increase wind power capacity.	Increase today's wind capacity by a factor of about 40. This would require 30 million hectares (74 million acres) of land (see Section 5.5).
Deforestation: Reverse deforestation in the tropics.	Provide economic alternatives to cutting forests for developing countries. Reduce demand for forest products in developed countries by supporting recycling and more efficient manufacturing.

be channeled into further developing carbon-free energy and other sectors of the green economy. Some 20 countries—including the United Kingdom, Sweden, India, Australia, and New Zealand—currently have carbon taxes. Some Canadian provinces, such as Quebec, have implemented carbon taxes. In the United States, carbon taxes are politically unpopular.

Cap-and-Trade

In a *cap-and-trade* system, a government allocates carbon "credits" to carbon producers, such as manufacturing companies. Fines result when emissions exceed the quota established for the credits. A polluting company may purchase more carbon credits from other cleaner companies that have extra carbon credits to sell. The limit on the overall number of carbon credits reduces the total carbon emissions of a country. All industrialized countries have some form of cap-and-trade system in place.

What Can You Do?

The challenge of coping with climate change can seem overwhelming to an individual. For many of us, almost all of our daily activities rely on energy from fossil fuels and emit CO_2 into the atmosphere. The average car emits about 5.5 metric tons of CO_2 each year. Heating, cooling, and electricity use in the average house in the United States produce about 11 metric tons of CO_2 each year. Only rare individuals are going to volunteer to go without driving, heating their homes, and using electricity, which have become necessities of modern life. In the green economy, all these activities will become *decarbonized,* meaning they will not result in CO_2 emissions.

Renewable and carbon-free energies, like wind and solar, will form the foundation of our energy needs.

Understanding your own personal carbon footprint provides useful insight. A **carbon footprint** is the amount of greenhouse gases (particularly CO_2) that an activity generates. Many of the steps required to reduce your personal carbon footprint call for lifestyle changes but not necessarily sacrifices **(Table 7.5)**. There are many online "carbon calculators" that allow you to estimate your CO_2 emissions and consider ways to reduce your carbon footprint.

Geoengineering

Because of the long atmospheric lifetime of CO_2, any increase in it is considered permanent as far as human societies are concerned. If the world were to magically stop emitting all carbon today, CO_2 concentrations would plateau immediately, but it would take much more than 10,000 years for natural carbon sinks to absorb the CO_2 out of the atmosphere and return atmospheric CO_2 to pre-industrial levels. If the world exceeds the 1,000 PgC limit, we may have to resort to geoengineering solutions. A growing number of scientists are convinced that geoengineering solutions are already unavoidable. Geoengineering (or *climate engineering*) is the deliberate, global-scale modification of Earth's environments to improve living conditions for people. There are two approaches to engineering the climate: *carbon dioxide removal* and *solar radiation management*. With the exception of planting real trees, each of these geoengineering schemes involves immense costs, serious environmental risks, and unpredictable effects on climate.

Table 7.5	**Carbon Footprint Reduction**
ACTIVITY	**REDUCING YOUR CARBON FOOTPRINT**
Reduce, reuse, recycle	Generating less waste and recycling reduce the amount of greenhouse gases emitted by resource extraction, manufacturing, transport, and disposal of materials.
Driving	Living closer to work, biking, walking, using public transportation, carpooling, and combining errands for fewer trips reduce the amount of fuel burned.
Home energy use	Insulating the attic and outside walls and installing LED lighting and efficient appliances reduce energy use.
Water use	Purifying and distributing water require energy. Using less water consumes less energy.
Renewable energy use	Using solar and wind energy decreases reliance on fossil fuels. Many local power companies allow users to select whether their power comes from conventional or renewable sources.
Diet	Raising livestock for meat is more carbon-intensive than growing plants for direct human consumption. Eating organic food reduces the use of fossil-fuel fertilizers. Eating locally produced food reduces the need to transport food long distances.
Flying	Flying is a carbon-intensive activity. Flying less reduces fuel use.
Political engagement	Climate-friendly legislation can be an effective way to manage greenhouse gas emissions. Vote for the climate!
Carbon offsets	Carbon offsets are investments in renewable, carbon-free energy sources such as wind energy and solar energy. These projects can reduce carbon emissions from conventional energy sources such as coal and oil by helping phase those out and replacing them with more projects on carbon-free energies.

Carbon Dioxide Removal

Techniques used to remove CO_2 from the atmosphere range from *forestation* (planting trees), to fertilizing the oceans with nutrients, to capturing CO_2 directly from the air by using "artificial trees" **(Figure 7.25)**. Some of the techniques, such as planting trees, have no risks. On the other hand, some techniques are inherently—and unpredictably—risky. For example, fertilizing the microscopic photosynthetic phytoplankton in the oceans with iron would stimulate their growth and pull CO_2 from the atmosphere and put it into the oceans—but no one knows what the effects would be on marine ecosystems. Similarly, it would take billions of "artificial trees" absorbing CO_2 to have a measurable effect. The cost of such a system would be astronomical, and it would take decades or longer to build such a system. Another method, carbon capture and storage (CCS), involves removing carbon emissions from coal-burning power plants before they enter the atmosphere. One major hurdle to these methods is that there is no known way to dispose of or use the immense amounts of captured and liquefied or powder-form CO_2 that would result from these methods. Scientists are working to economically convert captured waste CO_2 into useful products, like fuels, building materials, and plastics.

Solar Radiation Management

All solar radiation management techniques either propose to increase Earth's albedo or decrease the intensity of sunlight striking Earth from out in space. Earth's albedo could in theory be increased by injecting aerosols into the stratosphere with airplanes, mimicking the cooling effects of very large volcanic eruptions (see Section 7.2). Similarly, cloud seeding (see Section 4.7) could be used to make more reflective and cooling clouds. A more ambitious suggestion is to place a cloud of millions of small objects such as mirrors in an orbitally stable point between Earth and the Sun (called a *Lagrange point*). Such a solution would cool Earth by absorbing or reflecting sunlight for about 50 years. The costs and political details of such projects would be major hurdles to these schemes. Also, no one knows how the climate system would respond to such efforts.

The science is clear on one important point: The least expensive and least destructive path forward is to curb our greenhouse gas emissions. Engineering Earth's climate in response to increasing droughts, floods, disease, sea-level rise, and shifting agricultural zones would be far more costly and dangerous to society than curbing greenhouse gas emissions right now. Since the year 1976, every decade has been warmer than the previous one (see Figure 7.18). Atmospheric CO_2 concentrations and temperatures are still rising fast. But at the same time, green economies are quickly gaining momentum. It remains to be seen how this unintended planetary-scale experiment will play out.

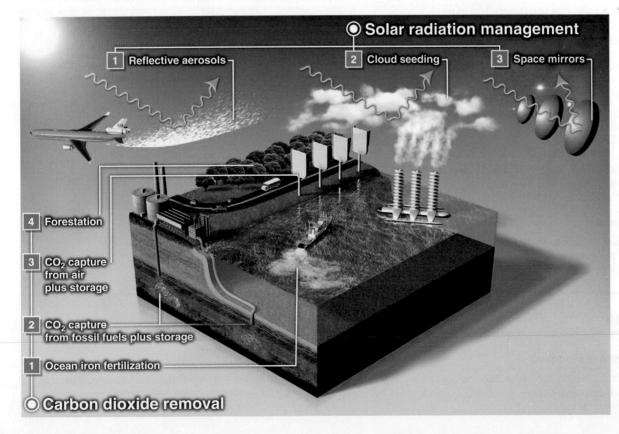

Figure 7.25 Geoengineering methods. The various geoengineering methods available today are illustrated here. All involve huge expense, and most involve considerable risk of uncontrolled changes to the climate system. *(Remik Ziemlinski/Climate Central)*

Chapter 7 **Study Guide**

Focus Points

The Human Sphere: Nuisance Flooding: Nuisance flooding events are increasing due to rising sea level.

7.1 The Climate System

- **Weather and climate:** Climate is the long-term average of the daily weather and the frequency of extreme events such as hurricanes, heat waves, or flooding rains.
- **The climate system:** The climate system is composed of the atmosphere, biosphere, lithosphere, hydrosphere, and cryosphere.
- **Temperature increase:** The average global atmospheric temperature has risen by 0.99°C (1.78°F) since 1880. Most of this warming is due to human activities.
- **Climate change:** Climate change occurs when the long-term trend of meteorological variables or weather extremes, such as the average temperature or the number of heat waves, changes.
- **Climate forcings and feedbacks:** Climate forcings, such as volcanic eruptions, are not influenced by the climate system; climate feedbacks, such as the permafrost–carbon feedback, change climate from within the climate system and are strongly influenced by climate.

7.2 Climate Trends, Cycles, and Anomalies

- **Modes of climate change:** Climate change occurs through long-term trends (Cenozoic cooling), repeating cycles (Milankovitch cycles), and unpredictable anomalies (volcanic eruptions).
- **The Holocene:** The Holocene (spanning the past 11,700 years) is Earth's current warm interglacial period. Climate has been unusually stable during the Holocene.
- **Rapid climate change:** Climate can rapidly switch between different states on a time scale of decades.
- **Reconstructing past climates:** Elements of Earth's past climates are recorded in various natural materials, including tree rings, glaciers, cave deposits, and ocean sediments.

7.3 Carbon and Climate

- **Carbon cycles:** Carbon is removed from the atmosphere, oceans, and biosphere and stored for many millions of years in the lithosphere in the long-term carbon cycle. Carbon moves relatively quickly among the atmosphere, biosphere, and oceans in the short-term carbon cycle.
- **Human activity and carbon cycling:** Each year, burning of fossil fuels and deforestation move up to 40 billion metric tons of carbon from the lithosphere and biosphere into the atmosphere.

7.4 Climate at the Crossroads

- **Carbon dioxide and temperature:** Atmospheric CO_2 concentrations and temperatures have risen and fallen together during the past 800,000 years.

- **Anthropogenic greenhouse effect:** Natural atmospheric CO_2 concentrations have not exceeded 300 ppm over the past 800,000 years. Because of human activities, atmospheric CO_2 concentrations are now 400 ppm and rising 2.5 ppm per year. As a result, atmospheric temperatures are rising.
- **Arctic amplification:** The ice–albedo positive feedback is warming the Arctic more than twice as quickly as the rest of the world.
- **Causes of the current warming trend:** No known natural climate forcing factor, such as volcanic activity or sunspot activity, can explain the current warming trend.
- **Effects of warming:** The negative aspects of warming far outweigh any positive aspects. Among the unwanted effects of climate change are sea-level rise, warmer and more acidic oceans, increased wildfires, and shifting agricultural zones.
- **Climate projections:** Projections based on computer modeling indicate that the mean global temperature will be between 1.8°C and 4.2°C (3.2°F and 7.6°F) warmer by 2100.

7.5 Geographic Perspectives: Fixing Climate: The 2-Degree Limit

- **2-degree goal:** Most climate scientists conclude that limiting atmospheric warming to less than 2°C (3.2°F) could prevent further dangerous climate change. To meet this goal, cumulative global carbon emissions must not exceed 1,000 PgC.
- **Cutting carbon:** There are many ways to cut carbon emissions, ranging from increasing wind and solar energies, to taxing carbon, to living closer to work.
- **Geoengineering:** Various geoengineering schemes are available to remove carbon from the atmosphere, such as building artificial "trees" and blocking sunlight with stratospheric aerosols. With the exception of simply planting more trees, each of them involves high expense and risk.

Key Terms

Arctic amplification, 213
carbon cycle, 208
carbon footprint, 222
Cenozoic era, 202
climate, 199
climate forcing factor, 200
general circulation model, 218
glacial period, 203
green economy, 221
Holocene epoch, 203
ice–albedo positive feedback, 200
interglacial period, 203
Intergovernmental Panel on Climate Change (IPCC), 218

Keeling curve, 212
Laurentide ice sheet, 207
Little Ice Age (LIA), 204
Medieval Warm Period (MWP), 205
Milankovitch cycles, 204
nuisance flooding, 198
ocean conveyor belt, 207
paleoclimate, 208
Quaternary period, 203
solar irradiance, 204
weather, 199
Younger Dryas, 206

Concept Review

The Human Sphere: Nuisance Flooding

1. What is nuisance flooding, and why is it occurring more commonly today than in the past? Where does it occur?

7.1 The Climate System

2. What is the difference between weather and climate?

3. How is *climate* defined?

4. Compare the definitions of *weather* and *climate*. Is one day of record heat an example of weather or climate? What about one warmer-than-average year? A decade of warmer-than-average years?

5. In addition to the atmosphere, what are the other parts of the climate system?

6. What happened to temperatures in the Northern Hemisphere during the Younger Dryas?

7. By how much has the average temperature of the lower atmosphere increased over the past 100 years?

8. Is the temperature increase over the past 100 years an example of weather or climate change? Explain.

9. If a severe storm strikes, can scientists definitely say it was caused by climate change? Explain.

10. What are climate forcing factors? How are they different from climate feedbacks? List examples of each.

11. Explain how ice cover at high latitudes can function as a positive feedback that destabilizes climate or as a negative feedback that stabilizes climate.

7.2 Climate Trends, Cycles, and Anomalies

12. What caused the Cenozoic cooling trend?

13. Describe the Quaternary ice age in terms of its timing and repeated climate cycles. Describe the climate of the Holocene Epoch.

14. What are Milankovitch cycles? With what kind of climate change pattern are they associated?

15. What are glacials and interglacials? About how long, on average, has each lasted during the past million years? What caused them: climate forcing or climate feedbacks?

16. Provide an example of a climate anomaly. Do climate forcings or climate feedbacks cause anomalies?

17. When did the Medieval Warm Period and Little Ice Age happen? What caused them? How did they affect human society?

18. What natural archives do paleoclimatologists examine to reconstruct ancient climates and environments?

7.3 Carbon and Climate

19. Where is most carbon on Earth stored? How does carbon enter and leave this long-term storage?

20. Compare and contrast the long-term carbon cycle with the short-term carbon cycle. Explain how carbon moves within each of these cycles.

21. What two main types of human activity are transferring carbon to the atmosphere? How many billions of metric tons of carbon are transferred to the atmosphere each year?

7.4 Climate at the Crossroads

22. What is the current rate of increase of CO_2 concentration in the atmosphere each year, in parts per million?

23. What is the Keeling curve? What does it show?

24. How do we know what prehistoric CO_2 concentrations were? Compare current CO_2 concentrations in the atmosphere with those over the past 800,000 years.

25. What concentration (in ppm) did natural atmospheric CO_2 not exceed over the past 800,000 years?

26. What is the current atmospheric CO_2 concentration (in ppm)? Where is this "extra" carbon coming from?

27. Describe the relationship between CO_2 and global atmospheric temperature.

28. Geographically, where is most of the current warming trend happening? Why is it happening there?

29. Are there any natural climate forcing factors that can explain the current warming trend?

30. What are some of the responses of Earth's physical systems to the current warming trend? What changes in Earth's physical systems are happening right now?

31. What is the IPCC? What range of atmospheric CO_2 and temperatures does the IPCC project for 2100?

32. What is a general circulation model? Why do general circulation models make a range of climate projections instead of just a single climate projection?

7.5 Geographic Perspectives: Fixing Climate: The 2-Degree Limit

33. Why does the IPCC state that the goal should be to limit warming to 2°C? What are scientists concerned about happening if the warming exceeds 2°C?

34. How many PgC can humans emit to the atmosphere in total and still remain under 2°C of warming?

35. How many PgC has human society thus far emitted in total? How much warming has that caused?

36. How, specifically, can society reduce its carbon emissions?

37. What are carbon taxes, and how do they work? What is cap-and-trade, and how does it work?

38. How, specifically, can individuals reduce their carbon emissions?

39. What is geoengineering? List and describe examples of geoengineering that may be used to change Earth's climate.

Critical-Thinking Questions

1. What differences and what similarities can you think of between the Greenland Norse and modern societies in the context of vulnerability to climate change?

2. Has reading this chapter altered your view on climate change? Explain.

3. Some scientists think we should begin ramping up geoengineering projects immediately. Do you agree or disagree with this view? Support your answer.

4. Why has the topic of climate change been politically controversial? What views might people with different backgrounds and interests take on this topic?

5. Read through Table 7.5 again. Do you find these individual approaches to addressing climate change agreeable or disagreeable? Specifically, do you think altering one's diet or plane travel is a reasonable response to the problem? Explain.

Test Yourself

Take this quiz to test your chapter knowledge.

1. True or false? A single year of record-breaking heat is a definite sign of climate change.

2. True or false? In just four days, human activity emits the same amount of CO_2 as natural emissions do over a whole year.

3. True or false? Most of Earth's carbon is stored in the lithosphere.

4. True or false? To safely stabilize climate, human carbon emissions must not exceed 2,000 PgC in total.

5. **True or false?** The ocean conveyor belt system caused the Cenozoic cooling trend.

6. **Multiple choice:** During the past 800,000 years, natural atmospheric CO_2 concentrations did not rise above

   a. 100 ppm.
   b. 200 ppm.
   c. 300 ppm.
   d. 400 ppm.

7. **Multiple choice:** Today's atmospheric CO_2 concentration is about

   a. 200 ppm.
   b. 300 ppm.
   c. 400 ppm.
   d. 500 ppm.

8. **Multiple choice:** Why are the oceans becoming more acidic?

   a. because they are warming
   b. because of ocean currents
   c. because organic activity is causing acidification
   d. because they are absorbing CO_2 from the atmosphere

9. **Fill in the blank:** The _____ is a graph of measurements that show increasing CO_2 concentrations in the atmosphere.

10. **Fill in the blank:** The study of ancient climate is called _____

Online Geographic Analysis

Tornado Analysis

In this exercise we analyze tornado activity in the United States.

Activity

Go to https://www.ncdc.noaa.gov/sotc/. This a National Oceanic and Atmospheric Administration (NOAA) web page. This "State of the Climate" page provides global climate data. Using the default settings on the page, click the "Submit" button.

1. **Scroll down and click and pull up the world map that shows extreme climate events worldwide. Compare extreme high temperature events to extreme low temperature events. Which do you see more of?**

2. **Where did the most extreme temperature event occur?**

Close the map and scroll back up to the selection fields. In the "Year" field select "2017." In the "Month" field select "annual." Click the "Submit" button.

3. **Scroll down to the table that lists the rankings of the warmest years. What 3 years were top ranked?**

4. **Only 1 year was not in the twenty-first century. Which year was it?**

Scroll back up to the selection field. In the "Report" field select "Global Snow and Ice." Keep the other fields the same. Click the "Submit" button.

5. **Scroll down to "Sea Ice Extent." Click the graph to enlarge it if you need to. What year experienced the smallest sea-ice extent?**

6. **On what date in 2016 did Arctic sea ice reach its smallest extent? Compared to the entire record, what was 2016 ranked in smallest minimum sea ice?**

Now go to https://data.giss.nasa.gov/gistemp/maps/. This NASA "GISS Surface Temperature Analysis" web page maps global temperatures. In the "Mean Period" field select "Annual (Dec-Nov)." In the "Time Interval" boxes select "2016" for both the "Begin" and "End" fields. Click the "Make Map" button.

7. **Scroll down to the surface winter temperature map. Where is the greatest warming taking place?**

8. **Why did that area experience the most warming?**

9. **Up to how much warming did that area experience, in both Celsius and Fahrenheit?**

10. **The data are given as "anomalies" in relationship to what?**

Picture This. *Your Turn*

Annual Carbon Dioxide Increase

The graph shows atmospheric concentrations of CO_2 measured at the Mauna Loa Observatory. The green line shows seasonal variation around the annual average (black line). Apply what you have learned in this chapter to answer the following questions.

1. Why does the global atmospheric CO_2 concentration fluctuate seasonally?

2. How has human activity changed the long-term carbon cycle to cause the rising trend shown in this graph?

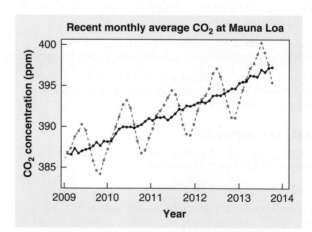

Recent monthly average CO_2 at Mauna Loa

For animations, interactive maps, videos, and more, visit www.saplinglearning.com.  **Sapling**Plus

Sunlight powers the biosphere. Almost all living organisms depend on solar energy. Plants and algae use sunlight directly to make food through photosynthesis. Other organisms either consume these photosynthesizers directly or consume other organisms. Part II explores the global geographic patterns of life on land and in the oceans, soil and water resources, and the human transformation of the biosphere.

CHAPTER 8

PATTERNS OF LIFE: Biogeography

Natural factors and human activities determine geographic patterns of life.

CHAPTER 9

CLIMATE AND LIFE: Biomes

Climate and human activity determine vegetation types across Earth's surface.

CHAPTER 10

OCEAN ECOSYSTEMS

Ocean ecosystems harbor a significant portion of Earth's biodiversity and are profoundly changed by human activity.

CHAPTER 11

SOIL AND WATER RESOURCES

Soil and water resources are essential for human well-being.

(Chapter 8: From: John Lund/Getty Images; Chapter 9: NASA/USGS; Chapter 10: Nature Picture Library/Alamy; Chapter 11: Aaron Joel Santos/ Getty Images)

Patterns of Life: Biogeography

Chapter Outline *and Learning Goals* ◎

8.1 Patterns of Biodiversity

◎ Identify and explain major geographic patterns of life on Earth.

8.2 Origins of Biodiversity

◎ Explain how evolution works and the geographic patterns that result from evolution.

8.3 Setting the Boundaries: Limiting Factors

◎ List and describe the factors that limit the geographic ranges of organisms.

8.4 Moving Around: Dispersal

◎ Explain why and how organisms disperse.

8.5 Starting Anew: Ecological Disturbance and Succession

◎ Describe the role of ecological disturbance and the return of life following disturbance.

8.6 Three Ways to Organize the Biosphere

◎ Describe three approaches to organizing the biosphere and how each works.

8.7 Geographic Perspectives: Journey of the Coconut

◎ Assess the relationship between people and the coconut palm and apply that knowledge to human relationships with other organisms.

The dragon blood tree (*Dracaena cinnabari*), shown here, grows only on the island of Socotra in the Arabian Sea. Socotra, which is part of Yemen, is a hot desert environment in which a lack of water and heat are the two most important *limiting factors* that organisms must cope with. The dragon blood tree's striking umbrella shape reduces heating from sun exposure and evaporation for the roots of the plant. New seedlings often establish only within the shade cover mature trees.

(John Lund/Getty Images)

To learn more about adaptations that organisms commonly employ in their environments, turn to Section 8.3.

THE HUMAN SPHERE   The New Biosphere

The biosphere today is not the same biosphere that existed just 200 years ago. The industrial era that began in the 1800s ushered in the age of mechanical engines and globetrotting people. As people traveled the world, they brought with them, intentionally and accidentally, innumerable species. **Species** are groups of individuals that naturally interact and can breed and produce fertile offspring. With the help of people, plants and animals, whose geographic ranges were formerly restricted by insurmountable barriers of oceans, mountains, and deserts, crossed these barriers with ease and colonized new areas. A **non-native** (also called *exotic* or *alien*) species is one that has been moved outside its original geographic range by people. Today, non-native species are on every continent. There are about 50,000 non-native species in the United States and Canada alone. This figure includes food crops and livestock.

Web Map

Non-native Zebra Mussels in North America

Available at
www.saplinglearning.com

The majority of non-native species cause no ecological harm, and many of them, such as crops, livestock, and honeybees, feed people and are economically essential. A small handful of non-native species, however, become destructive as they prey on or take resources from *native* species (those that originally inhabited the area). For example, the Nile perch (*Lates niloticus*) was intentionally brought into Lake Victoria in eastern Africa in the 1950s as a game fish to attract wealthy sport fishers from Europe and North America. Nile perch grow to nearly 2 m (6.5 ft) and can weigh 200 kg (440 lb). Because of their large size, they were also brought in as a food resource for local communities. Many native cichlid fish species in the lake have become extinct due to increasing fishing pressure, severe pollution in the lake, and predation from the Nile perch **(Figure 8.1)**.

Of all the U.S. states, Hawai'i has been most transformed by non-native species. Due to its extreme geographic isolation, the island chain has no native reptiles (such as snakes and lizards), no native amphibians (such as frogs and salamanders), no native parrots, no native ants, and only one native mammal—a bat. Today, however, Hawaiian landscapes are filled with non-natives, including escaped garden plants, wild pigs, piranhas, bass, trout, chickens, rats, mongooses, cats, snails, frogs, insects, cattle, deer, boa constrictors, and goats, among many others. These organisms could never naturally have crossed the vast distances of the Pacific Ocean to colonize the islands without the help of humans. Most of these non-native species are occupying habitat that has already been transformed by people and are not harming native species. Some of them, however, are threatening the islands' native species through predation and competition for resources.

Except in rare and limited circumstances, it is impossible to remove non-native species once they become established. Even if we wanted to, we cannot get rid of them. For better or worse, the global exchange of species has forever changed Earth's biosphere.

Figure 8.1 Nile perch. About 300 cichlid fish species native to Lake Victoria are near extinction or have become extinct in the past several decades. The three main factors to blame for their loss are increased fishing pressure, water pollution from local farming, and predation by Nile perch. *(Walter Astrada/AFP/Getty Images)*

8.1 Patterns of Biodiversity

◎ **Identify and explain major geographic patterns of life on Earth.**

Earth is covered by a continuous mantle of life. Everywhere, living creatures are rooting, swimming, drifting, flying, walking, crawling, climbing, and burrowing. Fern spores ride stratospheric winds. Bacteria reside in the depths of the oceans and kilometers deep within Earth's crust. Yet **biodiversity** (the number of living species in a specified region) is not haphazard across Earth's surface. The patterns of life on Earth reflect where a species first originated, the conditions that exist now, and the conditions that existed in the past.

Every species has a biogeographic story to tell. Why each species lives where it lives, what is currently preventing it from moving to other places, and what transpired in the past to bring it to its current geographic locations are all part of the story that biogeographers try to piece together to learn both about a given species' history and about Earth's history.

Biogeography is the study of the geography of life and how it changes through space and time. The roots of modern biogeography are often traced to the German explorer Alexander von Humboldt (1769–1859) and the English naturalist Alfred Russel Wallace (1823–1913). Here are some questions a biogeographer might ask:

- Why are there naturally no penguins in the Northern Hemisphere and no polar bears in the Southern Hemisphere?

- Why are the tropics rich with species, and why does biodiversity gradually decreases at higher latitudes?

- Why are some species geographically widespread and others restricted to isolated areas?

- How do terrestrial (land-based) species such as sunflowers reach remote places such as Hawai'i?

- How do people affect the geographic ranges of species?

- How are new species formed, and what causes species to go extinct?

- How do organisms respond to climate change?

The discipline of biogeography is closely tied to **ecology**, the study of the interactions between organisms and their environment. The **ecosystem** is a fundamental unit of ecology. Ecosystems include both living organisms within a community and the nonliving components of the environment in which the organisms live, such as energy, minerals, gases, and water.

Earth's Biodiversity

Scientists do not know exactly how many species there are on Earth. In all, scientists have named and identified 1.7 million different species. There are perhaps some 8 million readily seen organisms, such as plants, insects, and animals. But recent estimates conclude that when we include the microbes (microscopic organisms such as bacteria), the number of species on Earth increases to 100 billion or even 1 trillion.

Scientists group life-forms based on shared similarities among organisms and name those groups. For example, *botanists* (scientists who study plants) identify four major groups of plants. The mosses, or *bryophytes*, were the first land plants to evolve and they are the most primitive of the plant groups. All reproduce with spores and lack woody (or *vascular*) tissue. The flowering plants, or *angiosperms*, in contrast, are the most advanced of the plant groups—all angiosperms make flowers, seeds and fruit, and most attract pollinators with showy flowers. Angiosperms are also the most biodiverse of the plant groups **(Table 8.1)**.

Table 8.1	**The Four Plant Groups**		
SCIENTIFIC NAME	**EXAMPLES**	**GENERAL CHARACTERISTICS**	**ESTIMATED NUMBER OF SPECIES***
Bryophytes	Mosses and liverworts	No woody tissue; reproduce with spores	20,000
Pteridophytes	Ferns and horsetails	Possess woody tissue; reproduce with spores	13,000
Gymnosperms	Pine and spruce trees	Most produce cones; all produce seeds and reproduce with pollen	1,000
Angiosperms	Roses and lilies	Produce flowers that attract animal pollinators; produce seeds and fruit	352,000

*The species number estimates are approximations. Data are from www.theplantlist.org.

Animals are split into two major groups: the *vertebrates* (animals with backbones and a central spinal nerve column) and the *invertebrates* (animals without backbones). Most invertebrates are insects, but sea sponges, mollusks, and worms are also important in this group. Vertebrates are classed into five groups: fishes, amphibians, reptiles, birds, and mammals. **Table 8.2** lists the major characteristics of the vertebrate groups.

More than half (56.4%) of all known species are insects (**Figure 8.2**). The Coleoptera group (beetles) alone has nearly 400,000 species,

Table 8.2	The Five Vertebrate Groups			
GROUP	**EXAMPLES**	**GENERAL CHARACTERISTICS**	**NUMBER OF KNOWN SPECIES***	**BRANCH OF SCIENCE THAT STUDIES THE GROUP**
Fishes	Salmon and tuna	Gilled animals that live under water and lack limbs with digits	32,900	Ichthyology
Amphibians	Frogs and salamanders	Start out living in water and breathing with gills; adults breathe air through lungs and their permeable skin	7,302	Herpetology
Reptiles	Lizards and snakes	Mostly land dwelling; protected with scales; most lay soft-shelled eggs	10,038	Herpetology
Birds	Hummingbirds and parrots	Feathered; toothless beaked jaws; lay hard-shelled eggs; most can fly	10,425	Ornithology
Mammals	Whales and people	Mostly land dwelling; possess hair; feed their young with mammary glands	5,513	Mammalogy
			Total: 66,178	

* Species estimates do not include extinct species. Estimates are taken from the International Union of Conservation of Nature (IUCN) Red List of Threatened Species, 2014.

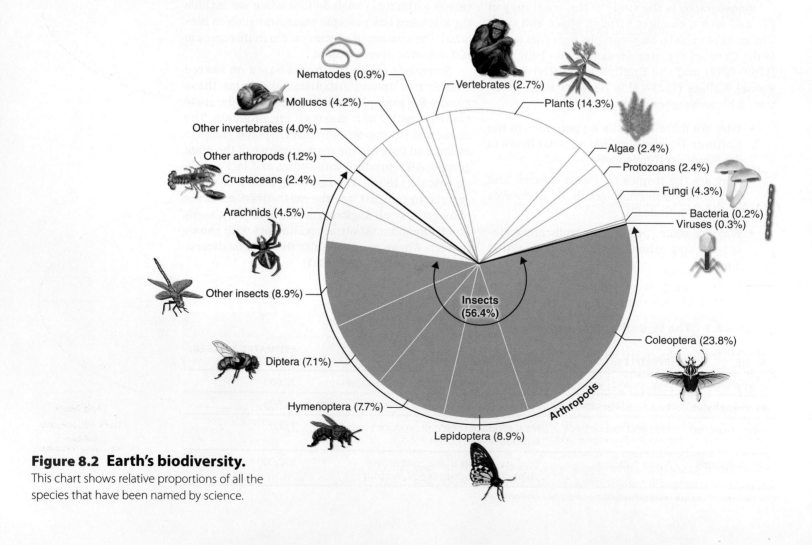

Figure 8.2 Earth's biodiversity.
This chart shows relative proportions of all the species that have been named by science.

almost 25% of all the known species of organisms! Microbes, such as protozoa, bacteria, and viruses, account for the greatest number of species, but they are very poorly known because they are very small and difficult to identify. Their true numbers are thus not represented in the pie chart in Figure 8.2.

Earth's biodiversity isn't permanent. Species can become **extinct**. Extinction is the permanent and global loss of the species. There have been five *mass extinction events*. During a mass extinction event, 75% or more of Earth's species become extinct. These extinction events have been caused by various factors, including asteroid impacts, severe global climate change, and volcanic activity. To the concern of many, currently humans appear to be causing what scientists are calling Earth's sixth mass extinction event. The International Union for Conservation of Nature (IUCN) reports that around 41% of all amphibians and 25% of all mammals are currently threatened with extinction primarily due to habitat fragmentation resulting from human activities.

Patterns of Biodiversity

As explained earlier in the chapter, biodiversity is the number of species in a given region, and species are groups of individuals that naturally interact and can breed and produce fertile offspring. Just as not all regions have the same species or the same number of species, not all regions are equally biodiverse.

Patterns across Latitude

Compared with higher latitudes, the tropics are bursting with species. Biodiversity is highest in the tropics and decreases toward the poles **(Figure 8.3)**. This *latitudinal biodiversity gradient* is Earth's most prominent small-scale biogeographic pattern.

Scientists do not know why most of the tropics and subtropics are more biodiverse than the mid- and high latitudes. Evidence from fossils indicates that this pattern has been in place for at least 300 million years. After many decades

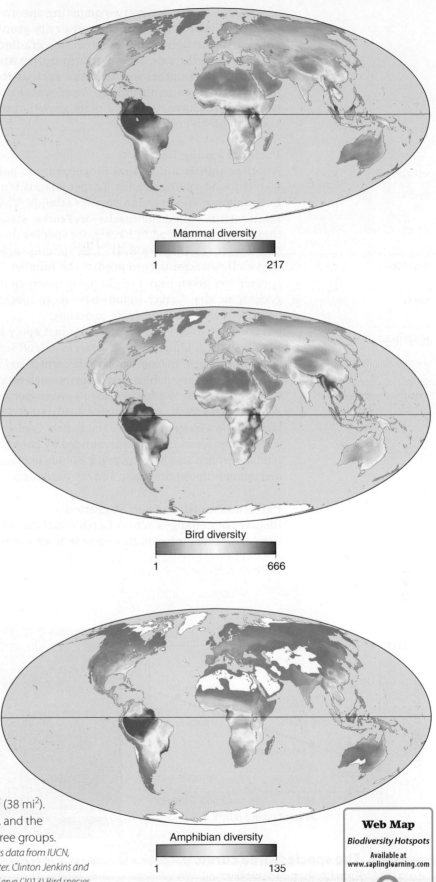

Mammal diversity

1   217

Bird diversity

1   666

Amphibian diversity

1   135

Figure 8.3 Biodiversity maps of mammals, birds, and amphibians. These three maps show the number of mammal, bird, and amphibian species per 100 km^2 (38 mi^2). The highest biodiversity is found in the tropics and subtropics, and the Amazon rainforest has the greatest number of species in all three groups.

(Top and bottom. Clinton Jenkins and BiodiversityMapping.org; based on species data from IUCN, available at http://www.iucnredlist.org/technical-documents/spatial-data; center. Clinton Jenkins and BiodiversityMapping.org; based on data from BirdLife International and NatureServe (2013) Bird species distribution maps of the world. BirdLife International, Cambridge, U.K. and NatureServe, Arlington, USA. Available at http://datazone.birdlife.org/species/requestdis)

Web Map
Biodiversity Hotspots
Available at
www.saplinglearning.com

of research, there are many competing theories to account for the latitudinal biodiversity gradient. The most significant factor in this gradient appears to be the rate of evapotranspiration and plant growth. In other words, areas such as the tropics with abundant sunlight, water, and warm temperatures have both high plant growth rates and high biodiversity.

Patterns among Islands

Another significant spatial biogeographic pattern is found among islands: Larger islands tend to have more species than smaller islands. This relationship, called the **species–area curve**, states that larger areas tend to have more species than smaller areas **(Figure 8.4)**. The species–area curve allows scientists to predict the number of species any given island might have, based on its geographic size. Larger islands have more species because they have more space to occupy.

The species–area curve does not just apply to islands in the traditional sense of pieces of land surrounded by water. In biogeography, the term *island* is used to refer to any habitat that is surrounded by an environment that is inhospitable to a given species. For instance, a freshwater pond that is surrounded by land can be a biogeographic island to fish. Cool and wet isolated mountaintops surrounded by hot lowland desert are also biogeographic islands to plants and animals intolerant of heat and dry conditions.

Patterns Resulting from Migration

Biodiversity changes across Earth's surface over time as species migrate in response to changing physical conditions. **Migration** is the seasonal movement of organisms from one place to another, usually for feeding or breeding. Many species migrate across latitudes (north and south) and up and down in elevation in response to changes in resource availability over time. Migration is a result of the "push" of dwindling resources in one region and the "pull" of increasing resources in another.

As organisms migrate, they create shifting cyclical patterns of biodiversity. For example, many species of whales and birds migrate to the Arctic to take advantage of the brief bounty of productivity in the Arctic spring and summer. Thus, Arctic biodiversity increases between June and August. Arctic biodiversity decreases after the organisms leave in fall. Likewise, biodiversity at high elevations in mountains decreases in winter as organisms migrate downslope to warmer elevations.

The physical demands of long-distance migration are considerable. Organisms expend great amounts of energy to travel long distances, and they must skillfully find their way using a variety of physical cues **(Figure 8.5A)**.

Scientists make great efforts to track the movements of animals to understand their ecological requirements. We are in the golden age of animal tracking science. Before the development of GPS technology, the migration patterns of many animals were poorly understood or completely unknown. Now, with ever-smaller and lighter GPS technology, accurately tracking large numbers of animals as they move around the globe is becoming possible **(Figure 8.5B** and **8.5C)**.

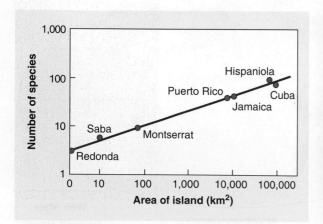

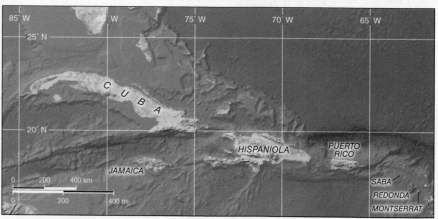

Figure 8.4 The species–area curve. This graph shows the number of reptile species for select islands in the Caribbean Sea. Cuba, as the largest island, has the greatest number of reptile species, as would be expected. Generally, larger Caribbean islands have more reptile species than smaller Caribbean islands. Note that the graph axes are ordered on a logarithmic scale to straighten the plotted line and make it easier to read.

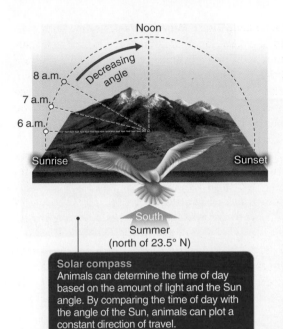

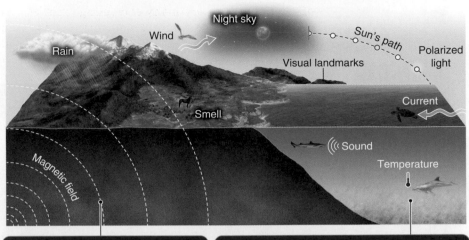

Solar compass
Animals can determine the time of day based on the amount of light and the Sun angle. By comparing the time of day with the angle of the Sun, animals can plot a constant direction of travel.

Magnetic compass
Many animal brains contain magnetite, a mineral that aids them in detecting Earth's magnetic field. Animals can determine latitude and direction by sensing the strength and the angle of the magnetic field.

Physical cues
Animals use a variety of physical cues to build mental maps they can use to navigate. Wind and ocean current direction, smell, climate, sound, temperature, the Moon and stars, and visual landmarks are physical cues animals use to choose their direction.

A

Plastic tags
GPS data loggers are too heavy for insects, so scientists use simple plastic labels to track them. Even this tiny plastic label is 2% of the weight of this monarch butterfly (*Danaus plexippus*). Scientists rely on people who find a tagged butterfly to return the tag by mail to the labeled address and indicate where and when the butterfly was captured.

Radio collars
Radio collars used on large mammals transmit the GPS coordinates of the moving animals to satellites continuously. In December 2011, a male offspring of this female gray wolf (*Canis lupus*), also radio-collared, was tracked as he entered California and became the first known wolf in that state since 1924.

PSATs
Pop-up satellite archival tags (PSATs) are used on large marine migratory animals such as sea turtles, seals, whales, and fish. After a set time, the tag detaches from the animal, floats to the surface, and transmits the data it has recorded to an orbiting satellite.

GPS rags
Birds and small fish are fitted with GPS archival tags that record data for a year or more. Tags record data such as changing light levels and day length. For birds, the tag is glued to the feathers and will fall off when the bird molts (sheds its feathers). Archival tags do not transmit information, so the animal must be recaptured and the tag must be removed in order for scientists to obtain data.

B

Figure 8.5 SCIENTIFIC INQUIRY: How do animals navigate, and how do scientists track their movement? (A) Animals use a variety of different methods to find their way as they migrate. (B) Scientists use different methods to track different animals. GPS is important in many, but not all, tracking methods. *(Top left. Will & Deni McIntyre/Getty Images; top right. Oregon Dept. of Fish & Wildife; bottom right. FLPA/Mark Newman/AGE Fotostock; bottom left. Eric Orbesen/NOAA)*

Video

Arctic Tern Migration

Available at
www.saplinglearning.com

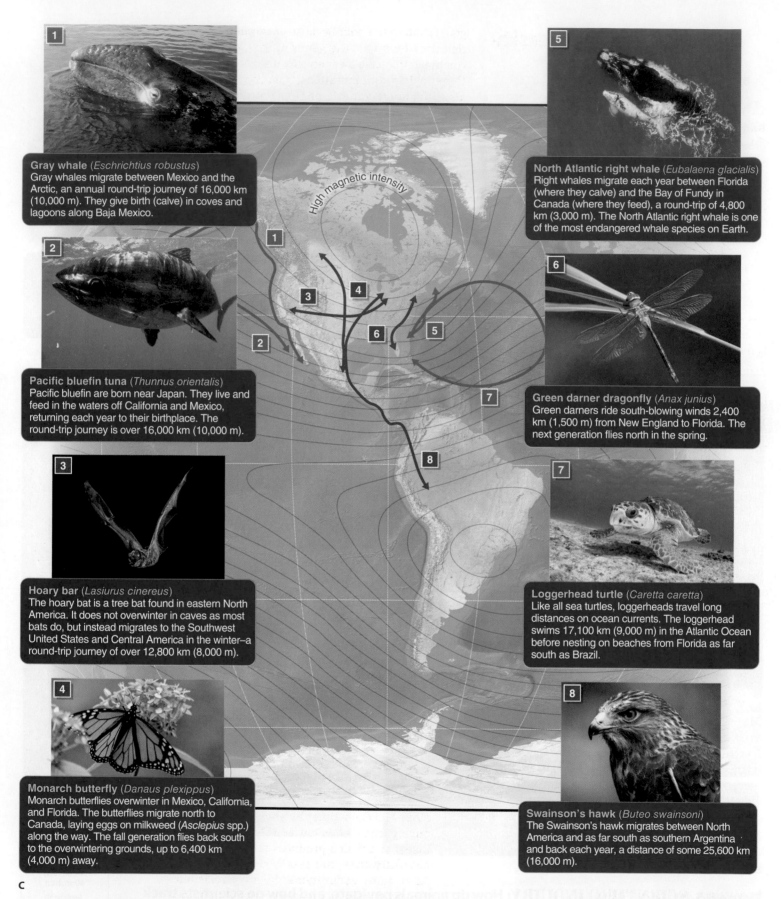

Gray whale (*Eschrichtius robustus*)
Gray whales migrate between Mexico and the Arctic, an annual round-trip journey of 16,000 km (10,000 m). They give birth (calve) in coves and lagoons along Baja Mexico.

Pacific bluefin tuna (*Thunnus orientalis*)
Pacific bluefin are born near Japan. They live and feed in the waters off California and Mexico, returning each year to their birthplace. The round-trip journey is over 16,000 km (10,000 m).

Hoary bar (*Lasiurus cinereus*)
The hoary bat is a tree bat found in eastern North America. It does not overwinter in caves as most bats do, but instead migrates to the Southwest United States and Central America in the winter—a round-trip journey of over 12,800 km (8,000 m).

Monarch butterfly (*Danaus plexippus*)
Monarch butterflies overwinter in Mexico, California, and Florida. The butterflies migrate north to Canada, laying eggs on milkweed (*Asclepius* spp.) along the way. The fall generation flies back south to the overwintering grounds, up to 6,400 km (4,000 m) away.

High magnetic intensity

North Atlantic right whale (*Eubalaena glacialis*)
Right whales migrate each year between Florida (where they calve) and the Bay of Fundy in Canada (where they feed), a round-trip of 4,800 km (3,000 m). The North Atlantic right whale is one of the most endangered whale species on Earth.

Green darner dragonfly (*Anax junius*)
Green darners ride south-blowing winds 2,400 km (1,500 m) from New England to Florida. The next generation flies north in the spring.

Loggerhead turtle (*Caretta caretta*)
Like all sea turtles, loggerheads travel long distances on ocean currents. The loggerhead swims 17,100 km (9,000 m) in the Atlantic Ocean before nesting on beaches from Florida as far south as Brazil.

Swainson's hawk (*Buteo swainsoni*)
The Swainson's hawk migrates between North America and as far south as southern Argentina and back each year, a distance of some 25,600 km (16,000 m).

C

Figure 8.5 (Continued). (C) The seasonal movements of eight migratory species are mapped here. *(1. VW Pics/Getty Images; 2. Michael Patrick O'Neill/ Science Source; 3. Merlin D. Tuttle/Science Source; 4. Gabriel Perez/Getty Images; 5. Peter Chadwick/Getty Images; 6. photography by Evy Lipowski/Getty Images; 7. Jim Abernethy/Getty Images; 8. garytog/Getty Images)*

8.2 Origins of Biodiversity

◎ **Explain how evolution works and the geographic patterns that result from evolution.**

All of Earth's biodiversity came about through the process of evolution. **Evolution** is genetically driven change in a population caused by selection pressures in the environment. A **population** is a group of organisms that interact and interbreed in the same geographic area. Through time, as genetic changes accumulate, speciation results. **Speciation** is the creation of new species through evolution. All species on Earth originated through the process of evolution and speciation.

Ideas about evolution were first developed by Charles Darwin (1809–1882), who observed biogeographic patterns among the finches of the Galápagos Islands, and by Alfred Russel Wallace, who observed birds in Malaysia and Indonesia. In 1859, Darwin and Wallace proposed a series of ideas that became known as the *theory of evolution*.

The genetic makeup of organisms is a result of physical and biological pressures imposed by their environment. Genetic traits (inherited traits) that help an organism reproduce tend to be preserved through the process of *natural selection*. Genetic traits that do *not* offer an advantage tend to be lost because organisms with them are less successful in surviving and reproducing than organisms without them.

Modern science knows much more about the cellular machinery that produces new species than Darwin and Wallace knew. Genetic traits are stored and inherited (passed on to the next generation) by organisms within their *genes*. Genes are arranged along strands of *DNA* (*deoxyribonucleic acid*), and these DNA strands occur in pairs called *chromosomes*. These three molecular elements—genes, DNA, and chromosomes—together form an organism's *genome*, and instruct cells how to replicate and function.

Every time a cell divides, its genetic instructions are copied into the new cell. If the DNA strands are not copied perfectly, genetic *mutations* result. Mutations occur due to internal copying errors or external factors that may harm DNA strands. Ultraviolet radiation and toxic chemicals, for example, can damage DNA and create mutations in the copying process. Over time small genetic mutations add up, and species can gradually change over time, in a process called *genetic drift*.

Evolution and Speciation

The genetic composition, or the genome, of all species changes through time. Most species experience continual genetic change from one generation to the next because of *sexual reproduction* (in which half the genes of each parent are combined to create a genetically unique offspring). Genetic drift and mutations from copying errors also introduce genetic changes to populations across generations. These genetic changes drive evolution in populations. If the genetic structure of organisms within populations never changed, there could be no evolution, and no new species would ever arise.

The role of genetic change in the process of evolution occurs three basic steps:

- First, in nature, more offspring are produced than the environment can support. Because of resource limitations, the environment cannot support an unlimited number of individuals within a population.

  Example: A single eucalyptus tree (*Eucalyptus* spp.) produces several thousand seeds per year. If all the seeds from all the trees in a eucalyptus forest grew into mature trees and reproduced, the number of eucalyptus trees would quickly overwhelm available resources. Therefore, only an extremely small fraction of all the seeds successfully germinates and grows into a mature tree.

- Second, the large number of offspring (in this case, seeds) produced through sexual reproduction results in a great amount of genetic variety in the population. Because of sexual reproduction, no two individuals are exactly alike. Some may be more fit for survival than others.

  Example: Eucalyptus leaves are pungent and toxic to repel insect attack. If a chance combination of genes or a genetic mutation in one eucalyptus among millions within a forest results in a slightly more toxic leaf that better wards off attacking insects, that individual may have a small advantage over others in the population, thanks to this adaptation.

- Third, through *natural selection*, individuals with beneficial genetic traits are more likely to reproduce and pass on those traits to the next generation. In this way, the beneficial traits become more common in the population.

  Example: Over time the slightly more toxic eucalyptus tree is better able to withstand insect attack and produces seeds more successfully than the less toxic individuals. After many generations, the more resistant tree type is slightly more successful than the others. The population (or forest) of eucalyptus will become increasingly composed of more toxic individuals.

Through these three steps, over time a new species of eucalyptus can evolve. At the same time, the insect populations that eat tree leaves experience exactly the same process. Through time, they too evolve and develop defenses that allow them to tolerate the increasing potency of toxins in the eucalyptus.

Patterns of Biodiversity from Evolution

As we discussed in Section 8.1, geographic patterns of Earth's biodiversity result from physical factors, such as latitude, island size, and migration. Another important factor that shapes the geography of life is the process of evolution. Convergent evolution can make unrelated organisms look similar and divergent evolution can make related organisms look very different.

Patterns of Convergence

Through the process of evolution, some unrelated organisms come to look strikingly alike. **Convergent evolution** is a process by which two or more unrelated organisms that experience similar environmental conditions evolve similar adaptations. No matter where they are located on Earth, all subtropical deserts, for example, impose similar selection pressures: intense sunlight, sparse vegetation cover, and persistent and severe moisture deficits. Because these environments select for the same traits, unrelated organisms in geographically isolated but similar desert environments may begin to look alike **(Figure 8.6)**.

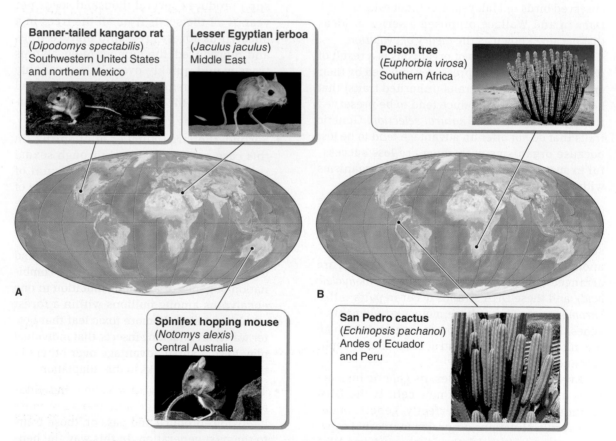

Figure 8.6 Evolutionary convergence. (A) Hopping, tufted tails, and nocturnal activity evolved in all three of these unrelated, geographically isolated mammals in response to their subtropical desert environments. Hopping is more efficient in sand than running; a flagged tail confuses desert predators, such as snakes; and nocturnal activity conserves water and helps the animals evade bird predators. (B) New World cacti and Old World euphorbs are genetically unrelated plants. However, they look alike because they have both evolved for life in the desert. Spines, photosynthetic stems rather than leaves, and a fluted body are all adaptations to deal with severe water deficits (see also Figure 8.12). *(A: Top left. Bob and Clara Calhoun/Photoshot; top right. Matthijs Kuijpers/Alamy; bottom. A.N.T. LIBRARY/Science Source; B: Top. age fotostock/Alamy; bottom. © Forest & Kim Starr, www.starrenvironmental.com)*

Patterns of Divergence

Another fascinating biogeographic pattern commonly seen on remote oceanic islands such as the Galápagos Islands of Ecuador or the Hawaiian Islands is dispersal and divergent evolution. **Dispersal** is the movement of an organism away from where it originated. Hawai'i, for example, is exceedingly remote from the mainland United States and was never connected to it. Native land organisms first colonized Hawai'i by dispersing thousands of miles across the Pacific Ocean long before humans ever arrived there. Organisms that managed to reach Hawai'i from the mainland became cut off from the mainland population from which they came. In a biological context, such isolation, called *reproductive isolation*, means that the two geographically separated populations are no longer interacting and that *gene flow* between them has stopped.

When populations become reproductively isolated, individuals within one population begin to diverge genetically from those in the other population through a process called **divergent evolution**. Those species that did disperse to Hawai'i long ago diverged genetically from their mainland ancestors for three reasons. First, they no longer interbred with and shared genes with those ancestors; second, there were unique selection pressures on each of the islands that were not present on the mainland; and third, random genetic copying errors resulted in genetic drift through time. **Figure 8.7** uses Hawaiian honeycreepers to illustrate the concept of divergent evolution.

8.3 Setting the Boundaries: Limiting Factors

◎ List and describe the factors that limit the geographic ranges of organisms.

No single species lives everywhere—and for good reason: No single species can adapt to the wide range of environmental conditions found across Earth's surface. Each species has an ecological **niche**, which consists of the living and nonliving resources and environmental conditions that the

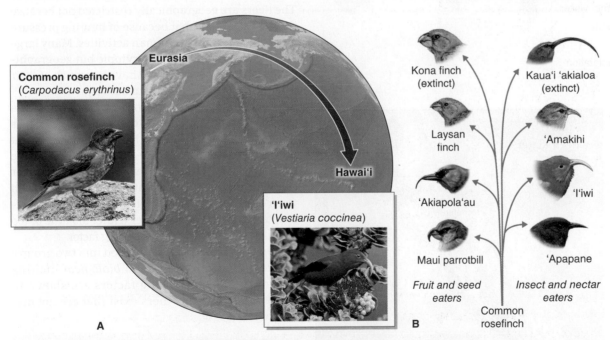

Figure 8.7 Divergent evolution in Hawai'i. (A) Genetic evidence indicates that about 5 million to 7 million years ago, a group of common rosefinches crossed 6,000 km (3,700 mi) of Pacific Ocean and colonized Hawai'i. (B) About 50 species of Hawaiian honeycreepers (of which 8 are shown here) evolved from a single ancestral species of common rosefinch. The beak shapes and sizes evolved and diverged as the birds began to occupy different environments. Those on the left eat fruit and seeds. Those on the right eat insects and flower nectar.

(A. Niko Pekonen/© NHPA/Photoshot; B. Cathy & Gordon Illg/Jaynes Gallery/DanitaDelimont.com/Danita Delimont Photography/USA)

species requires. A niche can be thought of as a way a species obtains food and resources, somewhat like its "job." Niches exist within a **habitat**, the physical environment in which a species lives.

Species with narrow ecological tolerances are called *stenotopic* (or *specialists*). Species with broad ecological tolerances are called *eurytopic* (or *generalists*). The peregrine falcon (*Falco peregrinus*), for example, is eurytopic and found on every continent **(Figure 8.8A)**. It is therefore said to have a *cosmopolitan* geographic range. The yellow-eared parrot (*Ognorhynchus icterotis*), on the other hand, is stenotopic, with a geographic range restricted to the Colombian Andes **(Figure 8.8B)**. Biogeographers describe the geographic range of a species

that is restricted to one area as **endemic**. The yellow-eared parrot, for example, is endemic to the Colombian Andes and is found nowhere else.

Why is the peregrine falcon cosmopolitan, but the yellow-eared parrot is endemic to the Colombian Andes? Generally speaking, species that have narrow ecological niches are stenotopic and usually geographically restricted, and species with broad niches are eurytopic and more likely to be geographically widespread. For example, the geographically restricted yellow-eared parrot has a narrow niche. It requires wax palm trees (*Ceroxylon* spp.), which grow in cloud forests (forests perpetually shrouded in clouds) of the Colombian Andes. The parrot eats the fruit of the palms and nests in their canopies. It can live only where the tree lives. In contrast, the geographically widespread peregrine falcon has a broad niche. It preys mostly on medium-sized birds, grabbing them out of the air, and builds nests in a variety of settings. As a result, it can live in many different habitats and climates, as long as there are birds to eat and places to nest.

One factor that complicates this generalization, however, is the influence of people. The endangered tiger (*Panthera tigris*), for example, is geographically restricted today into isolated pockets throughout southern Asia and far eastern Russia. The tigers are geographically restricted not because they are stenotopic but because of hunting pressure and habitat loss due to human activities. Many large vertebrate mammals are eurytopic but geographically restricted because they have lost much of their original habitat due to human activity.

Generally speaking, the narrower the niche of a species, the more limiting factors operate to restrict its geographic range. A **limiting factor** is any factor that prevents an organism from reaching its reproductive or geographic potential. For example, too little water can limit the growth of plants. Similarly, too much water can limit the growth of, or even kill, drought-adapted cacti. Therefore, water (either too little or too much of it) can function as a limiting factor.

Limiting factors can be divided into two groups: *physical* limiting factors and *biological* limiting factors. Six major limiting factors are shown in **Table 8.3**, but many others exist that are not discussed here.

- ■ Breeding summer visitor
- ■ Breeding resident
- ■ Winter visitor
- ■ Visitor during migration

A

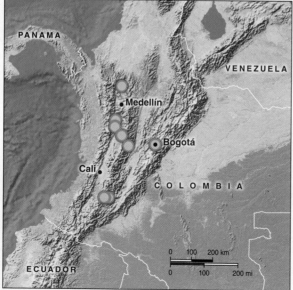

B

Figure 8.8 Range maps for the peregrine falcon and yellow-eared parrot. (A) The peregrine falcon is found on every continent. (B) The yellow-eared parrot is found only in the Colombian Andes. *(A. Gerard Soury/Getty Images; B. © Roland Seitre/naturepl.com)*

Table 8.3	Limiting Factors for Plants and Animals	
PHYSICAL LIMITING FACTORS	**BIOLOGICAL LIMITING FACTORS**
Light	Predation
Temperature	Competition
Water	Mutualism

Figure 8.9 Heliophytes and sciophytes. (A) This manzanita (*Arctostaphylos* spp.) displays many of the characteristics of heliophytes, including small leaves that are oriented edge-on to the Sun. (Note the hummingbird in the foreground for scale.) (B) The giant rhubarb (*Gunnera manicata*) has some of the largest leaves of any plant. Its leaves are up to 3 m (10 ft) across. The giant rhubarb's leaves have many sciophyte characteristics, including their large size and dark green color, which maximize the absorption of sunlight. *(A. Used with permission of www.laspilitas.com; B. Helen Hughes/MCT via Getty Images)*

A   **B**

Limiting factors result in stress for an organism and lower reproductive success. Reproductive success is important because it is the foundation of any species' long-term success. If a species does not reproduce, its fate is extinction.

Physical Limiting Factors

The physical limiting factors light, temperature, and water strongly influence the geographic range of both plants and animals.

Light

Some plants cannot grow in full sunlight or full shade, while others can tolerate a wide range of light conditions. Many plants can be categorized as either *heliophytes* that are adapted to strong sunlight or *sciophytes* that are adapted to low-light conditions. Heliophytes are intolerant of heavy shade. Their leaves tend to be small and may have a thick waxy coating that reduces moisture loss in intense sunlight. Many heliophytes also have a whitish or silvery appearance because of their small hairs that reflect sunlight. Some heliophyte leaves are oriented edge-on to the noontime sunlight to reduce its intensity **(Figure 8.9A)**. In contrast, the large, flat, dark leaves of sciophytes maximize their capture of dimmed light such as that found on a forest floor **(Figure 8.9B)**.

Temperature

Temperature strongly controls the geographic ranges of plants. Extreme high or low temperatures lead to reduced photosynthesis and mortality

in plants. Animals are also limited by temperature **(Figure 8.10)**.

Endothermic (or *warm-blooded*) organisms, such as birds and mammals, maintain a constant core body temperature that is usually much higher than the environmental temperature. The core

Figure 8.10 Temperature as a limiting factor. The geographic range of the eastern phoebe (*Sayornis phoebe*) in winter (blue and purple areas) coincides with areas where the average January temperature is −4°C (25°F) or warmer (south of the black line). North of this line, winter temperatures are too cold for the eastern phoebe. The bird moves north into the orange range of the map during summer. *(Don Johnston/Getty Images)*

body temperature of a warm-blooded organism is kept nearly constant by an internal metabolic process called *homeostasis*. Amphibians, insects, and most fishes and reptiles are *ectothermic* (or *cold-blooded*). Their body temperature is generally the same as the temperature of the surrounding environment. Although ectothermic animals cannot regulate their internal body temperature through homeostasis, they can regulate their internal temperature by moving to areas of suitable temperature or basking in the Sun.

As a means of avoiding environmental stresses, dormancy (a period of inactivity) is a common response of ectotherms to low temperatures. Frogs and insects, for example, often hide from low winter temperatures by going underground into mud or burrows. Some organisms, such as snails, insects, and reptiles, enter a state of *estivation* (summer dormancy) during hot or dry periods. This allows them to avoid water and heat stress. To avoid low winter temperatures and conserve energy when food is unavailable, many endothermic animals, such as most bears and bats, enter a state of *hibernation* (winter dormancy) in a protected environment during the winter. Endothermic animals that do not hibernate but instead remain active in cold winter temperatures, such as Arctic hares and many bird species, increase their metabolic rate and develop thick insulating layers of fat, fur, or feathers during winter.

Many mammals living in cold environments have shorter appendages (such as legs, ears, and tails) than related species in warm regions. This evolutionary pattern is called *Allen's rule* **(Figure 8.11)**.

Water

All organisms need liquid water to carry out their metabolic processes. Some organisms have evolved to exist on very little water, while others cannot survive without ample water. Plants, for example, can be grouped into three categories, based on their water requirements:

1. *Hydrophytes* are plants that require submersion in water. A water lily is an example of a hydrophyte.

2. *Mesophytes* are plants that can tolerate a broad range of soil moistures but are generally intolerant of prolonged drought or submersion. Most plants are mesophytes.

3. *Xerophytes* are plants that are adapted to environments with little available moisture. Cacti are xerophytes.

Plants have two main types of adaptations for coping with a lack of water: They *escape* water stress, or they *reduce* water stress. *Annual* plants (plants that go to seed and then die each year) escape water stress. Their seeds can remain dormant in the soil as a *seed bank* during long periods of drought and then germinate when conditions improve. Some seeds from the Sahara have germinated after 100 years of dormancy in the soil.

Perennial plants (plants that live longer than a year) must reduce water stress because they cannot escape it. The fishhook barrel cactus (*Ferocactus wislizeni*) is a xerophytic perennial that reduces water stress through a variety of adaptations **(Figure 8.12)**. Other perennial plants have a long *taproot* that reaches moisture deep in the ground. Taproots of mesquite (*Prosopis* spp.) have been measured at more than 50 m (160 ft) long, although the aboveground portion of this desert plant is no more than 3 m (10 ft) tall.

Like plants, animals either escape water stress or reduce water loss. Because animals are mobile, many of them escape dry conditions through migration. There are many examples of animals

Figure 8.11 Allen's rule. Allen's rule states that appendage length in mammals decreases as latitude increases. This anatomical variation helps animals maintain suitable body temperatures in these different climates. Among hares and rabbits in North America, the southernmost species have long appendages as a means to radiate body heat for cooling. Appendage length decreases to the north as a means of conserving heat. Fur length and thickness in these animals also increase as latitude increases.

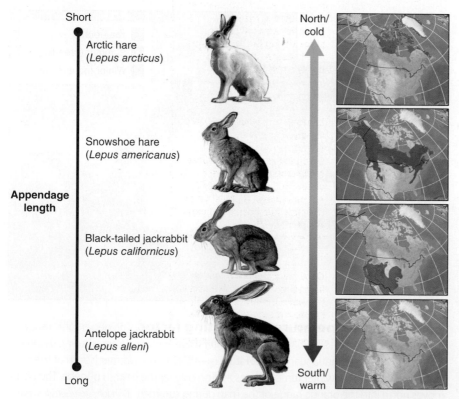

Short

Arctic hare
(*Lepus arcticus*)

Snowshoe hare
(*Lepus americanus*)

Appendage length

Black-tailed jackrabbit
(*Lepus californicus*)

Antelope jackrabbit
(*Lepus alleni*)

Long

North/cold

South/warm

that undertake seasonal migrations in response to changes in water availability. Large herds of animals throughout sub-Saharan Africa, such as wildebeests and zebras, follow the rainfall of the ITCZ and the plant growth (food) that comes with it.

Other animals have adaptations for reducing water stress. The kangaroo rats of the southwestern United States and northern Mexico (see Figure 8.6), for example, live in extremely arid desert climates. These animals never drink liquid water; instead, they derive water from the seeds and other dry foods they eat. Water is produced as the compounds in their food react and combine with oxygen in their bodies. Their other water-conserving adaptations include producing urine with high concentrations of *urea* relative to its water content, producing dry feces, and never sweating. Like many other desert animals, kangaroo rats are nocturnal, hiding away in their cool burrows during the day to avoid heat and daytime predators.

Biological Limiting Factors

Just as physical limiting factors such as light, temperature, and water influence the geographic range of a species, so too do the biological limiting factors of predation, competition, and mutualism.

Predation

Predation—the consumption of one organism by another—is one limiting factor that nearly all organisms must cope with. A lion that kills and eats an antelope is a predator. This type of predation, called *carnivory*, is lethal to prey. Not all predation is lethal carnivory, however. *Herbivory* is a form of what is usually non-lethal predation. An insect eating a plant and a giant panda (*Ailuropoda melanoleuca*) feeding on bamboo provide examples of non-lethal predation.

The geographic ranges of many species are determined by where their food sources live. Many insects are stenotopic and eat only one or a few species of plants. The monarch butterfly, for example,

CAM photosynthesis

Cacti and some other plants use crassulacean acid metabolism (CAM), a specialized photosynthetic process in which plants open their pores at night for photosynthesis. Opening pores during the day, as most plants do, would result in costly water loss.

Green stem

Large leaves would lose water rapidly. The leaves of cacti have evolved into spines, which also serve the purpose of warding off animals seeking water. Because its leaves do not produce food, the stem of the plant has adapted to carry out photosynthesis.

Waxy stem

The body of this cactus is its stem. Cactus stems have a thick waxy coating to prevent water loss.

Expandable stem

The stem of the cactus is ribbed to allow it to swell and store water.

Shallow root mat

Cacti have a shallow mat of fibrous roots (not visible here) that absorb rainwater quickly before it drains away into the porous sandy soil.

Figure 8.12 Barrel cactus adaptations. *(Craig K Lorenz/Getty Images)*

Figure 8.13 Monarchs and milkweed. The Rocky Mountains and southwestern deserts separate the monarch butterfly into a western population and an eastern population. The caterpillars of monarchs in both populations (see Figure 8.5C) eat only milkweed plants. The northern edge of the milkweed's geographic range is limited by low winter temperatures in Canada. The northern edge of the monarch's range is therefore bounded by the northern limit of milkweed. *(LuckyStep/Shutterstock)*

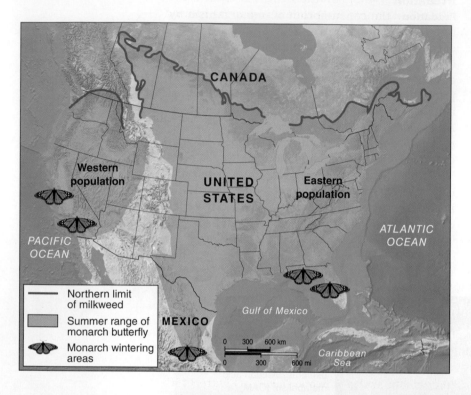

can live only where its *host plant*, milkweed (*Asclepias* species), lives, as shown in **Figure 8.13**.

Many mammals are eurytopic and have a broad dietary range. Humans are a good example of generalist eurytopic mammals. Stenotopic mammals, with a narrow dietary range, are less common. Two well-known examples are the giant panda, which eats only bamboo, and the koala (*Phascolarctos cinereus*), which eats only leaves from eucalyptus trees.

Native predators are unlikely to hunt their prey to extinction. Instead, a predator may reduce the population of a prey species until that prey species becomes too difficult to find, and the predator must find a new source of food. This process is called *prey switching*. As a mountain lion reduces a deer population, for example, deer become too difficult for the mountain lion to find. The mountain lion then prey switches and focuses on another prey species. Or the mountain lion may simply hunt in a different area, where deer are more abundant.

Native predators sometimes function as **keystone species**, species whose effects indirectly support many other species within an ecosystem. Keystone species play a crucial role in maintaining healthy ecosystems. The gray wolf (*Canis lupus*), for example, is a keystone species in Yellowstone National Park, but no one understood its importance to other species until wolves, which had been eradicated from the park, were reintroduced in 1995 **(Figure 8.14)**.

Figure 8.14 The gray wolf as a keystone species in Yellowstone National Park.
(Top left. jim kruger/Getty Images; top center. Tathoms/Shutterstock; top right. Ken M Johns/Getty Images; bottom left. William H Mullins/Getty Images; bottom right. National Park Service, Yellostone National Park)

Story Map
The Wolves of Yellowstone
Available at www.saplinglearning.com

Keystone species
Gray wolves were the primary predator of elk (*Cervus canadensis*) in Yellowstone National Park. After wolves were eradicated from the park in the 1930s, elk populations increased.

Wolf-kill ecology
When wolves have finished feeding on a large animal they have killed, grizzlies, coyotes (shown here), ravens, eagles, vultures, and many other scavengers also benefit from the kill. Wolf kills are an important component of the diets of these species.

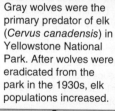

Riparian ecology
Elk browse on willow, cottonwood, and alder. When elk populations increased, these *riparian* (streamside) trees were eliminated from many environments.

Beavers
Beavers (*Castor canadensis*) eat these riparian trees and use them to build dams across streams, creating ponds. Beaver populations and their ponds declined with the loss of riparian trees.

Beaver-pond ecology
Beaver ponds create habitat for many other organisms. They slow stream erosion and create habitat for fish and wading birds. They increase willow growth, which attracts songbirds and songbird predators.

Wolf reintroduction
Wolves were gradually reintroduced beginning in 1995. Thereafter, elk populations in the park dropped from about 18,000 to about 9,000. Riparian plants returned, as did beavers and their ponds.

Competition

Competition is interaction between organisms that require the same resources. A *resource* is anything that an organism needs, including food, territory, and mates. All organisms must compete with other organisms for resources. Competition between individuals of the same species is *intraspecific competition*, and competition between different species is *interspecific competition*.

Plant species living in the same habitat compete with one another if their niches overlap. Hydrophytes do not compete with xerophytes because they occupy very different niches and habitats. However, xerophytes compete with other xerophytes when they occupy the same physical space. Plants compete with one another by shading each other or taking nutrients and moisture with their roots. Plants may also be equipped with chemicals that inhibit seed germination or growth in other plants. Such chemical competition is called *allelopathy*. Hardly anything grows beneath eucalyptus trees, for example, because their leaf litter contain chemicals that discourage the growth of other plants.

The degree of competition among organisms depends on how similar their niches are. Individuals of the same species have identical niches and compete for exactly the same resources. Very different species, such as wolves and beavers, can share the same habitat and yet not compete with one another for resources because they have very different niches. Wolves and coyotes, on the other hand, live in the same habitat and have similar niches. As a result, they compete for many of the same resources. Both compete for elk carcasses, for example. Mice and voles are also important to the diet of both species.

Animals may also compete indirectly. For example, the house cat (*Felis catus*) and the barn owl (*Tyto alba*) are competitors, although they rarely interact directly. The barn owl eats small rodents and hunts on the wing, mostly at night. House cats hunt small rodents from the ground day and night. Although the two predators rarely meet, they exploit the same resource.

Mutualism

The old saying that nature is "red in tooth and claw" suggests that predation and competition are the most common types of biological interactions. Yet there are just as many examples of benign interactions between species. A relationship between two species that benefits both of them is called **mutualism (Figure 8.15)**. Mutualism is a type of *symbiosis*, which refers to any close relationship between two species. More narrowly, mutualism refers a close relationship between two species that is mutually beneficial. Mutualism becomes a limiting factor when a species, without its mutualistic partner, is unlikely to reach its full reproductive potential.

Mutualism occurs in most ecological settings. In cases of *obligate mutualism*, neither party can survive without its partner. In *facultative mutualism*, both parties benefit from the relationship, but they do not need it to survive. In some cases of mutualism, such as the one shown in Figure 8.15C, one party is an obligate mutualist, while the other is a facultative mutualist.

A   B   C

Figure 8.15 Examples of mutualism. (A) Flowers receive pollination from the visiting insects and birds. The hummingbird benefits from the energy provided by flower nectar. The flowering plant benefits from pollination by the hummingbird. (B) Oxpecker birds (*Buphagus* species) remove bothersome insects, such as ticks, from large mammals in sub-Saharan Africa. The bird gets a meal, while the animal is cleaned. Here oxpeckers are working on a giraffe. (C) A clownfish escapes its predators by living in the stinging tentacles of a sea anemone. It also feeds on scraps from the anemone's fish meals. The clownfish keeps the anemone free of parasites and provides nutrient-rich feces for it to feed on. The clownfish is an obligate mutualist because it needs the anemone to survive. The anemone, on the other hand, is a facultative mutualist: It does better with clownfish mutualists but can survive without them. (*A. Rolf Nussbaumer/Getty Images; B. blickwinkel/Alamy; C. Richard Whitcombe/Shutterstock*)

8.4 Moving Around: Dispersal

◎ **Explain why and how organisms disperse.**

Recall that dispersal, the movement of organisms away from where they originated, allows organisms to reduce competition, to obtain more resources, and to respond to environmental change. Nearly all organisms disperse, some farther than others. It takes about 100 years for earthworms to disperse (unaided by people) 1 km (0.6 mi), while some birds, such as Arctic terns (*Sterna paradisaea*), can fly over 500 km (300 mi) in a single day.

Intra-range dispersal occurs when an organism disperses within the current geographic range of its species. *Extra-range dispersal* occurs when an organism disperses outside the current geographic range of its species. In most cases, extra-range dispersal is lethal to the disperser because the ecological conditions outside the current geographic range are inhospitable. Otherwise, the species would probably already occupy those regions. Successful extra-range dispersal is known as colonization. **Colonization** is the natural establishment of a population in a new geographic region without the help of people. Extra-range dispersal and colonization result in geographic range expansion, as we saw in Section 8.1.

Only a propagule can establish an extra-range population. A **propagule** is any unit, such as a seed, a mating pair, or a rooting branch, that is able to establish a new reproducing population. A single male or a group of males are not a propagule because they cannot start a new population.

Modes of dispersal can be grouped into two categories. Organisms dispersing under their own power, such as by flying, are *active dispersers*. Organisms that move about using outside forces, such as wind or ocean currents, or by hitching a ride on other organisms are *passive dispersers*. Most, but not all, animals are active dispersers. Barnacles and coral, for example, cast their eggs to the water, and the eggs then disperse passively on ocean currents. Except in the rare cases in which seeds are forcefully ejected from their fruits once they dry, all plants are passive dispersers; some plant dispersal strategies are shown in **Figure 8.16**.

Barriers to Dispersal

Many *biogeographic barriers*—objects or factors that restrict the dispersal of an organism—affect the geographic ranges of species. For example, species living on small, remote islands are surrounded by the formidable barrier of the ocean. Similarly, species living high on mountains in the Great Basin in the western United States dwell in virtual islands of suitable habitat surrounded by a barrier of inhospitable desert.

Biogeographic corridors, which allow unrestricted movement between habitats, are the opposite of barriers. The terms *barrier* and *corridor* are

Wind dispersal – anemochory	Animal dispersal – zoochory	Animal dispersal – zoochory	Water dispersal – hydrochory
Tumbleweeds (*Kali tragus*) detach from their roots and shake out seeds like a salt shaker as they roll across open fields. A single plant can release 50,000 seeds.	Acorns are cached (hidden) by squirrels, jays, and other animals. Many cached nuts are forgotten and successfully germinate. Blue jays (*Cyanocitta cristata*) have been recorded as caching some 4,500 acorns in a single season.	Colorful and nutritious fruit and seeds are eaten and passed through the digestive systems of animals. As the animals move about, they pass and disperse the seeds.	Coconuts (*Cocos nucifera*) float for months at sea. When they wash up on an island beach, freshwater rains trigger germination.

Figure 8.16 Dispersal strategies of plants.
(From left to right. Walter Meayers Edwards/Getty Images; Steve & Dave Maslowski/Getty Images; yuris/Shutterstock; M Swiet Productions/ Getty Images)

MODE OF DISPERSAL	DISPERSAL TERM
By wind	*Anemochore*
By water	*Hydrochore*
By animals	*Zoochore*

relative because what is an impassable barrier for one organism can be a corridor for another. For example, great white sharks (*Carcharodon carcharias*) frequently travel between California and Hawai'i. Seawater is a corridor to a shark; for a freshwater fish that is intolerant of salt water, however, seawater is an impassable barrier.

Barriers that allow certain types of organisms to disperse across them and not others are called *biogeographic filters* **(Figure 8.17)**. These filters include *climatic filters*, such as deserts, ice caps, and the heat of the tropics. Predation, competition, and lack of mutualistic partners may act as *biological filters*. The oaks and pines of the Northern Hemisphere, for example, have never crossed the equator to the Southern Hemisphere because of the severity of the competition (a biological filter) and the warmth in the tropics (a climatic filter). Likewise, only one bear species has successfully crossed the equator. The spectacled bear (*Tremarctos ornatus*) inhabits forests of the Andes as far south as northern Chile. All other true bears naturally live north of the equator. The heat of the tropics has presented an insurmountable barrier for them.

Small islands far away from the mainland continent are difficult for most organisms to reach. Successful dispersal to remote islands is therefore called *waif dispersal* because the organisms that reached the islands were presumably lost (*waif* means "abandoned") and far out of their normal ranges (see Figure 8.17). Waif dispersal is an unlikely event with a very low chance of success. Taken a step further, dispersal to small and extremely remote islands is so unlikely that it is like winning a lottery or a sweepstakes. Such long-distance dispersal to remote islands is called **sweepstakes dispersal**. Island chains (or archipelagos) aid in the dispersal of organisms (see Figure 8.17). They allow organisms to reach remote islands by "stepping" from one island to another rather than dispersing long distances. For this reason, they are called **stepping-stones**.

Biogeographic Realms

Because of natural barriers to dispersal, many regions of Earth's land surface are biologically distinct environments. **Biogeographic realms** are continental-scale regions that contain genetically similar groups of plants and animals. The realms result from the isolation of large continental regions from one another by physical barriers such as oceans, mountain ranges, and deserts. The geographic barriers between biogeographic realms prevent organisms from moving outside

Animation

Dispersing to Remote Islands

Available at
www.saplinglearning.com

Figure 8.17 Dispersing to remote islands.

Resistance to dispersal varies depending on distance, the severity of the barrier, and the type of organism dispersing. Here, species dispersing from Papua New Guinea or Australia to the remote island of New Caledonia must do so in one of three ways: via stepping-stone filters, waif dispersal, or sweepstakes dispersal. Species can disperse along a series of stepping-stone filters going past the Solomon Islands (shown with small yellow arrows). Successful waif dispersal directly from Australia, a distance of 1,250 km (780 mi) over open ocean (large yellow arrow), is very unlikely. Least likely of all, sweepstakes dispersal (red arrow) would occur directly from Papua New Guinea over open water, a distance of 1,700 km (1,060 mi). Note that the inset table shows the number of vertebrate species native to New Caledonia. All six of the mammals are flying bats. No amphibians have successfully dispersed because saltwater is an impassible barrier to them, and all of the reptiles that have reached the island are lizards. The lizards most likely clung to debris like tree trunks that washed ashore on New Caledonia from nearby islands like Vanuatu or from mainland Australia or they evolved on the island.

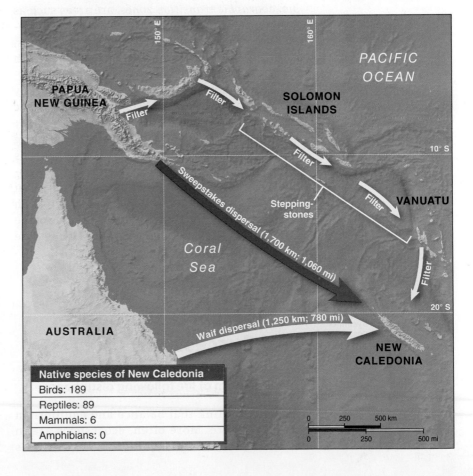

Native species of New Caledonia	
Birds:	189
Reptiles:	89
Mammals:	6
Amphibians:	0

Figure 8.18 Biogeographic realms. At a global scale, physical barriers isolate groups of organisms in one region from groups in another region. Over time, this isolation has resulted in eight biogeographic realms, each separated by a biogeographic transition zone. Each biogeographic realm is geographically isolated from the adjacent realm by deserts, oceans, mountain ranges, or deepwater channels.

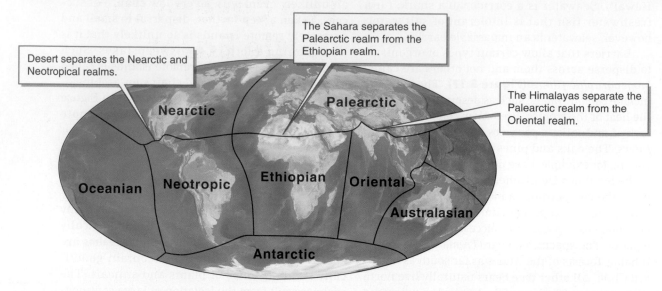

The Sahara separates the Palearctic realm from the Ethiopian realm.

Desert separates the Nearctic and Neotropical realms.

The Himalayas separate the Palearctic realm from the Oriental realm.

Nearctic

Palearctic

Oceanian   Neotropic   Ethiopian   Oriental

Australasian

Antarctic

the region in which they evolved, much as a fence keeps livestock from leaving a farm. **Figure 8.18** shows the names and locations of the world's eight biogeographic realms.

The edges of the biogeographic realms, called *biogeographic transition zones*, are areas such as deserts, mountains, and oceans that restrict movement of species and gene flow between regions, resulting in evolutionary divergence. The North American biogeographic realm is called the *Nearctic*. Most plants and animals there are genetically distinct from those of the *Neotropic* biogeographic region to the south.

Increasingly, as discussed in the opening Human Sphere section of this chapter, organisms are dispersing across these barriers with the assistance of people. The distinction between natural dispersal and human-assisted dispersal across an ocean barrier is not always clear, as the case of the cattle egret, discussed in the **Picture This** feature on the following page, illustrates.

8.5 Starting Anew: Ecological Disturbance and Succession

◎ **Describe the role of ecological disturbance and the return of life following disturbance.**

We tend to think that balance and equilibrium exist in nature. More often, ecosystems are in a constant state of disequilibrium. Ecosystems

change as they adjust to ecological disturbances through the process of ecological succession. An **ecological disturbance** is a sudden event that disrupts an ecosystem. **Ecological succession** is the step-by-step series of changes in an ecosystem that follows an ecological disturbance.

Fire in Ecosystems

Ecological disturbances range from the powerful winds of a tornado that uproot trees to massive volcanic eruptions that create an extensive barren landscape. Both *biotic* (living) forces and *abiotic* (nonliving) forces can create ecological disturbances. The removal of wolves from Yellowstone by people (see Figure 8.14) was a biotic disturbance, as was the introduction of the non-native Nile perch into Lake Victoria (see Figure 8.1).

Table 8.4 lists some examples of different types of ecological disturbances. In this section, we take a more detailed look at the role of one type

Table 8.4	**Types of Ecological Disturbances**
BIOTIC	**ABIOTIC**
Anthropogenic (such as logging or farming)	Fire
Loss of keystone species	Volcanic eruption
New non-native species	Wind
Insect outbreak	Landslide

Picture This Native or Non-Native? The Case of the Cattle Egret

(Bob Gibbons/Science Source)

Scientists know that the cattle egret (*Bubulcus ibis*) is native to Europe, Africa, and Asia. It is also found in the Americas, but it's not clear exactly how it got there. The bird first dispersed across the Atlantic Ocean to northern South America, perhaps as early as the 1870s. Prior to that it was not in the Americas. The map shows the documented expansion of the bird from northeastern South America in 1933 and outwards, both north and south. By the 1970s it had reached its full geographic range shown in the map. If it crossed the Atlantic Ocean to the Americas without the help of people it would be an example of a natural colonization event. If that happened the bird is native to the Americas.

Oddly, however, the cattle egret's crossing coincides with the increased movement of people across the Atlantic Ocean in the early 1900s, strongly suggesting that people transported the bird across the ocean. There is no known record

of anyone ever bringing the bird to the Americas. The cattle egret is eurytopic (a generalist) and is often found feeding on insects associated with cattle. Once it got the Americas, its successful range expansion was no doubt made possible by the expansion of cattle pasture. As cattle pasture habitat spread, so did the bird. For these reasons, many biogeographers argue that the cattle egret is a non-native species in the Americas.

Consider This

1. Given the balance of evidence presented in this short piece, do you think the cattle egret is native or non-native to the Americas? Support your answer.
2. Regardless of how the bird got to the Americas, is it important philosophically or ecologically whether the cattle egret is native or non-native? Explain your reasoning.

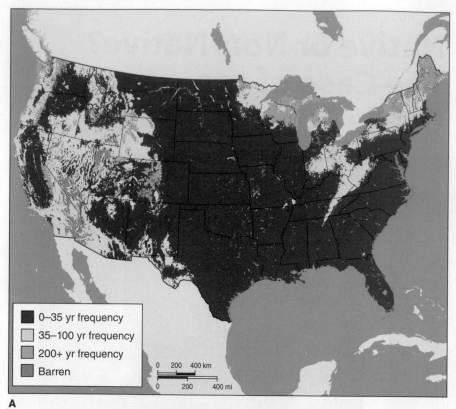

A

0–35 yr frequency

35–100 yr frequency

200+ yr frequency

Barren

| 0 | 200 | 400 km |
| 0 | 200 | 400 mi |

B

Crown sprouting

Crown sprouting allows a tree to sprout from the surviving root and trunk. These redwoods (*Sequoia sempervirens*) are crown sprouting after being burned.

Serotinous cones

Serotinous cones stay closed until heated by a fire. After the fire, they slowly open and release their seeds. These knobcone pines (*Pinus attenuata*) have serotinous cones.

Deep roots and seeds

Many plants of the North American prairies survive by sprouting from deep roots that are unharmed by fire or by germinating from seeds that survive in the soil.

1.5 m (5 ft)

3.0 m (10 ft)

Figure 8.19 Fire frequency and plant adaptations. (A) This map shows natural fire-return intervals for the United States. Fire burned in the red areas on average at least once every 35 years. The green areas experienced fire only once every 200 years, or longer. (B) Plants have many adaptations for surviving wildfires, including *crown sprouting*, *serotinous cones*, and deep roots and seeds. *(Left. Steve Norman, U.S. Forest Service; right. Paul Tomlins/Flower/AGE Fotostock)*

Web Map

Wildfires in the U.S.

Available at www.saplinglearning.com

Video

NASA Rainfall Data and Global Fire Weather

Available at www.saplinglearning.com

of disturbance—fire—in the landscapes of western North America.

Fire is a widespread, common, and important element in almost all natural ecosystems. Fire can be as important as water, light, or soil nutrients to the health of ecosystems. Wherever there is a dry and warm season and enough precipitation for vegetation to grow, wildfires will burn. Many North American ecosystems had natural *fire-return intervals* (the average number of years between fire events) of less than 35 years. Because wildfires burned so frequently, many species were well adapted to—and even benefited from—fire **(Figure 8.19)**.

Wildfires no longer burn as frequently as they once did. In the 1930s, to save property and trees with commercial value, the U.S. Forest Service began a "10 a.m." fire policy. The aim of this *fire suppression* campaign was to have any given wildfire contained by ten o'clock the morning after it was first reported. The *surface fires* (fires that burn gently on the forest floor and kill few trees) that once commonly occurred were extinguished **(Figure 8.20A)**. As a result, dead and living vegetation accumulated instead of being removed by frequent fires. Today many forests are overgrown and have very high *fuel loads* (material that can burn). When fuel loads are high, fires burn intensely hot, and plant adaptions to fire, such as crown sprouting and serotinous cones, are ineffective. In addition, catastrophic *canopy fires* (or *crown fire*) have become routine because of the high fuel loads. A *canopy fire* burns in the forest canopy (the uppermost layer of branches) and usually kills the trees **(Figure 8.20B)**.

Some trees, such as the lodgepole pine (*Pinus contorta*) and knobcone pine in the Sierra Nevada in California are adapted to crown fires. While the fire kills the tree, the tree's serotinous cones are tightly closed to protect the seeds inside from fire's heat. After the fire, the cones open, the seeds are released to the soil, and the new *saplings* (young trees) thrive in the nutrient-rich ashes of the burned landscape.

In the past few decades, the essential ecological role of fire has been increasingly recognized. In an attempt to address the problem of fuel buildup in western U.S. forests caused by a century of fire suppression, *prescribed burns* (in which fires are set intentionally under controlled conditions) and *mechanical thinning* operations (in which vegetation is removed by manual cutting) are sometimes employed to reduce fuel loads.

Ecological Succession: The Return of Life

Given enough time, life returns following fire or any other disturbance **(Figure 8.21)**. There are two types of ecological succession, depending on the severity of the disturbance. *Primary succession* occurs when life is completely removed from a landscape. Large volcanic eruptions and large landslides are disturbances that can trigger primary succession. In most cases, however, some life survives the disturbance, and ecosystems rebuild through the process of *secondary succession*.

Ecological succession takes place in stages and may require decades or centuries to complete. A **sere** is a stage of ecological succession that follows

Figure 8.20 Fire intensity. (A) When fires are frequent, fuel loads remain low and they burn as gentle surface fires, such as this one in the Sequoia and Kings Canyon national parks in California. (B) As a result of a century of fire suppression and resulting heavy fuel loads, the Waldo Canyon fire near Colorado Springs, Colorado, became an intense canopy fire that burned 74km² (29 mi²) in Pike National Forest in June 2012. It was the most destructive fire in Colorado history up to that time, forcing some 32,000 people out of their homes and destroying 346 homes. *(A. Eric Knapp, U.S. Forest Service; B. Karl Gehring/Getty Images)*

A

B

Figure 8.21 Disturbance and ecological succession. (A) In 1988, numerous fires burned 793,880 acres of Yellowstone National Park. This photo shows an area just after it was burned. (B) Twenty years later in 2008, lodgepole pine has regrown in the burned areas. *(A. National Park Service, Yellowstone National Park; B. © Katie LaSalle-Lowery)*

A.

B.

Figure 8.22 A model of ecological succession. The return of life after an ecological disturbance follows a sequence of vegetation seres. In this example, a farmer (far left) has cleared an oak–hickory forest (not shown), creating a bare field. After the farmer abandons the field, vegetation will return in a series of steps through the process of succession. First, a grass sere will develop. It is composed of pioneer r-selected plants. A perennial shrub sere follows, and that is followed by an oak sere. It may take a century or more for this process of succession to unfold and for the oak–hickory forest, composed mostly of K-selected species, to return (far right).

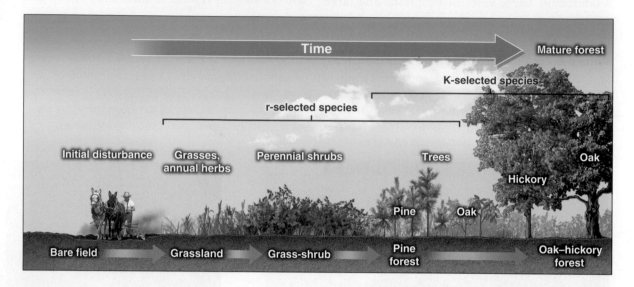

a disturbance. **Figure 8.22** illustrates a typical sequence of vegetation seres that would occur in the eastern United States. After a primary-disturbance event, the first organisms to move in usually are *pioneer species* or **r-selected species**. Species that are r-selected are generally short-lived and produce many offspring during their lifetime **(Table 8.5)**.

The return of the original ecosystem can take years to centuries, depending on the severity of the disturbance and the responsiveness of the ecosystem. Eventually, *climax species,* or **K-selected species**, become established. K-selected species are generally long-lived and produce few offspring during their lifetime (see Table 8.5). In the past, scientists promoted the *climax community model,* which assumed that a community would persist indefinitely in the absence of any further disturbance. In the example in Figure 8.22, the oak–hickory forest is the climax community.

The climax community model has been modified by a view that change and disturbance go on continually. Fire, volcanic eruptions, climate change, non-native species, and human transformation of ecosystems all create an ongoing cycle of disturbance in almost all ecosystems. Scientists now view communities as mosaics, with each community in a different stage of ecological succession and continually being reset by new disturbances.

Table 8.5 r-Selected and K-Selected Species Traits

r-SELECTED TRAITS	K-SELECTED TRAITS
Small body size	Large body size
Good dispersers	Poor dispersers
Short lifespan	Long lifespan
Early maturity	Late maturity
Produce many offspring	Produce few offspring
EXAMPLES	**EXAMPLES**
Insects	Elephants
Rodents	Humans
Grasses	Oaks
Dandelions	Redwoods

8.6 Three Ways to Organize the Biosphere

◎ **Describe three approaches to organizing the biosphere and how each works.**

The biosphere is the most complex of Earth's physical systems. To make sense of its great complexity, we take three approaches to structuring life on Earth: the trophic hierarchy, the taxonomic hierarchy, and the spatial hierarchy.

The Trophic Hierarchy

The first approach to structuring the biosphere is based on the movement of energy and matter through it, beginning with solar energy. Photosynthesis is the gateway through which solar energy enters the biosphere. Recall from Chapter 1, "The Geographer's Toolkit," that photosynthesis is a process by which plants, algae, and some bacteria convert the Sun's radiant energy to stored chemical energy in the form of sugars, as shown in **Figure 8.23**.

A **primary producer** is any organism that is able to convert sunlight to chemical energy through photosynthesis. On land, most primary producers are plants. In freshwater bodies and in the oceans, the primary producers are **phytoplankton,** microscopic algae and plantlike bacteria called *cyanobacteria*, which are suspended in the sunlit portions of water and also photosynthesize.

Photosynthesis does not truly produce energy. Instead, it converts energy from one form (radiant energy) to another (chemical energy). Of the total solar energy falling on a photosynthetic organism, less than 5% is converted to chemical energy through photosynthesis. This chemical energy is then transferred from the photosynthetic organisms to consumers. A **consumer** is any organism that cannot produce its own food through photosynthesis. As consumers eat primary producers, and as they are eaten in turn by other consumers, the solar energy made available by the primary producers flows through a *food chain* **(Figure 8.24)**.

Food chains are composed of **herbivores**, organisms that eat only primary producers; *carnivores*, organisms that eat only other consumers; and **omnivores**, organisms that eat both primary producers and other consumers. Herbivores such as grasshoppers are *primary consumers* because they eat plants directly. A mouse that eats a grasshopper is a *secondary consumer*, a snake that eats the mouse is a *tertiary consumer*, and so on.

When an organism dies, its tissues are recycled by scavengers, detritivores, and decomposers. The biosphere would have long ago run out of organic material to metabolize if not for the work of these organisms. *Scavengers* are animals such as coyotes and vultures that feed on *carrion* (dead animals). Many animals are at least part-time opportunistic scavengers. *Detritivores*, such as earthworms, fly larvae (maggots), and crickets, feed only on *detritus* (decomposing plant or animal remains). **Decomposers**, such as fungi and bacteria, are organisms that break down organic material into simple compounds that reenter the trophic system through plants.

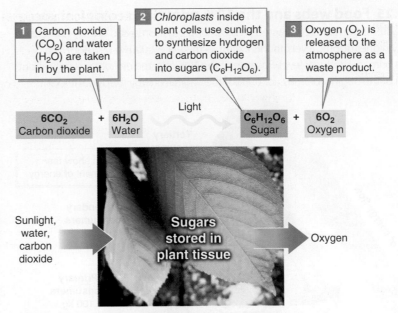

1 | Carbon dioxide (CO_2) and water (H_2O) are taken in by the plant.

2 | *Chloroplasts* inside plant cells use sunlight to synthesize hydrogen and carbon dioxide into sugars ($C_6H_{12}O_6$).

3 | Oxygen (O_2) is released to the atmosphere as a waste product.

$$\underset{\text{Carbon dioxide}}{6CO_2} + \underset{\text{Water}}{6H_2O} \xrightarrow{\text{Light}} \underset{\text{Sugar}}{C_6H_{12}O_6} + \underset{\text{Oxygen}}{6O_2}$$

Sunlight, water, carbon dioxide → Sugars stored in plant tissue → Oxygen

Figure 8.23 Photosynthesis. *(Bruce Gervais)*

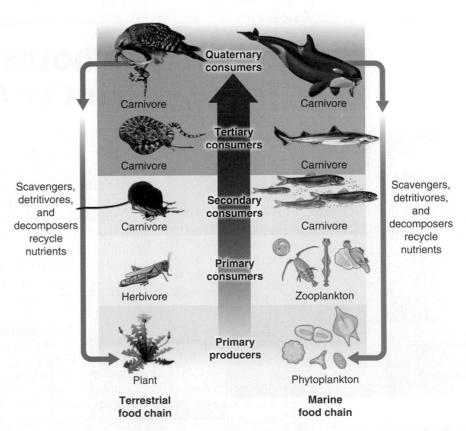

Figure 8.24 Food chains. Energy (red arrow) flows through a food chain from one organism to the next. Nutrients are recycled (green arrows) by scavengers, detritivores, and decomposers at all levels.

Figure 8.25 Food webs and the 10% rule. Flows of energy and matter in trophic systems form interconnected webs among organisms, and energy flows to successively higher trophic levels. In this example, there are 1,000 kg (2,204 lb) of biomass at the primary producer level. Because of the 10% rule, the owl obtains only about 1 kg (2.2 lb) of the original biomass, or 0.1% of the original energy fixed by the plants.

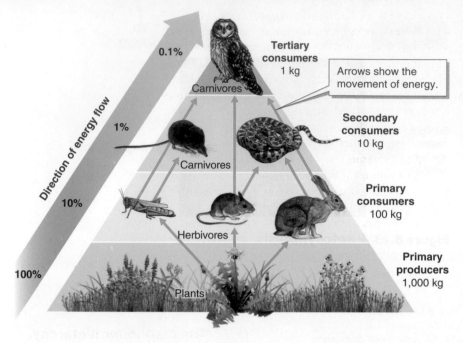

The food chain is a good basic explanation and schematic, but in nature, energy and matter do not flow in straight paths. Instead, the movements of energy and matter through the biosphere form complex systems of ecological interconnections called **food webs**. The different levels of an ecosystem through which energy and matter flow are called **trophic levels**.

Plants form the base of terrestrial trophic levels. Their chemical energy passes to consumers at higher trophic levels. With each higher trophic level, only about 10% of available energy is passed on. The remainder either is not absorbed by the consumer's body or is lost as heat. This phenomenon of energy loss is approximated as the *10% rule*. The 10% rule applies to **biomass** (the dried weight of living material) as well as to energy. Because of this loss of energy at successive trophic levels, biomass rapidly diminishes with each higher trophic level. Populations of top predators (such as wolves and owls) typically have very low total biomass compared with the trophic levels below them **(Figure 8.25)**.

Many organisms occupy more than one trophic level. Bears, for example, are scavengers when they eat carrion, herbivores when they eat berries,

Picture This A Photosynthetic Slug: Plant or Animal?

The eastern emerald elysia (*Elysia chlorotica*) is a 3 cm (1.2 in) sea slug. This solar-powered slug does something most animals do not: It converts sunlight to food through photosynthesis. The elysia lives on the East Coast of the United States, in shallow salt marshes and tidal pools such as those in the Chesapeake Bay. As the slug feeds on photosynthetic algae, it incorporates the algae's chloroplasts into its own gut cells. The elysia then use the chloroplasts to make sugars directly from sunlight, just as plants do. The elysia can go without eating algae for months, instead fueling itself with energy from sunlight. The elysia has incorporated the algal genes into its own DNA. Natural fusion of genes across such distantly related life-forms is extremely unusual.

Consider This

1. What trophic level(s) does the eastern emerald elysia occupy?
2. How would photosynthesis be an advantage to this animal?

(© Nicholas Curtis)

and carnivores when they eat insects. This makes them *omnivores*. Humans are omnivorous as well. The **Picture This** feature on the facing page describes an organism that defies conventional trophic categorization.

The Taxonomic Hierarchy

In addition to the trophic hierarchy, another way of structuring the biosphere is based on the genetic relationships among organisms. **Taxonomy** is the classification and naming of organisms based on their genetic similarities. Taxonomy has two main goals: (1) to assign a unique name to every known species and (2) to reveal genetic relationships among organisms.

Scientific Names

You may have noticed in this chapter that when we discuss a specific organism, we list its common name and its scientific name. The scientific name is given in parentheses. Scientists give every known species on Earth a unique scientific name that consists of a genus name and a species name.

Humans, for example, are *Homo sapiens*. *Homo* in Latin means "person," and *sapiens* means "wise." *Homo* is the genus name, and *sapiens* is the name for our species. It might help to think of the genus as a car make, such as "Honda." Just as there are many car models within a car make (such as "Honda Civic," "Honda Accord," and so on), there can be many species within a genus. This naming system, which was first developed by the Swedish naturalist Carl Linnaeus (1707–1778), is the *Linnaean binomial* (two-name) *classification system*. Note that the genus name is capitalized, and the species name is not. Also, Latin names are italicized, and the genus name (or its abbreviation) always accompanies the species name.

The taxonomic hierarchy consists of a series of nested groups. Each group belongs to another, larger group, in the same way folders are nested within one another in a computer or cell phone operating environment **(Figure 8.26)**. Each family has at least one genus, and each genus has at least one species.

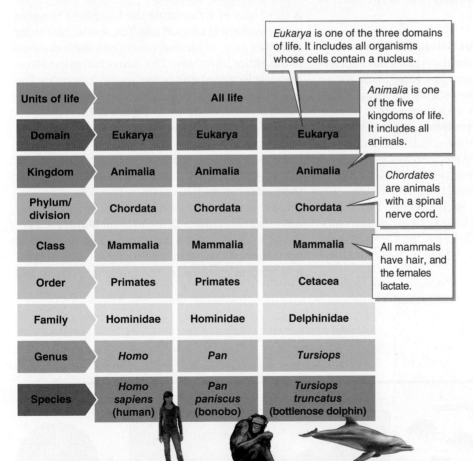

Eukarya is one of the three domains of life. It includes all organisms whose cells contain a nucleus.

Animalia is one of the five kingdoms of life. It includes all animals.

Chordates are animals with a spinal nerve cord.

All mammals have hair, and the females lactate.

Figure 8.26 The taxonomic hierarchy.

Humans, bonobos (similar to chimpanzees and the closest living relative to humans), and bottlenose dolphins are in the same domain, kingdom, phylum, and class. At the level of orders, they diverge. Dolphins belong to the order Cetacea, whose members are mammals adapted to marine environments. Humans and bonobos are both in the family Hominidae, but they are not in the same genus. Note that only the genus and species are italicized.

The taxonomic system uses *homologous traits* to group organisms that are genetically related to one another because they share a close common ancestor. For example, all birds have feathers, a characteristic that is unique to the class *Aves*. From ostriches to parrots to hawks—if it has feathers, it is a bird and is a member of *Aves*.

Many related organisms do not look alike. For example, a cow, a bat, and a whale do not look alike, but they are genetically similar. They all have hair, and they all feed their young milk from mammary glands. Only mammals have these features. Their differences in outward appearance result from divergent evolution (see Section 8.2).

Through convergent evolution (see Section 8.2), organisms sometimes look deceptively alike but have no close genetic ties. Such superficial resemblances between organisms are called *analogous traits* **(Figure 8.27)**. Analogous traits do not reflect genetic similarities and are never intentionally used in taxonomic groupings. It would be misleading to group such organisms together because such grouping would not reveal genetic relationships, which is one of the goals of taxonomy.

Common Names versus Scientific Names

Common names and scientific names have different uses. A common name is a local name for an organism. Common names are useful because they are familiar. A problem with them is that there is often more than one common name per species. The "black bear," for instance, is also called "North American black bear," "American black bear," and "cinnamon bear," depending where you live in North America. The Linnaean system assigns only one scientific name to each species. The scientific name for the black bear is *Ursus americanus*. *Ursus* means "bear" in Latin, and *americanus* describes where it lives. There is only one *Ursus americanus*.

A   **B**

Figure 8.27 Analogous traits. Birds (A) and bats (B) superficially look alike (in that they both have wings) and behave alike (in that they both fly), but they are only distantly related genetically. Birds are in the class Aves, and bats are in the class Mammalia. However, at the level of phylum, they both have backbones and so are vertebrates. At this taxonomic level, they are grouped together into Chordata. *(A. Tony Campbell/Shutterstock; B. Edwin Giesbers/Nature Picture Library/Getty Images)*

The Spatial Hierarchy

A third way of structuring the biosphere is based on the ecological units of life. The spatial hierarchy identifies units of life that range from the individual to the entire biosphere. The individual organism is the most localized unit of the spatial hierarchy. The whole biosphere is the largest unit in this categorization scheme, as shown in **Figure 8.28**.

Individuals make up populations of organisms, and populations form communities. A **community** consists of the populations of organisms interacting in a geographic area. Communities are named after their most conspicuous feature, based on their geographic setting, or based on the niche of the most prominent organisms living in them **(Figure 8.29)**.

As we have seen, a community and its physical environment constitute an ecosystem. The ecosystem concept emphasizes the interconnectivity and

Figure 8.28 The spatial hierarchy. The spatial hierarchy is scaled geographically from large-scale, localized phenomena through small-scale, continent-wide phenomena such as biomes. Biomes are extensive expanses of vegetation types that are determined by climate. The largest possible unit is the biosphere.

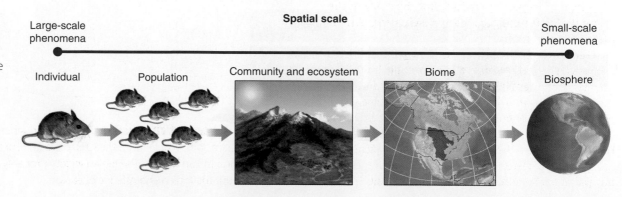

Spatial scale

Large-scale phenomena

Small-scale phenomena

Individual   Population   Community and ecosystem   Biome   Biosphere

A   **B**

Figure 8.29 Communities. (A) Populations of wildebeests, zebras, and antelope compose the *grazing community* on the grasslands in the Maasai Mara National Reserve in Kenya. This community is identified by the grazing niche of the large animals that dominate this landscape. (B) Sea urchins and sea stars make up this *tide pool community* along the rocky coast in Gwaii Haanas National Park, British Columbia. This community is identified by its geographic setting.   *(A. James Warwick/Getty Images; B. David Nunuk/Getty Images)*

interdependence of all facets of a landscape, both living and nonliving. Ecosystems are composed of the organisms living there, and the chemicals, nutrients, sunlight, and water found there.

The last category of the spatial hierarchy is the *biome*. Earth's biomes are geographically extensive. Only the biosphere is greater in spatial extent. Biomes are the focus of the next chapter.

GEOGRAPHIC PERSPECTIVES
8.7 Journey of the Coconut

◎ Assess the relationship between people and the coconut palm and apply that knowledge to human relationships with other organisms.

Many people associate the coconut (*Cocos nucifera*) with food or with a tropical vacation getaway. The coconut is certainly an important food, and it is an icon of tropical paradise. But the story of the coconut also illuminates many fascinating principles of biogeography.

Geographers are interested in where things are, why they are where they are, and how long they've been there. No one knows for certain where the coconut palm originated. Today, the tree has a *pantropical* distribution, meaning that it is found throughout the tropics. But where did it first evolve? How did it come to be spread across the tropics?

Genetic evidence suggests that the coconut palm arose some 11 million years ago in the

vicinity of the Malay Peninsula, Indonesia, the Philippines, and New Guinea. Wild coconuts were adopted by people as they began to cultivate the coconuts as a food source. Today there are two major varieties of the coconut palm, indicating that it was adopted in two different areas: somewhere in southern India, producing the *Indian variety*, and somewhere near Indonesia, producing the *Pacific variety* **(Figure 8.30)**.

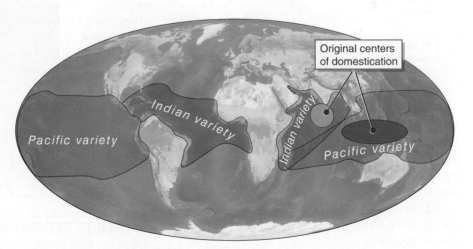

Figure 8.30 Coconut range map. The Pacific variety of coconut palm probably originated somewhere in southeastern Asia and spread throughout the tropical Pacific and Indian oceans. The Indian variety probably originated in India and spread throughout the Indian and Atlantic oceans.

Coconut Dispersal

But how did the coconut disperse across the tropical oceans from where it first arose? As we have learned in this chapter, plants disperse in a variety of ways, from being blown on the wind, to being transported in the digestive tracts of animals, to floating in water. As shown in **Figure 8.31**, coconuts are hydrochores and are well adapted to dispersing by floating in water.

If a coconut seed disperses north or south of about 26 degrees latitude, it cannot successfully establish a new population because of the limiting factors of low winter temperatures and dry air. The coconut is restricted to humid tropical regions and mostly to coastal regions.

Human Migration and Coconut Dispersal

As it turns out, the same traits that allow the coconut seed to survive long-distance travel on ocean currents also allow it to survive long-distance travel in outrigger canoes. The endosperm supports a young coconut palm as it first germinates and grows. The meat, oil, and water of the endosperm are also nourishing for people. During a 2,000-year period, people dispersed across the tropical Pacific Ocean from Taiwan to Papua New Guinea. Using the Pacific islands as stepping-stones, they dispersed all the way to Easter Island and the Hawaiian Islands. **Figure 8.32** provides a map of this remarkable migration event.

The coconut's stores of freshwater and nutrition made it an ideal food and source of water for ocean travelers. Archeologists know that migrating people brought the coconut with them because people and the coconut appeared on remote Pacific islands at the same time. The Hawaiian Islands, for example, were colonized about 1,100 years ago. The earliest middens (garbage heaps) in Hawai'i have coconut shells preserved in them. Before that time, there is no evidence of humans ever having set foot in Hawai'i, and no coconut had ever grown in the Hawaiian Islands naturally. The same is true for Easter Island (see Section 1.6).

When the first Europeans sailed to the western shores of tropical South America in the early 1500s, they found mature coconut palm groves there. Somehow, the coconut palm had dispersed to western South America. Whether Polynesian voyagers brought it there or the coconut drifted there on its own is still unknown.

Europeans also dispersed the coconut palm across the tropics. After 1500, Europeans began dispersing coconuts to regions where the plant already grew as well as to new regions **(Figure 8.33)**.

Artificial Selection and Domestication

The relationship between people and the coconut illustrates not only dispersal but also the important process of *artificial selection*. Long ago, people began selecting individual coconut palms with

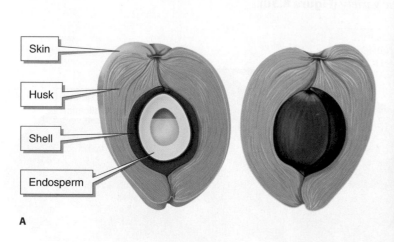

Figure 8.31 The coconut seed. (A) The coconut seed is surrounded by layers of waterproof barriers: an outer skin, a fibrous husk, and a shell. The inner *endosperm*, or "meat," contains sweet freshwater. (B) Coconuts can float in salt water for months and still germinate if washed up on a distant beach. *(Ethan Daniels/Shutterstock)*

Figure 8.32 Polynesian dispersal. Polynesians dispersed eastward across the Pacific Ocean in outrigger canoes (inset). The earliest dates of their arrival on Pacific islands are given. They brought with them several different types of plants and animals, including the coconut. *(Herbert Kawainui Kane/Getty Images)*

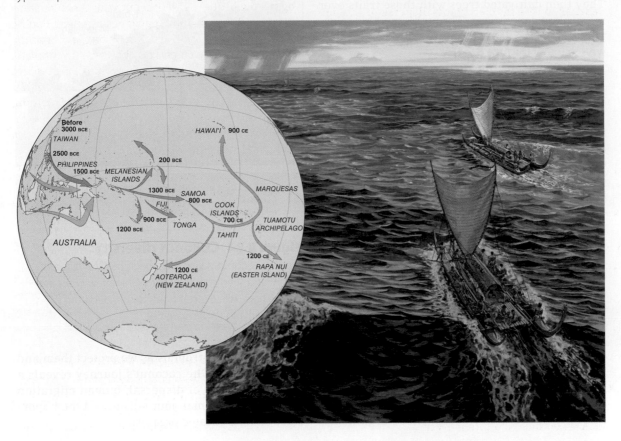

Figure 8.33 Coconut dispersal by Europeans.

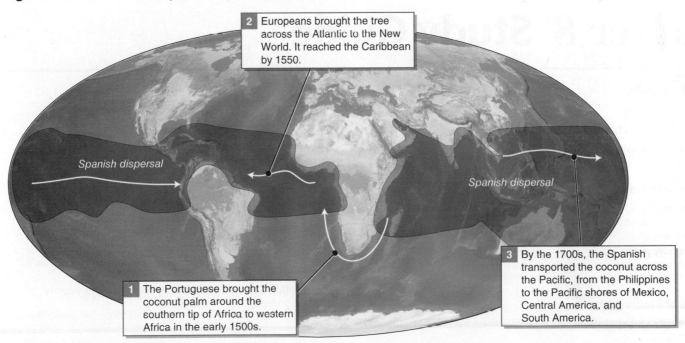

2 Europeans brought the tree across the Atlantic to the New World. It reached the Caribbean by 1550.

Spanish dispersal

Spanish dispersal

1 The Portuguese brought the coconut palm around the southern tip of Africa to western Africa in the early 1500s.

3 By the 1700s, the Spanish transported the coconut across the Pacific, from the Philippines to the Pacific shores of Mexico, Central America, and South America.

desirable traits, such as those with the sweetest and most nutritious meat and oil, those that did not grow too tall, and those with the best fiber. They then cultivated trees with those traits over many plant generations. Through time, this selection process created the 80 types of coconut palms we see today **(Figure 8.34)**.

Artificial selection *domesticated* the coconut. Domestication means to intentionally change the genetic makeup of an organism to suit human needs. Artificial selection works through many generations of selecting the qualities of an organism that are desirable. Today people rely on domesticated organisms for their survival. Nearly all of the foods we eat have been domesticated from wild ancestors. Grains, such as wheat, rice, and corn, supply over 50% of the world's caloric intake. These grains, as well as fruits, vegetables, and nuts in our grocery stores have all been created by people through artificial selection. Cows are domesticated from a now-extinct ancestor called the aurochs (*Bos primigenius*) from Europe. Chickens come from the red junglefowl (*Gallus gallus*) in Southeast Asia. All dog breeds (*Canis familiaris*), including chihuahuas and poodles, are descended from the gray wolf (*Canis lupus*) from Eurasia.

Humans and their domesticates, including the coconut palm, share a relationship of mutualism. Not only do we benefit from these organisms,

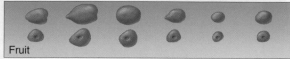

Figure 8.34 Coconut diversity. The many types of coconut palms have been bred for different height and seed characteristics. Only a few types are shown here. All coconut types descended from the two Indian and Pacific varieties.

they also benefit from us as we protect them and cultivate them. The coconut's journey reveals a fascinating story of dispersal, human migration history, plant–human mutualism, and the importance of domesticated organisms.

Chapter 8 Study Guide

Focus Points

The Human Sphere: The New Biosphere

- **Non-native organisms:** Non-native organisms, those transported outside their original range by people, have transformed the biosphere.

8.1 Patterns of Biodiversity

- **Biogeography:** Biogeographers study the geographic patterns of life to learn about Earth's physical systems and Earth's natural and human history.
- **Biodiversity patterns:** The tropics are the most biodiverse region on Earth. Large islands are more biodiverse than small islands. Biodiversity changes seasonally as animals migrate.

8.2 Origins of Biodiversity

- **Biodiversity and evolution:** Earth's biodiversity is the result of speciation.

- **Patterns resulting from evolution:** Biogeographic patterns of convergence and divergence result from the process of evolution.

8.3 Setting the Boundaries: Limiting Factors

- **Niche and geographic range:** Stenotopic species with narrow ecological niches usually have restricted geographic ranges; eurytopic species usually have broad geographic ranges.
- **Limiting factors:** Geographic ranges of species are determined by physical and biological limiting factors, such as light, water, competition, and predation.
- **Keystone species:** Keystone species play crucial roles in the functioning of ecosystems. Species with broad niches are usually geographically widespread.
- **Competition:** The degree of competition among organisms depends on the similarity of their niches.

8.4 Moving Around: Dispersal

- **Dispersal:** Dispersal allows organisms to reduce limiting factors such as competition, to obtain more resources, and to respond to environmental change.
- **Range expansion:** Geographic ranges expand when organisms successfully disperse outside the boundaries of their current range.
- **Dispersal modes:** Organisms may disperse under their own power (called active dispersal) or with the help of wind, water, or other organisms (called passive dispersal).

8.5 Starting Anew: Ecological Disturbance and Succession

- **Ecological disturbance:** Ecological disturbance is caused by both biotic and abiotic factors.
- **Fire ecology:** Plants are adapted to fire in many ways, including crown sprouting, serotinous cones, and roots and seeds protected by soil.
- **Fire suppression:** In North America, fires have been intentionally suppressed by people, resulting in an accumulation of fuel and subsequent catastrophic fires.
- **Ecological succession:** After a disturbance event, life returns to an area through ecological succession.

8.6 Three Ways to Organize the Biosphere

- **Organizing the biosphere:** The biosphere can be organized on the basis of flows of energy and matter, genetic similarities, or the spatial dimensions of ecological units of life.
- **The trophic hierarchy:** The trophic hierarchy is based on energy flow. Higher trophic levels have less available energy and biomass.
- **The taxonomic hierarchy:** The taxonomic hierarchy places organisms in nested groups according to their shared genetic relationships.
- **The spatial hierarchy:** Ecological units of life range from the individual to the entire biosphere.

8.7 Geographic Perspectives: Journey of the Coconut

- **Coconut dispersal:** The coconut seed is well adapted to disperse by floating on seawater as well as for dispersal by people.
- **Artificial selection:** The coconut palm illustrates the influences that people and their domesticated organisms have on one another.

Key Terms

biodiversity, 231
biogeographic realm, 247
biogeography, 231
biomass, 254
colonization, 246
community, 256
competition, 245
consumer, 253
convergent evolution, 238
decomposer, 253
dispersal, 239
divergent evolution, 239
ecological disturbance, 248
ecological succession, 248
ecology, 231
ecosystem, 231
endemic, 240
evolution, 237
extinct, 233
food web, 254
habitat, 240
herbivore, 253
keystone species, 244
K-selected species, 252

limiting factor, 240
migration, 234
mutualism, 245
niche, 239
non-native, 230
omnivore, 253
phytoplankton, 253
population, 237
predation, 243
primary producer, 253
propagule, 246
r-selected species, 252
sere, 251
speciation, 237
species, 230
species–area curve, 234
stepping-stone, 247
sweepstakes dispersal, 247
taxonomy, 255
trophic level, 254

Concept Review

The Human Sphere: The New Biosphere

1. What is the difference between a native species and a non-native species?

2. Why are non-native species sometimes destructive in their new habitats? What might be missing in those new habitats that was in their old habitats?

3. Are the majority of non-native species ecologically harmful? Explain.

8.1 Patterns of Biodiversity

4. Define *biogeography* and *biodiversity*. What are some questions a biogeographer might ask?

5. What are vertebrates? What are the five major groups of vertebrates?

6. What is a species? How many species have been counted? How many species may be on Earth?

7. Describe the generalized spatial and temporal patterns of biodiversity with respect to latitude, island size, and seasons.

8. Why do animals migrate? What cues do animals use to find their way during migration?

8.2 Origins of Biodiversity

9. What is a population? Provide biogeographic examples that result from evolution when populations are geographically isolated.

10. What is evolution? Describe the basic steps of the process and explain how it works.

11. What is convergent evolution? Describe the geographic circumstances under which it occurs and provide real-life examples of it.

12. What is divergent evolution? Describe the geographic circumstances under which it occurs and provide real-life examples of it.

8.3 Setting the Boundaries: Limiting Factors

13. Compare the terms *niche* and *habitat*. Why do species with a narrow niche usually have limited geographic ranges?

14. What does it mean to say that an organism is endemic to a place? What factors would cause an organism to be endemic?

15. What is a limiting factor? List examples of physical and biological limiting factors. Explain how each factor acts to limit the range of a species.

16. What is Allen's rule? How does it relate to limiting factors?

17. What adaptations do many cacti have to cope with life in the desert?

18. What is a keystone species? Give an example. How does the concept of a keystone species relate to limiting factors?

19. What is mutualism? Give examples of it. How can mutualism be a limiting factor for a given organism?

8.4 Moving Around: Dispersal

20. In the context of biogeography, what is dispersal? Why do organisms disperse?

21. By what means do plants and animals disperse?

22. What barriers present resistance to dispersal? Compare the relative ease or difficulty of dispersing to islands close to the mainland and far away from it.

23. What are biogeographic realms, and how do they form?

24. What is a colonization event? How does the cattle egret exemplify a native species in the Americas? How might it also exemplify a non-native species?

8.5 Starting Anew: Ecological Disturbance and Succession

25. What is ecological disturbance? What causes it?

26. Review the importance of fire to western North American forests. What problem has developed in western North American forests, and what is being done to address it?

27. What is ecological succession? What causes it? What are seres?

28. Compare the characteristics of r-selected species and those of K-selected species. Which first occupies disturbed environments?

8.6 Three Ways to Organize the Biosphere

29. What three hierarchies are used to organize the biosphere? Explain the basis of each.

30. What are trophic levels? Where do producers and consumers fit in? What is a tertiary consumer?

31. Compare a food chain with a food web. Which is more representative of the real world?

32. What is the 10% rule in relation to trophic levels?

33. Which types of consumers recycle dead organisms?

34. What are homologous traits? How are they used to create taxonomic groups?

35. Compare common names and scientific names. On what is each based? How is each type of name useful?

36. What is a community? How is the term defined? Give examples of a few communities.

8.7 Geographic Perspectives: Journey of the Coconut

37. Is *Cocos* the genus or the species name for the coconut palm?

38. What are the two basic varieties of coconut palm, and where did each originate?

39. The coconut palm is subject to two main modes of dispersal. What are they?

40. Where did Austronesians disperse the coconut? Where did Europeans disperse it?

41. What are the two primary limiting factors controlling the coconut's geographic range?

42. What is artificial selection? How does it relate to domestication? How does the coconut's history illustrate the process of artificial selection?

43. Describe the ecological relationship between humans and the coconut.

Critical-Thinking Questions

1. The theory of evolution has been a source of controversy within some religious groups since it was first developed. Do you find it controversial? Explain.

2. What trophic level(s) do you occupy? Why is it more efficient for people to eat as primary consumers rather than as secondary consumers?

3. By definition, when people move an organism to a new place, that organism is non-native. People brought the coconut to Hawai'i over 1,000 years ago. Because that was so long ago, do you think enough time has passed that we can consider this plant native to Hawai'i now? Explain.

4. Does evolutionary convergence (Section 8.2) result from analogous traits or homologous traits (Section 8.6)? Explain.

5. Compare natural selection (Section 8.2) with artificial selection (Section 8.7). How are they the same, and how are they different?

Test Yourself

Take this quiz to test your chapter knowledge.

1. True or false? When a species disperses on its own to a new location outside its geographic range, it is considered a native species.

2. True or false? Allen's rule states that there are more species in the tropics than at higher latitudes.

3. True or false? After ecological disturbance, ecosystems do not always return to the same climax community that existed prior to the disturbance.

4. True or false? Many species have more than one common name.

5. Multiple choice: What an organism does to obtain food is its
  a. niche.
  b. habitat.
  c. prey.
  d. community.

6. Multiple choice: Which of the following is not an evolutionary adaptation to fire by plants?
  a. serotinous cones
  b. crown sprouting
  c. deep roots
  d. fire-proof leaves

7. Multiple choice: Which of the following does fire suppression often lead to?
  a. a small fire
  b. a crown fire
  c. a surface fire
  d. a prescribed burn

8. **Multiple choice:** Which of the following is not a limiting factor?

   a. temperature
   b. light
   c. predation
   d. a scavenger

9. **Fill in the blank:** A(n) _____ is a stage of ecological succession following ecological disturbance.

10. **Fill in the blank:** _____ is the term used to describe an interaction between two species from which both species benefit.

Online Geographic Analysis

The Encyclopedia of Life (EOL)

We use the Encyclopedia of Life (EOL) web page (http://www.eol.org) in this exercise. The goal of the EOL web page is to place all species known to science into a single database and provide in-depth biological and biogeographic information on each. We will use this web page to examine the major characteristics of three species.

Activity

In the search field enter "polar bear." Using the "Brief Summary" information, answer the following questions.

1. **What is the polar bear's genus name? What is its species name?**

2. **What trophic level(s) does the polar bear occupy?**

3. **Is the polar bear stenotopic or eurytopic? Explain your choice.**

Now enter "koala" in the search field and answer the following questions.

4. **What are the genus and species of the koala?**

5. **Read the "Brief Summary" on the opening page. Is the koala a true bear? Why or why not?**

6. **Near the top of the page, click the "Detail" tab. Scroll down and read the content under the "Habitat and Ecology" heading. What do koalas eat? What trophic level(s) does the koala occupy?**

7. **Are koalas stenotopic or eurytopic? Explain your choice.**

Navigating the EOL page as before, enter "Great Basin bristlecone pine" in the search field and answer the following questions.

8. **What are the genus and species names of the Great Basin bristlecone pine? What is particularly distinctive about this plant?**

9. **Is this an r-selected species or a K-selected species? Explain your choice.**

10. **What is the growth rate of these trees? What causes this rate of growth?**

Picture This. *Your Turn*

Trophic-Level Feeding

The copper sharks (*Carcharhinus brachyurus*) shown in the photo are feeding on schooling sardines (*Sardinops sagax*) off the coast of South Africa. Sardines eat small animals called zooplankton (see Figure 8.25). The zooplankton eat microscopic plants called phytoplankton. Based on this information, answer the following questions. (Hint: Draw out this food chain with the phytoplankton at the primary producer level.)

1. What trophic level do the copper sharks occupy?

2. Given the 10% rule, how much energy from the primary producers do they receive?

(Alexander Safonov/Getty Images)

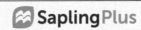

For animations, interactive maps, videos, and more, visit www.saplinglearning.com. **SaplingPlus**

Climate and Life: Biomes

Chapter Outline *and Learning Goals* ◎

This Landsat 8 image of New Zealand's Mount Taranaki was acquired July 3, 2014. The dark green circle-shaped area is protected temperate rainforest biome within the boundaries of Egmont National Park. The surrounding lighter green areas were once similarly forested but are now a mosaic of human-made pasture and farmland. For scale, the distance across Egmont National Park is about 19 km (12 mi).

(NASA/USGS)

To learn more about the temperate rainforest biome, go to Section 9.3.

THE HUMAN SPHERE Three Trillion Trees

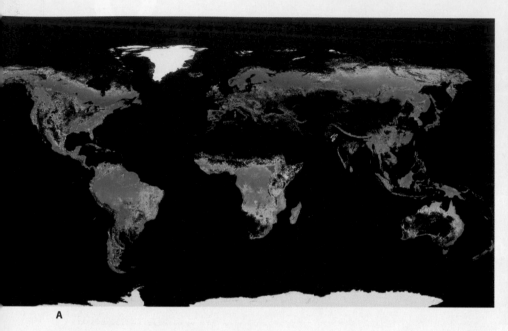
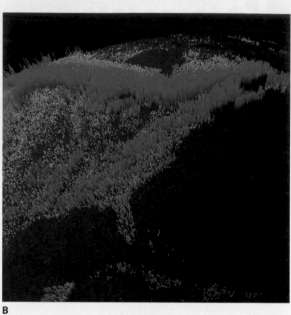

A

B

Figure 9.1 Earth's forest densities. (A) This map shows the density of forests worldwide. Yellow areas have the lowest tree density, and dark green areas have the highest. The highest densities are in northern forests. (B) Eastern North America is shown in greater detail. The heights of the green bars indicate forest density; taller bars represent denser forests. *(Republished with permission of SpringerNature, from Trillions of Trees, Rachel Ehrenberg, 525, 170–171, 2015, permission conveyed through Copyright Clearance Center, Inc. Source: T.W. Crowther et al. Nature 525, 201–205 (2015). Visualization: Jan Willem Tulp.)*

Nobody knows the exact number of trees on Earth, but recent research published in the journal *Nature* perhaps comes close to the mark: 3,040,000,000,000 (more than 3 trillion) trees. By using satellite remote sensing technology and surveys on the ground (called *ground-truthing*), scientists estimated the total number of trees and where the forests are densest. They found that about 43% (or 1.3 trillion) of all trees on Earth are in *tropical rainforests* and subtropical forests (the *tropical seasonal forests*). They also found that the greatest tree densities occur in the northern forests (the *boreal forest*) that grow across northern North America and Eurasia **(Figure 9.1)**. About 25% of all trees (750 billion) were found in these northern forests. Most of the rest of the trees grow in the mountains of the world (the *montane forests*).

In the same study, the researchers found that prior to the beginning of the *Agricultural Revolution* that began 10,000 years ago, there were perhaps 6 trillion trees on Earth. Since that time people have cut almost half of Earth's trees in the process of converting forests to farmland. Today about 15 billion trees are cut each year and not replaced.

Forests are critical elements of Earth's biosphere. They produce oxygen and transfer moisture from the soil to the air. Forests also prevent dangerous landslides by stabilizing slopes in mountainous regions. They cycle nutrients and filter and clean air and water, and they absorb carbon dioxide from the atmosphere and help check the pace of anthropogenic climate change.

9.1 Climates and Biomes

◎ **Explain the relationship between climate and vegetation structure.**

This chapter explores the geography of life in the context of regional climate types. The type of native vegetation found in any given area, whether it is forest or desert, reflects three fundamental physical limiting factors: temperature, water, and light. At the local level, other factors may be important, too, such as soil types and biological limiting factors like herbivory and nutrient supplies. In mountain ranges, slope steepness, slope *aspect* (the direction the slope is facing), and *microclimates* (the distinct climates of restricted areas) also determine the types of organisms that are found there.

Earth's land surface is covered by biomes. A **biome** is an extensive geographic region with relatively uniform vegetation structure. *Vegetation structure* refers to the type of vegetation that dominates a region, such as *closed-canopy forest* (dense) or open (sparse) grassland. Biomes are the second-largest units in the spatial hierarchy; only the biosphere is larger (see Section 8.6).

Every human alive lives in a biome. All land surfaces (except where there are ice sheets) are covered by the continuous mantle of biomes. Some biomes, such as a forest, are easy to recognize. Other biomes may be less easy to recognize because of how people have changed them. They often exist as small remnants of their former selves, such as natives trees mixed with non-native trees in an urban park. In many places natural biomes are completely gone, replaced by human-built landscapes such as farms and cities **(Figure 9.2)**.

Biome classifications are based on vegetation structure, which is determined mostly by climate. **Table 9.1** summarizes the terms describing vegetation structure that we use in this chapter.

Within each biome, plants take on a variety of *growth forms*. **Table 9.2** details these growth forms from largest to smallest. In addition to the names of these plant growth forms, this chapter includes a number of *botanical* (plant-related) terms that describe characteristics of plant growth and plant adaptations **(Table 9.3** on page 269**)**.

There is a spatial correspondence between climates and biomes. For this reason, classifying and categorizing climate types is useful when we are discussing biomes. One of the most widely used

Figure 9.2 New York's Central Park. A temperate deciduous forest biome once existed in New York City and its environs. The biome has been replaced by the urbanized city, but remnants of it are preserved in Central Park, where native trees are intermixed with non-native trees from around the world. In this photo, the fall colors of the forest light up the park. *(francois-roux/Getty Images)*

Table 9.1	Vegetation Structure
STRUCTURE	**DESCRIPTION**
Forest	Dominated by trees with a closed canopy
Woodland	Widely spaced trees with a grass understory
Shrubland	Continuous cover of shrubs
Grassland	Continuous cover of grasses
Scrubland	Widely spaced, dry-adapted shrubs
Desert	Sparse plant cover

Table 9.2	Plant Growth Forms
GROWTH FORM	**DESCRIPTION**
Tree	Tall, upright woody growth
Liana	Woody, climbing vine
Shrub	Many woody stems
Forb	Low, nonwoody plant other than grass
Epiphyte	Grows on surfaces of trees, not parasitic
Bryophyte	Member of the division that includes mosses

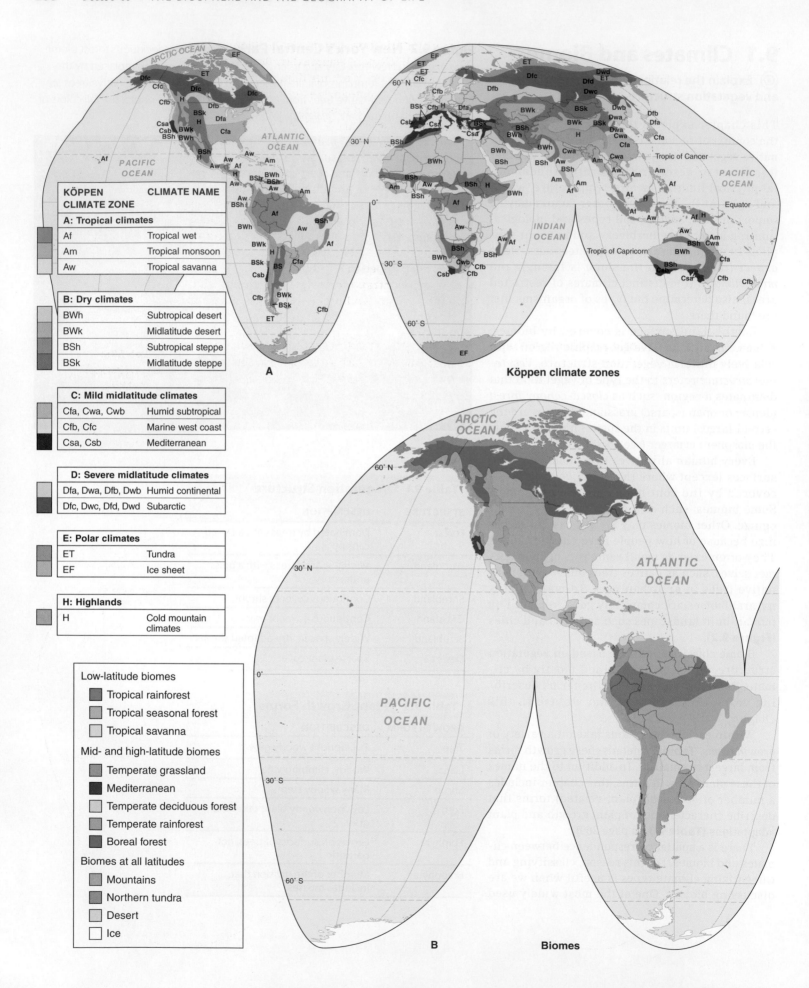

KÖPPEN CLIMATE ZONE	CLIMATE NAME
A: Tropical climates
Af	Tropical wet
Am	Tropical monsoon
Aw	Tropical savanna
B: Dry climates
BWh	Subtropical desert
BWk	Midlatitude desert
BSh	Subtropical steppe
BSk	Midlatitude steppe
C: Mild midlatitude climates
Cfa, Cwa, Cwb	Humid subtropical
Cfb, Cfc	Marine west coast
Csa, Csb	Mediterranean
D: Severe midlatitude climates
Dfa, Dwa, Dfb, Dwb	Humid continental
Dfc, Dwc, Dfd, Dwd	Subarctic
E: Polar climates
ET	Tundra
EF	Ice sheet
H: Highlands
H	Cold mountain climates

A

Köppen climate zones

Low-latitude biomes
- Tropical rainforest
- Tropical seasonal forest
- Tropical savanna

Mid- and high-latitude biomes
- Temperate grassland
- Mediterranean
- Temperate deciduous forest
- Temperate rainforest
- Boreal forest

Biomes at all latitudes
- Mountains
- Northern tundra
- Desert
- Ice

B

Biomes

Table 9.3	**Botanical Terms**
TERM	**DESCRIPTION**
Broad-leaved	Has wide, flat leaves; usually deciduous
Coniferous	Bears cones; usually needle-leaved and evergreen
Deciduous	Loses all leaves in one season, usually winter
Evergreen	Loses leaves gradually so it always has leaves
Herbaceous	Lacking woody tissue
Needle-leaved	Has narrow, needle-like leaves; usually evergreen
Sclerophyllous	Has hard, leathery, and waxy leaves
Woody	Grows rigid trunks and stems

climate classification systems is the **Köppen** (pronounced KUHR-pen) **climate classification system**, first published by Wladimir Köppen in 1884. The Köppen system categorizes climates based on the temperature and precipitation characteristics of a region. (Appendix V, "The Köppen Climate Classification System," provides further detail on the Köppen climate classification system.)

Because of the strong correspondence between vegetation structure and climate, the Köppen climate categories were originally based on the type of natural vegetation growing in an area. There are six major Köppen climate groups containing 25 climate zones. The climate zones and Earth's biomes are mapped in **Figure 9.3**.

We can portray biomes graphically by plotting temperature and precipitation. As shown in

Figure 9.3 Climate zone and biome maps. A comparison of the spatial distributions of (A) Köppen climate zones and (B) biomes shows that the two correspond roughly. Extreme events such as droughts and disturbances such as fire also determine the spatial distribution of vegetation. For this reason, biomes and Köppen climate zones do not overlap precisely.

Web Map
Climate and Biomes
Available at
www.saplinglearning.com

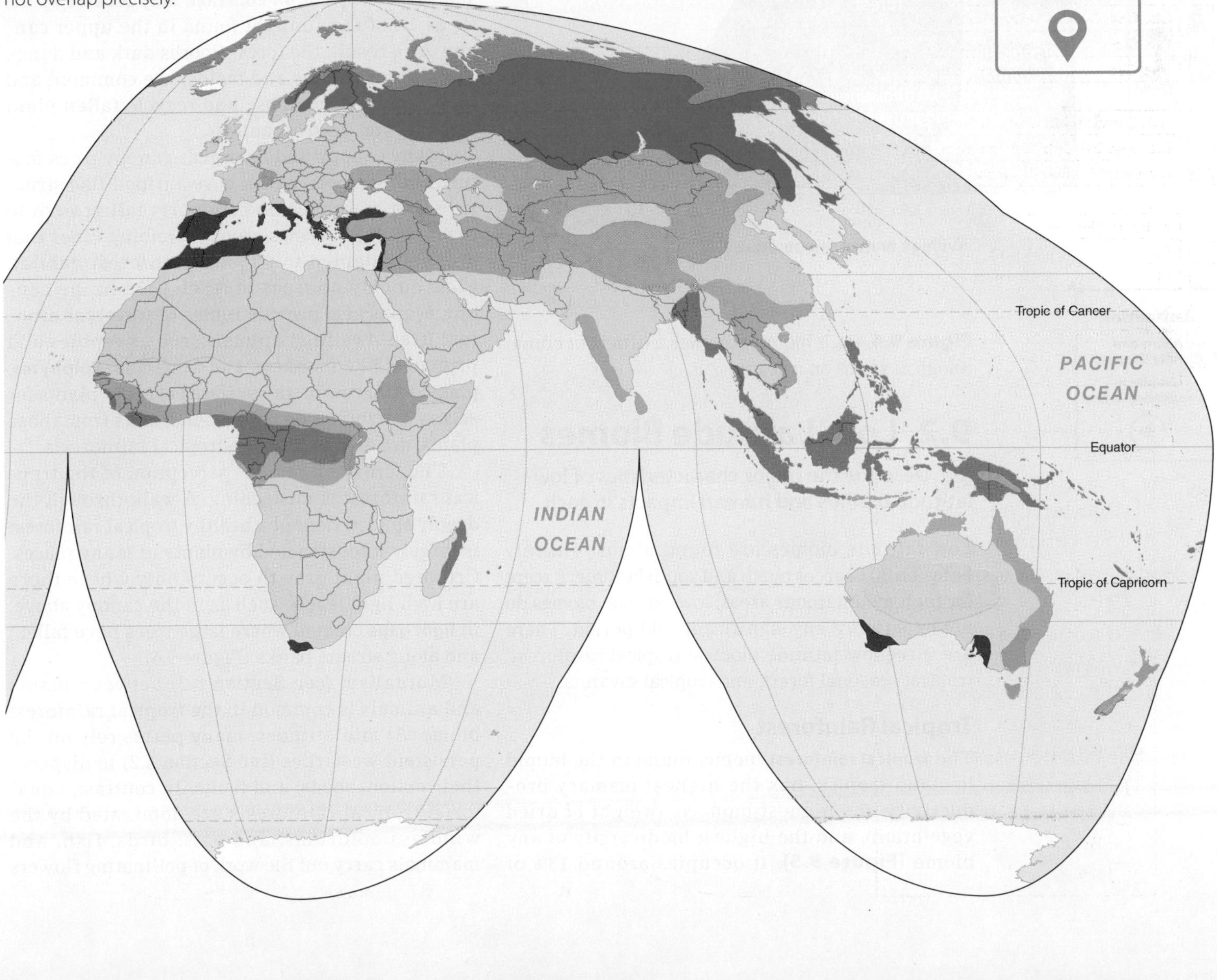

Figure 9.4 Biome and climate diagram. This diagram portrays the relationship between biomes and climates graphically. Tropical rainforests occupy the climates with the highest precipitation and temperature, represented by the ecological space at the top left. Tundra occupies the climates with the lowest precipitation and temperature, at the lower right, and subtropical desert occupies warm and dry climates, at the lower left.

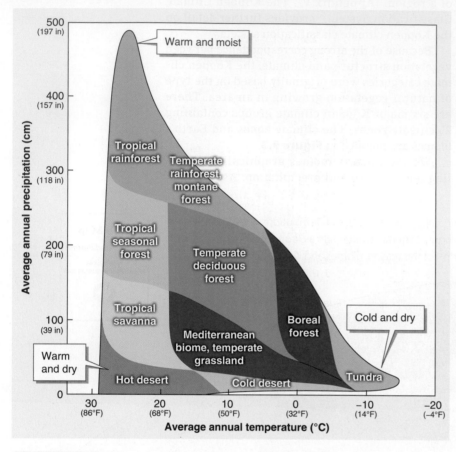

Animation

Biome and Climate Diagram

Available at
www.saplinglearning.com

Figure 9.4, each biome occupies a different climatological space on a graph.

9.2 Low-Latitude Biomes

◎ **Describe the major characteristics of low-latitude biomes and human impacts in each.**

Low-latitude biomes are found almost entirely between 30 degrees north and south latitude. Except for high mountainous areas, low-latitude biomes do not experience any significant cold period. There are three low-latitude biomes: tropical rainforest, tropical seasonal forest, and tropical savanna.

Tropical Rainforest

The **tropical rainforest** biome, found in the humid lowland tropics, has the highest primary productivity, the highest biomass (weight of dried vegetation), and the highest biodiversity of any biome **(Figure 9.5)**. It occupies around 13% of

Earth's land surface and accounts for about 40% of the world's biodiversity. A great variety of organisms in many different groups, including insects, birds, mammals, and amphibians, live in this biome. Worldwide, for example, this biome is home to about 250 species of *primates,* such as humans, gorillas, lemurs, and monkeys. All primates are mammals, and all have nails instead of claws, flexible hands and feet, good eyesight, and high intelligence. With the exception of humans and the Japanese macaque (*Macaca fuscata*), also known as the snow monkey, all primates live in tropical or subtropical areas. Most of them live in the tropical rainforests.

In the tropical rainforest biome, water is available in every month, and competition for light is a major limiting factor. The tropical rainforest biome develops a layered structure, with each layer adapted to the light conditions found there **(Figure 9.6** on page 272). The topmost layer, called the *emergent canopy*, consists of trees that protrude above the canopy into full sunlight. The shaded *forest floor*, in contrast, may receive as little as 1% of the sunlight found in the upper canopy. As a result, the forest floor is dark and damp. Fungi (mushrooms and molds) are common, and they rapidly decompose and recycle fallen plant material and dead organisms.

Many canopy and emergent-canopy trees feature **buttress roots**, which have a tripod-like structure that stabilizes and supports tall growth to reach the light. **Lianas**, woody climbing vines that are well adapted to tropical rainforest habitat, grow quickly up trees to reach light in the canopy. Lianas also provide routes of travel for arboreal (tree-dwelling) animals such as reptiles and primates, like monkeys and tamarins. **Epiphytes**, plants that grow on the surfaces of other plants for access to light but do not take nutrients from those plants, are also common in tropical rainforests.

The "crowded jungle" perception of the tropical rainforest is misleading. A walk through the deeply shaded floor of a healthy tropical rainforest is largely unobstructed by plants in many places. Crowded plant growth occurs only where there are high light levels, such as in the canopy above, in **light gaps** created where large trees have fallen, and along stream banks (Figure 9.6).

Mutualism (see Section 8.3) between plants and animals is common in the tropical rainforest biome. At midlatitudes, many plants rely on the persistent westerlies (see Section 5.2) to disperse their pollen, seeds, and fruits. In contrast, equatorial tropical rainforests are dominated by the windless doldrums, and bats, birds, fish, and mammals carry out the work of pollinating flowers

Figure 9.5 Tropical rainforest. (A) This photo of a tropical rainforest near Sandakan, Sabah, Malaysia, is of lianas growing up the trunks of tall trees to reach the light above. (B) This climate diagram for Sandakan shows that in tropical rainforests, the average annual temperature does not drop below 18°C (64°F), and an average of at least 250 cm (98 in) of precipitation falls each year. (C) Although tropical rainforest can occur within about 25 degrees north and south of the equator, this biome is found mainly within 10 degrees of the equator. About half of all tropical rainforest is in the Amazon Basin in South America. *(Mattias Klum/Getty Images)*

A

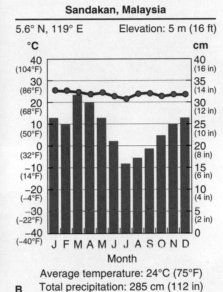

B

Sandakan, Malaysia

5.6° N, 119° E   Elevation: 5 m (16 ft)

Average temperature: 24°C (75°F)
Total precipitation: 285 cm (112 in)

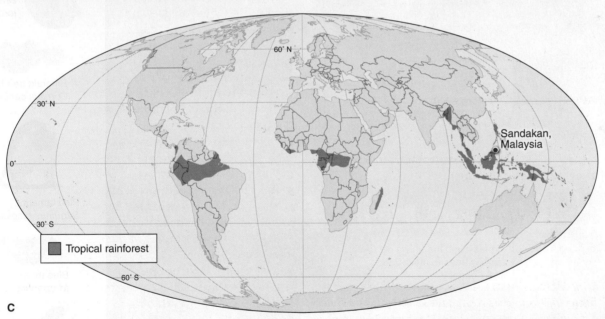

C

Tropical Rainforest	
Köppen climate zone	Tropical wet: Af
Vegetation	Broad-leaved evergreen trees, lianas, epiphytes
Notable features	ITCZ-dominated, no dry or cold season; layered forest structure and intense competition for light; very high biodiversity
Human footprint	Deforestation due to agriculture and livestock ranching, burning, and urbanization

Figure 9.6 Amazon rainforest layered structure. There are four layers of vegetation in many tropical rainforests: the forest floor, the understory, the canopy, and the emergent canopy. The Amazon rainforest's structure, the typical height of each canopy, and a few common species are illustrated here.

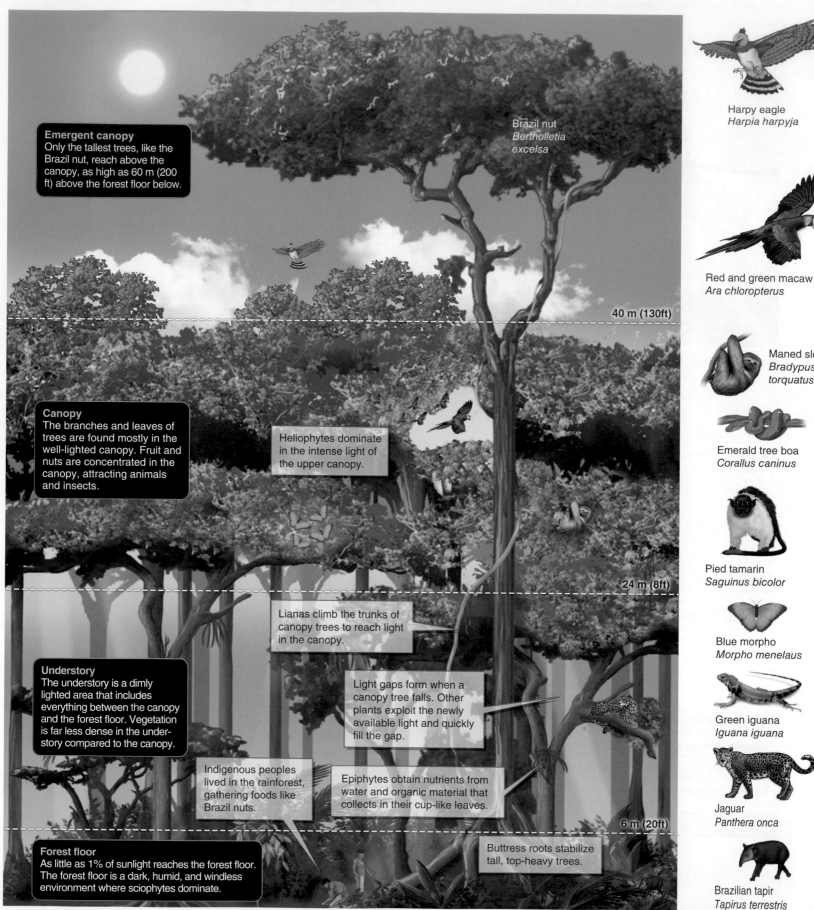

Emergent canopy
Only the tallest trees, like the Brazil nut, reach above the canopy, as high as 60 m (200 ft) above the forest floor below.

Brazil nut
Bertholletia excelsa

40 m (130ft)

Canopy
The branches and leaves of trees are found mostly in the well-lighted canopy. Fruit and nuts are concentrated in the canopy, attracting animals and insects.

Heliophytes dominate in the intense light of the upper canopy.

24 m (8ft)

Lianas climb the trunks of canopy trees to reach light in the canopy.

Understory
The understory is a dimly lighted area that includes everything between the canopy and the forest floor. Vegetation is far less dense in the understory compared to the canopy.

Light gaps form when a canopy tree falls. Other plants exploit the newly available light and quickly fill the gap.

Indigenous peoples lived in the rainforest, gathering foods like Brazil nuts.

Epiphytes obtain nutrients from water and organic material that collects in their cup-like leaves.

6 m (20ft)

Forest floor
As little as 1% of sunlight reaches the forest floor. The forest floor is a dark, humid, and windless environment where sciophytes dominate.

Buttress roots stabilize tall, top-heavy trees.

Harpy eagle
Harpia harpyja

Red and green macaw
Ara chloropterus

Maned sloth
Bradypus torquatus

Emerald tree boa
Corallus caninus

Pied tamarin
Saguinus bicolor

Blue morpho
Morpho menelaus

Green iguana
Iguana iguana

Jaguar
Panthera onca

Brazilian tapir
Tapirus terrestris

Picture This Lost Cities of the Amazon

As deforestation clears away the rainforest in Brazil, hundreds of mysterious human-built trench systems previously hidden by the forest are being revealed. The one shown here is near the city Rio Branco in western Brazil. Scientists call these trench systems *geoglyphs*. Some of the oldest geoglyphs were built as early as 2,000 years ago, and it is not clear why they were built. They indicate that large permanent settlements, or "lost cities," once existed in the rainforest. Most scientists in the past had assumed that there were few permanent human settlements in the Amazon. But now, based on the widespread occurrence of geoglyphs, some scientists think that at its peak, the Amazon lowland human population numbered as many as 10 million people.

Amazon rainforest soils are typically nutrient-poor and reddish in color (see Section 11.1). However, in many of the geoglyph locations, the soil, called *terra preta*, or "black earth," is uncharacteristically rich with organic matter and dark in color. The inhabitants of these ancient settlements built up the soil with charcoal, their own waste, animal bones, and food scraps. These modifications would have made the soil more suitable for growing crops to feed large populations.

(*Pulsar Imagens/Alamy*)

Consider This

1. Is *terra preta* an anthropogenic soil? Explain.

2. What does the information presented here indicate about the long-term human impacts on the forests of Brazil?

and dispersing seeds and fruit. This is why many tropical plants produce large, brightly colored flowers and nutritious seeds and fruits. In some tropical rainforests, up to 90% of plant species are pollinated and dispersed by animals.

Where there is volcanic activity, such as throughout Indonesia, tropical rainforest soils are young and nutrient-rich because they are continually replenished by volcanic ash. However, most tropical rainforest soils (called *oxisols*, see Section 11.1) in South America and Africa are poor in nutrients. *Soil weathering* (chemical disintegration) in the warm, humid climate and *leaching* by heavy rainfall move nutrients deep into the soil, out of the reach of plant roots. For this reason, decaying organic material is quickly taken up by

the shallow root systems of the forest vegetation rather than entering the soils.

Soils in the Amazon rainforest are particularly nutrient-poor due to leaching. People living in the Amazon long ago modified the soils to support farming. Those modified soils are still being discovered today, as the **Picture This** feature above discusses.

The Human Footprint in the Tropical Rainforest Biome

Natural resources are like the contents of full bank accounts: Spend all the money in one shopping spree, and it disappears. Carefully manage it, live off the interest, and it can last generations. Rainforests and other natural resources can be similarly spent or carefully managed. Today, undisturbed tropical rainforest is quickly being lost through deforestation. **Figure 9.7A** shows the typical sequence of events that cause deforestation in the Amazon rainforest. Encouragingly, the rate of loss of the Brazilian forest has decreased in recent years, due mostly to national and international pressure to save the Amazon rainforest **(Figure 9.7B)**. The important issue of habitat loss and fragmentation is further explored in the Geographic Perspectives, "The Value of Nature," at the end of this chapter.

For each geographic region, unique factors drive rainforest loss. As Figure 9.7A illustrates, hardwood logging, cattle ranching, and soybean and sugarcane cultivation are driving rainforest loss in the Amazon. In contrast, Indonesia is losing its rainforests largely to palm oil plantations. Orangutans have lost over 80% of their original habitat due to forest clearance for these plantations **(Figure 9.8)**.

The tropical rainforest biome is home to approximately 40% of all species on Earth, making this biome central to efforts to estimate the *global extinction rate*. The global extinction rate is in large part a reflection of the rate of deforestation in this biome. There are anywhere from 5 million to 8 million species worldwide. About 1% to 5% of those species are lost each decade, mostly because of habitat loss—and mostly in tropical rainforests.

Consumers of many of the products that are driving rainforest loss—such as palm oil,

Figure 9.7 Economic development of the Amazon.

(A) Deforestation in the Amazon rainforest often follows the sequence outlined here. Increasingly, the slash-and-burn step is skipped, as commercial logging is followed directly by forest burning and commercial-scale cattle ranching and agriculture. (B) This graph shows annual Brazilian forest loss between 1988 and 2017, in square kilometers per year (green bars). In 2017, approximately 6,600 km² (2,550 mi²) of rainforest was lost. The red trend line shows the percentage of remaining original forest compared to its extent in 1970. In 2017, about 81% of the Brazilian forest remained. *(Left. Stuart Wilson/Getty Images; center. © Julio Etchart/Alamy; right. Guido van der Werf, Vrije Universiteit, Amsterdam)*

1. Roads and logging

Logging companies or local political bodies build roads into remote, untouched forested regions. Widely spaced valuable hardwoods, such as teak and mahogany, are selectively removed from the forest. The remaining forest is left untouched.

2. Subsistence slash-and-burn agriculture

Logging roads make the forest accessible to poverty-stricken, small-scale, modern *subsistence* farmers who cut down the forests (*slash* them), burn the vegetation, and then plant crops. The nutrient-poor soils require that farmers move on after 3 to 4 years to another plot, where they must burn and plant again in a continuing *slash-and-burn* cycle.

3. Commercial ranching and agriculture

When subsistence slash-and-burn agriculture is no longer feasible, the lands are used to raise cattle. About 90% of all deforested areas in the Amazon are currently being grazed by cattle. Beef and leather are exported mainly to Asian, North American, and European markets. Soybeans and sugarcane are grown for export and to make *biofuels* for vehicles in Brazil.

A

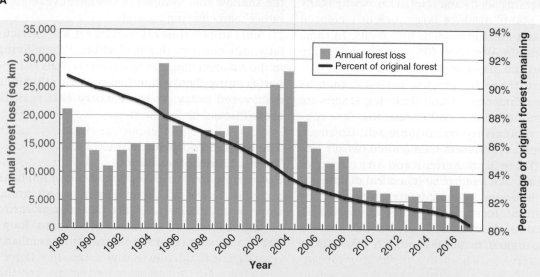

B

A   B   C

Figure 9.8 Orangutans and palm oil. (A) Orangutans (*Pongo pygmaeus* and *Pongo abelii*) are among the most intelligent land animals. Both species of orangutan are listed as "critically endangered," one step away from extinction in the wild. (B) This map shows their current range in the rainforests of Sumatra and Borneo. (C) Over 85% of the world's palm oil plantations are in Malaysia and Indonesia. Palm oil has a wide range of uses, from cosmetics to foods to lubricants and, increasingly, biofuels for cars and trucks. The oil is made from the seeds of the African oil palm (*Elaeis guineensis*).
(A. Guenter Guni/Getty Images; C. Michael Thirnbeck/Getty Images)

soybeans, beef, and hardwood lumber—are becoming increasingly aware of the connection between these products and the forests from which they come. To help consumers make informed choices, the European Union, for example, passed a law in 2014 requiring food products containing palm oil to be labeled as "palm oil" rather than something misleading, such as "vegetable oil." Likewise, a growing number of companies require that imports of palm oil be sourced from growers that meet stringent sustainability requirements, including the use of farming methods that do not cause deforestation. This allows them to label their palm oil product as "sustainably grown" in order to cater to consumers who are aware of and concerned about this problem. Roughly 45 million metric tons of palm oil are produced each year, and about 13% (6 million metric tons) is certified as sustainable. This percentage is small but growing.

Tropical Seasonal Forest

The **tropical seasonal forest** biome is often mistaken for tropical rainforest, but the tropical seasonal forest is distinguished from the rainforest biome by its winter season of reduced precipitation. The tropical seasonal forest is sometimes considered an *ecotone* (a transition between two biomes) for the tropical rainforest and the tropical savanna. The tropical seasonal forest biome is found in the warm lowland tropics bordering the tropical rainforest **(Figure 9.9** on the following page**)**. Many trees in this biome are broad-leaved and deciduous. A *deciduous tree* or shrub sheds its leaves, leaving bare branches. In the tropical seasonal forest biome, many trees shed their leaves in response to the winter dry season.

There are only three dominant layers in most tropical seasonal forests: canopy, shrubs, and forest floor. The canopy is lower and more open than that of the tropical rainforest, so more light reaches the forest floor. The diversity of insects, birds, mammals, reptiles, and amphibians is, in many cases, nearly as high as that of the tropical rainforest.

The Human Footprint in the Tropical Seasonal Forest Biome

Forests burn readily during the relatively dry season, making the tropical seasonal forest extremely vulnerable to fire. Increasing pressures from growing human populations, coupled with poverty, force people into the forest for subsistence farming. The soils of the tropical seasonal dry forest are better suited for crops and grazing than are the soils of tropical rainforests.

Tropical Savanna

Tropical seasonal forest transitions into tropical savanna, which is centered at about 25 degrees north and south latitude **(Figure 9.10** on page 277**)**. **Tropical savanna**, a woodland biome, featuring widely spaced trees with a continuous cover of grass, is characterized by wet summers and dry winters. The winter dry period is much drier and lasts much longer in tropical savanna than in tropical seasonal forest—as long as 7 months in some locations. In the drier portions of tropical savanna, where the winter dry season lasts 6 months or more, a type of tropical savanna called *thorn woodland* is found. In thorn woodland, which differs from more common tropical savanna, there are more shrubs than trees, and vegetation is tough and thorny in response to frequent fires, grazing pressure by animals, and the prolonged dry season.

Figure 9.9 Tropical seasonal forest. (A) This tropical seasonal forest at the Mayan archaeological site Tikal in eastern Guatemala, is located near Belmopan, Belize. (B) This climate diagram for Belmopan shows that the average annual temperatures in this biome are similar to those of the tropical rainforest biome, never dropping below 18°C (64°F). Precipitation averages roughly 150 to 275 cm (60 to 108 in) per year. (C) Most tropical seasonal forest lies within the tropics. In India, however, it was once found as far north as 30 degrees latitude (but has been converted to agriculture). *(Mlenny/Getty Images)*

A

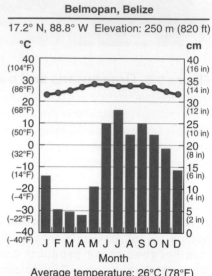

Belmopan, Belize

17.2° N, 88.8° W Elevation: 250 m (820 ft)

Average temperature: 26°C (78°F)
Total precipitation: 194 cm (76 in)

B

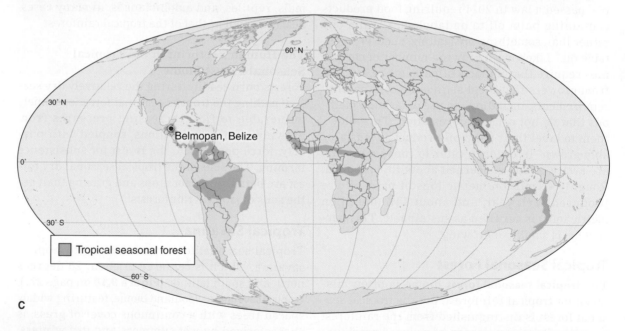

Belmopan, Belize

☐ Tropical seasonal forest

C

Tropical Seasonal Forest	
Köppen climate zone	Tropical monsoon: Am
Vegetation	Broad-leaved deciduous trees, shrubs
Notable features	ITCZ-dominated, some water stress in winter; vulnerable to winter fire; high biodiversity
Human footprint	Burning, mostly converted to agriculture

Figure 9.10 Tropical savanna. (A) An African savanna elephant (*Loxodonta africana*) roams the tropical savanna near the Okavango Delta, Botswana. (B) This climate diagram for Gaborone, Botswana indicates that in tropical savanna, average annual temperatures range from 15°C to 30°C (59°F to 86°F). (Note that Gaborone is in the Southern Hemisphere, so the summer rainy season occurs mainly in November through March.) Winters are warm. Annual average rainfall ranges from 40 to 175 cm (16 to 68 in). Thorn woodland may receive as little as 30 cm (12 in) of rainfall per year. (C) Tropical savanna has many local names, as labeled here. Half of all tropical savanna is found in Africa, where the biome covers about 65% of the continent. *(AfriPics.com/Alamy)*

A

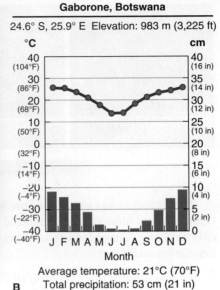

B

Gaborone, Botswana

24.6° S, 25.9° E Elevation: 983 m (3,225 ft)

Month

Average temperature: 21°C (70°F)

Total precipitation: 53 cm (21 in)

C

Tropical Savanna	
Köppen climate zones	Tropical savanna: Aw; subtropical steppe: BSh
Vegetation	Broad-leaved deciduous trees, woody and thorny shrubs, grass and forb understory
Notable features	Water stress in winter, summer rain from the ITCZ; frequent fires; plant adaptations to dry season and fire; large grazing animals (in Africa)
Human footprint	Overgrazing by livestock leading to desertification; hunting and poaching of large animals

Figure 9.11 African savanna grazing sequence. Throughout Africa's savanna biome, different herbivorous mammals eat different parts of plants at different times. This variation allows large populations of many species to coexist in the same region without exhausting the resources. As the summer rains diminish and the vegetation dries, the animals migrate to a different area with new growth. This diagram shows a common progression of animals that takes places over weeks and months on the savanna. *(1. James Hager/robertharding/Getty Images; 2. Jim Richardson/Getty Images; 3. Vadim Petrakov/Shutterstock; 4. Martin Harvey/Getty Images)*

1. Water buffalo

After monsoon rains, water buffalo (*Bubalus bubalis*) arrive first, eat leaves and tall river grasses, and then move on to areas where there is new growth.

2. Zebras

The plains zebra (*Equus quagga*) follow the water buffalo. Zebras eat the shortened grasses and new grass growth stimulated by the trampling and grazing of water buffalo.

3. Wildebeests

Wildebeests (*Connochaetes gnou* and *C. taurinus*) eat taller grasses that zebras do not eat.

4. Topi

Topi (*Damaliscus korrigum*), eland (*Taurotragus oryx*), and Thompson's gazelles (*Eudorcas thomsonii*) eat herb and grass growth stimulated by earlier animal activity.

Time →

The tropical savanna has low biomass and relatively low biodiversity compared to the tropical rainforest. Three factors strongly influence its vegetation structure:

1. *Seasonally intense rainfall:* The rainfall arrives with the ITCZ in summer, all within a few months, and often in the form of thunderstorms that produce heavy downpours. Deciduous plants grow during the rainy season and are dormant during the dry season.

2. *Fire:* During the warm winter dry season, abundant fuel and dry conditions favor frequent wildfires. Many savanna plants are adapted to survive fire.

3. *Grazing pressure:* More than 90 species of grazing *ungulates* (hoofed mammals) roam the African tropical savanna.

Although the tropical savanna has low overall biomass, it is able to support the world's largest and densest grazing animal community because the animals do not all graze at the same time **(Figure 9.11)**.

The Human Footprint in the Tropical Savanna Biome

One prevalent threat to the tropical savanna biome is overgrazing by livestock, particularly cattle, sheep, and goats. Increased burning by people to stimulate grass growth for livestock is also degrading savanna woodlands and threatening the habitat of native grazing ungulates. In addition, populations of these native animals have declined greatly as a result of hunting for trophies, elephant ivory, and rhinoceros horn **(Figure 9.12)**. Many large game preserves and parks have been

Figure 9.12 Saving large animals. (A) Hunting pressures have reduced the once-widespread populations of the black rhino (*Diceros bicornis*) in Kenya. In 2011, a subspecies of the black rhino, called the western black rhino, was officially declared extinct. The remaining black rhino populations are critically endangered. (B) Wildlife wardens are armed to protect animals from poachers. *(A. Martin Harvey/Getty Images; B. Raffaele Meucci/AGE Fotostock)*

A

B

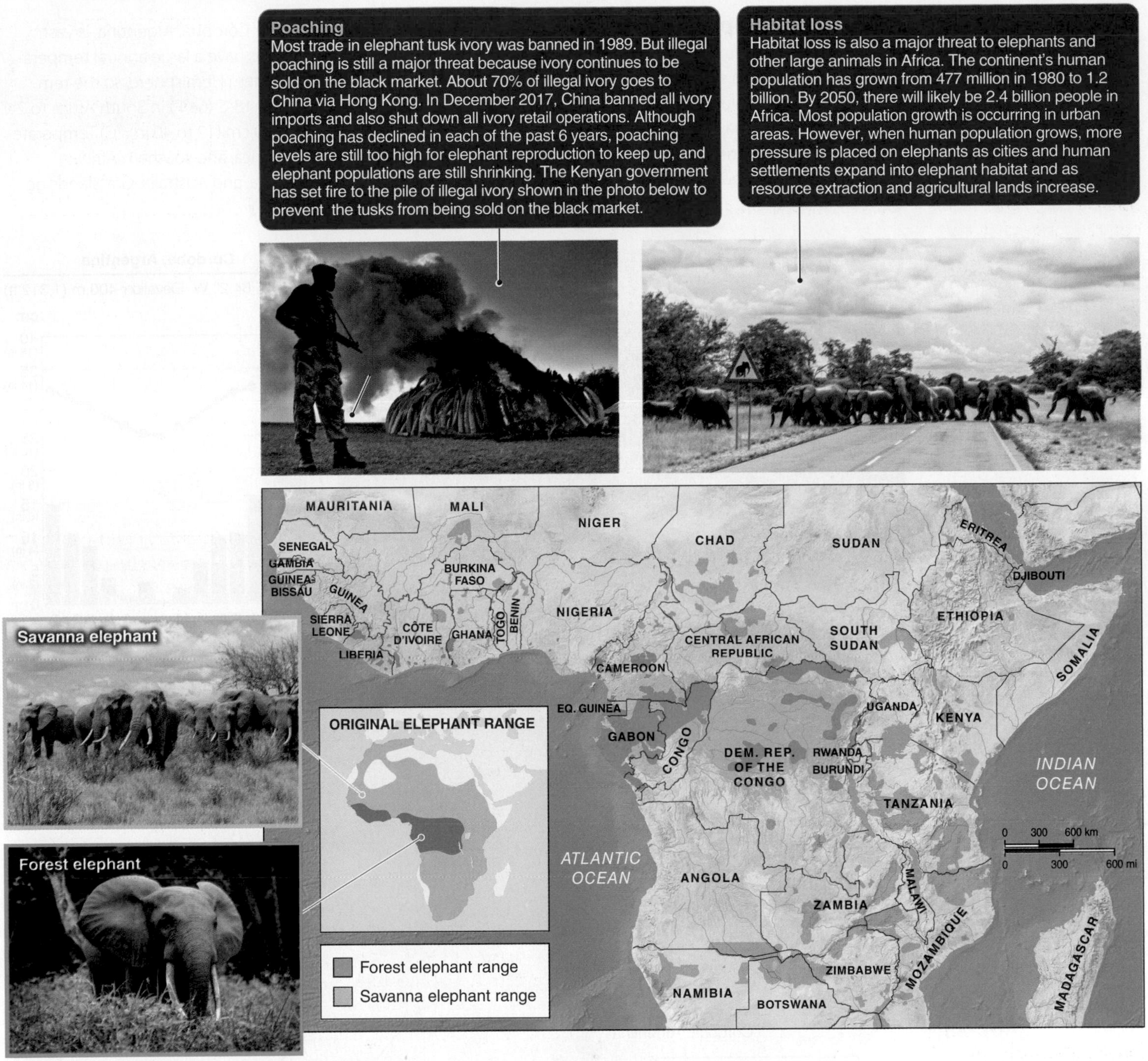

Poaching
Most trade in elephant tusk ivory was banned in 1989. But illegal poaching is still a major threat because ivory continues to be sold on the black market. About 70% of illegal ivory goes to China via Hong Kong. In December 2017, China banned all ivory imports and also shut down all ivory retail operations. Although poaching has declined in each of the past 6 years, poaching levels are still too high for elephant reproduction to keep up, and elephant populations are still shrinking. The Kenyan government has set fire to the pile of illegal ivory shown in the photo below to prevent the tusks from being sold on the black market.

Habitat loss
Habitat loss is also a major threat to elephants and other large animals in Africa. The continent's human population has grown from 477 million in 1980 to 1.2 billion. By 2050, there will likely be 2.4 billion people in Africa. Most population growth is occurring in urban areas. However, when human population grows, more pressure is placed on elephants as cities and human settlements expand into elephant habitat and as resource extraction and agricultural lands increase.

Figure 9.13 SCIENTIFIC INQUIRY: What are the biggest threats to African elephants? Based on genetic evidence, scientists recognize two species of African elephant, the African savanna elephant (*Loxodonta africana*) and the smaller African forest elephant (*Loxodonta cyclotis*). Both species are considered vulnerable to extinction in the wild, according to the International Union for Conservation of Nature (IUCN). There were originally about 27 million African elephants living throughout the continent (inset map). The period 2014–2016 saw the most severe decline in African elephant populations in the past quarter century: In 2014, there were 700,000 African elephants in the wild, and by 2017, the numbers had dropped to roughly 415,000, a 40% decline from 2014. The two most prominent factors threatening African elephants, illegal poaching (*hunting*) for ivory and habitat loss, are explained here.   *(Top left. Carl De Souza/AFP/Getty Images; top right. SoopySue/Getty Images; bottom left-up. Nigel Pavitt/Getty Images; bottom left-down. Education Images/UIG via Getty Images)*

established throughout Africa to protect large animals. These parks also generate revenue from *ecotourists*, travelers who seek to visit natural places.

Increasingly, conflicts between African elephants and people have arisen in parts of Africa as the human population has increased and expanded into elephant territory. Today, most African elephant populations are in steep decline as a result of poaching and habitat loss **(Figure 9.13)**.

Figure 9.14 Temperate grassland. (A) The pampas grassland biome near Córdoba, Argentina, is vast. (B) This climate diagram for Córdoba indicates that most temperate grassland regions have a large annual temperature range, with warm summers and cold winters. (Note that Córdoba is in the Southern Hemisphere, so the temperature is coolest in June through August.) Average annual temperatures vary from 18°C (64°F) in South Africa to 2°C (35°F) or less in Canada and Eurasia. Precipitation in grasslands ranges from 30 to 100 cm (12 to 40 in). (C) Temperate grasslands occur extensively in interior North America, Eurasia, southern South America, and southern Africa. There are also scattered temperate grasslands in Madagascar, New Zealand, California, and Australia. Grasslands go by many local names, as shown on this map. *(Eduardo Pucheta Photo/Alamy)*

A

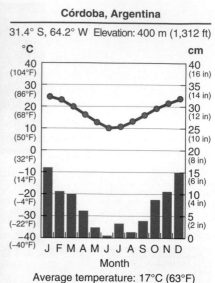

Córdoba, Argentina

31.4° S, 64.2° W Elevation: 400 m (1,312 ft)

Month

Average temperature: 17°C (63°F)
Total precipitation: 87 cm (34 in)

B

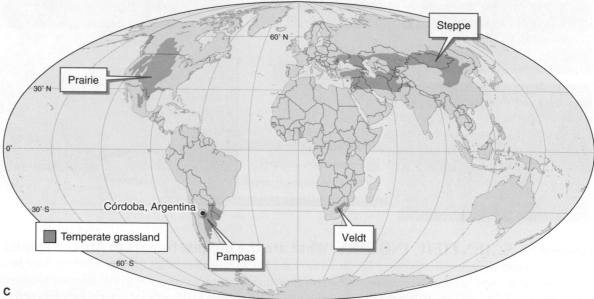

C

Temperate Grassland	
Köppen climate zones	Subtropical steppe: BSh; Midlatitude steppe: BSk
Vegetation	Continuous cover of grasses and forbs
Notable features	Aridity caused by interior locations and the subtropical high; strongly continental climate; moisture deficits in most or all months; too dry for forest development; frequent fires
Human footprint	Mostly converted to agriculture and rangeland

9.3 Midlatitude and High-Latitude Biomes

◎ **Describe the major characteristics of mid- and high-latitude biomes and human impacts in each.**

In this section, we continue traveling higher in latitude as we survey Earth's biomes. The midlatitudes, or *temperate* latitudes, are the transitional regions between warm subtropical air and cold polar air. Temperate biomes experience a large annual temperature range if they are far from the moderating effects of the oceans; they have a smaller annual temperature range if they are near the coast. The five biomes found at middle and high latitudes are temperate grassland, the Mediterranean biome, temperate deciduous forest, temperate rainforest, and boreal forest.

Temperate Grassland

The **temperate grassland** biome is largely dominated by grasses. It is characterized by significant moisture deficits for most of the year, natural fires, and grazing herbivores—all factors that keep trees from becoming established in most temperate grasslands. Temperate grasslands are found mostly between 30 and 60 degrees latitude in continental interiors. They go by several local names: *prairie* in North America, *pampas* in South America, *steppe* in Eurasia, and *veldt* in South Africa **(Figure 9.14)**.

About 90% of the plant biomass in temperate grasslands is grasses, but grasses comprise only about 20% of the species diversity. Forbs (non-grass herbs) such as milkweed (*Asclepias* spp.), purple coneflower (*Echinacea purpurea*), and black-eyed Susan (*Rudbeckia* spp.) greatly increase the plant biodiversity in grasslands in North America. Most of the living portions of grassland vegetation is found in the roots of grasses, which may have a biomass some three times greater than the aboveground portions of the plant.

Before its conversion to agriculture, the temperate grassland of North America was dominated by three prairie types: *tallgrass prairie* in the eastern Great Plains, *mixed-grass prairie* in the central Great Plains, and *short-grass prairie* in the western Great Plains **(Figure 9.15)**. The prominent grass species found in North America were buffalo grass (*Buchloe dactyloides*), little bluestem (*Schizachyrium scoparium*), and switchgrass (*Panicum virgatum*).

Figure 9.15 Map of prairie types of North America. The height of grasses decreases westward, reflecting the increasing aridity in interior North America. The grassland–desert ecotone lies on the western margin of the short-grass prairie.

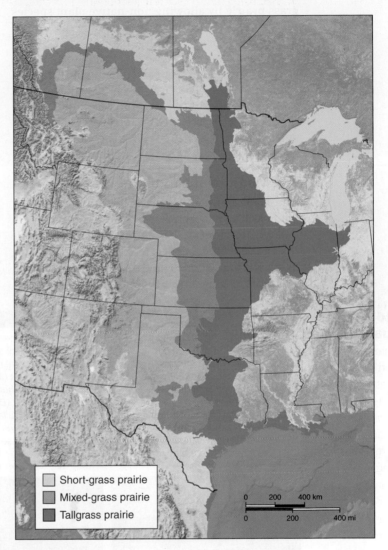

Short-grass prairie
Mixed-grass prairie
Tallgrass prairie

0   200   400 km
0   200   400 mi

The Human Footprint in the Temperate Grassland Biome

The geographic extent of the temperate grassland biome has been greatly reduced. This biome is a desirable place for people to live, farm, and raise livestock. Roughly 80% of the original North American prairie is now either developed or being used for commercial agriculture and cattle ranching. Many animal species populations collapsed with the loss of these grasslands in the North American grassland ecosystem during the early twentieth century. Prairie dogs (*Cynomys* spp.)

were important keystone species that are discussed further in **Figure 9.16**.

In areas where temperate grassland has not been cultivated for crops, extensive grazing by cattle has altered this biome. Cattle grazing has changed nearly all of the North American short-grass prairie by allowing non-native plant species to establish themselves and become dominant. Where cattle are grazed intensively, some 90% of the plant species are non-native. Light to moderate cattle grazing, however, can simulate the effects of grazing by the now-scarce bison, to which grasses are adapted, and may therefore be beneficial for the prairie ecosystem. Grassland restoration ecologists are increasingly seeing the value of managed cattle grazing as a stand-in for native grazers.

The Mediterranean Biome

The **Mediterranean biome** is the smallest of the biomes. It is characterized by hot, dry summers and winter rainfall. Three physical factors control the vegetation structure of the Mediterranean biome: (1) an extended summer dry season lasting 5 months or longer, (2) frequent wildfires, and (3) low soil nutrients and soil organic matter. Many plants in this biome are adapted to a pronounced dry season and fire. About half the area of this biome is found in the Mediterranean region of southern Europe and northern Africa **(Figure 9.17)**.

The Mediterranean biome has high biodiversity and high endemism, meaning that many species are geographically restricted to it and found nowhere else. The fynbos of South Africa and the chaparral of California, for example, each have about 6,000 different plant species, many of them unique to those areas. The Mediterranean region has more than 7,000. About half of all plant species in the Mediterranen biome are annuals (plants that live for a year or less and perpetuate themselves by seeds). A few perennials (plants that live longer than a year) are summer deciduous. Many plants of this biome have **sclerophyllous** (pronounced skler-AH-fuhlis) leaves—hard, leathery, waxy leaves adapted to reduce water loss and herbivory.

Many plants of the Mediterranean biome display fire-adapted traits, such as serotinous (fire-adapted) cones, crown sprouting, and thick bark (see Section 8.5). The ecological health of the

Figure 9.16 Prairie dog mutualism.

Prairie dogs are keystone species in North American temperate grasslands. They are mutualistic partners of many grassland organisms, including bison and pronghorn. Prairie dog burrowing improves plant growth in grasslands by aerating the soil and allowing water to better infiltrate the ground. The animals also add nitrogen to the soil with their fecal pellets. *(Left. Merilee Phillips/ Getty Images; center. Werner & Kerstin Layer Naturfoto/Getty Images; right. Danita Delimont/ Getty Images)*

Bison

Bison (*Bison bison*) prefer to forage near prairie dog colonies, or "towns," where the prairie dogs' activities improve plant growth. Bison keep the grass around prairie dog colonies short, which gives the prairie dogs a clear view of approaching predators such as birds of prey, foxes, and ferrets.

Prairie dogs

There are five species of prairie dogs in North America. All of them provide habitat or food for many other species, including:
- Black-footed ferret (*Mustela nigripes*)
- Burrowing owl (*Athene cunicularia*)
- Ferruginous hawk (*Buteo regalis*)
- Golden eagle (*Aquila chrysaetos*)
- Swift fox (*Vulpes velox*)
- Grasshopper mouse (*Onychomys leucogaster*)

Pronghorn

Pronghorn (*Antilocapra americana*) feed on forbs in preference to grasses, which provide less nutritional benefit. Like bison, they find good forage near prairie dog colonies and help to keep vegetation near the colonies short.

Mutualism   Mutualism

Conservation status

Between 10 million and 75 million bison originally occupied the grasslands of North America. By 1884, the bison was almost extinct as a result of exploitation by European immigrants. Due to conservation efforts, there are about 15,000 wild-ranging bison in North America today. About 500,000 are raised in captivity for food.

Conservation status

There were originally several billion prairie dogs in North America. Farming and intentional extermination reduced their numbers. By the turn of the twenty-first century, they were at a fraction of their former numbers, and currently all five species are at risk due to human activities.

Conservation status

Pronghorn may originally have been more numerous than bison. By 1920, there were about 13,000 left. Thanks to careful management, they have increased to nearly 1 million.

Figure 9.17 The Mediterranean biome. (A) This photo taken near Tunis, Tunisia, shows the shrubland vegetation structure of the Mediterranean biome. The trees seen here are drought- and fire-adapted Aleppo pine (*Pinus halepensis*). (B) This climate diagram for Tunis illustrates an average annual temperatures in the Mediterranean biome range from 5°C to 20°C (41°F to 68°F), and annual precipitation totals average 50 to 120 cm (19 to 47 in). (C) The Mediterranean biome is centered at about 35 degrees latitude, and it is geographically isolated on the western margins of five different continents. With the exception of the Mediterranean region itself, all of these locations have cold ocean currents offshore that inhibit evaporation of seawater and increase aridity. The Mediterranean biome goes by several different local names, as shown on this map. *(Fletcher & Baylis/Science Source)*

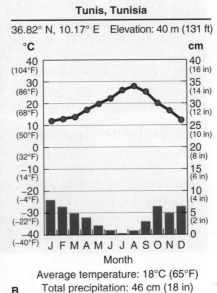

A

B

Tunis, Tunisia

36.82° N, 10.17° E   Elevation: 40 m (131 ft)

Month

Average temperature: 18°C (65°F)
Total precipitation: 46 cm (18 in)

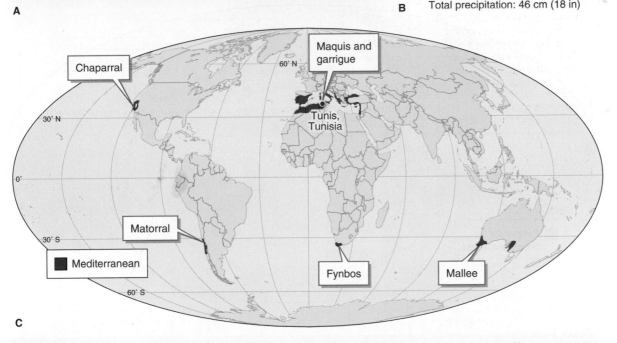

C

Mediterranean Biome	
Köppen climate zones	Mediterranean: Csa, Csb
Vegetation	Few trees; sclerophyllous shrubs; adapted to fire
Notable features	Aridity caused by the subtropical high and cold offshore ocean currents; mild winters, extended summer dry season; frequent fires; high biodiversity
Human footprint	Agriculture, non-native species, human development, fire suppression

Figure 9.18 Temperate deciduous forest. (A) This temperate deciduous forest is in Cherokee National Forest, near Chattanooga, Tennessee. (B) This climate diagram for Chattanooga shows an average annual temperatures in the temperate deciduous forest biome vary from 2°C to 20°C (35°F to 68°F). Average annual precipitation ranges from 50 to 250 cm (20 to 98 in). (C) Temperate deciduous forest is or was located mainly in eastern North America and western Europe, and it was also formerly found in eastern Asia. To a lesser extent, it is found at midlatitudes in South America, Australia, and New Zealand (not shown here). *(Raymond Gehman/Getty Images)*

A

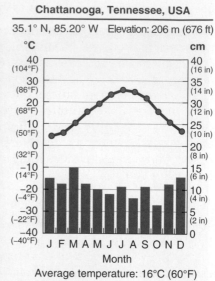

B

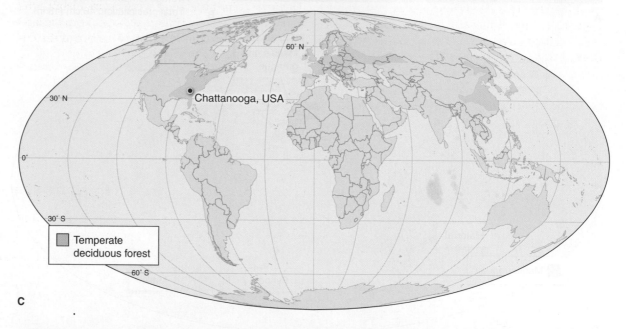

C

Temperate Deciduous Forests	
Köppen climate zones	Humid subtropical: Cfa, Cwa, Cwb; humid continental: Dfa, Dwa, Dfb, Dwb
Vegetation	Broad-leaved deciduous trees
Notable features	Winter storms caused by the subpolar low and summer thunderstorms; cold winters, warm summers; fall colors
Human footprint	Extensively converted to agriculture and urbanized settlement in China and Europe

Mediterranean biome is dependent on fire. Fires cycle nutrients back into the soil and stimulate germination of seeds and vegetation regrowth.

The Human Footprint in the Mediterranean Biome

The main anthropogenic agents of change in the Mediterranean biome are agriculture and urban development, overgrazing of livestock (particularly by sheep and goats around the Mediterranean Sea), fire suppression, and non-native plants. California has been particularly affected in this last regard, as more than 1,000 non-native plants have established populations there. The native perennial grasses of California have been almost completely replaced by agriculture and non-native annual species, mostly from the Mediterranean region. Many plants in the Mediterranean region, in turn, are from California.

Temperate Deciduous Forest

Temperate forests are the result of the precipitation caused by the subpolar low at midlatitudes. There are two kinds of temperate forest: temperate deciduous forest and temperate rainforest.

The **temperate deciduous forest** biome is dominated by trees that shed their leaves in winter in response to low temperatures. Examples of such trees include oak (*Quercus*), maple (*Acer*), elm (*Ulmus*), and beech (*Fagus*). This biome occurs at midlatitudes where the annual temperature range is large and winters bring below-freezing temperatures. As the map in **Figure 9.18** shows, this biome is located mainly in the Northern Hemisphere because there is more land at midlatitudes there than in the Southern Hemisphere.

The deciduous trees in these forests lose their leaves through *abscission*, a process triggered by changes in light and moisture conditions in the fall. Leaves often turn bright colors as they lose their green chlorophyll, revealing brighter-colored *anthocyanin* pigments beneath.

The Human Footprint in the Temperate Deciduous Forest Biome

In the temperate deciduous forest biome, nearly all of the **primary forest**—forest that has never been significantly altered by people—has been lost to agriculture and human settlement. The soils of the temperate deciduous forest are fertile and well suited for farming once the forest has been cleared. Logging for hardwood lumber has also reduced the extent of temperate deciduous forests.

Today, little of this biome remains in the British Isles, much of Europe, and eastern China. Instead, these regions support extensive agricultural systems. The forests in these areas were cleared beginning some 8,000 years ago. Human activities keep the land under agricultural production even though the climate is suitable for the return of forest **(Figure 9.19)**.

Figure 9.19 Agricultural fields in County Kerry, Ireland. Ireland, like most of the rest of northern Europe, has a climate that supports temperate deciduous forest. Throughout Europe, forests were cleared several thousand years ago for farming and livestock grazing. This entirely anthropogenic landscape is maintained by people. If left alone, it would develop into oak and pine forest through ecological succession over many centuries. *(AIMSTOCK/Getty Images)*

Forests that have been cleared or disturbed and regrown are called **secondary forests**. Secondary forests usually have less biodiversity than primary forests. In some areas, the temperate deciduous forest has rebounded after having once been cut. In eastern North America, for example, deciduous forests are more geographically extensive today than they were a century and a half ago because many farmers abandoned their farms and the forest grew back through plant succession (see Figure 8.22). In addition, the forests today are less damaged by acid rain than they were a century ago. Legislation such as the Clean Air Act of the 1970s (see Section 2.4) is responsible for this positive change.

Temperate Rainforest

Compared with temperate deciduous forest, temperate rainforest is geographically limited, and far fewer people ever see it. The **temperate rainforest** biome occurs where annual precipitation is high and temperatures are mild. It is characterized by large trees that form a dense canopy. A dense understory layer of vegetation lies beneath the trees. Epiphytic species, such as ferns, lichens, and bryophytes (mosses and their relatives), thrive in the canopy. In the Northern Hemisphere, dominant tree species include western red cedar (*Thuja plicata*), Sitka spruce (*Picea sitchensis*), and western hemlock (*Tsuga heterophylla*). The California redwood (*Sequoia sempervirens*) dominates the temperate rainforest found in northern California's coast. The California redwood holds the world record as the tallest tree: The Hyperion tree stands 115.9 m (380.3 ft) tall.

Many temperate rainforests are located along the western coasts of continents, where abundant precipitation, fog, and high humidity occur. Evergreen needle-leaved trees dominate this biome in the Northern Hemisphere. In the Southern Hemisphere, forests are composed of evergreen broad-leaved trees, such as the southern beech (*Nothofagus* spp.), podocarps (*Podocarpus* spp.), and eucalyptus (*Eucalyptus* spp.). As the climate diagram in **Figure 9.20** shows, the temperate rainforest receives more precipitation, and has a more moderate maritime climate, than the temperate deciduous forest.

The Human Footprint in the Temperate Rainforest Biome

The future of the temperate rainforest biome has become a divisive political issue, as different sides fight either to use it for profit or to preserve it. California redwoods, for example, make prized outdoor decking and furniture because of the rot-resistant qualities of the wood. Logging that began in the late 1800s and continued through much of the twentieth century cleared large expanses of old-growth groves of redwoods. Today, only about 5% of the original temperate rainforest in California, Oregon, and Washington has remained unlogged. In British Columbia, about 50% remains, and Alaska has about 90% of its original temperate rainforest, most of it in the Tongass National Forest, which has largely been protected from logging. Worldwide, about half of the temperate rainforest has been cut. In many areas, temperate rainforest is actively managed as a timber resource.

Boreal Forest

The **boreal forest** (called *taiga* in Eurasia) is a cold coniferous forest biome found at high latitudes in the Northern Hemisphere. The boreal forest is among the largest biomes, comprising about one-fourth of all forested land on Earth. This biome occurs in continental interiors where low winter temperatures and a short summer growing season are typical. As **Figure 9.21** on page 288 shows, the boreal forest is not found in the Southern Hemisphere due to the lack of interior continental climates at high latitudes there. Boreal forest vegetation is dominated by *coniferous*, or cone-bearing, trees, most of which are needle-leaved and evergreen. Dominant coniferous trees of this biome include pine (*Pinus*), spruce (*Picea*), fir (*Abies*), and larch (*Larix*). An understory of mosses, lichens, and herbaceous plants lies beneath the canopy.

Fire can be an important factor in the boreal forest during the summer. The fire-return interval ranges from a few decades in the southern portions of the boreal forest to over 1,000 years at the northern edge of the forest.

The Human Footprint in the Boreal Forest Biome

The boreal forest is the most sparsely human populated forested region on Earth. Vast tracts through Eurasia and North America have remained mostly unaltered by people. This is quickly changing, however, as world demand grows for forest products, such as paper and lumber, and minerals, such as petroleum. Recent open-pit mining in Alberta's tar sand deposits has generated considerable conflict between the energy industry's political representatives and those seeking cleaner fuels with fewer environmental impacts in the United States and Canada. The boreal forest is also one of the fastest-warming biomes due to anthropogenic climate change. On top of that, the warmer atmosphere has decreased snowfall and increased fire activity and intensity in the boreal forest as well.

Figure 9.20 Temperate rainforest. (A) This temperate rainforest is near Vancouver, British Columbia, Canada. Because of the abundant precipitation and mild temperatures, immense trees dominate temperate rainforests if they have not been logged. (B) This climate diagram for Vancouver, shows that summers are cool, and winters are mild. Average annual temperatures in the temperate rainforest biome range from about 3°C to 20°C (37°F to 68°F). Average annual rainfall ranges from 170 to 350 cm (67 to 138 in), and precipitation falls in all months. (C) The temperate rainforest biome is scattered throughout midlatitudes where precipitation is sufficient to support it. The largest intact temperate rainforests are along the coasts of western North America, southern Chile, and southeastern Australia. *(Steve Ogle/Getty Images)*

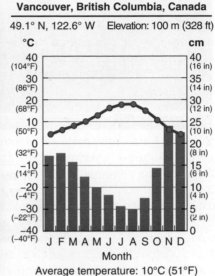

A

B

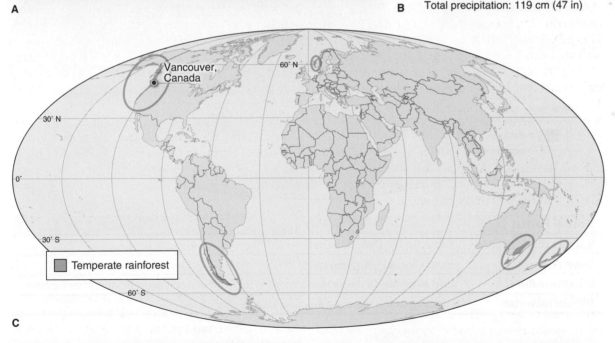

C

Temperate Rainforest	
Köppen climate zones	Marine west coast: Cfb, Cfc
Vegetation	Needle-leaved and broad-leaved evergreen trees and layered forest structure; epiphytes, bryophytes, and ferns
Notable features	Coastal locations and orographic precipitation; abundant precipitation brought by the subpolar low; mild winters and cool summers
Human footprint	Logging

Figure 9.21 Boreal forest. (A) This boreal forest is near Fort Smith, Northwest Territories, Canada. Conifers are tall and columnar to maximize the sunlight, which is always near the horizon, regardless of the time of day. (B) This climate diagram for Fort Smith shows an average annual temperatures in the boreal forest biome range from –5°C to 3°C (23°F to 37°F). Annual precipitation averages 40 to 200 cm (16 to 79 in). (C) The boreal forest extends across North America and Eurasia. It is centered at about 60 degrees north latitude.

(© Garry Foote)

Conifers are tall and columnar to maximize exposure to sunlight, which is always near the horizon.

A

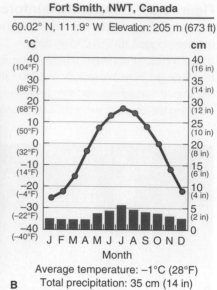

Fort Smith, NWT, Canada

60.02° N, 111.9° W   Elevation: 205 m (673 ft)

Average temperature: –1°C (28°F)
Total precipitation: 35 cm (14 in)

B

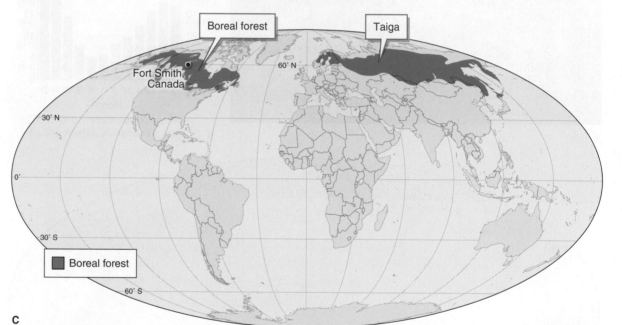

Boreal forest

Taiga

Fort Smith, Canada

Boreal forest

C

Boreal Forest	
Köppen climate zones	Subarctic: Dfc, Dwc, Dfd, Dwd
Vegetation	Needle-leaved evergreen trees
Notable features	Cold winters and a short summer growing season; found only in the Northern Hemisphere
Human footprint	Logging, oil and mineral extraction, climate change and fire

9.4 Biomes Found at All Latitudes

◎ **Describe the major characteristics of the biomes found at all latitudes and human impacts in each.**

Three biomes occur across a wide range of latitudes. We cover the biomes in this section using a moisture gradient, from wet to dry, rather than organizing according to latitude. We start with the montane forest biome, move into the relatively dry tundra, and conclude with the desert biome.

Montane Forest

The **montane forest** biome occurs where orographic lifting increases precipitation on the windward side of a mountain range (see Section 4.4). The

A

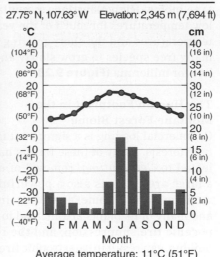

Creel, Chihuahua, Mexico

27.75° N, 107.63° W   Elevation: 2,345 m (7,694 ft)

Month

Average temperature: 11°C (51°F)
Total precipitation: 74 cm (29 in)

B

Figure 9.22
Montane forest.

(A) This montane forest is near Creel, Chihuahua, Mexico. (B) This climate diagram for Creel shows that the average annual temperatures vary greatly within the montane forest biome, depending on elevation and latitude. Average annual precipitation also varies significantly by latitude, with some areas receiving over 250 cm (98 in). As a rule, precipitation is always sufficient to support trees. (C) Montane forest is found in North America from the highlands of Mexico to southern Alaska. Other significant montane forests are found in the Andes in South America, the eastern African highlands, the European Alps, and the mountains of Asia.

(© Sebastian Rauprich)

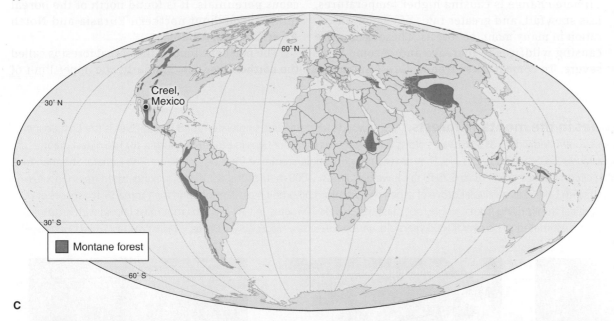

Creel, Mexico

Montane forest

C

Montane Forest	
Köppen climate zone	Cold mountain climates: H
Vegetation	Needle-leaved evergreen trees in the Northern Hemisphere, broad-leaved evergreen trees in the Southern Hemisphere
Notable features	Cool temperatures; cold winters at midlatitudes; low seasonality in the tropics
Human footprint	Logging, fire suppression, climate change, and fire

vegetation of the montane forest biome is needle-leaved in the Northern Hemisphere and broad-leaved in the Southern Hemisphere. Because it is found at a wide range of latitudes and elevations, the montane forest is one of the most climatologically diverse biomes **(Figure 9.22)**.

The montane forests of the Northern Hemisphere are dominated by species in the pine family, such as spruce, larch, pine, and fir. The pines do not occur in the Southern Hemisphere, and montane forests there are dominated by different tree species, depending on the location. Evergreen broad-leaved eucalyptus trees, for example, dominate Australia's montane forests. New Zealand's montane forests are dominated by evergreen broad-leaved southern beeches, and podocarps, as are Patagonia's in southern South America.

The world's oldest trees are found in the montane biome. High elevations at midlatitudes have low temperatures for much of the year. These low temperatures and the short growing season cause many tree species to grow slowly and allow them live for millennia **(Figure 9.23)**.

The Human Footprint in the Montane Forest Biome

Commercial logging is a significant factor in montane forests. Many of these forests have been subjected to *clear-cutting*: removal of all trees over a large area. Fire is also a powerful force in the montane forest biome, as we saw in Section 8.5, and many of its plants are adapted to fire. Anthropogenic fire suppression, and the resulting fuel buildup, can result in catastrophic fires that change the forest structure and species composition. Climate change is causing higher temperatures, less snowfall, and greater rates of evapotranspiration in many montane forests. These factors are causing wildfires to increase and become more severe. To combat these trends, forest managers use prescribed burns and selective cutting to reduce accumulations of fuel. Where these techniques are applied, they can be costly but effective.

Tundra

The **tundra** biome occurs at any elevation or latitude where it is too cold for trees to grow. There are three types of tundra: alpine tundra, found at high elevations; northern tundra, found at high latitudes in the Northern Hemisphere; and *Antarctic tundra*, found only in limited coastal areas in Antarctica. **Alpine tundra** is a cold, treeless high-elevation biome whose vegetation consists mainly of shrubs and herbaceous perennials. Alpine tundra occurs in high mountainous areas, including the Canadian Rockies, the Andes, and the Himalayas. **Northern tundra** is a cold, treeless high-latitude biome, also dominated by herbaceous perennials. It is found north of the boreal forest throughout northern Eurasia and North America **(Figure 9.24)**.

The northern limit of the boreal forest is called the **northern tree line**. Similarly, the upper limit of

Figure 9.23 Notable trees in the montane forests. (A) The world's longest-lived, single-stemmed tree species is the bristlecone pine (*Pinus longaeva*). The oldest-known individual, at 5,067 years old, is living in the White Mountains in eastern California. (B) The oldest giant sequoias (*Sequoiadendron giganteum*) are over 3,000 years old. They are the heaviest single-stemmed trees on the planet. The General Sherman tree, shown here, is 31.3 m (102 ft) in circumference and 83.8 m (275 ft) tall. It weighs almost 2,000 metric tons. (C) These quaking aspen trees (*Populus tremuloides*) are found near Utah's Fish Lake. This particular grove of aspen trees is informally known as "Pando" (Latin for "I spread"). It consists of more than 40,000 genetically identical trunks rising from a single root mass that covers 106 acres. In total, Pando's root mass (along with its trunks) weighs over 6,000 metric tons and is estimated to be over 80,000 years old.   *(A. Rob Blakers/Getty Images; B. Bruce Gervais; C. Diane Cook, Len Jenshel/Getty Images)*

A   B   C

Figure 9.24 Tundra. (A) Purple saxifrage (*Saxifraga oppositifolia*) is typical of the low-growing herbaceous plants in the tundra biome. This plant is in full bloom near Resolute, Nunavut, Canada. (B) This climate diagram for Resolute shows that the average annual temperatures in tundra range from –15°C to –5°C (–5°F to 23°F). Average annual precipitation varies considerably with latitude but is generally less than 100 cm (39 in). (C) This map shows the extent of the northern tundra (in purple) and mountainous areas with potential alpine tundra (in gray). The Antarctic tundra is too geographically limited to be visible on this map. *(All Canada Photos/Alamy)*

A

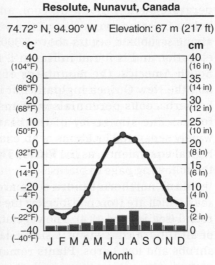

Resolute, Nunavut, Canada

74.72° N, 94.90° W   Elevation: 67 m (217 ft)

Month

Average temperature: –16°C (4°F)
Total precipitation: 16 cm (6 in)

B

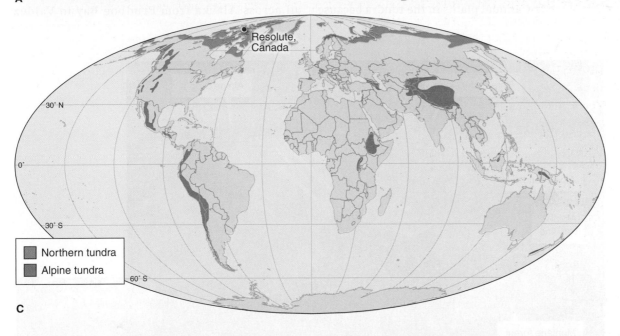

Resolute, Canada

■ Northern tundra
■ Alpine tundra

C

Tundra	
Köppen climate zone	Tundra and cold mountain climates: ET, H
Vegetation	Small shrubs and herbs, no trees
Notable features	Very cold winters, short summer growing season; too cold for trees; occurs at high latitudes and high elevations; soils often permanently frozen
Human footprint	Mineral extraction, ecosystems shifts, and permafrost thawing due to climate change

the montane forest is the **alpine tree line** (or *timberline*) **(Figure 9.25)**.

The northern tundra and alpine tundra at midlatitudes have large annual temperature ranges. In contrast, alpine tundra within the tropics, called *tropical alpine scrubland*, experience little annual temperature variation. Instead, these areas experience diurnal temperature swings: Days are cool, and nights are below freezing every day of the year. Tropical alpine scrubland occurs above 3,300 m (10,560 ft) elevation and is found in Hawai'i, the Andes of South America, the mountains of East Africa, and the New Guinea highlands. It is dominated by herbaceous perennials with rosette growth forms. The similarity of the plants in these widely separated locations is an example of ecological equivalence, as the **Picture This** feature on the following page explains.

Many northern tundra soils are *permafrost soils*, which are frozen just below the surface year-round (see Section 18.4). The vegetation structure of all tundra consists of a single layer of small shrubs and low herbs. Plants remain low to the ground, where it is warmer and where they can be protected by an insulating layer of snow in winter. There are few annual plants in the tundra because the growing season is too short and cold for most plants to complete their life cycle and set seed. The biomass of standing tundra vegetation is low, as is the diversity of species. Many of the same species grow throughout the tundra across broad geographic regions.

An important feature of the northern tundra is the number of migratory animals it receives each summer. Although its summers are brief, they have an abundance of food resources because the daylight hours are very long. North of the Arctic Circle, the Sun does not set (see Section 3.5), and biological productivity experiences a brief burst during this time. Many bird species use the northern tundra for summer breeding because of the relatively small number of predators found there and the plentiful food supply for their nestlings.

The Human Footprint in the Tundra Biome

Human impacts in the northern tundra are occurring mostly through road building and resource extraction and transportation. Oil and gas exploration in the Prudhoe Bay region on the North Slope of Alaska has had an impact on the tundra there. The Trans-Alaska pipeline transports oil across Alaska from Prudhoe Bay to Valdez

Figure 9.25 Alpine tree line, Alberta, Canada. The alpine tree line is visible in this photo in Banff National Park, Alberta. Harsh conditions stunt tree growth and prevent trees from growing above this elevation. *(© Daniel Mosquin)*

Picture This Ecological Equivalence

(A. Rich Reid/Getty Images; B. Prisma by Dukas Presseagentur GmbH/Alamy; C. Photo © Marjn van den Brink)

All of Earth's biomes reflect the response of the animal and plant life to regional climate. Through evolution, different, unrelated groups of plants and animals within similar climates begin to resemble one another. Thus, widely separated but similar climates produce similar plant and animal adaptations. This biome-level resemblance is called *ecological equivalence*.

Although tropical alpine scrublands in different regions are structurally alike, their plant species are genetically unrelated. In this example, different groups of plants share the structure that works best in the Köppen H climate zone in the tropics. Tropical high-elevation settings receive intense ultraviolet radiation. Many tropical alpine plants are adapted to reduce the effects of UV exposure. Many have evolved parabolic *rosettes*, a growth form in which all the leaves emerge from one location on the plant. Reflective silver hairs, also common in tropical alpine plants, reflect UV radiation to protect the plant from its harmful effects. Shown here are the Haleakalā silversword (*Argyroxiphium sandwicense*) of Hawai'i (A), espeletia (*Espeletia pycnophylla*) of the Andes in Ecuador (B), and giant lobelia (*Lobelia deckenii*) on Mount Kilimanjaro in Tanzania (C).

Consider This

1. Are there examples of ecological equivalence among the other biomes in this chapter? Give examples.
2. How does ecological equivalence illustrate convergent evolution (see Section 8.2)?

(Figure 9.26). The pipeline blocks the movement of migrating animals and also has been involved in oil spills.

Climate change is having a growing impact on tundra permafrost soils as well. The Arctic is the fastest-warming region on Earth, and tundra ecosystems are rapidly changing as their permafrost soils thaw. Permafrost soils are rich in carbon from organic remains. As they thaw, they emit methane and carbon to the atmosphere, contributing to the greenhouse effect (see Section 18.4).

Desert

The largest of the biomes in area, **desert** covers nearly 30% of Earth's land surface and is found on every continent. The desert biome features chronic moisture deficits and sparse, dry-adapted vegetation. Most deserts receive less than 25 cm (10 in) of annual precipitation. The single largest desert is the Sahara of northern Africa, covering almost 10 million km² (3.9 million mi²) **(Figure 9.27)**.

The three broad groups of deserts are hot deserts (which are located beneath the subtropical high), rain shadow deserts (on the leeward sides of mountain ranges), and cold deserts (found at high latitudes and high elevations).

1. *Hot deserts,* such as the Namib Desert in Namibia and the Sahara in Northern Africa, exist because of the descending air of the subtropical high in the vicinity of 30 degrees latitude north and south.

2. *Rain-shadow deserts* form on the leeward sides of mountain ranges due to adiabatic heating (see Section 4.4). The Gobi Desert, for example, is found throughout Mongolia in the rain shadow of the Himalayas, the Pamirs, and the Altai Mountains. Its

location in the continental interior and its high elevation and resulting low temperatures also contribute to the Gobi's aridity.

3. *Cold deserts* are found at the poles and at high elevations. Polar regions are deserts because cold air has low water vapor content. The McMurdo Dry Valleys of Antarctica, for example, are exceedingly cold and arid. Two major cold deserts in Asia are the Taklamakan Desert in China and the Gobi Desert Mongolia.

In many deserts, a combination of factors together cause aridity. For example, the Atacama Desert of Chile averages 0.4 cm (0.15 in) of precipitation per year and is among the driest locations on Earth. It lies in the rain shadow of the Andes and is under the influence of the subtropical high. In addition, a cold offshore current inhibits evaporation and rainfall from the Pacific Ocean.

In all deserts, plants and animals exhibit a wide range of physiological and behavioral responses to the scarcity of water, including nocturnal activity in animals, deep roots to reach water, and germination from seeds in plants. (See Figure 8.12 for more plant adaptations to the desert.)

The Human Footprint in the Desert Biome

Deserts are largely off-limits to permanent human settlement because of the lack of available water. As a result, human impacts have been relatively light in this biome. There are places, however, where large populations live in the desert biome. In the United States, for example, Las Vegas, Nevada, and Phoenix, Arizona, are located in deserts. These cities import water, mostly from the Colorado River,

Video

The Heat is On: Desert Tortoises and Survival

Available at www.saplinglearning.com

▶

Figure 9.26 Trans-Alaska oil pipeline.

The Trans-Alaska pipeline carries oil from Prudhoe Bay on the North Slope of Alaska south to the port city of Valdez. From Valdez it is transferred onto oil tankers and brought to oil refineries in the United States. The oil is heated so that it will flow through the pipeline. Because the oil is heated, the pipeline must be elevated above the tundra so that it does not thaw the permafrost, which would result in a buckled, broken, and leaky pipeline.
(sarkophoto/Getty Images)

Figure 9.27 Desert. (A) The Hoggar Mountains near Salah, Algeria are located in the Sahara. (B) This diagram summarizes Salah's desert climate. (C) The largest and driest deserts are found beneath the subtropical high, centered at about 30 degrees latitude north and south, but deserts are found at other latitudes as well. *(muha04/Getty Images)*

A

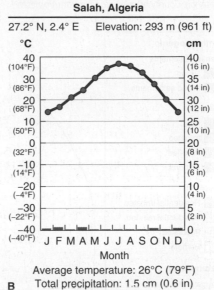

Salah, Algeria

27.2° N, 2.4° E   Elevation: 293 m (961 ft)

Month

Average temperature: 26°C (79°F)
Total precipitation: 1.5 cm (0.6 in)

B

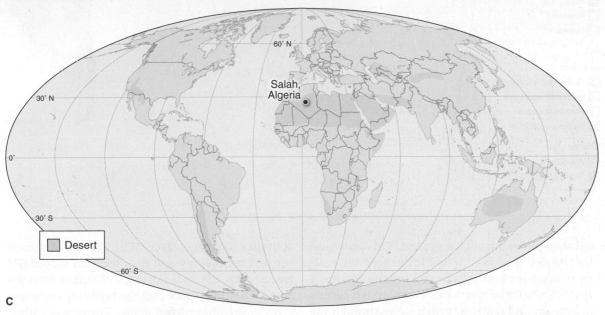

C

Desert	
Köppen climate zones	Subtropical desert: BWh; midlatitude desert: BWk; ice sheet: EF
Vegetation	Xerophytes
Notable features	Severe moisture deficits in most or all months; very low biomass
Human footprint	Water diversion from rivers, off-road vehicles, solar power facilities, livestock grazing

Figure 9.28 Anthropogenic biomes map. A GIS was used to map the different types of human activities that are affecting natural biomes. In some regions, such as in India and eastern China, the original natural biomes no longer exist. In other regions, such as northern Canada, the natural biomes are intact. *(Erle C. Ellis, University of Maryland, Baltimore County http://ecotope.org/anthromes/v2)*

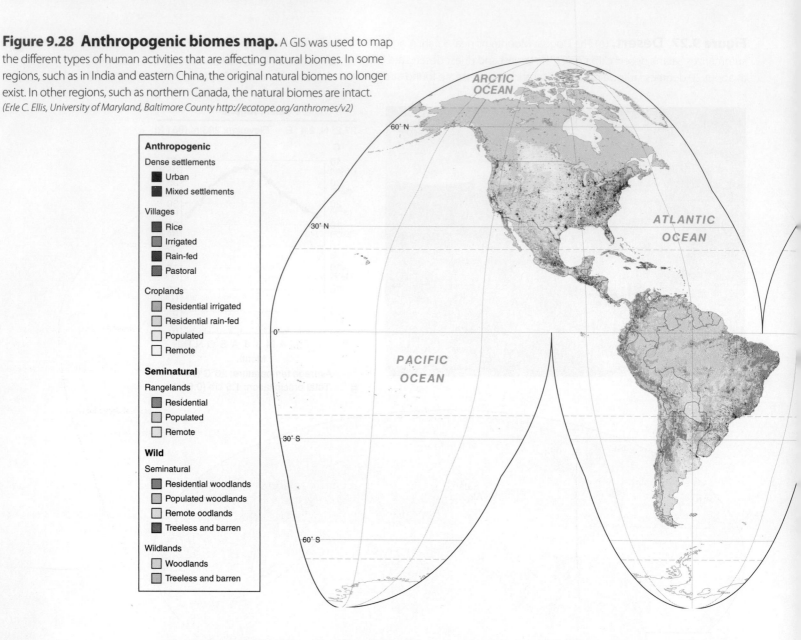

Anthropogenic

Dense settlements
- Urban
- Mixed settlements

Villages
- Rice
- Irrigated
- Rain-fed
- Pastoral

Croplands
- Residential irrigated
- Residential rain-fed
- Populated
- Remote

Seminatural

Rangelands
- Residential
- Populated
- Remote

Wild

Seminatural
- Residential woodlands
- Populated woodlands
- Remote oodlands
- Treeless and barren

Wildlands
- Woodlands
- Treeless and barren

and they pump it from the ground. This water use has reduced stream flow and influenced the ecology of many streams and riparian (streamside) areas significantly. Other rivers that flow through deserts include the Indus River, which flows through the deserts of eastern Pakistan and western India, and the Nile River, which flows through the easternmost Sahara. These rivers support large desert populations. People have lived sustainably in these desert areas for millennia.

GEOGRAPHIC PERSPECTIVES
9.5 The Value of Nature

◎ **Assess the connection between the well-being of people and the well-being of nature.**

As we have seen throughout this chapter, all of Earth's biomes have been changed by human activities to some degree. With the aid of remote sensing technology and the powerful tool of GIS (see Section 1.5), detailed maps of human land use across Earth's surface and the resulting *anthropogenic biomes* have been made. These maps illustrate how natural biomes have been changed by human activity and highlight areas where natural biomes remain **(Figure 9.28)**.

Habitat fragmentation is the division and reduction of natural habitat into smaller pieces by human activity. In his book *The Song of the Dodo*, David Quammen compares the process of habitat fragmentation to cutting up a Persian rug: A whole Persian rug functions as a beautiful rug, but if the rug is cut into pieces, it becomes worthless. Our tour of biomes in this chapter has revealed that habitat fragmentation is reducing Earth's natural biomes and ecosystems into ever-smaller pieces that are surrounded by anthropogenic landscapes.

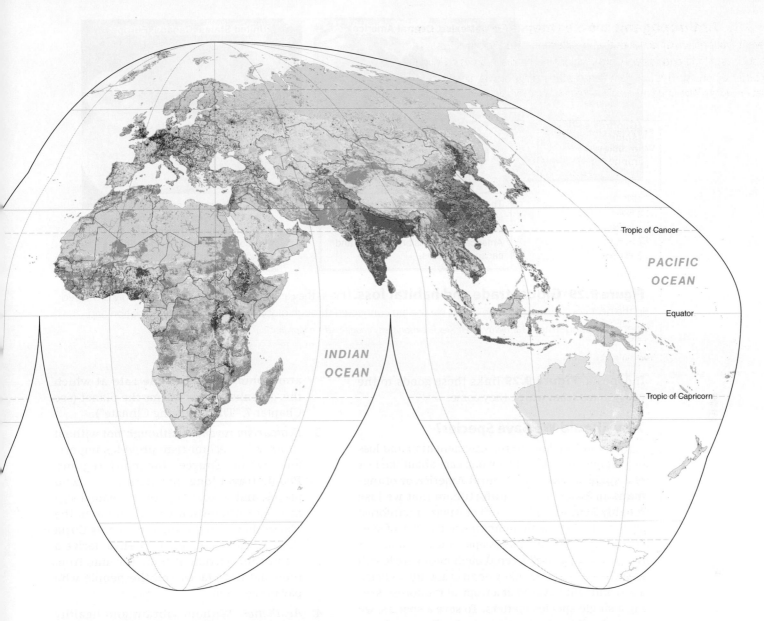

India, Java, and much of eastern China are bright blue and purple in Figure 9.28, indicating the presence of villages with large human populations engaged in irrigated and rain-fed agriculture. The seasonal tropical forest and tropical savanna biomes that would occur naturally in India are completely gone, as are China's temperate deciduous forests. The slopes of the Himalayas, once montane forest, are now rice-growing agricultural settlements and dense mixed settlements.

Food production accounts for a large proportion of human land use. The most widespread human land use is *rangelands*, where domesticated livestock graze. Rangelands are followed by *croplands*, the second most widespread human land use.

Habitat and Species Loss

The materials that surround us—homes, phones, books, clothes, the materials used to make our roads and buildings—without exception, were at one point resources found in nature. Like any other species, people require resources and food and space. The needs of the 7.6 billion people who are alive in the world lead to the transformation of Earth's surface from natural biomes to mosaics of human land uses.

Habitat loss is now the most significant factor causing species endangerment and extinction. One-third of the species threatened with extinction today are threatened because their habitat is being degraded or is disappearing altogether. The materials people use and the foods people eat come to us through a global network of trade. Many of the goods and materials North Americans and Europeans enjoy originate overseas, where their extraction or production often takes an environmental toll. Coffee, for example, comes from Brazil, beef from Argentina, cacao (chocolate) from Central America, and palm oil from

The Geoffroy's spider monkey (*Ateles geoffroyi*) is critically threatened in Mexico and Central America by expansion of coffee and cacao plantations.

Demand for coffee and chocolate in the United States, Canada, and Europe drives habitat loss in Mexico and Central America and threatens the spider monkey.

Figure 9.29 Global trade and habitat loss. Unless they are locally produced or responsibly produced in the host country, many of the goods we use are connected through global trade to habitat degradation and species endangerment far away. *(Left. Alex Robinson/Getty Images; right. happydancing/Shutterstock)*

Indonesia. **Figure 9.29** links these goods to the ecosystems from which they come.

Why Should We Save Species?

Why should we lament the fragmentation and loss of a biome? Why should anyone care about the loss of a spider monkey in Central America or orangutans in Borneo? It is unfortunate that we lose roughly 7,000 km² (2,700 mi²) of Amazon rainforest each year, along with an unknown number of species never before seen by people. But does it matter in a practical sense? Even though most people will never see a spider monkey or an orangutan outside a zoo and will never visit a tropical rainforest, saving a single species matters. To save a species, we must save the ecosystem in which it lives, and saving ecosystems helps humans in four direct ways:

1. *Ecosystem services:* Natural ecosystems and the species in them provide humans with clean air and drinking water, fertile soil, filtration of pollutants along coastal zones, buffers against hurricane storm surges, pollination of food crops, and compounds for new medicines. These benefits are services because people need them for their health and sustenance.

2. *Climate stabilization:* Saving ecosystems forestalls climate change. Whereas deforestation releases the carbon stored in trees and the soil to the atmosphere, healthy forests absorb carbon dioxide from the atmosphere and slow the rate at which this greenhouse gas warms the planet (see Chapter 7, "The Changing Climate").

3. *Ecotourism revenue:* Although not without controversy, ecotourism provides important revenue sources for many regions. People travel long distances to see wild places, and in so doing, they create a significant revenue source for people in the places they visit. Countries such as Costa Rica, Ecuador, Nepal, and Kenya derive a significant portion of their revenue from their natural places and the people who pay to see them.

4. *Aesthetics:* Without vibrant and healthy nature, the world would be gritty, boring, and monotonous. Ecosystems and their species do more than clean the air and water and fill our stomachs: They feed our spirits. Without the complete Persian rug of life, the world is less interesting and nurturing for people.

Humans are connected to and supported by the biosphere. The well-being of our lives matters, and so does the well-being of human societies. Saving species and their habitats preserves ecosystem services, curbs climate change, generates revenue through ecotourism, and preserves a richer and more interesting world. The health and well-being of people are inextricably connected to the health and well-being of the biosphere.

Web Map

Global Deforestation

Available at
www.saplinglearning.com

Chapter 9 Study Guide

Focus Points

9.1 Climates and Biomes

- **Biomes:** Biomes are regions that share uniform vegetation structures. Biome locations are determined mostly by climate.
- **Köppen climate classification system:** The Köppen climate classification system is used to classify different climate types.

9.2 Low-Latitude Biomes

- **Three biomes:** The three low-latitude biomes are tropical rainforest, tropical seasonal forest, and tropical savanna. Tropical rainforest is the most biodiverse of all biomes.
- **Climate:** Low-latitude biomes have warm climates, and they are either wet all year (tropical rainforest) or have a winter dry season.
- **Vegetation:** Plant adaptations in low-latitude biomes vary, from the complex layered structure of the tropical rainforest to the fire-adapted plants of the tropical savanna.
- **Human footprint:** Roads, logging, farming, and livestock ranching have had the greatest impacts in low-latitude biomes.

9.3 Midlatitude and High-Latitude Biomes

- **Five biomes:** The five midlatitude and high-latitude biomes are temperate grassland, the Mediterranean biome, temperate deciduous forest, temperate rainforest, and boreal forest.
- **Climate:** Winters are mild in the Mediterranean and temperate rainforest biomes; winters are cold in the other biomes at mid- and high latitudes. Precipitation is lowest in temperate grasslands and highest in temperate rainforests.
- **Vegetation:** Vegetation structure at midlatitudes varies widely, from the dry Mediterranean biome to temperate rainforest, which is home to the world's largest trees. Fire occurs in all midlatitude biomes.
- **Human footprint:** Human impacts includes non-native species and logging in forested biomes and agriculture and livestock operations in grasslands.

9.4 Biomes Found at All Latitudes

- **Three biomes:** Montane forest, tundra, and desert are found at all latitudes. Some montane forests support large trees. Alpine tundra occurs at high elevations, and northern tundra occurs at high latitudes, where it is too cold for trees to grow. Desert has low biomass and dry-adapted vegetation.

- **Climate:** The climates of these biomes vary by latitude. The montane forest biome has sufficient moisture to support forests and is cool. Tundra is always cold. Deserts have perennial moisture deficits and may be hot or cold.
- **Vegetation:** Vegetation in these three biomes ranges from thickly wooded montane forest to the shrubs and herbs of the treeless tundra to the sparsely vegetated desert.
- **Human footprint:** Human impacts include logging in montane forest, mineral extraction in tundra, and water diversions from desert streams.

9.5 Geographic Perspectives: The Value of Nature

- **Habitat fragmentation:** Habitat fragmentation is the most significant driver of species extinctions today.
- **International trade:** International trade is a driving force in habitat fragmentation and degradation of natural biomes.
- **Value of natural biomes:** Natural biomes provide humans with resources, services, aesthetic value, and climate stabilization.

Key Terms

alpine tree line, 292
alpine tundra, 290
biome, 267
boreal forest, 286
buttress roots, 270
desert, 294
epiphyte, 270
habitat fragmentation, 296
Köppen climate classification system, 269
liana, 270
light gaps, 270
Mediterranean biome, 282

montane forest, 288
northern tree line, 290
northern tundra, 290
primary forest, 285
sclerophyllous, 282
secondary forest, 286
temperate deciduous forest, 285
temperate grassland, 281
temperate rainforest, 286
tropical rainforest, 270
tropical savanna, 275
tropical seasonal forest, 275
tundra, 290

Concept Review

The Human Sphere: Three Trillion Trees

1. When combined, which two biomes have the greatest number of trees? In which biome is forest density greatest?

9.1 Climates and Biomes

2. What is a biome? What physical factors determine biome type?

3. What is vegetation structure? Plant growth form? Give examples of each.

4. List four examples of botanical terms used in this chapter.

5. Briefly describe the Köppen climate classification system. What are the six major climate types used in this system?

6. Describe the spatial relationship between biomes and climate.

7. For each biome in this chapter, describe where it occurs, its major physical characteristics in terms of biodiversity, its structure, climate, the prevalence of fire, and important limiting factors.

8. Discuss the primary impacts humans have on each biome.

9.2 Low-Latitude Biomes

9. Why does the Amazon tropical rainforest have poor soils, while the soils of the tropical rainforest in Indonesia are rich?

10. What percentage of Earth's species is estimated to live in the tropical rainforest biome?

11. How do poor soils, farming, and deforestation relate to one another in the tropical rainforest?

12. Why is tropical seasonal forest particularly vulnerable to human activities?

13. What other regional names does the tropical savanna go by?

14. What are the three factors that largely shape the tropical savanna biome?

15. What is a grazing sequence, and how does it allow many large animals to use the same areas in the same season?

9.3 Midlatitude and High-Latitude Biomes

16. List the various local names for the temperate grassland.

17. What three prairie types are found in the North America?

18. Describe the ecological role of the prairie dog. Specify its relationships with bison and pronghorn.

19. List the various local names for the Mediterranean biome.

20. How does the Mediterranean biome's biodiversity compare with that of other temperate biomes?

21. What is unusual about the climate of the Mediterranean biome?

22. Why does fire often occur in the Mediterranean biome?

23. Define *deciduous*.

24. What is the most prominent human impact in the temperate deciduous forest biome?

25. What name is used for the boreal forest in Eurasia?

26. Describe the climate and vegetation of the boreal forest.

9.4 Biomes Found at All Latitudes

27. Why is it that forests occur on the windward slopes of mountains but are much less common on the leeward slopes?

28. What is the difference between northern tundra and alpine tundra?

29. What is an alpine tree line? What is a northern tree line?

30. What is tropical alpine scrubland, and where is it found? How do the life-forms found there illustrate convergent evolution?

31. Compare the biomass of the desert with the biomass of a tropical rainforest.

32. What is the main limiting factor in deserts?

9.5 Geographic Perspectives: The Value of Nature

33. Give examples of the main types of human land uses that have modified natural biomes.

34. What is the main driver of species extinctions today?

35. Why should we save a spider monkey species in a remote tropical seasonal forest in Mexico? Of what practical use is the spider monkey to people?

Critical-Thinking Questions

1. On Figure 9.2 find the biome you live in. Referring to Figure 9.27, can you determine the leading category of human land use in your biome now?

2. Many people in Brazil claim that countries such as the United States have no business telling Brazil not to cut its rainforests because most wealthy countries (such as the United States) cut their forests early in their histories. Do you agree or disagree with this argument? Explain.

3. Most tropical rainforest canopy trees have a flat, umbrella-like structure. Most boreal forest trees have a narrow, pole-like structure. In the context of solar altitude (the height of the noontime Sun), explain the structure of these trees.

4. Non-native species are often most successful when they are transported to the same or a similar biome. They are less successful if they are transported to a different biome. For example, many non-native plants in California came from the Mediterranean region. Why do you think these plants have been able to successfully establish in California?

5. In addition to the reasons discussed in Section 9.5, what are other reasons to save habitat and species?

Test Yourself

Take this quiz to test your chapter knowledge.

1. **True or false?** The Amazon rainforest has nutrient-poor soils.

2. **True or false?** Trees in temperate rainforests and tropical rainforests are mostly evergreen.

3. **True or false?** The geographic distribution of the boreal forest, tundra, tropical alpine scrubland, and tropical savanna all are determined by low temperatures.

4. **True or false?** Of all the biomes, deserts have the lowest biomass, and tropical rainforests have the highest biomass.

5. **True or false?** At small spatial scales, climate is the most important determinant of vegetation structure.

6. **Multiple choice:** Which of the following biomes is characterized by summer dry season, winter rain, and frequent fire?
   a. Mediterranean   c. tropical rainforest
   b. tundra   d. tropical savanna

7. **Multiple choice:** Which of the following biomes is treeless due to chronic moisture deficits, grazing, and fire?
   a. grassland   c. desert
   b. tundra   d. tropical alpine scrubland

8. **Multiple choice:** Today, the most important agent of extinction is
   a. exotic species.   c. habitat loss.
   b. hunting.   d. climate change.

9. **Fill in the blank:** _____ marks the uppermost elevations of trees in mountains.

10. **Fill in the blank:** A(n) _____ is a plant that lives on other plants.

Online Geographic Analysis

Anthropogenic Biomes and Change in Tree Cover
This exercise focuses on present-day human land uses and forest change.

Activity
Go to http://ecotope.org/anthromes/v1/maps/a2000/. This interactive web map shows anthropogenic biomes (or *anthromes*). We will use this web page to examine land use patterns on Earth's surface. Examine the map located on this interactive web page and and answer these questions.

1. **Zoom in to the map until India fills the view screen. What are three primary types of land uses in India?**

2. **Refer to your book and Figure 9.2. What were the primary biomes that occurred in India?**

Now zoom out, pan over to northern South America, and zoom until the Amazon rainforest biome fills the view.

3. What are the three major types of land use or vegetated surfaces in this biome now?

4. What is the largest urbanized region you can find in the Amazon region? You might need to zoom a bit to see the map more clearly.

Zoom out and pan northward until your map view is centered on and filled with the United States.

5. How would you compare land uses in the eastern half of the United States to land uses in the western half?

6. Zoom in to the region where you are currently living. What land use types are common in your area?

Enter the URL http://www.globalforestwatch.org/ and click the "Map" button at the top of the page. This interactive web map has many layers of data. The default settings have two active layers: "Tree cover gain," shown in blue, and "Tree cover loss," shown in purple. Pan until North America fills your view screen. Turn the blue "Tree cover gains" layer on and off by clicking it's "eye" icon in the left-side menu. Do the same for the "Tree cover losses" layer.

7. In North America, which has been greater overall: tree cover gains or losses?

8. Where in North America have tree cover gains been greatest?

9. Where in North America have tree cover losses been greatest?

Zoom out as far as the map allows and turn the blue "Tree cover gains" layer on and off by clicking it's "eye" icon in the left-side menu. Do the same for the "Tree cover losses" layer.

10. List the three countries where tree cover loss has been greatest.

11. List three countries where tree cover has increased.

Picture This. *Your Turn*

Name That Biome

Match each of the biomes listed in the table on the right to the appropriate photo. Use each biome name only once. Hints for each biome are provided.

BIOME	PRECIPITATION	TEMPERATURE
1	Summer wet, winter dry	Always warm
2	High year-round	Mild
3	Low, mostly in summer	Very low
4	Low	Hot summers, cold winters

Northern tundra

A

Temperate rainforest

B

Temperate grassland

C

Tropical savanna

D

(A. John Warburton-Lee/Getty Images; B. Micha Pawlitzki/Getty Images; C. Mark Cosslett/Getty Images; D. Tetra Images/Getty Images)

For animations, interactive maps, videos, and more, visit www.saplinglearning.com. SaplingPlus

Ocean Ecosystems

Chapter Outline *and Learning Goals* ◎

10.1 The Physical Oceans

◎ List the major physical features of the oceans.

10.2 Life on the Continental Margins

◎ Describe the ecosystems along continental margins and human influences on those ecosystems.

10.3 Life in Polar Waters

◎ Explain how polar marine ecosystems function.

10.4 Life in Open Waters

◎ Describe the geographic patterns of life in the open ocean.

10.5 Geographic Perspectives: The Problem with Plastic

◎ Assess the environmental impact of and remedies for plastic pollution in the oceans.

This photo, taken in the Cayman Islands, a part of the British West Indies in the Caribbean Sea, is of the silhouette of the critically endangered elkhorn coral (*Acropora palmata*). Corals look like plants, but they are animals that are related to jellies. They build hard reef structures in clean coastal waters in the tropics and subtropics.

(Nature Picture Library/Alamy)

To learn more about coral reefs, see Section 10.2.

THE HUMAN SPHERE Coastal Dead Zones

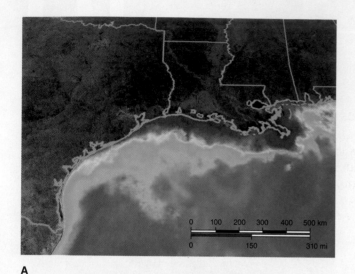

A

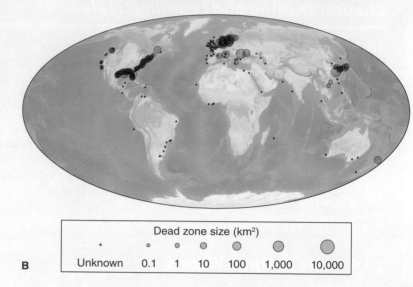

B

Figure 10.1 Coastal dead zones. (A) Here a phytoplankton bloom and subsequent dead zone occur along the coast of the Gulf States. The area in red is a continuous dead zone stretching hundreds of kilometers. (B) On this map, red circles indicate the locations and sizes of coastal dead zones. The largest circles represent dead zones 10,000 km² (3,860 mi²) in extent or greater. Dead zones of unknown size are shown with black dots. Note that there are fewer dead zones in the Southern Hemisphere because there is less agricultural activity there. *(Goddard SVS/NASA)*

Coastal dead zones are areas in coastal waters where there is too little dissolved oxygen to support most forms of marine life. Coastal dead zones are a natural phenomenon, but human activity has increased both their numbers and their geographic extent in recent decades. Dead zones occur around the world, but more occur in the Northern Hemisphere—where there are more people creating more sewage and farming runoff—than in the Southern Hemisphere. Dead zones are closely tied to agriculture and form or increase in size as fertilizers, such as nitrogen and phosphorus, applied in excess to agricultural fields, are carried into streams by rainfall runoff. Each year the Mississippi River carries about 1.7 million tons of fertilizers to the Gulf of Mexico. These fertilizers stimulate the growth of **phytoplankton** (microscopic photosynthetic algae and other microorganisms), creating phytoplankton "blooms," which lead to coastal dead zones **(Figure 10.1)**.

A phytoplankton bloom attracts zooplankton primary consumers. The zooplankton (plankton consisting of small animals and immature stages of larger animals) release waste, which settles to the seafloor, where microbes consume it. In the process, the microbes remove dissolved oxygen from the water, creating *hypoxic* (very low oxygen) and *anoxic* (oxygen-free) dead zones. Hypoxic conditions occur naturally, but human activity creates more of them and enhances naturally occurring ones. Most marine life either leaves the area of a dead zone or asphyxiates and dies if it cannot escape.

10.1 The Physical Oceans

◎ **List the major physical features of the oceans.**

Physical geography explores all of Earth's physical systems. The oceans are a particularly important realm of physical geography. Consider these facts:

1. All the organisms in terrestrial environments, from rainforests to savanna to tundra, make up only about 1% of Earth's total biomass. The rest is in the oceans.

2. Almost all (99%) of the biosphere is in the oceans, which are immense and deep. The oceans cover 71% of Earth's surface, and they are on average 4 km (2.5 mi) deep **(Table 10.1)**. Living creatures can occupy all of this space. On land, in contrast, life is restricted to a thin shell: mostly less

than a few meters deep in the soil or a few hundred meters above the land surface.

3. Because of their size and because they are composed of water, the oceans regulate the atmosphere's chemical composition and modify Earth's climate and weather systems.

4. The oceans, not the rainforests, are the true lungs of the planet. About 70% of the oxygen in the atmosphere comes from photosynthetic phytoplankton in the open oceans.

5. There would be no life on Earth without the oceans. Life began in the oceans, and without the water evaporated from the oceans and precipitated onto land, there could be no life on land.

6. Nearly half the world's human population depends directly on the oceans as a primary source of food.

Table 10.1	Physical Characteristics of the Oceans	
CHARACTERISTIC	SIZE (METRIC)	SIZE (U.S. CUSTOMARY)
Total area	331,441,932 km^2	127,970,392 mi^2
Total volume	1,303,155,354 km^3	312,643,596 mi^3
Average depth	4 km	2.5 mi
Greatest depth	10,916 m	35,814 ft
Longest mountain range	16,000 km	9,920 mi

Earth's Five Oceans

The terms *sea*, *gulf*, *bay*, *sound*, and *strait* are used in reference to large bodies of salt water. **Figure 10.2** provides an example of each of these water bodies, using the Gulf of Mexico region. The term *ocean* is used to refer to the five major water-filled basins on Earth's surface, shown in

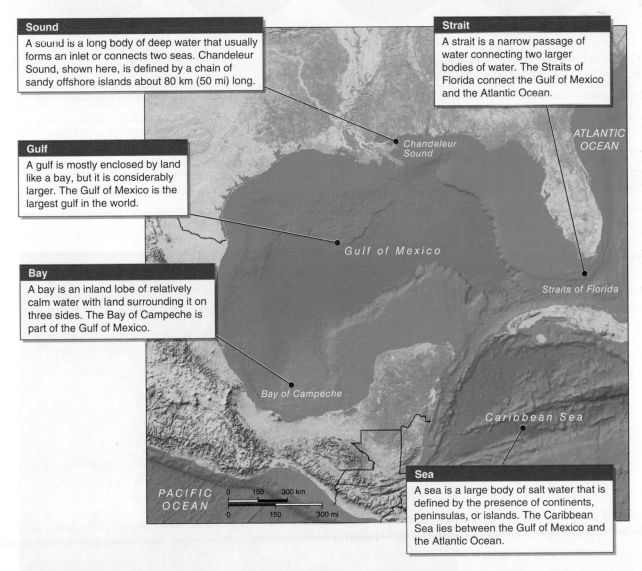

Sound
A sound is a long body of deep water that usually forms an inlet or connects two seas. Chandeleur Sound, shown here, is defined by a chain of sandy offshore islands about 80 km (50 mi) long.

Strait
A strait is a narrow passage of water connecting two larger bodies of water. The Straits of Florida connect the Gulf of Mexico and the Atlantic Ocean.

Gulf
A gulf is mostly enclosed by land like a bay, but it is considerably larger. The Gulf of Mexico is the largest gulf in the world.

Bay
A bay is an inland lobe of relatively calm water with land surrounding it on three sides. The Bay of Campeche is part of the Gulf of Mexico.

Sea
A sea is a large body of salt water that is defined by the presence of continents, peninsulas, or islands. The Caribbean Sea lies between the Gulf of Mexico and the Atlantic Ocean.

Chandeleur Sound
ATLANTIC OCEAN
Gulf of Mexico
Straits of Florida
Bay of Campeche
Caribbean Sea
PACIFIC OCEAN

0   150   300 km
0   150   300 mi

Figure 10.2 Names for large bodies of water. The most common geographic names for ocean features are applied here to the Gulf of Mexico region.

Figure 10.3. Throughout the remainder of this chapter, we use the term *sea* and *ocean* interchangeably to refer to the world's five oceans.

Seafloor Topography

Imagine the ocean basins without water. What would the exposed seafloor look like? The features of the seafloor include major mountain ranges, extensive plains, and deep trench systems **(Figure 10.4)**. The process of measuring the depth and topographic features beneath the surface of a body of water is called **bathymetry** (see also Figure 10.7).

If the oceans were dried up, you would see **continental shelves** (the shallow, gently sloping areas of seafloor near continental margins). You would see steeper *continental slopes* forming the transition between the continental shelves and the endless flat stretches of the deep seafloor.

Figure 10.3 The five ocean basins and their major characteristics.

Pacific Ocean   Atlantic Ocean   Indian Ocean   Southern Ocean   Arctic Ocean

OCEAN BASIN	AREA (KM²)	AREA (MI²)	PERCENTAGE OF TOTAL OCEAN AREA	NOTABLE FEATURE
Pacific	152,617,159	58,925,790	46	The largest and oldest ocean basin; includes the Mariana Trench, Earth's deepest water
Atlantic	81,705,396	31,546,616	25	Includes a prominent submerged mountain range called the Mid-Atlantic Ridge
Indian	67,469,539	26,050,123	20	Mostly located in the Southern Hemisphere
Southern	20,973,318	8,097,840	6	Surrounds Antarctica
Arctic	8,676,520	3,350,021	3	Mostly ice-covered

Figure 10.4 Bathymetry of the North Atlantic.
The bathymetric features of the oceans include the shallow and sloping continental shelves, the continental slopes, the flat abyssal plains, mid-ocean ridges, and deep-sea trenches. The vertical scale of the ocean depths is greatly exaggerated here. *(NOAA National Environmental Satellite and Information Service)*

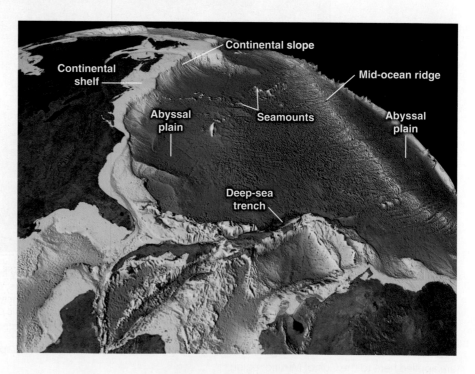

The **abyssal plains** are large flat areas on the ocean floor at depths between 4,000 and 6,000 m (13,000 and 20,000 ft).

From these plains rise the world's longest mountain ranges, called **mid-ocean ridges**, which are submarine mountain systems stretching the length of ocean basins from pole to pole. Isolated **seamounts** (submarine mountains) rise from the seafloor. Many of these are flat-topped inactive volcanoes that once protruded above sea level. For example, islands like Hawai'i are enormous mountains protruding from the seafloor, with only their tops above the surface of the ocean. **Deep-sea trenches** are long, narrow valleys on the seafloor that are the deepest parts of the oceans. Some of these trenches drop more than 6 km (4 mi) vertically.

Layers of the Ocean

Sunlight powers the biosphere, both on land and in the oceans. Only the ocean's surface layers absorb light and are heated directly by sunlight. The depth to which light can penetrate seawater depends on the time of day, the weather above the sea surface, and water clarity. In murky water containing *suspended sediments*, light barely penetrates 10 m (33 ft). Different wavelengths of light penetrate seawater to different depths. As **Figure 10.5** shows, blue wavelengths of light penetrate deepest.

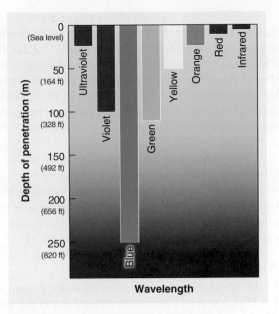

Figure 10.5 Depth of penetration of solar radiation. In clear seawater, relatively long wavelengths of insolation, such as infrared, red, orange, and yellow, are absorbed in the first 50 m (164 ft) of water. Shortwave UV radiation is also absorbed near the water surface. Green and violet light penetrate just over 100 m (328 ft), and blue light penetrates as much as 250 m (820 ft).

The ocean's surface temperatures vary considerably with latitude. Because of the direct sunlight in the tropics and the largely cloud-free skies of the subtropical high, tropical and subtropical waters receive strong solar heating. Higher-latitude waters receive far less solar heating and are much colder, as shown in **Figure 10.6**.

Figure 10.6 Average annual sea surface temperatures. Latitudinal differences in sea-surface temperature, like differences in air temperature, are a result of latitudinal differences in solar heating. The Arctic Ocean's surface is largely frozen throughout most of the year, and in the tropics, water temperatures can reach 30°C (86°F) or more. The three inset graphs show how water temperature changes with depth at different latitudes.

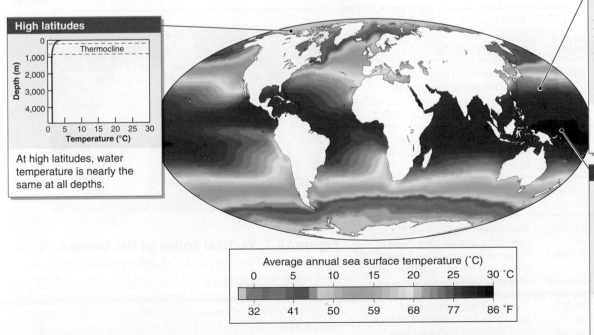
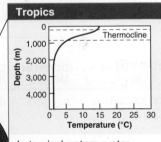
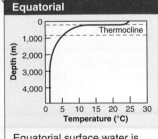

In the tropics, starting at a depth of about 100 m (328 ft), the water temperature quickly declines with depth until about 500 to 1,000 m (1,640 to 3,280 ft). This transitional zone of temperature decline is the **thermocline**. The depth of the thermocline varies by latitude and by ocean (see Figure 10.5). Below 2,500 m (8,200 ft) at all latitudes, seawater temperatures stabilize to a temperature of 0°C to 2°C (32°F to 35.6°F).

The oceans can be divided into five vertical layers, or zones, that are distinguished by light penetration, water temperature, and seafloor features. The topmost layer, the **epipelagic zone**, is the relatively warm and sunlit surface of the ocean. It extends down to 200 m (650 ft) **(Figure 10.7)**.

Water Pressure

At the ocean's surface, there is no water pressure. Yet if you were to dive down to 10 m (33 ft), you would experience 1 kg/cm^2 (14.7 psi) of pressure. For every 10 m in depth, water pressure increases approximately 1 bar (or 1 atmosphere), which is 1 kg/cm^2 (see Section 2.2). Sperm whales dive as deep as 3,000 m (10,000 ft). The pressure at those depths is on the order of 300 kg/cm^2 (4,450 psi). This means that every 4 cm^2 (0.62 in^2) of the animal's body has over 1 metric ton pushing against it. The enormous pressure in the deep ocean creates a formidable barrier to scientists attempting to explore life there. It also presents a host of challenges for organisms, such as sperm whales and elephant seals, that live at the surface and dive to great depths to feed.

Ocean Salinity

Ocean **salinity**, the concentration of dissolved minerals in water, is important for many reasons. Life in the oceans must adapt to it, and humans

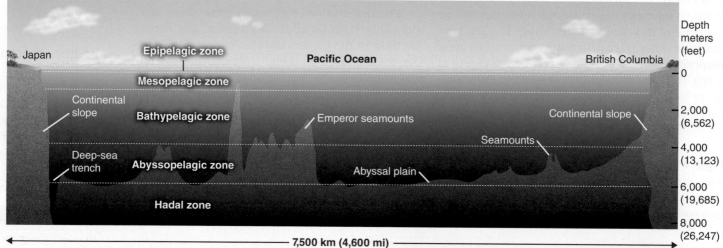

7,500 km (4,600 mi)

ZONE	DEPTH IN METERS	DEPTH IN FEET	CHARACTERISTICS
Epipelagic zone	0 to 200	0 to 650	The sunlit surface region. Also called the *photic zone*. In the clearest of seawater, only about 1% of sunlight reaches 100 m (328 ft).
Mesopelagic zone	200 to 1,000	650 to 3,300	Also called the *twilight zone*. Illumination is so dim as to be barely perceivable.
Bathypelagic zone	1,000 to 4,000	3,300 to 13,000	Also called the *aphotic zone*. No light penetrates to this depth.
Abyssopelagic zone	4,000 to 6,000	13,000 to 20,000	The seafloor at these depths is called the *abyssal plain*. Most of the seafloor consists of flat abyssal plains that cover more than 50% of Earth's surface.
Hadal zone	6,000 to 11,000	20,000 to 36,000	The deepest ocean extends to more than 10,000 meters. Deep-sea trenches are in the hadal zone.

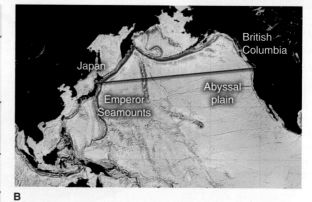

Figure 10.7 Vertical zones of the oceans. (A) This figure sketches the overall seafloor bathymetry along a transect (line) between Japan and British Columbia. The vertical dimension has been stretched about 200 times to make the seafloor features visible. (B) This map shows seafloor bathymetry and the transect followed between Japan and British Columbia. *(David Sandwell, Scripps Institution of Oceanography)*

cannot drink untreated seawater because of its salinity. Variations in salinity also drive global ocean circulation, which in turn affect global climate (see Section 7.2).

Why are the oceans salty? As rivers flow over the continents, they dissolve minerals from the rocks and transport those dissolved minerals to the ocean. Each year, freshwater rivers deposit about 2.5 billion tons of dissolved minerals, including salt, in the oceans. Other factors that raise the salinity of the oceans are evaporation and the formation of sea ice. Evaporated water (water vapor) and sea ice have no salt in them. During the evaporation and freezing processes, salts remain in the seawater and increase its salinity.

These factors that raise ocean salinity are counterbalanced by factors that decrease its salinity. Although streams carry salt, their freshwater is far less salty than the oceans. The continuous input of freshwater from streams in coastal areas, precipitation of rain and snow, the melting of sea ice in polar regions, and glaciers flowing into the ocean all decrease the salinity of seawater.

On longer time scales, shell-building organisms, chemical reactions with rocks, chemical precipitation, and the recycling of Earth's crust deep into its interior through plate tectonics (see Section 13.2) remove salt from the oceans at the same rate at which it is deposited by streams. As a result of all these factors, the salinity of the oceans, although it varies regionally, remains constant.

The average ocean salinity is 3.5% by volume (35 g/L [4.6 oz/gal]). Ocean salinity varies geographically from about 1% to 4.1%. There are two main reasons for this geographic variation: (1) Salinity increases where evaporation is high, and (2) salinity decreases where rivers discharge their freshwater or where there are persistent heavy rains. The subtropics have the highest evaporation rates and, therefore, the highest salinity values **(Figure 10.8)**.

By volume, most of the material dissolved in ocean water is chlorine and sodium, but every

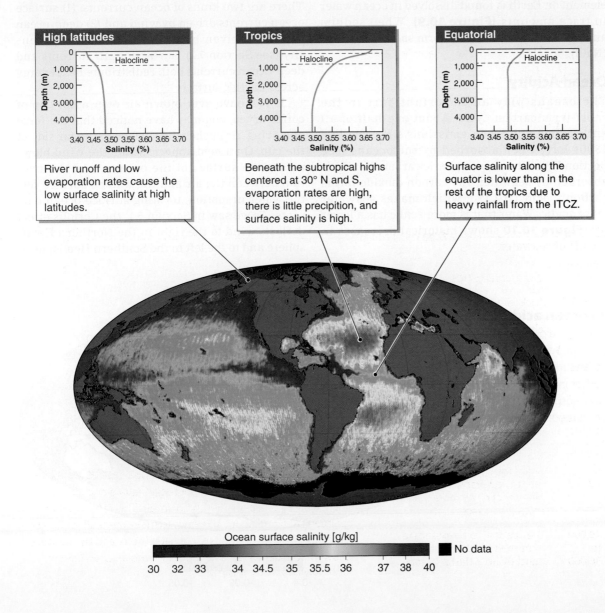

High latitudes

River runoff and low evaporation rates cause the low surface salinity at high latitudes.

Tropics

Beneath the subtropical highs centered at 30° N and S, evaporation rates are high, there is little precipitation, and surface salinity is high.

Equatorial

Surface salinity along the equator is lower than in the rest of the tropics due to heavy rainfall from the ITCZ.

Ocean surface salinity [g/kg]

30   32   33   34   34.5   35   35.5   36   37   38   40

No data

Figure 10.8 Variation in ocean salinity. The map shows global variation in salinity at the ocean's surface in grams per kilogram, as measured by the NASA's Aquarius satellite in 2011. Purple areas have the lowest salinity and orange areas have the highest salinity. The graphs show variations in salinity with depth as the percentage of salt by volume. The halocline is the transitional zone of salinity change between the ocean surface and the deep sea. At a depth of about 3,500 m (11,500 ft), salinity is about 3.5% at all latitudes. *(NASA/GSFC/JPL-Caltech)*

Video

Aquarius: One Year Observing the Salty Seas

Available at www.saplinglearning.com

Figure 10.9 Materials dissolved in seawater. Ocean water contains many dissolved materials.

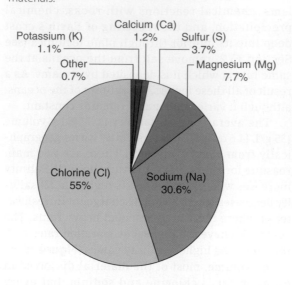

element on Earth is found dissolved in ocean water in trace amounts **(Figure 10.9)**. When sodium and chloride combine, they form sodium chloride (NaCl), or salt.

Ocean Acidity

The oceans play an important part in the short-term carbon cycle. About one-half of all anthropogenic carbon emissions since the late 1800s have been absorbed by the oceans (see Section 7.3). When atmospheric carbon dioxide is absorbed by seawater, the carbon dioxide forms carbonic acid (H_2CO_3), which makes seawater more acidic. Water that is more acidic has a lower pH. **Figure 10.10** shows historical decreases in the pH of seawater.

The chemistry of this ocean acidification is well understood. However, scientists are less certain how it will affect marine organisms in the coming decades. They are concerned that shell-building organisms, such as corals, mollusks, and microscopic plantlike organisms called *foraminifera*, will be affected by the increasing acidity of seawater. Acidic water contains fewer available calcium ions, and a decrease in calcium ions could inhibit the ability of these animals to build their calcium carbonate shells.

Many scientists are concerned that the current rate of ocean acidification is faster than what has occurred naturally at any time during the past 300 million years. Past ocean acidification has been linked to global mass extinctions of shell-building organisms.

Surface Ocean Currents

The atmosphere and the oceans are both fluids. Therefore, they flow. Just as there is wind in the atmosphere, there are currents in the oceans. There are two kinds of ocean currents: (1) surface ocean currents driven by wind and (2) deep-ocean currents driven by differences in water density (see Section 7.2). Surface ocean currents and deep-ocean currents both redistribute heat energy across Earth's surface.

If you have ever blown air over a hot cup of coffee or tea, you may have noticed that the force of moving air pushes the surface to the far side of the cup. On a hemispheric scale, the wind blows across the surface of the oceans, and the friction between the air and the water's surface sets the water in motion, forming surface ocean currents. As we saw in Section 5.1, the Coriolis effect deflects wind to the right in the Northern Hemisphere and to the left in the Southern Hemisphere,

Figure 10.10 Changes in ocean acidity.

This map shows estimated changes in the pH of ocean surface water from the 1700s to recent. Seawater pH has decreased in all ocean basins. Ocean surface pH varies geographically because of differences in the activities of organisms, the concentrations of atmospheric CO_2, and the effects of ocean currents, stream runoff, and precipitation amounts.

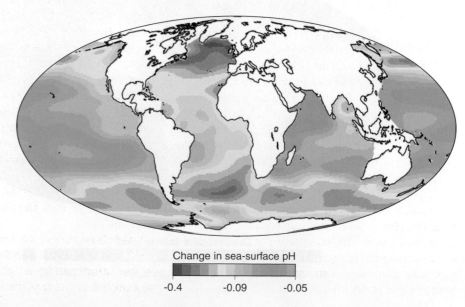

Change in sea-surface pH

-0.4   -0.09   -0.05

Figure 10.11 Surface ocean currents. Red arrows represent warm currents, and blue arrows represent cold currents. The five gyres and other major surface ocean currents are labeled.

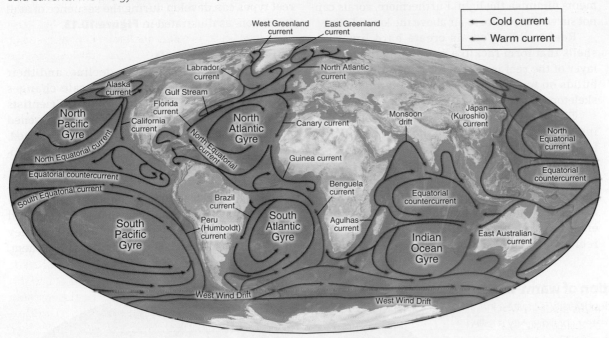

creating anticyclonic systems. It deflects moving water in the same way, creating large circular ocean currents called **gyres (Figure 10.11)**.

One notable result of ocean gyres is that they corral plastic and other trash in the oceans, concentrating it into vast garbage patches. The Geographic Perspectives, "The Problem with Plastic," at the end of this chapter explores this problem further.

10.2 Life on the Continental Margins

◎ Describe the ecosystems along continental margins and human influences on those ecosystems.

Life in the oceans is not grouped into biomes, as it is on land. Marine biological interactions are instead referred to as *ecosystems* (living organisms within a community and the nonliving components, such as sunlight and water) and, more locally, as *communities* of interacting organisms.

Our tour of life in the oceans begins at the margins of continents, where people interact intensively with marine ecosystems that are readily accessible to be seen, enjoyed, used, and influenced by people. The ecosystems at the margins of continents include coral reefs, mangrove forests, seagrass meadows, estuaries, kelp forests, and beaches and rocky shores.

Coral Reefs

Coral reefs, often called the rainforests of the sea, are one of nature's greatest displays of life. Coral reefs are the most biodiverse of the marine ecosystems. Coral reefs occupy only 0.1% of the area near the ocean's surface, about 286,000 km² (110,000 mi²), but they are used by approximately 4,000 species of marine fishes at some point in their life cycles. Scientists are still learning about coral reefs. Only about 10% of coral reef species have been identified.

There are two broad types of corals: cold-water corals and warm-water corals. Cold-water corals live at all latitudes and at great depths in the ocean. This section focuses on warm-water corals, about which we know much more. Warm-water corals live only near the sea surface in coastal regions at latitudes less than 30 degrees north and south **(Figure 10.12** on the following page**)**.

Corals are animals called *polyps* (pronounced PAW-lips), which are related to jellies. There are about 800 known species of hard corals. Most corals share an obligate mutualism with algae, called *zooxanthellae* (pronounced ZOO-zan-thel-ee): The coral polyps filter nutrients from seawater and make them available to photosynthetic zooxanthellae living within the corals' tissues. The zooxanthellae, in turn, provide the polyps with carbohydrates and fats.

Corals require clean, well-lighted water between 18°C and 30°C (64°F and 86°F). They are intolerant of low salinity and low light levels. Therefore, they are absent at the mouths of large rivers, such the

Amazon River or the great rivers of Southeast Asia, where the water is too fresh, and suspended sediments diminish the light. Furthermore, corals cannot survive exposure to air above the low-tide line.

Reef-building corals create hard limestone shells that form rocklike reefs. Only the topmost layer of the reef is alive; each generation of corals builds on top of the preceding generation's hard skeletons, adding to the reef over time.

There are three kinds of coral reefs: fringing reefs, barrier reefs, and atolls. A **fringing reef** forms near and parallel to a coastline. A **barrier reef** runs parallel to a coastline and forms a deep-water lagoon behind it. An **atoll** is a ring of coral reefs with an interior lagoon that forms around a sinking volcano. **Lagoons** are fully or partly enclosed stretches of salt water formed by a coral reef or sand spit. Charles Darwin was the first to recognize that atolls develop as volcanoes gradually sink beneath the ocean's surface. All three reef types can develop during the sequence of atoll formation, as illustrated in **Figure 10.13**.

Threats to Coral Reefs

Worldwide, coral reefs are in decline, and their decline points to deeper, global-scale changes occurring in Earth's physical systems. Scientists estimate that 90% of coral reefs will be threatened by 2030, and by 2050, all reefs could be threatened. As **Figure 10.14** shows, fewer than half the world's coral reefs are currently healthy.

The first sign of trouble for a reef is coral bleaching. **Coral bleaching** is the loss of coloration in corals that occurs when they have been stressed or have died. When corals become stressed, they expel the zooxanthellae algae living within them.

Figure 10.12 Distribution of warm-water coral reefs. (A) Coral species diversity is greatest in Southeast Asia. The area of greatest coral diversity is called the Coral Triangle. (B) Royal gramma (*Gramma loreto*) swim among lettuce corals in this coral reef in the Phoenix Islands of the Republic of Kiribati. *(Paul Nicklen/National Geographic Creative)*

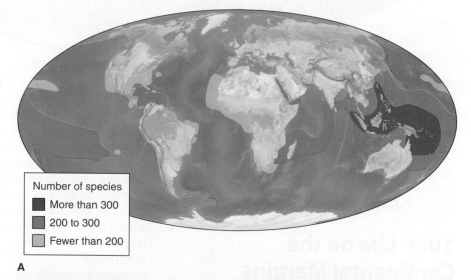

Number of species
- More than 300
- 200 to 300
- Fewer than 200

A

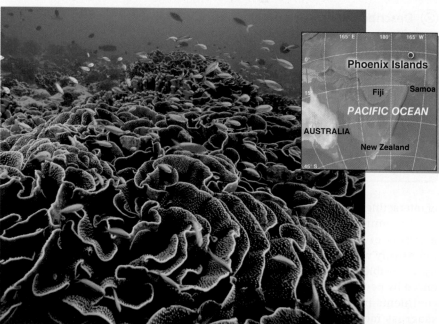

B

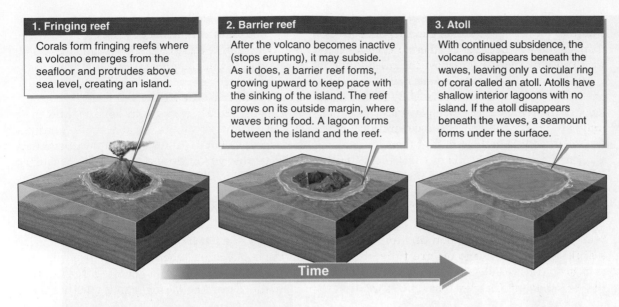

1. **Fringing reef**

Corals form fringing reefs where a volcano emerges from the seafloor and protrudes above sea level, creating an island.

2. **Barrier reef**

After the volcano becomes inactive (stops erupting), it may subside. As it does, a barrier reef forms, growing upward to keep pace with the sinking of the island. The reef grows on its outside margin, where waves bring food. A lagoon forms between the island and the reef.

3. **Atoll**

With continued subsidence, the volcano disappears beneath the waves, leaving only a circular ring of coral called an atoll. Atolls have shallow interior lagoons with no island. If the atoll disappears beneath the waves, a seamount forms under the surface.

Time

Figure 10.13 Three kinds of coral reefs.
A fringing reef, a barrier reef, and an atoll may develop in sequence around a sinking volcanic island. Fringing reefs also form near mainland coasts. Barrier reefs form along mainland coasts as well, but are separated from the coast by open water.

Animation

Three Kinds of Coral Reefs

Available at www.saplinglearning.com

The corals then starve because they are unable to obtain enough food. Most of the color of corals comes from their algal partners. Once the algae are gone, the flesh of the coral polyps appears white. If the environmental stress is not relieved, the coral polyps will die. The reef will soon become covered with a different type of algae that does not live within the polyp's tissues but instead blankets it and blocks sunlight (**Figure 10.15** on the following page).

There are two kinds of bleaching events: local and global. Local bleaching events are caused by local factors, such as pollution. Global bleaching events are caused by warm seawater forming throughout the tropics during El Niño events. The first-ever global bleaching occurred in 1998. Then another one occurred in 2010. The most recent one lasted from late 2015 to late 2017. Global coral bleaching events are increasing because strong El Niño events (see Section 5.4) are increasing, perhaps as a result of climate change. **Table 10.2** summarizes important local and global factors that stress corals and trigger bleaching.

Threatened
(could be lost
in 40 years)
20%

Critical
(will probably be
lost in 20 years)
15%

Healthy
46%

Already lost
19%

Figure 10.14 Coral reef status.
At present, only 46% of the world's warm-water coral reefs are considered healthy.

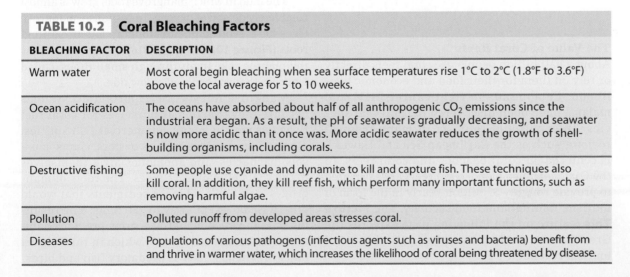

TABLE 10.2	**Coral Bleaching Factors**
BLEACHING FACTOR	**DESCRIPTION**
Warm water	Most coral begin bleaching when sea surface temperatures rise 1°C to 2°C (1.8°F to 3.6°F) above the local average for 5 to 10 weeks.
Ocean acidification	The oceans have absorbed about half of all anthropogenic CO_2 emissions since the industrial era began. As a result, the pH of seawater is gradually decreasing, and seawater is now more acidic than it once was. More acidic seawater reduces the growth of shell-building organisms, including corals.
Destructive fishing	Some people use cyanide and dynamite to kill and capture fish. These techniques also kill coral. In addition, they kill reef fish, which perform many important functions, such as removing harmful algae.
Pollution	Polluted runoff from developed areas stresses coral.
Diseases	Populations of various pathogens (infectious agents such as viruses and bacteria) benefit from and thrive in warmer water, which increases the likelihood of coral being threatened by disease.

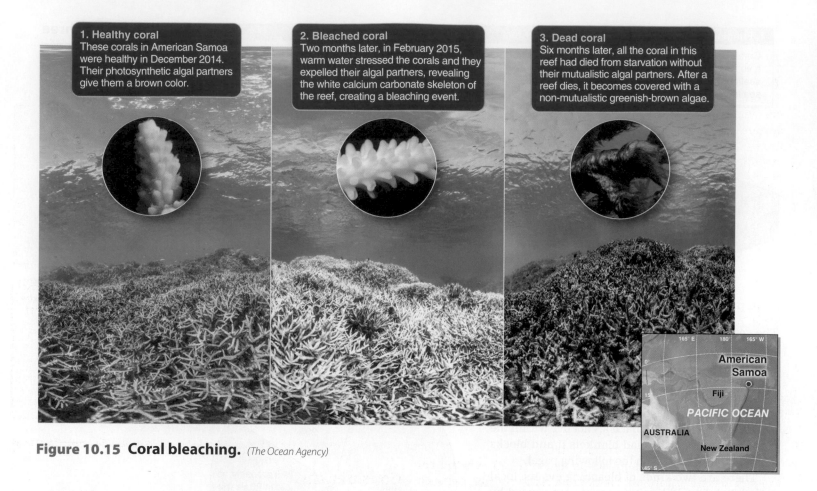

1. Healthy coral
These corals in American Samoa were healthy in December 2014. Their photosynthetic algal partners give them a brown color.

2. Bleached coral
Two months later, in February 2015, warm water stressed the corals and they expelled their algal partners, revealing the white calcium carbonate skeleton of the reef, creating a bleaching event.

3. Dead coral
Six months later, all the coral in this reef had died from starvation without their mutualistic algal partners. After a reef dies, it becomes covered with a non-mutualistic greenish-brown algae.

American Samoa
Fiji
PACIFIC OCEAN
AUSTRALIA
New Zealand

Figure 10.15 Coral bleaching. *(The Ocean Agency)*

Many coral species have the ability to survive bleaching events and recover in about a decade if the external stresses cease. In an unusual case, in the Phoenix Islands coral colonies suffered widespread bleaching and mortality in 2009 and again in 2012. In 2015, researchers were surprised to see that the health of the reefs there had rebounded. It remains to be seen how quickly, if at all, corals will be able to adapt to rising ocean temperatures in the coming decades.

The Value of Coral Reefs

About a half billion people use coral reefs directly or rely on them for their food and economic livelihood. Compounds to fight cancer, HIV, and malaria have been discovered in coral reef species. International tourism in many tropical regions, such as the Caribbean Sea and Hawai'i, is centered on coral beaches and coral reefs. In the Caribbean Sea alone, coral reefs are estimated to provide roughly $5 billion yearly in the form of tourism, fishing, and food security. The **Picture This** feature on the following page explores the Great Barrier Reef of Australia in the context of tourism revenue.

Mangrove Forests

Mangrove forests are ecosystems dominated by saltwater-tolerant coastal shrubs and trees. They are found in the tropics and subtropics. *Mangroves,* unique halophytic (salt-tolerant) plants, occupy, or once occupied, nearly all tropical coastal waters. Where the water is warm enough, their range extends far into the subtropics of North America, Brazil, and eastern Australia **(Figure 10.16)**.

The mud in which mangrove roots grow is almost completely devoid of oxygen. In response, mangrove trees have evolved pneumatophore (air-breathing) roots **(Figure 10.17** on page 316**)**. These roots allow the plant to take in air through small tubes, called *lenticels,* that are exposed at low tide.

Mangrove forests are important ecosystems because they function as nurseries for coral reef fishes as well as many commercial fish species. Marine invertebrates such as oysters, clams, mussels, barnacles, and anemones affix themselves to the roots of mangrove trees. The roots of mangroves trap and hold loose sediments that would otherwise be swept away by currents. The resulting mud in which mangroves grow provides habitat for crabs and sea worms, which in turn attract other organisms, such as predatory fish and birds.

Picture This **The Great Barrier Reef**

The Great Barrier Reef in northeastern Australia is the largest coral reef system in the world. It is a labyrinth of nearly 3,000 reef formations. The reef stretches more than 2,600 km (1,600 mi), is up to 60 km (40 mi) wide in some places, and covers some 350,000 km² (135,135 mi²) in area (red area of the inset map). This satellite image shows only the southern portions of the reef, off the central Queensland coast.

The Great Barrier Reef possesses some of Earth's richest biodiversity. It supports 350 to 400 species of corals, 1,500 fish species, 4,000 species of mollusks, some 240 species of birds, 30 whale species, and 6 sea turtle species.

In response to widespread declines in fish populations and poor reef health, the Great Barrier Reef Marine Park was established in 2004. Fishing was banned in 32% of the reef's area. Only 2 years after the protections were put in place, the reef's fish biomass had doubled, and the reefs had recovered in the protected areas. About 2 million tourists visit the reef each year, and the park generates over $3.4 billion from ecotourism annually. Australia has now enacted the Reef 2050 Plan, which will guide protection efforts and long-term sustainable uses of the Great Barrier Reef into the middle of this century.

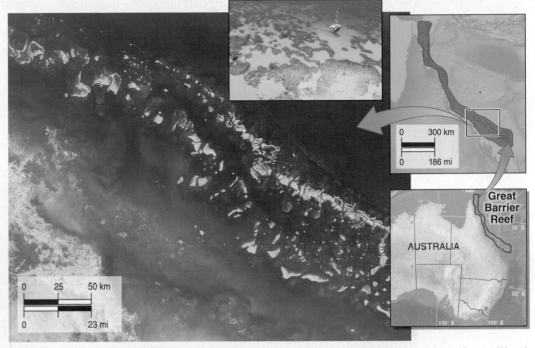

(Image courtesy NASA/GSFC/LaRC/JPL, MISR Team; Inset: Regien Paassen/Alamy)

Consider This

1. Weigh the pros and cons of restricting or banning fishing (or hunting or collecting) from a natural area to allow wildlife populations to recover.
2. Have you ever been an ecotourist? If so, where?

Mangrove species

1   3   5   9   13   17   21   26   36   41   66

Figure 10.16 Mangrove forests. (A) Mangrove forests are found mostly between 30 degrees north and south latitude along coastal shorelines. Mangrove tree species diversity is greatest in Southeast Asia. (B) Mangroves on Nusa Lembongan Island, near Bali, Indonesia. Mangrove trees are the only trees that can grow immersed in salt water. They have several adaptations to cope with high salinity, including the ability to exude salt from pores on their leaves. *(Jason Edwards/Getty Images)*

As **Figure 10.18** details, there are many threats to mangrove forests. The single most important force of degradation for mangrove forests is shrimp aquaculture (farming).

Half the world's original mangrove forests are now gone. About 1% of the remaining mangrove forest area, or 142,000 ha (350,000 acres), is lost each year. At this rate, mangrove forests outside protected reserves will be gone by the end of the century.

Mangroves are a renewable economic resource for people, and they provide a defense against dangerous surges of water caused by hurricanes or tsunamis (see The Human Sphere in Chapter 15, "Geohazards: Volcanoes and Earthquakes"). Coastlines with healthy mangrove forests experience

Figure 10.17 Mangrove roots. Mangroves have broad stilt-like roots that stand above low tide, provide support for the plants in loose sediments, and allow the plants to exchange gases with the atmosphere. A juvenile lemon shark (*Negaprion brevirostris*) swims among mangrove roots in the Bimini Islands of the Bahamas. *(Brian J. Skerry/Getty Images)*

Figure 10.18 Threats to mangrove forests. The table lists the factors that are detrimental to coastal mangrove forests. The pie chart shows where the remaining mangrove forests are located. Asia and Africa together account for 63% of the world's remaining mangrove forests.

THREATS TO MANGROVE FORESTS
Large-scale conversion to shrimp aquaculture
Conversion for tourism
Coastal pollution
Clearance for agriculture
Urbanization
Natural disasters such as hurricanes

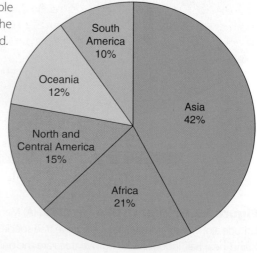

considerably more fishery resources and less flooding and erosion compared with regions where the mangrove forests are gone. Protecting mangroves goes hand-in-hand with protecting the livelihoods and health of people and the habitats of species.

Seagrass Meadows

Seagrass meadows, like mangrove forests, are an often-overlooked marine ecosystem. **Seagrass meadows** are shallow coastal ecosystems dominated by flowering plants that resemble grasses. They are important natural fish nurseries. Although they are most common in tropical waters, seagrass meadows are found in temperate waters as well **(Figure 10.19)**.

The plant species we call seagrasses are not true grasses. The seagrasses comprise about 60 different species within four families of flowering plants, all of which look like grasses, with flat and narrow leaf blades. Common seagrasses include eelgrass (*Zostera* spp.), turtle grass (*Thalassia testudinum*), and manatee grass (*Syringodium filiforme*).

Seagrass meadows provide critical habitat for many species of marine life in their juvenile stages of development. Some animals, such as green sea turtles, manatees (*Trichechus* spp.), and the dugong (*Dugong dugon*), graze on seagrasses. The plants trap sediment, which reduces suspended sediments and improves coastal water quality. Their roots stabilize sediments in coastal areas, reducing coastal erosion.

Several anthropogenic forces, summarized in **Table 10.3**, are causing losses of seagrass meadows globally. Like coral reefs and mangrove forests, seagrass meadows contribute most to local economies when they are preserved rather than converted to other uses. In the Mediterranean Sea and elsewhere, effective efforts to save and restore seagrass meadows have focused on stemming pollution from stream runoff in coastal areas (particularly phosphorus and nitrogen from agricultural fields), establishing protected areas, and replanting seagrass meadows that have diminished or have been lost.

TABLE 10.3 Threats to Seagrass Meadows
Coastal pollution
Dredging (deepening of ports and harbors)
Bottom trawling (dragging fishing nets along the seafloor)
Aquaculture
Beach development

Figure 10.19 Seagrass meadows. (A) This map shows the distribution of seagrass meadows. Like that of corals and mangroves, seagrass species richness is greatest in Southeast Asia and northern Australia. (B) A green sea turtle (*Chelonia mydas*) hovers over turtle grass (*Thalassia testudinum*) in Hol Chan Marine Reserve in Belize. A single acre of seagrass can support 40,000 fish and 50 million invertebrates such as clams, burrowing worms, sea stars, and conches. (© *Brian J. Skerry/Getty Images*)

A

B

Estuaries

An **estuary** is a brackish-water ecosystem at the mouth of a river that is influenced by tides. *Brackish water*—water that is salty but less salty than ocean water—forms where fresh river water and ocean water mix. Salinity in estuaries varies considerably, depending on streamflow levels and tidal levels.

Many large estuaries are about 10,000 years old. They were formed as the great continental ice sheets in the Northern Hemisphere (see Section 7.2) began melting 15,000 years ago. The meltwater drained into the oceans and caused the global sea level to rise about 85 m (280 ft).

Coastal estuaries formed as low-lying coastal areas flooded as the global sea level rose. In many areas, *rias* were formed where coastal river valleys were flooded. Any non-glaciated drowned river valley is called a ria. San Francisco Bay, Rio de la Plata, the mouth of the Amazon River, Puget Sound in Washington State, and Chesapeake Bay **(Figure 10.20)** are some of the world's largest rias.

In the tropics, mangrove forests colonized these drowned river environments. At middle and high latitudes, a variety of salt-tolerant herbaceous plants dominate estuarine environments. Common estuary plants include cordgrass (*Spartina alterniflora*), pickleweed (*Salicornia* spp.), and needlerush (*Juncus roemerianus*).

As rivers flow into estuaries, they bring in organic-rich clays and silts. These sediments are deposited and accumulate into *mudflats*, which are exposed twice daily during low tide. All estuaries have deep deposits of mud that are rich in organic material. Primary productivity is high in estuaries, as plant growth is fueled by the nutrient-rich sediments.

Estuaries are important habitats for many kinds of organisms, particularly for the juvenile stages of many fish species. Some 75% of the commercial fish catch in the United States consists of fish that depend on estuaries for at least part of their life cycle. Many bird and mammal species also rely on estuary habitat. Migratory birds use estuaries as rest and feeding stops on their long-distance migrations.

Because estuaries are flat and have calm waters, they make ideal locations for human settlement. Seventy percent of the world's largest cities, including London, Shanghai, Buenos Aires, Hong Kong, Boston, and New York City, are located on estuaries (or what were once estuaries). People expand settlements onto estuaries by first filling them in with debris, such as soil or even garbage, then expanding the cities onto the newly created surface.

Because of the heavy human influence on estuaries, they are the world's most endangered marine ecosystems. **Table 10.4** summarizes the threats to estuaries.

A

B

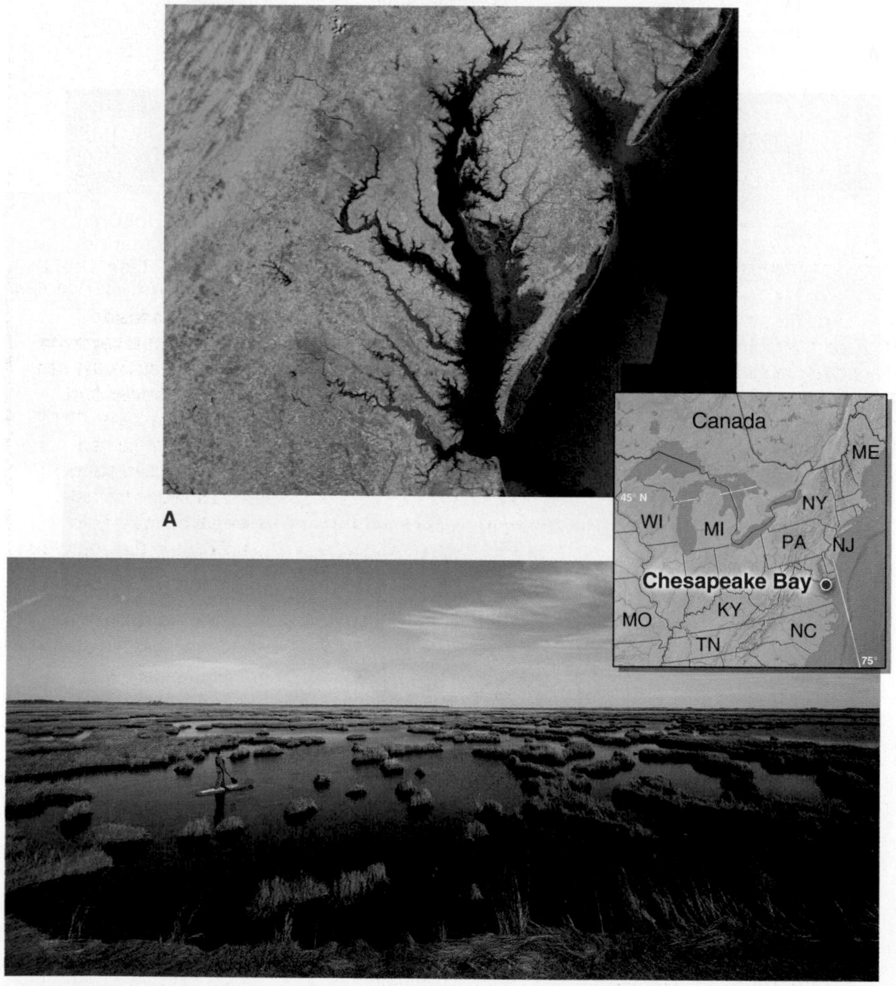

Figure 10.20 Chesapeake Bay. (A) At top is a Landsat image of Chesapeake Bay. This estuary was a river valley when sea levels were lower, about 15,000 years ago. (B) Most of the original wetlands of Chesapeake Bay have been developed and are gone. Some, such as these wetlands in the Blackwater National Wildlife Refuge in Maryland, have been preserved.   *(A. U.S. Geological Survey, NASA; B. Trevor Clark/Getty Images)*

TABLE 10.4	**Threats to Estuaries**
Heavy industry	
Dredging, infilling, and housing development	
Coastal pollution	
Seawalls	

Estuaries provide many economic benefits to people. They not only provide shelter for juvenile commercial fish species (such as herring and salmon) and production of shellfish such as oysters and crustaceans like crabs. They also provide recreational activities like kayaking and sport fishing. Estuaries also absorb wave energy from large storms, such as hurricanes, significantly reducing coastal erosion and property loss. Furthermore, they filter pollutants from rivers before they enter the sea, significantly improving coastal water quality.

Kelp Forests

Kelp forests are marine ecosystems in temperate and polar waters where the water temperature does not exceed 20°C (68°F) **(Figure 10.21)**. As their name implies, these ecosystems are dominated by large algae called *kelp*. There are three groups of kelp: green kelp (*Chlorophyta*), tan kelp (*Phaeophyta*), and red kelp (*Rhodophyta*). Only tan kelp form kelp forests.

A

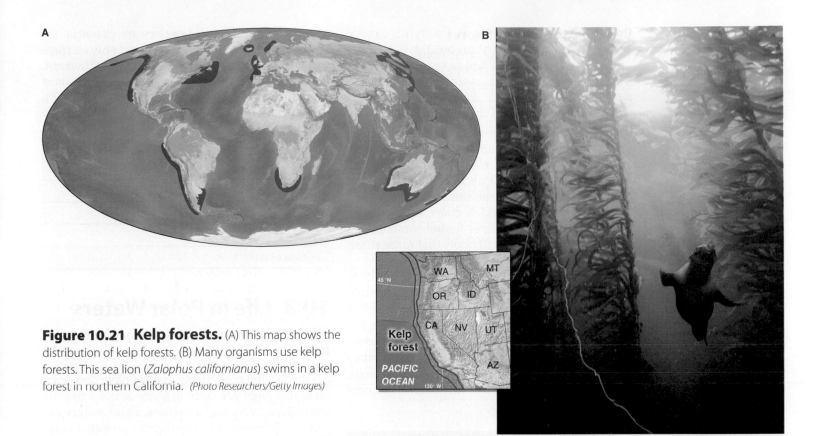

Figure 10.21 Kelp forests. (A) This map shows the distribution of kelp forests. (B) Many organisms use kelp forests. This sea lion (*Zalophus californianus*) swims in a kelp forest in northern California. *(Photo Researchers/Getty Images)*

Limited to zones with adequate light, kelp forests grow only in water no deeper than 50 m (165 ft). The kelp rely on *holdfasts*, which resemble roots, but function only to anchor the algae to the rocks. Many kelp grow quickly; some tan kelp can grow 1 m (3.3 ft) per day.

Fish communities, sea urchins, anemones, sea stars, mollusks, and crustaceans inhabit kelp forests. Kelp produce toxins to deter most grazers. Sea urchins, however, do eat kelp, and if their populations increase, kelp forests can suffer steep declines. **Figure 10.22** explores the Pacific sea otter's role as a keystone species in kelp forests.

Like other coastal ecosystems, kelp forests are vulnerable to threats posed by human activities **(Table 10.5)**. Fortunately, the rapid growth rates of kelp allow them to recover rapidly if degradation pressure stops.

TABLE 10.5	**Threats to Kelp Forests**
Pollution and sediment runoff from streams	
Ocean warming caused by climate change and El Niño	
Removal of keystone species, such as sea otters	
Non-native sea urchins	
Harvesting of kelp for food, food additives, and other products	

1. Keystone species

Each day, a Pacific sea otter (*Enhydra lutris*) eats about 25% of its body weight in food. Much of its diet is purple sea urchins (*Strongylocentrotus* species).

2. Overhunting

In the 1800s, the Pacific sea otter disappeared throughout most of its range in the northern Pacific Ocean due to hunting for its fur. By 1900, it was nearly extinct. In 1911, it was legally protected.

3. Ecosystem changes

A *trophic cascade* occurs when a keystone species, such as the sea otter, is removed and other species are affected. Purple sea urchins (shown below) feed on kelp holdfasts, killing the whole kelp plant. When the sea otter was removed, sea urchin populations grew. As a result, kelp forests and the species that depend on them suffered.

4. Reintroduction

Pacific sea otter reintroductions (return to a former geographic range) have contributed to the recovery of many Pacific kelp forests. Although the Pacific sea otter has been returned to some of its former range, its population numbers are still low.

Figure 10.22 The Pacific sea otter as a keystone species.
(1. Marc Shandro/Getty Images; 2. Gulf of Maine Cod Project, NOAA National Marine Sanctuaries; Courtesy of National Archives; 3. Mark Conlin/Getty Images)

Beaches and Rocky Shores

The coastal ecosystems most familiar to people may be beaches and rocky shores. Most of us have strolled along a beach barefoot or gazed at the horizon from a rocky shore. Because they are exposed to the force of crashing waves and wind, beaches and rocky shores are high-energy environments. As a result, many organisms in these environments are either anchored to the rocks (such as barnacles, sea urchins, sea stars, and kelp) or hidden within the sand or rock crevices (such as clams, periwinkles, crabs, and octopus). Rocky coastlines sometimes harbor communities that live in **tide pools,** still pools that form at low tide **(Figure 10.23)**.

In general, biodiversity is not particularly high in beach and rocky shore environments, but many organisms live only in these habitats or rely on them during at least part of their life cycle. Shorebirds such as gulls, sandpipers, terns, and pelicans are common on beaches and rocky shores. Marine mammals such as sea lions, seals, and sea otters are found in these environments

as well. Warm tropical beaches are essential for sea turtle survival because they nest only on tropical and subtropical beaches. Human settlement, pollution runoff, and erosion of sand are the most significant threats to beach and rocky shoreline ecosystems **(Table 10.6)**.

TABLE 10.6 **Threats to Beaches and Rocky Shores**
Conversion to vacation resorts and construction
Pollution runoff
Fishing pressure
Erosion of sand

10.3 Life in Polar Waters

◎ **Explain how polar marine ecosystems function.**

All the marine ecosystems that we have described up to this point are close to shore and are heavily influenced by human activities. Polar waters are close to the edges of the continents as well, but far fewer people live near them and influence them.

The Importance of Phytoplankton

At the base of the food web in the polar oceans are the microscopic photosynthetic algae and other microorganisms collectively called phytoplankton. All oceans contain phytoplankton. Even though ocean phytoplankton account for only 1% of the world's photosynthetic biomass, they are very important, producing 70% of the oxygen in Earth's atmosphere.

Marine ecosystems rely on phytoplankton to convert solar energy to the chemical energy that all other marine organisms require. (Hydrothermal and geothermal vent ecosystems are the exception, as we will see in Section 10.4.) The key to primary productivity in ocean water is **upwelling**: the circulation of water from the seafloor to the ocean surface. Upwelling brings nutrients from the seafloor sediments to the sunlit epipelagic zone, where phytoplankton reside. The more upwelling that occurs, the better the phytoplankton are fertilized and grow, and the more productive the ecosystem.

Phytoplankton in Polar Waters

Polar waters have no thermocline, meaning that the surface water is as cold as the water at depth (see Figure 10.6). As a result, deep, cold water can readily mix to the surface and stimulate phytoplankton "blooms." Where surface water is warm in the tropics and subtropics, the deep cold water is relatively dense. This greater density inhibits the upwelling of nutrients. Upwelling is certainly

Figure 10.23 Tide pools. Tide pools harbor unique communities composed of sea anemones, crabs, octopuses, kelp, and many other organisms. Most tide pools are easy to reach and explore at low tide. Here purple and pink starfish (*Pisaster ochraceus*) cling to a rock at low tide in Sechelt, British Columbia. *(Julian Nieman/Alamy)*

not restricted to polar waters, but because of the geographically widespread occurrence of upwelling in polar waters, they are among the most biologically productive oceans in the world **(Figure 10.24)**.

In contrast to polar terrestrial ecosystems, polar marine ecosystems can respond rapidly to spring warming because phytoplankton, with a life span of a few days, grow as soon as temperature and light levels are suitable. Land plants at high latitudes, however, must wait for the snow to melt and ground temperatures to increase. Thus, there is far less available energy in polar terrestrial ecosystems and far less biomass on land. Antarctica, for instance, is barren and has little biomass, while the seas just offshore are teeming with life supported by phytoplankton blooms.

Many migratory birds, as well as mammals such as whales, make their way to the high latitudes each summer to gorge on the brief but bountiful productivity of the polar waters. During the dark polar winter, phytoplankton activity shuts down, and many species migrate to more productive latitudes.

10.4 Life in Open Waters

◎ **Describe the geographic patterns of life in the open ocean.**

Because the oceans are so large and so deep, **pelagic** (open-ocean) ecosystems are the most geographically extensive ecosystems on the planet. *Open waters* (also called the *high seas*) cover about 50% of Earth's surface in total. Our knowledge of the open ocean, however, is still in its infancy. Scientists are working to understand what lives there, how pelagic ecosystems function, and how people are influencing them.

In this section of the chapter we discuss two geographic approaches to examining the open waters: (1) tracing the daily movements of organisms vertically through layers of the ocean and (2) following the periodic movements of organisms horizontally across the ocean's surface.

Daily Vertical Migrations

Much of the open ocean's surface is a biological desert by day, but it is full of life at night, when organisms migrate upward from the safety of the dark depths to feed at the surface. These daily vertical migrations, or "commutes," represent the largest synchronized animal movements on Earth. All types of organisms rise to the surface each night, including zooplankton, jellies, fish, and squid. Some organisms, such as jellies, take hours to reach the surface, spending more time migrating than feeding.

Figure 10.24 Phytoplankton productivity. Phytoplankton blooms are shown in green, yellow, orange, and red over the oceans. Blue areas have the lowest phytoplankton activity, and red areas have the highest. Note the phytoplankton activity at high latitudes. *(A. SeaWiFS Project, NASA/Goddard Space Flight Center, and DigitalGlobe™; B. SeaWiFS Project, NASA/Goddard Space Flight Center, and DigitalGlobe™)*

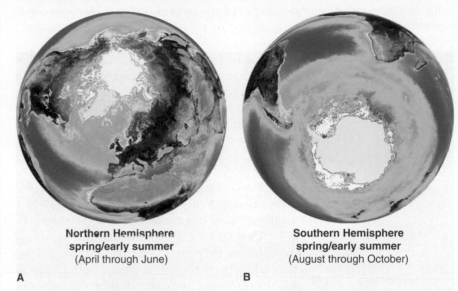

Northern Hemisphere spring/early summer
(April through June)

A

Southern Hemisphere spring/early summer
(August through October)

B

Many other organisms make a "reverse commute" to feed. In other words, they live at the lighted ocean surface but dive into the depths for their food **(Figure 10.25** on the following page**)**.

Trans-Ocean Migrations

During the past decade, thanks to new GPS tagging and tracking technology, scientists have begun to piece together where pelagic migratory organisms go. Scientists are using a variety of electronic tagging techniques on migratory species to reveal where they migrate, when they migrate, and why they migrate. **Figure 10.26** on the following page traces the migrations of two species, the great white shark and the bluefin tuna.

Life in the Deep

No light reaches below the mesopelagic zone, which ends at a depth of 1,000 m (3,280 ft) (see Figure 10.6). Below the surface epipelagic zone, the open ocean is pitch black and perpetually just above freezing. Pressures at these depths are extremely high because of the immense weight of water above.

There is light at these depths, but it doesn't come from the surface. The vast majority (90%) of organisms in the deep sea make their own light. Some living organisms are capable of **bioluminescence**, the production of light through chemical means. Bioluminescence, used to attract mates and prey and to hide and escape from predators, has evolved independently among unrelated groups of deep-sea organisms at least 40 times **(Figure 10.27** on page 323**)**.

Figure 10.25 Deep-diving organisms. (A) The strange-looking ocean sunfish (*Mola mola*) lives at the ocean surface and dives about 200 m (660 ft) deep to eat jellies. Note the GPS data recorder tag just behind the top fin of the fish. (B) Sperm whales (*Physeter macrocephalus*) dive as deep as 3,000 m (10,000 ft)—3 km straight down—to feed on giant squid. Like dolphins and bats, sperm whales locate their prey in the dark using sound, or echolocation. *(A. © Mike Johnson; B. Reinhard Dirscherl/ullstein bild via Getty Images)*

A

B

Figure 10.26 Pelagic animal migrations. An understanding of the movement patterns of marine organisms is an important conservation tool for protecting populations. (A) Great white sharks (*Carcharodon carcharias*) routinely travel between California and Hawai'i and congregate in between at a location scientists call the shark café, probably to feed. The yellow dots represent the locations of several dozen tagged sharks over a period of a few years. (B) Bluefin tuna (*Thunnus thynnus*), like many other pelagic organisms, routinely cross ocean basins. The dots represent the locations of a single 15 kg (33 lb) bluefin that crossed the North Pacific between California and Japan three times in less than 2 years. *(Photos by Michael O'Neill/Science Source)*

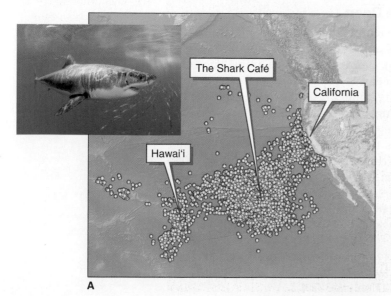

A

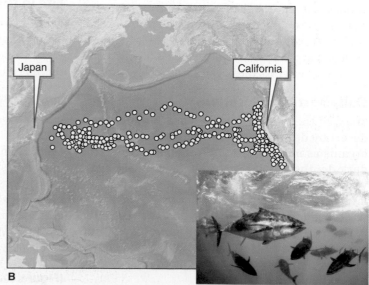

B

Scientific study of life in the deep ocean is a particularly challenging endeavor. For example, scientists have only established within the past decade that the bristlemouth fish (the genus *Cyclothone*) living in the dark mesopelagic zone is the most abundant vertebrate on Earth. Only the length of a finger, this fish exists in staggering numbers. Scientists estimate that there are hundreds of *trillions* of them in the world's oceans. The tremendous pressures of the deep have created a formidable barrier to scientific investigation, and staffed missions are very expensive. Robotic submersibles and trawls, which are far less costly, have been among the most important sources of scientific information on deep-ocean ecosystems **(Figure 10.28)**.

Biological Islands: Seamounts and Hydrothermal Vents

The abyssal plains are relative biological deserts. Rising from these plains are rich biological islands in the form of seamounts and hydrothermal vents. Collectively, seamount ecosystems are larger in area than any terrestrial biome. Currently 33,452 of them have been identified in the world's oceans, constituting almost 5% of the seafloor. Together, they make up an area approximately the same size as the continent of South America. These ecosystems,

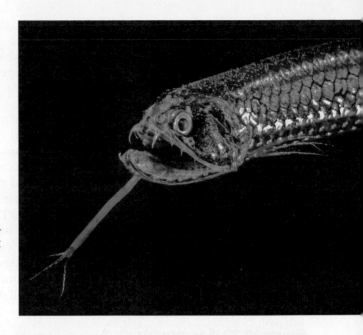

Figure 10.27 Adaptations to the deep sea. The black-belly dragonfish (*Stomias atriventer*) has many traits that are common among deep-sea fishes: large eyes for detecting bioluminescence in extremely low-light conditions, a bioluminescent lure to attract prey, and a large jaw with cage-like teeth to trap prey. Silver scales reflect the faint light and reduce the fish's silhouette, lowering its visibility to predators and prey. *(© Dave Wrobel)*

however, are largely inaccessible to scientific study and barely understood by scientists **(Figure 10.29** on the following page**)**.

Like winds around a high mountain, ocean currents flow over and around large seamounts. These currents may accelerate and become turbulent,

Trawl sampling
Nets, called trawl nets, that are dragged along behind vessels are the mainstay of research on life in the deep. *Benthic trawls* capture samples from the ocean floor. *Pelagic trawls* capture samples from the middle of the water column. *Plankton nets* collect plankton-sized organisms in a fine net with a mesh size of 335 µm (0.01 in).
Advantages: Trawling makes it possible to sample large areas of the ocean.
Limitations: Some animals swim out of the nets and avoid capture. Soft-bodied animals such as jellies are damaged beyond recognition.

Remotely Operated Vehicle (ROV) sampling
ROVs such as the *Hercules*, shown here, are equipped with instruments for exploration of the deep. *Slurp guns* vacuum up specimens. *Mechanical arms* reach out and grab samples from the seafloor.
Advantages: ROVs can be used to target specific habitats for study and collect live specimens for study.
Limitations: The sampling area is limited, and ROVs are more expensive to operate than trawls.

Figure 10.28 SCIENTIFIC INQUIRY: How do scientists explore life in the deep? The deep oceans are the least understood of Earth's ecosystems, but evolving submersible technology is improving scientists' understanding of deep-sea ecosystems. Each approach to scientific exploration of the deep has advantages and disadvantages. *(Top left. Image courtesy of Lewis and Clark 2001, NOAA/OER; top right. NOAA Ocean Explorer; bottom. Solvin Zankl/Alamy)*

Remotely Operated Vehicle (ROV) photo documentation
Photos of organisms taken from ROVs are invaluable in documenting marine life in its natural ecological setting. This humpback anglerfish (*Melanocetus johnsonii*) spends its life suspended in total darkness and near-freezing temperatures at a depth of 3,200 m (10,500 ft). It uses its bioluminescent lure to attract prey.
Advantages: Photos of deep-sea organisms in their natural habitat provide an ecological context and show how they behave.
Limitations: Identification of unknown species can be difficult using photography alone. A physical specimen is necessary for genetic analysis.

stirring up nutrients from the seafloor and bringing them to the sunlit surface. The nutrients stimulate phytoplankton blooms, which support other organisms. Many migratory species, including sharks, tuna, whales, and seabirds, seek out and congregate above seamounts.

Like seamounts, hydrothermal vent communities are rich biological islands on the seafloor. **Hydrothermal vent communities** are unique ecosystems found at *hydrothermal vents*, volcanic hot springs that emit mineral-rich water. Seawater circulates through fissures in the crust and becomes heated to over 400°C (750°F) before emerging from the seafloor. The water becomes a stew of acids, sulfides, methane, and carbon dioxide as rocks dissolve in it. Specialized bacteria metabolize these chemicals by the process of *chemosynthesis*, and other organisms feed on those bacteria. These hydrothermal vent communities get their energy from geothermal energy rather than solar energy. If the Sun were to stop shining, these organisms would probably not be affected, while the rest of the biosphere that is dependent on solar energy would disappear. The locations of some hydrothermal vents are shown in **Figure 10.30**.

Reaping the Bounty: Industrial Fishing

All marine ecosystems, whether on the continental margins or in the open ocean, support fisheries of one type or another. A **fishery** is a region where fish are caught for human consumption. The livelihoods of about a half billion people, mostly in economically developing countries, are dependent on small-scale, local, and *sustainable fishing*. In sustainable fishing, the amount of fish taken is equal to or less than the reproductive rate of the fish population.

Figure 10.29 Seamounts. (A) This map shows the distribution of the thousands of mapped seamounts. All of them are volcanic in origin. (B) A sonar image of the Kawio Barat seamount in the Celebes Sea, south of the Philippines. The vertical heights of this seamount are color coded. From its base to its peak, it is about 3,000 m (10,000 ft) high. *(Courtesy Christopher Small, Lamont Doherty Earth Observatory, Columbia University)*

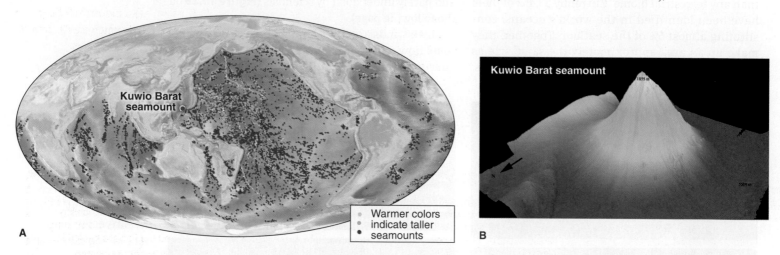

Figure 10.30 Hydrothermal vents. (A) This map shows the locations of the 200 known hydrothermal vents. Like seamounts, hydrothermal vents are all volcanic. (B) This hydrothermal vent was photographed at a depth of 2,250 m (7,500 ft) near Vancouver, Canada, on the Juan de Fuca Ridge. As mineral-rich water escapes from the seafloor, the minerals precipitate out, forming chimneys called *black smokers*. More than 600 new animal species have been discovered at hydrothermal vents, including the tube worms (*Riftia pachyptila*) seen here. *(© Verena Tunnicliffe, University of Victoria)*

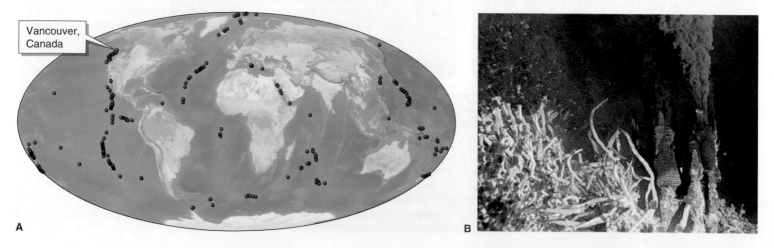

Unsustainable fishing, in contrast, occurs when fish are taken out of a fishery faster than the fish population can reproduce, causing the population to decline. Most *industrial fishing* (fishing done one a large scale for commercial profit) is unsustainable. Due to unsustainable fishing, about 30% of the world's fisheries have *collapsed*. We define a collapsed fishery here as a fishery with a fish population that is just 10% of its former population.

A growing human population, coupled with advances in industrial fishing technology in the past several decades, are to blame for global fishery declines. The global fish catch peaked in 1989, at 90 million tons, and has since leveled off or declined. Fish catches are down because there are fewer fish in the oceans now than in the past. The decline is not due to conservation efforts; there are now more fishing boats plying the oceans than ever. Industrial fishing activity now covers 55% of the world's ocean area **(Figure 10.31A)**. This is an area four times larger than the land area used for agriculture.

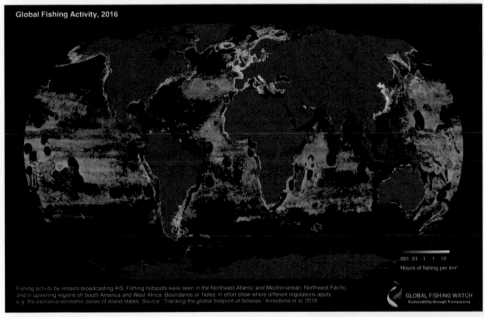

Global Fishing Activity, 2016

.001 .01 .1 1 10
Hours of fishing per km²

Fishing activity by vessels broadcasting AIS. Fishing hotspots were seen in the Northeast Atlantic and Mediterranean, Northwest Pacific, and in upwelling regions off South America and West Africa. Boundaries or 'holes' in effort show where different regulations apply e.g. the exclusive economic zones of island states. *Source 'Tracking the global footprint of fisheries.' Kroodsma et al, 2018*

GLOBAL FISHING WATCH
Sustainability through Transparency

A

Figure 10.31 Industrial fishing.
(A) This map shows the intensity of fishing per square kilometer in 2016 (as hours fished per year). The most intensively fished areas show up in yellow and green in coastal Europe and Southeast Asia. The dark circular regions are marine protected areas in which fishing is banned. Some 70,000 vessels were tracked to make this map. More than 75% of them were industrial fishing boats more than 36 m (118 ft) in length. (B) The four major industrial fishing methods and solutions to bycatch are illustrated here. *(David Kroodsma/Global Fishing Watch)*

1. Drift net fishing (or gillnetting)

Huge factory ships use sonar and airplanes to locate fish. Each ship then launches 20 to 50 small, fast boats to set out drift nets spanning thousands of miles in length. The nets are set out for several days. Fish are caught in the nets by their gills.

Addressing bycatch:
Drift nets can be outfitted with sonar devices that whales and dolphins can hear, allowing them to avoid the nets. Restricting drift nets to minimum distances from regions where birds breed can help minimize bird bycatch.

2. Trawling

Trawling nets can be as large as a football field. *Bottom trawling* involves dragging nets on the seafloor, raking in seafloor organisms indiscriminately.

Addressing bycatch:
Turtle-excluding devices (TEDs) are small doorways that allow many sea turtles to escape the net while keeping desired fish. Bottom trawling has the highest bycatch of any fishing method and should be minimized or, many scientists believe, abolished.

3. Longlining

Long-lining employs a central line up to 80 km (50 mi) long with thousands of baited hooks dangling from it. The depth of the central line can be set specifically for the targeted fish species.

Addressing bycatch:
Setting the line at specific depths and using specially designed hooks reduces bycatch of sea turtles, albatross, and other unwanted species.

4. Seining

Seining involves using a net that surrounds schooling fish. The net is drawn shut, and the fish are hauled in.

Addressing bycatch:
Seining has high dolphin bycatch, because dolphins are attracted to the schooling fish. Lowering the lip of the net so the dolphins can escape is one way to reduce dolphin kills.

B

Drift net fishing, trawling, longline fishing, and seining are four methods commonly employed in industrial fishing. **Drift net fishing** involves suspending large nets in the upper reaches of the ocean. **Trawling** involves dragging nets through the water column or along the seafloor. **Longline fishing** employs thousands of baited hooks on lines up to 80 km (50 mi) in length. **Seining** involves encircling fish and trapping them in a large net **(Figure 10.31B** on page 325**)**. All these fishing methods invariably kill other marine species besides the intended ones, a phenomenon called **bycatch**. These organisms are usually thrown back to sea dead. Bycatch is a serious problem that threatens sea turtles, albatross, whales, dolphins, and many species of fish.

Approximately one-third of the world's open-water shark species (more than 100 species) are considered *endangered* or *critically endangered*, according to the International Union for Conservation of Nature (IUCN), due to forces such as habitat degradation and unsustainable fishing. Because sharks are keystone species in many marine ecosystems, healthy functioning of those ecosystems is compromised when shark numbers are reduced. The IUCN estimates that around 75 million sharks are killed every year for just their fins, which are used to make shark fin soup **(Figure 10.32)**.

Figure 10.32
Drying shark fins.
Thousands of shark fins have been laid out to dry on this rooftop in Hong Kong, the capital of the shark fin industry. In the past, the fins were dried mostly on sidewalks. In recent years, rooftops have become the preferred location to dry the fins to avoid increasing scrutiny of this destructive practice. *(Antony Dickson/AFP/Getty Images)*

GEOGRAPHIC PERSPECTIVES
10.5 The Problem with Plastic

◎ **Assess the environmental impact of and remedies for plastic pollution in the oceans.**

Plastic has transformed human society. Plastic is cheaper than most natural materials, and it often performs better. Nearly everything we use is composed at least partly of plastic: electronics (cell phones and computers), packaging, facial scrubs, toothbrushes, credit cards, bags, bottles, car parts, combs and hairbrushes, clothing—we simply could not do without plastic. So how is plastic relevant to physical geography or the oceans? The answer has to do with these statistics:

1. Each year, about 311 million metric tons of plastic are made. That is about 42 kg (92 lb) per year for each person on Earth.

2. In all, we've made 8.3 billion metric tons of plastic, a weight equal to 40 million adult blue whales. Half of all that was made in just the past 13 years.

3. No commonly used plastic is *biodegradable*. Because it does not biodegrade, it lasts for centuries—or longer—in the environment.

4. Europe recycles 30% of its plastic. China recycles 25%. The United States recycles 9%.

5. Worldwide, 79% of plastic goes to the landfill, and 12% is incinerated.

What Happens to Plastic in the Oceans?

About 2% to 4% of all plastic made each year (5 to 13 million metric tons) finds its way to the oceans. Most of the plastic entering the oceans comes from streams flowing into the sea and from beaches. Plastic trash in the streets of towns and cities is washed into storm drains, then into rivers, then into the oceans.

About 70% of all plastic that enters the oceans eventually breaks apart into smaller and smaller pieces and gradually sinks. Scientists believe that through time these *microplastics* are being transported long distances by deep-ocean currents and accumulating as plastic sediment in the world's ocean basins. That plastic will persist for thousands to tens of thousands years or longer. Large accumulations of plastic presumably lie somewhere on the seafloor, but they have not yet been found.

The 30% of plastic that remains suspended in the sunlit epipelagic zone is carried on the ocean gyres. Eventually, that plastic collects in the center of the gyre, in what is often referred to as a garbage patch. **The Great Pacific Garbage Patch**, a region of concentrated plastic litter formed by the North Pacific Gyre, is the largest and best known of these collections. It is made up of an estimated 80,000 metric tons totaling 1.8 trillion pieces of plastic trash. This suspended plastic ranges in size from massive clumps of fishing nets that weigh tons to minuscule bits of microplastics invisible to the human eye. Most of the pieces are less than 1 cm (0.4 in) in length and cannot be seen from a boat. It is possible to catch them by towing a fine sieve (net) through the water **(Figure 10.33)**.

The plastic in garbage patches is undetectable to satellites or aerial photography because most of it is suspended just beneath the water surface. However, ocean current patterns are well known, and computer simulations reveal where the plastic debris concentrations are greatest. Transported by ocean currents, suspended plastic finds its way to the most remote areas. There, waves wash it onto beaches, where it collects and breaks down further **(Figure 10.34)**.

Figure 10.33 Suspended plastic. Charles Moore, one of the first discoverers of the plastic in the oceans, holds a jar containing small bits of plastic captured by a towed sieve. Note that the plastic is suspended in the water, not floating at the surface. *(Courtesy of Algalita Marine Research Institute)*

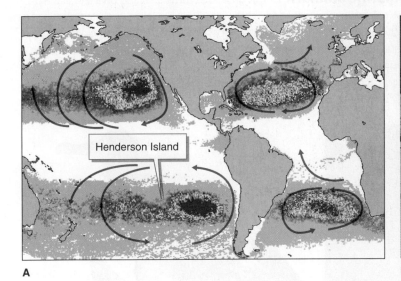

A

B

Figure 10.34   Plastic in the ocean gyres. (A) This map shows the circulation of the ocean gyres (blue arrows) in the Atlantic and Pacific oceans and the probable locations of concentrated plastic debris modeled by computer simulations of ocean current patterns. Red areas show the greatest concentration of plastic debris. Each of the regions of plastic concentrated in the gyres covers an area roughly equivalent to the size of the lower 48 United States. (B) This beach is on uninhabited Henderson Island (a World Heritage Site), which is only 37.3 km^2 (14.4 mi^2). Henderson is located on the easternmost edge of the plastic debris field in the South Pacific Gyre. Henderson's beaches have the highest reported plastic density in the world: Researchers found 671 plastic items for every square meter (11 square feet) of beach. *(A. Nikolai Maximenko, International Pacific Research Center, SOEST, University of Hawaii; B. USFWS photo by Susan White)*

Story Map

The Problem with Plastic

Available at
www.saplinglearning.com

How Does Plastic in the Oceans Pose a Problem?

Suspended plastic provides a surface onto which many forms of life cling, creating what's called the *plastisphere*. Some of this life can be harmful. For example, bacteria in the genus *Vibrio*, which cause cholera and other human health ailments, have been found on ocean plastic. Researchers suspect that such pathogens are able to persist longer and travel farther as they cling to plastic particles adrift in the ocean.

Suspended plastic particles in the oceans also attract and *adsorb* (bind to their surface) toxic chemicals that have entered ocean waters (such as *PCBs* and *DDEs* from industrial activity and pesticides from agricultural spraying). Plastics have been found to carry concentrations of PCBs and DDEs 1 million times those of seawater. The toxins enter organisms that eat the plastic. Scientists have documented ingestion of microscopic bits of plastic by organisms such as barnacles, krill (small shrimplike organisms), and fish that filter-feed on plankton. Toxins from the plastic particles could work their way through almost all marine trophic levels and even into terrestrial trophic systems, eventually potentially finding their way to our dinner plates. It is not yet known, however, the extent to which these toxic chemicals are making their way through marine food webs.

Plastic does more obvious harm to marine life when animals eat it or become entangled in it **(Figure 10.35)**. Scientists estimate that 100,000 marine mammals and sea turtles are killed by plastic each year, mostly through entanglement.

A 2018 study took 1-liter samples of sea ice from five locations across the Arctic and found plastic in each sample. One sample had over 12,000 bits of plastic in it. Plastic debris in our waters is not limited to the oceans. The Great Lakes contain high concentrations of microscopic plastic particles as well. They come from facial cleansers that contain tiny plastic beads and from plastic fibers in our clothes. Plastic micro-fibers dislodge when we wash our clothes. They are so small they pass through water treatment facilities and then are discharged with the treated water into the rivers. Microplastics have even been found in sea ice in the Arctic ocean.

Video

A Rescue Mission to Save Sea Birds

Available at www.saplinglearning.com

Figure 10.35 Plastic threatens marine animals. (A) This green sea turtle is dining on a jelly, a main food for sea turtles. To sea turtles, translucent plastic bags suspended in the water look like jellies. Ingested plastic can kill a sea turtle by blocking its digestive tract. (B) This adolescent Laysan albatross (*Phoebastria immutabilis*) from Kure Atoll probably died of starvation as its parents fed it a diet of plastic. Its stomach was filled with 340 g (12 oz) of trash. Plastic ingestion is a common cause of mortality for albatross nestlings. Worldwide, seabird mortality from eating plastic is estimated to be about 1 million birds per year. Fish mortality related to plastic is unknown. *(A. Ai Angel Gentel/Getty Images; B. David Littschwager/National Geographic Creative)*

A

B

Fixing the Problem

The problem with plastic in the oceans resembles other situations in which a harmful agent released into the environment by people has led to subsequent problems in Earth's physical and biological systems. Toxic air pollutants (Section 2.4), CFCs in the ozonosphere (Section 2.5), and carbon in the atmosphere (Section 7.3) are three examples we have examined in this book.

There is too much plastic being made to clean it up after it has entered the environment, and doing away with plastic altogether would be impossible. What can society do? There is no single solution. An important step toward addressing this problem is making the problem widely known. As **Figure 10.36** demonstrates, a social media campaign to inform the public is under way.

What can an individual do? Reduce, reuse, and recycle. Many of the plastic items that enter the oceans are designed to be one-use items, such as water bottles, grocery bags, and utensils.

These items are used for only a few days, hours, or even minutes. They then spend millennia in landfills or in the environment. About half of all plastic produced is made to be disposed of after a single use.

To address this issue, *closed-loop systems* need to be developed. In closed-loop systems, plastics are designed to be reused and recycled indefinitely. Many plastics made now (such as PVC, polyurethane, and polystyrene foam) have properties that make them difficult or impossible to recycle. Similarly, the many different types of plastics must be separated before they can be recycled. But separating many plastics—for example, the different types of plastics found in toothbrushes and electronic devices—is virtually impossible.

Another important step is to create *bioplastics* that biodegrade in the environment. Bioplastics, which are made from renewable biomass materials, such as corn, bacteria, and vegetable oils, are

Just because you can't see it, doesn't mean it isn't there

Figure 10.36 Plastic pollution awareness. This message about plastic pollution in the oceans captures the viewer's attention and conveys the information at a glance. Many major cities, such as Los Angeles, and even states, such as Hawai'i, are restricting the use of plastic, such as bans on plastic bags in grocery stores. Such measures have passed because the public supports them, and public support can only come through public awareness.

(© 2011 Ferdi Rizkiyanto)

becoming increasingly available. The more consumers demand them, the more they will be made available at a price equal to that of traditional petroleum-based plastics. Reusable and biologically based plastics are two important solutions to this problem.

Modern society needs plastic, and it needs healthy ecosystems. These two needs are not necessarily mutually incompatible. The way plastic is perceived and manufactured will certainly change as more people become aware of this growing ecological problem in the oceans.

Chapter 10 Study Guide

Focus Points

10.1 The Physical Oceans

- **Five oceans:** The Pacific Ocean is the largest and oldest of the five ocean basins.
- **Layers:** The oceans can be divided into five vertical layers. Sunlight enters only the uppermost layer.
- **Surface ocean currents:** Surface ocean currents are driven by wind. The Coriolis force deflects the moving water, creating large circular ocean currents called gyres.
- **Seafloor features:** Seafloor topography includes continental shelves, continental slopes, abyssal plains, major mountain ranges, and deep-sea trenches.

10.2 Life on the Continental Margins

- **Ecosystem types:** Marine ecosystems along continental margins range from tropical coral reefs to kelp forests, rocky shorelines, and beaches.
- **Human footprint:** All coastal marine ecosystems are influenced by human activity, particularly pollution, overfishing, habitat loss, and non-native species.
- **Fish nurseries:** Mangroves, seagrass meadows, and estuaries provide important shelter for young fish.
- **Conservation:** Coastal ecosystems provide long-term, sustainable benefits for people when they are preserved rather than converted to other uses.

10.3 Life in Polar Waters

- **Human footprint:** Polar marine ecosystems are relatively unaffected by people.
- **Food webs:** Phytoplankton form the base of the food web in marine ecosystems.
- **Seasonal productivity:** Polar waters are highly productive during spring and summer.

10.4 Life in Open Waters

- **Migration:** Many marine organisms migrate vertically between ocean surface waters and deeper layers. Others migrate horizontally across ocean basins.

- **Bioluminescence:** Many deep-water organisms produce light to find mates, capture prey, and avoid predators.
- **Biodiversity:** Seamounts and hydrothermal vents support species-rich ecosystems on the deep seafloor.
- **Fishing:** Industrial fishing employs drift nets, trawling, longlines, and seining to catch seafood. Bycatch is a serious threat to ocean life.

10.5 Geographic Perspectives: The Problem with Plastic

- **Path to the oceans:** Plastic enters the oceans mostly in stream runoff from towns and cities and from beaches.
- **Gyres and beaches:** Marine plastic is concentrated by ocean gyres and deposited on beaches.
- **Problems:** Marine plastic carries high concentrations of industrial pollutants absorbed from seawater. Wildlife eat toxic plastic and become entangled in plastic.
- **Fixing the problem:** Solving this problem requires changing the way we use and make plastic.

Key Terms

abyssal plain, 307	hydrothermal vent community, 324
atoll, 312	kelp forest, 318
barrier reef, 312	lagoon, 312
bathymetry, 306	longline fishing, 326
bioluminescence, 321	mangrove forest, 314
bycatch, 326	mid-ocean ridge, 307
continental shelf, 306	pelagic, 321
coral bleaching, 312	phytoplankton, 304
deep-sea trench, 307	salinity, 308
drift net fishing, 326	seagrass meadow, 317
epipelagic zone, 308	seamount, 307
estuary, 317	seining, 326
fishery, 324	thermocline, 308
fringing reef, 312	tide pool, 320
Great Pacific Garbage Patch, 327	trawling, 326
gyre, 311	upwelling, 320

Concept Review

The Human Sphere: Coastal Dead Zones

1. What is a coastal dead zone? How does it form? How have people influenced dead zones in recent decades?

10.1 The Physical Oceans

2. How many oceans are there? What are their names? Which is the largest and oldest?

3. How deep does light penetrate in the oceans? Which color penetrates deepest?

4. At what rate does pressure increase with depth in the oceans?

5. Compare the surface temperatures of tropical oceans with those of polar oceans. Compare the deep-water temperatures (below 2,500 m [8,200ft]) of tropical and polar oceans.

6. Why is seawater salty? Describe the pattern of ocean surface salinity from the equator to the poles and explain why it changes.

7. What drives surface ocean currents? What are gyres? Why do gyres flow in large circular loops?

8. List and describe the major topographic features of the seafloor.

10.2 Life on the Continental Margins

9. What are the two major types of corals? Which type of coral builds reefs in the tropics?

10. Describe the overall biodiversity of coral reefs.

11. Describe the mutualistic relationship between corals and algae.

12. What are the three kinds of coral reefs? Use each to describe the process of atoll formation.

13. Which coral reef system is the largest in the world? Where is it located?

14. What are the major threats to corals today?

15. What is coral bleaching? Is it permanent?

16. Describe the three steps that lead to reef mortality.

17. What is a mangrove forest? Where are mangrove forests found? Why are they important?

18. What is the most significant threat to mangrove forests?

19. What are seagrass meadows? Where are they found? Why are they important?

20. What are estuaries? How does brackish water relate to them?

21. How were many estuaries formed? Why are estuaries important? Why are so many major cities located on estuaries?

22. What is a kelp forest? Where are kelp forests found? What are the threats to kelp forests?

10.3 Life in Polar Waters

23. What are phytoplankton? Why are they essential to nearly all marine ecosystems?

24. Why do so many organisms migrate to polar waters in spring and summer? Give examples of organisms that do this.

10.4 Life in Open Waters

25. Why do so many organisms undergo a daily vertical migration? Where do they migrate to and from?

26. What compels organisms to migrate across ocean basins?

27. What and where is the shark café? What species of shark goes there, and why?

28. What is bioluminescence? Where in the oceans do organisms use it? What functions does it provide?

29. How do scientists sample life from the deep sea?

30. What are seamounts, and why do they have high biodiversity?

31. What are hydrothermal vents? Describe the life found at them. What is extremely unusual about the life-forms at these vents?

32. What are the four types of industrial fishing in wide use? What is bycatch?

33. How does soup relate to the endangerment of sharks?

10.5 Geographic Perspectives: The Problem with Plastic

34. How does plastic enter the oceans?

35. What is the relationship between ocean gyres and ocean garbage patches?

36. Why does plastic become more toxic the longer it resides in the ocean?

37. How does plastic pose a threat to marine life? To people?

38. What are some of the solutions to the problem of plastic in the oceans?

Critical-Thinking Questions

1. What is the connection between the burning of fossil fuels and ocean acidity? In this regard, what role have the oceans played in slowing atmospheric warming?

2. What food served in a local seafood restaurant may be connected directly to the loss of mangrove forests in the tropics?

3. Trace the connection between upwelling and ocean productivity. Why does upwelling occur more readily in cold polar waters? Explain why tropical ocean water has low phytoplankton productivity.

4. The Monterey Bay Aquarium in California publishes a *Seafood Watch Pocket Guide* that details how consumers can protect marine ecosystems by avoiding seafood caught using ecologically destructive fishing techniques. In your region, what are the "best choices" of seafood species? What species should be avoided?

5. Can you think of any solutions to the plastic problem not proposed in the Geographic Perspectives?

Test Yourself

Take this quiz to test your chapter knowledge.

1. **True or false?** The wind creates ocean gyres.

2. **True or false?** Ocean salinity is greatest along the equator.

3. **True or false?** Trawls, longlines, drift nets, and seines are used in industrial fishing.

4. **True or false?** In the open ocean, many organisms migrate to the epipelagic zone at night.

5. **True or false?** Kelp forests occur at all latitudes in coastal areas.

6. **Multiple choice:** Which of the following is the most biodiverse ecosystem?

a.   mangrove forest   c.   estuary
b.   coral reef   d.   seagrass meadow

7. **Multiple choice:** Why do seamounts have high biodiversity?

a.   because they are warm
b.   because they are near the lighted surface
c.   because there is nutrient upwelling around them
d.   because of volcanic minerals in seawater near them

8. **Multiple choice:** If you live in a coastal city in the midlatitudes, chances are that your city is sited on what was once

a.   an estuary   c.   a mangrove forest
b.   a coral reef   d.   a kelp forest

9. **Fill in the blank:** Unwanted organisms, such as birds and turtles, caught by industrial fishing methods are called _____.

10. **Fill in the blank:** The concentrated region of plastic in the subtropical Pacific Ocean is the _____.

Online Geographic Analysis

Dead Zones and Wildlife Movement Data

In this exercise we will use near-real-time data to examine dead zones in the Gulf of Mexico and marine wildlife tracking information.

Activity

Go to https://service.ncddc.noaa.gov/website/Hypoxia/viewer.htm. This interactive web map shows dissolved oxygen levels in the Gulf of Mexico along much of the U.S. coastline.

1. Pan around and zoom in to the area or areas with the lowest dissolved oxygen levels. Which states have the lowest off-shore oxygen levels?

2. Click the map (but not a red station dot) where the lowest levels are seen. What are the dissolved oxygen levels there? (Units are given in milligrams per liter [mg/L].)

3. When were the data acquired?

4. Bottom-dwelling organisms, such as crabs and oysters, require a minimum of 1 to 6 mg/L of dissolved oxygen. Fish require 4 to 15 mg/L. Are the levels you found safe for bottom-dwelling animals? Are they safe for fish?

5. The contours of the map were drawn using data from sampling stations on ships. Those stations are shown as red dots. Click a red station dot closest to or within the lowest dissolved oxygen levels on the map. What are the dissolved oxygen levels that this ship sampled? What was the depth of seafloor (called "Bottom Depth") where the sample was taken?

Next go to the web page www.gtopp.org. This page provides maps of tracked marine wildlife, ranging from sharks, to elephant seals, to whales. In the lower right of the page there is a small map labeled "Real Time Data Server." Click that map and in the new page scroll down until you see the "Deployment Information" list of animals.

6. Select any animal from this list. What is the name of the animal you selected?

7. Click the "ptt number" for the animal you selected. How long has this animal been tracked?

8. Describe where your animal has migrated during the time it's been tracked. Refer to place names such as California, Alaska, Hawai'i, etc.

9. Next, examine the map that shows the animal's migration and sea surface temperatures. Do you see any relationship between where the animal has migrated and the surface temperature of the water? Explain.

10. Using the map's legend, estimate the water temperature range the animal has experienced during its movements. (Range is the difference between the highest and the lowest.)

Picture This. *Your Turn*

Marine Ecosystems

Nine marine ecosystems are marked on the globe. Use the hint given for each letter to identify the correct ecosystem. Choose from the following numbered list of terms, using each term only once.

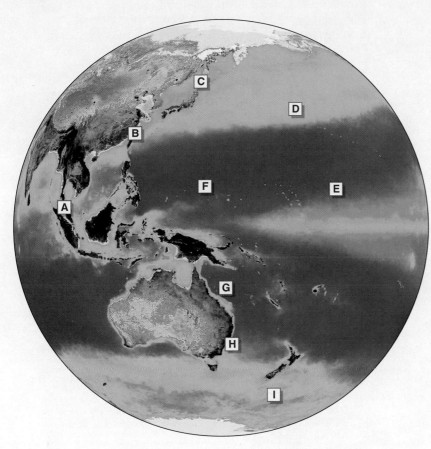

1. Coral reef

2. Estuary

3. Hydrothermal vent

4. Kelp forest

5. Mangrove forest

6. Open waters

7. Polar waters

8. Seagrass meadow

9. Seamount

Hints:

A   Chemosynthetic ecosystems

B   Daily vertical migrations

C   Dominated by algae

D   High seasonal primary productivity

E   Most species-rich marine ecosystem

F   Most threatened ecosystem on Earth

G   Preferred by manatees and dugongs

H   Submerged atolls

I   Threatened by shrimp farming

(SeaWiFS Project, NASA/Goddard Space Flight Center, and DigitalGlobe™)

For animations, interactive maps, videos, and more, visit www.saplinglearning.com.  Sapling Plus

Soil and Water Resources

Chapter Outline *and Learning Goals* ◎

11.1 The Living Veneer: Soils

◎ Explain how soils form and erode and why soils are important to people.

11.2 Surface Water Resources

◎ Describe human use of freshwater and the factors that affect water availability.

11.3 The Hidden Hydrosphere: Groundwater

◎ Describe different features of aquifers and explain how water moves through the ground.

11.4 Problems Associated with Groundwater

◎ Describe groundwater problems that result from human activities.

11.5 Geographic Perspectives: Water Resources under Pressure

◎ Assess the importance of water resources to human societies.

This photo from the northern island of Luzon in the Philippines illustrates how local farmers have created *terraced* hillsides that hold standing water in which to grow rice. This form of agriculture allows people to create standing pools of water on steep slopes while also preventing the erosion of valuable topsoil. The original forests that once grew here were eliminated centuries ago to build these terraces.

(Aaron Joel Santos/Getty Images)

To learn more about terracing, see Figure 11.13.

THE HUMAN SPHERE The Collapse of the Maya

The collapse of the powerful Maya civilization is a reminder of the importance of soil and water resources to human societies. Maya society stretched from Mexico's Yucatán Peninsula to as far south as Honduras and El Salvador (**Figure 11.1**). Complex Maya societies existed for over 3,000 years, from approximately 2000 BCE to 1100 CE. The Maya developed one of the most technologically advanced and successful pre industrial civilizations in the world. They had fully developed writing, mathematics, architecture, calendar, and astronomical systems. At its peak, the Maya population was perhaps 19 million people or more. By 1100 CE, their civilization had disintegrated, their populations had collapsed, and their great inland cities were abandoned.

Many theories have been proposed about the demise of the Maya. Wasted natural resources, as a result of perpetual warfare between neighboring Maya city states, drought, overpopulation, and environmental change from deforestation are most often cited as plausible explanations. Recent research published in the journal *Science* established that Maya populations grew when the climate became wetter and shrank in times of drought. This finding indicates that water resources played a key role in the collapse of Maya civilization.

The region in which the Maya lived is composed mostly of limestone bedrock. Water dissolves limestone, creating underground stream systems. As a result, there were few surface streams, and the Maya had little available surface water. A series of droughts (prolonged periods of water shortage) between 1020 and 1100 CE brought severe water scarcity to the Maya. Furthermore, the tropical soils of the Maya region are low in fertility. Scientists think that drought and the resulting reduction of plant cover probably further reduced the fertility of these soils. A lack of soil and water resources played a pivotal role in the collapse of one of the world's greatest pre industrial societies.

Figure 11.1 A Maya ruin. The Mask Temple in Lamanai, Belize, like most other Maya works, was carved out of limestone. Limestone bedrock probably played a central role in the demise of the Maya. *(Alex Robinson/Getty Images)*

11.1 The Living Veneer: Soils

◎ Explain how soils form and erode and why soils are important to people.

Soils are a fundamentally important natural resource to people. **Soil** is the layer of sediment closest to Earth's surface into which plant roots extend and that has been modified by organisms and water. Although the average thickness of Earth's soils is only about 1 m (3.3 ft), soils play an incredibly important role in human lives and biological systems. The gateway for solar energy entering the terrestrial biosphere, as we have seen, is plant photosynthesis, and almost all plants (including our food crops) are rooted in soil. Without soil, there would be no plants and little life on land.

Pedogenesis, the process of soil formation, occurs through the weathering of rocks, the activities of organisms, the movement of rainwater, and time. **Weathering** is the process by which solid rock is dissolved and broken apart into smaller fragments. Soils form as weathering breaks down bedrock into smaller fragments. **Bedrock** is rock that is structurally part of and connected to Earth's crust. Any loose Earth material that covers bedrock, including soil, is generally called **regolith (Figure 11.2)**. The term *soil*, in contrast, refers to only the uppermost portion of regolith that has been modified by organisms.

Soil Characteristics

Most soils, when exposed in a cross section *or soil profile*, have several distinct layers called soil horizons. **Soil horizons** are horizontal zones within the soil that are identified by their different physical and chemical properties. Weathering and the activities of organisms create soil horizons. Rainwater dissolves and carries nutrients downward through the process of **leaching**. In areas of heavy rainfall, leaching may deplete the topsoil of nutrients. This has happened in the Amazon rainforest. **Eluviation**, the process in which rainwater carries soil particles downward, is also a significant factor that forms soil horizons. The process in which soil particles are deposited (after being transported downward through eluviation) is called **illuviation**. Leaching, eluviation, and illuviation play important roles in the development of soil horizons, as shown in **Figure 11.3** on the following page.

Most soils contain components from all four of Earth's major physical systems: rock fragments and minerals (lithosphere), water (hydrosphere), air (atmosphere), and organic material (biosphere). Although no two soils are alike, most soils are composed chiefly, by volume (but not by weight), of rock fragments, minerals, and air.

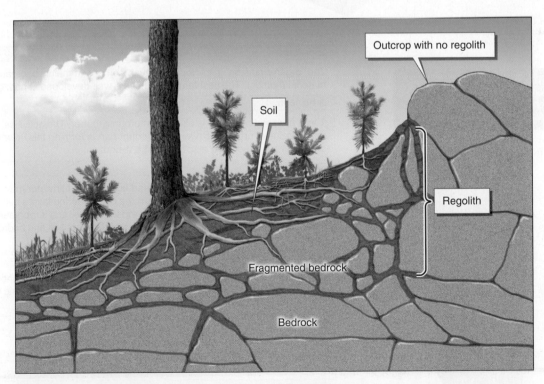

Figure 11.2 Regolith. Regolith is composed of inorganic fragments ranging from large rocks to tiny grains of sand, as well as organic material found in soil.

Figure 11.3 Soil and soil horizon development.
(A) Through weathering and biological activity, bare bedrock forms soils over time. (B) Eluviation and leaching form soil horizons. (C) There are six main categories of soil horizons, arranged vertically. Few soils have all six categories, and some soils have none. *(1–4. Bruce Gervais)*

1. Bare bedrock
This granite surface was recently covered by a glacier. It has no soil cover.

2. Weathering
Over centuries, weathering weakens and breaks the solid bedrock apart.

3. Soil development
As weathering continues, small plants colonize the area as soil begins forming.

4. Developed soil
Organic matter in the soil increases as trees become established. The activities of earthworms, burrowing insects, and rodents mix the soil.

Time

A

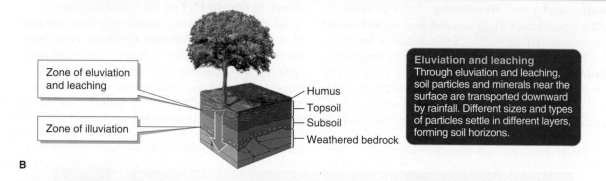

Zone of eluviation and leaching

Zone of illuviation

Humus
Topsoil
Subsoil
Weathered bedrock

Eluviation and leaching
Through eluviation and leaching, soil particles and minerals near the surface are transported downward by rainfall. Different sizes and types of particles settle in different layers, forming soil horizons.

B

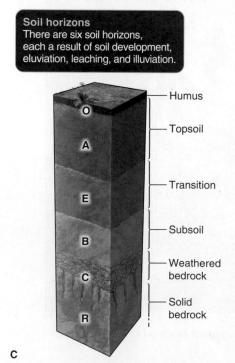

Soil horizons
There are six soil horizons, each a result of soil development, eluviation, leaching, and illuviation.

Humus — O
Topsoil — A
Transition — E
Subsoil — B
Weathered bedrock — C
Solid bedrock — R

C

O horizon
The "organic horizon" is the topmost layer, composed mainly of organic material (humus). The best farmlands have a well-developed O horizon.

A horizon
The A horizon (or topsoil) is also rich in organic matter. The activities and remains of organisms play an important part in forming this layer of soil. Most plant roots are restricted to the A horizon.

E horizon
The E horizon is formed by eluviation and leaching as rainwater moves dissolved chemicals and small clay particles deeper to the layers below. The E horizon is lighter in color than the topsoil above, and it represents a transition zone between the topsoil layers and the subsoil and bedrock below.

B horizon
Transported clays and dissolved chemicals are deposited through the process of illuviation in the B horizon, or subsoil. High concentrations of clay, aluminum, and iron are typically found in this horizon.

C horizon
The C horizon is composed of unconsolidated weathered rock sandwiched between the base of the soil horizons above and the bedrock below. There is little to no organic material in the C horizon.

R horizon
The R horizon (for rock) consists mostly of solid, unweathered bedrock.

Soil is often described by its *texture,* which is determined by the relative amounts of sand, silt, and clay composing it. *Sand, silt,* and *clay* are inorganic soil particles of three size classes **(Figure 11.4)**. Soil textures can be summarized graphically in a *soil texture triangle* **(Figure 11.5)**.

Loam, which has a mineral content of about 40% sand, 40% silt, and 20% clay, contains large amounts of organic material. Farmers value loam because it retains and transmits moisture and nutrients that are easily accessible to plants. Loam is the ideal soil texture for growing most crops. In contrast, *sandy soils* and *rocky soils* are dominated by mineral particles and have less organic content. Large *pores* (or air spaces) between particles allow water to drain from these soils rapidly, and nutrients are quickly leached out of reach of plants. As a result, sandy and rocky soils make poor agricultural soils. Clay and silt soils are also unsuitable for many plants but for the opposite reason: Clay and silt particles are extremely small and impede water flow.

Soil Formation Factors

Many different factors determine what types of soils form in any given place and how fast they form. Recall that soils form as weathering breaks down bedrock into smaller fragments. Both physical weathering and chemical weathering are particularly important factors in soil development. *Physical weathering* is the physical breakdown of rocks into smaller fragments without alteration of their chemical makeup. *Chemical weathering* is the process of changing minerals in a rock through chemical reactions involving water. (See Section 16.1 for more about these weathering processes.)

Climate

Temperature, rainfall, and even the types of organisms living in and affecting the soil are a function of climate. Soil develops most quickly where chemical weathering predominates, such as in the lowland tropics, where it is always warm and moist. Soil develops more slowly in cold climates and in arid climates. As **Figure 11.6** on the following page shows, soil types and soil development coincide with the broad latitudinal climate zones determined by global atmospheric pressure systems (see Section 5.2).

Parent Material

A soil's *parent material* is the rock and mineral matter that is weathered into particles of sand, silt, and clay. Parent material is composed of both the

Figure 11.4 Sand, silt, and clay. The relative sizes of sand, silt, and clay particles are shown here. For comparison, human hair is about 0.1 mm or less in diameter.

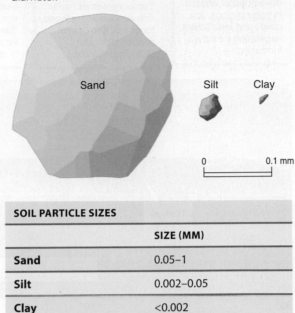

SOIL PARTICLE SIZES	
	SIZE (MM)
Sand	0.05–1
Silt	0.002–0.05
Clay	<0.002

Figure 11.5 Soil texture triangle. Soil texture is determined by a soil's relative proportions of sand, silt, and clay. To read this diagram, follow any soil type up and to the left. For example, find 60% along the bottom (sand) line and follow the diagonal line within the triangle up and to the left. Stop at the horizontal line for 30% clay. This soil type is "sandy clay loam," with 60% sand, 30% clay, and 10% silt.

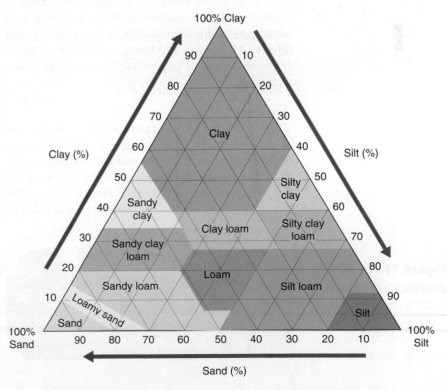

Figure 11.6 Soils and climate zones. Generally, the character of soils reflects climate patterns.

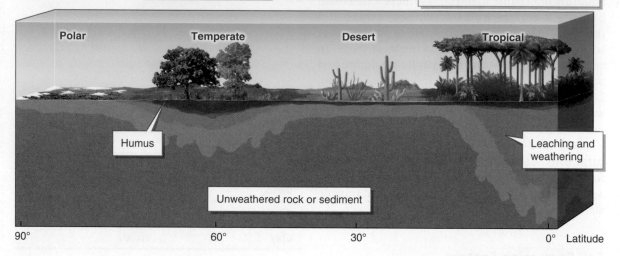

Polar high: Cold	Subpolar low: Cold and moist	Subtropical high: Warm and dry	ITCZ: Warm and wet
Soils are poorly developed or absent in polar regions. Ice cover and insufficient vegetation inhibit soil formation.	Temperate soils have rich and well-developed topsoil due to ample precipitation and biological activity.	Subtropical desert soils are poorly developed due to sparse vegetation and a lack of available moisture.	Tropical soils are deeply weathered due to large amounts of rainfall and warm temperatures that enhance chemical weathering. Rainwater leaches nutrients out of reach of plants. As a result, many tropical soils are nutrient-poor.

weathered bedrock beneath the soil and the sediments that have been transported to the site by streams, glaciers, or wind. Parent material plays an important part in the chemical composition of the soil, the texture of the soil, and the rate at which soil can form. Different rocks vary in hardness. Granite, for example, is a type of rock that is more resistant to weathering than sandstone, and sandstone is much harder than soft volcanic ash **(Figure 11.7)**. (You will learn more about all these types of rocks in Chapter 14, "Building the Crust with Rocks.")

Organisms

"Dirt" is what we track into the house. Soil is the living medium in which plants spread their roots, organisms live, and we grow our food. Healthy soil is a teeming metropolis of worms, insects, and microbes that churn and digest organic material. Without the effects of biological activity, broken bedrock would

be sterile and composed of gravel and sand, like the regolith found on the Moon or Mars.

The topmost layer of soil, the O horizon (see Figure 11.3C), is composed primarily of partially decomposed vegetation such as leaf litter. Just below the O horizon is the A horizon, also called *topsoil*. The A horizon is thin; its average depth worldwide ranges from about 5 to 20 cm (2 to 8 in).

Soil horizons O and A are strongly influenced by the activities of organisms **(Figure 11.8)**. Seeds and plant roots are concentrated in these horizons. Microbes, nematodes, earthworms, beetles, and other insects all digest, *bioturbate* (mix), and build soil. Bioturbation redistributes nutrients and particles throughout the A horizon. Wastes from these organisms and from larger animals also contribute to the chemical and physical structure of these horizons. Soil organisms create **humus** (pronounced HYOO-mes), the brownish-black mixture of organic material that makes up the bulk of the O and A horizons by volume.

On a local level, people also can play an important role in soil development. There are many examples of organic-rich anthropogenic soils, including the *terra preta* that was developed by indigenous cultures in the Amazon rainforest (see Section 9.2), *plaggen soils* that were developed throughout Europe during medieval times, and modern soils developed by composting food scraps **(Figure 11.9)**.

Figure 11.7 Soils and parent material. Bedrock that is resistant to weathering forms soils more slowly, and the resulting soils are thinner.

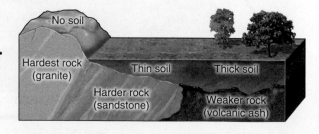

Figure 11.8 Living soils. The O and A horizons of healthy soils are alive with organisms whose activities contribute to soil formation and health.

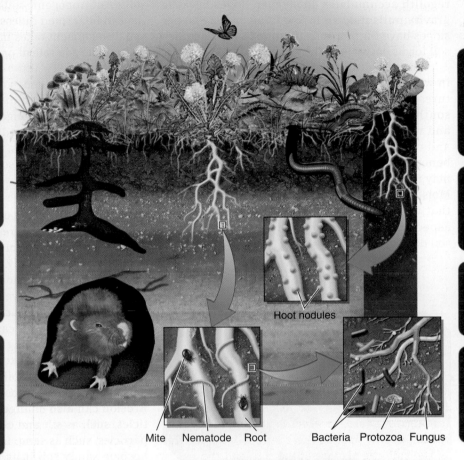

Invertebrates
The activities of mites, nematodes, larvae of moths and butterflies, centipedes, and others all help build soil.

Fungi
The majority of fungi exists below ground in the form of thread-like filaments called *mycelia*. These bind the soil together and facilitate in nutrient transport.

Plant roots
Plant roots pry soil apart and provide channels for water and air to penetrate soil. Roots decompose and add organic material.

Nitrogen-fixing bacteria
Some plants, such as those in the bean family, harbor mutualist bacteria in their root nodules. These bacteria convert or "fix" unusable nitrogen from the air (N_2) into ammonia (NH_3), which plants are able to use.

Plant litter
Within forests the surface of soil is hidden beneath a layer of plant litter. The litter serves to moderate the temperature and moisture of the soil and nutrients from it adds to soil fertility.

Earthworms
Earthworms take in soil and digest the organic material. They deposit their waste and undigested soil particles as they move through the soil.

Mammals
Mammals burrow and dig in soil, mixing and aerating it. Their waste also add significantly to soil nutrients.

Microbes
Bacteria and protozoa are essential to soil formation. They decompose the organic material in soil, breaking it down so that plants can use it.

Root nodules

Mite   Nematode   Root   Bacteria   Protozoa   Fungus

Figure 11.9 Anthropogenic soil. The soil shown here is a result of San Francisco's mandatory composting ordinance. The city requires all residents and restaurants to compost food scraps and paper food wrappers. Farmers often call this composted soil "black gold," referring to its color and its richness for growing crops. (*© Jose Luis Villegas/ Sacramento Bee/ZUMAPRESS.com*)

Topography and Moisture

The steepness of a slope is directly related to the depth of soil that can develop on the slope. Less regolith accumulates on steeper slopes because gravity pulls it down to low-lying areas. Steeper slopes have thinner soils **(Figure 11.10)**, and thick soil forms only where the ground is level.

Within a climate zone, *aspect* (the direction in which a slope is facing) determines soil moisture. In the Northern Hemisphere, for example, south-facing slopes are usually relatively dry and warm, while those facing north are moister and cooler. Similarly, **groundwater** (water found beneath Earth's surface in sediments and rocks) may rise to the surface as springs in some areas. Moisture is a key determinant of the type of soil that will form in a given area. Factors such as aspect and groundwater can strongly influence soil development.

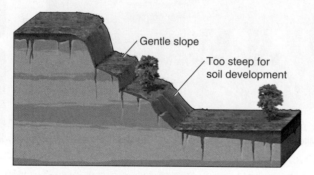

Figure 11.10 Soil and slope. Flat land develops thicker soil than gentle slopes. Soils cannot form where topography is too steep.

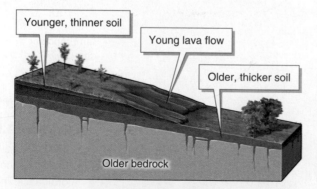

Figure 11.11 Soil and time. Older exposures of bedrock have had more time to develop thicker soils than new rock exposures, such as those created by a relatively recent lava flow.

Time

Over time, inputs of new inorganic material as well as biological activity both build and renew soils. Anthropogenic soils made through composting can be formed in less than a year, but natural soil development takes much longer. Natural soils accumulate at a rate of only 2.5 cm (1 in) per 200 to 1,000 years.

As we have noted, soil formation takes longer in cold climates than in warm climates. In addition, it takes much longer to form soil from bare hard rock than from sediments such as sand. As **Figure 11.11** illustrates, older surfaces are generally covered by thicker soils.

Soil Erosion

The National Academy of Sciences estimates that the world's croplands are losing soil 10 to 40 times faster than soil can form naturally. Worldwide, an area of cropland roughly the size of Indiana (about 10 million ha [25 million acres]) is significantly degraded or lost every year.

Soil erosion can occur rapidly in a single heavy downpour, or it can happen gradually and imperceptibly over long periods. Either way, soil is lost if soil erosion exceeds the rate of soil formation. Valuable topsoil is the first to be eroded away because of its position at the soil surface. Erosion can also change soil texture. Finer particles, such as silt and clay, erode before heavier particles, such as sand. Loam, for example, can be become sandy soil if its lighter silt and clay are blown away.

The most important factor in soil loss and degradation is erosion by water runoff and wind. Plant roots anchor soil. When plant cover is removed, the soil is subject to erosion by flowing water and the wind. Deforestation, overgrazing, fire, and plowing are a few of the factors that remove anchoring plant roots. Farm fields that are left bare for extended periods without protection by a cover of crops or crop residue are also prone to erosion **(Figure 11.12)**. In addition, *compaction* of soil layers by heavy machinery can decrease the pores in soil into which water can seep, leading to greater surface runoff and faster erosion. Other important causes of soil loss include new settlements, expansion of cities onto farmland, accumulations of salt through *soil salinization,* and poisoning with pesticides and herbicides.

When steep slopes are farmed, severe soil erosion can occur. To reduce erosion on slopes, farmers can employ techniques such as *contour plowing* and *terracing* **(Figure 11.13)**.

Figure 11.12 Soil erosion in Georgia. Western Georgia's Providence Canyon (also called Georgia's Little Grand Canyon) largely did not exist before 1800. The early settlers used poor farming practices, allowing runoff to flow unimpeded over plowed fields, which triggered soil erosion that led to the formation of gullies, then deep ravines, and eventually canyons 50 m (160 ft) deep. The canyons are still deepening by erosion. Tourists are drawn to the area today to hike in the canyons. *(Anita Patterson Peppers/Shutterstock)*

Figure 11.13 Agriculture on slopes. (A) This photo shows contour plowing in Montgomery County, Iowa. This farming method involves creating furrows that follow a line of equal elevation on a slope to prevent water from collecting in small channels that run downslope and erode the soil. The green strips of vegetation catch sediments and further reduce soil loss. (B) The Banaue Rice Terraces in the Philippines are shown in this photo. Terracing is a farming method used on very steep slopes. Rice has been sustainably grown in the Philippines using this method for over 2,000 years. (C) Most of Haiti's hills are severely eroded due to farming on steep slopes without protective contouring or terracing. The hills shown here were once forested. The native vegetation was cleared for farming or cut for fuel. Erosion then washed away the soil, leaving the hills permanently damaged and largely without cover. *(A. Photo by Tim McCabe, USDA Natural Resources Conservation Service; B. Travel Ink/Getty Images; C. James P. Blair/Getty Images)*

Naming Soils: Soil Taxonomy

Many localities are named after the color of their dominant soils: the Yellow River (Huang He) of China, the Red River between Oklahoma and Texas, the White Sands of New Mexico, the Painted Desert of Arizona, and the Kemet, or "black soils," of the Nile Delta in Egypt.

Soil color reveals the soil's history and the conditions under which it formed, as well as its composition. Soil scientists commonly use the *Munsell Soil Color Book*, shown in **Figure 11.14**, in identifying soil types in the field based on their colors.

Scientists use the Linnaean taxonomic system to name organisms and the Köppen system to classify climates. Similarly, scientists use several different systems to classify soils. In Canada, for example, the *Canadian System of Soil Classification* is used. One soil classification system that is widely employed in the United States, called **soil taxonomy**, arranges soils into 12 *soil orders*, as shown in **Figure 11.15**.

Figure 11.14 Identifying soils by color. Scientists use the *Munsell Soil Color Book*, a page of which is shown here, to identify soils. There is a Munsell Chart app as well. This table lists the biomes in which certain soil colors are often found. *(© J. Kelley, SoilScience.info)*

SOIL COLOR	CHARACTERISTICS	BIOME
Black or dark brown	High organic content	Temperate grasslands
Brown	Iron oxides with high organic content	Temperate deciduous forest
Red or orange	Strong chemical weathering of iron and aluminum	Tropical rainforest
Blue/green-gray	Persistently saturated soils	Estuaries
Gray	Heavy leaching of iron	Boreal forest

The Importance of Soils to People

Soils are essential to life and to humans as they form the medium in which we grow almost all of our food. This section explores some of the other services that soils provide for people.

Medicines

Scientists have learned much from the chemicals that soil bacteria produce. Among the first naturally occurring antibiotics used in the treatment of human diseases was tyrothricin (or gramicidin), which was isolated from soil bacteria in 1939. The Actinobacteria, a widespread group of soil bacteria, have been used to make more than 20 different antibiotic compounds. The antibiotics actinomycin, neomycin, and streptomycin are in widespread commercial use today and have saved countless human lives. They were all derived from soils.

Mitigation of Climate Change

Soils contain enormous amounts of organic material and, therefore, carbon. Terrestrial *carbon stocks* (the amount of carbon) in the soil are greater than those in living biomass (plants). The boreal forest contains by far the greatest amount of carbon in its soils. **Table 11.1** lists the global carbon stocks in various vegetation types, sorted by soil carbon amounts.

Table 11.1	Soil Carbon Stocks		
VEGETATION TYPE	**SOIL CARBON (PG C)***	**PLANT CARBON (PG C)**	**TOTAL CARBON (PG C)**
Boreal forest	471	88	559
Temperate grasslands	295	9	304
Tropical savanna	264	66	330
Wetlands	225	15	240
Tropical rainforest	216	212	428
Desert	191	8	199
Cropland	128	3	131
Tundra	121	6	127
Temperate forest	100	59	159

*1 Pg C (petagram of carbon) = 10^{15} g of carbon (2.2 trillion lb).

When carbon is exposed to the air, it *oxidizes* (reacts with oxygen) to form the greenhouse gases carbon dioxide and methane. Removal of vegetation cover that holds soil in place can result in oxidation of soil carbon. Scientists estimate that greenhouse gas production from human-caused soil changes have produced about 20% of the climate warming of the past century (see also Section 7.4).

Figure 11.15 Soil taxonomy. (A) The soil taxonomy classification system recognizes 12 soil orders, based on soil color, texture, and chemical makeup. The 12 soil orders display a wide range of colors and horizons as a result of the different processes from which they form. Here the soil orders are arranged according to the extent of formation and latitude. (B) Climate is often an important factor in determining the geographic distribution of soil orders. This soil map bears many similarities to the map of Köppen climate zones (see Figure 9.3). In many areas, however, there is no spatial correspondence between soils and climate zones because many other factors, such as parent material and topography, determine soil types. Andisols, for example, are a product more of their parent material (volcanic ash) than of climate. *(All photos courtesy of USDA Natural Resources Conservation Service)*

SOILS THAT ARE STILL DEVELOPING

Andisols	Entisols	Histosols	Inceptisols
Soils that develop from volcanic ash. Fertile and suitable for agriculture. Easily eroded if not protected by vegetation.	Poorly developed soils with no horizon development. Often found in new deposits from rivers, glaciers, or sand dunes.	Soils that form in wetlands or places with deep forest litter with very high organic content. No horizon development.	Young soils with poorly developed horizons.

LOW AND MIDDLE LATITUDE SOILS

Aridisols	Oxisols	Ultisols	Vertisols
Desert soils with little organic matter and rapid drainage.	Heavily weathered soils rich in iron and aluminum minerals. Low nutrient content. Form in tropical regions with high rainfall. Found at tropical and subtropical latitudes.	Old, deeply weathered soils. Low in nutrients, high in clays. Common in the southeastern United States.	Clay-dominated soils that shrink and crack when dry and swell when wet. Low organic content.

MIDDLE AND HIGH LATITUDE SOILS

Alfisols	Gelisols	Mollisols	Spodosols
Soils that develop in humid climates. Found under forests. B horizon is grayish-brown, with clay accumulation. Little color change with depth.	Soils that show evidence of disturbance by frost. They occur where the ground is permanently frozen (permafrost).	Humus-rich soils. Thick, dark A horizon with a soft texture. High nutrient content, among the most fertile soils in the world. Form within grasslands at midlatitudes. Well-developed horizons.	Acidic, nutrient-poor soils. Often with bleached E horizon. Found under coniferous forests in snowy climates.

A

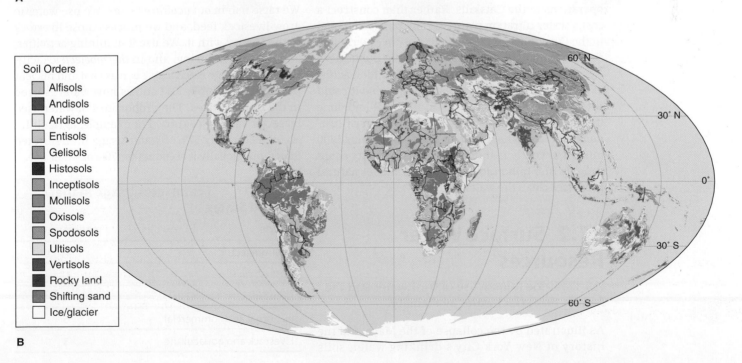

Soil Orders
- Alfisols
- Andisols
- Aridisols
- Entisols
- Gelisols
- Histosols
- Inceptisols
- Mollisols
- Oxisols
- Spodosols
- Ultisols
- Vertisols
- Rocky land
- Shifting sand
- Ice/glacier

B

Figure 11.16 New York City drinking water. The Ashokan Reservoir, in the Catskill Mountains of southeastern New York, provides the city of New York with drinking water. It is one of 19 reservoirs and three lakes that provide water for the city. *(AP Photo/Mike Groll)*

Water Purification

Up until about the 1980s, New York City's water supply, located in the Catskill Mountains, was filtered by vegetation and soils. New York's water was nicknamed "the champagne of drinking water" in reference to its purity. The city's water supply began to be polluted in the 1980s, however, due to increasing construction of vacation homes near streams, building of new roads, and poor farming and livestock operations in the Catskills. Rather than construct a costly water filtration plant, the city purchased land in the Catskills through *eminent domain* (the right of the government to take private land for public use). New York allowed some 830,000 ha (2 million acres) of temperate deciduous forest to regrow so it could filter and clean the water naturally, as it had done in the past **(Figure 11.16)**. Building a water filtration plant would have cost up to $8 billion (plus $300 million in yearly maintenance costs); the city spent roughly $1.5 billion to buy the land and let natural processes clean the water.

11.2 Surface Water Resources

◎ **Describe human use of freshwater and the factors that affect water availability.**

As illustrated by the collapse of the Maya and the history of New York City's drinking water, soils and water are two closely tied natural resources that people require. Soil is the medium that holds and provides nutrients to crop plants and natural vegetation, and water gives soil life. This section begins by exploring surface freshwater, such as the water that occurs in lakes and streams. It then examines groundwater—water that is stored in the ground within sediments and rocks.

Every drop of water that people drink and every bite of food that we eat (except seafood from the oceans) depends on the availability of freshwater. Most of Earth's water is in the oceans, unusable in its saline (salty) state **(Figure 11.17)**. The vast majority of Earth's freshwater is locked up in ice sheets and inaccessible. Only a small proportion of the planet's water is liquid freshwater, and only a small proportion of *that* is *surface water*, freshwater that is accessible to people at Earth's surface in streams and lakes.

In many regions there is not enough freshwater to meet the needs of people. This predicament often compels people to turn to the ocean for water. To render seawater usable for human consumption, the dissolved minerals in the water must be removed through **desalination** techniques that are both expensive and potentially environmentally harmful. The **Picture This** feature on page 348 explores this interesting topic further.

Water Use

People use water for nearly everything. We generate electricity from it (hydroelectricity). We water our crops with it through *irrigation*. We use it in our homes, drinking it, bathing in it, washing clothes and dishes in it, and watering lawns and gardens with it. We raise fish in *aquaculture* farms. We use water to grow livestock feed, and we process those livestock in the factory with it. We use it in mining activities, and in manufacturing, and in *thermoelectric* facilities that generate electricity, such as coal and nuclear power plants. **Table 11.2** shows how water is used in the United States. The proportions of water used by sector vary by country. For example, agriculture accounts for only 37% of water usage in the United States, but globally it accounts for 70% of water use.

Table 11.2 Freshwater Usage in the United States

USAGE	PERCENTAGE
Thermoelectric	41
Agriculture	37
Domestic	13
Industrial and commercial	6
Livestock and aquaculture	3

Water Scarcity

The health of any natural ecosystem or human-built environment depends in large part on the water that's available. The *climatic water balance* for any given location is a result of inputs and outputs of water. Inputs of water to an area include precipitation and streamflow. Water outputs include water lost through evapotranspiration—evaporation and transpiration from plants. Most water is lost through evaporation because of high temperatures, but windiness and low humidity also play important roles in how fast water is lost to the atmosphere. When water outputs exceed water inputs, water scarcity results.

The geographer Charles W. Thornthwaite (1899–1963) established a widely used method of determining water scarcity in the soil within the context of crops and natural vegetation. The Thornthwaite system considers three key variables: (1) the amount of precipitation, (2) actual

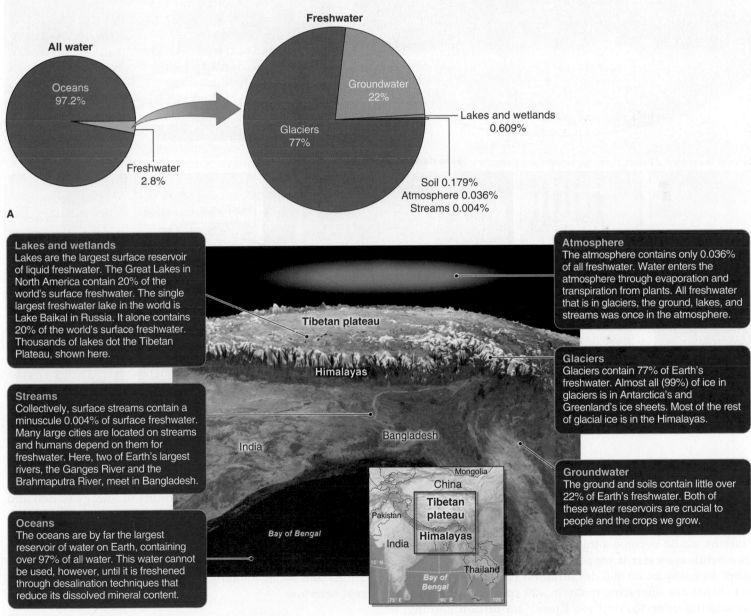

Figure 11.17 Earth's water. (A) Oceans contain 97.2% of Earth's water. The rest of the planet's water is freshwater, stored mostly in glaciers. (Note that the proportions of glaciers and groundwater have been rounded. As a result, the freshwater proportions do not add up to 100%.) (B) Satellite data are georeferenced onto a digital elevation model (DEM), and the vertical height of the topography is exaggerated 50 times in this image of Earth's surface and freshwater reservoirs in South Asia. *(NASA/Goddard Space Flight Center Scientific Visualization Studio)*

Picture This Desalination

Locations of major desalination plants

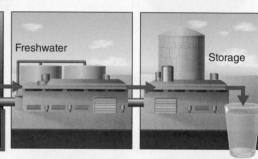

The Jebel Ali desalination plant in Dubai, United Arab Emirates. It is one of the world's largest desalination plants. It can produce 818 million liters (216 million gallons) of water a day.

How desalination plants work

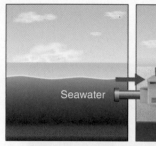
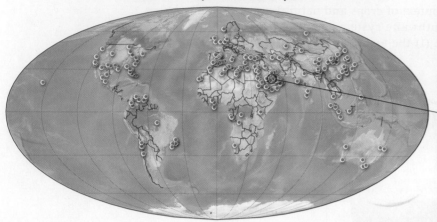
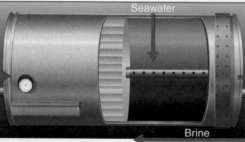

1. Water intake
Intake pipes take in millions of gallons of seawater.

2. Filtration
Water is first coarsely filtered to remove sediment and large debris.

3. Desalination
There are many methods to remove the salt. This diagram shows *reverse osmosis* where seawater is pumped through fine filters that filter out minerals.

4. Waste brine
The resulting waste brine is mixed with seawater to reduce its salinity before it is returned to the sea. However, the water is still very salty and potentially ecologically harmful.

5. Post-treatment
The treated freshwater is chlorinated and minerals are added to improve taste.

6. Delivery
The treated water enters the municipal water supply to be mixed with other water sources and used.

(Stanislav71/Shutterstock)

"Water, water every where, Nor any drop to drink" lamented Samuel Taylor Coleridge in his poem *Rime of the Ancient Mariner*. If only we could drink seawater, water resource limitations would become a thing of the past. And for a cost, we *can* drink seawater if we remove its salt through desalination. As many as 20,000 desalination plants (many of them on ships) are operating today in 120 countries, collectively producing over 13.5 billion liters (3.5 billion gallons) of freshwater each day.

Desalination plants can cost billions of dollars to construct and hundreds of millions of dollars each year to operate and maintain. Each of the technologies used to freshen salt water uses large amounts of energy to boil and evaporate water or push water through membranes. It's far cheaper to pump water up from the ground if it is available, to recycle water, or to simply use less water through conservation. But for many coastal cities with growing populations and water demands, there is no other option but to treat seawater.

Consider This

1. Looking at the map, where are desalination plants densest?
2. Why do you think most desalination plants are located in those areas?

evapotranspiration, and (3) *potential evapotranspiration*, the amount of water that would evaporate and be transpired if it were available. Whenever potential evapotranspiration exceeds precipitation, there is a natural water deficit.

Hot deserts, for example, have few permanent bodies of water and little vegetation cover. The potential for evapotranspiration in deserts is high, but actual evapotranspiration is low due to the lack of plants and surface water. Deserts are therefore in perpetual water budget deficits. Conversely, tropical rainforests have equally high potential evaporation rates, but because they receive high rainfall amounts each month, they experience only brief water deficits or none at all.

In other biomes, such as tropical deciduous forest or tropical savanna, water deficits are normal for only a few summer months of the year. When water deficits persist longer than normal as a result of a lack of precipitation, the result is a **drought**: a prolonged period of water shortage. The *Palmer Drought Severity Index* is a practical application of the Thornthwaite system. It provides a measure of dryness based mostly on potential evapotranspiration. This index is useful because it indicates the extent of drought or water surplus occurring in an area **(Figure 11.18)**.

Droughts are extremely costly to society in a variety of ways, including lost hydroelectricity output, reduced crop yields, and stress on municipal water systems. Temperatures that are rising as a result of climate change put additional stress on water resources. We explore this important topic in the Geographic Perspectives, "Water Resources under Pressure" in Section 11.5.

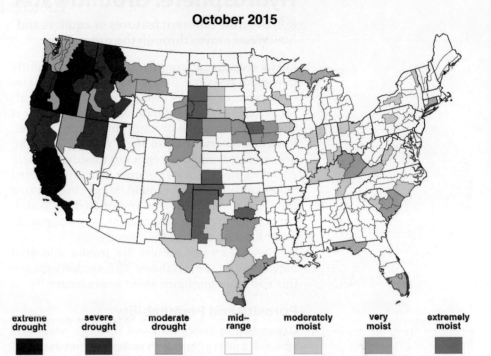

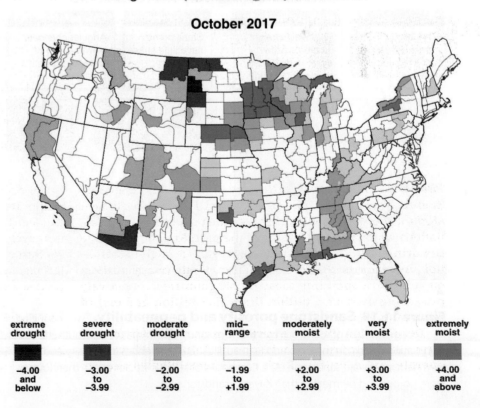

Figure 11.18 Tracking drought in the West. The Palmer Drought Severity Index uses zero as a normal baseline value. Negative values indicate water deficits (shown in orange, red, and maroon); positive values indicate water surpluses (shown in light and dark green). In October 2015, much of the western United States was gripped by unprecedented "extreme drought" (top). The severe nature of the drought was due to low precipitation and record high temperatures that raised the potential evapotranspiration rates. By 2017, one of the worst droughts in the West's history had ended (bottom). *(NOAA)*

Web Map

Palmer Drought Severity Index

Available at
www.saplinglearning.com

11.3 The Hidden Hydrosphere: Groundwater

◎ Describe different features of aquifers and how water moves through the ground.

Surface water in streams and lakes is a vitally important resource, but its availability fluctuates over weeks or months. Groundwater acts as a buffer in times of short-term drought. Groundwater is slower to respond to drought than surface water, but after several back-to-back years of drought, groundwater supplies begin diminishing as well. About half of the water people use in the United States comes from the ground. Many cities located far from permanent sources of surface water, particularly those in arid regions, rely on groundwater for most or all of their needs.

How does water get into the ground, and what happens to it once it is there? This section explores this and other questions about groundwater.

Porosity and Permeability

Groundwater flows into and through pores (air spaces) underground. **Porosity** is a measure of the available air space within soil, sediments, or rocks. The porosity of the ground is expressed as a percentage. If, for example, half of a sandy soil consists of open pores, then its porosity is 50%. In most regions, significant stores of water do not exist much deeper than 0.8 km (0.5 mi) from the surface. At greater depths, the pressure from the weight of the ground compacts the pores, leaving little room for water.

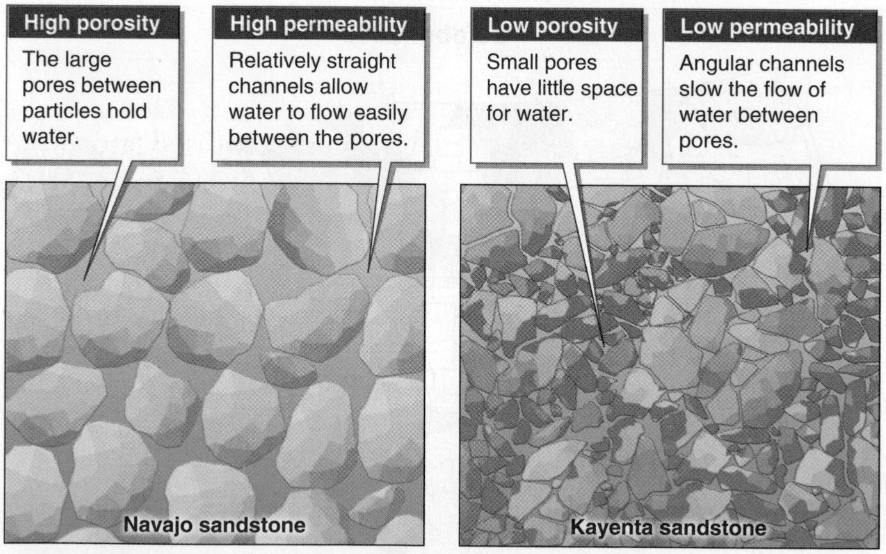

The rate at which water flows through pores within the ground varies from several centimeters to several meters per day. The ease with which water can flow through soil, sediments, or rocks is called **permeability**. The permeability of a material depends on the size, number, and configuration of pores within it. Permeability is high where there are many adjacent pores, creating straight paths for water flow.

Sand has both high porosity (it can absorb water) and high permeability (water cannot pass easily through it). In contrast, clay has high porosity but low permeability. Clay has many pores, but they are very narrow because clay particles are flat and platelike in shape. Like narrow streets in a city that restrict the flow of cars, the small and narrow connections between pores in clay restrict the flow of water. Different rock types have different porosities and permeabilities **(Figure 11.19)**. Most rocks have no porosity and no permeability and therefore contain no water.

During a rainstorm or snowmelt, water on the ground surface is pulled downward by gravity and enters the soil in the process of **infiltration**. How fast water infiltrates the soil, the *infiltration rate*, depends on the permeability of the soil. Soils with high permeability have high *infiltration rates*. After water enters the soil through infiltration, it moves around soil particles through narrow, meandering channels in the process of **percolation** (Latin for "to trickle through"). Whereas infiltration is gravity driven, water movement through percolation is a result of water's cohesive and adhesive properties (see Section 4.2 on the properties of water).

Aquifers

An **aquifer** is any sediment or rock with pores that contain water. Aquifers store and transmit water, and they are an important water resource for many people living in arid regions. **Figure 11.20** shows the global distribution of major and minor aquifer systems. An **aquiclude**, in contrast, is a sediment or rock layer that lacks pores and cannot contain water. Aquicludes have low or no porosity or permeability. They greatly limit or altogether prevent water movement.

Most aquifers are *unconfined aquifers*, meaning that rainwater can move into them directly from the surface. When an aquiclude separates an aquifer from the surface, the aquifer is a *confined aquifer* **(Figure 11.21)**.

High porosity	High permeability	Low porosity	Low permeability
The large pores between particles hold water.	Relatively straight channels allow water to flow easily between the pores.	Small pores have little space for water.	Angular channels slow the flow of water between pores.

Navajo sandstone

Kayenta sandstone

A

B

Figure 11.19 Sandstone porosity and permeability. Two magnified thin sections (slices) of rock from two different sandstone formations in southern Utah's Zion National Park, on the Colorado Plateau, illustrate differences in porosity and permeability. Navajo sandstone (A) is composed of larger particles and is therefore more porous and permeable than Kayenta sandstone (B).

Figure 11.20 World map of aquifers. Regions in blue have the greatest amount of groundwater.

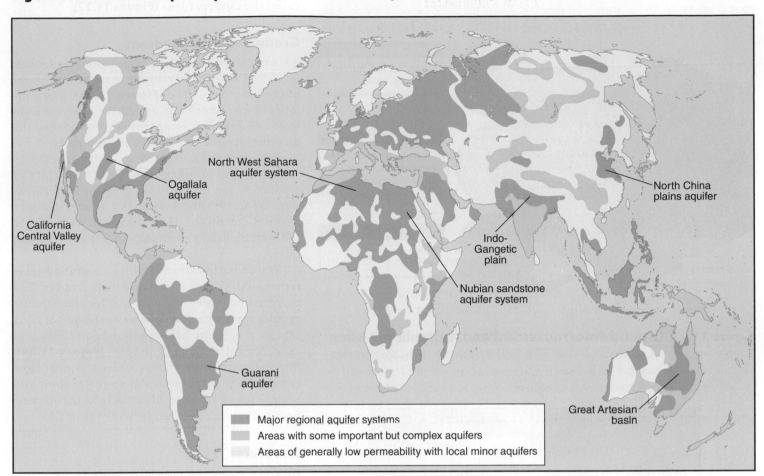

North West Sahara
aquifer system

Ogallala
aquifer

North China
plains aquifer

California
Central Valley
aquifer

Indo-
Gangetic
plain

Nubian sandstone
aquifer system

Guarani
aquifer

Great Artesian
basin

Major regional aquifer systems
Areas with some important but complex aquifers
Areas of generally low permeability with local minor aquifers

What would happen if you were to go out into your backyard with a shovel and dig a deep hole to the aquifer below? You would first dig through the aquifer's **zone of aeration**, the layer of the ground that is not permanently saturated. You might dig a few meters or a few hundred meters before you began hitting water. The depth of water-saturated ground depends on the depth of the water table. The **water table** is the top surface of the aquifer's **zone of saturation**, the layer of the ground usually saturated with water. You would not reach the water table abruptly. Instead, the ground would gradually become wetter and wetter as you approached the zone of saturation. Eventually, the hole might fill with a mud slurry just above the water table.

The ground becomes gradually wetter as you dig because water migrates up from the zone of saturation into the pore spaces in the zone of aeration because of water's cohesive and adhesive properties. The gradual transition between

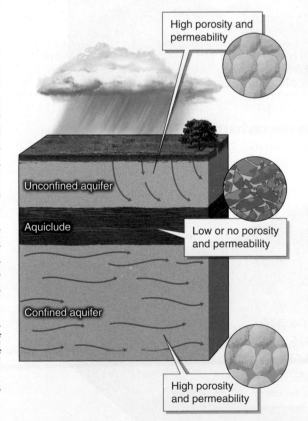

High porosity and
permeability

Unconfined aquifer

Aquiclude

Low or no porosity
and permeability

Confined aquifer

High porosity
and permeability

**Figure 11.21
Confined and
unconfined
aquifers.** Rainwater
moves into an unconfined
aquifer directly from
Earth's surface. A confined
aquifer is separated from
the surface by a layer
of impermeable rock or
sediments (an aquiclude).

Figure 11.22 Groundwater zones. The water table lies at the top of the zone of saturation. The capillary fringe forms a gradual transition between the zone of aeration and the zone of saturation.

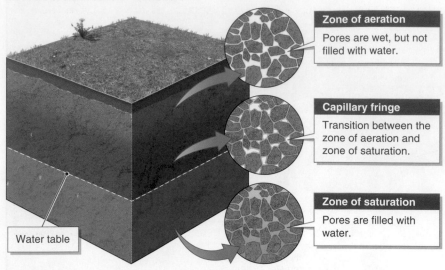

Zone of aeration
Pores are wet, but not filled with water.

Capillary fringe
Transition between the zone of aeration and zone of saturation.

Zone of saturation
Pores are filled with water.

Water table

Figure 11.23 Groundwater movement and hydraulic pressure. In this aquifer, water flows from high elevation to low elevation, or h_1 to h_2, and from high pressure to low pressure, or P_1 to P_2. There is more hydraulic pressure (the force exerted by the weight of water) at P_1 because it has a higher column of water above it than does P_2.

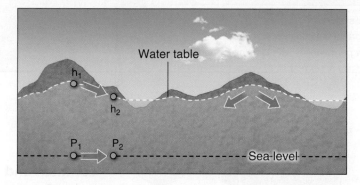

Water table

h_1
h_2

P_1 P_2
Sea-level

Figure 11.24 Groundwater recharge and discharge. Groundwater recharge areas are usually higher in elevation than groundwater discharge areas. Pictured here is a gaining stream; groundwater discharge is supplying the stream with water.

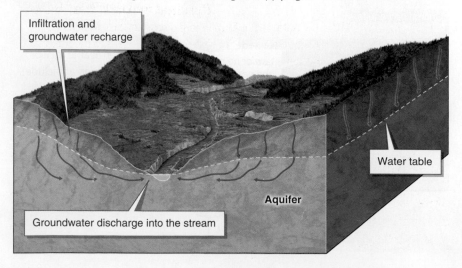

Infiltration and groundwater recharge

Water table

Aquifer

Groundwater discharge into the stream

the zone of aeration and the zone of saturation is called the **capillary fringe (Figure 11.22)**.

Groundwater Movement

An unconfined aquifer's water table is seldom level. The water table roughly parallels the height of the ground surface **(Figure 11.23)**. The contours of the water table determine differences in *hydraulic pressure* (water pressure), and thus the direction of water movement, in the aquifer.

Why does the water table follow surface topography instead of being level like the surface of a pond? When rain falls on areas of high elevation, the infiltrating water has a greater distance to travel downward. If no more rain were ever to fall, gravity would eventually create a level water table.

Water infiltration results in **groundwater recharge**, the entry of water into an aquifer. The movement of water out of an aquifer and onto the ground surface is **groundwater discharge**. Streams that gain water by having groundwater flow into them are called *gaining streams* **(Figure 11.24)**. Streams that flow over areas where the water table is very deep may lose water to the ground and become *losing streams*. Eventually losing streams dry up if rainfall does not replenish their flow.

The Water Table

How far below the ground surface is the water table? The water table may lie at the ground surface, forming a wetland, lake, or stream, or it may lie hundreds of meters below the surface. In most cases, bodies of surface water, such as ponds or streams, represent the height of the water table. In arid regions, the water table typically lies far below the surface of the ground because of the lack of precipitation and infiltration to fill the pores. As a result, there are few permanent surface streams or water bodies in most arid regions. Streams in these regions are often *ephemeral* or *exotic*: They flow only after sudden heavy precipitation events add water to the stream channel faster than it can infiltrate the ground.

At any given location, the water table is not fixed. It fluctuates because of seasonal changes in precipitation, water withdrawals from aquifers by people, and long-term changes in climate **(Figure 11.25)**.

Localized impermeable layers of rock or sediment, called *discontinuous aquicludes*, sometimes form **perched water tables**—localized water tables that lie above the *regional water table* **(Figure 11.26)**.

Perched water tables sometimes create springs. A **spring** arises where hydraulic pressure (the force exerted by the weight of water) pushes groundwater onto the surface. Springs may form

Figure 11.25 The varying height of the water table. Permanent lakes and streams indicate the local height of the water table. (A) In the Um El Ma Oasis in the Sahara, in southwestern Libya, an oasis forms where a depression in the sand dunes dips below the water table. Although almost no water falls from the sky, the climate was once wetter. The water from that wetter period is stored in the aquifer just below the sand surface. (B) In humid regions, drought can lower the water table. Standing bodies of water can disappear if the water table drops. *(Frans Lemmens/Getty Images)*

A

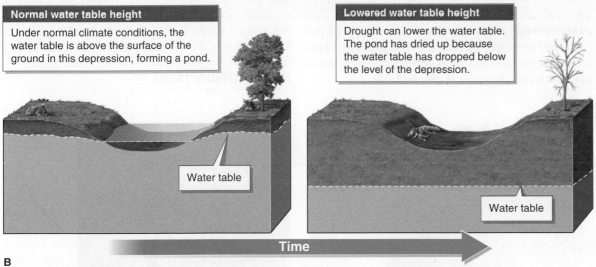

Normal water table height

Under normal climate conditions, the water table is above the surface of the ground in this depression, forming a pond.

Lowered water table height

Drought can lower the water table. The pond has dried up because the water table has dropped below the level of the depression.

Time

B

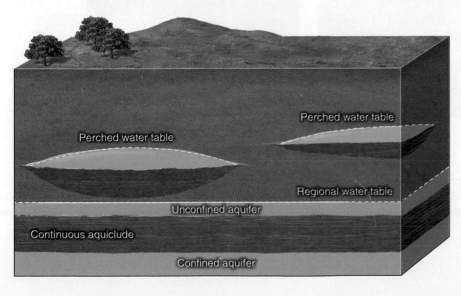

Figure 11.26 Perched water table. Perched water tables form where discontinuous aquicludes prevent water from flowing downward to the regional water table.

along cliff faces and hillsides where water collects above a small, localized aquiclude, as shown in **Figure 11.27**.

Hydraulic Pressure and the Potentiometric Surface

Have you ever wondered why water forcefully gushes out of your home's faucets or garden hose? The hydraulic pressure pushing the water out is created by gravity. The water supply is fed to our homes from a higher elevation, and the pull of gravity forces the water through the pipes that lead into our homes and faucets. In some places, the water supply flows from a reservoir located at an elevation higher than the region it supplies; in other places, water is pumped mechanically up into a water tower to raise it above the region it supplies. In either situation, the weight of the water creates pressure that pushes the water through pipes under pressure.

Water gushes from our faucets because our homes lie below the potentiometric surface created by the elevation of the water supply. The **potentiometric surface** is the elevation to which hydraulic pressure pushes water in pipes or wells. Structures higher than the potentiometric surface have no water pressure **(Figure 11.28)**.

Wells are holes dug or drilled by people to get water from the ground. A well drilled down only as far as the zone of aeration will not produce water unless the water table rises to the well during the rainy or snowmelt season. When a well is drilled downward into the zone of saturation, it fills with water up to the height of the water table. The tops of most wells lie above the potentiometric surface of the aquifer, so pumps (or buckets) must be used to lift the water up out of the well.

An **artesian well** is a well that has been drilled through an aquiclude into a confined aquifer below. Water gushes out of some artesian wells with no pumping required because they are below the potentiometric surface of the recharge area of the aquifer. Artesian wells may form where sedimentary rocks (see Section 14.3) are tilted and permeable and impermeable layers of rock intersect the ground surface, forming a confined aquifer. Rainwater flows into the permeable layers where the layers are exposed at the surface. Because the recharge area is higher in elevation than the aquifer, the recharge area creates hydraulic pressure that pushes water through the aquifer, much as a water tower pushes water through the pipes of a building **(Figure 11.29)**.

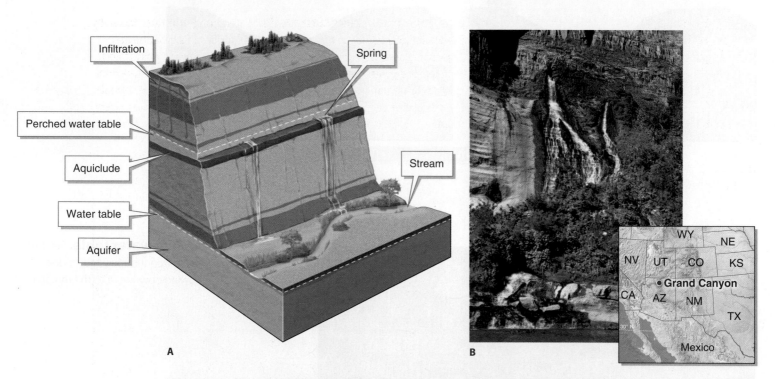

A

B

Figure 11.27 Grand Canyon spring. (A) Water infiltrates permeable sedimentary rocks and collects above a localized aquiclude, forming a perched water table. A spring results where the water emerges from the side of a cliff. (B) Vasey's Paradise is a spring that flows into the Colorado River in Coconino County, Arizona. The Grand Canyon has many such springs that form where localized aquicludes create perched water tables that abut canyon walls.

(UniversalImagesGroup/Getty Images)

Figure 11.28 The potentiometric surface. (A) The potentiometric surface can be visualized as a dome with a surface that drops in all directions with distance away from the water supply. The house farthest from this water tower will have poor water pressure because it is at the level of the potentiometric surface. (B) Water tanks in New York City provide water pressure for the buildings on which they sit. Modern skyscrapers use mechanical pumps, rather than water tanks, to provide water pressure. *(John Cairns/Alamy)*

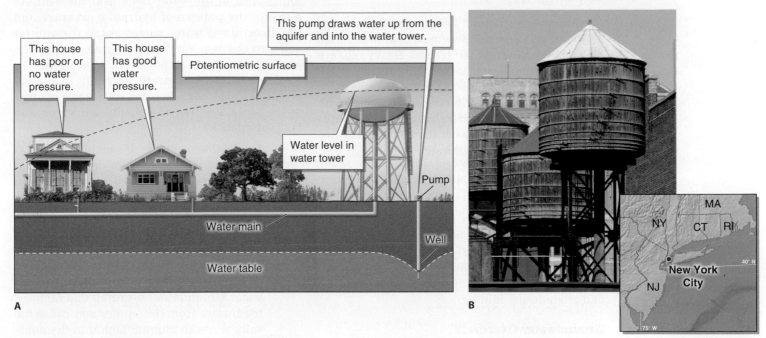

This pump draws water up from the aquifer and into the water tower.

This house has poor or no water pressure.

This house has good water pressure.

Potentiometric surface

Water level in water tower

Pump

Water main

Well

Water table

A

B

Figure 11.29 Artesian wells. Water gushes from artesian wells that are located below the potentiometric surface of a groundwater recharge area. Water must be pumped up from an artesian well located above the potentiometric surface. Vendome Well (inset), in south-central Oklahoma, is an artesian well that gushes 9,500 L (2,500 gal) of naturally saline water each minute. *(Courtesy of Butch Bridges, www.oklahomahistory.net)*

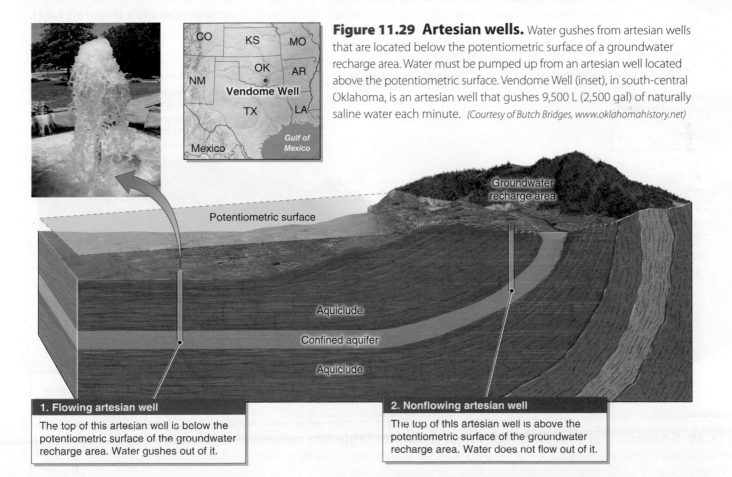

Potentiometric surface

Groundwater recharge area

Aquiclude

Confined aquifer

Aquiclude

1. Flowing artesian well
The top of this artesian well is below the potentiometric surface of the groundwater recharge area. Water gushes out of it.

2. Nonflowing artesian well
The top of this artesian well is above the potentiometric surface of the groundwater recharge area. Water does not flow out of it.

11.4 Problems Associated with Groundwater

◎ **Describe groundwater problems that result from human activities.**

Groundwater is essential for human needs, particularly in arid regions, where most or all water used by people comes from the ground. Heavy reliance on groundwater can create conflicts among neighboring communities drawing from the same aquifer. There are two major problems associated with human use of groundwater resources: (1) withdrawal of groundwater faster than it is replenished and (2) pollution of groundwater.

Too Much Too Fast: Hydrologic Imbalance

The height of the water table in an unconfined aquifer is the result of a balance between groundwater recharge (inputs) and groundwater discharge (outputs). Human activities can disrupt this hydrologic balance by groundwater overdraft and groundwater mining.

Groundwater Overdraft

Groundwater overdraft is the removal of water from an aquifer faster than the aquifer is recharged at the site of a well. Groundwater overdraft often forms a **cone of depression**: a cone-shaped lowering of the water table around the well from which water is being removed **(Figure 11.30)**. Overdraft and the resulting cone of depression change the topography of the water table near the well. As a result, the pattern of hydraulic pressure and the direction of water movement in the aquifer can also change. Water can migrate toward the lowered water table at the cone of depression.

Groundwater overdraft can result in several problems, including a lowered water table at other wells, contamination of wells by salt water, and land subsidence:

1. *Lowered water table:* If neighboring wells are not deep, the cone of depression can lower the water table and cause them to go dry (see Figure 11.30).

2. *Saltwater intrusion:* In coastal regions, a freshwater aquifer may lie on top of salty groundwater. Freshwater is relatively less dense and light compared to salt water, so the freshwater remains above the salt water. Groundwater overdraft that removes freshwater from the aquifer may cause the salty water to migrate higher in the aquifer and contaminate wells. This process is called **saltwater intrusion (Figure 11.31)**.

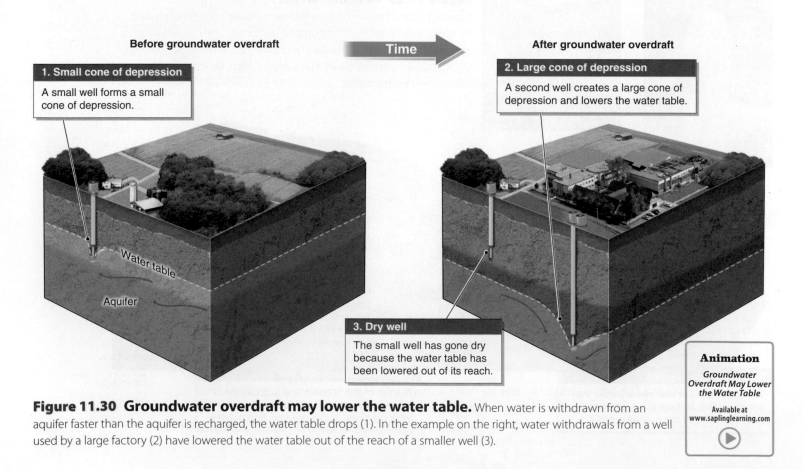

Before groundwater overdraft

1. Small cone of depression
A small well forms a small cone of depression.

Time

After groundwater overdraft

2. Large cone of depression
A second well creates a large cone of depression and lowers the water table.

Water table

Aquifer

3. Dry well
The small well has gone dry because the water table has been lowered out of its reach.

Animation
Groundwater Overdraft May Lower the Water Table
Available at
www.saplinglearning.com
▶

Figure 11.30 Groundwater overdraft may lower the water table. When water is withdrawn from an aquifer faster than the aquifer is recharged, the water table drops (1). In the example on the right, water withdrawals from a well used by a large factory (2) have lowered the water table out of the reach of a smaller well (3).

Saltwater contamination of a well is permanent and irreversible.

3. *Land subsidence:* The pressure of the water in the pores of sediments keeps sediment particles apart while the ground is saturated. When water is removed, these pores can collapse under the weight of the sediments. When this happens, the elevation of the land surface drops as the sediments are compacted **(Figure 11.32A** on the following page**)**. In most cases, once the pores in the aquifer collapse, they can no longer hold water, and the aquifer is lost permanently. Near Mendota, in the San Joaquin Valley, California, the land elevation has dropped more than 8.5 m (28 ft) due to groundwater overdraft. In some cases of land subsidence, large surface fissures (or cracks) can open up **(Figure 11.32B** on the following page**)**.

Groundwater Mining

Groundwater mining is the process of extracting groundwater from areas where there is little to no groundwater recharge. In areas where there is no recharge, once groundwater has been mined, it is gone permanently.

North America's largest aquifer, the Ogallala Aquifer (also called the *High Plains Aquifer*), lies beneath the Great Plains **(Figure 11.33A** on page 359**)**. About 200,000 wells tap into the Ogallala Aquifer. Most of them are connected to *center-pivot* irrigation systems used for agricultural fields **(Figure 11.33B** on page 359**)**. Agriculture in the Great Plains generates some $20 billion in revenue each year, or about one-ninth of the total U.S. agricultural production. The agricultural productivity in the region is dependent on groundwater taken from the Ogallala Aquifer.

The water in the Ogallala is **fossil groundwater**: water that entered the aquifer long ago and is no longer being replenished. Most of it came from the now-melted Laurentide ice sheet that covered much of North America during the most recent glacial period (see Section 7.2). In the southern portions of the Ogallala, the climate is too arid to

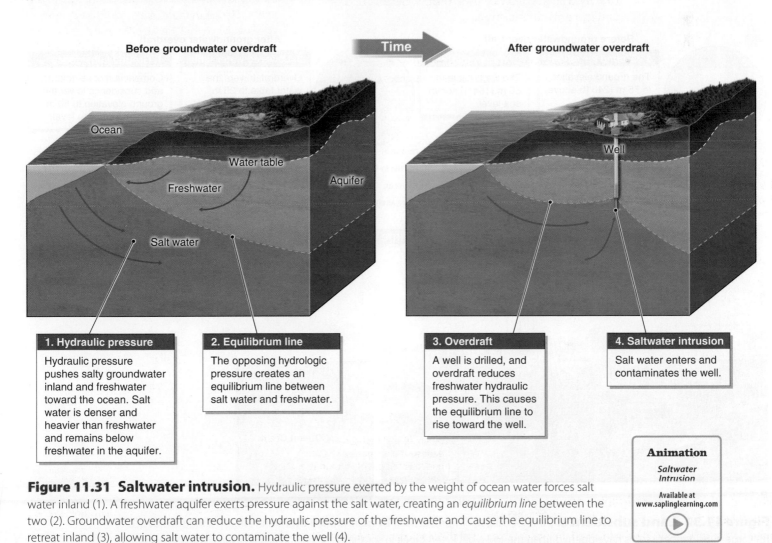

Before groundwater overdraft — Time → **After groundwater overdraft**

Ocean
Water table
Freshwater
Aquifer
Salt water
Well

1. Hydraulic pressure
Hydraulic pressure pushes salty groundwater inland and freshwater toward the ocean. Salt water is denser and heavier than freshwater and remains below freshwater in the aquifer.

2. Equilibrium line
The opposing hydrologic pressure creates an equilibrium line between salt water and freshwater.

3. Overdraft
A well is drilled, and overdraft reduces freshwater hydraulic pressure. This causes the equilibrium line to rise toward the well.

4. Saltwater intrusion
Salt water enters and contaminates the well.

Animation
Saltwater Intrusion
Available at www.saplinglearning.com

Figure 11.31 Saltwater intrusion. Hydraulic pressure exerted by the weight of ocean water forces salt water inland (1). A freshwater aquifer exerts pressure against the salt water, creating an *equilibrium line* between the two (2). Groundwater overdraft can reduce the hydraulic pressure of the freshwater and cause the equilibrium line to retreat inland (3), allowing salt water to contaminate the well (4).

recharge the aquifer. There, farmers and ranchers are mining the groundwater, and the water table is dropping across the southern stretches of the aquifer—by more than 1.6 m (5 ft) each year in some places **(Figure 11.34)**. The deeper the water table becomes, the more energy and money is required to pump water because it has a longer way to go to reach the surface.

Aquifers in arid regions, where there are no permanent surface water bodies, are always composed of fossil groundwater (also see Figure 11.25A). Many countries face looming problems with the loss of groundwater, including China, India, Pakistan, and most countries in the Middle East. About one-fourth of India's food is grown using groundwater that is not being replaced. Saudi Arabia, in the hyperarid Arabian Desert, has developed an agricultural economy dependent on mining fossil groundwater, as highlighted in the **Picture This** feature on page 360.

Groundwater Pollution

Common sources of groundwater pollution include leaks from landfills (or dumps), septic systems, and gas station tanks; agricultural chemicals; animal sewage from factory farms; and mining activities. The process of mining natural gas from shale rock is also linked to groundwater pollution (see Section 14.5).

In some areas, toxic chemicals are (or were) intentionally dumped on the ground or injected into aquifers from wells in an effort to dispose of them cheaply. In the 1950s and 1960s, in California the large utility company Pacific Gas and Electric (PG&E) dumped hexavalent chromium into a collecting pool. From there, the chemical seeped into the ground and migrated several miles through the aquifer. The wells in the nearby town of Hinkley, California, gradually became contaminated, and rates of various types of cancers skyrocketed in the small town. Many of the residents became ill and died. Although PG&E denied wrongdoing, the

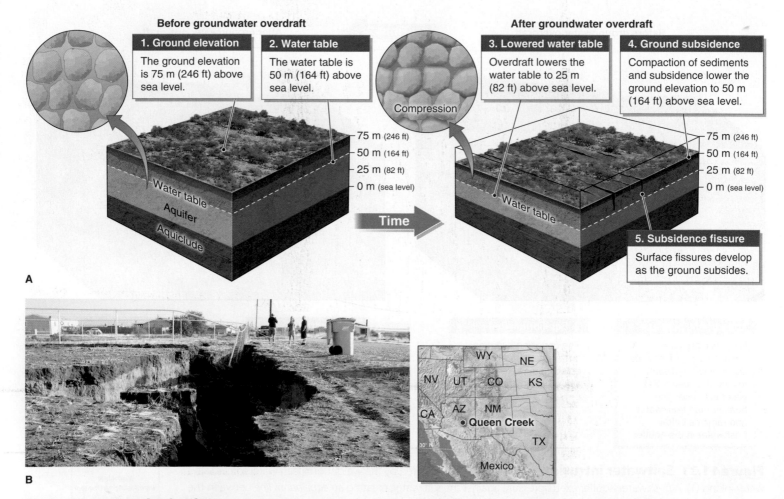

Figure 11.32 Land subsidence. (A) The collapse of pores caused by groundwater overdraft results in land subsidence. (B) Large subsidence fissures have opened up in the town of Queen Creek in southern Arizona. *(Arizona Department of Water Resources)*

Figure 11.33 The Ogallala Aquifer.

(A) The Ogallala Aquifer stretches from South Dakota to northern Texas. Its zone of saturation is thickest in Nebraska, where it is about 300 m (1,000 ft) thick. (B) This aerial photo shows the circular crop patterns resulting from the use of center-pivot irrigation systems near the Texas Panhandle. *(Robert S. Ogilvie)*

Figure 11.34 Water table changes in the Ogallala Aquifer.

The water table has dropped more than 46 m (150 ft) in historic times in some areas, mostly in northern Texas and western Kansas.

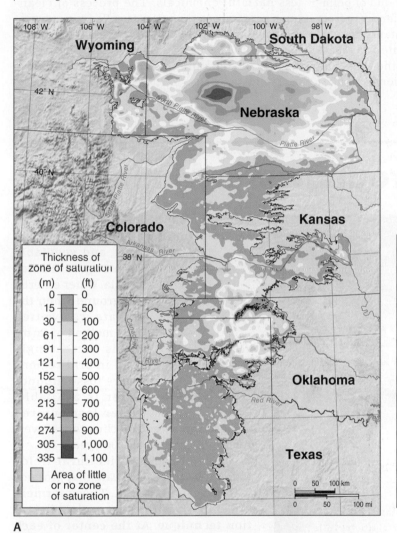

A

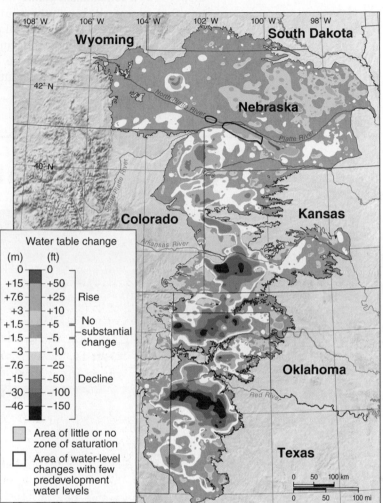

B

Video

Water-level Change in the High Plains Aquifer System

Available at www.saplinglearning.com

company was held accountable in court, thanks to the efforts of the environmental activist Erin Brockovich. Environmental regulations are now in place to stop (or at least reduce) such practices.

Once pollutants have entered an aquifer, they disperse as a **contaminant plume**: a cloud of pollution that migrates through the aquifer away from its source **(Figure 11.35)**. Any pollutants, natural or anthropogenic, that enter the aquifer stay there for centuries or longer. Therefore, the loss of an aquifer through groundwater pollution is, in most cases, permanent.

Various techniques have been tried for cleaning up polluted groundwater. These techniques include the use of bacteria to digest the contaminants and the use of chemicals to react with the contaminants, changing them into other, less harmful chemicals. The process of cleaning a contaminated aquifer is called **groundwater remediation (Figure 11.36)**. In most instances, however, the best option is to stop the "upstream" sources of contaminants and permanently close contaminated wells.

Picture This Groundwater Mining

This Landsat 8 satellite image shows center-pivot irrigation plots for grain, fruit, and vegetable farming in Saudi Arabia. Water entered the ground when the climate was wetter. This method of farming is increasingly used in arid regions where fossil groundwater is available. Water last entered the ground when the climate was wetter, roughly 5,000 to 9,000 years ago.

These desert agricultural fields (shown in green) indicate that groundwater is being mined. These farmers are employing the center-pivot irrigation technique. At the center of each green circle is a well pipe and a diesel pump that pulls water up from the aquifer below. Water is pumped from as deep as 1 km (0.6 mi) beneath the surface. The water moves through a tubular arm called a *gantry* that stretches across the circle's radius (inset). The gantry slowly sweeps around the field. The diameter of each circle is about 1 km (0.6 mi). The water will run out and the fields will return to desert in about 40 years or less, depending on how fast water is taken from the ground.

(MAP: NASA Earth Observatory. image created by Robert Simmon and Jesse Allen, using Landsat data provided by the United States Geological Survey.)

(© Matt Green)

Consider This
1. Why do circular agricultural fields always indicate that groundwater is being used?
2. Why is farming in the Arabian Desert unsustainable?

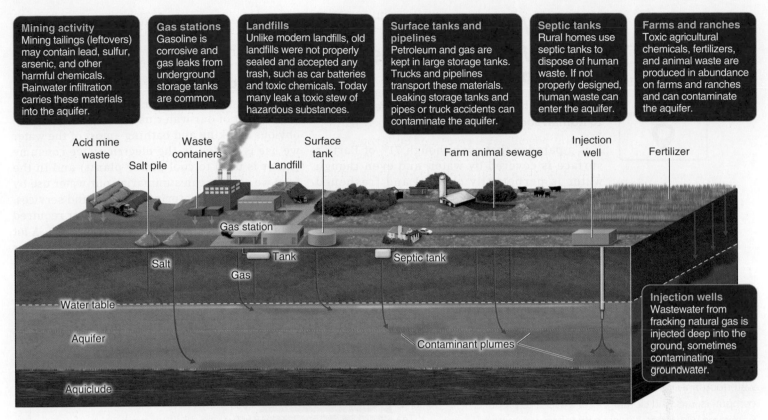

| **Mining activity** Mining tailings (leftovers) may contain lead, sulfur, arsenic, and other harmful chemicals. Rainwater infiltration carries these materials into the aquifer. | **Gas stations** Gasoline is corrosive and gas leaks from underground storage tanks are common. | **Landfills** Unlike modern landfills, old landfills were not properly sealed and accepted any trash, such as car batteries and toxic chemicals. Today many leak a toxic stew of hazardous substances. | **Surface tanks and pipelines** Petroleum and gas are kept in large storage tanks. Trucks and pipelines transport these materials. Leaking storage tanks and pipes or truck accidents can contaminate the aquifer. | **Septic tanks** Rural homes use septic tanks to dispose of human waste. If not properly designed, human waste can enter the aquifer. | **Farms and ranches** Toxic agricultural chemicals, fertilizers, and animal waste are produced in abundance on farms and ranches and can contaminate the aquifer. |

Injection wells Wastewater from fracking natural gas is injected deep into the ground, sometimes contaminating groundwater.

Figure 11.35 Aquifer contaminant sources. Sources of groundwater contaminant plumes (in light orange) come from a variety of human activities. Contaminant plumes migrate vertically depending on their density. For example, salty water sinks to the bottom of the aquifer, and gasoline rises to the top. The contaminants also migrate in the direction of water flow within an aquifer.

Figure 11.36
SCIENTIFIC INQUIRY: How is contaminated groundwater cleaned?

In 1980, the U.S. Environmental Protection Agency Superfund Program was established to clean up the nation's worst toxic contamination. Over 80% of EPA Superfund sites are areas of contaminated groundwater. The method of cleanup used depends on the type of contaminant, the physical circumstances at the site, and the cost. Generally, contaminated water is either pumped from the ground and treated or treated while in the ground, a method called in *situ remediation*.

Pump and treat
Extraction wells are placed in the path of the migrating contaminant plume, and the contaminated water is pumped from the aquifer. Once at the surface, the contaminated water is treated with filtration, chemical additives, or bacteria. A combination of all these methods may be used. Once cleaned, the treated water is discharged into the local watershed or sewer system.

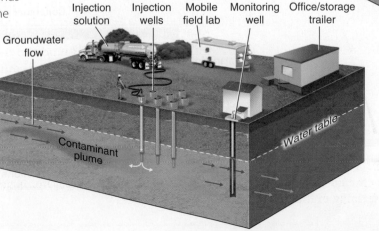

In situ remediation
In situ remediation involves the injection of chemicals, such as hydrogen peroxide (H_2O_2), or bacteria into the contaminated aquifer. The chemicals or bacteria break down the contaminants into safer compounds.

GEOGRAPHIC PERSPECTIVES
11.5 Water Resources under Pressure

◎ **Assess the importance of water resources to human societies.**

It is a paradox that even though 71% of Earth's surface is covered by water, and even though water is perpetually cleaned and freshened through the hydrologic cycle (see Section 4.1), freshwater is a resource that many people lack. There is a limit to how quickly water can be cleaned in the hydrologic cycle. As the global human population and its economic affluence grow, water—both on the surface and in the ground—is becoming an increasingly important and scarce resource.

Water Footprints

How much water people use may come as a surprise. Most of our water use comes not from the obvious drinking and bathing—most of the water we use is hidden in the electricity we consume (water is used to cool power plants) and in the food we eat. We can summarize our water use by creating a water footprint for items and services. A **water footprint** is the amount of water required to produce a specific item, food, or service. A lot of water goes into producing food and consumer goods **(Figure 11.37)**.

Figure 11.37 Water footprints of foods and consumer goods. (A) This graph shows the water footprint of common foods in liters per kilogram. Gallons per pound are in parentheses. Data are from waterfootprint.org. (B) Manufactured goods, such as computers, cars, and clothing, all take water to produce. Compared with foods, it is less easy to see why material goods require water, but many of them have substantial water footprints.

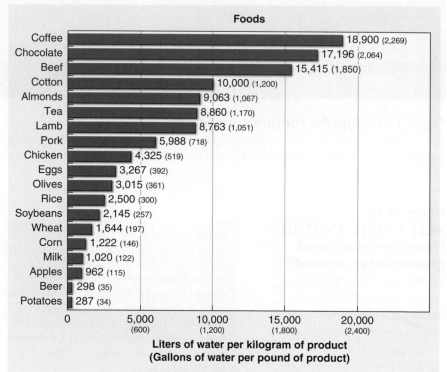

A

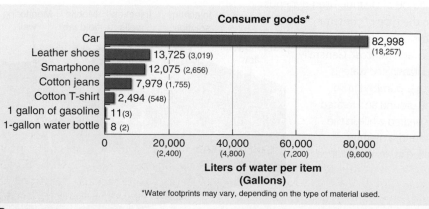

B

Why does producing a single kilogram of beef require 15,415 liters of water? Most of the water footprint of beef is a result of watering the crops that cattle eat. In addition, water is used to process and prepare the beef in factories for consumers.

When food is produced in a suitable climate, much of the water used to make it falls directly from the sky. Chocolate, for example, comes from the seeds of the cacao tree (*Theobroma cacao*). Plantations of cacao trees are located in tropical regions where the climate naturally meets the plant's water needs. This applies to coffee as well. In contrast, some foods, such as beef in Texas, are grown or raised in semiarid or arid lands. In the case of Texas-raised beef, the water needed does not fall from the sky and must instead be pumped from the Ogallala Aquifer or diverted from surface streams.

Many services require water as well. Golf courses, for example, have an inordinately large water footprint. Their close-cut turf grass stores little water and requires uninterrupted irrigation. There are more than 16,000 golf courses in the United States. The average water consumption of all U.S. golf courses is roughly 2.9 trillion L (766 billion gal) per year. Golf courses in the desert use even more water **(Figure 11.38)**.

Out of economic necessity, many golf course managers are adapting to the realities of increasing water scarcity. About 1,000 golf courses in the United States are now using, at least in part, recycled or reclaimed water (wastewater from the golf facility and treated sewage water). They are also planting more drought-tolerant varieties of turfgrass to reduce their water footprint.

The Global Reach of Virtual Water

Water resources have a borderless, global dimension. Water consumption is not restricted by political boundaries because many of the goods people use and the foods people eat come from somewhere else, often very far away. Imported foods and goods are sometimes more common than domestic products. Chinese toys, New Zealand apples, Chilean grapes and wines, California almonds, Japanese electronics and cars, and midwestern beef all exact a water toll on their country or region of origin. The hidden flow of water embedded in these goods and foods is called **virtual water.** As shown in Figure 11.37, a pair of cotton jeans, for example, has a water footprint of 7,979 L (2,108 gal); this water footprint includes virtual water accumulated during the

Figure 11.38 A desert golf course. The average 18-hole golf course in the United States uses about 500,000 L (132,000 gal) of water per day. In the summer, desert golf courses, such as the one shown here in Palm Springs, California, use up to 3 million L (800,000 gal) of water each day, the same amount a family of four uses in five years. There are 123 golf courses in Palm Springs alone, using 369 million L (98.4 million gal) of water every day during summer. *(JenniferPhotographyImaging/Getty Images)*

steps of manufacturing and transporting the jeans across international borders. As goods such as jeans and cars and food move around the planet, immense amounts of virtual water travels with them **(Figure 11.39)**.

The average annual global flow of virtual water is about 908 trillion L (240 trillion gal). Of this, 92% was in the form of agricultural crops or products derived from crops (such as paper and cotton). The average worldwide *per capita* (per person) water use is 3,794 L (1,002 gal) each day. This figure includes water used directly, such as in bathing, and virtual water associated with commodities consumed. In the United States and Canada, the average per capita water use is roughly twice the worldwide average.

Earth's physical systems, such as soils and water, and the people in them are interconnected. Countries that receive imports have impacts on those exporting countries. The economies of exporting countries benefit—although sometimes at the expense of the environment. For example, diversions of water from the Aral Sea are used to grow cotton in the desert of Uzbekistan. These diversions are causing the Aral Sea to shrink and its wetland ecosystems to disappear, leaving behind toxic dust that has poisoned local residents. The cotton grown in Uzbekistan using the diverted water is then exported somewhere else, such as China, to manufacture goods such as jeans and T-shirts. These items are then exported to world markets such as Japan, the United States, Canada, and Europe. Although virtual water is unseen, its effects link a simple cotton T-shirt with the loss of the entire inland ecosystem.

The Future of Water

The human population has just passed the 7.6 billion mark and continues to grow. This single statistic underpins all environmental issues today, but it is particularly relevant to the scarcity of water resources. Global population numbers are an important part of the story, but they are not the only part. Higher levels of economic development are usually associated with higher water demands, and unequal political power in a geographic region can also exacerbate water shortages for politically disadvantaged populations.

In wealthy nations such as the United States and Canada, water shortages range from an inconvenience to causing serious economic and ecological harm. In poorer countries, such as Mozambique or Chad, water resources can be a matter of life and death. Worldwide, 844 million people (or about 1 in 9) do not have access to safe drinking water. About 40% of those people live in sub-Saharan Africa. By some estimates, half the world's hospital beds are filled with people who are sick from drinking contaminated water.

Economic development for poverty-stricken countries is absolutely a good thing. It improves access to clean drinking water, medicines, education, health, and a whole range of other social benefits. Economic development also creates more efficient use of water and possibilities for recycling water. However, affluence also increases demand for material goods. These material goods put more pressure on water resources because water is needed to make them. Human population growth, economic development, and climate change are three interwoven global forces that make freshwater an increasingly precious natural resource.

Figure 11.39 Virtual water movement. Japan exports roughly 1.5 million automobiles each year to the United States. Meanwhile, the United States exports roughly 150,000 metric tons of rice to Japan each year. When these products are converted to virtual water, it is clear that more than just material goods are being exchanged. Not all of the virtual water used to produce these cars came from Japan, however. Many of the materials used to make an automobile, such as rubber and metals, were themselves imported to Japan from somewhere else. So water is only passing through Japan to the United States as virtual water in the form of automobile components. *(Left. Ken Shimizu/AFP/Getty Images; right. Ken James/Bloomberg via Getty Images)*

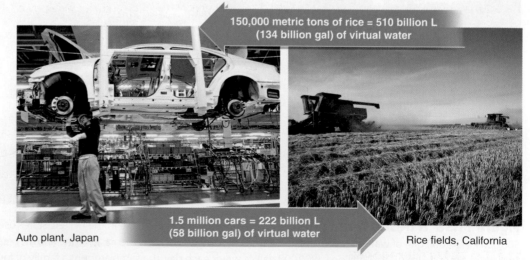

150,000 metric tons of rice = 510 billion L (134 billion gal) of virtual water

1.5 million cars = 222 billion L (58 billion gal) of virtual water

Auto plant, Japan

Rice fields, California

Chapter 11 Study Guide

Focus Points

11.1 The Living Veneer: Soils

- **Soil composition:** By volume, soils are composed mostly of weathered fragments of rocks and minerals. Healthy soils are filled with living organisms and their organic remains.
- **Soils and climate:** Soils develop most quickly where it is warm and wet and most slowly where it is cold and dry. On a small (broad) geographic scale, climate is an important soil-forming factor.
- **Anthropogenic soils:** People can form soils intentionally by composting.
- **Soils and time:** In nature, a few centimeters of soil take centuries or more to form.
- **Soil erosion:** Human activity is causing soil erosion to happen faster than soils form naturally.
- **Soil taxonomy:** The soil taxonomy classification system groups soils into 12 orders, based mainly on horizon development and color.
- **Importance of soils:** Soils provide food and other ecosystem services, such as new medicines, climate change mitigation, and water purification.

11.2 Surface Water Resources

- **Water proportions:** Only 2.8% of all water on Earth is fresh, and only a small fraction of that occurs in lakes and streams.
- **Water uses:** Agriculture and power production account for 78% of freshwater use in the United States.
- **Water scarcity:** The Palmer Drought Severity Index provides a measure of water deficits and ranks the severity of drought.

11.3 The Hidden Hydrosphere: Groundwater

- **Groundwater:** About half the water used in the United States comes from the ground.
- **Permeability:** Water flows through rocks or sediments at a rate of centimeters to meters per day, depending on the permeability of those materials.
- **Surface water:** Bodies of surface water, such as lakes and streams, occur where the water table reaches Earth's surface.
- **Water table fluctuations:** The water table fluctuates over time as a result of seasonal changes in precipitation, long-term climate change, and withdrawals of groundwater by people.

- **Wells:** The tops of most wells lie above the potentiometric surface of the groundwater recharge area, and water must be pumped out of them. Water gushes out of some artesian wells.

11.4 Problems Associated with Groundwater

- **Groundwater overdraft and mining:** Groundwater overdraft and groundwater mining withdraw water from aquifers more rapidly than it can be replenished.
- **Groundwater pollution:** Contaminants that enter groundwater by infiltration come from a variety of sources, including gasoline tanks, sewage, and landfills.

11.5 Geographic Perspectives: Water Resources under Pressure

- **Water footprints:** Producing food crops and material goods, such as computers and clothing, requires water.
- **Virtual water:** Global virtual water flow exceeds 900 trillion L (237 trillion gal) each year.

Key Terms

aquiclude, 350
aquifer, 350
artesian well, 354
bedrock, 337
capillary fringe, 352
cone of depression, 356
contaminant plume, 360
desalination, 346
drought, 349
eluviation, 337
fossil groundwater, 357
groundwater, 342
groundwater discharge, 352
groundwater mining, 357
groundwater overdraft, 356
groundwater recharge, 352
groundwater remediation, 360
humus, 340
illuviation, 337
infiltration, 350
leaching, 337

loam, 339
pedogenesis, 337
perched water table, 352
percolation, 350
permeability, 350
porosity, 350
potentiometric surface, 354
regolith, 337
saltwater intrusion, 356
soil, 337
soil horizon, 337
soil taxonomy, 344
spring, 352
virtual water, 363
water footprint, 362
water table, 351
weathering, 337
well, 354
zone of aeration, 351
zone of saturation, 351

Concept Review

The Human Sphere: The Collapse of the Maya

1. When and where did the Maya civilization flourish?

2. Describe the link between the collapse of the Maya civilization and their soil and water resources.

11.1 The Living Veneer: Soils

3. What is pedogenesis?

4. What are soil horizons? How do they form? How many are there? Do all soils have each of them?

5. Describe the major characteristics of each of the soil horizons.

6. What are the four major components of soils?

7. On a small (broad) geographic scale, what is the single most important factor that determines what kinds of soils will form? Explain why this factor is important.

8. What other soil-forming factors are there? Briefly explain each one.

9. What factors contribute to soil loss and soil degradation? Which of these is the most important?

10. Compare tropical soils, subtropical desert soils, and temperate soils with regard to degree of weathering and leaching as well as topsoil depth.

11. What is contour plowing? What is terracing? How do they reduce soil erosion?

12. What are the 12 soil orders, and what characteristics are used to identify them?

13. What overall information can be gained from the color of soil? For example, what does a blue-green soil signify?

14. What kinds of ecosystem services do soils provide?

15. Why did New York City choose to buy and preserve temperate deciduous forest land rather than build a water filtration plant?

16. How are soils linked to climate change?

11.2 Surface Water Resources

17. What percentage of Earth's freshwater is stored in ice? In the ground? In lakes? In streams?

18. What is desalination? Where is this process used, and how does it work?

19. What is a drought?

20. What is the Palmer Drought Severity Index, and how does it help scientists determine the degree of drought for a given region?

11.3 The Hidden Hydrosphere: Groundwater

21. How does water get into the ground?

22. Compare the terms *porosity* and *permeability*. What do they relate to, and how are they different? Discuss these terms in the context of sand and clay.

23. Compare the terms *infiltration* and *percolation*. How are they different?

24. Compare the terms *aquiclude* and *aquifer*. Where does each of them occur, and how are they different?

25. Compare a confined aquifer with an unconfined aquifer. What confines an aquifer?

26. Compare the zone of aeration with the zone of saturation. What is the capillary fringe, and where does it occur in relationship to the zones of aeration and saturation?

27. How does water enter an aquifer? How does water leave an aquifer?

28. What does a lake or stream tell us about the depth of the water table?

29. What is a perched water table?

30. What is a spring, and how does it form? Is it an area of recharge or an area of discharge for an aquifer? Explain.

31. What is the potentiometric surface? If a building is above the potentiometric surface of its water supply, will water flow from its faucets? Explain.

32. What is a well? How do some artesian wells differ from other wells?

33. Using the concepts of the potentiometric surface and the area of groundwater recharge, explain how water can gush from an artesian well.

11.4 Problems Associated with Groundwater

34. In the context of recharge and discharge, what is groundwater overdraft?

35. What is groundwater mining?

36. Discuss the problems resulting from groundwater overdraft. What are they, and what processes cause each of them?

37. What is fossil groundwater? Can you relate this term to groundwater mining?

38. How do aquifers become polluted?

39. What is a contaminant plume? What cleanup options are available to deal with contamination?

11.5 Geographic Perspectives: Water Resources under Pressure

40. What is a water footprint? Why does producing one kilogram of beef require 15,415 liters of water?

41. Why do golf courses, particularly those in the desert, have unusually large water footprints?

42. What is virtual water? Give examples.

43. Where are water resources most stressed today?

Critical-Thinking Questions

1. Examine the soil taxonomy map in Figure 11.15 and find where you live. What soil order(s) occurs where you live? What other methods could you use to determine the soil type where you live?

2. Section 11.1 outlines the ecosystem services provided by soils. Can you think of any additional services that soils provide?

3. How can a commodity consumed in one country (for example, a T-shirt) could affect a far-away region's water supply?

4. What is the relationship between the level of economic development and access to clean and safe freshwater?

5. Do you think luxury services such as golf courses are justifiable, given their large water footprint? Does your answer depend on whether you golf or not? Are there other similar luxuries you could argue as being unjustifiable? If so, what are they?

Test Yourself

Take this quiz to test your chapter knowledge.

1. True or false? Loam is composed of about 40% sand, 40% silt, and 20% clay.

2. True or false? Aquifers form mainly in porous sediments or rocks.

3. True or false? Groundwater overdraft can increase the height of the water table.

4. True or false? The water footprint for a kilogram of wheat is greater than the water footprint for a kilogram of beef.

5. Multiple choice: Which of the following soils would most likely be found in a tropical rainforest?
a. alfisol
b. gelisol
c. oxisol
d. vertisol

6. Multiple choice: Which of the following biomes has the greatest stores of soil carbon?

a. tropical rainforest
b. boreal forest
c. temperate grassland
d. wetland

7. Multiple choice: Which of the following problems is not associated with groundwater overdraft?

a. subsidence
b. dry wells
c. surface fissures
d. lowered potentiometric surface

8. Fill in the blank: _____ occurs when a well becomes contaminated with salt water.

9. Fill in the blank: The unseen water used to make a product is called _____ .

10. Fill in the blank: The process of soil formation is called _____ .

Online Geographic Analysis

Calculating Your Water Footprint

In this exercise you will evaluate your personal water footprint.

Activity

Go to http://www.watercalculator.org, an interactive web page where you can enter your water usage information and have the site calculate your personal water use. Answer the questions in the water calculator to obtain an estimate of your personal water use. Once you have done that, answer the following questions.

1. What is your personal water footprint, in gallons per day?

2. Is your water usage higher or lower than the average water usage?

3. Scroll down to see a bar chart of three categories of water use: "Indoor Water," "Outdoor Water," and "Virtual Water." In which of these three categories is your water footprint greatest?

4. Move your cursor over each bar (or tap the bar) to get more information about that specific category. Which three activities and uses account for most of your water use?

5. Scroll down through your results until you find your highest category of water use. (For example, it may be "Shopping Habits.") Click or tap the "Tips" button to the right. What are some suggestions the site provides for reducing your water use in this category?

6. Scroll down through your results until you see the "Diet" category. Click or tap it to modify that information. Compare the water usage by "Vegan," "Vegetarian," and "Meat Eater" categories. As you change these categories, how does the water footprint change? Which is the highest, and which is the lowest?

7. Select the "Tips" button for the "Diet" category. What are some suggestions the site provides for reducing your water use in this category?

8. How can line-drying your clothes rather than using a clothes dryer save water? Hint: Select the "Tips" button in "Laundry."

9. What is a "greywater" system? How does it save water?

10. What is a "rain barrel," and how does it save water?

Picture This. *Your Turn*

Groundwater Features

Use the following terms to fill in the boxes on the diagram. Use each term only once.

1. Aquiclude

2. Artesian well

3. Confined aquifer

4. Perched water table

5. Potentiometric surface

6. Recharge area

7. Unconfined aquifer

8. Water table

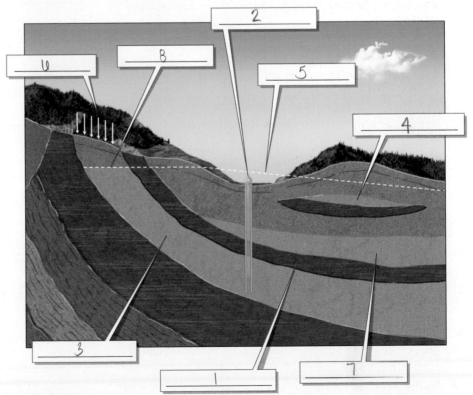

PART III
TECTONIC SYSTEMS: Building the Lithosphere

Earth's internal heat energy drives the movement of lithospheric plates that move entire continents, lift mountains, build volcanoes, and generate powerful earthquakes. Part III explores the role of plate tectonics in building and shaping Earth's surface.

CHAPTER 12
Earth History, Earth Interior

Earth and the solar system formed 4.6 billion years ago, and life began soon after. The planet's hot interior is structured in layers and protects Earth's life.

CHAPTER 13
Drifting Continents: Plate Tectonics

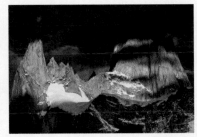

Earth's internal heat energy drives the movement of lithospheric plates, creating mountains, valleys, and other features.

CHAPTER 14
Building the Crust with Rocks

Plate tectonics drives the rock cycle that creates and transforms rocks.

CHAPTER 15
Geohazards: Volcanoes and Earthquakes

Volcanoes and earthquakes shape the crust's surface and are significant hazards for people.

Earth History, Earth Interior

Chapter Outline *and Learning Goals* ◎

12.1 Earth Formation

◎ Explain the origin of Earth, its atmosphere, and oceans and summarize when major life-forms first arose.

12.2 Deep History: Geologic Time

◎ Describe the major divisions of geologic time and explain how the age of ancient Earth material is determined.

12.3 Anatomy of a Planet: Earth's Internal Structure

◎ Describe the planet's internal structure and understand the importance of lithospheric plate movement.

12.4 Geographic Perspectives: Earth's Heat and the Biosphere

◎ Assess the connection between Earth's internal heat and life on Earth.

This incandescent lava is flowing out of Kilauea Volcano on the Big Island of Hawai'i. The Hawaiian Islands are built entirely of lava flows such as this. Lava forms as melted rock called magma forms deep within Earth's interior and spills onto the surface through a volcano.

(© Andrew Hara)

To learn more about the composition and structure of Earth's interior, turn to Section 12.3.

THE HUMAN SPHERE The Anthropocene

Figure 12.1 Tokyo, Japan. Tokyo is the largest urban area in the world. The city proper has 13.2 million people. When the surrounding metropolitan region is also included the total population is 38 million people. Like many other regions on Earth, this city and the area surrounding it have been completely transformed by people. *(ngkaki/Getty Images)*

Nearly all of the world's major rivers have been dammed, and about half the world's freshwater runoff is used by people. Nearly 40% of Earth's land surface not covered by glaciers is being used to grow food for people. People and domesticated animals, such as cows and sheep, comprise about 90% of all mammal biomass. About 10,000 years ago, the percentage of mammal biomass made up of people and domesticated animals was only 0.1%. Globally, people move more sediment and rock using bulldozers and other earth-moving equipment than is moved naturally by weathering and erosion. Earth has experienced five *mass extinction events,* a geologic event when 75% of all species go extinct. Given the rate of human-caused habitat loss and species extinctions today, most scientists are convinced that Earth is now in the midst of a sixth mass extinction event.

The extent of these anthropogenic changes to Earth's physical environments are significant enough to warrant the designation of a new geologic epoch in Earth history. This new epoch, called the **Anthropocene** (pronounced AN-thruh-pa-seen), or the "Age of Humans," is the age of human transformation of Earth's biological and physical systems **(Figure 12.1)**.

Although the term *Anthropocene* is in widespread use, this new epoch has not yet been officially recognized as a formal geologic period. Before that can happen, scientists first need to agree on when it began. Some argue that it began 10,000 years ago, when early farming in what is today the Middle East became common. Others believe the advent of the Industrial Revolution, around 1800, marks the beginning of the new era. At that time, humans began employing fossil fuels to run machines, and, in turn, the chemistry of the atmosphere and oceans began changing. Still others argue that the Anthropocene's beginning should be set at 1950, when the rate of world population and human environmental impacts began greatly accelerating. Although scientists do not yet agree on when the Anthropocene began, no one doubts that the current Age of Humans is one of the most significant periods in Earth's history.

12.1 Earth Formation

◎ **Explain the origin of Earth, its atmosphere, and oceans and summarize when major life-forms first arose.**

Earth is one of eight planets in our *solar system*. Earth's solar system includes the Sun and all objects orbiting the Sun, including hundreds of thousands of rocky *asteroids* and icy *comets,* as well as the eight planets. It is not the only solar system in existence, however. As of April 2018, using powerful telescopes and other means of detection, astronomers have discovered 2,800 different solar systems and 3,760 planets within those solar systems. About three new planets are discovered every week.

Planets orbit stars, and stars make up galaxies. Earth is in the *Milky Way* galaxy. When the night sky is viewed away from city lights and under clear skies, the Milky Way is a visible band of stars **(Figure 12.2)**. Galaxies are organized, rotating masses of hundreds of billions of stars. Scientists estimate that there are between 200 billion and 400 billion stars in the Milky Way and at least that many planets.

Galaxies are immensely large, and the distances between them are vast. The Milky Way, for example, is 120,000 *light-years* across; that is, it takes light 120,000 years to cross the galaxy from end to end. Light travels at 300,000 km (186,000 mi) per second. The Andromeda galaxy, the closest large galaxy to Earth, is 2.5 million light-years away. Some of the most distant galaxies are more than 13 billion light-years away from Earth. In other words, it has taken more than 13 billion years for their light to reach Earth. We cannot see such distant objects without powerful telescopes. When we look up into the night sky without telescopes, the stars we see are relatively close to Earth. Their light has traveled for decades to a few thousand years before reaching our eyes.

The Milky Way and Andromeda galaxies are two among many hundreds of billions of galaxies in the *universe*. The universe began about 13.8 billion years ago. This age estimate is based on astronomer Edwin Hubble's observation that galaxies are all moving away from Earth and away from one another. From this observation, Hubble surmised that the universe is expanding, an idea called the *expanding universe theory.* From the rate of their movement, Hubble and other astronomers calculated that all the galaxies would have been at a single point about 13.8 billion years ago. At that time, that single point rapidly expanded in the *Big Bang.* Since then, the universe has continued expanding.

Figure 12.2 The Milky Way. From the vantage point of Earth, the disk structure of the Milky Way is revealed in the band of stars we can see at night if we are away from artificial lighting. When stars are very far away and densely packed together, they appear "milky." All of the points of lights we can see in the night sky are stars relatively close to Earth. This photo was taken near Bardenas, Spain. *(Inigo Cia/Getty Images)*

How the Solar System Formed

Our solar system and everything in it, including Earth, formed from an immense cloud of dust and gas that were the remnants of a star that exploded long ago. A neighboring star is thought to have exploded in a *supernova,* creating a shockwave through space that squeezed a portion of the dust and gas cloud, causing it to collapse about 5 billion years ago. As the cloud began to collapse, it also began to spin. Our Sun formed as this cloud of dust and gas, called a *nebula,* coalesced (came together) by mutual gravitational attraction and collapsed into a smaller mass of gas called a *protostar.* As the mass of this *protostar* increased, its internal pressure and temperature rose. Eventually, as more and more material accumulated and the pressure increased, the object's interior temperature reached about 10 million °C (18 million °F). At this temperature, hydrogen atoms begin fusing to form helium atoms. The fusion of these atomic nuclei (called *nuclear fusion*) produced enormous amounts of heat and light and formed the Sun.

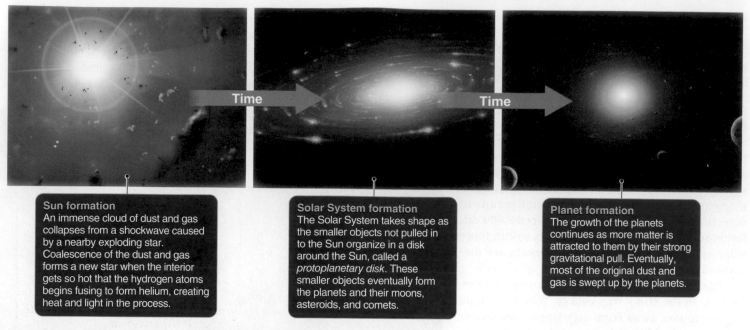

Sun formation
An immense cloud of dust and gas collapses from a shockwave caused by a nearby exploding star. Coalescence of the dust and gas forms a new star when the interior gets so hot that the hydrogen atoms begins fusing to form helium, creating heat and light in the process.

Solar System formation
The Solar System takes shape as the smaller objects not pulled in to the Sun organize in a disk around the Sun, called a *protoplanetary disk*. These smaller objects eventually form the planets and their moons, asteroids, and comets.

Planet formation
The growth of the planets continues as more matter is attracted to them by their strong gravitational pull. Eventually, most of the original dust and gas is swept up by the planets.

Figure 12.3 Solar system and planet formation. *(SPL/Science Source)*

As it formed, the Sun captured 99.8% of all the mass in the cloud of dust and gas that became the solar system. The remaining 0.2% of material accreted (formed by gradual accumulation) into smaller bodies that were not massive enough to trigger nuclear fusion, resulting in the formation of planets, asteroids, and other objects of the solar system **(Figure 12.3)**.

The mass of the Sun is 333,000 times the mass of Earth, and about 1 million Earths would fit inside the Sun. The mass of the Sun exerts a strong gravitational pull that keeps the planets locked in their respective orbits along a flat plane **(Figure 12.4)**. Earth is the third closest planet to the Sun; Venus and Mars are its two nearest neighbors.

Most of the planets in the solar system have orbiting moons. Saturn has 53 moons. The leading hypothesis to account for the formation of Earth's Moon, called the *giant-impact hypothesis*, states that another planet in the solar system collided with Earth about 4.5 billion years ago. That other planet, called *Theia*, is thought to have been about two-thirds the size of Earth. The impact was so violent that much of Theia melted and was incorporated into Earth. A significant portion of the remnants of the collision vaporized and dispersed away in space. The material that didn't escape Earth's gravity eventually coalesced to form the Moon. Many scientists suspect that the debris field from the impact may have even formed a ring system around Earth, similar to Saturn's ring system, before it eventually formed the Moon.

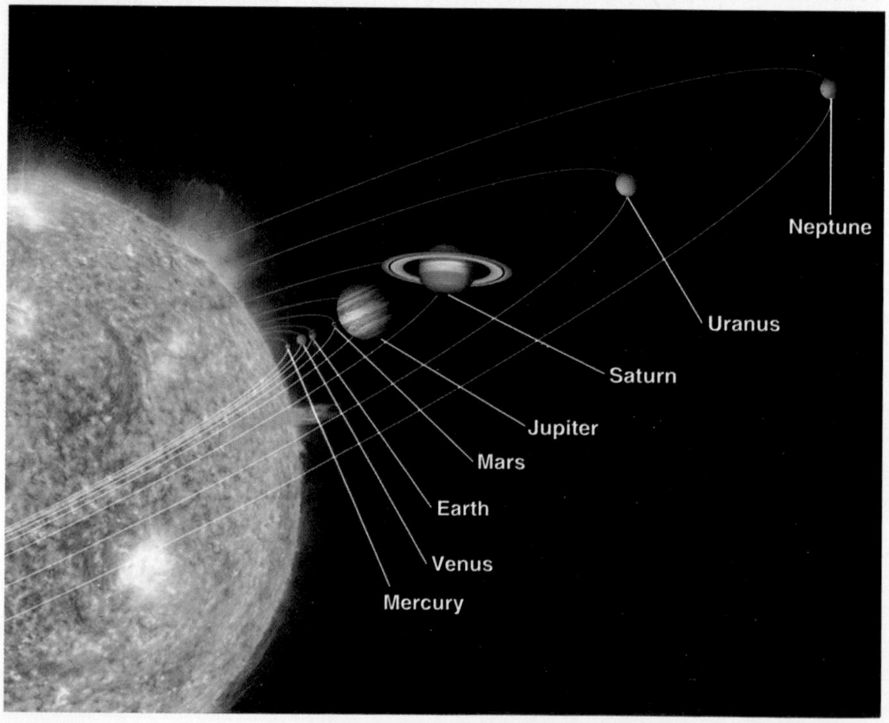

Neptune

Uranus

Saturn

Jupiter

Mars

Earth

Venus

Mercury

Figure 12.4 The solar system. The solar system consists of the Sun and eight planets, as well as many smaller objects. The planets and the Sun are scaled to show their relative sizes in this illustration, but distances between planets are not to scale; the planets are much farther apart than this illustration indicates. The orbital paths that the planets take around the Sun are shown here as white lines. *(NASA/JPL-Caltech/T. Pyle (SSC))*

From space, Earth looks like a perfect sphere, equal in all dimensions. Yet it is not. Earth is actually an *oblate spheroid*, a slightly flattened sphere. This slightly flattened shape is produced by centrifugal force that results from the planet's rotation.

Earth has been pounded by large rocky bodies called *asteroids* throughout its history. During the *Late Heavy Bombardment*, which took place between 4.1 and 3.8 billion years ago, large impact collisions were common. So many asteroids struck Earth that heat from the friction of their impacts kept Earth's surface molten for several million years. Eventually, most of the objects in Earth's orbit were swept up by these collisions. Earth's molten surface then cooled and hardened into the rigid outermost portion of Earth, called the **crust**.

Today, about 15 metric tons of rock and dust from space enter Earth's atmosphere each year. Most debris particles entering Earth's atmosphere are no bigger than grains of sand. Occasionally, large objects do hit Earth and leave their mark in the form of an impact crater. Most impact craters on Earth are erased by erosion, but some remain **(Figure 12.5)**.

How the Atmosphere and Oceans Formed

Earth's atmosphere formed from gases emitted by volcanoes. A **volcano** is a mountain or hill formed by eruptions of lava and rock fragments. While Earth's crust was forming, extensive volcanic activity pumped huge quantities of gases such as hydrogen, water vapor, carbon dioxide, hydrogen sulfide, and nitrogen from Earth's interior to form the early atmosphere.

Scientists are less certain about the origin of Earth's water. Evidence from ancient *zircon crystals* that contain the chemical signature of water indicates that Earth had extensive liquid water as early as 4.4 billion years ago. One idea is that water vapor from volcanic emissions condensed out of the atmosphere and collected in the low-lying areas of the crust, forming the oceans. A contending theory is that icy comets from space delivered water to Earth. Certainly both processes played a role in forming Earth's water, but at present the volcanic eruptions theory is the leading theory to account for the majority of Earth's water.

There are two great reservoirs of water on Earth. The first is on Earth's surface, mostly in the form of oceans and ice sheets. Groundwater (described in Section 11.3) is also part of this reservoir of water. Experimental evidence, however, indicates that another reservoir of water may exist deep within Earth's interior. The interior of Earth

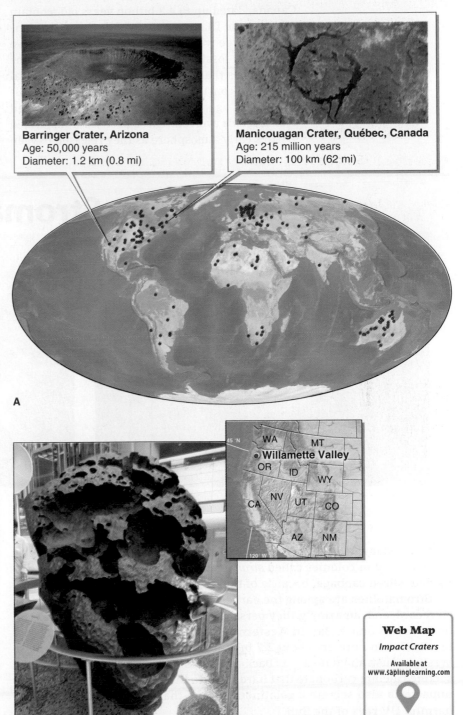

Figure 12.5 Impact craters. (A) The locations of the major surviving impact craters on Earth's surface are shown as red dots. Barringer Crater (or Meteor Crater) in Arizona and Manicouagan Crater in Quebec are highlighted. Manicouagan Crater is one of the oldest surviving impact craters. (B) The Willamette meteorite, displayed in the American Museum of Natural History in New York, is the largest meteorite ever found in the United States. It is 3.05 m (10 ft) tall. The 15.5-ton meteorite, composed of iron and nickel, is more than 1 billion years old. It entered the atmosphere over the Willamette Valley in Oregon at an estimated 64,400 km/h (40,000 mph). Intense heat caused by friction with the atmosphere pitted its surface. *(A: left. Imagestate Media Partners Limited - Impact Photos/Alamy, A: right. NASA image courtesy NASA/GSFC/LaRC/JPL, MISR Team; B. Bruce Gervais)*

Barringer Crater, Arizona
Age: 50,000 years
Diameter: 1.2 km (0.8 mi)

Manicouagan Crater, Québec, Canada
Age: 215 million years
Diameter: 100 km (62 mi)

A

Willamette Valley

B

Web Map

Impact Craters

Available at
www.saplinglearning.com

is an environment of great heat and great pressure. Common rocks found near Earth's surface, such as *olivine*, are dry. Deep inside the planet, however, olivine is heated to high temperatures and crushed under enormous pressures, which transform it into *wadsleyite* and *ringwoodite*. These rocks can incorporate up to 3% of their weight in water. This suggests that Earth's interior may hold as much water as the oceans.

The timing of Earth's earliest start of life is similarly a subject of debate. Fossils from Quebec, Canada, dated up to 4.3 billion years old, contain microscopic filaments and tubes that many scientists agree were Earth's very first life-forms. These tiny fossils are thought to be the remains of microbes that lived in deep-sea hydrothermal vents and consumed iron. Later, around 3.5 billion years ago, photosynthetic bacteria called **cyanobacteria** evolved. These bacteria transformed Earth's early atmosphere as they released oxygen as a waste product of photosynthesis (the conversion of sunlight energy to chemical energy). They are still found on Earth today, as the **Picture This** feature below illustrates.

When cyanobacteria evolved 3.5 billion years ago, the oxygen they produced was pulled out of the atmosphere as it combined with minerals in rocks. After that process began to slow, oxygen began to accumulate in the atmosphere about 2.4 billion years ago, in what is called *The Great Oxygenation Event*. Oxygen levels at that time were only 1% to 4%, but even such small amounts were significant as Earth had never had oxygen in its atmosphere before. (Oxygen makes up about 21% of Earth's atmosphere today.) Around 600 million years ago, oxygen levels reached about 15%, high enough to form a significant ozonosphere. The ozonosphere blocked the harmful UV rays of the Sun (see Section 2.3), making it possible for life to leave the protective cover of the oceans and move onto land.

Picture This Stromatolites

1.6 billion-year-old fossil stromatolites

A

Living stromatolites

B

AUSTRALIA

Shark Bay

Timor Sea

Great Australian Bight

(A. Courtesy Eve Stueeken/ University of St Andrews; B. BIOSPHOTO/Alamy)

The curved layers of the rock shown in photo A are fossilized cyanobacteria that lived in colonies called *stromatolites*. Stromatolite fossils are nicknamed "sliced cabbage" because of their appearance.

Stromatolites are among the earliest forms of life preserved in the fossil record, and, amazingly, they persist to this day. Living stromatolites can be found in Shark Bay, in Western Australia, as shown in photo B. The fossils shown here are about 2.7 billion years old. Stromatolites found in Western Australia date as far back as 3.5 billion years. The cyanobacteria secrete calcium carbonate that hardens into the colony structure. The cyanobacteria also secrete a gelatinous mucus that protects them from the harmful UV rays of the Sun.

Consider This

1. How did stromatolites make it possible for life to colonize land?

2. Areas with stromatolite colonies are found only in shallow water. Why do you think this is the case?

Table 12.1

LIFE-FORM	FIRST APPEARANCE
Homo sapiens	200 kya*
Flowering plants	160 mya*
Birds	160 mya
Mammals	200 mya
Dinosaurs	240 mya
Reptiles	320 mya
Amphibians	370 mya
Insects	416 mya
Land plants	475 mya
Fish	530 mya
Multicellular life	2 bya*

* kya = thousand years ago; mya = million years ago; bya = billion years ago

The Rise of Multicellular Life

Multicellular life first arose as early as 2 billion years ago. Before that, all life consisted of single-celled microbes such as cyanobacteria. Only in the last half billion years did life branch out into the forms we are familiar with today **(Table 12.1)**.

Dinosaurs appeared in the fossil record roughly 240 million years ago and then went extinct 66 million years ago, when a large asteroid hit the Yucatán Peninsula in Mexico. The earliest mammals appeared about 200 million years ago, but mammals did not flourish until after the dinosaurs became extinct. The earliest evidence of anatomically modern humans dates back to about 200,000 years ago. In the next section, we discuss how these events fit into Earth's history.

12.2 Deep History: Geologic Time

◎ Describe the major divisions of geologic time and explain how the age of ancient Earth material is determined.

Earth is 4.6 billion years old. Using the *geologic time scale*, we divide (and subdivide) this immense span of Earth history into time periods called *eons, eras, periods*, and *epochs*. **Figure 12.6A** on the following page illustrates Earth's four eons. The divisions of time are based on major geologic events, such as mass extinction events. The five mass extinctions in Earth's history were caused by many factors, including periods of intense volcanic activity, global climate change, and asteroid impacts. Episodes of rapid evolution of new species also provide a basis for defining new divisions of time. Using these great events in Earth's history, the four eons are subdivided into eras, periods, and epochs **(Figure 12.6B** on the following page**)**. The immensity of geologic time and relative timing of the events are nearly impossible to comprehend. One can gain a sense of perspective of Earth's history when geologic time is compressed into a single calendar year **(Figure 12.6C** on the following page**)**.

Slow and Gradual Change: Uniformitarianism

From our human perspective, the physical world appears unchanging. Yet this is not the case. The processes that shape Earth are similar to tree growth: It is nearly impossible to watch a tree grow, yet we know that even the largest trees were once very small. Similarly, the Grand Canyon was formed over millions of years as the land slowly rose and the Colorado River slowly carved into the rocks, eventually creating a mile-deep canyon. Most physical and biological systems that we see today resulted from small and gradual changes that accumulated over long periods of time. **Uniformitarianism** is the principle that the same gradual and nearly imperceptible processes operating now have operated in the past. One way to think about uniformitarianism is to consider the phrase "The present is the key to the past." The same processes that are active today were also active in the past, but they are usually too slow and gradual to perceive on human time scales.

Uniformitarianism underpins almost all the tectonic and erosional processes that we describe in the rest of this book. This key concept is also central to understanding the process of biological evolution that resulted in Earth's biodiversity. Through the accumulation of gradual changes, over the past 4.3 billion years the first single-celled microbes differentiated into every single living organism that is alive today, as well as every organism that that has ever lived.

Not all changes on Earth have been gradual. As we have discussed in this chapter, sudden catastrophic events also play important roles in Earth history. Collisions with asteroids, for example, have changed the course of history for the biosphere. After the dinosaurs became extinct, most likely due to an asteroid impact 66 million years ago, mammals evolved and filled the niches vacated by the dinosaurs. Catastrophic events occur intermittently, but uniformitarianism is constant.

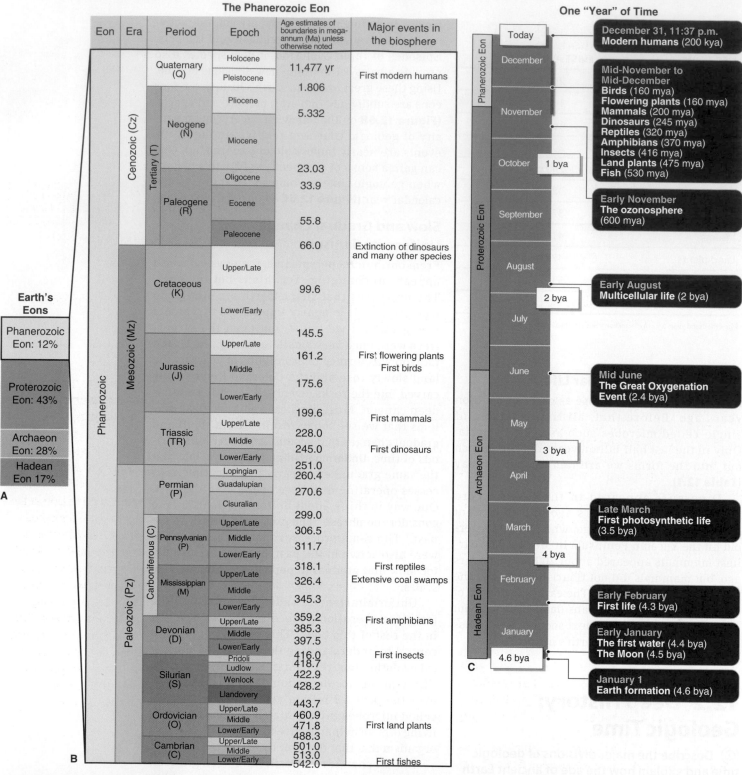

Figure 12.6 Geologic time. (A) Collectively Earth's first three eons, the Hadean Eon, Archaean Eon, and the Proterozoic Eon, constitute 88% of Earth's history. The top of the diagram shows the present day, and the lower parts of the diagram show increasingly older times. Our current eon, the Phanerozoic Eon, comprises 12% of Earth's history. Nearly all major evolutionary events of Earth history have happened in the Phanerozoic. (B) The Phanerozoic Eon, shown in detail here, is subdivided into numerous eras, periods, and epochs. We live in the Holocene Epoch, which is nested in the Quaternary Period, which in turn is nested the Cenozoic Era of the Phanerozoic Eon. Note that the Anthropocene Epoch is not included above the Holocene epoch because scientists have not yet formally recognized it as a new epoch. (C) In this graphic, Earth's entire history is compressed into a single year of time, with Earth and the solar system forming on January 1 (at the bottom of the graphic) and the present day being midnight on December 31. Each month on the calendar represents about 383 million years, and each second is about 145 years. Using this perspective, we can see that most life-forms we are familiar with arose only recently, no earlier than mid-"November" (kya = thousand years ago; mya = million years ago; bya = billion years ago).

Story Map

Geologic Time

Available at
www.saplinglearning.com

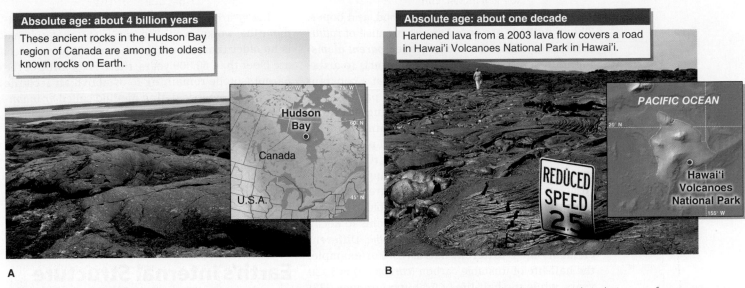

Absolute age: about 4 billion years

These ancient rocks in the Hudson Bay region of Canada are among the oldest known rocks on Earth.

Absolute age: about one decade

Hardened lava from a 2003 lava flow covers a road in Hawai'i Volcanoes National Park in Hawai'i.

A

B

Figure 12.7 Absolute age. The ancient rocks near Hudson Bay, Canada (A), have both an absolute age and a relative age far greater than the newly formed rocks encasing this road sign in Hawai'i (B). *(A. Jonathan O'Neil, University of Ottawa; B. G. Brad Lewis/Getty Images)*

How Do Scientists Date Earth Materials?

How do we know the ages of ancient events and materials? There are two ways to evaluate the age of an object or event. **Relative age** compares the age of one object or event with the age of another, without specifying how old either object is. **Absolute age** is specified in years before the present **(Figure 12.7)**.

Relative age accounts for the order of events. For example, if two trees are growing side by side and one is smaller, we might reasonably conclude that the smaller tree is younger. We may not know how old it is in absolute terms but only that it is younger relative to the larger tree.

This concept can be applied to rock layers in Earth's crust. A sequence of rock layers forms as layers of sediments are deposited, one after another, and eventually harden into sedimentary rock (a process described in more detail in Section 14.3). The **superposition principle** states that in such a sequence of rock layers, the oldest rocks are at the bottom, and the youngest rocks are at the top **(Figure 12.8)**.

Absolute age, which is given in actual numbers of years, is determined using various dating techniques. For example, tree-ring analysis (called *dendrochronology*) provides absolute ages of trees. Most trees create one new growth ring each year. Counting the growth rings in a cut tree allows an investigator to determine the absolute age of a tree.

Another way absolute ages can be determined is through **radiometric dating**, which involves using unstable atoms in materials to assign ages to those materials. Radiometric dating methods, which have been in wide use since the 1950s, provide a means by which we can determine the absolute ages of

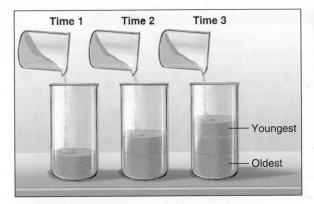

Time 1   Time 2   Time 3

Youngest

Oldest

A

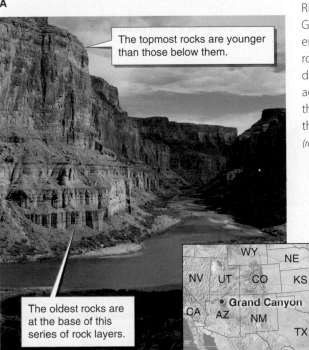

The topmost rocks are younger than those below them.

The oldest rocks are at the base of this series of rock layers.

WY NE

NV UT CO KS

CA AZ • Grand Canyon

NM

TX

Mexico

B

Figure 12.8 Relative age and the superposition principle. (A) The layers of sand poured into this glass represent sedimentary rock layers that have formed from sand over time. The layer at the bottom was deposited first and is older than the top layer. (B) The Colorado River cut the mile-deep Grand Canyon in northern Arizona, exposing rock layers that were first deposited 1.7 billion years ago. Rocks at the top of the canyon are younger than those below. *(robertharding/Alamy)*

ancient materials such as rocks, wood, and bones. These methods are based on the premise of *radioactive decay*: Unstable atoms (called *parent atoms*) found in some elements decay (convert) to a stable element (called *daughter atoms*) at a constant rate through time. For example, the parent atom *carbon-14* decays into the stable daughter atom *nitrogen-14*. Young objects that contain an unstable atom have high proportions of that parent atom because there has not been enough time for the parent form to decay to the stable, daughter form. In very old objects, the unstable parent atom is largely gone, having decayed to its stable form. The time it takes for half of the parent atoms to decay to the daughter atoms is called a *half-life*. Different elements have different half-lives. For example, the half-life of unstable carbon (*carbon-14*) is 5,730 years, while the half-life of uranium (*uranium*-238) is 4.5 billion years. **Figure 12.9** discusses the *radiocarbon dating* method.

Radiocarbon dating is a radiometric dating technique that works on once-living organic material that is no older than about 60,000 years. When objects are older than 60,000 years, there is too little of the parent isotope remaining to establish an accurate date. A radiometric dating method called *uranium-lead dating* can be used to assign absolute ages to Earth's oldest rocks and minerals. This technique measures the decay of unstable uranium-238 atoms to stable lead atoms. Using it, scientists have been able to date the oldest minerals on Earth to 4.4 billion years before the present. Those minerals are found in rocks from the Jack Hills in Western Australia.

12.3 Anatomy of a Planet: Earth's Internal Structure

◎ Describe the planet's internal structure and understand the importance of lithospheric plate movement.

Generations have been intrigued by the mystery of what lies beneath Earth's surface. In his *Divine Comedy*, which he completed in 1321, Dante Alighieri imagined Earth's interior as composed of nine concentric shells, each named after a human mortal sin, and a lake of ice at the center.

Earth's Hot Interior

The modern scientific understanding of Earth's interior does not include mortal sin or a frozen core. We do know, however, that Earth's interior is hot and under enormous pressure due to the weight of overlying rock. At Earth's center, the pressure is some 3,600,000 times greater than that found on the surface.

As depth increases, so does temperature; this pattern is called the *geothermal gradient*. How quickly the temperature increases with depth depends on the geographic location on Earth. In deep gold mines in South Africa, temperatures reach almost 54°C (130°F) only 3.5 km (2 mi) below the surface. Near volcanic activity the temperature increases far more rapidly. On average, at 40 km (24 mi) below the surface, the temperature climbs to about 500°C (930°F). The temperature at the core of the planet approaches 6,000°C (10,832°F)— nearly as hot as the Sun's surface.

Earth's interior heat is an increasingly important source of carbon-free energy that many countries are beginning to develop. The **Picture This** feature on the following page discusses renewable geothermal energy.

Why is Earth's interior hot? Two factors have created and continue to Create a hot Earth interior: radioactivity and friction. The most important source of Earth's internal heat is radioactive decay

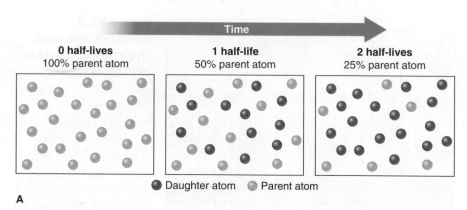

Time →

| 0 half-lives | 1 half-life | 2 half-lives |
| 100% parent atom | 50% parent atom | 25% parent atom |

● Daughter atom   ● Parent atom

A

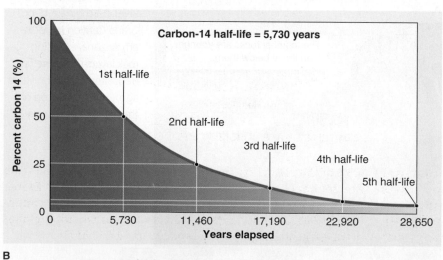

Carbon-14 half-life = 5,730 years

1st half-life
2nd half-life
3rd half-life
4th half-life
5th half-life

Percent carbon 14 (%)

100
50
25
0

0   5,730   11,460   17,190   22,920   28,650
Years elapsed

B

Figure 12.9 Radioactive half-life. (A) These three panels illustrate the conversion of parent atoms after two half-lives. After one half-life, 50% of the parent atoms are present. After two half-lives, 25% of the parent atoms are present. (B) The amount of carbon-14 diminishes during each consecutive half-life. After 5,730 years (one half-life), half of the parent carbon-14 atoms are still present. After 28,650 years (five half-lives) only 3% of the original parent carbon-14 atoms are still present.

of unstable elements, mainly potassium, uranium, and thorium. These elements give off heat during the decay process.

In addition to radioactivity, friction has played and continues to play a role in heating Earth's interior. When the Moon first formed, it was only about 24,300 km (15,000 mi) away from Earth (compared with today's average distance of 384,399 km [238,854 mi]). Because it was so close, the Moon's gravitational pull exerted enormous *lunar tidal forces*, which distorted Earth's shape and created heat from the friction of the movement of Earth's interior. Although it is farther away today, the Moon still exerts tidal forces on Earth, distorting its shape, and creating a small amount of heating through friction.

As we have seen, Earth formed through the accretion of dust and gas. The denser elements, such as iron and nickel, settled toward the middle of the planet beneath the less dense materials, such as silicon and magnesium. This early settling of materials by density also resulted in movement that generated friction and, therefore, heat.

How Do Scientists Know What Is Inside Earth?

The average density of Earth is about 5.5 g/cm^3, a figure calculated by dividing the mass of the planet by its volume. The average density of the crust's rocks, however, is less than this (about 2.92 g/cm^3), indicating that Earth's interior must be denser than its surface.

Picture This Using Earth's Heat

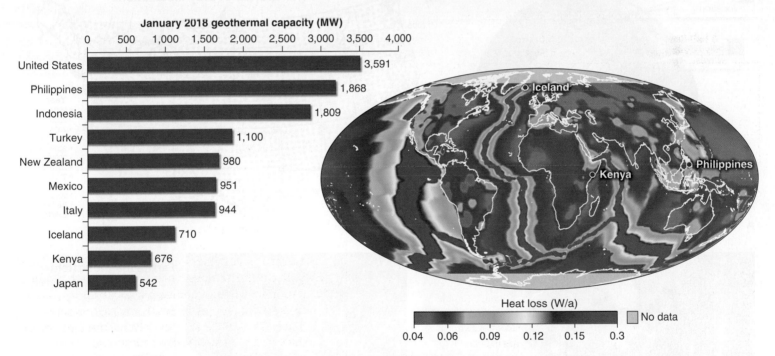

January 2018 geothermal capacity (MW)

United States	3,591
Philippines	1,868
Indonesia	1,809
Turkey	1,100
New Zealand	980
Mexico	951
Italy	944
Iceland	710
Kenya	676
Japan	542

Heat loss (W/a)   □ No data

0.04   0.06   0.09   0.12   0.15   0.3

This map portrays the thermal energy that Earth's interior emits through its crust in watts per square meter. Heat loss is greatest where the crust is thinnest (red areas), mostly along the plate boundaries. The graph data, in megawatts (MW), show the top 10 countries converting geothermal energy into electricity as of January 2018.

Like wind and solar energy, geothermal energy is sustainable, renewable, and carbon-free, and it has a low environmental impact. Kenya has an installed geothermal power capacity of 676 MW, meeting a little over half of its total electricity demand. The Philippines and Iceland generate 30% of their electricity from geothermal energy. Only 0.3% of the electricity in the United States is generated from Earth's heat. In absolute terms, however, the United States is the leader, having an installed geothermal power capacity of 3,591 megawatts (MW). That's 25% of the total global installed geothermal power capacity of 14,060 MW.

Consider This

1. Which of the three countries labeled on the map has the greatest geothermal energy capacity?

2. Which geographic region generates most of the geothermal power in the United States?

Most of what we know about Earth's deep interior has come from analysis of the behavior of **seismic waves**—energy released by earthquakes that travels through Earth's interior. An **earthquake** is a sudden shaking of the ground caused by movements of Earth's crust. We also have other direct and indirect evidence related to Earth's internal structure **(Figure 12.10)**.

Earth's Layers

Earth's atmosphere, oceans, and interior are arranged in layers. The heaviest and most dense layers and the greatest pressures are found at Earth's center. Outward from the center, the density of rocks and the pressure decrease. The atmosphere and oceans are a continuum with the solid Earth. Each layer is made of matter in solid,

Seismic waves

A powerful earthquake sends out seismic waves that travel through the planet. As seismic waves pass through materials of differing densities in the planet's interior, they are refracted (that is, bent and change speed).

Measuring seismic waves

Data about the paths of seismic waves help scientists understand the composition of Earth's deep interior. The EarthScope program, funded by the U.S. National Science Foundation, used more than 400 seismometers (instruments that detect seismic waves) to develop a detailed understanding of Earth's interior, down to 90 km (56 mi) beneath the continental United States. The EarthScope seismometer network was deployed in sections across the continental United States between 2004 and 2013 in steps, as mapped here.

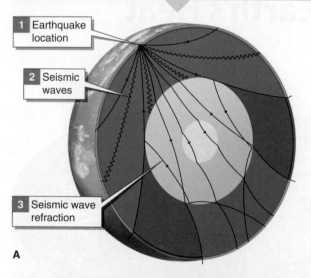

1 Earthquake location
2 Seismic waves
3 Seismic wave refraction

A

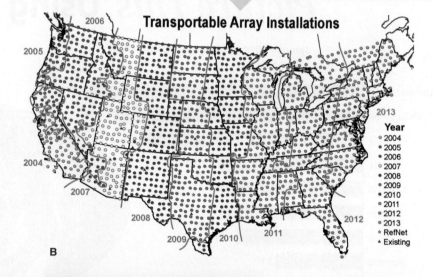

Transportable Array Installations

2006
2005
2004
2007
2008
2009
2010
2011
2012
2013

Year
○ 2004
● 2005
● 2006
○ 2007
● 2008
● 2009
● 2010
● 2011
● 2012
● 2013
★ RefNet
▲ Existing

B

Volcanic rock samples

Scientists sample and analyze lava that travels from Earth's interior to its surface through volcanoes. The chemistry of lava reflects the chemistry of Earth's interior.

C

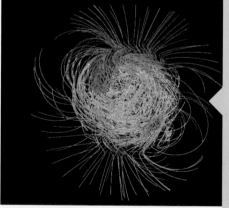

D

Earth's magnetic field

Magnetic fields are generated by circulating liquid iron. The presence of Earth's magnetic field indicates that the planet has a dynamic liquid metal core. The lines on this illustration represent alignment of Earth's magnetic field. The magnetic field flows from out of the South Pole, around the planet, and back into the North Pole. The blue lines show where the magnetic field is flowing out at the South Pole, and orange lines show where it is entering at the North Pole.

Figure 12.10 SCIENTIFIC INQUIRY: How do scientists study Earth's hidden interior? An understanding of Earth's internal composition and structure is essential to understanding geologic phenomena. Volcanic eruptions and earthquakes, for example, are directly related to the composition and movement of Earth's interior. *(B. Incorporated Research Institutions for Seismology; C. U.S. Dept. of Interior, U.S. Geological Survey; D. G. Glatzmaier, LANL/P. Roberts, UCLA/Science Source)*

liquid, or gaseous states. It is no accident that the layers are arranged this way—from most dense to least dense. As Earth accreted from dust and gas during the formation of the solar system, the densest materials settled deepest, and the least dense materials, like gases and water, rose to the surface. You may have observed the same thing happen when a glass of unfiltered orange juice is left undisturbed for several hours: The heavy pulp settles to the bottom. The atmosphere's gas molecules are the least dense and lightest of Earth's materials, and they therefore rest on top of the denser and heavier liquid oceans and the solid Earth. **Figure 12.11** illustrates Earth's interior layers.

The Inner Core

The center of Earth is 6,371 km (3,959 mi) below sea level. The solid **inner core**, which extends from Earth's center to about 5,150 km (3,200 mi) below Earth's surface, is a mixture of dense elements, mostly iron and nickel. Inner core temperatures approach 6,000°C (10,800°F). These temperatures would melt the inner core if not for the extremely high pressure found there. The extreme pressure in the inner core keeps the inner core solid. Atoms in molecules are forced closer together under great pressure. As a result, to break molecular bonds and melt rocks in the inner core, temperatures must be very high; temperatures need not be as high to melt rocks at or near Earth's surface, where there is much less pressure.

Animation

Earth's Interior Structure

Available at www.saplinglearning.com

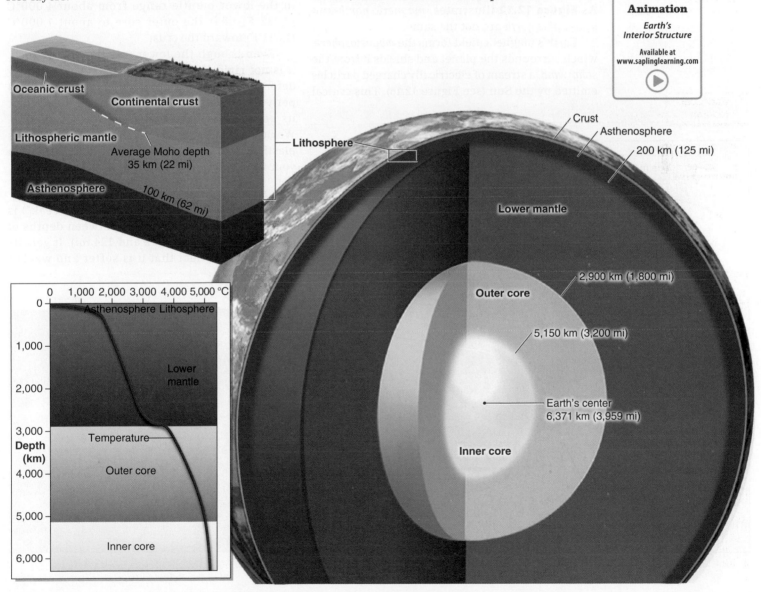

Figure 12.11 Earth's interior structure. The layers of Earth's interior are distinguished by the density of their rock material and by their chemical makeup. The top inset diagram shows the detail of the lithosphere. The thickness of the crust varies; the depths shown here are average depths below sea level. To increase their visibility, the lithospheric mantle and the crust are not drawn to relative scales. The bottom inset diagram shows how temperature increases with depth.

The Outer Core

The **outer core**, which surrounds the inner core, is composed of a liquid alloy of iron and nickel. Temperatures in the outer core begin around 4,000°C (7,232°F) and increase toward the inner core. The outer core extends to about 2,900 km (1,800 mi) below Earth's surface. The outer core is liquid because pressures within it are less than the pressure in the inner core; lower pressure allows the outer core to melt and flow. This circulating liquid metal generates electrical currents and creates Earth's *magnetic field*. Together, the inner and outer core make up about 15% of Earth's volume.

People and migratory animals use the magnetic field as a navigational aid (see Figure 8.5.). As **Figure 12.12** illustrates, *magnetic north* and *geographic north* are not the same.

Earth's magnetic field forms the *magnetosphere*, which surrounds the planet and shields it from the *solar wind*, a stream of electrically charged particles emitted by the Sun (see Figure 12.16). This critical function is explored further in the Geographic Perspectives, "Earth's Heat and the Biosphere," at the end of this chapter. Because the magnetic field enters the planet near the poles, the solar wind is able to reach the uppermost atmosphere there, causing it to light up, creating auroras, or northern and southern lights (see Section 2.3).

The Lower Mantle

The **lower mantle** is the layer of heated and slowly deforming solid rock that lies between the base of the crust and the outer core. The lower limit of the lower mantle lies 2,900 km (1,800 mi) below the surface; it extends upward to approximately 200 km (125 mi) below the surface. Temperatures in the lower mantle range from about 4,000°C (7,232°F) near the outer core to about 1,000°C (1,832°F) toward the crust.

Even though the lower mantle is solid rock, it is not rigid and unmovable. Instead, it slowly deforms and flows at a rate of about 15 cm (6 in) per year. When a solid is able to deform and flow, its behavior is described as *plastic* (or *ductile*). Warm (but unmelted) candle wax or beeswax is plastic in that it can be squeezed and shaped with your fingers once it is warmed.

The Asthenosphere

The **asthenosphere** (from the Greek for "weak") is the layer of the mantle found between depths of about 100 and 200 km (62 and 124 mi). It gets its name from the fact that it is softer and weaker

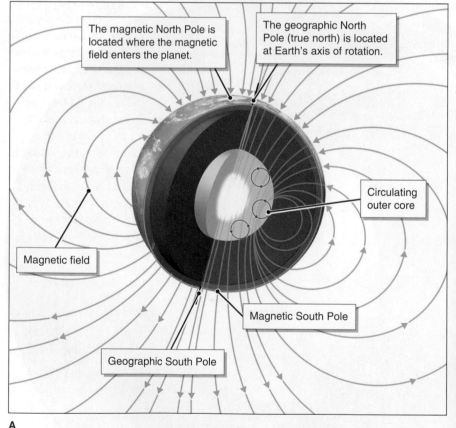

A

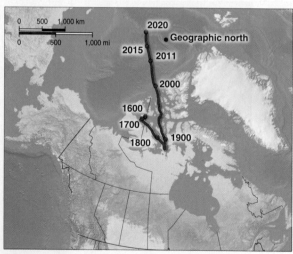

B

Figure 12.12 Geographic north and magnetic north. (A) The geographic North Pole (or true north) and the magnetic North Pole are located at different positions in the Northern Hemisphere. The magnetic north or south is found where compass needles no longer point in a single direction but instead spin chaotically. (B) Whereas the geographic North Pole is permanently fixed, the magnetic North Pole moves in unpredictable directions and speeds. Currently, it is moving about 55 km (34 mi) per year. In the past century, the location of the magnetic North Pole has migrated over 2,000 km (1,242 mi). This map tracks the position of the magnetic North Pole from 1600 to 2015. Its projected position for 2020 is plotted as well. If the magnetic North Pole continues at its present trajectory and speed, it will be in northern Russia by the end of this century.

than the lower mantle beneath it. Like the lower mantle, this layer of rock is solid but can flow under pressure. There is considerably less pressure in the asthenosphere than in the lower mantle below it. As a result, although the rocks here are still in a solid state, they are nearer to melting and are consequently weak and easily deformed.

The asthenosphere and the *lithospheric mantle* (discussed next) together make up the *upper mantle*. The lower mantle and upper mantle together make up the *mantle*, which accounts for about 84% of Earth's volume. The mantle is composed of *silicate* rocks (made mostly of minerals that contain silica) that are rich in iron and magnesium.

The Lithosphere

The **lithosphere** is relatively strong and consists of Earth's rigid crust and the rigid lithospheric mantle beneath it, extending to a depth of about 100 km (62 mi) on average (see Figure 12.11). Unlike the asthenosphere and lower mantle, the lithosphere is not plastic and does not deform and flow. Instead, when subjected to stresses from the moving asthenosphere beneath, it cracks and breaks, forming *lithospheric plates*. The movement of these plates drives earthquake and volcanic activity, which we explore in Chapter 15, "Geohazards: Volcanoes and Earthquakes."

The outermost portion of the lithosphere, the crust, is the part of Earth we walk on and that the atmosphere and oceans rest on. The crust is not melted by Earth's internal heat because it is in contact with the atmosphere and oceans, which are relatively cold. Although Earth's crust makes up about 1% of Earth's volume, it is very important to people. It is home. It is where we grow food, extract minerals and energy resources such as petroleum, and build our houses and cities.

There are two types of crust: continental crust and oceanic crust (see Figure 12.11). **Continental crust** makes up the continents. It is composed mainly of **granite**, a silica-rich rock made up of coarse grains. **Oceanic crust** lies beneath the oceans and is composed mainly of **basalt**, a dark, heavy, fine-grained volcanic rock.

Granite forms from magma that cools deep in the crust. **Magma** is melted rock that is below the surface of the crust. Chemically, granite is composed chiefly of silica, aluminum, potassium, calcium, and sodium. It is light in both weight (2.7 g/cm^3) and color compared with basalt, which weighs about 3 g/cm^3. But not all continental rocks are granite. Rocks found on the continents vary more in age and chemistry than do those in oceanic crust, and they come from different formative processes than the rocks found on the seafloor.

Basalt, in contrast, is formed from lava. **Lava** is magma that spills onto the surface of Earth's

crust. Chemically, basalt is composed mostly of compounds of silica that are high in iron and magnesium. When melted, basalt has low *viscosity*, which means it is runny and flows easily.

But how thick is the crust? The Croatian scientist Andrija Mohorovičić (1857–1936) identified the crust–mantle boundary by studying seismic waves and finding a point where rock density rapidly increases. That point, named the *Mohorovičić discontinuity*, or the **Moho**, after Mohorovičić, who formalized the concept in 1909, is the boundary that separates the crust from the lithospheric mantle below. The depth of the Moho is about 35 km (22 mi) on average, but it varies depending on the type of crust it lies beneath.

The Moho is deepest beneath continental crust and shallowest beneath oceanic crust. Beneath the oceans the Moho is found at about 7 km (4 mi) in depth. Beneath the continents it is found at about 70 km (40 mi) in depth, but the depth varies with continental topography **(Figure 12.13)**. The Moho is deepest beneath high mountains

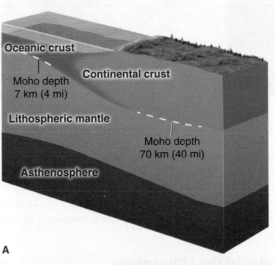

A

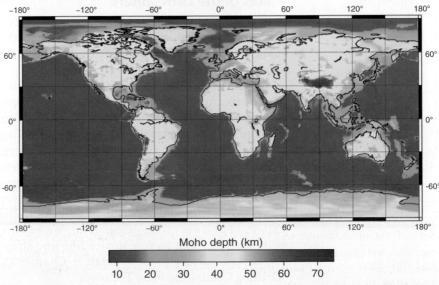

B

Figure 12.13 Depth of the Moho. (A) The Moho follows the contours of the crust's surface but in reverse. It is deepest where there are high mountains and most shallow beneath the oceans. (B) This map shows that the thickest crust (shown in red) is found beneath the Tibetan Plateau and Himalayas in Eurasia and the Andes in South America. Ocean crust (shown mostly in blue) is the thinnest.

Figure 12.14 Lithospheric plates. (A) The San Andreas Fault runs down the length of western central and southern California. It defines the boundary between the North American plate and the Pacific plate. Relative to the North American plate, the Pacific plate is moving north. This photo was taken in southern California looking southeast from an altitude of about 1,680 m (5,500 ft). (B) There are 14 major lithospheric plates as well as several minor plates. Black triangles point toward the direction of movement. For example, the Nazca plate is moving eastward into the South American plate. Red boundaries show where plates are moving apart, and black-line boundaries show where plates are moving laterally past each other. *(© David Lynch)*

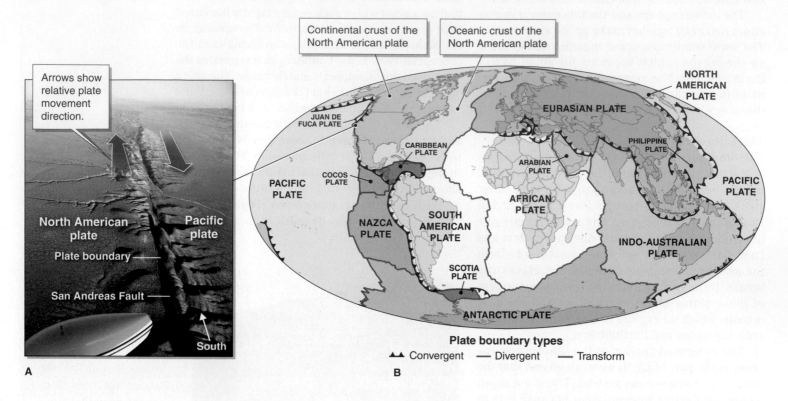

and much less deep in areas where the continental crust is being stretched by tectonic processes. Overall, the thickness of Earth's crust is 0.01% to 1% of the radius of the planet—about the same as the thickness of an apple peel compared to the rest of the apple.

Plates of the Lithosphere

The lithosphere is broken into pieces, or *plates*, each of which moves over Earth's surface as the mantle beneath it slowly circulates. **Figure 12.14** shows Earth's major lithospheric plates.

As the lithospheric plates move in the process of plate tectonics, they buckle, bend, and warp into mountains, split into valleys, and break, causing earthquakes. **Plate tectonics** is a theory addressing the origin, movement, and recycling of lithospheric plates and the landforms that result. (If there were no plate tectonics, Earth's surface would become flat because weathering and erosion would wear down mountains, and sediments would fill in the valleys (a process called *denudation*). We explore plate tectonics and its role in building and shaping Earth's surface in more detail in Chapter 13, "Drifting Continents: Plate Tectonics."

GEOGRAPHIC PERSPECTIVES
12.4 Earth's Heat and the Biosphere

◎ Assess the connection between Earth's internal heat and life on Earth.

What would happen if Earth's interior cooled? How does Earth's internal heat support its biosphere? We cannot cool down Earth's interior to see what happens. We can, however, learn from other planets. If Earth's interior cooled, the planet would become a lifeless body, much like Mars or the Moon today.

Lessons from Mars

Today, as far as we can tell, Mars is barren, bitter cold, and devoid of life. There is water on Mars, but it is permanently frozen at both poles and hidden deep within the Martian soils. The average temperature on Mars at midlatitudes is about −45°C (−50°F). Atmospheric pressure is only a hundredth of Earth's—so low that a person could not survive for a full minute without a pressurized suit. In fact, in terms of atmospheric pressure and temperature, a typical day on Mars might be similar to a typical day 120 km (75 mi) or so above Earth's surface—where one would find little breathable air, bitter cold, and an utterly inhospitable environment.

Earth stands in sharp contrast to Mars. Whereas Mars appears reddish-beige from space, Earth is blue and white, due to its immense liquid water oceans and the clouds in its oxygen-rich atmosphere **(Figure 12.15)**. But we now know that Mars, too, once had oceans and a thick atmosphere.

What caused Earth and Mars to take such very different paths early in the history of the solar system? Why did Mars lose its water and atmosphere? Why was Earth's retained? The answer lies just below our feet.

Earth's Outer Core and Life

Our tour of the internal Earth in this chapter revealed the layered structure of the planet. Recall that Earth's outer core is made predominantly of liquid iron and nickel and that it generates a magnetic field as it circulates deep within Earth's interior. Also recall that Earth's magnetic field forms a shell-like **magnetosphere** that envelops the planet and protects it from the solar wind **(Figure 12.16)**.

Sudden increases in the solar wind create problems for people on Earth, including cancer, interruptions in satellite and cell-phone communications, and drought. Because of the damaging effects of the solar wind on organisms, life on land probably would not be possible without the magnetic field. Life would reside only within the protective water of the oceans.

The magnetic field plays a second important role in making Earth a habitable planet. Scientists know that Mars once had a stronger magnetic field than it does today. Evidence for this conclusion lies in rocks found on Mars and on Earth that have become magnetized as they cooled from molten lava. Satellites with *magnetometers* (instruments that detect magnetism) orbiting Mars can detect this magnetism in surface rocks. On Earth, new rocks still become magnetized today. On Mars, only rocks several billion years old or older are magnetized; newer rocks are not. Several billion years ago, the Martian magnetic field shut down, and the rocks on Mars stopped becoming magnetized as they cooled.

Scientists do not know why Mars lost its magnetic field. One possibility is that the interior of Mars cooled enough that its liquid metal core stopped circulating and creating a magnetic field. Another possibility is that Mars was hit by a series of very large asteroids, which heated the surface, disrupting the core-to-surface temperature

Figure 12.15 Earth and Mars. The true colors and relative sizes of Earth and Mars are shown here. Robotic missions to Mars and satellite imaging of the surface have revealed evidence that Mars once had extensive liquid oceans and rivers like those of Earth. It also had a thick atmosphere like Earth's. So different was ancient Mars from what we see today that scientists are hopeful of finding evidence that life once existed there. *(NASA/JPL)*

Figure 12.16 The magnetosphere. The white lines represent the solar wind, which is deflected by Earth's magnetosphere, shown as a blue elongated bubble around Earth.

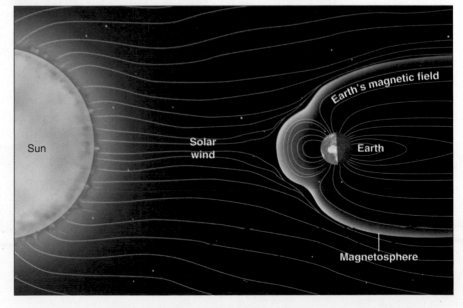

contrast and the core's ability to generate a magnetic field. Once the magnetic field of Mars was weakened, the atmosphere was exposed to the solar wind. Mars's atmosphere and oceans were then stripped away by the charged particles of the solar wind, leaving the planet the cold and sterile place it is today.

If Earth lost its magnetic field, over many millions of years its atmosphere and oceans would also be stripped away by the solar wind. Because Earth's internal heat is maintained, the planet generates the magnetic field that prevents this from happening.

The Global Thermostat

Earth's internal heat makes the planet hospitable to life in a third way—through plate tectonics. The movement of the lithospheric plates creates a global thermostat that keeps Earth's atmospheric temperature in a range suitable for life. Because Earth's interior is hot, the mantle and asthenosphere slowly move. This in turn moves the plates of the lithosphere, creating plate tectonics.

Scientists believe that plate tectonics has played a central role in keeping Earth from getting too hot like Venus or too cold like Mars in two ways: by building mountain ranges that weather and erode and by creating volcanic activity. Like a thermostat in a house that keeps the temperature stable, **weathering** (the process by which solid

rock is dissolved and broken apart into smaller fragments) and volcanic activity keep the atmosphere's overall temperature stable in a range that is favorable for life.

Weathering of Mountain Ranges Cools Earth

Earth's internal heat drives movement of the plates in the crust, and this builds mountains. As mountain ranges weather and erode, carbon dioxide (CO_2) in the atmosphere is transferred to the ocean basins. This happens because CO_2 from the atmosphere binds with materials in the weathered rocks. Streams transport and deposit the carbon-containing weathered materials in the oceans. There they form carbon-rich sediments and rocks (or *carbonates*) on the seafloor **(Figure 12.17)**.

Recall from Chapter 7, "The Changing Climate," that carbon dioxide is a greenhouse gas that strongly affects the atmosphere's temperature. When CO_2 in the atmosphere increases, Earth warms; when it decreases, Earth cools. If the weathering and erosion process occurred in isolation over geologic time, the atmosphere's CO_2 would be drawn down as carbon became locked away in the ocean basins. Earth would eventually become too cold to support life. However, because of volcanic activity, this doesn't happen.

Volcanoes Warm Earth

While the weathering and erosion of mountains are removing CO_2 from the atmosphere, volcanoes are adding CO_2 to the atmosphere. Volcanic activity is driven by Earth's heat and plate movement. The carbonate rocks on the seafloor formed by weathering and erosion of mountains are subducted (drawn down) along the margins of the continents and deep into the mantle (see Section 13.3). The resulting volcanic eruptions along the continental margins return the carbon from the rocks to the atmosphere, where it recombines with oxygen to form CO_2. There is therefore a balance of CO_2 leaving the atmosphere through weathering and CO_2 entering the atmosphere from volcanoes (see Figure 12.17).

Several periods in Earth's ancient history—the most recent occurring 650 million years ago—were so cold that the entire planet was covered in ice. This most likely happened because of weathering uplifted mountain ranges. During these *Snowball Earth* periods, the absence of liquid water slowed the process of weathering. The removal of CO_2 from the atmosphere consequently also slowed. Volcanoes, continually emitting CO_2, eventually warmed the atmosphere and brought Earth out of these cold periods. Likewise, were it not for the effects of weathering, volcanoes would continually

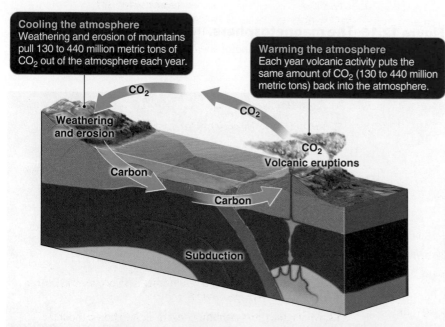

Cooling the atmosphere
Weathering and erosion of mountains pull 130 to 440 million metric tons of CO_2 out of the atmosphere each year.

Warming the atmosphere
Each year volcanic activity puts the same amount of CO_2 (130 to 440 million metric tons) back into the atmosphere.

CO_2

CO_2

Weathering and erosion

CO_2
Volcanic eruptions

Carbon

Carbon

Subduction

Figure 12.17 Global thermostat. Each year, weathering removes about the same amount of CO_2 that volcanoes put into the atmosphere, creating a long-term equilibrium of CO_2 in the atmosphere.

emit CO_2, and the atmosphere would accumulate CO_2 and become inhospitably hot.

Were it not for its hot interior, Earth would be as lifeless and barren as the Moon. Earth's internal heat generates the magnetic field. The magnetic field shields life from the harmful solar wind and prevents the atmosphere from being stripped away. Internal heat also drives the movement of the lithospheric plates, which builds mountains and creates volcanic activity. Together volcanic emissions and the effects of weathering keep the atmosphere's temperature in a range suitable for life.

Chapter 12 **Study Guide**

Focus Points

12.1 Earth Formation

- **Formation of the solar system:** The solar system, including Earth, was formed through the accretion of dust and gas.
- **Formation of the atmosphere and oceans:** Earth's atmosphere and oceans formed from gases emitted by volcanoes early in Earth's history.

12.2 Deep History: Geologic Time

- **The geologic time scale:** Earth is 4.6 billion years old. Its history is divided into epochs, periods, eras, and eons, based on major geologic events that occurred in the past.
- **Uniformitarianism:** Earth's physical systems operate mostly under gradual changes that accumulated over geologic time.
- **Absolute and relative age:** Relative age accounts for the order of events. Absolute age is the specific age of an object.

12.3 Anatomy of a Planet: Earth's Internal Structure

- **Earth's interior:** Earth's interior is hot and under enormous pressure.
- **Earth's interior layers:** Earth is composed of an inner core and an outer core, the lower mantle, and the lithosphere, which includes the crust.
- **Types of crust:** There are two types of crust: oceanic and continental. Oceanic crust is thin, and continental crust is thick.

- **Plate tectonics:** Convection in the mantle drives the movement of lithospheric plates and creates surface topography and volcanic activity.

12.4 Geographic Perspectives: Earth's Heat and the Biosphere

- Earth's internal heat energy protects Earth's atmosphere from the solar wind, and it generates plate tectonics, which stabilizes Earth's temperature.

Key Terms

absolute age, 379
Anthropocene, 372
asthenosphere, 384
basalt, 385
continental crust, 385
crust, 375
cyanobacteria, 376
earthquake, 382
granite, 385
inner core, 383
lava, 385
lithosphere, 385
lower mantle, 384

magma, 385
magnetosphere, 387
Moho, 385
oceanic crust, 385
outer core, 384
plate tectonics, 386
radiometric dating, 379
relative age, 379
seismic wave, 382
superposition principle, 379
uniformitarianism, 377
volcano, 375
weathering, 388

Concept Review

The Human Sphere: The Anthropocene

1. What is the Anthropocene? Do scientists agree about when it started? Explain.

12.1 Earth Formation

2. What is a solar system? How many planets are in Earth's solar system?

3. What is a galaxy? What are galaxies composed of? What is the name of the galaxy to which Earth belongs?

4. What units are used to measure the size of a galaxy? How big is the galaxy to which Earth belongs?

5. What is the universe? What was the Big Bang, and when did it happen?

6. Describe the sequence of events that formed Earth's solar system.

7. How did Earth's Moon form? When did it form?

8. How did the atmosphere and oceans form?

9. When did life on Earth first evolve? What kind of life was it? What did it consume?

12.2 Deep History: Geologic Time

10. What is the geologic time scale? What marks the beginning or end of a division of time?

11. How old is Earth?

12. Review some of the major events that have occurred in Earth's history and when they occurred.

13. What is uniformitarianism? Use it to explain how a mountain range forms.

14. Compare and contrast relative age with absolute age. How is each determined?

15. Explain what a half-life is and how it is applied to radiometric dating.

12.3 Anatomy of a Planet: Earth's Internal Structure

16. Why is Earth's interior hot?

17. What evidence is used to determine Earth's internal structure?

18. List the layers of Earth's interior, starting from the core and moving outward to the crust.

19. What generates Earth's magnetic field? What does the magnetic field deflect?

20. What is the difference between the magnetic North Pole and the geographic North Pole?

21. What is the Moho?

22. What are the two kinds of crust, and where is each of them found?

23. Where on the planet is the crust thickest? Where is it thinnest?

24. What is plate tectonics?

25. Why is the lithosphere broken into plates?

12.4 Geographic Perspectives: Earth's Heat and the Biosphere

26. Explain the connection between the loss of the Martian magnetic field and the loss of that planet's early atmosphere and oceans.

27. In what two ways is Earth's outer core, made of liquid iron and nickel, essential for life?

28. What is the "global thermostat"? Explain how it works.

29. How is plate tectonics essential to the functioning of the global thermostat?

Critical-Thinking Questions

1. What observations would you need to make to determine the approximate depth of the Moho where you live?

2. What physical features can you identify where you live that were formed under the principle of uniformitarianism?

3. New evidence shows that most stars have orbiting planets and that planets are more common than stars. Do you think that there might be life on other planets besides Earth? What conditions would be necessary to support life?

4. How do human activities affect the global thermostat?

Test Yourself

Take this quiz to test your chapter knowledge.

1. True or false? Earth is 4.6 billion years old.

2. True or false? Deep drill holes reveal the internal structure of Earth.

3. True or false? Oceanic crust is composed of basalt and is heavier than continental crust.

4. True or false? Earth's atmosphere and oceans originated from volcanic gases.

5. True or false? The magnetic North Pole is also called geographic north.

6. Multiple choice: Which of the following generates Earth's magnetic field?

a. the lithosphere
b. the crust
c. the outer core
d. the inner core

7. Multiple choice: Which of the following is an example of the superposition principle?

a. mountain building over long stretches of geologic time
b. Earth's interior heat
c. layered sediments with the oldest layers at the bottom
d. the process of plate movement

8. Multiple choice: Which of the following lists the interior layers of Earth in the correct order, from the surface downward?

a. lithosphere, asthenosphere, lower mantle, inner core
b. asthenosphere, lower mantle, lithosphere, inner core
c. crust, lithospheric mantle, lower mantle, asthenosphere
d. crust, lower mantle, asthenosphere, outer core

9. Fill in the blank: The principle of _____ is based on the accumulation of slow and gradual changes over geologic time.

10. Fill in the blank: Radiometric dating provides a(n) _____ for materials.

Online Geographic Analysis

Exploring Events in Earth History

EarthViewer allows you to explore events in Earth history.

Activity

Go to http://www.hhmi.org/biointeractive/earthviewer. Scroll down and click "Launch Online Version." Watch the brief tutorial or dismiss it. Next, select the "Ancient Earth" tab at the bottom on the far left. Then click the "Charts" button, which is also at the bottom of the window. Select "Oxygen" and click the "Charts" button again to minimize the menu. Grab and drag the silver "needle" on the left to experiment with moving it vertically through time. Notice how oxygen levels change through time. The geologic time is displayed at the top left of the window.

1. At what point did oxygen levels permanently rise above 2%? What was this event called?

2. What was the peak oxygen level value?

3. When did the peak oxygen level occur?

Close out the oxygen chart. In the "View" menu at bottom, activate "Biological Events" and "Geological Events." Drag the silver needle down to the bottom of the screen and then move slowly up through time to view major events.

4. When did water first form?

5. How old are the oldest rocks, and where are they found?

6. When did Earth's plates begin moving in the process of plate tectonics?

7. What was the "Cambrian explosion"? When was it, and what happened?

8. When did the first land vertebrates occur? What type of organism were they?

9. What was *Pangaea*, and when did it occur?

10. When did the genus *Homo* first occur? Where? When did "anatomically modern" humans show up?

Picture This. *Your Turn*

Earth in Cross Section

Label each part of Earth's interior described in this chapter. Put one star next to layers that are plastic and deformable. Put two stars next to the layer that generates the magnetic field. You should use each of these terms only once.

1. Asthenosphere

2. Continental crust

3. Inner core

4. Lithosphere

5. Lower mantle

6. Moho

7. Oceanic crust

8. Outer core

9. Lithospheric mantle

Continent   Ocean

Drifting Continents: Plate Tectonics

Chapter Outline *and Learning Goals* ◎

Morning light strikes Mount Fitz Roy in Patagonia. Standing at 3,405 m (11,171 ft) and 49 degrees south latitude, Mount Fitz Roy straddles the border of Chile and Argentina and is part of the southern Andes mountain range. The southern Andes are a volcanic mountain range formed where the Antarctic lithospheric plate is subducting beneath the South American plate.

(DPK-Photo/Alamy)

To learn more about subduction and the landforms it produces, turn to Section 13.3.

THE HUMAN SPHERE

Life on Earth's Shifting Crust

For better or for worse, human lives and global economies are closely connected to Earth's physical processes. This is particularly true when it comes to plate tectonics. Recall from Chapter 12, "Earth History, Earth Interior," that Earth's internal heat energy drives the movement of the lithospheric plates. The movement of the plates lifts mountains, builds volcanoes, and generates powerful earthquakes. This plate movement often negatively affects humans. Earthquakes, which are a result of plate movement, range from minor annoyances to major disasters. Small tremors can shake the windows and cause chandeliers to gently sway. Large earthquakes can reduce cities to rubble and generate enormous ocean waves, called tsunamis, which have the potential to destroy coastal areas (see the Human Sphere section in Chapter 15, "Geohazards: Volcanoes and Earthquakes").

Volcanic eruptions similarly influence the lives of people. Some volcanic eruptions are catastrophic events that bury whole cities with ash and lava. Most, however, go unnoticed or are inconveniences. In April and May 2010, the eruption of Eyjafjallajökull volcano in Iceland caused the cancellation of some 100,000 airline flights, at considerable expense. The next year, Grímsvötn, another Icelandic volcano **(Figure 13.1)**, grounded thousands more flights.

Movement of the crust also benefits people in subtle ways that don't make headlines. There really is no direct benefit to earthquakes, but volcanoes are essential elements of many human societies. Volcanic eruptions and their lava flows build new land. The land surface of many island nations, such as Japan, Indonesia, and much of New Zealand, was built by volcanoes. Volcanoes also, indirectly, help to feed the world: Ash from erupting volcanoes settles to the ground and creates some of the world's most fertile agricultural soils.

Figure 13.1 Volcanic ash from Grímsvötn.
On May 22, 2011, Grímsvötn sent plumes of ash into the air that traveled over northern Europe, causing the cancellation of thousands of flights. Aircraft engines fail when they take in volcanic ash, so planes must be rerouted or grounded until the ash clears.   (© Sigurlaug Linnet/AFP/Getty Images)

13.1 Continental Drift: Wegener's Theory

◎ **Compare the theory of continental drift with the theory of plate tectonics.**

Some 300 million years ago, Earth would hardly have been recognizable to us from space. At that time, Earth had only one ocean and one *super-continent*, called **Pangaea** (meaning "whole land"), which was a single large landmass. The ocean that surrounded Pangaea, called the *Panthalassic Ocean,* was much larger than the Pacific Ocean. Eventually the Pangaea landmass was *rifted* (split apart) as a result of convection in the asthenosphere beneath it. The transition from Pangaea to today's continental configuration occurred in two major steps: (1) the opening of the North Atlantic Ocean and the *Tethys Sea* and (2) the formation of the Atlantic Ocean **(Figure 13.2)**.

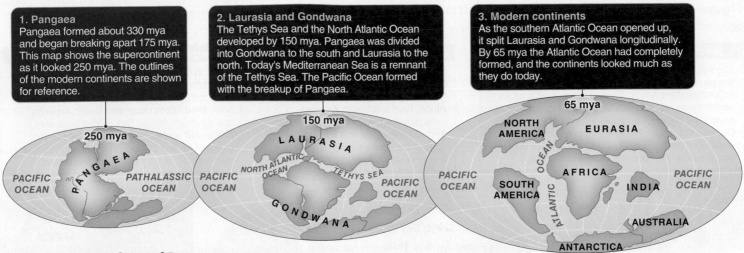

1. Pangaea
Pangaea formed about 330 mya and began breaking apart 175 mya. This map shows the supercontinent as it looked 250 mya. The outlines of the modern continents are shown for reference.

2. Laurasia and Gondwana
The Tethys Sea and the North Atlantic Ocean developed by 150 mya. Pangaea was divided into Gondwana to the south and Laurasia to the north. Today's Mediterranean Sea is a remnant of the Tethys Sea. The Pacific Ocean formed with the breakup of Pangaea.

3. Modern continents
As the southern Atlantic Ocean opened up, it split Laurasia and Gondwana longitudinally. By 65 mya the Atlantic Ocean had completely formed, and the continents looked much as they do today.

Figure 13.2 Breakup of Pangaea.

The opening of the North Atlantic Ocean and the Tethys Sea, about 150 million years ago, created two large landmasses: Laurasia to the north and Gondwana to the south. **Laurasia** consisted of the landmasses that would become North America, Greenland, and Eurasia. **Gondwana** consisted of the landmasses that would become South America, Australia, Africa, India, and Antarctica. By about 65 million years ago the Atlantic was fully opened. As the Atlantic Ocean basin developed, the shapes of the continents took their modern form (see Figure 13.2).

Pangaea is only the most recent supercontinent. *Rodinia* formed about 1 billion years ago. There were other supercontinents before Rodinia, each of which broke apart and rearranged into a new supercontinent. The details about these ancient supercontinents are poorly known.

It took a long time and much scientific evidence for scientists to accept that Earth's crust is dynamic and always on the move. As recently as the early 1960s, almost all scientists believed that the crust was fixed and immovable. After all, they thought, how could entire continents made of solid rock move thousands of miles? As we will see, the growing body of evidence supporting the idea of a dynamic lithosphere eventually became indisputable.

The first person to propose a theory of continental movement was the German meteorologist Alfred Lothar Wegener (1880–1930; his name is pronounced VEG-en-er). Wegener's theory of **continental drift** proposed that continents move slowly across Earth's surface. One piece of evidence that motivated Wegener to pursue his theory was the jigsaw-puzzle fit between the east coast of South America and the west coast of Africa **(Figure 13.3)**.

Wegener was not the first to be struck by how the continental outlines fit together, but he was the first to assemble a formal theory proposing that the continents shift. Wegener supported his theory using evidence from matching rock types on separated continents, deposits of glacial gravels, and fossil remains of organisms **(Figure 13.4** on the following page**)**.

Wegener formally proposed his theory in 1912 and published it in his book *The Origins of Continents and Oceans*. Wegener's theory was dismissed by his professional colleagues because no

Animation
Breakup of Pangaea
Available at
www.saplinglearning.com
▶

Story Map
Breakup of Pangaea
Available at
www.saplinglearning.com
📍

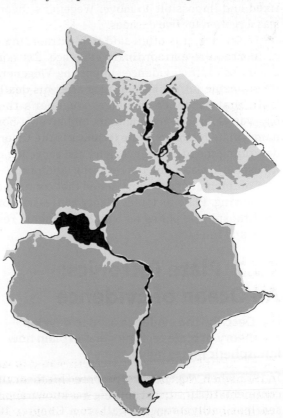

**Figure 13.3
Evidence for continental drift: The fit of South America and Africa.** If the Atlantic Ocean were closed, North America and South America would fit tightly against Europe, Greenland, and Africa.

Figure 13.4 Evidence for continental drift: Rock types and fossils. Evidence for movement of the continents includes the geographic patterns of rock types, fossils, and glacial gravels.

1. Matching mountains and rock types	2. Glacial gravels	3. The geography of fossils
The Appalachian Mountains in the eastern United States were once connected to the Atlas Mountains in northwestern Africa. Eastern Greenland's rocks match those found in Scotland and Norway. In other words, the rocks types on either side of the Atlantic Ocean are the same in age and composition and were once part of a single continuous mountain range before the continents were split apart.	Similarities among the ancient *tillites* (rocks formed from glacial gravel deposits) in southern Africa, southern South America, India, southern Australia, and Antarctica suggest that those continents were once joined and covered with ice.	Geographically separated fossils of now-extinct organisms suggest formerly connected continents. The extinct *Mesosaurus*, shown here, lived in both southern South America and southern Africa. This reptile lived only in freshwater and could not have crossed the ocean. The distribution of fossils suggests that the two continents were once connected.

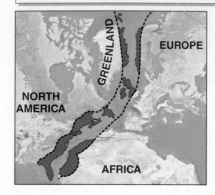
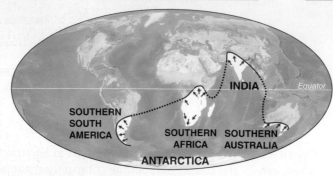
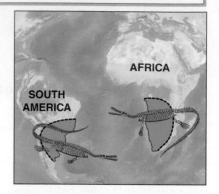

forces sufficiently powerful to move continents such great distances were known.

Wegener's biography reveals that scientific revolutions are not neat and tidy. Established views are hard to change. At the time, scientists rejected Wegener's idea because it conflicted with their *paradigm* (conceptual model) of the crust being fixed and impossible to move. Wegener's theory was forgotten for five decades.

In science, it is often said that extraordinary claims require extraordinary evidence. Extraordinary scientific evidence supporting Wegener's theory came only later, decades after his death. With that evidence in place, Wegener's theory was embraced by scientists and given a new name: plate tectonics. **Plate tectonics** is the theory describing the origin, movement, and recycling of lithospheric plates and the resulting landforms.

Continental drift theory stated that continents are moving. Plate tectonics theory addresses how the lithospheric plates on which continents rest move and interact.

13.2 Plate Tectonics: An Ocean of Evidence

◎ Describe the evidence used to develop the theory of plate tectonics and explain how lithospheric plates move.

In 1912, when Wegener first proposed his theory of continental drift, almost nothing was known about seafloor bathymetry. (Recall from Chapter 10,

"Ocean Ecosystems," that bathymetry is the measurement of depth and topographic features beneath the surface of a body of water.) Following World War II, emerging seafloor research methods using sonar provided new evidence that led to the development of the theory of plate tectonics.

One important research tool was the ship *Glomar Challenger*. The *Challenger* was able to drill 1.6 km (1 mi) long samples from seafloor rock in water 5 km (3 mi) deep. Researchers aboard this vessel sampled rock materials in the Atlantic Ocean along a *transect* (line) that crossed the mid-ocean ridges (see Section 10.1). In addition to the material the *Challenger* obtained, scientific data about the rock found at mid-ocean ridges began to pour in from other sources. Scientists working within the paradigm that the continents are fixed and unmoving could not make sense of this new information about mid-ocean ridges.

Harry Hess is considered one of the founders of the theory of plate tectonics. In 1962, he formalized the theory of *seafloor spreading*, which eventually developed into the broader theory of plate tectonics in 1967. Hess used data on the ages of seafloor rocks and on patterns of magnetism in those rocks to propose that new oceanic crust forms at mid-ocean ridges and then moves apart. Hess's theory and additional scientific evidence (as illustrated in **Figure 13.5**) led not only to a consensus among scientists that the lithospheric plates are in motion but also to a complex model showing how it is possible for them to move.

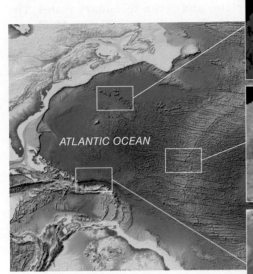

ATLANTIC OCEAN

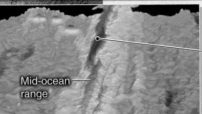

Seamounts
Flat-topped, submerged mountains called seamounts were discovered using sonar (see Section 10.4). Seamounts were found to be predominantly volcanic in origin. This seamount is approximately 5 km (3 mi) in diameter at the base.
Significance: The flat tops of many seamounts indicate that they were once subject to erosion at the sea surface before they were carried to deep water by plate movement.

Mid-ocean ridges
Bathymetric mapping with sonar in the 1950s revealed mid-ocean ridges.
Significance: Sampling of rocks from these mountain ranges revealed that they are volcanic in origin. These volcanic mountain ranges occur where new oceanic crust is being formed on the seafloor.

Mid-ocean range

Deep-sea trenches
Deep-sea trenches are located on the edges of continents and island chains (see Section 10.1). The Puerto Rico Trench is the deepest location in the Caribbean Sea. Florida is located in the upper right.
Significance: Deep-sea trenches reveal the locations where oceanic crust dives deep into the mantle in the process of subduction.

Florida

Puerto Rico

Normal magnetism
At Time 1 (T₁), new rocks cool from lava at a mid-ocean ridge and record the direction of magnetic north.

Mid-ocean ridge

Normal polarity

T₁

Magma

Time

Reversed magnetism
At Time 2 (T₂), new rocks record the new direction of magnetic north after a magnetic reversal.

Reversed polarity

T₁   T₂   T₁

Time

Normal magnetism
At Time 3 (T₃), new rocks record the direction of magnetic north after another magnetic reversal.

Normal polarity

T₁ T₂ T₃ T₂ T₁

Ages of seafloor rocks
The age of the seafloor increases away from a mid-ocean ridge (black lines). Few seafloor rocks are older than 150 million years (areas in blue and purple).
Significance: This pattern of older crust farther from mid-ocean ridges indicates that the crust forms at mid-ocean ridges and then moves away from them.

Seafloor magnetic reversals
Every 200,000 years, on average, for unknown reasons, Earth's magnetic field flips: North becomes south, and south becomes north. As lava hardens into rock, iron particles in the lava record the location of magnetic north like a compass. Samples of seafloor rocks reveal a matching pattern of magnetic reversals in rocks that are located at equal distances in opposite directions from mid-ocean ridges.
Significance: These matching magnetic stripes on both sides of mid-ocean ridges can be explained only if oceanic crust is moving away from mid-ocean ridges through time.

Mid-ocean ridge

Figure 13.5 SCIENTIFIC INQUIRY: What evidence supports plate tectonics theory? A wide range of data and observations, gathered from the ocean basins with many scientific tools, supports the theory of plate tectonics.
(Above left. NCEI/NOAA; top right. Gulf of Alaska 2004. NOAA, WHOI, and the Alvin Group; center right. SPL/Science Source; bottom right. Woods Hole Science Center, U.S. Geological Survey; below left. NOAA/NGDC)

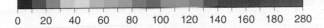

Age of oceanic crust (millions of years)

0   20   40   60   80   100   120   140   160   180   280

Animation
Seafloor Magnetic Reversals
Available at
www.saplinglearning.com

According to this model, new lithosphere forms at mid-ocean ridges, and old lithosphere is recycled deep into the mantle through **subduction**, a process in which oceanic lithosphere of one plate bends and dives into the mantle beneath another plate.

Plate Tectonics: A Fractured Crust

Earth's lithosphere is fractured into 14 major plates and many smaller fragments, all of which move independently of one another across the surface of the planet. The places where lithospheric plates meet are **plate boundaries**. The large plates are called *primary plates,* and smaller plates are called *secondary plates.* There are seven primary plates and seven secondary plates. The largest plate is the Pacific plate **(Figure 13.6)**.

The plates lie atop and are set in motion by the convecting mantle. The plates are not floating on a sea of molten rock. With the exception of a few isolated pockets of melted rock, the rocks of the asthenosphere are solid. The rocks of the asthenosphere are hot enough that they would melt if they were at the surface, but because they are deep in the Earth, they are under great pressure and, as a result, do not melt. Instead, the asthenosphere

Figure 13.6 Plates of the lithosphere. The 14 major lithospheric plates are shown here. Black triangles are used to show where one plate is moving toward another; the triangles point in the direction the plate is moving. For example, the Nazca plate is moving eastward toward the South American plate. At each transform boundary, plates move laterally past one another.

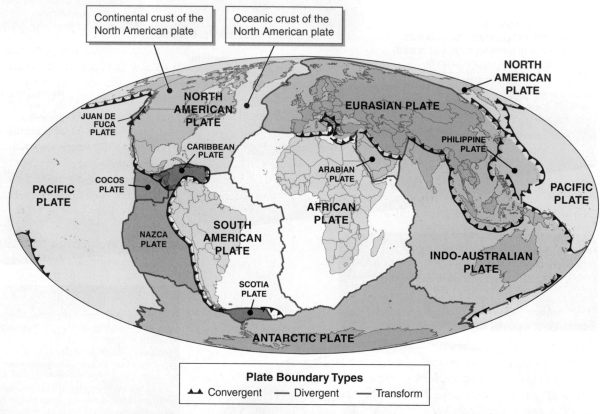

Web Map

Plate Boundary Types

Available at
www.saplinglearning.com

Plate Names	
PRIMARY PLATES	**SECONDARY PLATES**
African plate	Arabian plate
Antarctic plate	Caribbean plate
Eurasian plate	Cocos plate
Indo-Australian plate	Juan de Fuca plate
North American plate	Nazca plate
Pacific plate	Philippine plate
South American plate	Scotia plate

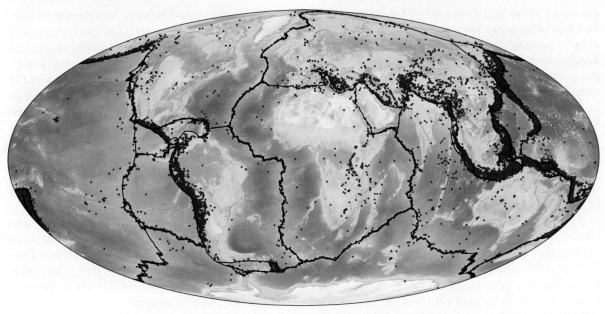

· Earthquake   —— Plate boundaries

Figure 13.7 Global earthquake patterns. This map shows the global geographic pattern of earthquakes. Although some earthquakes do occur far from plate boundaries, the majority of them occur on plate boundaries.

Web Map

Earthquakes and Plate Boundaries

Available at
www.saplinglearning.com

is solid rock that is plastic (also called *ductile*) and can deform and slowly flow, like warm (but not melted) wax (see Section 12.3).

The majority of the 14 lithospheric plates are composed of both continental and oceanic crust. Notice in Figure 13.6 that some plate boundaries follow the outlines of the continents and some do not. Geographers distinguish between active continental margins (borders) and passive continental margins. An **active continental margin** follows a plate boundary. A **passive continental margin** has a broad sloping continental shelf and does not coincide with a plate boundary. The west coasts of North America and South America, for example, are active continental margins. In contrast, their east coasts are passive continental margins.

How Do We Know Where the Plate Boundaries Are?

The geographic pattern of earthquakes reveals the locations of plate boundaries. After World War II, scientists began placing extensive networks of seismometers around the ocean basins to measure and record earthquake activity. To their astonishment, they found that earthquakes are not randomly distributed, as they had anticipated. Instead, as shown in **Figure 13.7**, earthquakes occur in a geographic pattern that delineates the outlines of the lithospheric plates.

How Do the Plates Move?

When the theory of plate tectonics was first proposed, it was assumed that the underlying mantle grabbed and moved the overlying lithospheric plates by friction as the mantle flowed and convected. Such models could not explain the

directions of plate movement, however, because it is impossible to arrange mantle convection patterns in a way that accounts for the direction of observed plate movements. For that reason, scientists modified the theory by concluding that there must be additional forces moving the lithospheric plates.

Today, plate movement is thought to be a result of three factors: ridge push, mantle drag, and slab pull **(Figure 13.8)**. **Ridge push** is the process by which magma rising along a mid-ocean ridge lifts oceanic lithosphere and forms a slope. The higher

Video

What Drives Plate Tectonics?

Available at
www.saplinglearning.com

Figure 13.8 Subducting lithosphere and slab pull.

About 90% of plate movement may be caused by slab pull, where the lithosphere is subducting into the mantle. Slab pull can even work against and override the effects of mantle drag. The process of slab pull is similar to a long chain sliding off the edge of a tabletop, where the chain gains momentum as gravity pulls it downward.

Animation

Subducting Lithosphere and Slab Pull

Available at
www.saplinglearning.com

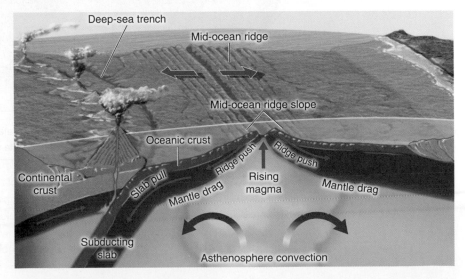

Deep-sea trench
Mid-ocean ridge
Mid-ocean ridge slope
Oceanic crust
Continental crust
Ridge push
Ridge push
Rising magma
Slab pull
Mantle drag
Mantle drag
Subducting slab
Asthenosphere convection

elevation of the mid-ocean ridge causes gravity to pull the plate away from the ridge center. **Mantle drag** is caused by convection within the asthenosphere. As the asthenosphere moves, friction drags the lithosphere above it. In the process of **slab pull**, the weight of the subducting portion of a plate (a *slab*) accelerates plate movement by pulling the plate deeper into the mantle.

How Fast Do the Plates Move?

Two kinds of plate velocity can be measured: relative and absolute. *Relative plate velocity* is the speed of a plate in relation to the speed of another plate, and *absolute plate velocity* is the speed of a plate in relation to a fixed object, such as the center of Earth.

Imagine riding your bike at 10 km/h (6 mph) past someone who is standing still on the sidewalk. To that person, your speed is 10 km/h. That is your *absolute* velocity. If another bicyclist passes you traveling 12 km/h (7.5 mph), that rider's absolute velocity is 12 km/h (to the person standing on the sidewalk). But from your perspective on your bicycle, that rider is traveling only 2 km/h (1.2 mph)—which is that rider's *relative* velocity.

Similarly, if two lithospheric plates each have an absolute velocity of 5 cm (2 in) per year, and they are moving away from each other, their relative velocity to each other is 10 cm (4 in) per year. If they are

traveling in the same direction, however, their relative velocity is 0 cm per year because they are both moving in the same direction at the same speed, like two bicyclists traveling together at the same speed.

Satellite GPS technology is so accurate today that even the smallest movements in the plates can be measured with precision. Today, plate movement can be measured with millimeter-level accuracy using orbiting satellites and GPS technologies. **Figure 13.9** shows that the Pacific and Nazca plates near Easter Island are the fastest-moving plates, each having a relative plate velocity of 18 cm (7 in) per year.

Because the plates are moving, they are drifting off the geographic coordinates that were established in the late 1800s. The **Picture This** feature on the following page discusses how Australia's geographic coordinates are updated to keep up with plate movement.

Why Is the Theory of Plate Tectonics Important?

Plate tectonics is one of the most important scientific theories developed in the twentieth century because it explains a wide range of geophysical phenomena. Plate tectonics is described as a *unifying theory* because it gathers together seemingly unrelated phenomena and puts them together

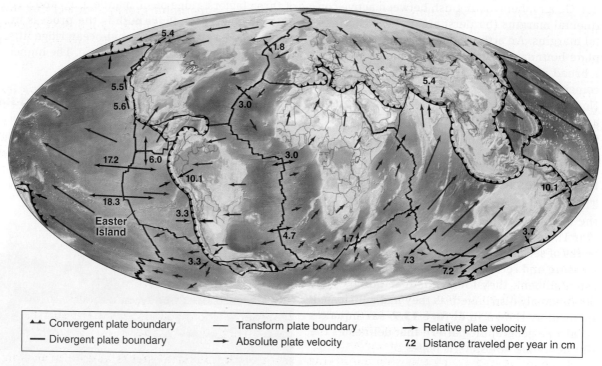

| ▲▲ Convergent plate boundary | —— Transform plate boundary | ⟶ Relative plate velocity |
| —— Divergent plate boundary | ⟶ Absolute plate velocity | **7.2** Distance traveled per year in cm |

Figure 13.9 Plate movement velocity. Each of the 14 lithospheric plates moves independently and at a different speed. Red arrows show absolute plate velocities; black arrows show relative plate velocities. Longer arrows indicate faster movement. The numbers are distances traveled each year, in centimeters. Symbols for different types of plate boundaries are given below the map: Triangles show where plates converge. Red lines show where plates diverge. Thin black lines show where plates slip past one another sideways at transform plate boundaries.

Picture This Chasing Australia

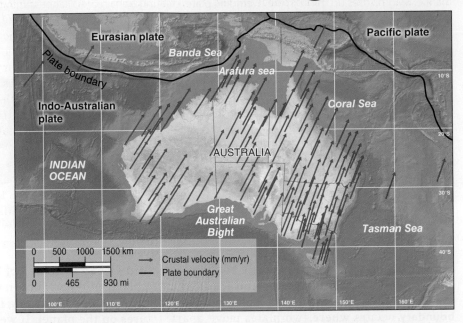

The Indo-Australian plate is among the lithosphere's fastest-moving plates. Australia, on the Indo-Australian plate, is moving at an average absolute rate of about 6.9 cm (2.7 in) per year. (In comparison, the North American plate moves at an absolute rate of 2.5 cm [1 in] each year.) This rapid movement of Australia makes its geographic grid coordinate system increasingly inaccurate every passing year. Imagine that a stake is driven into the ground on the Indo-Australian plate. When it is hammered in, its precise latitude and longitude coordinates are measured. After only 1 year, it will be almost 7 cm away from the initial coordinate; after a decade, it will be nearly 70 cm (27 in) off the coordinate. Like the stake, the entire Australian continent is moving. Consequently, officials have to periodically reset the geographic coordinates of the country to keep up with its movement.

In 1994, a correction of 200 m (656 ft) had to be applied. Then between 1994 and 2016 Australia moved another 1.5 m (5 ft) to the north. On January 1, 2017, officials adjusted Australia's geographic grid to 1.8 m (5.9 ft) north, where Australia is predicted to be on January 1, 2020. Australia's geographic grid system will therefore not be precisely accurate until January 2020. And, shortly after, it will move past that location, and the process will start over.

Consider This

1. Observe the map and the arrow legend. Longer arrows represent faster movement. Is all of Australia moving at the same speed? Explain.
2. What practical aspects of life would Australia's movement disrupt if its grid coordinate system weren't reset? (Hint: see Section 1.4.)

---

under the same umbrella of explanation. The formation and geographic locations of volcanoes, earthquake patterns, rock types and ages, island formation, lake formation, undersea mountain ranges, deep-sea trenches, continental mountain ranges, the shapes and outlines of continents, and the geographic ranges of many plants and animals can all be explain by this single unifying theory. Plate tectonics even plays a central role in scientists' understanding of why Earth's climate changes over time periods of millions of years. The next section explores how this theory explains the formation of all mountain ranges.

13.3 Plate Boundary Landforms

◎ Describe the three types of plate boundaries and the landforms that result from each plate boundary type.

Most of the physical stress and deformation of Earth's crust occur at the boundaries of lithospheric plates. The way the plates move in relationship to one another determines the extent of stress and deformation—and, therefore, the resulting landforms. There are three types

Figure 13.10 Three types of plate boundaries. (A) In this divergent plate boundary scenario, two plates move apart as new oceanic crust forms at a mid-ocean ridge. (B) In this convergent plate boundary scenario, one plate subducts beneath another, forming a deep-sea trench. The subducting plate is recycled deep into the mantle. Magma rises up through the overriding plate, creating a chain of volcanoes. (C) A transform plate boundary forms where two plates slide past each other.

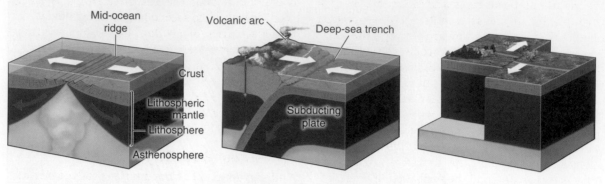

A. Divergent plate boundary   **B.** Convergent plate boundary   **C.** Transform plate boundary

of plate boundaries: divergent plate boundaries, convergent plate boundaries, and transform plate boundaries **(Figure 13.10)**. A **divergent plate boundary** occurs where two plates move apart. A **convergent plate boundary** occurs where two plates move toward each other. A **transform plate boundary** occurs where one plate slips laterally past another.

Divergent Plate Boundaries

Telltale landforms result when two plates move apart at divergent plate boundaries. These include mid-ocean ridges (in the oceans) and rift valleys (on continents), which we describe in detail in the material that follows.

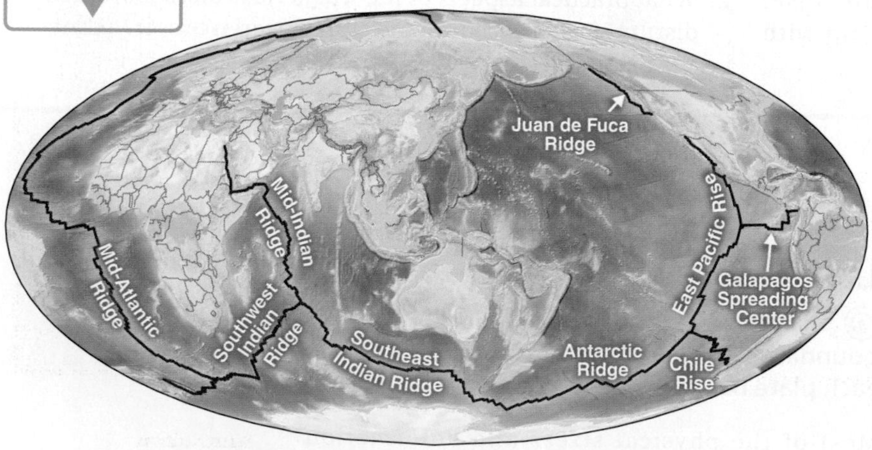

Figure 13.11 Mid-ocean ridges. Diverging plates create mid-ocean ridges (shown with a red line) that run nearly from pole to pole. These underwater mountain ranges are far larger than any mountain range on dry land.

Mid-Ocean Ridges

Although they are hidden beneath the ocean, mid-ocean ridges are among the largest surface features on the planet. If we could see through the waters of the oceans, we would see the mountain ranges of mid-ocean ridges stretching out beyond the horizon, running the entire length of the ocean basins **(Figure 13.11)**.

Along the mid-ocean ridges, the mantle convects (flows upward) to the base of the lithosphere. The pressure deep in the mantle is much greater than the pressure near the crust. As mantle material migrates upward, the pressure on it decreases. As the pressure decreases, the melting point of the rock is lowered, and the rock melts into magma. This process of melting hot mantle through changes in pressure is called **decompression melting**. It occurs as mantle material is brought to a shallow depth in the lithosphere.

Magma has a lower density—and is thus more buoyant—than the solid rock of the crust that surrounds it. As a result, magma rises upward. It may cool beneath the crust's surface. If it is extruded (pushed out) onto the seafloor, however, it forms submarine volcanoes that make up the mid-ocean ridges.

This rising magma creates the ridge push effect described in the previous section and forms new oceanic crust. Gravity pulls the plates at the mid-ocean ridge farther apart, allowing more magma to rise into the crust from below and continue the process (see Figure 13.8). This process is called *seafloor spreading*. The seafloor spreading model explains why the youngest seafloor is always found along the mid-ocean ridges, where it has recently formed, and it explains how ocean basins grow and become wider.

Rifts

A **rift** is a region where continental crust is stretching and splitting. This happens as the asthenosphere beneath pulls the crust apart, just as it does at a mid-ocean ridge. As the crust is split apart, the mantle material rises and melts into magma through decompression melting. The buoyant magma then migrates upward through the crust to create volcanoes. This process results in a **rift valley**: a long valley with volcanoes formed by rifting of continental crust. Because they form closed basins, rift valleys often fill with freshwater to form very deep lakes. As rift valleys continue to open and deepen, they may be flooded with seawater, creating an inland sea and, with more time, a new ocean basin. **Figure 13.12A** illustrates the sequence of rifting and the formation of a rift valley and a new ocean basin.

Rift valley lakes are among the largest bodies of freshwater in the world. Lake Tanganyika in the East African Rift System is 1,470 m (4,820 ft) deep **(Figure 13.12B)**. Because of its great depth, it is the world's second largest lake by volume of water (though not by surface area). Lake Malawi, at 706 m (2,316 ft) deep, is the world's sixth largest lake by volume. Also on the East African Rift System, Lake Victoria is the second largest lake in the world by surface area (68,800 km^2 or 26,600 mi^2). Only Lake Superior in Michigan is larger in surface area.

The world's largest lake by volume is Lake Baikal, in eastern Russia—also a result of rifting. It is 1,642 m (5,387 ft) deep and contains 20% of Earth's surface freshwater. It is also thought to be the oldest lake in the world, at 25 million years old.

Convergent Plate Boundaries

While divergent plate boundaries produce important landforms, perhaps the most remarkable and dangerous tectonic features on Earth are made where plates converge.

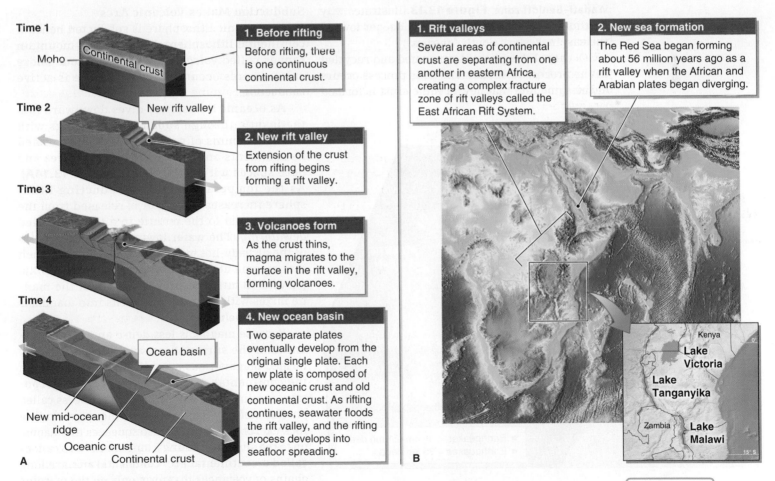

Time 1

Moho — Continental crust

1. Before rifting
Before rifting, there is one continuous continental crust.

Time 2

New rift valley

2. New rift valley
Extension of the crust from rifting begins forming a rift valley.

Time 3

3. Volcanoes form
As the crust thins, magma migrates to the surface in the rift valley, forming volcanoes.

Time 4

Ocean basin

4. New ocean basin
Two separate plates eventually develop from the original single plate. Each new plate is composed of new oceanic crust and old continental crust. As rifting continues, seawater floods the rift valley, and the rifting process develops into seafloor spreading.

New mid-ocean ridge
Oceanic crust
Continental crust

A

1. Rift valleys
Several areas of continental crust are separating from one another in eastern Africa, creating a complex fracture zone of rift valleys called the East African Rift System.

2. New sea formation
The Red Sea began forming about 56 million years ago as a rift valley when the African and Arabian plates began diverging.

B

Kenya
Lake Victoria
Lake Tanganyika
Zambia
Lake Malawi
15° S

Animation
Rifting
Available at www.saplinglearning.com

Figure 13.12 Rifting. (A) Rifting of a continent can eventually lead to the development of a new ocean basin. (B) The East African Rift System has been opening at a rate of about 3 mm (0.12 in) per year over the past 40 million years. The rifting started in the north, near the Red Sea, and has been working its way southward, like a big zipper unzipping the margin of the African continent. Lakes Victoria, Tanganyika, and Malawi are deep rift valleys that have filled with water (inset map). *(NOAA/NGDC)*

Subduction

We live on a planet rife with natural hazards. Among the most deadly of those natural hazards are the powerful and explosive volcanoes that form along convergent plate boundaries where subduction is occurring (see Figure 13.10B).

As plates move past one other during subduction, they generate earthquakes, some of which can be very powerful. The *driving force* that moves the plates is mantle convection fueled by geothermal heat. The *resisting force* that inhibits plate movement is friction. The plates become locked together through friction. When the driving force exceeds the resisting force, the plates suddenly become unlocked and move, releasing energy that radiates out as seismic waves (see Section 15.3), and an earthquake occurs. The locations of earthquakes follow the profiles of subducting plates. Thus, subduction zones produce a sloping pattern of increasingly deep earthquake centers, or *foci* (pronounced FOHS-eye; plural of *focus*), called a **Wadati–Benioff zone. Figure 13.13** illustrates why earthquakes become progressively deeper toward the east in western South America.

Oceanic lithosphere is consumed and recycled in the process of subduction. This process occurs at the same rate that new oceanic crust is formed at mid-ocean ridges. Almost all oceanic crust is younger than 150 million years because of this recycling process (see Figure 13.5).

Continental lithosphere, in contrast to oceanic lithosphere, is not subducted because it is less dense than oceanic lithosphere and thus too buoyant. Because of this difference in density, the continents have been free of subduction for most of Earth's history. Therefore, the oldest and most tectonically stable portions of the lithosphere are found on continents in areas called *continental shields*. The oldest materials from a continent are 4.4-billion-year-old zircon crystals found in continental rocks from Western Australia. Based on the dating of these materials, the Australian Shield must have formed just after early Earth's surface cooled from a molten state 4.6 billion years ago (see also Section 12.1). Scientists are still uncertain about the details of how the continents first formed.

Subduction Makes Volcanic Arcs

Where oceanic lithosphere is subducted beneath continental lithosphere, long volcanic mountain ranges called *volcanic arcs* are created. These volcanic arcs occur only where there is active subduction.

As oceanic lithosphere dives downward into the mantle through subduction, it brings with it large amounts of seawater that has saturated its sediments and infiltrated any fractures and pore spaces within the rocks **(Figure 13.14A)**. As the temperature of the subducting lithosphere increases, the water is released from the oceanic crust to the mantle in a process called *dewatering*. The water lowers the melting point of the already-hot mantle rock, causing it to melt into magma at depths between 100 and 125 km (62 and 78 mi). The process in which the mantle mixes with seawater and melts into magma is called **flux melting**.

Because magma is less dense and more buoyant than the surrounding solid rock, the newly formed magma rises up, melting its way through the continental crust. As it does so, it incorporates continental crust material in a process called *partial melting*. The weight of the surrounding rock exerts pressure and also squeezes the magma up to the surface, forming a long chain of volcanoes called a continental arc. **Continental arcs** are long chains of volcanoes that form only on the margins of continents where subduction is occurring.

The subducting lithosphere has to reach the asthenosphere, at a depth of at least 100 km (62 mi) and 700°C (1,300°F), before it can melt into magma.

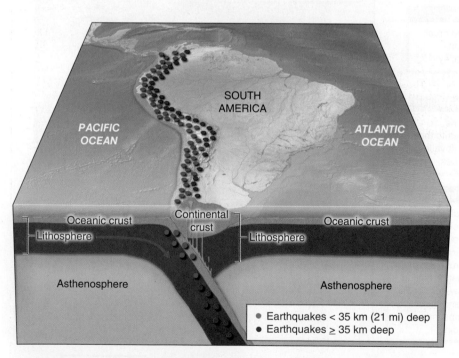

Figure 13.13 Wadati–Benioff zones. The Nazca plate is subducting beneath the South American plate, creating a pattern of progressively deeper earthquakes toward the east. Note that the thickness of the lithosphere in this drawing is greatly exaggerated in relationship to the size of South America.

Because the subducting crust has to reach a depth of 100 km, the distance of the continental arc from the deep-sea trench depends on the dip angle of the subducting lithosphere. On average, subducting plates dip at an angle of about 45 degrees as they dive into the mantle. As a result, the continental arc lies about 100 km (62 mi) from the deep-sea trench (see Figure 13.14A). If the dip angle of the subducting plate is less (say 30 degrees), the continental arc lies about 200 km (124 mi) from the deep-sea trench. If the angle is greater, the continental arc is closer to the trench.

As oceanic lithosphere is subducted beneath continental lithosphere, a deep-sea trench forms where the oceanic lithosphere bends downward into the mantle. During subduction, sediments and rocks on the seafloor are scraped up against the edge of the overriding continent, forming a folded mass of sediments and rock known as an **accretionary prism**. All subduction zones have deep-sea trenches and accretionary prisms, and these features never form anywhere else **(Figure 13.14B)**.

Subduction can also occur where the oceanic lithosphere of two plates converges. In the case of oceanic–oceanic convergence, a similar sequence of events occurs: Oceanic lithosphere of one plate subducts beneath oceanic lithosphere of another plate, resulting in an arc of volcanoes. The denser plate subducts beneath the slightly less dense plate. The difference, however, is that the region of subduction is submerged beneath the ocean, and the volcanoes form on the seafloor; magma produced through flux melting forms volcanoes that eventually rise above sea level to form an island arc. An **island arc** is a chain of volcanic islands formed where oceanic lithosphere of one plate is subducting beneath oceanic lithosphere of

Figure 13.14 Continental arcs. (A) Deep-sea trenches, accretionary prisms, and continental arcs form where oceanic lithosphere is subducting beneath continental lithosphere. (B) The Cascade Range in the Pacific Northwest of the United States and southern Canada is a continental arc. It is the result of the Juan de Fuca plate subducting beneath the North American plate; the black line with triangles shows the plate boundary and the direction of plate movement. Each red triangle is an active volcano. Mount Rainier and Mount Saint Helens (labeled) are two particularly active volcanoes in the Cascade Range.

Animation

Subduction

Available at
www.saplinglearning.com

▶

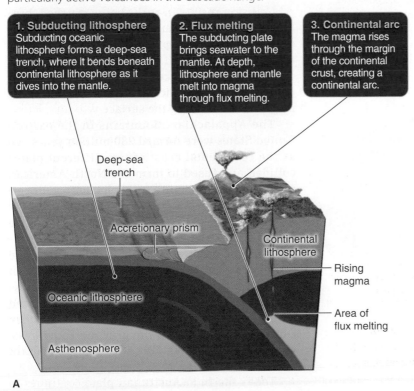

1. **Subducting lithosphere** Subducting oceanic lithosphere forms a deep-sea trench, where it bends beneath continental lithosphere as it dives into the mantle.

2. **Flux melting** The subducting plate brings seawater to the mantle. At depth, lithosphere and mantle melt into magma through flux melting.

3. **Continental arc** The magma rises through the margin of the continental crust, creating a continental arc.

Deep-sea trench

Accretionary prism

Continental lithosphere

Rising magma

Oceanic lithosphere

Area of flux melting

Asthenosphere

A

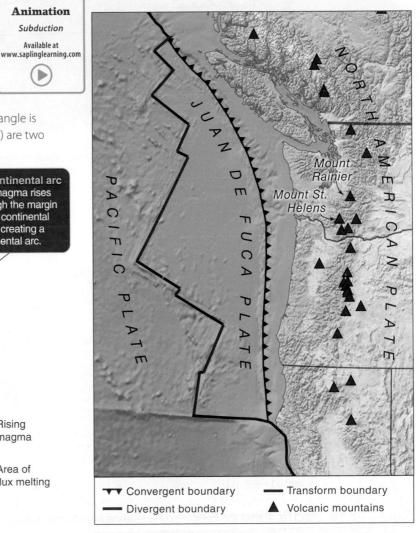

B

▼▼ Convergent boundary   —— Transform boundary
—— Divergent boundary   ▲ Volcanic mountains

another plate **(Figure 13.15)**. Examples of island arcs include Tonga, the western Aleutian Islands of Alaska, northeastern Japan, East Timor, the Philippines, and most of Indonesia.

Volcanic arcs, formed by both continental–oceanic and oceanic–oceanic subduction, occur around the perimeter of the Pacific Ocean, from Chile to Mexico to California, north to Alaska, and south again to Japan, the Philippines, and Indonesia, all the way to Antarctica. Collectively,

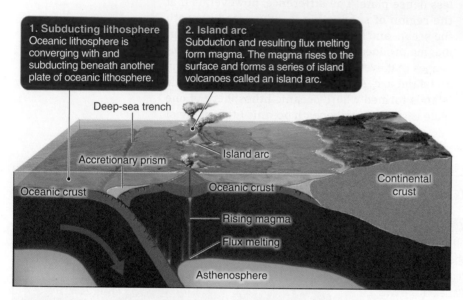

Figure 13.15 Island arcs. In this scenario, two oceanic plates are colliding. The relatively dense one subducts. As it sinks into the mantle, it is heated and melts through flux melting. The resulting buoyant magma rises through the overlying oceanic lithosphere and forms an island arc.

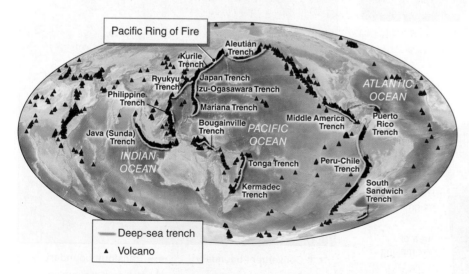

Figure 13.16 Pacific Ring of Fire. The Pacific Ring of Fire spans some 40,000 km (25,000 mi) and encompasses over 75% of the world's active volcanoes. With the exception of the Puerto Rico Trench, all deep-sea trench systems are found in the Pacific Ring of Fire.

these volcanic mountain ranges make up the **Pacific Ring of Fire**: a zone of volcanically active mountain chains resulting from subduction on the margins of the Pacific Ocean **(Figure 13.16)**.

Collision

Portions of the Himalayas—Earth's highest mountain range—are composed of rocks that were once seafloor sediments. Fossils of ancient marine (ocean) species, such as whales, corals, and shelled mollusks, are found preserved in rocks high in the Himalayas. This remarkable fact makes sense in the context of plate tectonics.

In our discussion of Figure 13.6, we noted that lithospheric plates are composed of both oceanic and continental crust. As a plate with oceanic crust continues to subduct, eventually all of its heavy oceanic crust subducts into the mantle, but the portion of that same plate that includes continental crust never subducts because it is too light and buoyant. In this scenario for a lithospheric plate that is composed of both oceanic and continental crust, once the oceanic crust is subducted, the continental crust attached to the plate collides with the continental crust of an opposing plate. **Collision** occurs where the continental crust of two plates converges. Because continental crust is too buoyant to subduct, collision forms mountain belts as the crust is heaved upward. As illustrated in **Figure 13.17**, collision can follow the closing of an ocean basin.

Collision results in rocks being crushed, broken, folded, and chemically transformed by the enormous heat and pressure involved. Mountains formed through collision are not volcanically active. On the contrary, collision sutures plates together and thickens the crust, preventing magma from reaching the surface.

The Appalachian Mountains in the eastern United States were formed 280 million years ago as the continental crust of two different plates collided and fused to form the North American plate. The Himalayas are still being formed by collision between the Indo-Australian plate and the Eurasian plate. The Indian landmass was once part of Gondwana and located south of the equator. It moved north with the Indo-Australian plate at a very fast absolute rate of about 15 cm (6 in) per year and collided with the Eurasian landmass **(Figure 13.18)**. As it moved northward, it plowed seafloor sediments and rocks (and the fossils they contained) upward into the growing Himalayas, which explains why today's Himalayas contain marine fossils.

Today the Indo-Australian plate continues to push northward and slide horizontally beneath the Eurasian plate. In all, it has moved 1,900 km (1,200 mi)

Figure 13.17 Continental–continental convergent plate boundary.

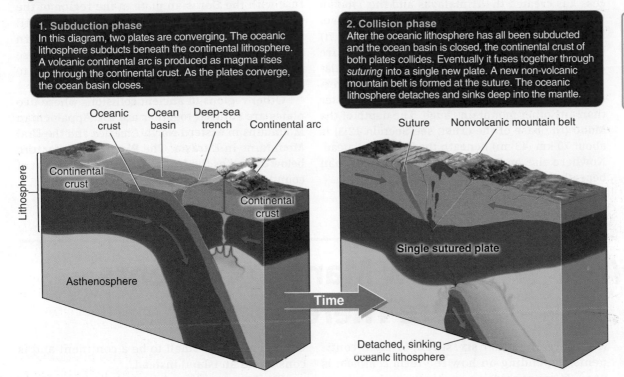

1. Subduction phase
In this diagram, two plates are converging. The oceanic lithosphere subducts beneath the continental lithosphere. A volcanic continental arc is produced as magma rises up through the continental crust. As the plates converge, the ocean basin closes.

2. Collision phase
After the oceanic lithosphere has all been subducted and the ocean basin is closed, the continental crust of both plates collides. Eventually it fuses together through *suturing* into a single new plate. A new non-volcanic mountain belt is formed at the suture. The oceanic lithosphere detaches and sinks deep into the mantle.

Animation
Collision
Available at
www.saplinglearning.com

Figure 13.18 The collision of India and Eurasia.

(A) This figure draws Eurasia in its present location, but 55 million years ago it was closer to the equator (shown with the dotted line). As a result, although not apparent here, India began colliding with Eurasia about 55 million years ago. (B) This digital elevation model shows the height of the Tibetan Plateau in false color. Red areas are over 5,000 m (16,400 ft), and white/gray areas are over 6,500 m (21,300 ft). *(Courtesy Chris Duncan)*

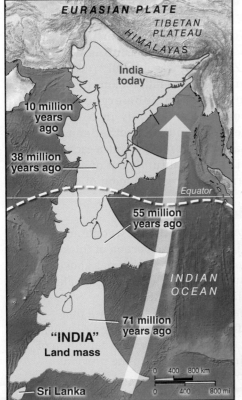

A

B

northward beneath the Eurasian plate. This process has created the Himalayas and the Tibetan Plateau, landforms that are rising at a net rate of about 5 mm (0.2 in) every year, or 50 cm (1.6 ft) every century. After the effects of erosion are accounted for, the western portions of the Himalayas in Pakistan are rising even faster—at an average rate of up to 10 mm (0.39 in) per year, faster than any other mountain range. The depth of the Moho (the base of the crust; see Section 12.3) is about 70 km (43 mi) beneath the Tibetan Plateau. Nowhere else on Earth is the crust thicker than beneath the Himalayas and the Tibetan Plateau.

Eventually, the Indo-Australian plate will fuse with the Eurasian plate in the region of the Himalayas, the growth of the Himalayas will stop, and then this majestic range will be worn down flat through weathering and erosion. These events will take several hundred million years to unfold.

Other regions of ancient collisions where two plates are sutured together are the Appalachian Mountains in eastern North America and the Ural Mountains in Eurasia. The **Picture This** feature below discusses the Urals further in the context of counting continents.

Picture This **How Many Continents Are There?**

(A) There are either six, seven, or eight continents, depending on how the term *continent* is used. If we consider a continent to be a large landmass surrounded by water, there are six continents: North America, South America, Eurasia, Africa, Australia, and Antarctica.

Greenland is too small to be a continent and is considered an island instead.

(B) Human geographers factoring in political regions, however, divide Eurasia into two continents: Europe and Asia. The boundary between Europe and Asia is traced along an ancient collision zone: the Ural Mountains. The Urals stretch 2,500 km (1,550 mi) from northern Kazakhstan through Russia to the Arctic Ocean. West of the Urals is Europe, and East of them is Asia.

(C) Now many scientists are recognizing a newly discovered eighth continent: *Zealandia*. In 2017, Zealandia was formally recognized as a new continent. It is 5 million km^2 (2 million mi^2) in extent and 94% submerged. Zealandia rises above sea level only in New Zealand, New Caledonia, and a few other small islands. Zealandia is a thick mass of silica-rich continental crust that rises abruptly out of oceanic crust. It was formerly a large fragment of Gondwana.

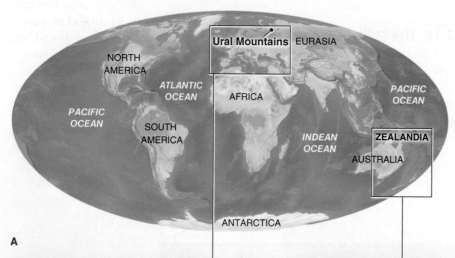

A

B

C

Consider This

1. One common definition of a continent is that it is an extensive portion of thickened continental crust that protrudes above surrounding oceanic crust. Does Zealandia meet this definition? Explain.
2. Does each of the two landmasses that result from dividing Europe and Asia using the Ural Mountains meet the definition of a continent, as defined in the previous question? Explain.

Accreted Terranes

Some continents have an odd assortment of mismatched crust types, called accreted terranes, along their margins. An **accreted terrane** is a mass of crust that is transported by plate movement and fused onto the margin of a continent. (An *accretionary prism*, discussed earlier, is a type of accreted terrane when it is exposed as dry land by tectonic uplift.) Accreted terranes result from a combination of subduction of oceanic lithosphere and collision of landmasses that are smaller than continents. Terranes can travel thousands of miles across ocean basins before they encounter a continent. (Note that *terrane* refers to a piece of land added to continental crust, while *terrain* refers to the topography or surface features of a landscape. The two terms are pronounced the same.)

India is not considered an accreted terrane because the Indian landmass before it joined to Eurasia was large enough to be a continent. However, as the Indo-Australian plate pushed northward into the Eurasian plate, it bulldozed several preexisting terranes ahead of it. Thus, the Himalayas are composed of mismatched pieces of volcanic rocks that formed during the early subduction phase, accreted terranes, and the Indian continent.

Accreted terranes are found worldwide, but the largest accreted terrane group is in western North America. From Alaska to Mexico, terranes that have traveled across the Pacific Ocean basin have piled up on the western edge of the North American continent. As **Figure 13.19** shows, these terranes have significantly increased the size of the continent.

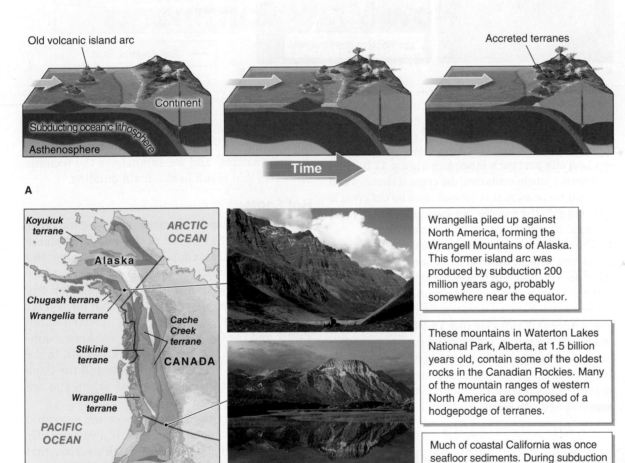

Figure 13.19
Accreted terranes.
(A) Pieces of continental crust and island arcs are carried along on oceanic crust toward a continent. The oceanic crust subducts, forming active volcanoes, but the relatively buoyant terranes do not. Instead, they are sheared off and fused onto the continent's margin. (B) Through this process, many different terranes have been added to the North American landmass at various times in the past. *(Top. Andrew Greene, Hawaii Pacific University; center. Don Johnston/Getty Images; bottom. © David K. Lynch)*

Wrangellia piled up against North America, forming the Wrangell Mountains of Alaska. This former island arc was produced by subduction 200 million years ago, probably somewhere near the equator.

These mountains in Waterton Lakes National Park, Alberta, at 1.5 billion years old, contain some of the oldest rocks in the Canadian Rockies. Many of the mountain ranges of western North America are composed of a hodgepodge of terranes.

Much of coastal California was once seafloor sediments. During subduction about 150 million years old, these sediments were scraped up into an accretionary prism of piled and folded sediments at a deep-sea trench (see Figure 13.15). Some of those sediments, like those shown here, located just north of San Francisco in Marin County, were transformed into a type of rock formed from seafloor sediments. The layered and folded nature of the rocks is clearly visible.

Figure 13.20 Transform plate boundary. Los Angeles sits on the Pacific plate, and San Francisco sits on the North American plate. In California, the two plates are slipping past each other along a transform plate boundary, so that Los Angeles is moving toward San Francisco. The relative velocity of the two plates is 5 cm (2 in) per year. At the present rate of movement, the two cities will meet in about 17 million years. *(A. EROS Center, U.S. Geological Survey; B. NASA/JPL)*

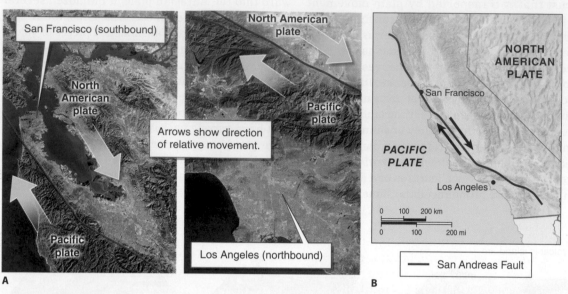

Transform Plate Boundaries

A transform plate boundary exists where two plates slip laterally past each other (see Figure 13.10C). At transform plate boundaries, the crust is sheared and torn, but no new crust is formed, and no old crust is recycled into the mantle. No significant mountains are built, and there are no volcanic eruptions.

Transform plate boundaries are most common on the seafloor at mid-ocean ridges. They seldom occur on dry land, but where they do, they are conspicuous. Few transform plate boundaries are more conspicuous or more notorious than California's San Andreas Fault **(Figure 13.20)**. The Pacific and North American plates grind past one another along this transform plate boundary, creating hundreds of earthquakes each day (most too small to be felt). Occasionally, this plate movement creates powerful and dangerous earthquakes.

13.4 Hot Spots, Folding and Faulting, and Mountain Building

◎ **Describe landforms that result from tectonic processes away from plate boundaries.**

Two tectonic processes do not fit into the three types of plate boundary interactions discussed in the previous section but are nevertheless important in building landforms: hot spots and folding and faulting. In this section we describe those processes, and we summarize the tectonic processes that result in mountain building.

Hot Spots

So far in this chapter, we have discussed the types of volcanoes that occur at the edges of plates where divergence or subduction is occurring. But not all volcanoes occur on plate boundaries. There are some 30,000 islands in the Pacific Ocean, and most of them are volcanic islands that formed far from plate boundaries. Hawai'i, Easter Island, Tahiti, and the Galápagos Islands are a few examples of non-plate boundary volcanic islands. In the Atlantic Ocean, the Canary Islands and the Cape Verde Islands are also volcanic in origin and are not directly on plate boundaries.

Volcanic islands such as these were created by geologic hot spots. A **hot spot** is a volcanically active and isolated location on Earth's surface that is caused by a vertical column of hot rock, called a **mantle plume**, that extends down as far as Earth's outer core **(Figure 13.21)**.

Mantle plumes are not well understood. They are thought to remain mostly stationary, with rock moving upward through them at a rate of several centimeters per year. The rock within the plumes is not liquid; rather, it is slowly deforming solid rock. As rock rises within a mantle plume, it encounters less pressure, and in the asthenosphere it melts through decompression melting. The resulting buoyant magma melts its way through the overlying lithosphere, and once on the surface, it forms a volcano. As a lithospheric plate moves over a hot spot, a line of volcanoes may form **(Figure 13.22)**.

If you look carefully at a bathymetric map of the world's ocean basins, you will find many linear ridges that look as if a giant knife made long incisions in the crust. Volcanic islands and submerged seamounts are often associated with these linear ridges. Stationary hot spots melt through the crust as the plates move over them, creating these lines of inactive volcanoes, called *hot spot tracks* **(Figure 13.23** on the following page**)**.

An active continental hot spot persists beneath the North American plate at the location of Yellowstone National Park. This hot spot has produced a number of volcanic eruptions, including one of Earth's largest known eruptions 640,000 years ago (see Section 15.2). It also produced a string of volcanic landforms called *calderas* (collapsed volcanoes) **(Figure 13.24** on the following page**)**. Some scientists think that the magma body that resides just below Yellowstone National Park could erupt again someday.

Figure 13.21 Island hot spot. A column of hot solid rock slowly rises and melts through decompression melting. The resulting buoyant magma melts upward through the lithosphere. In oceanic crust, as shown here, it creates underwater lava flows that build oceanic islands once they rise above sea level.

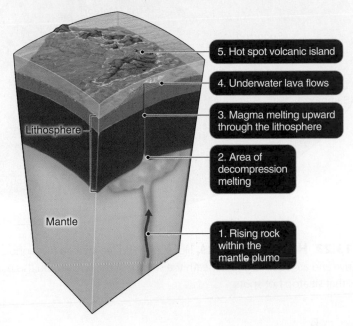

5. Hot spot volcanic island

4. Underwater lava flows

3. Magma melting upward through the lithosphere

2. Area of decompression melting

1. Rising rock within the mantle plume

Lithosphere

Mantle

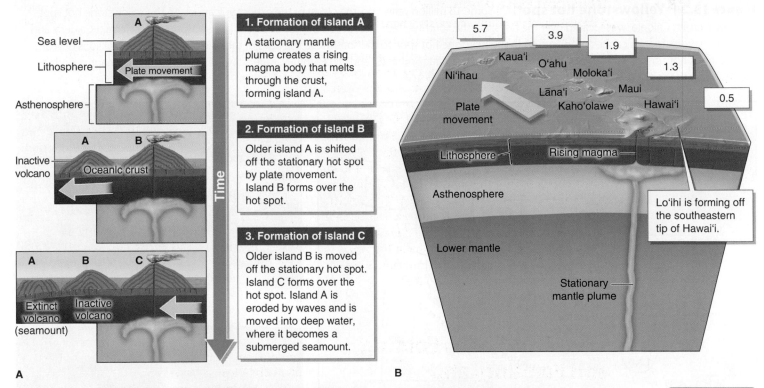

1. **Formation of island A**

A stationary mantle plume creates a rising magma body that melts through the crust, forming island A.

2. **Formation of island B**

Older island A is shifted off the stationary hot spot by plate movement. Island B forms over the hot spot.

3. **Formation of island C**

Older island B is moved off the stationary hot spot. Island C forms over the hot spot. Island A is eroded by waves and is moved into deep water, where it becomes a submerged seamount.

Sea level
Lithosphere
Plate movement
Asthenosphere
Inactive volcano
Oceanic crust
Extinct volcano (seamount)
Inactive volcano
Time

5.7   3.9   1.9   1.3   0.5
Kaua'i   O'ahu   Moloka'i
Ni'ihau   Lāna'i   Maui
Plate movement   Kaho'olawe   Hawai'i
Lithosphere   Rising magma
Asthenosphere
Lo'ihi is forming off the southeastern tip of Hawai'i.
Lower mantle
Stationary mantle plume

A   **B**

Figure 13.22 Archipelago formation at a hot spot. (A) The process of lithosphere moving over a stationary hot spot creates a chain of volcanic islands, or an *archipelago*. An old volcano that has moved off its hot spot becomes extinct, erodes, and diminishes in size. Eventually, inactive volcanoes are moved away from their hot spots into deeper water, where they become flat-topped seamounts. The lithosphere in hot spot regions is heated and rises upward, creating a relatively shallow ocean. (B) The Hawaiian Islands were formed by a stationary hot spot that is still active. As the Pacific plate moves over the hot spot, new islands are formed, and old islands are moved into deeper water. The maximum ages of the islands are given in millions of years. A new island, named Lo'ihi, is forming and will rise above the sea in about 10,000 years.

Animation

Archipelago Formation at a Hot Spot

Available at www.saplinglearning.com

▶

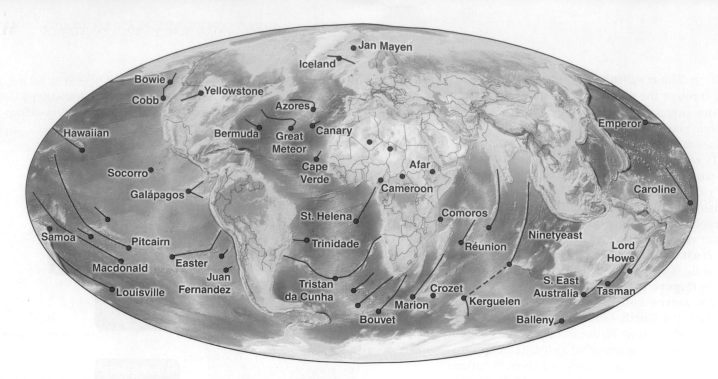

Figure 13.23 Hot spot tracks. Hot spot tracks are located throughout the world's ocean basins. Most of their volcanic mountains are no longer active and are submerged beneath the surface of the oceans. On this map, hot spot tracks are shown in red. Red dots show the locations of volcanoes that sit atop hot spots.

Figure 13.24 Yellowstone hot spot. (A) Some 16 million years ago, the oldest of the extinct volcanoes on the Yellowstone hot spot track, McDermitt Caldera, was located over the hot spot. The southwestern direction of plate movement has transported McDermitt Caldera, and the other now-extinct volcanoes produced by the Yellowstone hot spot, to the southwest. The ages of the calderas are given in millions of years. Yellowstone's hot springs and geysers, such as the Grand Prismatic Spring (B) and Old Faithful (C), are active results of the active magma body that resides beneath the park. *(B. Justin Reznick/Getty Images; C. U.S. Geological Survey)*

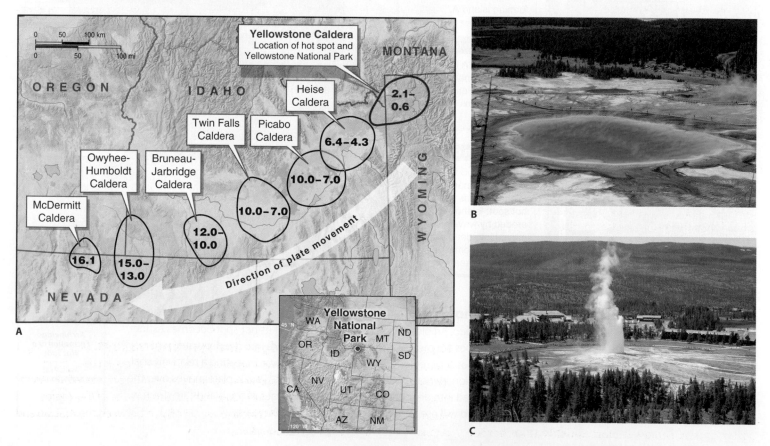

Figure 13.25 Folding. (A) These folded volcanic rocks are located in Hamersley Gorge, Karijini National Park, Australia. The distance across this photo is only a few tens of meters. (B) A satellite image of kilometers-long folds in sedimentary rocks in the Béchar Basin, northwestern Algeria. These rocks were folded in a continental collision between the African and Eurasian plates beginning about 65 million years ago. The ground distance of this satellite image is approximately 100 km (60 mi) across. *(A. Auscape/Getty Images; B. Image courtesy of the Image Science & Analysis Laboratory, NASA Johnson Space Center)*

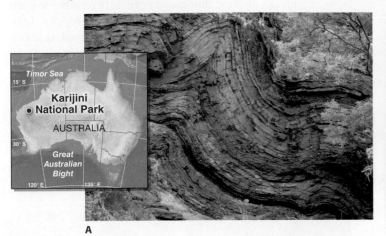

A

B

Bending and Breaking: Folding and Faulting

When hit with a hammer, rocks break into smaller fragments because they are brittle. When they are under pressure and heat, however, rocks in many cases deform and fold (bend) before they fault (break). A **fold** is a wrinkle in the crust that results from deformation caused by geologic stress, and a **fault** is a fracture or a break in the crust where movement and earthquakes occur.

Folding and faulting can occur anywhere on Earth's surface. Folding tends to occur most often where two plates are converging, particularly in regions of subduction and collision. Faulting, too, occurs most often near plate boundaries, but many regions far away from plate boundaries are also faulted. Folding and faulting are particularly important elements of lithospheric plate movement because each can produce prominent surface landforms.

Folded Landforms

Folding, which can occur anyplace where *compressional forces* are pushing the crust together, has created many mountain ranges. Folds may also occur at just about any spatial scale, from centimeters to kilometers, and in many different rock types, as shown in **Figure 13.25**.

The two major types of folds are anticlines and synclines. An **anticline** is a fold in the crust with an archlike ridge, and a **syncline** is a fold in the crust with a U-shaped dip. Generally, when synclines and anticlines occur across a wide region, ridges form on the anticlines and valleys form on the synclines **(Figure 13.26)**.

When a folded surface is weathered and eroded, softer rock is removed more quickly than harder

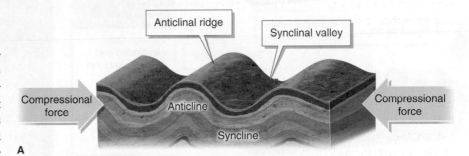

A

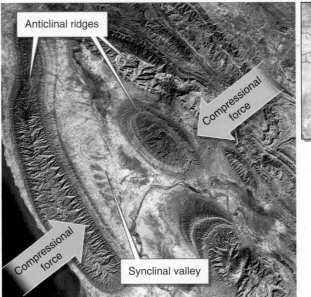

B

Figure 13.26 Anticlinal ridges and synclinal valleys. (A) Like a rug left rumpled on the floor when its two ends have been pushed toward each other, Earth's crust deforms into synclinal and anticlinal folds when it is compressed. (B) An anticlinal ridge and a synclinal valley are visible in this satellite image of the Zagros Mountains of western Iran. These folds are a result of collision between the Arabian and Eurasian plates. *(EROS Center, U.S. Geological Survey)*

rock, leaving the more resistant rock to form a ridge. This process can result in an *inverted topography*, with synclinal ridges and anticlinal valleys **(Figure 13.27)**.

Block Landforms

Faulting results when rocks can deform no further by folding and they break. At the moment that this happens, the energy stored in the rocks is released and travels through the crust as seismic waves that shake the ground in an earthquake. (Faulting and earthquakes are covered in greater detail in Section 15.3.)

Through time, continued faulting can create blocks of crust that move vertically relative to each other. These blocks are called *fault blocks*. Like folds, fault blocks occur across many spatial scales and in a variety of tectonic settings. There are many fault-block mountain ranges in the world, and some of them have dramatic relief **(Figure 13.28)**.

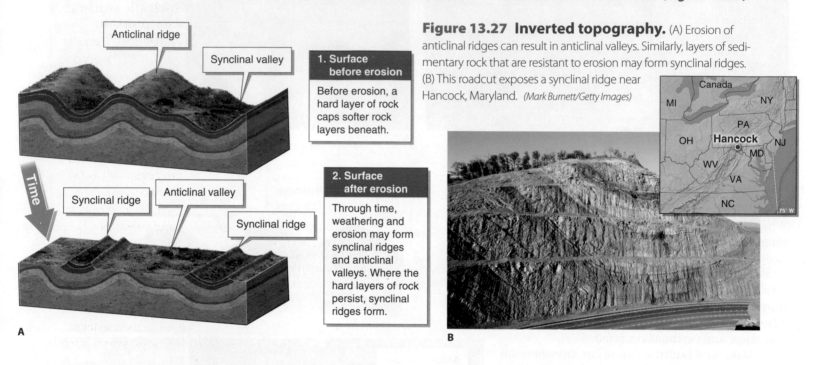

Figure 13.27 Inverted topography. (A) Erosion of anticlinal ridges can result in anticlinal valleys. Similarly, layers of sedimentary rock that are resistant to erosion may form synclinal ridges. (B) This roadcut exposes a synclinal ridge near Hancock, Maryland. *(Mark Burnett/Getty Images)*

Figure 13.28 The Teton Range, Wyoming.
(A) This diagram illustrates the fault-block system that formed the Teton Range. The range formed when two opposing 65 km (40 mi) long fault blocks were lifted and lowered in relationship to one another, beginning about 13 million years ago. Erosion by streams and glaciers subsequently cut into the higher block, removing most of the overlying rocks and creating the jagged mountain topography seen today. (B) Grand Teton, standing at 4,197 m (13,770 ft) and visible here, is the highest peak in the Teton Range. *(© John Wang/ Photodisc/Getty Images)*

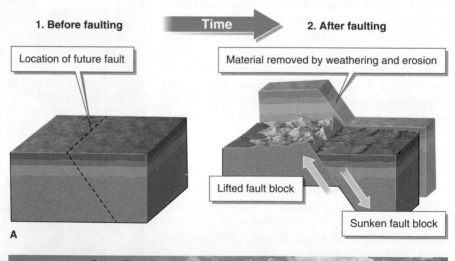

Orogenesis: Tectonic Settings of Mountains

The building of mountain ranges by any tectonic process is called **orogenesis** (from the Greek *oros*, "mountain," and *genesis*, "origin"). Most mountains are grouped together to form linear ranges. These ranges, called **orogenic belts**, form most commonly along plate boundaries, particularly in areas of collision and subduction. Away from plate boundaries, isolated mountains and orogenic belts form in areas of rifting, hot spot tracks, and folding and faulting. **Figure 13.29** summarizes the tectonic settings and processes covered in this chapter and their roles in orogenesis.

TECTONIC SETTING	VOLCANICALLY ACTIVE MOUNTAINS?
Divergent Plate Boundaries	
Seafloor spreading	Yes
Rifting	Yes
Convergent Plate Boundaries	
Subduction	Yes
Collision	No
Accretion of terranes	No
Hot spots	Yes
Folding and Faulting	Sometimes (depending on the tectonic setting)

A

Figure 13.29 Tectonic settings of mountain ranges. (A) This table summarizes the tectonic settings of mountain ranges and specifies whether each setting produces volcanically active mountains. (B) Earth's 10 longest terrestrial mountain ranges are labeled here. Red and orange areas show the highest surface elevations. Note that the Antarctica insert map is not to scale. The accompanying table orders the ranges by their length and indicates their tectonic settings. *(Inset. British Antarctic Survey)*

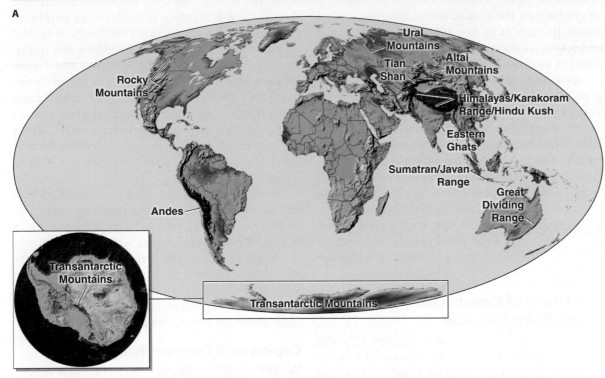

RANGE	LOCATION	LENGTH (KM)	LENGTH (MI)	TECTONIC PROCESS
Andes	South America	7,300	4,526	Subduction
Rocky Mountains	North America	6,000	3,720	Folding and faulting
Himalayas/Karakoram Range/ Hindu Kush	Asia	4,000	2,480	Collision
Great Dividing Range	Australia	3,600	2,232	Collision
Transantarctic Mountains	Antarctica	3,500	2,170	Rifting
Sumatran/Javan Range	Sumatra, Java	2,900	1,798	Subduction
Tian Shan	China	2,300	1,426	Collision
Eastern Ghats	India	2,100	1,302	Collision
Altai Mountains	Asia	2,000	1,240	Collision
Ural Mountains	Russia	2,000	1,240	Collision

B

GEOGRAPHIC PERSPECTIVES
13.5 The Tibetan Plateau, Ice, and People

◎ Assess the importance of glaciers on the Tibetan Plateau for southern Asia and explain how people are adapting to climate change there.

As we have seen throughout *Living Physical Geography,* people and Earth's physical environments are closely intertwined and strongly influence one another. The relationship between people and the process of plate tectonics is no exception. The Tibetan Plateau is a region that was built by a collision between the Eurasian plate and the Indo-Australian plate (see Figure 13.18 on page 407). The plateau and its surrounding mountains cover some 2.9 million km^2 (1.1 million mi^2), or almost half the area of the continental United States. Because of its size and its average elevation of 4,500 m (14,800 ft), the Tibetan Plateau is often referred to as the "Roof the World."

As discussed in Chapter 7, "The Changing Climate," the long cooling trend that began 55 million years ago was caused by weathering and erosion of the Himalayas and surrounding mountains. This weathering and erosion transferred carbon from the atmosphere to long-term storage in sea-floor sediments. Carbon dioxide levels dropped from about 1,000 ppm (parts per million) to the recent preindustrial levels of 300 ppm and lower. This process is ongoing, but for the time being, the weathering of the Himalayas is causing far less carbon to be stored than human activities are emitting.

The Effects of Climate Change on the Tibetan Plateau

In the past, the Tibetan Plateau changed the climate; today, anthropogenic climate change is rapidly changing the plateau. Over the past 150 years, people have been burning fossil fuels and releasing the greenhouse gas CO_2 to the atmosphere much faster than natural processes such as weathering (and the ocean waters and plants) can remove it from the atmosphere. Fossil fuel combustion has resulted in a 30% increase in atmospheric CO_2, from roughly 280 parts per million (ppm) in pre-industrial times to today's 408 ppm and rising at about 2 ppm per year (see Section 7.4).

The atmosphere gets warmer as the concentration of atmospheric CO_2 increases. Satellite data show that, as a result of the ice–albedo positive feedback, the Tibetan Plateau and its mountains are warming up to three times faster than the global average.

The Tibetan Plateau and the Himalayas and other surrounding mountains contain the largest volume of ice outside the polar regions. The plateau is home to some 49,500 glaciers, most of which are in the Himalayas. But as the atmosphere warms, these glaciers are quickly shrinking and retreating upslope, where it is cooler. Between 2000 and 2016, conservative estimates put the total annual Himalayan ice loss at 16 billion tons per year. According to the European Geoscience Union, the Tibetan Plateau as a whole could lose 70% of its glaciers by the end of the century due to anthropogenic climate change.

Ice and Water, Food, and Energy Security

The loss of Himalayan ice is connected to water, food, and energy security throughout southern Asia. Many of Earth's largest rivers originate from glaciers high on the Tibetan Plateau and its surrounding mountains, including the Indus, Ganges, Irrawaddy, Mekong, Yangtze, and Yellow **(Figure 13.30)**. Roughly 20% of the world's population (1.4 billion people) get their water from these rivers. This water is essential to all aspects of human life, providing drinking water, water for household needs, fishing, navigation, hydroelectric power generation, and water for crop irrigation.

Climate change can be a two-edged sword. In the short term, warming temperatures have helped crops in the region, increasing the yields of rice, apricot, apple, lentil, wheat, and barley in the foothills around the plateau. In addition, rapidly melting glaciers are now bringing a surplus of water to downstream areas. In the long term, however, this supply will be threatened as the glaciers from which the rivers originate disappear.

Coping with Climate Change: Water Megaprojects

In response to the situation in the region of the Tibetan Plateau, several countries, particularly China and India, are planning large-scale *megaprojects* on the region's rivers to store and divert water from the streams running off the plateau. The Mekong, Salween, and Brahmaputra are just a few of the rivers that are slated for dozens of new water megaprojects. The large dams and reservoirs that are planned will both address the problem of disappearing glaciers by storing water for release all year long and generate clean, carbon-free electricity to meet the CO_2 emissions targets agreed to in the 2015 Paris Agreement

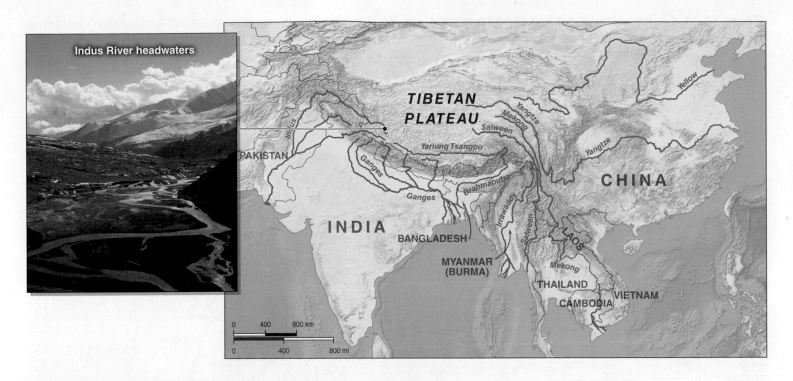

Figure 13.30 **Asian rivers.** The major streams that originate on the Tibetan Plateau are shown here. The headwaters of the Indus are shown in the inset photo. Glaciers are faintly visible in the background mountains. *(TAO Images Limited/Alamy)*

(see Section 7.5). They also will boost the economies of the region and facilitate navigation along the waterways.

China alone has 80 major dams planned, under construction, or completed along the streams that flow off the plateau. One of these projects, the Shuangjiangkou Dam, is on the Dadu River that flows into the Yangtze River. When finished, it will be the tallest dam in the world, standing at 312 m (1,024 ft) tall, only 12 meters shy of the height of the Eiffel Tower. Along the Mekong, a series of 26 dam projects are planned or being built. **Figure 13.31** on the following page shows megaprojects slated for the upper reaches of the Yangtze, Mekong, and Salween rivers.

China is also constructing history's most ambitious water-transfer project—the South–North Water Transfer Project. In addition to building dozens of new dams, the project will divert the flow of several rivers originating on the plateau northward to the capital city Beijing, where water is increasingly scarce because of population growth and reduced rainfall due to climate change. Construction of the South–North Water Transfer Project is now well under way, and China aims to complete this massive project by 2050.

Opposition to these megaprojects runs deep because they have significant environmental and social consequences. Most of the major rivers of Asia flow through several countries. The Mekong, for example, winds through six countries, and 95 ethnic groups dwell along its length. International conflicts arise when water projects upstream in one country result in less water for people downstream in another country. These huge projects also flood preexisting wildlife habitat, cultural heritage sites, and valleys were people live.

Furthermore, many scientists are deeply concerned about earthquake-induced dam failure and the catastrophic consequences that would follow for people living downstream. The Tibetan Plateau, which is one of the world's most active collision zones, is riddled with geologic faults and already prone to earthquakes (see Figure 13.7 on page 399). Compounding the danger, large reservoirs are known to trigger earthquakes as the weight of their water forcibly enters rocks and lubricates underlying geologic faults.

Just as the Tibetan Plateau was formed by the collision between two continental plates, a similar collision of sorts is forming between people and the changing climate. People need water, food, and energy sources, all of which are increasingly threatened by climate change. As glaciers disappear in the Tibetan Plateau region, water megaprojects may help pick up the slack and provide security—but not without environmental cost and risk.

Figure 13.31 Chinese hydropower megaprojects. This map shows the hydropower megaprojects either built, under construction, under consideration, or proposed along the headwaters of the Yangtze, Mekong, and Salween rivers and their tributaries in China. The names of the projects and the hydroelectric capacities, in megawatts (MW), are shown at left. *(Tibetan Plateau Blog)*

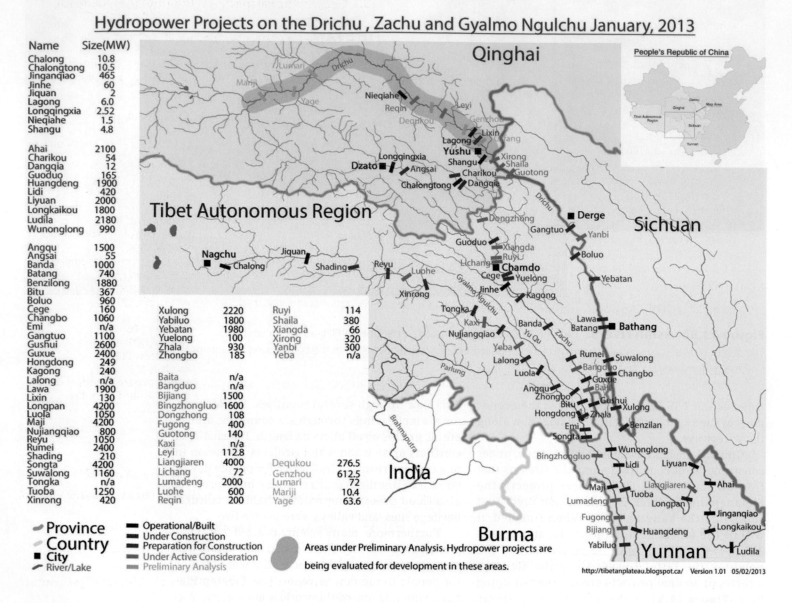

Hydropower Projects on the Drichu , Zachu and Gyalmo Ngulchu January, 2013

Name	Size(MW)
Chalong	10.8
Chalongtong	10.5
Jinganqiao	465
Jinhe	60
Jiquan	2
Lagong	6.0
Longqingxia	2.52
Nieqiahe	1.5
Shangu	4.8
Ahai	2100
Charikou	54
Dangqia	12
Guoduo	165
Huangdeng	1900
Lidi	420
Liyuan	2000
Longkaikou	1800
Ludila	2180
Wunonglong	990
Angqu	1500
Angsai	55
Banda	1000
Batang	740
Benzilong	1880
Bitu	367
Boluo	960
Cege	160
Changbo	1060
Emi	n/a
Gangtuo	1100
Gushui	2600
Guxue	2400
Hongdong	249
Kagong	240
Lalong	n/a
Lawa	1900
Lixin	130
Longpan	4200
Luola	1050
Maji	4200
Nujiangqiao	800
Reyu	1050
Rumei	2400
Shading	210
Songta	4200
Suwalong	1160
Tongka	n/a
Tuoba	1250
Xinrong	420

Xulong	2220	Ruyi	114
Yabiluo	1800	Shaila	380
Yebatan	1980	Xiangda	66
Yuelong	100	Xirong	320
Zhala	930	Yanbi	300
Zhongbo	185	Yeba	n/a
Baita	n/a		
Bangduo	n/a		
Bijiang	1500		
Bingzhongluo	1600		
Dongzhong	108	Dequkou	276.5
Fugong	400	Genzhou	612.5
Guotong	140	Lumari	72
Kaxi	n/a	Mariji	10.4
Leyi	112.8	Yage	63.6
Liangjiaren	4000		
Lichang	72		
Lumadeng	2000		
Luohe	600		
Reqin	200		

Legend:
- ⌇ Province
- ⌇ Country
- ■ City
- ⌇ River/Lake
- ▬ Operational/Built
- ▬ Under Construction
- ▬ Preparation for Construction
- ▬ Under Active Consideration
- ▬ Preliminary Analysis
- ⬤ Areas under Preliminary Analysis. Hydropower projects are being evaluated for development in these areas.

http://tibetanplateau.blogspot.ca/   Version 1.01   05/02/2013

Chapter 13 Study Guide

Focus Points

13.1 Continental Drift: Wegener's Theory

- **Breakup of Pangaea:** The supercontinent Pangaea split into Laurasia and Gondwana about 150 million years ago. The opening of the Atlantic Ocean led to the modern configuration of the contents.
- **Continental drift:** Alfred Wegener was the first to formally propose the theory of continental drift, which states that continents move.

13.2 Plate Tectonics: An Ocean of Evidence

- **Plate tectonics theory:** Plate tectonics theory addresses why and how lithospheric plates move.
- **Oceanographic evidence:** Many forms of evidence from the ocean basins, including seafloor bathymetry, rock sample ages, and patterns of magnetization, led to the development and acceptance of the theory of plate tectonics.
- **Moving plates:** Plate tectonics theory states that there are 14 major lithospheric plates that move independently of one another across the surface of the planet.
- **Plate boundaries:** Earthquake activity reveals plate boundary locations.

- **How plates move:** Ridge push, mantle drag, and slab pull move the lithospheric plates.
- **Plate velocity:** The fastest relative plate movement is 18 cm (7 in) per year.
- **Unifying theory:** Plate tectonics is a unifying theory that helps explain many seemingly separate physical phenomena.

13.3 Plate Boundary Landforms

- **Plate boundaries:** There are three types of plate boundaries: divergent plate boundaries, convergent plate boundaries, and transform plate boundaries.
- **Plate boundary landforms:** Landforms created by plate movements include deep-sea trenches, continental arcs, island arcs, and non-volcanic mountain ranges.
- **Rifting:** Continental rifting splits a single continental landmass, creating a volcanic rift valley. Rifting can lead to new plate margins and a new ocean basin.
- **Plate recycling:** Subduction recycles oceanic lithosphere into the mantle.
- **Buoyant continental crust:** Continental lithosphere does not subduct. Because of this, Earth's oldest crust is continental crust.
- **Accreted terranes:** Much of western North America is accreted terranes.

13.4 Hot Spots, Folding and Faulting, and Mountain Building

- **Hot spots:** Hot spots exist where stationary mantle plumes melt through the crust. Hot spots beneath oceanic crust create oceanic islands such as Hawai'i.
- **Folding and faulting:** Earth's crust folds and faults in response to tectonic stresses. Folding produces anticlines and synclines, and faulting produces earthquakes.

- **Orogenesis:** Many mountain ranges form in orogenic belts along plate boundaries.

13.5 Geographic Perspectives: The Tibetan Plateau, Ice, and People

- **The Tibetan Plateau and people:** About 1.4 billion people living in Asia rely on rivers that originate from dwindling glaciers on the Tibetan Plateau. Water megaprojects are rapidly being undertaken to provide water, food, and energy security.

Key Terms

accreted terrane, 409
accretionary prism, 405
active continental margin, 399
anticline, 413
collision, 406
continental arc, 404
continental drift, 395
convergent plate boundary, 402
decompression melting, 402
divergent plate boundary, 402
fault, 413
flux melting, 404
fold, 413
Gondwana, 395
hot spot, 410
island arc, 405
Laurasia, 395

mantle drag, 400
mantle plume, 410
orogenesis, 415
orogenic belt, 415
Pacific Ring of Fire, 406
Pangaea, 394
passive continental margin, 399
plate boundary, 398
plate tectonics, 396
ridge push, 399
rift, 403
rift valley, 403
slab pull, 400
subduction, 398
syncline, 413
transform plate boundary, 402
Wadati–Benioff zone, 404

Concept Review

The Human Sphere: Life on Earth's Shifting Crust

1. How are plate movements harmful to people? How are they beneficial?

13.1 Continental Drift: Wegener's Theory

2. What was Pangaea, and when did it exist?

3. Which of today's continents made up Laurasia, and which ones made up Gondwana?

4. What theory did Alfred Wegener develop? In what year? On what evidence did he base his theory?

5. Was Wegener's theory accepted by his colleagues during his lifetime? Why or why not?

13.2 Plate Tectonics: An Ocean of Evidence

6. What is plate tectonics theory? How is it the same as, and how is it different from, continental drift theory?

7. What geophysical evidence was used to develop plate tectonics theory?

8. What is a lithospheric plate, and how many are there?

9. What is the difference between an active continental margin and a passive continental margin?

10. How are the plate boundaries determined?

11. Why do the plates move?

12. Using relative velocity, how fast do the plates move?

13. What does it mean to say that plate tectonics is a unifying theory?

13.3 Plate Boundary Landforms

14. Describe the movements of lithospheric plates at divergent, convergent, and transform plate boundaries.

15. What is a mid-ocean ridge? Where do mid-ocean ridges form? How do they form?

16. What is rifting? Where does it occur? What happens to a continent when rifting occurs on it?

17. How does rifting relate to the formation of new oceanic crust and new ocean basins?

18. How does rifting create a new plate boundary?

19. What is subduction? At what type of plate boundary does it occur?

20. Describe the relative ages of oceanic crust and continental crust and explain why they are different.

21. What is a Wadati–Benioff zone, and how does it provide supporting evidence for subduction?

22. What is the Pacific Ring of Fire? What kinds of mountains form along the Pacific Ring of Fire?

23. What is collision? What kind of mountains result from collision? Give a real-world example.

24. What are accreted terranes? Where do they come from, and why do they pile up against continental margins? Why do they not subduct?

25. What is a transform plate boundary? What kind of landforms do transform plate boundaries produce?

13.4 Hot Spots, Folding and Faulting, and Mountain Building

26. What is a hot spot? What physical features on Earth are produced by hot spots?

27. What is a hot spot track? Explain how hot spot tracks are formed.

28. How are rocks folded? What is the difference between folding and faulting?

29. Compare anticlines and synclines. What is an inverted topography, and how does it form?

30. What is a fault block? What do fault blocks create?

31. What is orogenesis? Review the major mountain ranges of the world and the tectonic formation of each.

13.5 Geographic Perspectives: The Tibetan Plateau, Ice, and People

32. When and how did the Tibetan Plateau and its mountains form?

33. What does it mean that the Tibetan Plateau is a climate driver? In what three ways does the plateau modify climate?

34. What is currently happening to the plateau's glaciers? Why?

35. What are the connections among geologic uplift, glaciers, rivers, and feeding people?

Critical-Thinking Questions

1. What is the nearest plate boundary to where you live? What kind is it? Refer to Figure 13.6 for help.

2. What do rift valley lakes, volcanoes, and earthquakes have in common?

3. The rate of crust formation at mid-ocean ridges must be exactly equal to the rate of crust recycling at subduction zones. Is this statement true or false? Explain.

4. What mountains are nearest where you live? How were they formed? If you do not know, what kind of questions would you ask to determine how they were formed? Refer to Figure 13.29 for help.

5. Most hot spot tracks are straight, but some curve sharply. How would you explain a curved hot spot track?

Test Yourself

Take this quiz to test your chapter knowledge.

1. True or false? Gondwana consisted of South America, North America, and Eurasia.

2. True or false? Continental crust does not subduct.

3. True or false? The Pacific Ring of Fire is a result of rifting.

4. True or false? The Andes were formed by diverging plates.

5. True or false? Earth's oldest crust is found on the ocean floor.

6. Multiple choice: Which of the following was formed by a hot spot?

   a. the Himalayas
   b. the Yellowstone Caldera
   c. the East African Rift system
   d. the Teton Range

7. Multiple choice: Which of the following is not related to the Tibetan Plateau and climate change?

   a. changing volcanic activity
   b. changing atmospheric CO_2 concentrations
   c. changing flow patterns of the westerlies
   d. changing midlatitude albedo

8. Multiple choice: Which of the following cannot occur where two plates converge?

   a. earthquakes
   b. volcanoes
   c. rift valleys
   d. mountains

9. Fill in the blank: A(n) _____ forms when oceanic crust subducts beneath other oceanic crust.

10. Fill in the blank: _____ is thought to be responsible for most of the movement of subducting oceanic crust.

Online Geographic Analysis

Monitoring Volcanic Ash in Alaska

In this section we analyze volcanic ash hazards in Alaska.

Activity

The Alaska Volcano Observatory interactive web page allows you to explore volcanic activity in Alaska. Go to https://www.avo.alaska.edu/activity/index.php and answer the following questions.

1. What volcanic island chain shown on this map extends westward from Alaska? You may have to zoom in or out to activate the island chain name.

2. Explain why these volcanoes are here and why they are aligned in an arc. You may have to refer to your reading in this chapter.

3. Based on your answer to the previous question, would you expect there to be a deep-sea trench near these islands? Where specifically? Explain.

4. What does each triangle represent?

5. What do the colors of the triangles represent?

6. Click on a red or orange triangle. If those aren't available, click on a yellow or green one. You should now see more detail about the volcano you choose. What is the name of the volcano you chose? Given its color, what is its current activity level?

7. In the same pop-up window, click on the "Description" link below the name of the volcano. What are the geographic coordinates of this volcano? What is its elevation in feet?

8. What is the name of the nearest town and how far away from the volcano is it?

9. Reread the Human Sphere section at the beginning of this chapter. Why does NOAA monitor these volcanoes so closely?

10. Refer to the discussion about great circle routes in Chapter 1, "The Geographer's Toolkit." Aircraft flying between what geographic areas should be concerned about activity of volcanoes in this area?

Picture This. *Your Turn*

Tectonic Landforms and Features

Fill in the blanks on the diagram. You may need to use some of the following terms more than once.

1. Asthenosphere
2. Continental crust
3. Deep-sea trench
4. Earthquakes
5. Lithosphere
6. Mantle drag
7. Mid-ocean ridge
8. Oceanic crust
9. Ridge push
10. Slab pull
11. Subducting slab
12. Subduction zone
13. Volcanoes

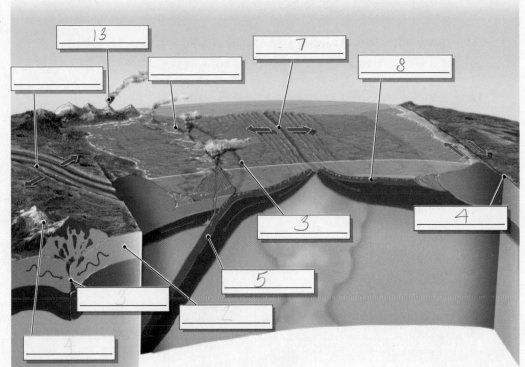

For animations, interactive maps, videos, and more, visit www.saplinglearning.com.   Sapling Plus

Building the Crust with Rocks

Chapter Outline *and Learning Goals* ◎

14.1 Minerals and Rocks: Building Earth's Crust

◎ List the main mineral classes that make up rocks and discuss how rocks are formed, transformed, and recycled.

14.2 Cooling the Inferno: Igneous Rocks

◎ Describe the different groups of igneous rocks and where they form.

14.3 Layers of Time: Sedimentary Rocks

◎ Describe the three groups of sedimentary rocks, where they form, and how they provide information about Earth's history.

14.4 Pressure and Heat: Metamorphic Rocks

◎ Describe where metamorphic rocks form and how they are categorized.

14.5 Geographic Perspectives: Fracking for Shale Gas

◎ Assess the pros and cons of extracting natural gas trapped in shale through hydraulic fracturing.

A hiker examines the layers of rock in the Coyote Buttes area of the Vermilion Cliffs National Monument. These *sandstone* rocks are composed of sand that was once part of an extensive system of sand dunes in a desert that existed roughly 190 million years ago. Through time the sand grains were compressed and cemented together to form layers of sedimentary rock.

(Carles Zamorano Cabello/Alamy)

Section 14.3 discusses in more detail how sedimentary rocks form.

THE HUMAN SPHERE **People and Rocks**

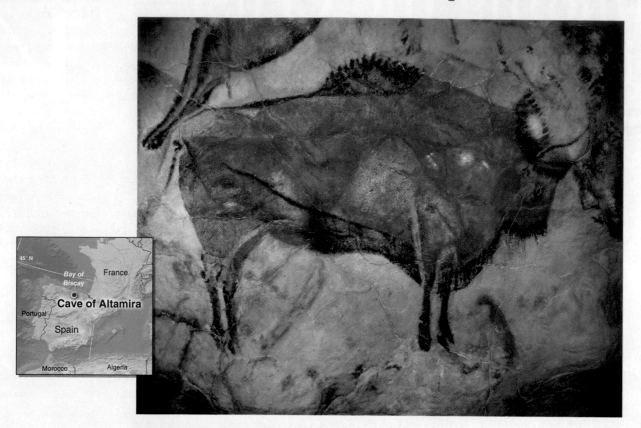

Figure 14.1 Cave of Altamira. This mural of a now-extinct species of bison was found in the Cave of Altamira, located near Santander, Spain. Some of the images in the cave were made about 35,600 years ago and are among the earliest-known cave art in Europe. Stone tools were used to grind various Earth materials, including iron oxides and ochre from rocks, to make the paints. *(Heritage Images/Getty Images)*

Our success as a species is directly linked to Earth's rocks. As early as 3.2 million years ago in eastern Africa, our early hominid ancestors began using rocks as tools. Those first tools were not used for hunting. Instead, they were used for breaking animal bones for their marrow and cutting and scraping meat from animal carcasses.

Through time, as humans evolved, so did their use of rocks and an increasingly wide range of Earth materials. In more recent prehistoric times, Paleolithic humans used materials derived from Earth's crust in a wide range of applications, from shaping rocks into spear tips to using pigments to portray the landscapes and animals around them in cave paintings **(Figure 14.1)**.

Earth materials are still prominent in the lives of humans today. Copper and gold are used in electronics manufacturing.

Diamonds and rubies are carefully cut and polished to make jewelry. Uranium, a mineral found in many rocks in Earth's crust, is used in applications ranging from nuclear energy to atomic weaponry. Pencil lead (graphite), electronic circuitry (silicon and quartz), wallboard and concrete (gypsum), baby powder (talc), plant fertilizer (sulfur), soap (borax), and toothpaste (calcite) are all produced from minerals and rocks derived from Earth's crust. Coal has been among the most economically important rocks for the past century and a half. Coal fueled the Industrial Revolution and continues to power a large portion of the modern global economy. And now we are transitioning to renewable energy technologies that also require earth minerals; for example, photovoltaic solar power (see Section 3.6) uses silicon.

14.1 Minerals and Rocks: Building Earth's Crust

◎ **List the main mineral classes that make up rocks and discuss how rocks are formed, transformed, and recycled.**

Minerals are the fundamental building blocks of Earth's crust. Both the ground we walk on and the crust beneath the oceans are composed of rock that is made up of minerals. **Minerals** are naturally occurring, crystalline, solid chemical elements or compounds with a uniform chemical composition. A **rock** is a solid mass composed of one or more types of minerals. Minerals are *abiogenic* (not made by organisms). Thus, naturally forming ice is a mineral. Common table salt is also a mineral, but sugar is not. Table sugar is crystalline, but it is an organic compound made by living plants, and it is therefore not a mineral. Petroleum is not a mineral because it is liquid.

Web Map

Mining

Available at
www.saplinglearning.com

Scientists have identified more than 4,600 different minerals and continue to discover dozens of new ones each year. Some minerals are very rare and form only where conditions are exactly right. Precious gemstones and metals, including diamond, gold, silver, ruby, opal, sapphire, and jasper, are minerals.

All minerals form through the process of **crystallization**, which occurs when atoms or molecules come together in an orderly patterned structure called a *crystal*. Although crystallization can occur on Earth's surface—as it does where salt forms, for example—the process of crystallization is most common in subterranean (underground) cavities where water circulates and *precipitates* (deposits) minerals.

Minerals are *homogenous* in their makeup; any part of a mineral sampled, no matter the size, has the same composition. Quartz, for example, is a mineral composed of silica, which is another name for the silicon dioxide (SiO_2) molecule. The volcanic rock obsidian is also chemically uniform, and it is also composed mostly of silica. Obsidian is not a mineral, however, because it is noncrystalline; its atoms are arranged randomly, and so it is referred to as *glassy*. Because the atoms within minerals are arranged in a regular, repeating structure, the mineral crystals have geometric shapes with angular and flat surfaces. As **Figure 14.2** shows, the surfaces of crystals can reflect the arrangement of interlocking atoms and molecules that make up the mineral.

Mineral Classes in Rocks

Although there are thousands of minerals, just a few dozen of them make up most rocks in Earth's crust. The lion's share of the crust (74%) is made up of just two elements: oxygen and silicon (**Figure 14.3**).

All of the minerals found in Earth's crust can be categorized into silicate minerals and non-silicate minerals. Within these two large groupings, minerals are further subdivided into *classes*, based on their chemical composition. The silicate minerals are a class of minerals composed mainly of silica. They are by far the most common class of mineral, accounting for 95% of the minerals found in Earth's crust. The molecules in the silicate minerals each contain one silicon atom (Si) that is bonded with four oxygen (O) atoms to form silicate (SiO_4). The non-silicate minerals do not contain

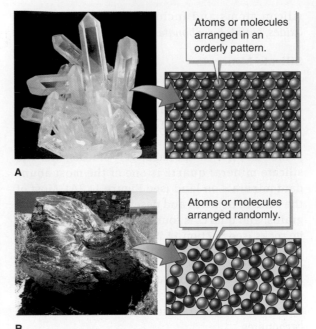

Figure 14.2 Crystalline and noncrystalline materials. (A) Like all other minerals, quartz crystals have a geometric arrangement of atoms. The naturally formed flat edges (or faces) of quartz crystals reflect this atomic orderliness. (B) Like quartz, the volcanic rock obsidian is chemically uniform, but it does not have geometric arrangement of atoms. It is therefore not a mineral. *(A. De Agostini Picture Library/ Getty Images; B. Bruce Gervais)*

A Atoms or molecules arranged in an orderly pattern.

B Atoms or molecules arranged randomly.

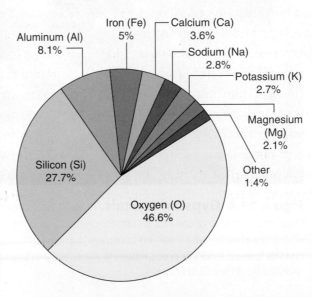

Aluminum (Al) 8.1%
Iron (Fe) 5%
Calcium (Ca) 3.6%
Sodium (Na) 2.8%
Potassium (K) 2.7%
Magnesium (Mg) 2.1%
Other 1.4%
Silicon (Si) 27.7%
Oxygen (O) 46.6%

Figure 14.3 Elements of the crust. This pie chart shows the relative percentages of the main elements that make up Earth's crust.

silicon atoms and include the mineral classes *oxides*, *sulfides*, *carbonates*, and *halides*.

Silicate Minerals

Rocks composed mainly of silica are structurally strong and relatively resistant to weathering and erosion. The silicate mineral *feldspar* is the most common type of mineral in Earth's crust. Oceanic crust is the most extensive type of crust on the planet, and it is composed largely of feldspars. The silicate mineral quartz is one of the most abundant minerals on land (see Figure 14.2A). Most of the sand in beaches and deserts is made of quartz grains, and quartz is found in abundance in granite rocks (see Section 14.2).

Non-silicate Minerals

The non-silicate minerals include many different classes. The most common non-silicate mineral classes are the halides, oxides, sulfides, and carbonates.

The *salt minerals*, or halides, make up the largest class of non-silicate minerals. The salt minerals form when halogen compounds, such as chlorine and iodine, bond with

metallic elements. A common salt mineral is *sodium chloride* ($NaCl$), also called halite or rock salt. Sodium chloride usually forms as salt water evaporates.

Oxide minerals, or oxides, form when other chemical elements combine with oxygen. Iron combined with oxygen, for example, forms the mineral *hematite*. Hematite gives many rocks a reddish color.

Sulfide minerals, or sulfides, form when sulfur bonds with other metallic elements such as copper, iron, zinc, or lead. Pyrite (FeS_2) (also called "fool's gold") is a common sulfide mineral. *Gypsum* ($CaSO_4 \bullet 2H_2O$), another common sulfide, forms when sulfur and oxygen combine with calcium and other elements **(Figure 14.4)**.

Carbonate minerals, or carbonates, form when carbon combines with other elements, particularly oxygen. *Calcite* is the most common type of mineral in the carbonate mineral class. One of the most common carbonate rocks is limestone, composed mostly of calcium carbonate ($CaCO_3$).

Figure 14.4 Gypsum crystals. These gypsum crystals in Mexico's Cueva de los Cristales ("Cave of the Crystals") are some of the world's largest crystals. (Note the people in the photo for scale.) This cave was inundated with groundwater until a mining company pumped the water out, revealing the large crystals within. Because these crystals formed through precipitation in water, they will not grow further as long as the cave is kept dry. *(Carsten Peter/ Speleoresearch & Films/Getty Images)*

Rocks and Outcrops

Recall from earlier in this section that rocks are composed of one or more types of minerals. A few rocks are composed of single minerals. Sodium chloride, for example, forms rock salt, and silicon dioxide forms obsidian. But most rocks are assemblages of many minerals interlocked and cemented together.

Earth's crust is composed of rocks and covered by sediments derived from rocks. **Sediments** are accumulations of small fragments of rock and organic material that are not cemented together. Rock fragments, or *clasts*, are formed by weathering of larger blocks of rock. Clasts are transported by flowing water, ice, or wind and deposited in layers of loose sediments. The sand on a beach, for example, is deposited by flowing rivers and the action of coastal waves. Over most of Earth's surface, sediments form a veneer (a thin surface) that covers the underlying *bedrock* (rock that is structurally part of Earth's crust).

Exposed areas of bedrock, called **outcrops**, dominate the landscape in tectonically and volcanically active mountainous regions, but there are few outcrops in the flat interior regions of continents, such as the Great Plains of North America. In these flatter areas, the bedrock in most places is buried deep beneath sediments. Most outcrops are found in mountainous areas, in stream canyons, on rocky coastlines, and where roads have been cut into hills or mountains **(Figure 14.5)**.

The Rock Cycle

There are many kinds of rocks, from soft to hard, light to dark, and in all colors. Some rocks look like human hair, and some are so filled with pockets of air that they float. The wide array of rocks can be broken into three major categories, or *families*: igneous rocks, sedimentary rocks, and metamorphic rocks. **Igneous rocks** form when magma or lava cools. **Sedimentary rocks** form through cementation and compaction of sediments. **Metamorphic rocks** form when heat and pressure are applied to preexisting rocks.

In a way, scientists view rocks as they would the life of an organism: Rocks experience a cycle of beginning, middle, and end. They are "born" in the furnace of Earth's heat. They undergo a series of transitions during their "lifetimes," eventually returning to the molten state where they began or being recycled deep into the mantle. The **rock cycle** is a conceptual model that describes the formation and transformation of rocks in the crust. In the rock cycle, rocks are formed, transformed from one type to another, and then re-melted or recycled into the mantle. Rocks first form when magma cools and solidifies (or freezes) into rock from its molten (liquid) state, either deep within

Figure 14.5
Roadcut outcrops.
A roadcut near Golden, Colorado, has created an outcrop that reveals layers of sedimentary rock. *(Francois Gohier/Science Source)*

Earth's crust or as lava extrudes (pushes out) onto the crust's surface from a volcanic vent **(Figure 14.6)**.

Rocks are subject to *weathering*, a process by which they are reduced to fragments of varying size or are chemically broken down. The resulting fragments are transported downslope by flowing water, wind, and glaciers and deposited as sediments. Through the processes of compaction and cementation, loose sediments may experience

lithification (transformation into rock). As sediments accumulate and their weight compresses the particles, compaction occurs. **Cementation** is a process in which minerals, such as calcite, fill the spaces between sediment particles and bind them together to form sedimentary rock. Most sedimentary rocks are formed through a combination of compaction and cementation **(Figure 14.7)**.

In the next phase in the rock cycle, sedimentary rock can be moved deep into the crust through

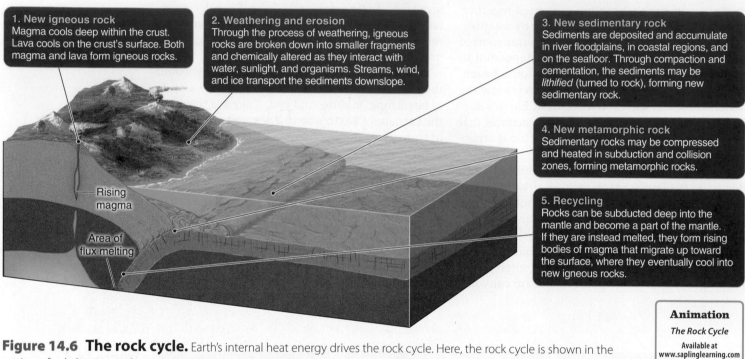

1. New igneous rock
Magma cools deep within the crust. Lava cools on the crust's surface. Both magma and lava form igneous rocks.

2. Weathering and erosion
Through the process of weathering, igneous rocks are broken down into smaller fragments and chemically altered as they interact with water, sunlight, and organisms. Streams, wind, and ice transport the sediments downslope.

3. New sedimentary rock
Sediments are deposited and accumulate in river floodplains, in coastal regions, and on the seafloor. Through compaction and cementation, the sediments may be *lithified* (turned to rock), forming new sedimentary rock.

4. New metamorphic rock
Sedimentary rocks may be compressed and heated in subduction and collision zones, forming metamorphic rocks.

5. Recycling
Rocks can be subducted deep into the mantle and become a part of the mantle. If they are instead melted, they form rising bodies of magma that migrate up toward the surface, where they eventually cool into new igneous rocks.

Rising magma

Area of flux melting

Animation
The Rock Cycle
Available at
www.saplinglearning.com

Figure 14.6 The rock cycle. Earth's internal heat energy drives the rock cycle. Here, the rock cycle is shown in the setting of subduction and a continental arc.

Figure 14.7 Lithification. As sediments are transported out to sea by a stream, they accumulate on the seafloor. Their weight compacts the sediments, and the growth of minerals in the remaining pore spaces cements the particles together, forming new sedimentary rock.

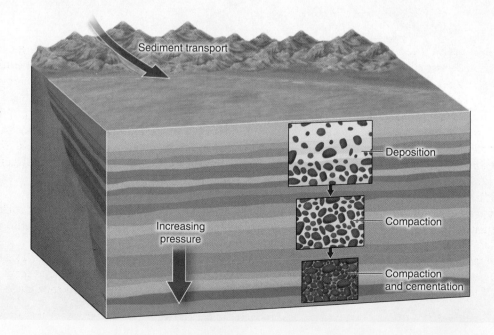

Sediment transport

Deposition

Compaction

Compaction and cementation

Increasing pressure

tectonic collision or subduction (see Section 13.3). There it may be compressed and heated under enormous pressure. Through compression and heating, any rock can become a new metamorphic rock. If a rock is heated enough, it melts into magma. Thus, through the rock cycle, rocks may be transformed from igneous rock, to sedimentary rock, to metamorphic rock, and back to igneous rock.

This conceptual model of the rock cycle is a good start, but it falls short of capturing the complexity of the process of rock formation and transformation. Rock may take any of several transformational pathways, depending on the tectonic setting in which it forms. In Figure 14.6, if the metamorphic rock were to be exposed at the crust's surface, it would be weathered and eroded into sediments. Those sediments could then lithify into sedimentary rock. In another example, seafloor igneous rocks form from cooled magma and lava at mid-ocean ridges and are then transported across the ocean basin and recycled into the mantle by subduction (see Section 13.3), without ever undergoing weathering.

Interestingly, a completely new type of rock has begun forming in the past few decades—an anthropogenic rock called *plastiglomerate*. The **Picture This** feature below discusses this rock further.

Picture This A New Rock: Plastiglomerate

A *(Patricia L. Corcoran and GSA Today)*

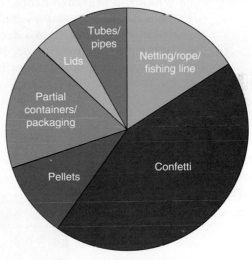

B

About a decade ago, while walking on a beach on the Big Island of Hawai'i, researchers Patricia Corcoran and Charles Moore found an unusual "rock" that they had never seen before. This rock was composed of a mass of sand, pebbles, and small rocks locked tightly together by melted plastic. They called the new stone "plastiglomerate" because it is a conglomerate (collection) of materials welded together by melted plastic (A).

Plastiglomerate forms wherever a large mass of plastic melts and mixes with debris. Forest fires, lava flows, and campfires all forge plastiglomerate. In addition, this anthropogenic rock commonly forms on beaches where plastic regularly washes ashore (see Section 10.5) and people burn the plastic in an effort to get rid of it. As the plastic melts, it encases and cements together any materials in contact with it, including rocks, shells, wood, and other partially melted

or unmelted plastic garbage. Most plastiglomerate rocks have recognizable plastic junk in them, such as toothbrushes, cigarette lighters, disposable utensils, or ropes. The pie chart (B) shows the proportions of recognizable plastic items found in a 2013 study of plastiglomerate on Kamilo Beach on the island of Hawai'i. As long as it is not melted, plastiglomerate is extremely durable and will persist for hundreds of thousands of years or longer.

Consider This

1. In what family of rocks (igneous, sedimentary, or metamorphic) would you place plastiglomerate? Explain your choice.
2. Where in Figure 14.6 would you insert a description of the process that forms plastiglomerate? Does it fit into the traditional rock cycle model?

14.2 Cooling the Inferno: Igneous Rocks

◎ **Describe the different groups of igneous rocks and where they form.**

As you might recall from Section 12.3, the asthenosphere (part of the upper mantle) is not in a molten state. Under surface conditions, it would be hot enough to melt, but because of the great pressures it is under, it does not melt. Instead, the asthenosphere is composed of solid rock that slowly deforms and flows.

Magma and lava are molten materials that originate from either the solid asthenosphere or the lithospheric mantle. The word *igneous* is derived from the Latin word *ignis*, which means "fire." As discussed in Section 13.3, the asthenosphere melts when pressure on it decreases as it is brought to a shallow depth in the lithosphere. When the pressure drops, the asthenosphere melts through decompression melting. (The term *decompression* refers to a decrease in pressure.) Decompression melting occurs in three tectonic settings: at mid-ocean ridges, in rift valleys, and at hot spots. Mantle material also melts into magma

in subduction zones, where water reduces the melting point of the mantle, a process called *flux melting*. Subduction zones are the fourth setting for igneous rock formation.

Decompression melting and flux melting both result in the production of magma, which contributes to building the crust when it cools and solidifies into new igneous rock. **Figure 14.8** summarizes the tectonic settings in which melting of mantle material into magma occurs. (See Sections 13.3 and 13.4 for further details about these tectonic settings.)

Once magma forms, it rises up into the crust because it is less dense and more buoyant than the lithospheric mantle and also because pressure from the rocks squeezes it upward. Magma may cool either within the crust or on the crust's surface. Rock that cools from lava on the crust's surface forms **extrusive igneous rock** (or *volcanic rock*). **Intrusive igneous rock** (or *magmatic rock*) forms as magma deep underground cools and solidifies.

Igneous Rock Formations

At the surface, common extrusive igneous landforms include solidified lava flows and volcanoes. As magma moves upward through the

Figure 14.8 Tectonic settings for igneous rock formation. Igneous rocks are formed through decompression melting or flux melting in four tectonic settings.

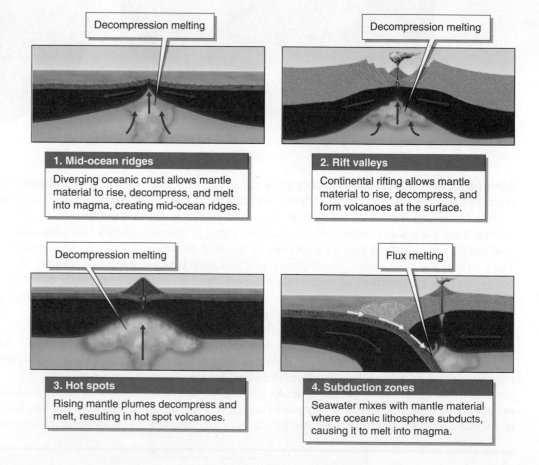

Decompression melting

Decompression melting

1. Mid-ocean ridges

Diverging oceanic crust allows mantle material to rise, decompress, and melt into magma, creating mid-ocean ridges.

2. Rift valleys

Continental rifting allows mantle material to rise, decompress, and form volcanoes at the surface.

Decompression melting

Flux melting

3. Hot spots

Rising mantle plumes decompress and melt, resulting in hot spot volcanoes.

4. Subduction zones

Seawater mixes with mantle material where oceanic lithosphere subducts, causing it to melt into magma.

surrounding rock (called *country rock*) and solid-ifies deep beneath the surface, several different intrusive igneous features may form, including batholiths, plutons, sills, dikes, and laccoliths. A **batholith** is a dome-shaped body of intrusive igne-ous rock hundreds of kilometers in extent, formed by the movement and fusion of numerous plutons. A **pluton**, like a batholith, is a dome-shaped igneous rock mass. Plutons are typically a few kilometers in diameter and are chemically dis-tinct from neighboring plutons. A **sill** is a hori-zontal sheet of igneous rock that has cooled from magma injected between layers of preexisting

rock. A **dike** is similar to a sill but forms as a ver-tical sheet of igneous rock. A **laccolith** is a shal-low, dome-shaped igneous rock body. These intrusive igneous rock formations are hidden beneath Earth's surface until they are exposed by **exhumation**—the removal of overlying rock and sediment at the surface that exposes deeper rocks. This process takes millions of years and involves geologic uplift that exposes rocks to ero-sive forces. **Figure 14.9** illustrates these intru-sive igneous features.

The exhumation of intrusive igneous rock features sometimes creates spectacular surface

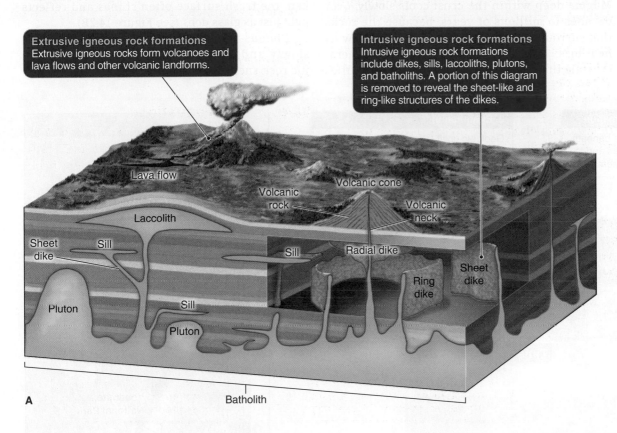

Extrusive igneous rock formations
Extrusive igneous rocks form volcanoes and lava flows and other volcanic landforms.

Intrusive igneous rock formations
Intrusive igneous rock formations include dikes, sills, laccoliths, plutons, and batholiths. A portion of this diagram is removed to reveal the sheet-like and ring-like structures of the dikes.

Lava flow
Laccolith
Sheet dike
Sill
Pluton
Sill
Pluton
Volcanic rock
Volcanic cone
Volcanic neck
Sill
Radial dike
Ring dike
Sheet dike

A   Batholith

Figure 14.9 Igneous rock formations.

(A) Extrusive igneous land-forms include lava flows and volcanic mountains. Intrusive igneous landforms remain buried unless erosion exposes them. (B) Pine Valley Mountain (left) in southern Utah is the one of the largest laccoliths in the United States, measuring about 35 km (20 mi) across. This laccolith formed between layers of sedimentary rock that have since been eroded away. Shiprock (right) is a prominent landform in northwestern New Mexico. A large volcano once existed there. After the volcano became extinct, its outer layers of rock were eroded away, exposing the relatively resistant volcanic neck and a dike.
(Left: Utah Geological Society; right: Nature/UIG/Getty Images)

Laccolith

WA
MT
OR
ID
WY
NV
UT
CO
CA
Pine Valley Mountain
NM
AZ

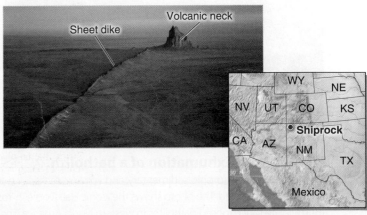

Sheet dike
Volcanic neck

WY
NE
NV
UT
CO
KS
CA
AZ
Shiprock
NM
TX
Mexico

B

landforms. **Figure 14.10** illustrates the exhumation of the Sierra Nevada batholith in California and shows other exhumed batholiths in the western United States.

Igneous Rock Groups

Just as criminals leave behind evidence of their crimes, rocks always leave behind evidence about the conditions that formed them. Scientists, like detectives, can examine the evidence and reconstruct the conditions under which a rock formed. The presence (or absence) of mineral crystals and the resulting rock texture reveal much about the conditions under which an igneous rock formed. Magma deep within the crust cools slowly, over decades to millions of years, because the rocks that surround the magma insulate it and slow its heat loss. This long span of time gives mineral crystals time to develop as the magma solidifies into rock. These slowly cooled rocks often have a rough texture and visible crystals, which may be several centimeters across or larger.

Three groups of igneous rock are based on the characteristics of crystals within the rock: *phaneritic*, *aphanitic*, and *glassy*. Coarse-grained rocks with large crystals that formed slowly deep in the crust are called phaneritic rocks. These rocks sparkle as large crystal faces reflect light. Aphanitic igneous rocks cool more quickly closer to the surface, and as a result they have smaller crystals that cannot be seen without the aid of magnification. Glassy rocks have no orderly crystalline arrangement of atoms because they cool rapidly at the surface. When a glassy rock is broken, the fresh surface often shines and reflects light just as glass does (see Figure 14.2B).

Plutons and batholiths (see Figure 14.8) cool slowly and are usually composed of phaneritic rock such as *granite* or *diorite*. Laccoliths and

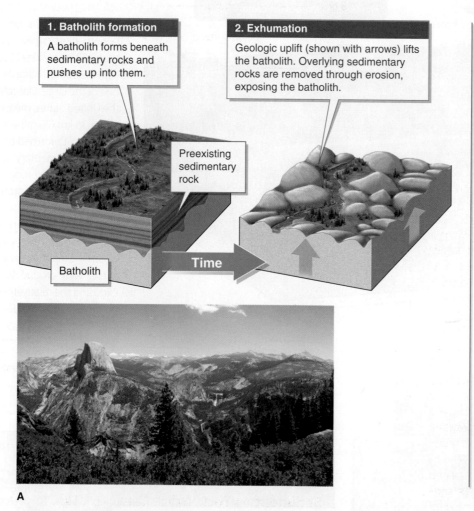

1. Batholith formation
A batholith forms beneath sedimentary rocks and pushes up into them.

2. Exhumation
Geologic uplift (shown with arrows) lifts the batholith. Overlying sedimentary rocks are removed through erosion, exposing the batholith.

Preexisting sedimentary rock

Batholith

Time

A

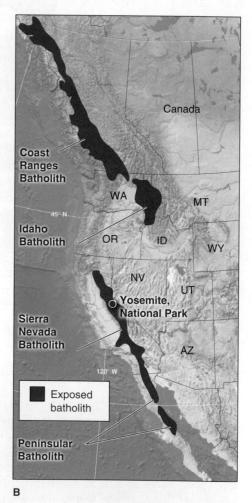

Canada

Coast Ranges Batholith

WA
45° N
MT

Idaho Batholith

OR
ID
WY

NV

Yosemite, National Park

UT

Sierra Nevada Batholith

AZ

120° W

■ Exposed batholith

Peninsular Batholith

B

Figure 14.10 **Exhumation of a batholith.** (A) The Sierra Nevada mountain range in California is an exhumed batholith. Many kilometers of metamorphic, sedimentary, and volcanic rocks once covered the Sierra Nevada. These older rocks were removed by erosion as the batholith was uplifted. The photo shows the iconic Half Dome in Yosemite National Park, one of many plutons that protrude up from the batholith. (B) There are four exposed batholiths in western North America. The largest of them is the Coast Range Batholith in western Canada. *(Bruce Gervais)*

dikes, on the other hand, cool more rapidly nearer the surface and are typically composed of aphanitic rocks, such as *rhyolite* or *andesite*. Some lava flows cool so quickly that crystals do not have time to form. Such flows may be composed of glassy rocks such as *obsidian* or *pumice*.

In addition to using the rock texture, scientists also use the silica content to describe igneous rocks **(Figure 14.11)**. The amount of silica in magma depends largely on the amount of country rock melted and incorporated into the magma as it rises through the crust. *Felsic* igneous rocks (also called *granitic* rocks) are composed of about 70% silica. They are light in color and contain high concentrations of silica, aluminum, potassium, and sodium. They are also less dense than mafic rocks. Granite, for example, is composed mostly of the light-colored minerals quartz and feldspar. *Mafic* igneous rocks (also called *basaltic* rocks), in contrast, contain only about 50% silica. Mafic rocks such as basalt are relatively high in magnesium, iron, and olivine, and they are denser than felsic rocks. (Section 15.1 discusses the silica content of lava in the context of volcanic landforms and hazards.)

Figure 14.11 Classifying igneous rocks. Igneous rocks can be organized by their texture, silica content, density, and color. Felsic rocks (such as granite), which have a high silica content, tend to be lighter in color, while mafic rocks (such as basalt), with less silica, are darker. *(Top left. Arterra Picture Library/Alamy; top center. A.B. JOYCE/Science Source; top right. Science Stock Photography/Science Source; bottom left. Joel Arem/Getty Images; bottom center. Breck P. Kent; bottom right. Julie S. Woodhouse/Alamy)*

14.3 Layers of Time: Sedimentary Rocks

◎ Describe the three groups of sedimentary rocks, where they form, and how they provide information about Earth's history.

Each layer of sedimentary rock is like a page from the book of Earth's history. Throughout time, sediments have been deposited, compacted, and cemented together into sedimentary rock. As each "page," or layer of sediment, is created, it records certain aspects of the environments existing at the time of its formation. Important events in Earth's history, such as the evolution and extinction of species, natural climate change, and the opening and closing of oceans, are all recorded in sedimentary rocks.

The desert in today's southwestern United States, for example, was once a large, shallow sea and then later a series of large freshwater lakes that dried up and re-formed many times. Rivers carried dissolved and fragmented rock material into these water bodies and deposited it as sediments. Many of these sediments were rich in hematite (or other iron oxides). Later, the sediments lithified and were then exposed through erosion, in many places forming dramatic and brightly colored scenery **(Figure 14.12)**.

Figure 14.12 Sedimentary rock, Bryce Canyon, Utah. The sediments that formed these rocks were laid down beginning about 63 million years ago over a period of about 23 million years by a system of streams and freshwater lakes. The oldest sediments are at the bottom (according to the superposition principle; see Section 12.2). Hematite gives the sediments their vibrant orange color. *(Bruce Gervais)*

Sedimentary Rock Groups

There are three groups of sedimentary rock: clastic sedimentary rock, organic sedimentary rock, and chemical sedimentary rock. These three groups are distinguished by the kinds of sediments of which they are composed, the environment in which they form, and the processes that form them **(Table 14.1)**. **Clastic sedimentary rock** is composed of broken pieces of other rocks; the sizes of those pieces determine the kind of clastic sedimentary rock that forms (Table 14.1). **Organic sedimentary rock** is composed mostly of organic material derived from ancient organisms or their shells. **Chemical sedimentary rock** forms as dissolved minerals precipitate out and as water evaporates from sedimentary deposits.

Table 14.1 Sedimentary Rock Types

1. Clastic sedimentary rocks (Composed of rock and mineral fragments)

ROCK TYPE	FORMED FROM
Claystone	Clay
Shale	Clay
Siltstone	Silt
Mudstone	Silt and clay
Sandstone	Sand
Conglomerate	Cobbles, sand, silt, and clay

2. Organic sedimentary rocks (Composed of the remains of organisms)

ROCK TYPE	FORMED FROM
Fossiliferous limestone	Coral and shell fragments
Chalk	Marine phytoplankton shells
Bituminous coal	Peat

3. Chemical sedimentary rocks (Composed of precipitated and evaporated mineral deposits)

ROCK TYPE	FORMED FROM
Rock salt (halite)	Evaporated water
Gypsum	Evaporated seawater
Limestone and tufa	Precipitated calcium carbonate

Sedimentary rocks form only in *depositional environments*, which are places where sediments accumulate. The sedimentary rock formations we see in today's landscapes reflect the environments in which their sediments accumulated and lithified long ago. Any place where flowing water, wind, or ice no longer carry their load of sediments can be a depositional environment. The most common depositional environments include river floodplains, estuaries and lakes, desert dunes, and the seafloor in coastal and offshore regions **(Figure 14.13)**.

The processes that form sedimentary rocks operate under the principle of uniformitarianism (see Section 12.2). These slow and gradual processes are difficult to observe on human time scales but nonetheless operate in the world around us today.

The Three Most Common Sedimentary Rocks

Although 96% of Earth's crust is composed of igneous and metamorphic rocks, most of the rocks exposed on the crust's surface are sedimentary rocks **(Table 14.2)**. About 99% of those sedimentary rocks are shale, sandstone, or limestone. **Shale** is a clastic sedimentary rock formed from clay-sized particles; 45% of all sedimentary rocks are shale. **Sandstone**, which makes up 32% of all sedimentary rocks, is also a clastic sedimentary rock; it is composed chiefly of quartz sand grains. Making up 22% of all sedimentary rocks, **limestone**, which is composed of at least 50% calcite, can be either a chemical sedimentary rock or an organic sedimentary rock.

Table 14.2 Rock Proportions and Exposures

	PROPORTION OF THE CRUST	PROPORTION EXPOSED AT THE SURFACE
Igneous and metamorphic rocks	96%	25%
Sedimentary rocks	4%	75%

Economically Significant Sedimentary Rocks

Human society is built on natural resources. Most natural resources are either mined or grown; minerals and rocks, for example, must be mined. Among the sedimentary rocks, coal, shale, and some chemical sedimentary rocks are particularly important from an economic standpoint.

The Industrial Rock: Coal

The most economically valuable sedimentary rock is coal. **Coal** is an organic rock that is formed from the remains of terrestrial wetland forests. Coal fueled the industrial era beginning in the early 1800s. The two main types of coal are bituminous coal and anthracite coal. *Bituminous coal* is an organic sedimentary rock formed from peat, and *anthracite coal* is a metamorphic rock (see Table 14.3 on page 441) that forms as bituminous coal is subjected to great heat and pressure.

Coal forms from ancient deposits of **peat**, a brownish-black, heavy soil found in wetlands made up of the partially decomposed remains of plants. When peat is buried deeply beneath other

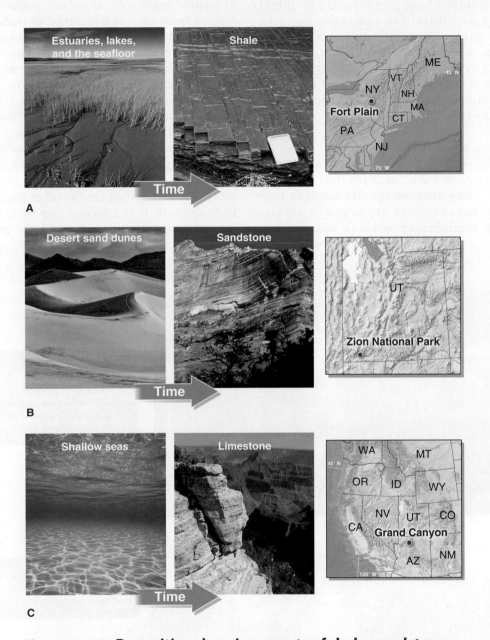

Figure 14.13 Depositional environments of shale, sandstone, and limestone. (A) Deposits of clay in estuaries, in lakes, and on the seafloor lithify into shale. This shale, called Utica shale, near Fort Plain, New York, is 450 million years old. (B) Sand dunes form in desert and coastal beach environments. These lithified dunes, which formed about 170 million years ago, are part of the Navajo Sandstone formation in Zion National Park, Utah. (C) Most limestone forms as minerals dissolved in water precipitate out as solids. These limestone rocks, which formed about 260 million years ago, are part of the Kaibab Formation in the Grand Canyon in northern Arizona. *(A. Ron Erwin/Getty Images; Michael C. Rygel via Wikimedia Commons; B. Witold Skrypczak/Getty Images; Bruce Gervais; C. Tobias Bernhard/Getty Images; Bruce Gervais)*

sediments, it is compacted and heated and can lithify into coal **(Figure 14.14)**.

Coal still plays a central role in today's world economies. About 30% of the electricity generated in the United States is produced by burning coal, and worldwide coal accounts for 40% of electricity generation. These proportions are quickly dropping as natural gas and renewable technologies such as wind and solar become more prevalent. Many types of coal mining are environmentally destructive, particularly the widespread practice of coal mountaintop removal in West Virginia, described in the **Picture This** feature on the following page.

Petroleum and Natural Gas

Petroleum (or *oil*) and natural methane gas are not minerals or rocks because they are liquid and gas. They deserve special mention here, however, because they are found in association with sedimentary rocks. In addition, they are an economically vital material resource for humanity.

Petroleum is a *hydrocarbon*, which means it is a liquid composed of chains or rings of molecules made of hydrogen and carbon atoms. Like coal, petroleum is derived from the organic remains of ancient organisms. However, whereas coal is formed from ancient terrestrial wetland forests, petroleum is formed from the remains of ancient marine zooplankton and phytoplankton.

As much as modern society depends on coal, petroleum is at least as important to our everyday lives, accounting for about 25% of the U.S. energy supply. On average, each person in the United States uses about 70 L (20 gal) of petroleum each day.

Sedimentary rocks are often associated with deposits of petroleum and natural gas. Shale can form a porous reservoir, where natural gas and petroleum accumulate. On the other hand, if shale is nonporous, it can form a *cap rock*, trapping petroleum and natural gas beneath the ground.

1. Ancient coal forests
"Coal" forests existed over 300 million years ago. These forests partially decomposed to form peat. Millions of years of deep burial and the resulting pressure and heat transformed the peat to coal.

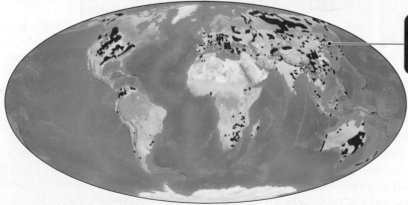

2. Coal deposits worldwide
Coal deposits, which are shown on this map, are remnants of these ancient coal forests.

Figure 14.14 Coal formation and distribution. Coal is sometimes referred to as "buried sunshine" because it is solar energy stored in plants that grew long ago. Ancient terrestrial forests that lived some 300 million years ago partially decomposed to form peat and then lithified into coal. *(New York Public Library/Science Source)*

Picture This Moving Mountains

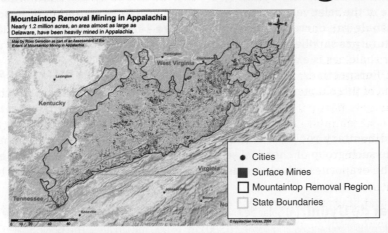

Mountaintop Removal Mining in Appalachia
Nearly 1.2 million acres, an area almost as large as Delaware, have been heavily mined in Appalachia.

Map by Ross Geredien as part of an Assessment of the Extent of Mountaintop Mining in Appalachia

- Cities
- Surface Mines
- Mountaintop Removal Region
- State Boundaries

© Appalachian Voices, 2009

This small hill and its trees were spared (after many lawsuits) to save a small historic cemetery located on the hill.

Forested area before mining

Mined area

Web Map

Coal Mining Scars in Appalachia

Available at
www.saplinglearning.com

(Top. © Appalachian Voices 2009; bottom. Ami Vitale/National Geographic Creative)

Coal obtained through *mountaintop removal* (*MTR*) is arguably the most environmentally destructive form of energy. MTR is a coal mining technique used when coal deposits are too close to the surface to be safely mined underground. Surface vegetation, soil, and rock are stripped away to extract the coal deposits beneath, leaving lifeless, toxic moonscapes that were once verdant forested mountains (shown in the photo). Today, MTR is employed mostly in Virginia, West Virginia, Kentucky, and Tennessee. In all, about 4,856 km² (1,875 mi²) of temperate deciduous forest have been destroyed by coal mountaintop removal in the eastern United States.

MTR also causes significant water pollution when mining companies deposit the removed soil and rock, called *overburden*, in adjacent valleys. More than 1,100 km (700 mi) of streams in the Appalachian Mountains now lie beneath overburden. The overburden leaches (emits) toxic trace elements

such as selenium, lead, nickel, and cadmium. These elements enter stream systems, where they poison the organisms that live in the streams and the people who drink water from them. The elevated health risks to people living near MTR are well documented and include a wide range of ailments, including lung cancer and birth defects.

Consider This

1. *Reclamation* is the process of replanting areas mined by MTR after the coal is mined out. How do you think the landscapes that have been reclaimed differ from those that existed before MTR? Explain.
2. One of the arguments for MTR coal mining is that it provides jobs. Weigh the benefits of the jobs it provides against the problems that MTR creates. Which do you think is more important?

Some porous shale holds natural gas within its pore spaces. Such "shale gas" is currently playing a major role in the North American energy supply. The past decade has seen a boom in shale gas production. Because of the relatively new process called "fracking," shale gas currently provides about 67% of the natural gas supplies in the United States. Fracking for shale gas is explored further in the Geographic Perspectives, "Fracking for Shale Gas" at the end of this chapter.

Evaporites

Some chemical sedimentary rocks have major economic value. One such group of chemical sedimentary rocks is the evaporites, which are used in many applications, ranging from lithium ion batteries (lithium) to construction materials (gypsum) to seasonings (table salt). An **evaporite** is a deposit of one or more minerals resulting from the repeated evaporation of water from a basin. **Figure 14.15** highlights one particularly important evaporite deposit, Salar de Uyuni in Bolivia.

Windows to the Past: Fossils

Sedimentary rocks reveal much about Earth's ancient life and environments; it is in these rocks that fossils are preserved. **Fossils** are the remains or the impressions of organisms preserved in sedimentary rock. **Figure 14.16** explores the use of fossils to study Earth's ancient history.

Figure 14.15 Salar de Uyuni. (A) Salar de Uyuni in Bolivia is the largest exposed evaporite deposit in the world, measuring 200 km (124 mi) across and some 10,500 km² (4,000 mi²) in area. The polygonal cracks (left) in the sediments were formed by repeated swelling and shrinking due to wetting and drying. Lithium, borax, and rock salt (or table salt) are some of the minerals mined and processed here (right). The deposit is estimated to contain 10 billion tons of salt as well as 42% of the world's lithium supply. (B) Evaporites accumulate as runoff pools into shallow lakes which evaporate repeatedly over time.   *(Left. Mike Theiss/Getty Images; right. Christian Kober/robertharding/ Getty Images)*

A

1. Stream inflow

A local thunderstorm creates a stream that flows into Salar de Uyuni. The stream carries dissolved minerals in its water.

2. Evaporite deposition

Because Salar de Uyuni is in a desert, the stream stops flowing once the storm has passed, and the water in the basin evaporates. The dissolved minerals in the water are left behind, forming evaporites.

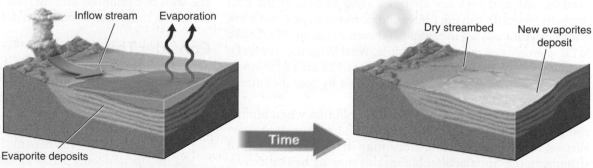

B

The history of life
Although the record of Earth's life preserved by fossils is incomplete because fossil preservation is rare, fossils in sedimentary rocks have provided most of what we know about Earth's biological history. According to the superposition principle (see Section 12.2), the lowest sedimentary rock layers contain the oldest fossils, and the youngest fossils are at top. As different fossil types disappear and appear through time, the processes of extinction and evolution are revealed.

Only species A is still living. Species B and C are extinct.

Modern surface

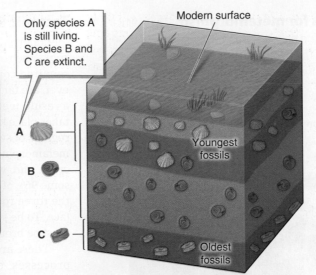

A

B

C

Youngest fossils

Oldest fossils

Figure 14.16
SCIENTIFIC INQUIRY:
What information do fossils provide about the past? Most of what scientists know about Earth's ancient past has been learned by studying fossils. *(Center left. Mark Taylor/Warren Photographic/Getty Images; center right. Rainbow Forest Museum, NPS; bottom. © Christos Kotsiopoulos)*

Environmental change
Fossils reveal that Earth's environments are always changing. This 225-million-year-old fossilized tree trunk is in the Petrified Forest Nation Park in northern Arizona. After the tree fell, it was washed down a stream and buried in silt and volcanic ash. Over time, silica and other minerals replaced the wood, leading to its fossilization. Today's American southwestern desert was once a verdant forest of giant trees (shown in the art illustration).

Dating the past
Fossils tell us the ages of rocks and past events, such as volcanic eruptions. *Index fossils* provide markers that allow scientists to determine the ages of sedimentary rocks at a glance. The ages of index fossils are already known, so they can be used as a quick reference to determine the ages of rocks in which they are found. These are now-extinct 240-million-year-old ammonites from Ligourio, Greece. Ammonite fossils make good index fossils because they occur worldwide, they are easy to identify, and their absolute ages have been determined.

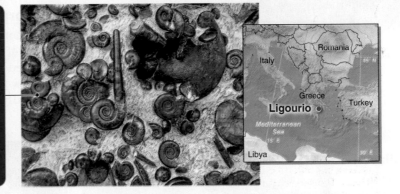

14.4 Pressure and Heat: Metamorphic Rocks

◎ **Explain where metamorphic rocks form and how they are categorized.**

Caterpillars undergo *metamorphosis* and become butterflies. Rocks in the crust undergo *metamorphism* and become new rocks. All rocks found in the crust can undergo the process of metamorphism. Even metamorphic rocks can be *metamorphosed* more than once. The result of metamorphism is a physical, mineralogical, or chemical transformation of the **protolith**—the parent, or original, rock.

The process of metamorphism involves either enormous pressures or high temperatures or both. Heated and chemically reactive fluids such as water can play an important role in metamorphism as well. The thermal range for metamorphism is roughly between 100°C and 900°C (200°F and 1,600°F). There is an upward limit because if rock is heated too much while it is near the surface, it will melt into magma and then cool and solidify into igneous rock.

Figure 14.17 Tectonic settings for metamorphism. These illustrations show the differences of heating and pressure found within the crust in (A) subduction zones and (B) collision zones. The degree of heating or pressure depends on the depth of the rocks, the compressional forces, and how close the rocks are to magma bodies.

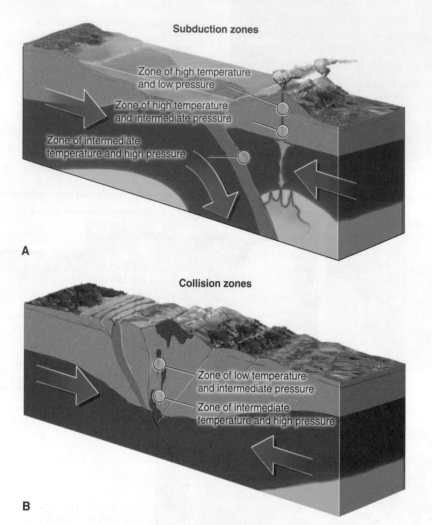

Tectonic Settings of Metamorphism

Metamorphism can occur at the surface where lava cooks the rocks with which it comes in contact. The required conditions for rock metamorphosis, however, are far more common deep within the crust. As a result, metamorphism is a process that normally takes place kilometers deep in the crust, primarily in regions of subduction and collision. Therefore, most metamorphic rocks are hidden away deep within the crust, and, while they and igneous rocks constitute some 96% of the crust, they are the least common of the three rock families exposed on the crust's surface. To be seen and walked on, metamorphic rocks must be brought to the surface through exhumation.

There are two broad categories of metamorphic processes: contact metamorphism and regional metamorphism. *Contact metamorphism* occurs when rock comes into contact with and is heated by magma. *Regional metamorphism* occurs where the crust is being compressed, such as in a collision zone between two convergent continental plates.

The many variables involved in the metamorphism process determine the characteristics of the rock that ultimately forms. These variables include the degree of heating, the degree of compression, the length of time under these conditions, the original chemical constitution of the protolith, and the presence or absence of hot and chemically reactive fluids such as water. **Figure 14.17** illustrates the different degrees of heating and pressure in the context of subduction zones and collision zones.

Metamorphic Rock Groups

All metamorphic rocks fit into one of two groups on the basis of their physical structure: foliated and nonfoliated **(Figure 14.18)**. *Foliated* metamorphic rocks show flat or wavy banding patterns that may superficially resemble sedimentary layers but are

Figure 14.18
Foliated and nonfoliated metamorphic rocks. (A) These metamorphic rocks near Geiranger, Norway, a type called *gneiss* (pronounced NICE), show a conspicuous banded foliation pattern. (B) Marble is a nonfoliated metamorphic rock that is prized by sculptors for its fine grain and ease of carving. Shown here is a detail of Michelangelo's *David*, carved from marble.
(A. SPL/Science Source; B. Fratelli Alinari IDEA S.p.A./Getty Images)

A

B

unrelated to them in how they were formed. Foliation is caused by compression forces. As the rock is squeezed under great pressure, minerals in the rock may become flattened and aligned, creating a grain to the rock called *cleavage*. The flattened minerals form a grain that runs perpendicular to the direction of compression. With further compression, foliation develops. Foliated metamorphic rocks typically form where there is regional metamorphism and the rocks are being squeezed.

Nonfoliated metamorphic rocks form where pressure and heat are uniform. They have little or no structured grain pattern or mineral alignment and usually lack any banded or layered appearance. Nonfoliated metamorphic rocks can form both under conditions of contact metamorphism, where rocks are being heated by a body of magma, and where rocks are being squeezed in environments of regional metamorphism.

Foliated metamorphic rock types are determined by grain size. Nonfoliated metamorphic rock types are classified by their chemical composition. The metamorphic rocks shown in **Table 14.3** represent only a small fraction of the types of metamorphic

Table 14.3 Metamorphic Rock Classification

	ROCK NAME	GRAIN SIZE	PROTOLITH	TEMPERATURE REQUIRED	DESCRIPTION
Foliated metamorphic rocks (classified by grain size)	**Slate**	Small	Shale or mudstone	Low	Dark in color; mineral grains are too small to see with naked eye. Used in building materials such as roofing tiles.
	Schist	Intermediate	Slate	Intermediate	Minerals are medium grained and visible with naked eye. Used in outdoor paving or indoor flooring.
	Gneiss	Large	Schist	High	Coarse grained and typically banded or wavy, with dark and light layers of minerals. Used in flooring, building facades, and gravestones.

	ROCK NAME	PRIMARY MINERAL	PROTOLITH		DESCRIPTION
Nonfoliated metamorphic rocks (classified by chemical composition)	**Marble**	Calcite	Limestone or chalk		Hard rock that may be banded. Often used in sculptures, building facades, and flooring.
	Quartzite	Quartz	Sandstone		Very hard rock, highly resistant to weathering. Commonly used in sandpaper and sandblasting.
	Anthracite Coal	Carbon	Bituminous coal		Hard, black, and shiny, with a high carbon content.

(Slate and schist. Tyler Boyes/Shutterstock; gneiss, marble, quartzite. Breck P. Kent; anthracite: Gary Ombler/Getty Images)

rocks found in nature, but they are among the most common or economically important metamorphic rocks exposed at the crust's surface.

GEOGRAPHIC PERSPECTIVES
14.5 Fracking for Shale Gas

◎ Assess the pros and cons of extracting natural gas trapped in shale through hydraulic fracturing.

The past decade has seen an energy revolution in fossil fuels. In 2008, the price of petroleum was at a high of $157 per barrel (adjusted for inflation). Since then petroleum prices have been on a decreasing trend, dropping as low as $30 a barrel in 2016. The price of petroleum has declined for one primary reason: fracking.

Fracking, short for *hydraulic fracturing*, is a procedure in which water, sand, and chemicals are pumped under high pressure into shale bedrock to extract natural gas and petroleum trapped in the pores of the shale. This technique was first developed in the 1940s, but it was originally an expensive way to extract petroleum and gas from the ground. Advances in technology over the past decade have reduced the cost and transformed fracking into a lucrative way to extract natural gas and petroleum from shale. Fracking has become common in many regions worldwide, and the technique has dramatically increased the amount of available natural gas and petroleum. As a result, prices for petroleum have dropped.

According to the U.S. Energy Information Administration (EIA), in the year 2000 there were 26,000 wells in the United States producing natural gas through fracking. Cumulatively, these wells produced 3.6 billion cubic feet of natural gas per day, 7% of the nation's total. By 2015, the number of fracking wells in the United States had grown to 300,000. Each day these wells produce 53 billion cubic feet of natural gas and account for a total of 67% of all natural gas and 50% of all petroleum produced in the United States.

There are about 187 trillion m^3 (6.6 quadrillion ft^3) of natural gas locked in shale worldwide. China, Argentina, the United States, and Mexico have the largest shale gas reserves in the world. Within the United States, most of the natural gas produced comes from just a handful of shale formations (**Figure 14.19**).

Figure 14.19 Shale gas map for North America. The geographic extent of shale formations from which shale gas is potentially recoverable by fracking is shown in dark orange. The shale gas formations that have produced most of the natural gas in the United States are labeled. Yearly natural gas extraction in billions of cubic feet per day (cf/day) are given for each labeled location.

Story Map

Fracking Tour

Available at
www.saplinglearning.com

Bakken
(2 billion cf/day)

Barnett
(5 million cf/day)

Marcellus
(13 billion cf/day)

Permian Basin
(10.4 billion cf/day)

Eagle Ford
(6 billion cf/day)

0   300   600 km

0   300   600 mi

How Fracking Works

Shale, as discussed in Section 14.3, is a fine-grained sedimentary rock formed from clay. Although the pore spaces in shale are small, they often contain natural gas. Conventional gas extraction techniques are ineffective in shale because the gas flows out of the formation too slowly to be of economic use.

In the hydraulic fracturing process, to speed up the flow of gas from a well drilled in shale, water, sand, and a cocktail of nontoxic and toxic chemicals are injected at high pressures into the well to fracture the rock. Fracturing increases the permeability of the rock and lets petroleum and gas escape up the wellhead so that it can be captured and used. **Figure 14.20** diagrams the process of fracking.

Effects of Fracking

The procedure sounds straightforward: We get more natural gas and petroleum from shale by injecting water, sand, and chemicals into it. Several potentially serious problems result from this method of gas extraction, however. One problem is where to put all the contaminated water that the process generates.

Fracking Fluid Disposal

Each fracking operation at each of the 300,000 wells requires roughly 26 million L (7 million gal) of water. A single well may be fracked several times before it is shut down. Millions of liters of toxic and radioactive wastewater, called *fracking fluid*, are left over after a fracking operation is complete. This wastewater is collected at the fracking site and must be safely disposed of.

Why is the water toxic and radioactive? Chemicals such as benzene and xylene, two known carcinogens, are sometimes added to water during the fracking process to help extract the gas from the shale. In addition, the water absorbs natural radium from the rock, becoming radioactive in the process.

The fracking industry has been grappling with how to dispose of the toxic fracking fluid it produces. Fracking fluid has been discharged into rivers and sent to wastewater treatment plants, neither of which can handle its volume and contaminants. So far, the best solution seems to be to inject it back underground under high pressure.

Unfortunately, injecting fracking fluid back in the ground under high pressure can cause faults (cracks) in the rocks to widen and slip, creating an earthquake. There are now about 3,200 active fracking-related wastewater injection wells in Oklahoma alone. Prior to 2009 Oklahoma had almost no earthquakes of magnitude 3.0 or

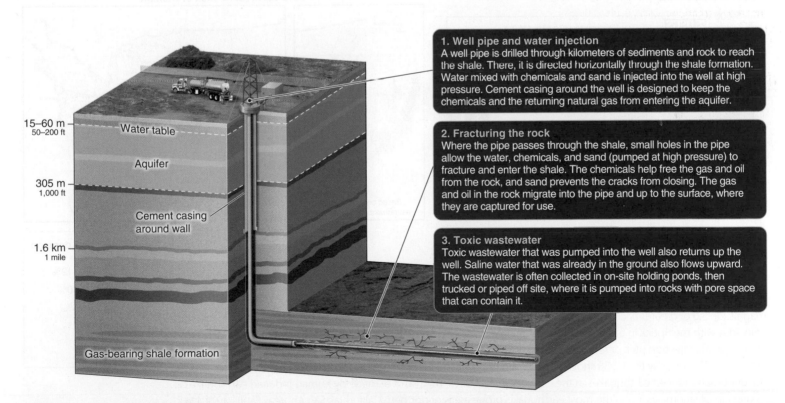

Figure 14.20 How fracking is done. Hydraulic fracturing involves freeing natural gas from within shale by injecting water under high pressure and cracking the shale.

greater. Since then, the number of such earthquakes each year has increased dramatically (**Figure 14.21**).

Drinking Water Contamination

Injecting fracking fluid back into the ground can cause an even bigger problem than earthquakes: It can contaminate groundwater supplies. Opponents of fracking claim that it contaminates drinking water supplies in two ways. First, the toxic fracking fluid finds its way to surface streams and drinking water supplies through accidental trucking spills or leaks from pipelines and containment ponds (**Figure 14.22A**). In addition, opponents argue that fracking threatens drinking water supplies with methane gas contamination (**Figure 14.22B**). Methane has contaminated drinking water in Colorado, Ohio, New York, Pennsylvania, Texas, and West Virginia. Samples from 60 water wells near fracking operations in northeastern Pennsylvania and upstate New York found methane concentrations 17 times higher than average in drinking water supplies. Where it occurs, methane contamination of water by fracking is most likely to be caused by failure of the cement casing placed around the well where it passes through groundwater (see Figure 14.20).

Air Pollution and Climate Change

When burned, natural gas produces less air pollution and carbon dioxide than coal, and it has been hailed by the natural gas industry as a clean alternative to coal. Unfortunately, scientific analysis has found that up to 9% of the natural gas being extracted from fracking wells in Utah leaks directly into the atmosphere. NOAA's recent measurements of methane emissions from petroleum and gas fields by have shown that the Bakken natural gas field leaks 275,000 tons of methane each year, and the Barnett field leaks nearly twice that amount. As a greenhouse gas, methane is 21 times more potent than carbon dioxide, molecule for molecule. Because of these leaks, scientists suspect that natural gas obtained through fracking is a more carbon-intensive fuel (in other words, it causes more global warming) than even coal. Between 2002 and 2014, U.S. methane natural gas emissions increased more than 30%. Although scientists are not sure where the "mystery methane" came from, several studies have linked the source of the methane to fracking activities in North America.

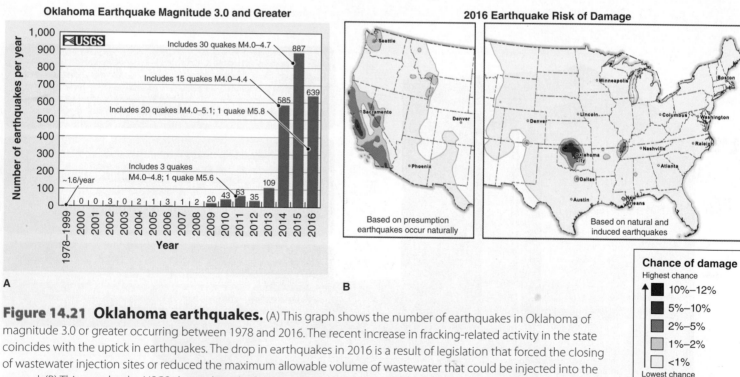

Figure 14.21 Oklahoma earthquakes. (A) This graph shows the number of earthquakes in Oklahoma of magnitude 3.0 or greater occurring between 1978 and 2016. The recent increase in fracking-related activity in the state coincides with the uptick in earthquakes. The drop in earthquakes in 2016 is a result of legislation that forced the closing of wastewater injection sites or reduced the maximum allowable volume of wastewater that could be injected into the ground. (B) This map by the USGS shows the risk of natural earthquakes on the West Coast of the United States and the risk of human-induced earthquakes in the central United States. The risk of damage by human-induced earthquakes is as great in Oklahoma as it is in the most earthquake-prone areas of northern California's San Andreas Fault. (A. USGS-NEIC ComCat & Oklahoma Geological Survey; B. USGS)

The Pros and Cons of Fracking

In some areas, the perceived risks of fracking outweigh its perceived benefits. Countries that have banned it include Scotland, Wales, France, Germany, and Bulgaria. New York, Maryland, and Vermont have banned fracking. Numerous counties across the United States have also prohibited the technique, including five in California. Four of Canada's provinces have banned the procedure as well.

Despite the downsides, we all enjoy the lowered cost of energy that fracking brings, and fracking proceeds apace in most of the United States, Canada, and around the world. Some governments, such as South Africa's, have even lifted bans on fracking and are forging ahead with it. The pros and cons of fracking for the United States are summarized in **Figure 14.23** on the following page.

People need energy. From the standpoint of climate change, natural gas may be a preferable option to relatively dirty coal, assuming the wells don't leak methane. Unfortunately, all fossil fuels, natural gas included, contribute to climate change. Looking ahead several decades, it remains to be seen whether the current boom in natural gas development will provide a relatively clean source of energy while renewable energy sources are developed or if it is simply a continuation of society's dependence on fossil fuels.

A

B

Figure 14.22 Water contamination by fracking. (A) Hazardous fracking fluid is stored on-site in plastic-lined open containment ponds like this one. (B) The groundwater in some areas near fracking operations has become so contaminated with methane that it is possible to set the water on fire as it comes out of the faucet. *(A. Mladen Antonov/AFP/Getty Images; B. Melanie Stetson Freeman/Christian Science Monitor/Getty Images)*

Figure 14.23
The fracking debate. The table summarizes the pros and cons of fracking. *(Left. Mladen Antonov/AFP/Getty Images; right. Justin Sullivan/Getty Images)*

Fracking Pros and Cons	
REASONS FOR	**REASONS AGAINST**
It keeps fossil fuel and gasoline prices low.	Natural gas leaks and heavy machinery that runs on polluting diesel fuel put pollution and greenhouse gases into the atmosphere, contributing to climate change.
It buys time while renewable energy technologies are developed.	Cheap natural gas competes with renewable energy markets, slowing the development of renewable energy technology.
It provides jobs, which is good for the economy.	It causes surface water and groundwater pollution and earthquakes at some locations.
Natural gas produces less air pollution and greenhouse gases than coal and reduces use of coal.	It causes large amounts of methane to be released into the atmosphere both accidentally and deliberately.
It reduces dependence on foreign petroleum.	It creates air pollution around drilling sites.

Chapter 14 Study Guide

Focus Points

14.1 Minerals and Rocks: Building Earth's Crust

- **Composition of rocks:** A rock is composed of one or more minerals.
- **Rock families:** All rocks can be grouped into the igneous, sedimentary, and metamorphic rock families.
- **The rock cycle:** Rocks are formed, transformed, and recycled into the mantle through the rock cycle.

14.2 Cooling the Inferno: Igneous Rocks

- **Melting:** Mantle material melts into magma by means of decompression melting and flux melting.

- **Tectonic settings:** Igneous rocks form in hot spots, mid-ocean ridges, rift valleys, and subduction zones.
- **Intrusive igneous rocks:** Intrusive igneous rocks cool slowly, allowing time for the growth of crystals.
- **Extrusive igneous rocks:** Extrusive igneous rocks cool relatively quickly and have smaller or no crystals.

14.3 Layers of Time: Sedimentary Rocks

- **Prevalence of sedimentary rocks:** Sedimentary rocks cover 75% of the crust's surface, but they make up only 4% of the crust.
- **Sedimentary rock types:** The three types of sedimentary rocks are clastic, organic, and chemical sedimentary rocks.
- **Coal:** Coal is widely used as an energy source, but its mining causes environmental problems.

- **Evaporites:** Evaporites are an economically important type of chemical sedimentary rock.
- **Sedimentary rocks and Earth history:** Fossils in sedimentary rock reveal ancient life and ancient environments on Earth.

14.4 Pressure and Heat: Metamorphic Rocks

- **Metamorphism:** Rock of any type can experience metamorphism if it is heated, subjected to high pressures, or both.
- **Tectonic settings of metamorphism:** Most metamorphism occurs at convergent plate boundaries.

14.5 Geographic Perspectives: Fracking for Shale Gas

- **Fracking and fuel supply:** Shale gas obtained through fracking accounts for 67% of natural gas production and 50% of petroleum production in the United States.
- **Pros and cons:** Fracking for petroleum and natural gas provides cheap energy, but it can pollute the environment.

Key Terms

batholith, 431
cementation, 428
chemical sedimentary rock, 434
clastic sedimentary rock, 434
coal, 434
crystallization, 425
dike, 431
evaporite, 438
exhumation, 431
extrusive igneous rock, 430
fossil, 438
fracking, 442
igneous rock, 427
intrusive igneous rock, 430
laccolith, 431
limestone, 434

lithification, 428
metamorphic rock, 427
mineral, 425
organic sedimentary rock, 434
outcrop, 427
peat, 435
pluton, 431
protolith, 439
rock, 425
rock cycle, 427
sandstone, 434
sediment, 427
sedimentary rock, 427
shale, 434
sill, 431

Concept Review

The Human Sphere: People and Rocks

1. What roles have minerals and rocks played in human history?

14.1 Minerals and Rocks: Building Earth's Crust

2. What is a mineral? Provide some examples.

3. What are the major mineral classes? Which is most common?

4. What are rocks composed of?

5. Within the context of the rock cycle, compare igneous, sedimentary, and metamorphic rocks.

6. Through what process are rocks recycled into the mantle?

7. Where does the energy needed to drive the rock cycle come from?

8. How does plate tectonics relate to the rock cycle?

14.2 Cooling the Inferno: Igneous Rocks

9. Describe the tectonic settings in which igneous rocks form.

10. In what two ways does mantle material melt into magma in the crust?

11. What is an igneous rock? How is an intrusive igneous rock different from an extrusive igneous rock?

12. Give examples of igneous rock formations found beneath the surface. How are these subsurface features exposed at the surface through geologic time?

13. Compare phaneritic igneous rocks with aphanitic igneous rocks with respect to where each forms and the size of the crystals within each. Give examples of each.

14. Provide examples of igneous rock types.

15. Compare felsic igneous rocks with mafic igneous rocks with respect to their silica content.

14.3 Layers of Time: Sedimentary Rocks

16. What is a sedimentary rock? How are sedimentary rocks formed?

17. What percentage of Earth's crust is sedimentary rock? What percentage of the exposed crust is sedimentary rock?

18. Compare clastic, organic, and chemical sedimentary rocks in terms of how each is formed.

19. Give examples of depositional environments where sedimentary rocks form.

20. Provide examples of sedimentary rock types.

21. Coal is both a sedimentary rock and a metamorphic rock. Explain how this is true. Describe how coal forms.

22. Describe petroleum and natural gas in the context of their importance to people.

23. What is mountaintop removal mining? Where is it practiced? How is it environmentally destructive?

24. How is salt formed? What kind of sedimentary rock is salt?

25. What are fossils? What information do fossils give us about Earth's history and life?

14.4 Pressure and Heat: Metamorphic Rocks

26. What is metamorphism? List two tectonic environments in which metamorphism occurs.

27. What are the two categories of metamorphic rocks? How do they differ in appearance?

28. What is a protolith?

29. List several examples of metamorphic rocks and their protoliths.

14.5 Geographic Perspectives: Fracking for Shale Gas

30. What is fracking? Describe the steps in the process of fracking.

31. What are the roles of water and chemicals in fracking?

32. As of 2015, how many petroleum and gas wells using the fracking technique were in operation in the United States?

33. What is fracking fluid, and why is it contaminated with toxins and radioactivity?

34. List the pros and cons of fracking.

Critical-Thinking Questions

1. Using only your own knowledge, draw a mental map connecting your house or campus to the nearest outcrop of rock. How far do you have to go?

2. Do all rocks—those of the oceans and those of the continents—experience recycling in the rock cycle? Explain.

3. Without exhumation, would we be able to find metamorphic rocks at Earth's surface? Explain.

4. How do shale and coal relate to climate change?

5. Weigh the pros and cons of fracking. If a big oil and gas company proposed fracking near where you live, do you think you would support it or oppose it? If it is already going on in your area, do you think it has had a positive or negative influence on the region?

Test Yourself

Take this quiz to test your chapter knowledge.

1. **True or false?** An outcrop is exposed bedrock.

2. **True or false?** If Earth's interior were cold, there would be no rock cycle.

3. **True or false?** Most of Earth's crust is composed of sedimentary rocks.

4. **True or false?** Clay lithifies into shale.

5. **True or false?** Sandstone is an example of an organic sedimentary rock.

6. **Multiple choice:** Which of the following rocks is formed only with great heating and compression?
   a. marble
   b. sandstone
   c. granite
   d. shale

7. **Multiple choice:** Which of the following is the most common sedimentary rock type?
   a. limestone
   b. sandstone
   c. shale
   d. coal

8. **Multiple choice:** Which of the following is not an intrusive igneous feature?
   a. batholith
   b. dike
   c. sill
   d. fossil

9. **Fill in the blank:** _____ is the removal of overlying material through erosion to expose igneous rock formations.

10. **Fill in the blank:** The process of turning sediments into rock is called _____.

Online Geographic Analysis

Earthquake Activity in Oklahoma

In this online interactive exercise you will explore earthquake activity in Oklahoma and examine satellite imagery for drill rigs.

Activity

Go to https://earthquake.usgs.gov/earthquakes/, click the "Latest Earthquakes" link, and answer the following questions.

1. **You may need to zoom in closer. Examine Oklahoma. The default map shows earthquakes that have occurred within the past day. Has an earthquake(s) occurred in Oklahoma within the past day? If yes, proceed to Question 2. If not, proceed to Question 3.**

2. **Click one of the dots signifying an earthquake. What was the magnitude (M)? What was the depth?**

3. **In the upper right of the page, click the settings icon (which looks like a gear). From the drop-down menu select "30 Days, Magnitude 2.5+ U.S." Also, make sure the "hamburger" icon in the upper right is activated so that the list of all earthquakes appears on the left margin of the page. Now reexamine Oklahoma and compare it to neighboring states. Does Oklahoma have more or fewer earthquakes compared to its neighbors?**

4. **Zoom out and pan over to California. How does Oklahoma's earthquake activity compare with California's? Based on visual inspection, which state would you say has more earthquakes?**

5. **What do the different colors of the earthquake dots represent? Click a few different colors and compare the information about them to answer this question.**

6. **What do the different sizes of the earthquake dots represent?**

7. **Zoom so that you can see only Oklahoma in the window view. Now scan the list of earthquakes in the list on the left of the screen. What is the deepest earthquake, and what is the shallowest earthquake?**

8. **Click the settings icon in the upper right, scroll down, and activate the "U.S. Faults" option. California has hundreds of faults and earthquakes. Oklahoma has virtually no faults but lots of earthquakes. Can you explain why this is?**

9. **Click the settings icon in the upper right, scroll down, and activate the "Satellite" map view. Find a cluster of earthquakes and zoom in as close as you can get. Are there drill rigs near the earthquakes? (A drill rig is recognizable as a single dead-end road leading to a small rectangular, gray clearing with machinery on it.)**

10. **Do the same for the strongest earthquakes you can find in Oklahoma. Are there drill rigs near those?**

Picture This. *Your Turn*

Rock Identification

Identify the major rock types pictured here. Write the answer in the space provided under each photo. Choose from the following list of terms and use each term only once.

1. coal

2. extrusive igneous rock

3. gneiss

4. granite

5. marble

6. sandstone

This rock cools directly from lava.

This rock forms under great heat and pressure and has a banded structure.

This rock forms layers and is composed of cemented sediments.

This rock cools deep within the ground directly from magma.

This is an organic sedimentary rock.

This rock has formed under great heat and pressure. Its protolith is limestone.

(Top row. Stocktrek Images/Richard Roscoe/Getty Images; Siim Sepp/ Alamy; center row. Bruce Gervais; © Siim Sepp/Shutterstock.com; bottom row. li Jingwang/Getty Images; Grant Faint/Getty Images)

For animations, interactive maps, videos, and more, visit www.saplinglearning.com.   **SaplingPlus**

Geohazards: Volcanoes and Earthquakes

Chapter Outline *and Learning Goals* ◎

15.1 About Volcanoes

◎ Describe the four main types of volcanoes and major landforms associated with each.

15.2 Pele's Power: Volcanic Hazards

◎ Explain the hazards volcanoes pose and how volcanoes are ranked.

15.3 Tectonic Hazards: Faults and Earthquakes

◎ Explain what causes earthquakes and which geographic areas are most at risk.

15.4 Unstable Crust: Seismic Waves

◎ Describe the types of seismic waves produced by earthquakes, how earthquakes are ranked, and what can be done to reduce human vulnerability to earthquakes.

15.5 Geographic Perspectives: The World's Deadliest Volcano

◎ Assess the links between large volcanic eruptions, Earth's physical systems, and people.

This ancient fresco, found in the town of Pompeii, Italy, depicts the Greek god Heracles. A catastrophic eruption from nearby Mount Vesuvius buried Pompeii in volcanic ash in 79 CE. Those unfortunate victims who didn't escape were buried alive. The ash perfectly preserved the city and all its art, including delicate wall murals such as the one shown here.

(Peter Horree/Alamy)

To learn more about Pompeii, turn to page 465.

THE HUMAN SPHERE Killer Waves

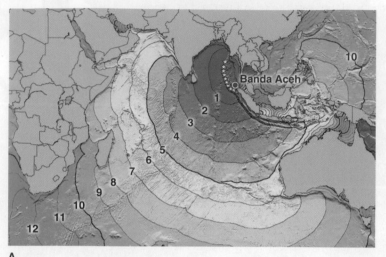

A

Figure 15.1 Banda Aceh. (A) This isochron (*isochrons* are lines of equal time) map shows the first-arrival time of waves that radiated out from the Indonesian earthquake on December 26, 2004. The first waves reached Australia after 5 hours and Africa after 7 hours. (B) A French military helicopter surveys the destruction in Banda Aceh, Indonesia, on January 14, 2005. There the tsunami waves towered 24 m (80 ft) and surged across the city at about 60 km/h (35 mph), far faster than a person can run. (C) Today, the Museum Tsunami Aceh displays the names of many of the 170,000 victims who died in Banda Aceh. on the day of the tsunami. (A. NOAA; B. Joel Saget/AFP/Getty Images; C. Florian Kopp/Getty Images)

Web Map
Tsunami Hazards
Available at
www.saplinglearning.com

B

C

Just before 8 a.m. on December 26, 2004, the seafloor off the coast of the island of Sumatra, in Indonesia, was thrust upward 5 m (16 ft) in a magnitude 9.1 earthquake. This earthquake was the third strongest in recorded history. The movement of the seafloor heaved an estimated 30 km^3 (7.2 mi^3) of seawater upward, creating a series of waves that radiated across the Indian Ocean. Such large ocean waves triggered by an earthquake or other natural disturbance of the ocean floor are called **tsunamis**.

In the open ocean, the waves traveled at nearly the speed of a jetliner (800 km/h [500 mph]), but they went largely undetected because they had a wavelength (the distance between wave crests) of hundreds of kilometers. For this reason, the thousands of boats in the Indian Ocean did not detect the waves as they passed underneath.

As the waves approached shallow water, however, the height of the waves surging ashore grew to 30 m (100 ft) in some regions. The waves devastated coastal areas along the Indian Ocean, particularly in regions nearest the earthquake. Most of the city of Banda Aceh, on the island of Sumatra (**Figure 15.1**), was destroyed.

In response to this catastrophe, the Indian Ocean Tsunami Warning System, similar to a warning system already active in the Pacific Ocean, was developed and activated in June 2006. Cell-phone users can access a free app that is connected to the detection system and provides real-time data and warnings. It is hoped that with this system in place, catastrophic loss of life can be avoided. This chapter focuses on Earth's two most dangerous geologic hazards, or **geohazards** (natural hazards that threaten people): volcanoes and earthquakes.

15.1 About Volcanoes

◎ Describe the four main types of volcanoes and major landforms associated with each.

Plate tectonics theory provides the framework to understand why volcanoes occur where they do. All volcanoes are found where the mantle is melted into **magma**, melted rock below the surface of the crust. Magma forms at or near mid-ocean ridges and continental rifts (divergent plate boundaries), subduction zones (convergent plate boundaries), and hot spots (see Sections 13.3 and 13.4). More than 60% of the Pacific Ocean's margins (borders), totaling some 40,000 km (25,000 mi), are subduction zones with active and potentially dangerous volcanoes. **Active volcanoes**—volcanoes that have erupted in the past 10,000 years and could erupt again—pose the greatest geohazard. Volcanoes that have not erupted for 10,000 years or more but could awaken again are considered *dormant* or *inactive*. An **extinct volcano** is one that has not erupted for tens of thousands of years and can never erupt again.

Four Types of Volcanoes

Volcanoes are surface landforms created by accumulations of **lava** (which is magma that spills onto the surface of Earth's crust) and other materials they emit over time. Volcanoes range in size from small hills of solidified lava to immense mountains. Although they take on many shapes and sizes, most volcanoes can be categorized, from largest to smallest, as shield volcanoes, stratovolcanoes, lava domes, or cinder cones.

A **shield volcano** is a broad, domed volcano formed from innumerable layers of fluid basaltic lava laid down over tens to hundreds of thousands of years (**Figure 15.2**). Shield volcanoes are slowly built as magma travels up from the **magma chamber** (the reservoir of magma beneath a volcano) and through the **volcanic vent** (the conduit through which magma moves and reaches the surface). A **volcanic crater**, a bowl-shaped volcanic depression, may form at the summit of a volcano. **Flank eruptions** happen where lava pours out the sides (or flanks) of a volcano through side vents, and they play an important role in building shield volcanoes.

Shield volcanoes are the largest volcanoes on Earth. In fact, they can be so large that they look like broad, gently sloped horizons and may be difficult to identify as volcanoes from the ground.

Stratovolcanoes differ from shield volcanoes in that they are smaller, steeper sided, and composed of lava and volcanic ash. A **stratovolcano**, or *composite volcano*, is a large, potentially explosive, cone-shaped volcano composed of alternating layers of lava and pyroclasts. **Pyroclasts** (meaning "fire clasts"), or *pyroclastic materials*, encompass any fragmented solid material that is ejected from a volcano. Pyroclasts range in size from volcanic **ash**—pulverized rock particles and solidified droplets of lava that form a fine powder—to large boulders. Stratovolcanoes form over millennia as lava flows and pyroclasts accumulate layer by layer. Eventually, if stratovolcanoes do not

Video

Kīlauea Volcano — Fissure 8 Flow: From Vent to Sea

Available at
www.saplinglearning.com

▶

Animation

Shield Volcano Formation

Available at
www.saplinglearning.com

▶

Figure 15.2 Shield volcano structure. (A) Shield volcanoes are built of innumberable layers of basaltic lava flows. (B) Mauna Kea, on Hawai'i, is among Earth's largest shield volcanoes and has a typical shield volcano profile. The island of Hawai'i, formed on a hot spot (see Section 13.4), is made up of five shield volcanoes that have joined together. Mauna Kea stands at 4,207 m (13,803 ft) and is the highest of the five volcanoes on Hawai'i. *(Peter French/Getty Images)*

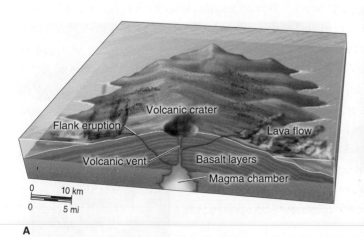

A

B

blow themselves apart with violent eruptions, their profiles can tower over landscapes and can been seen from hundreds of kilometers away **(Figure 15.3)**.

Lava domes (or *plug domes* or *volcanic domes*) are dome-shaped volcanoes that form when thick lava that cannot easily flow piles up around a volcanic vent and solidifies into a domed structure **(Figure 15.4)**. Lava domes are far smaller than stratovolcanoes, typically only a few hundred meters high. A lava dome usually forms from a single eruptive episode over a period of weeks to a few years. Lava domes may form on the flanks of shield volcanoes and stratovolcanoes; they may also cap shield volcano and stratovolcano summit craters.

Cinder cones are small, cone-shaped volcanoes consisting of pyroclasts (often called *cinders* or *scoria* when associated with cinder cones) that

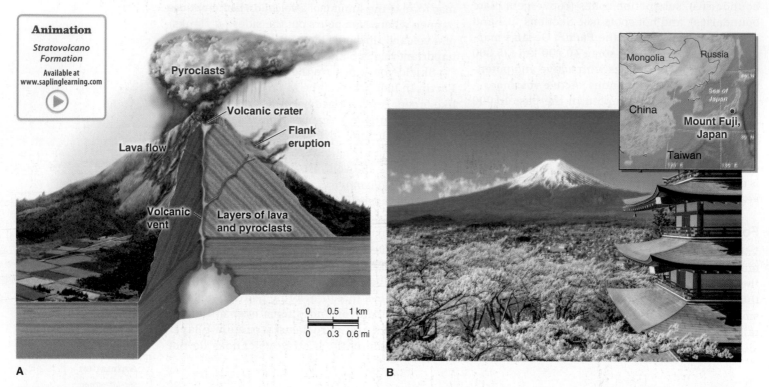

Animation

Stratovolcano Formation

Available at www.saplinglearning.com

Figure 15.3 Stratovolcano structure. (A) The interior structure of a stratovolcano consists of a volcanic vent surrounded by alternating layers of lava flows and pyroclasts. (B) Mount Fuji, an active stratovolcano, has a symmetrical conical profile typical of stratovolcanoes. It reaches a height of 3,775 m (12,387 ft). *(Ikunl/Getty Images)*

Figure 15.4 Lava dome structure.

(A) A lava dome forms as thick lava solidifies into a rounded protrusion, typically no more than a few hundred meters in diameter. (B) The photograph shows one of the many lava domes that make up the Mono Craters region in eastern California. *(RGB Ventures/SuperStock/Alamy)*

Video

Time-lapse Images of Mount St. Helens Dome Growth

Available at www.saplinglearning.com

settle at the **angle of repose**: the steepest angle at which loose sediments can settle. The steepness of the slope of a cinder cone ranges from 25 to 35 degrees, depending on the size of the pyroclasts that were ejected during their formation. Cinder cones erupt for a few years or decades and then become extinct. Most cinder cones are smaller than 400 m (1,300 ft) and symmetrically cone-shaped **(Figure 15.5)**.

Flows of lava also form during the ejection of pyroclasts that make up the cinder cone. Because it is heavier than the pyroclasts, the lava moves beneath the accumulated heap of pyroclasts. Cinder cones can form in any volcanic setting but are particularly common on the slopes of shield volcanoes.

Of the four types of volcanoes, cinder cones are the smallest. **Figure 15.6** illustrates and compares the differences in the sizes and shapes of the four volcano types.

Volcanic Materials and Landforms

Volcanoes build Earth's crust as they pour lava onto the surface which solidifies into volcanic rocks (or extrusive igneous rocks). Stratovolcanoes and shield volcanoes produce a variety of physical materials, such as ash, rock debris, and lava flows. They also create landforms ranging in size from small volcanic craters to immense lava fields covering thousands of square kilometers. Here we discuss three categories of volcanic products: lavas, pyroclasts and gases, and volcanic landforms.

Flowing Molten Rock: Lava

Lava is one of the most conspicuous products of volcanic activity. Lava comes only from volcanoes or volcanic *fissures* in the ground. Lava flows range from fast-moving sheets of basaltic lava to blocky, glowing boulders that slowly push and tumble across a landscape. Lava also forms

Figure 15.5 Cinder cone structure. (A) Cinder cones consist of pyroclastic material that has settled out in a cone near a volcanic vent. The larger, heavier material settles close to the vent, and the smaller, lighter material settles farther away. (B) The southern end of Lake Turkana in Kenya is the site of several cinder cones. This one is 1 km (0.6 mi) in diameter and 220 m (700 ft) tall, and its crater is 187 m (600 ft) deep. *(Gallo Images/Alamy)*

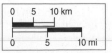

A

B

Figure 15.6 Volcano sizes. A typical cinder cone, a lava dome, Mount Fuji stratovolcano, and the Big Island of Hawai'i (composed of five fused shield volcanoes) are drawn to scale to show their relative sizes. Much of Hawai'i is submerged beneath the ocean, and the immense size of its shield volcanoes is hidden beneath the water.

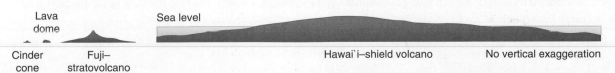

Lava dome

Sea level

Cinder cone   Fuji–stratovolcano   Hawai'i–shield volcano   No vertical exaggeration

0   5   10 km

0   5   10 mi

cohesive masses of molten rock that, when thick enough, can plug a volcanic vent and create lava domes.

The *viscosity* of lava refers to its thickness and resistance to flowing. The higher the lava's viscosity, the more resistant it is to flowing. The viscosity of lava depends on many factors, including its temperature, gas content, crystal content, and silica (SiO_2) content. Silica plays an important role in determining lava viscosity because it forms long chains of molecules that bind the lava together.

Three types of lava are classified according to their silica content and temperature: mafic lava, intermediate lava, and felsic lava. **Mafic lava** has a temperature of about 1,000°C to 1,200°C (1,800°F

to 2,200°F) and a silica content of 50% or less **(Figure 15.7)**. These features mean that mafic lava has a low viscosity, and, as a result, it flows easily.

Some of the most spectacular lava formations are created where mafic lava flows over the surface of a volcano. **Figure 15.8** shows some of these mafic lava formations.

Intermediate lava has a temperature of about 800°C to 1,000°C (1,500°F to 1,800°F), a silica content between 50% and 70%, and a medium viscosity (see Figure 15.7). *Andesite lava,* often called *blocky* lava because of its blocky texture as it moves downslope, is one type of intermediate lava. Stratovolcanoes are composed mostly of andesite and pyroclasts.

Figure 15.7 Three types of lava. (A) Mafic lava forms basalt lava flows. Basalt flows have a low viscosity and flow in streams or sheets downslope. These flows can eventually form shield volcanoes. In this photo, a *volcanologist* (a scientist who studies volcanoes) is sampling mafic lava in Hawai'i Volcanoes National Park, on the island of Hawai'i. (B) Intermediate lava resists movement because it has more silica and is more viscous than mafic lava. Intermediate lava forms andesite flows that solidify into andesite. Stratovolcanoes are composed primarily of andesite and pyroclasts. This intermediate lava, being photographed on Mount Etna, Sicily, has a blocky consistency. (C) Felsic lava has the highest silica content and moves sluggishly downslope as rhyolite flows that solidify into rhyolite. Felsic lava formed the lava dome in the volcanic vent of Mount St. Helens, a stratovolcano in the state of Washington, which is shown here.   *(A. David R. Frazier/Science Source; B. © Tom Pfeiffer/www.volcanodiscovery.com; C. USGS/photo by John S. Pallister)*

Of the three lava types, **felsic lava** has the lowest temperature range—about 650°C to 800°C (1,200°F to 1,500°F)—and the highest silica content—70% or more. The resulting high viscosity restricts felsic lava's ability to flow. Lava domes, which may block volcanic vents, are composed of viscous felsic lava that forms rhyolite rocks (see Figure 15.7).

What determines the amount of silica in magma? Two main factors determine its silica content: the makeup of the solid mantle material from which the magma first melted and the type of rock (called *country rock*) the magma passes through on its way to the surface of the crust. For example, as magma migrates through granitic crust in a subduction zone, it partially melts the surrounding granite through which it is passing (see Section 14.2). Granite is high in silica and is mixed into the magma, creating felsic magma. On the other hand, magma migrating up through basalt oceanic crust, as it does at a hot spot, becomes mafic lava that spills from a volcano.

Ejected into the Air: Pyroclasts and Gases

Explosive volcanic eruptions produce pyroclasts that settle to the ground to form layers of ash and debris, collectively known as *tephra*. Pyroclasts have a wide range of sizes, shapes, consistencies, and means of formation. Some common types of pyroclasts are described here.

During explosive eruptions, volcanoes spray droplets of lava high into the atmosphere. These tiny droplets solidify in the air as they cool, forming fine, powdery fragments of *volcanic ash*. Powdery volcanic ash is also formed in powerful eruptions when existing rock from the volcano is shattered, pulverized, and blown aloft during the eruption.

A volcano's eruption column consists mostly of ash, but it is also mixed with other pyroclasts of larger sizes. Both lapilli and pumice, for example, are formed as ejected lava fragments cool while in the air. **Lapilli** (pronounced la-PILL-eye) are rounded marble- to golf ball–sized fragments of lava that solidify while airborne

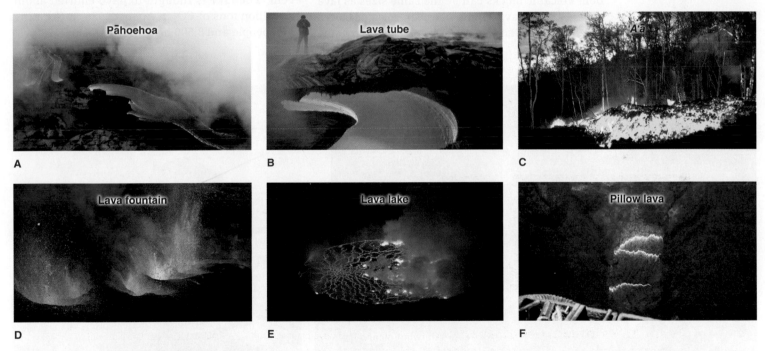

Figure 15.8 Mafic lava formations. (A) When mafic lava flows in smooth streams or sheets over the surface, it is called *pāhoehoe* (pronounced pa-HOY-HOY). (B) Pāhoehoe often flows beneath the surface through a channel called a *lava tube*. (C) When mafic lava takes on a blocky, rough surface, it is called *'a'ā* (pronounced AH-a). (D) Sometimes it is ejected forcefully to form a *lava fountain*. (E) In some instances, the volcanic vent fills with a pool of lava, creating a *lava lake*. There are only about five lava lakes in the world. The number of them varies as they fill and drain through time. (F) Underwater, mafic lava forms rounded mounds called *pillow lava*. *(A. Iordache Gabriel/Alamy; B. Images & Volcans/Science Source; C. Stephen & Donna O'Meara/Volcano Watch Int'l/Getty Images; D. Arctic-Images/Getty Images; E. Stocktrek Images/Richard Roscoe/Getty Images; F. NOAA and NSF)*

(Figure 15.9A). **Pumice** (pronounced PUM-iss) is a lightweight, porous rock with at least 50% air content. Pumice forms from silica-rich lava, frothy with gas bubbles, that has cooled rapidly, either while airborne or under water. The air spaces that the bubbles occupied are preserved as the lava solidifies. These pore spaces within pumice rock are called *vesicles*. Pumice size varies from small pebbles to large boulders **(Figure 15.9B)**. Pumice rock floats on water, and island volcanoes sometimes disgorge huge amounts of pebble-sized pumice into the oceans, forming sheets of floating pumice debris called pumice rafts that can cover large swaths of ocean surface. Eventually these *pumice rafts* either wash up on the beaches of a shoreline or, after many months, become waterlogged and sink to the seafloor.

A **lava bomb** is a streamlined or spherical fragment of lava ejected from a volcano that cools and hardens as it is moving through the air. Lava bombs are generally potato to soccer ball sized, but they can be the size of a minivan or even larger. A **volcanic block** is a fragment of rock that is torn from the volcano's cone and ejected during an explosive eruption. Volcanic blocks can be the same sizes as lava bombs. Unlike lava bombs, however, volcanic blocks are not in a molten state as they are ejected from the volcano; rather, they are solid pieces of the volcano itself that are forcibly blown out of the volcano. As a result, they have angular sides, in contrast to the more rounded shape of lava bombs **(Figure 15.10)**.

By volume, about 8% of most magma is gas. *Volcanic gas* is not a pyroclastic material, but as gas forcefully exits a volcano, it blasts lava and rock debris into the air, generating pyroclasts. Gas in magma expands as the magma migrates toward the surface of the crust, where there is less pressure. At the surface, the gases in magma expand rapidly, creating explosions.

The main gases emitted by volcanoes are water vapor (H_2O), carbon dioxide (CO_2), sulfur dioxide (SO_2), and hydrogen sulfide (H_2S). Aside from their role in generating pyroclasts, volcanic gases are not usually lethal to people because they are usually diluted before they reach populated areas. Where volcanic gases are concentrated, however, they can be lethal. An example of this occurred near Lake Nyos, in the western African country Cameroon, in 1986. Lake Nyos is located on an inactive volcano. A magma chamber below the lake leaks CO_2 into the lake. Occasionally, the CO_2 is released from the lake in a sudden *outgassing* event. The lake is thought to have emitted about 1.6 million tons of CO_2 in August 1986, suffocating 1,700 people and 3,500 head of livestock.

Figure 15.9 Lapilli and pumice. (A) The small stones in the top photo are lapilli formed by an eruption on the flank of Darwin Volcano on the Galápagos Islands. The lower photo is of a polished cross section of a lapillus. The rings in the lapillus formed as the lava fragment traveled through the air and molten debris stuck to it, much the way hailstones grow in a thunderstorm. (B) A geography student easily lifts a large, air-filled pumice boulder near Mono Lake, California. *(A. © David K. Lynch; B. Bruce Gervais)*

After the Lava Cools

Volcanic landforms are among the most spectacular landforms on the planet. Some notable volcanic landforms, in addition to volcanic mountains, are columnar jointing, large igneous provinces, and calderas.

As basaltic lava cools and hardens into rock, cracks and weak planes in the rock, called **joints**, develop. A geometric pattern called **columnar jointing**, shown in **Figure 15.11**, sometimes forms when angular columns result from joint formation in the lava during cooling.

Figure 15.10 Lava bombs and volcanic blocks. (A) This lava bomb was found in the Pinacate Volcanic Field in northwestern Sonora, Mexico. Its typical streamlined shape is the result of airflow around it as it cooled in flight. (B) The flanks of Licancabur Volcano, near the town of San Pedro de Atacama in central Chile, are littered with volcanic blocks that once made up the volcano but were torn from it during eruptions. *(A. © Peter L. Kresan; B. Paul Harris/Getty Images)*

Figure 15.11 Columnar jointing.

Most of Staffa, an island in the Inner Hebrides of northwestern Scotland, is composed of mafic lava that slowly cooled 55 million years ago, allowing time for columnar jointing to form. Here, the joints mainly sit perpendicular to the cooling surface.

(Image Source/Getty Images)

Large igneous provinces (LIPs) are accumulations of basaltic lava that cover extensive geographic areas. If you have ever driven through eastern Washington and Oregon, you have driven over the Columbia Plateau. The rocks of the Columbia Plateau superficially resemble sedimentary rocks with horizontal layers, but they are *flood basalts*, lava flows that poured onto the crust over a period of several million years. The Columbia Plateau flows formed between 17 million and 6 million years ago and created a large igneous province. There are several dozen large igneous provinces around the world **(Figure 15.12)**.

After a volcanic eruption, the emptied magma chamber can collapse, forming a large circular depression called a **caldera** (from the Spanish word for "cauldron"). The process of caldera formation is illustrated in **Figure 15.13**. Calderas usually have flat bases and steep slopes and can be many kilometers in diameter. Although calderas can be mistaken for meteor impact craters, the two can be differentiated because each leaves distinct evidence (see the **Picture This** on the following page).

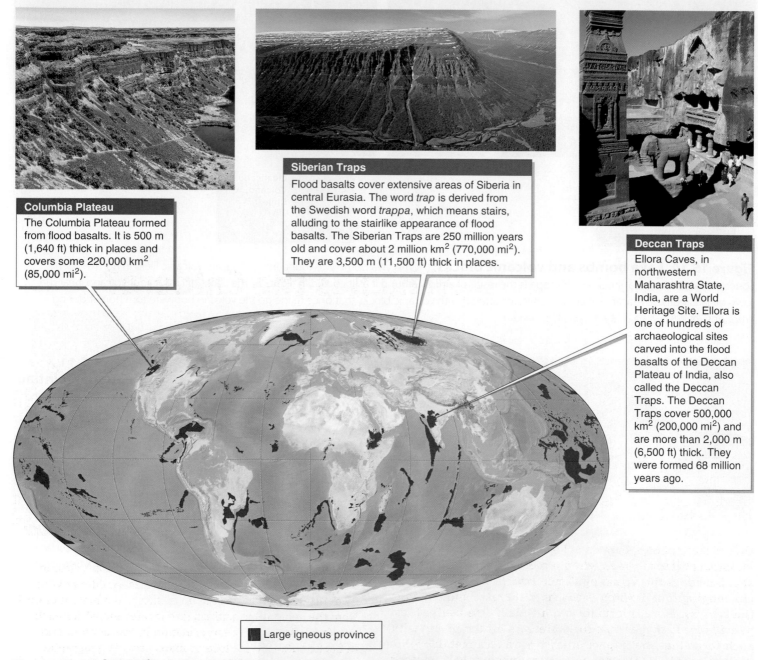

Columbia Plateau

The Columbia Plateau formed from flood basalts. It is 500 m (1,640 ft) thick in places and covers some 220,000 km^2 (85,000 mi^2).

Siberian Traps

Flood basalts cover extensive areas of Siberia in central Eurasia. The word *trap* is derived from the Swedish word *trappa*, which means stairs, alluding to the stairlike appearance of flood basalts. The Siberian Traps are 250 million years old and cover about 2 million km^2 (770,000 mi^2). They are 3,500 m (11,500 ft) thick in places.

Deccan Traps

Ellora Caves, in northwestern Maharashtra State, India, are a World Heritage Site. Ellora is one of hundreds of archaeological sites carved into the flood basalts of the Deccan Plateau of India, also called the Deccan Traps. The Deccan Traps cover 500,000 km^2 (200,000 mi^2) and are more than 2,000 m (6,500 ft) thick. They were formed 68 million years ago.

Large igneous province

Figure 15.12 Large igneous provinces. All of the world's large igneous provinces formed where mantle plumes created geologic hot spots. Most of the eruptions that created these provinces caused global climate change, and some even caused global mass extinction events when they rapidly elevated atmospheric CO_2 levels. *(Left. © Peter L. Kresan; center. Zastavkin/Getty Images; right. Tony Waltham/robertharding/Getty Images)*

Figure 15.13
Quilotoa caldera, Ecuador. (A) A caldera forms as a magma chamber partially empties and collapses. (B) A lake fills the caldera of Quilotoa in Ecuador, which was formed about 8,000 years ago when a major volcanic eruption caused the collapse of the volcano's magma chamber. The caldera gradually filled with rainwater to form this caldera lake. The lake's greenish color is due to dissolved minerals in the water. *(Hemis/Superstock)*

1. Lava extrusion
A magma chamber extrudes lava.

2. Gas expansion
Gases expand and escape explosively.

3. Magma chamber collapse
The partially emptied magma chamber collapses, forming a caldera.

Time

A

B

Honduras
Nicaragua
Panama
PACIFIC OCEAN
Colombia
Quilotoa Caldera, Ecuador
90° W

Animation
Caldera Formation
Available at
www.saplinglearning.com

Picture This **Which Is the Caldera?**

(Randy Olson/Getty Images)

Wolf Creek Crater
AUSTRALIA
15° S
30
45° S   120°   135°   150° W

(WaterFrame/Alamy)

O'ahu, Hawai'i
20° N
PACIFIC OCEAN
155° W

One of these photos shows a volcanic caldera, and one shows an impact crater formed when a meteor struck Earth long ago. Based on the visual evidence from these photos, it is challenging to tell which is the impact crater and which is the caldera. More information is needed. One useful form of evidence is *shatter cone* rock. Meteors hit the planet with such force that the impact energy produces metamorphic shatter cones. Shatter cones are produced only at meteor impact sites. They are not visible in either of these photos.

Consider This

1. If you found a large crater-shaped landform in volcanic rock, could you be 100% certain that it is a caldera? What evidence would you need to be certain it is a meteor crater?

2. Note the geographic settings (see locator maps) for each landform. Based on the information in this chapter and in Section 13.4, do the locator maps provide geographic information that could help you decide which landform is the caldera?

15.2 Pele's Power: Volcanic Hazards

◎ **Explain the hazards volcanoes pose and how volcanoes are ranked.**

In Hawaiian myth, Pele is the volcano goddess. She is said to reside in the summit caldera of Kilauea on the island of Hawai'i. Pele embodies the many facets of volcanoes, ranging from life-sustaining benevolence to destructive malevolence. Volcanoes shape Earth's crust. They can pour cubic kilometers of lava onto Earth's surface to build new islands and landmasses, and they provide nutrient-rich soils that plants thrive in. Volcanoes can also be powerfully explosive and cause catastrophic loss of human life. In this section, we turn to Pele's malevolent side and examine the main geohazards that volcanoes present.

Two Kinds of Eruptions: Effusive and Explosive

Shield volcanoes, such as those found on Hawai'i, present little threat to human life. Shield volcanoes have nonexplosive **effusive eruptions**. Mafic lava from shield volcanoes usually flows slowly downhill and can be easily avoided.

Stratovolcanoes may produce effusive outpourings of mafic lava like shield volcanoes, but they also produce intermediate and felsic lava. As a result, they are potentially serious geohazards because they produce **explosive eruptions**.

The primary difference between effusive eruptions and explosive eruptions is explained by the silica content of the magma. As discussed in Section 15.1, silica causes the viscosity of lava to increase. In thin, low-viscosity lava, gases that form within the lava can escape. In thicker magmas with higher silica content, gases become trapped in the magma, building up to erupt in a potentially explosive way **(Figure 15.14)**.

Explosive eruptions send rock, ash, and volcanic gases high into the troposphere, or even into the stratosphere. In the troposphere, rain washes the volcanic material out in a few days or weeks. There is no rainfall in the stratosphere, however, so once ash and sulfur gases enter the stratosphere, they can remain suspended there for 5 years or more. These materials can encircle the globe and cause climate cooling for a few years (see Section 7.2). **Figure 15.15** is a photo of an explosive volcanic eruption in which the force of the expanding gases and collapsing magma chamber sent ash billowing high into the atmosphere.

Figure 15.14 Explosive eruptions. The way gas forms and escapes from a bottle of soda is similar to the way gases form and escape from magma. When thick felsic magma traps the gases, the volcano can become dangerously explosive. This process can unfold over the course of hours to days, depending on the size of the volcano, and can result in a catastrophic explosive eruption. *(Bruce Gervais)*

2. Gases leave solution
Quickly removing the cap removes the pressure that holds the CO_2 in solution in the soda. With less pressure, the H_2CO_3 separate to form CO_2 and H_2O. The CO_2 produces gas bubbles, which are lighter than the liquid and come rushing out.

1. Gases in solution
Sodas like this root beer have CO_2 added to them to make them fizzy. Then they are sealed under pressure in the factory. Most of the CO_2 in the soda is not a gas but instead dissolves in *solution*, forming *carbonic acid* (H_2CO_3). Carbonic acid forms by combining water and carbon dioxide:
$$CO_2 + H_2O \longleftrightarrow H_2CO_3$$

A

3. Explosive eruption
Whether a volcanic eruption is violent or not depends in large part on how much gas is trapped within the magma and released. High-viscosity (silica-rich) felsic magma traps more gases, creating more explosive eruptions.

2. Gases leave solution
When magma is pushed upward through the vent, pressure on the magma is reduced, and the gases escape solution and form gas bubbles. This makes the magma more buoyant, causing it to rise faster, further reducing pressure, which causes more gas to form.

1. Gases in solution
Like the contents of a soda bottle, deep magma is under great pressure. Its gases are therefore in solution, particularly H_2O, CO_2, and SO_2 (sulfur dioxide).

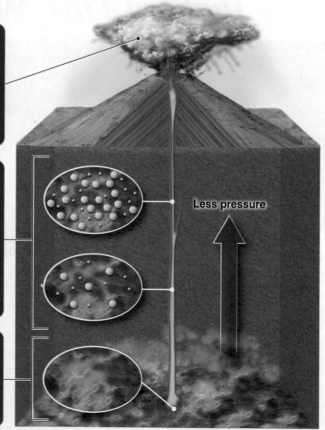

Less pressure

B

Figure 15.15 Volcanic ash cloud. This photo of the Soufrière Hills volcano on the island of Montserrat in the Caribbean Sea was taken March 24, 2010, by a passenger on a commercial aircraft. The volcano sent ash 12,500 m (40,000 ft) into the atmosphere. *(Mary Jo Penkala/Solent News & Photo/Sipa Press/Newscom)*

Ranking Volcanic Eruption Strength

The **Volcanic Explosivity Index (VEI)** ranks volcanic eruption magnitude based on the amount of material a volcano ejects during an eruption. A VEI 5 eruption emits more than 1 km³ (0.24 mi³) of pyroclastic material into the atmosphere, and a VEI 6 eruption emits more than 10 km³ (2.4 mi³). During the past 10,000 years, there have been about 50 VEI 6 eruptions. The eruption of Tambora, described in the Geographic Perspectives at the end of the chapter, has been the only VEI 7 eruption in historic times. **Figure 15.16** compares large historical eruptions to the colossal prehistoric eruption of the Yellowstone caldera, 640,000 years ago.

The Two Greatest Threats: Lahars and Pyroclastic Flows

Stratovolcanoes are among the most dangerous geohazards on the planet. Seventy percent of the world's active land volcanoes are located in the Ring of Fire, and the majority of them are stratovolcanoes. Lava is not the biggest threat these volcanoes pose because stratovolcano lava usually flows slowly, and people can escape. The two greatest volcanic hazards are lahars and pyroclastic flows. Together, they account for about half of the volcano-related deaths in any given year.

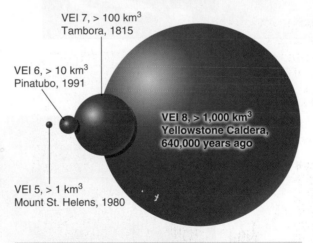

VEI 7, > 100 km³
Tambora, 1815

VEI 6, > 10 km³
Pinatubo, 1991

VEI 8, > 1,000 km³
Yellowstone Caldera,
640,000 years ago

VEI 5, > 1 km³
Mount St. Helens, 1980

Volcanic Explosivity Index

VEI	PLUME HEIGHT (KM)	FREQUENCY ON EARTH
0	< 0.1	Daily
1	0.1 to 1	Daily
2	1 to 5	Weekly
3	3 to 15	Yearly
4	10 to 25	Every 10 yrs
5	> 25	Every 50 yrs
6	> 25	Every 100 yrs
7	> 25	Every 1,000 yrs
8	> 25	Every 10,000 to 100,000 yrs

Figure 15.16 The Volcanic Explosivity Index.

The eruption of Mount St. Helens in 1980 (VEI 5) emitted about 1 km³ of pulverized rock and volcanic ash into the atmosphere. In contrast, the most recent major Yellowstone eruption (VEI 8), which occurred about 640,000 years ago, ejected 1,000 times more material. The table included here indicates how often eruption magnitudes occur and the height of the eruption plume. VEI 8 events like the Yellowstone eruption, for example, occur only once every 10,000 to 100,000 years and send ash material greater than 25 km (15 mi) into the atmosphere.

Figure 15.17 A lahar. Plymouth, the capital city of the island Montserrat, was destroyed by a lahar from the Soufrière Hills volcano in August 1997. This 2010 photo shows the town buried in 12 m (39 ft) of ash and mud, with the volcano steaming in the background. About 1,500 people were evacuated from Plymouth just before the 1997 eruption. The town was abandoned after it was buried. The Soufrière Hills volcano continues to be very active today, routinely forcing evacuations from the southern part of the island. *(© Bernhard Edmaier/Science Source)*

Torrents of Mud: Lahars

A **lahar** (a Javanese word that means "mudflow" or "debris flow") is a mudflow that results when a snow-capped stratovolcano erupts. During an eruption, the snow melts. A lahar is a thick slurry of mud, ash, and other debris that moves rapidly down the volcano's flank. Lahars can travel tens of kilometers down the slopes of volcanoes and into the flatlands below, where people may reside. Lahars are not hot. Their danger lies in the fact that they move quickly and can engulf whole villages in minutes. **Figure 15.17** shows a lahar that entombed Plymouth, the former capital city of Montserrat, an island in the Caribbean Sea.

Blazing Clouds: Pyroclastic Flows

Pyroclastic flows (also called *nuées ardentes*, meaning "blazing clouds") are quick-moving avalanches of gas and ash. Pyroclastic flows are some of the greatest volcanic hazards because they can travel at speeds up to 700 km/h (450 mph) and can be as hot as 500°C (930°F). These avalanches may even glow orange from their intense heat. The largest flows can travel hundreds of kilometers from the volcanic vent. **Figure 15.18** shows a small pyroclastic flow.

Lahars and pyroclastic flows are by far the most significant geohazards volcanoes present, but they are not the only ones. Volcanoes can also produce large earthquakes, dangerous lava flows, and smothering ashfalls. The **Picture This** on the following page explores an unusual and unfortunate volcanic event that happened in Italy many centuries ago.

Figure 15.18 Pyroclastic flows. Mount Merapi, on the island of Java, is the most active volcano in Indonesia. Here a pyroclastic flow roils down Merapi's flank during a relatively small 2010 eruption. *(Beawiharta/Reuters/Newscom)*

Picture This The Pompeii Disaster

(Bettmann/Getty Images)

In 79 CE, on the morning of August 24, a series of earthquakes shook the region near the Italian city of Pompeii. At about 1 p.m., a menacing black ash cloud billowed up 25 km (15 mi) and shrouded Pompeii (and the nearby city of Herculaneum) in blackness. Eruptions continued for a week. As many as 16,000 people may have died, crushed under the weight of ash as rooftops collapsed or asphyxiated as they were buried alive. The town was entombed beneath 6 m (18 ft) of ash.

In 1749, mysterious terra cotta roof tiles were found beneath farm fields where a canal was being dug, hinting at a lost city beneath. But it was not until the late 1880s that archaeologists began to excavate the ash to reveal the ruins of Pompeii beneath (see the opening photo of this chapter). As they were digging, they found many mysterious cavities in the ash.

When these cavities were injected with plaster, shapes of people were revealed.

Mount Vesuvius, which was responsible for the destruction of Pompeii, is still alive and active. It last erupted for a period of 31 years from 1913 to 1944. Since then, it has been silent. Fully aware of the risk posed by the volcano, the Italian government has offered up to 30,000 euros (US$40,000) to each of the 500,000 people living within the "red zone," the area within 8 km (5 mi) of the volcano, to move farther away. Most have declined this offer.

Consider This

1. Examine the inset map that shows ash fall depths. Which way was the wind blowing when Mount Vesuvius erupted?
2. Why is Vesuvius a far greater geohazard today than at any other time in the past?

Can Scientists Predict Volcanic Eruptions?

Because volcanoes can be such a serious geohazard, predicting their eruptions would save many lives. Volcanic eruptions are not easy to predict. In November 2017, 70,000 people were evacuated from the red zone of Mount Agung in Indonesia because of scientists' warnings of an imminent eruption. The evacuees soon began moving back to their homes, however, because the volcano did not erupt. If a volcano gives clear warning signs scientists can sometimes accurately predict eruptions. The monitoring of Mount St. Helens in Washington State is a good example of the process of monitoring warning signs and successfully anticipating an eruption, as illustrated in **Figure 15.19**.

Figure 15.19
SCIENTIFIC INQUIRY: Can scientists predict dangerous volcanic eruptions?

Careful monitoring of Soufrière Hills volcano in August 1997 allowed officials to evacuate Plymouth, Montserrat (see Figure 15.17), and avoid catastrophic loss of life. Likewise, rumblings from Mount St. Helens allowed scientists to predict its eruption in 1980 and warn people to get out of harm's way. This figure details the evidence that scientists use to ascertain the likelihood of an eruption. *(1. USGS, photo by Thomas Casadevall; 2 and 3. U.S. Geological Survey, photo by Lyn Topinka; 4. U. S. Geological Survey, photo by P. W. Lipman; bottom right. U.S. Geological Survey, Volcano Hazards Program, photo by Mike Doukas)*

1. Scientists take gas samples to understand how magma is moving beneath the ground.

2. Scientists measure surface cracks. Widening of cracks could indicate that magma is rising up through the magma chamber.

3. Seismic stations on the volcano measure earthquake activity.

4. Scientists measure surface swelling from a distance.

The May 1980 eruption of Mount St. Helens was not a surprise. Scientists collected data and closely monitored the volcano before its eruption.

15.3 Tectonic Hazards: Faults and Earthquakes

◎ **Explain what causes earthquakes and which geographic areas are most at risk.**

Video

March 11, 2011 Japan Earthquake and Tsunami

Available at www.saplinglearning.com

▶

On Friday, March 11, 2011, seismographs around the world began detecting one of the largest earthquakes in recorded history, now called the 2011 Tōhoku earthquake. The shaking began at 2:46 p.m. local time. The earthquake was calculated at magnitude 9.0, a colossal event. There are more than 1 million detectable earthquakes on the planet each year, and this single 9.0 event released more energy than all of the others for the year combined. Only four other recorded earthquakes have been larger. The earthquake was 32 km (20 mi) deep and 128 km (80 mi) from Sendai, on the island of Honshu, Japan. The aftershocks that followed for weeks were as powerful as magnitude 7.2.

The damage caused by the earthquake and its aftershocks was made much worse by a tsunami that reduced the low-lying coastal regions in its path to ruins. (See the Human Sphere section at the beginning of this chapter to read more about tsunamis.) Over a million buildings were destroyed by the earthquake and subsequent tsunamis in Japan, totaling US$230 billion dollars in damage. More than 24,000 people died. To make matters even worse, although the Fukushima Daiichi nuclear power plant had withstood the shaking, it was not designed to be flooded by salt water from a tsunami. After it was flooded, its cooling generators failed and, consequently, three of its six reactors suffered critical meltdowns, leaking into the atmosphere radiation that eventually traveled across the Northern Hemisphere. Bringing the reactors under control and stopping their radiation leaks have been among the greatest challenges Japan has faced in the wake of this earthquake. As of 2018, radiation continued to leak from the Fukushima Daiichi nuclear power plant's three reactors into the Pacific Ocean, with no end in sight.

Faulting and Earthquakes

Earthquakes are as dangerous as volcanoes—and they can be even more devastating. The 2011 Tōhoku earthquake, like all other earthquakes, occurred when stresses acting on Earth's crust broke the crust along a geologic fault, which is a fracture in the crust where movement and earthquakes occur. As the crust slipped along the fault, the energy that had accumulated over decades or hundreds of years was all released within a matter of minutes. The energy traveled outward through the Earth as seismic waves, resulting in an earthquake.

Three Types of Faults

The Tōhoku earthquake occurred on a type of fault called a reverse fault. There are three basic types of faults: normal faults, reverse faults, and strike-slip faults **(Figure 15.20** on the following page**)**. A **normal fault** is a result of *tensional* (extension) *force* as two pieces of Earth's crust, called *fault blocks*, are pulled apart. As a result, one fault block slips downward in relation to the other fault block. A **reverse fault** results from *compressional force* that pushes one block upward in relation to another block. Under certain circumstances, reverse faults are also called *thrust faults*. A **strike-slip fault** occurs where one block moves horizontally in relation to another block as a result of *shearing* (lateral) *force*.

Normal and reverse faults create a **fault scarp**, or cliff face, that results from the vertical movement of the fault blocks. Strike-slip faults create little up or down block movement. Where strike-slip faults cross orchards, streams, roads, sidewalks, fence lines, and other linear features, those features may be *offset* by fault movement. *Left-lateral* strike-slip faults occur when, from the perspective of either block, the opposite block moves to the left. *Right-lateral* strike-slip faults, as shown in **Figure 15.21** on the following page, occur when the opposite block moves to the right.

Fault scarps indicate that a normal or reverse fault is at work, and offset features indicate that a strike-slip fault is present. Much of the western United States, including California and Nevada, have many fault systems with all three fault types, as shown in **Figure 15.22** on page 469.

How Do Faults Generate Earthquakes?

When subjected to geologic stresses, fault blocks can creep smoothly past one another. But more often, friction between them causes them to stick together, and stress energy builds up in the crust. Eventually, the geologic stress exceeds the friction, the crust breaks (either along a preexisting fault or along a new fault), and the blocks move. As each block moves, the built-up stress energy is released and travels through the crust as seismic waves, resulting in an earthquake.

Elastic-rebound theory describes how fault blocks bend, break, and rebound to a new position as they move in relation to one another. The blocks may become stuck again from friction and then slip again in this *stick-slip* process. The **focus** (or hypocenter) is the location of initial crust movement along a fault during an earthquake. Most earthquake *foci* (plural of focus; pronounced FOHS-eye)

Figure 15.20 Three fault types. Faults occur where breakage and slippage happen in the crust. The direction of force and the resulting block movement determine the type of fault.

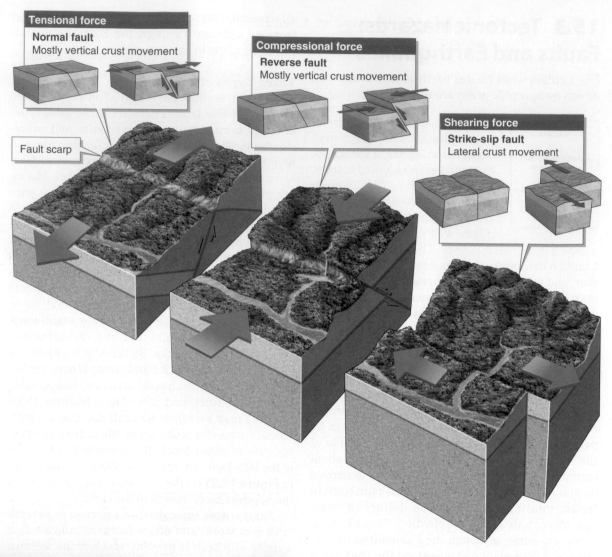

Tensional force
Normal fault
Mostly vertical crust movement

Fault scarp

Compressional force
Reverse fault
Mostly vertical crust movement

Shearing force
Strike-slip fault
Lateral crust movement

Figure 15.21 Right-lateral strike-slip fault. On September 4, 2010, the magnitude 7.1 Canterbury earthquake struck South Island, New Zealand. The tire tracks on this dirt road once connected. This fault is a right-lateral strike-slip fault because the opposite side moved to the right. *(© Kate Pedley Photography)*

Papua New Guinea

AUSTRALIA

Coral Sea

Tasman Sea

South Island, New Zealand

occur at depths less than 50 km (30 mi). Foci in subduction zones can be deeper than 600 km (370 mi). The **epicenter** is the location on the ground's surface immediately above the focus and is usually the area of greatest shaking **(Figure 15.23)**.

What Are Foreshocks and Aftershocks?
Small **foreshock** earthquakes sometimes precede large earthquakes. Foreshocks may be caused by smaller cracks developing as the deformed and stressed crust is about to fail. Consider the bending stick analogy used in Figure 15.23. As stresses are applied and the stick bends, small breakages and splinters of wood may form; these are like the foreshocks that precede an earthquake. Foreshocks may indicate that the stick is about to break—in other words, that the fault is about to slip. The breaking of the stick represents the main earthquake that occurs when the fault slips.

Very commonly, especially with large earthquakes, smaller earthquakes called **aftershocks** follow the main shock. Aftershocks occur as

Figure 15.22 Fault map of California and Nevada. (A) The North American and Pacific plates are fractured by many fault systems in the western United States. (B) In the Great Basin Desert of Nevada, the crust is being rifted and stretched, creating a series of normal faults oriented north–south and resulting in *horst-and-graben topography*. The fault blocks have rotated slightly as the crust has been stretched. Portions of the blocks form grabens (valleys), and portions of them form horsts (mountain ranges), as illustrated here. The photograph shows Nevada's snow-capped Wheeler Peak, part of one of the many mountain ranges in Nevada produced by a rotated and tilted block. *(Bruce Gervais)*

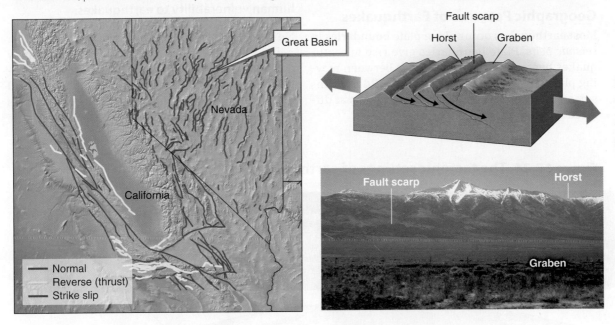

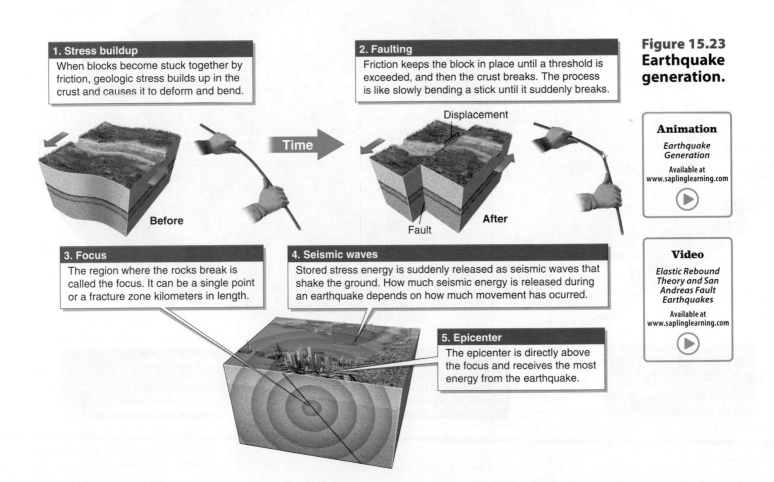

Figure 15.23 Earthquake generation.

1. Stress buildup
When blocks become stuck together by friction, geologic stress builds up in the crust and causes it to deform and bend.

2. Faulting
Friction keeps the block in place until a threshold is exceeded, and then the crust breaks. The process is like slowly bending a stick until it suddenly breaks.

3. Focus
The region where the rocks break is called the focus. It can be a single point or a fracture zone kilometers in length.

4. Seismic waves
Stored stress energy is suddenly released as seismic waves that shake the ground. How much seismic energy is released during an earthquake depends on how much movement has ocurred.

5. Epicenter
The epicenter is directly above the focus and receives the most energy from the earthquake.

Animation
Earthquake Generation
Available at www.saplinglearning.com

Video
Elastic Rebound Theory and San Andreas Fault Earthquakes
Available at www.saplinglearning.com

the crust is settling into its new position after it has been moved. Most aftershocks are much smaller than the main earthquake and occur on the same fault as the initial earthquake. Occasionally, aftershocks occur on different faults nearby.

Geographic Patterns of Earthquakes

Most earthquakes occur along plate boundaries in *seismic belts*. Plate boundaries give rise to earthquakes because of the interactions between moving plates that occur there. **Figure 15.24** explains some major characteristics of earthquakes at different types of plate boundaries.

Web Map

Earthquake Hazards

Available at
www.saplinglearning.com.

15.4 Unstable Crust: Seismic Waves

◉ Describe the types of seismic waves produced by earthquakes, how earthquakes are ranked, and what can be done to reduce human vulnerability to earthquakes.

No two earthquakes are exactly alike. After people have been in an earthquake, they may describe "rolling" or "up-and-down" or "sideways" movement. Earthquakes generate several different types of seismic waves. The movements people experience depend on the dominant type

Figure 15.24 The tectonic settings of earthquakes.

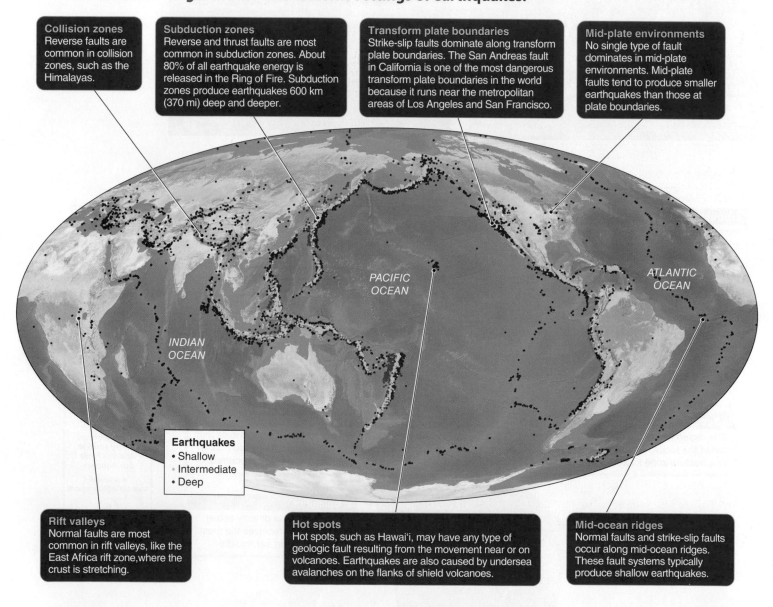

Collision zones
Reverse faults are common in collision zones, such as the Himalayas.

Subduction zones
Reverse and thrust faults are most common in subduction zones. About 80% of all earthquake energy is released in the Ring of Fire. Subduction zones produce earthquakes 600 km (370 mi) deep and deeper.

Transform plate boundaries
Strike-slip faults dominate along transform plate boundaries. The San Andreas fault in California is one of the most dangerous transform plate boundaries in the world because it runs near the metropolitan areas of Los Angeles and San Francisco.

Mid-plate environments
No single type of fault dominates in mid-plate environments. Mid-plate faults tend to produce smaller earthquakes than those at plate boundaries.

PACIFIC OCEAN

ATLANTIC OCEAN

INDIAN OCEAN

Earthquakes
• Shallow
• Intermediate
• Deep

Rift valleys
Normal faults are most common in rift valleys, like the East Africa rift zone, where the crust is stretching.

Hot spots
Hot spots, such as Hawai'i, may have any type of geologic fault resulting from the movement near or on volcanoes. Earthquakes are also caused by undersea avalanches on the flanks of shield volcanoes.

Mid-ocean ridges
Normal faults and strike-slip faults occur along mid-ocean ridges. These fault systems typically produce shallow earthquakes.

of seismic waves passing through the ground beneath them and the type of ground underfoot. Seismic waves can be categorized by where they travel and how they move through the crust. Seismic waves can be grouped into two categories based on where they travel: through Earth's interior or exclusively through the crust. *P waves* (or *primary waves*) and *S waves* (or *secondary waves*) are called *body waves* because they travel through Earth's interior. *R waves* (named after physicist Lord Rayleigh) and *L waves* (named after physicist Augustus Love) are called *surface waves* because

they move only through the crust. These wave types are illustrated in **Figure 15.25**.

Detecting Earthquakes

The instruments used to detect, measure, and record ground shaking are **seismographs** (or *seismometers*). Before the digital era, a seismograph consisted of a swinging pendulum that recorded ground shaking. Modern electronic seismographs generate an electrical signal to measure ground shaking and are far more sensitive and accurate than traditional pendulum

Figure 15.25 Seismic wave types.

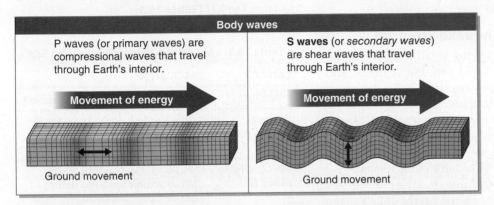

Animation

Seismic Wave Types

Available at
www.saplinglearning.com

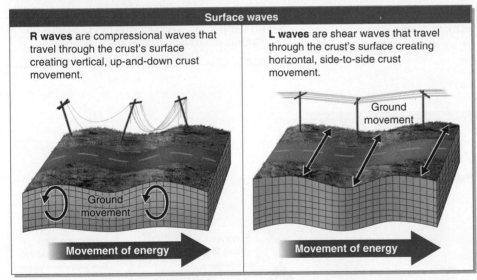

Categories of Seismic Waves
CATEGORIES BASED ON WHERE WAVES TRAVEL:
Body waves can pass through the "body," or interior, of Earth and through the crust.
Surface waves can move only through Earth's crust. They do not move through the interior.
CATEGORIES BASED ON HOW WAVES MOVE:
Compressional waves produce movement that goes back and forth in a direction parallel to the direction of the traveling waves.
Shear waves move back and forth, perpendicular to the direction the waves are traveling.

seismographs. Seismographs are anchored to the ground and record Earth movement digitally. They can also draw the seismic waves on paper in real time. P waves travel the fastest and are the first to arrive after an earthquake. They are soon followed by S waves. L waves and R waves, which arrive last, produce the greatest shaking **(Figure 15.26)**.

The seismic waves detected at the earthquake focus always reach the epicenter first and usually shake the ground at the epicenter the most. Ground shaking typically decreases with distance away from the epicenter because the crust absorbs seismic wave energy.

Ranking of Earthquake Strength

Most earthquakes are too small to be felt by people. Only seismographs can detect them. Many earthquakes that do strongly shake the ground occur in remote areas and are harmless to people. Very rarely, a massive earthquake, such as the Tōhoku earthquake, occurs near a populated region, causing catastrophic loss of life and major structural damage to the built environment.

The amount of ground shaking caused by an earthquake depends on the earthquake's magnitude, the distance from its focus, and the composition of the ground being shaken. Two measures are used to characterize an earthquake's strength: *intensity* and *magnitude*.

Earthquake Intensity

Earthquake intensity is determined by the amount of damage an earthquake causes to physical structures. The *Mercalli intensity scale* (or *Mercalli scale*) was developed in 1902 by Italian scientist Giuseppe Mercalli as a means to estimate the intensity of shaking. No instruments are used to rank earthquakes on the Mercalli scale; instead, the scale is subjectively based on the observed damage done to structures. Later, the Mercalli intensity scale was developed into the **Modified Mercalli Intensity (MMI) Scale**. In this system, earthquakes are ranked using Roman numerals, ranging from I to XII **(Table 15.1)**.

Table 15.1	Modified Mercalli Intensity Scale
CATEGORY	**DESCRIPTION**
I–III: Slight	Not felt, barely noticeable movement.
IV–VI: Moderate to strong	Dishes can be broken. Easily felt by those near the epicenter.
VII–IX: Very strong to violent	Difficult to stand upright. Poorly built structures are badly damaged. Considerable damage to well-built structures may occur.
X–XII: Intense to cataclysmic	Poorly built structures collapse. All buildings may be destroyed, and rivers may be rerouted.

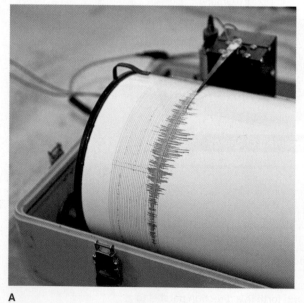

A

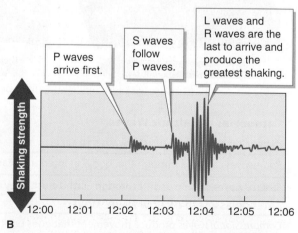

P waves arrive first.

S waves follow P waves.

L waves and R waves are the last to arrive and produce the greatest shaking.

Shaking strength

12:00   12:01   12:02   12:03   12:04   12:05   12:06

B

Figure 15.26 Seismograph and seismic wave sequence. (A) A seismograph uses a stationary magnet and a wire coil to generate an electronic signal. Greater Earth movement creates a stronger voltage that moves the needle more. (B) This seismograph recording shows a typical sequence of seismic wave types over 1-minute increments. *(Zephyr/Science Source)*

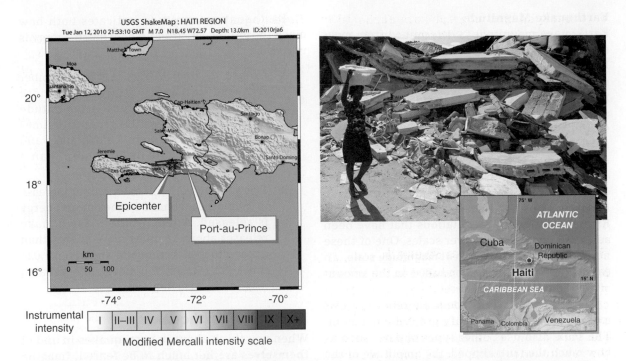

USGS ShakeMap : HAITI REGION
Tue Jan 12, 2010 21:53:10 GMT M 7.0 N18.45 W72.57 Depth: 13.0km ID:2010rja6

| Instrumental intensity | I | II–III | IV | V | VI | VII | VIII | IX | X+ |

Modified Mercalli intensity scale

Figure 15.27 Intensity rankings for the 2010 Haiti earthquake. Color is used in a continuous gradation to denote the intensity of ground shaking during the January 12, 2010, Haiti earthquake. Red areas experienced the greatest shaking. Port-au-Prince, which was close to the epicenter, experienced an intensity of VIII, "very strong to violent." Many structures collapsed on people (right). The death toll for this event is estimated to have been up to 316,000. *(Left. Macmillan Learning; right. Thony Belizaire/AFP/Getty Images)*

No single MMI value is assigned to any given earthquake. Instead, locations progressively farther away from the epicenter experience less shaking as the seismic waves dissipate with distance, so each location is given its own MMI value. **Figure 15.27** provides an MMI map of the 2010 Haiti earthquake.

The distance seismic waves travel through the crust depends in large part on the integrity of the crust. In the western United States, for example, many faults separate sections of the crust, and seismic waves do not travel as far as they would if the crust were not fractured. In the eastern United States, seismic waves tend to travel greater distances because there are fewer faults.

Another factor that influences the intensity of an earthquake is the composition of the ground. Loose, wet sediments deposited by rivers or human-made landfills are susceptible to liquefaction. **Liquefaction** is the transformation of solid sediments into an unstable slurry by ground shaking. Buildings resting on top of sediments may sink during liquefaction, as **Figure 15.28** shows, unless their supporting piles are anchored in more stable ground, such as bedrock.

Figure 15.28 Liquefaction. This building sank into soil that liquefied during the magnitude 7.4 Kocaeli (Izmit) earthquake in Turkey on August 17, 1999. *(Ali Kabas/Getty Images)*

Earthquake Magnitude

Earthquake magnitude is determined from measurements of ground movement using seismographs. More ground movement creates higher-magnitude earthquakes. Each earthquake is given a single magnitude number that indicates the maximum shaking at the epicenter. Scientists can calculate earthquake magnitude from any seismograph on Earth if its distance from the epicenter is correctly established.

In 1935, the American geologist Charles Richter developed the *Richter scale* to quantify earthquake magnitudes using seismographic measurements. Richter's system had limitations that have been addressed by several newer scales. One of these newer scales is the **moment magnitude scale**, an earthquake-ranking system based on the amount of ground movement produced.

The moment magnitude scale relies on seismographic data to quantify ground movement. The scale also uses other types of data, such as how much the fault slipped, the amplitude of the ground movement (its up-and-down and back-and-forth extent), and the physical characteristics of the rocks at the epicenter. It takes several weeks to collect data and calculate the moment magnitude because scientists have to go out and inspect the ground for indications of the extent of movement. Although there is no upper limit to the moment magnitude scale, no earthquake exceeding magnitude 10.0 has ever been recorded. The 2010 Haiti earthquake (Figure 15.27) was a magnitude 7 event. The strongest earthquake ever recorded, which occurred in Chile in 1960, was a magnitude 9.5.

Earthquake magnitude indicates both how much the ground shakes and how much energy is released:

1. *Ground shaking:* With each whole-number increase in magnitude, 10 times more ground movement occurs. A magnitude 5 earthquake shakes the ground 10 times more than a magnitude 4 earthquake and 100 times (10×10, or 10^2) more than a magnitude 3 earthquake.

2. *Energy released:* With each unit of increase in magnitude, about 32 times more energy is released. A magnitude 5 earthquake releases about 32 times more energy than a magnitude 4 earthquake and about 1,024 times (32×32, or 32^2) more energy than a magnitude 3 earthquake.

Living with Earthquakes

When you think about it, earthquakes in and of themselves are not much to be feared. Imagine that you are picnicking in an open, grassy field when a strong earthquake occurs. What would happen? You would first experience up-and-down movement with the arrival of P waves, then you would experience side-to-side movement with the following S waves. Then the R waves and L waves would move the ground up and down and sideways more vigorously. Your drinks would spill, and you might be tossed into the air a few inches or higher. The sensation would be disorienting and exhilarating or terrifying, depending on your perspective. But you would probably not get hurt.

Table 15.2	Top 10 Most Deadly Earthquakes			
RANK	**DATE**	**LOCATION**	**MAGNITUDE**	**FATALITIES**
1	Jan. 23, 1556	Shaanxi, China	8*	830,000
2	Jul. 28, 1976	Hebei, China	7.8	700,000
3	**Jan. 12, 2010**	**Haiti**	**7**	**316,000**
4	Dec. 16, 1920	Ningxia, China	7.8*	273,400
5	**Dec. 26, 2004**	**Sumatra, Indonesia**	**9**	**230,200**
6	Sep. 25, 1303	Shanxi, China	8*	200,000
7	Dec. 28, 1908	Messina, Italy	7.1*	123,000
8	Oct. 6, 1948	Turkmenistan	7.3	110,000
9	Sep. 1, 1923	Kanto, Japan	7.9*	105,400
10	**Oct. 8, 2005**	**Muzaffarabad, Pakistan**	**7.6**	**87,400**

* Earthquakes occurring before 1935 occurred before Richter Scale development. Their magnitude is based on estimates of damage.

The same strong earthquake occurring in a populated area, however, could bring death to thousands by causing structures to collapse, bridges to fail, bricks and glass to rain down from above, and gas mains to burst into flames. Large earthquakes near urbanized areas are extremely dangerous. On average, each year about 10,000 people die in earthquakes or earthquake-triggered events, such as tsunamis and landslides. **Table 15.2** lists the 10 deadliest earthquakes in recorded human history. Notice that three of them (shown in bold) have occurred in the twenty-first century.

Except in the instances where tsunamis are generated and devastate low-lying coastal areas, ground movement during earthquakes is not directly the cause of fatalities; the structures that fail during ground movement are the cause of fatalities. **Figure 15.29** illustrates examples of the effects of earthquakes on structures.

Figure 15.29
Earthquake damage. (A) Strong L-wave shearing caused this freeway overpass to collapse in Kobe, Japan, in 1995 during a magnitude 6.9 earthquake. (B) Intense R-wave shaking lasted 3 minutes in February 2010 during the 8.8 magnitude earthquake near Santiago, Chile. The overpass structure failed, overturning these cars.
(A. AFP/Getty Images; B. AP Photo/David Lillo)

Earthquakes are as old as Earth's crust itself. As long as Earth's mantle moves the crust's plates, there will be earthquakes. So we have to learn to live with them. But what are our options?

Saving Lives

Because many lives are lost when structures fail during earthquakes, engineers have redesigned structures to better withstand ground shaking. In the United States, building codes require that

Table 15.3	Earthquake Preparedness
HEAVY ITEMS	Secure unstable heavy items, such as bookshelves, to walls and check for other objects that could become hazards during shaking.
SAFE PLACES	Identify safe places indoors and outdoors that you can reach quickly.
SHUTOFFS	Learn and make sure other family members know how to turn off gas, electricity, and water to your home. Gas leaks are common sources of fires after an earthquake has struck.
SURVIVAL KIT	Keep a survival kit in a safe place. It should include a flashlight, a radio, batteries, a first-aid kit, emergency food and water, a nonelectric can opener, essential medicines, and shoes.

engineers build structures in accordance with the seismic risk for the region. Older structures must be *retrofitted* with steel supports to make them safer. These building codes have made earthquakes far less hazardous than they once were. Unfortunately, many buildings in the United States are not retrofitted because of the backlog in retrofitting projects and because of budgetary constraints. In Los Angeles some 15,000 buildings have been identified as being vulnerable to collapse and are in immediate need of retrofitting. In San Francisco, nearly 4,000 apartment buildings housing more than 100,000 people also require retrofitting as quickly as possible. Even worse, many countries do not even have building codes requiring retrofitting, and their residents are ever vulnerable to the collapse of buildings in earthquakes.

Another effective means of saving lives is earthquake warnings. Electrons in copper wire travel far faster than seismic waves in Earth's crust. Thus, after an earthquake occurs, an automated system of alerts can be broadcast electronically. After the 2011 Tōhoku earthquake, for example, people living in Tokyo, 370 km (230 mi) from the epicenter, had 80 seconds to shut off gas

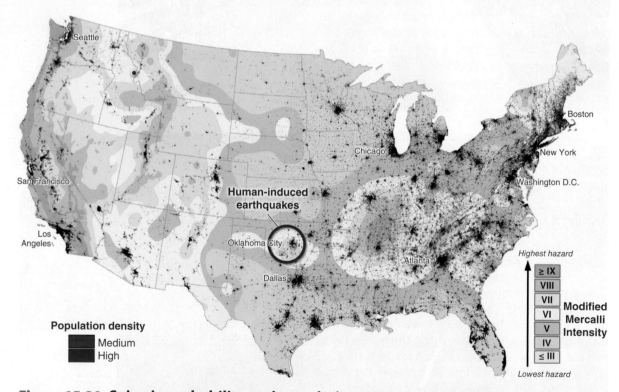

Figure 15.30 Seismic probability and population map. The earthquake hazard is greatest where the seismic probability and the number of people are both high. These conditions occur mostly on the West Coast, from Los Angeles to Seattle. Note that the earthquakes in Oklahoma are human-induced, caused by fracking activities (see Section 14.5). Using this map, scientists have estimated that about half of the population in the lower 48 U.S. states (143 million people) is exposed to potentially damaging earthquakes. California, Washington, and Utah have the most people at risk. *(USGS)*

mains, stop trains, seek shelter, and leave buildings. These actions saved many lives. The USGS is developing an earthquake warning system in southern California. This system could give downtown Los Angeles up to 50 seconds of warning time if a major earthquake occurred along the nearby San Andreas Fault.

Responsibility for earthquake safety also rests with every individual who lives where earthquakes are common. **Table 15.3** lists some of the important ways individuals can prepare themselves for an earthquake.

Predicting Earthquakes

Scientists cannot predict earthquakes. Many seismologists believe we will never be able to predict earthquakes because their precise timing and location are largely random. Seismologists are much better at determining long-term *seismic probabilities* than at making short-term predictions. The seismic probability of an area is determined by considering many factors, including the history of large earthquakes, the types of faults present, and how active those faults have been in recorded history.

For example, they know that a 6.7 earthquake has a 99% probability of happening on the San Andreas Fault in California within the next three decades, and they know at what places along the fault the probability of such an event is highest. But it is not possible to predict exactly where and when such an earthquake will occur. **Figure 15.30** shows a map that shows both seismic probability and population in the United States.

GEOGRAPHIC PERSPECTIVES
15.5 The World's Deadliest Volcano

◎ **Assess the links between large volcanic eruptions, Earth's physical systems, and people.**

The Gothic horror novel *Frankenstein* and physical geography are linked in a surprising way. The novel's author, Mary Shelley, was vacationing on Lake Geneva in Switzerland in the summer of 1816. The weather was uncharacteristically cold, gloomy, and stormy, so her vacation was spent confined indoors with her husband and friends. They held a ghost story–writing contest to see who could best express how he or she felt about the miserable situation. Shelley won, and *Frankenstein* was born. Little did Shelley know that her inspiration was due to the eruption of Tambora, a volcano 12,300 km (7,600 mi) away

on the island of Sumbawa, east of Java, the previous year.

Tambora (elevation 2,850 m [9,350 ft]) is part of the Sunda Arc, the volcanic island arc that makes up part of Indonesia. Subduction of the Indo-Australian plate beneath the Eurasian plate formed the Sunda Arc, which is the most volcanically active and dangerous section of the Ring of Fire. Three particularly notable volcanoes in the Sunda Arc are Krakatau, Toba, and Tambora **(Figure 15.31)**.

Krakatau's most recent big eruption was in 1883. It was a VEI 6 eruption that killed more than 36,000 people. Toba last erupted about 73,000 years ago. Its eruption was one of the largest known volcanic eruptions in Earth's history. Scientists estimate that it had a magnitude of VEI 8 and believe that it caused significant climate change worldwide as ash veiled the Sun and cooled the planet for as much as several centuries. Genetic evidence indicates that humans nearly went extinct because of its eruption. Tambora's most recent big eruption, in 1815, holds the dubious distinction of causing the greatest known human death toll of any volcanic eruption.

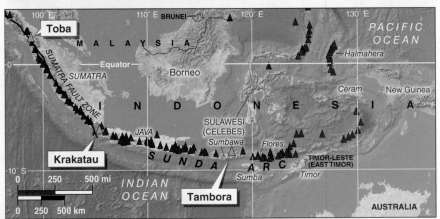

Figure 15.31 Sunda Arc volcanoes. Krakatau and Tambora are two of Indonesia's approximately 150 active volcanoes (red triangles). Toba is considered extinct because it has not erupted in over 10,000 years. The caldera of Tambora, which was formed in 1815, is 7 km (4.5 mi) across and approximately 1,240 m (3,970 ft) deep. *(AP Photo/KOMPAS Images, Iwan Setiyawan)*

The Waking Giant

In 1812, after centuries of sleep, Tambora awoke with a series of blasts that could be heard over hundreds of miles. These massive detonations culminated in an 1815 eruption of colossal proportions.

At about 7 p.m. on April 10, 1815, a VEI 7 eruption occurred that could be heard as far away as South Asia, some 2,600 km (1,600 mi) away. Similar eruptions thundered during the night and into the next day. Mount Tambora was blown apart by the force of these blasts and was reduced in height by almost 1.5 km (1 mi). A column of ash punched into the stratosphere and reached an altitude of 43 km (27 mi), nearly to the base of the mesosphere. Thick ash rained down to the north **(Figure 15.32)**. Up to 100,000 people died in the immediate vicinity of Tambora, both directly in pyroclastic flows and indirectly in tsunamis, as well as from starvation resulting from crop failure.

Today, Tambora continues to show signs of life. In August 2011, the mountain erupted small amounts of ash into the atmosphere. Earthquakes are common nearby, indicating shifting magma and gas in the magma chamber. Surprisingly, few of the 3,000 or so people living on the flanks of the volcano today even know about the 1815 eruption and the loss of life that occurred. Some 130 million people now live on the nearby island of Java. A similar eruption today would bring catastrophic loss of life. Scientists are carefully monitoring Tambora for signs of reawakening.

Tambora's Wide Reach

A VEI 7 stratovolcano eruption puts more than 100 km³ (24 mi³) of ash, as well as an estimated 400 million tons of sulfur gases, into the atmosphere. When sulfur combines with water, it creates droplets of sulfuric acid. These droplets can remain suspended in the stratosphere for several years. The ash and sulfuric acid droplets reflect and absorb solar radiation in the stratosphere and cool Earth's surface (see Section 7.2). Climatologists have coined the term *volcanic winter* to describe the cooling effects of large volcanic eruptions.

The Year without a Summer

The year 1816 was nicknamed the Year without a Summer or Eighteen-Hundred-and-Froze-to-Death because it was unusually cold in both eastern North America and northern Europe **(Figure 15.33)**.

North America, Europe, Argentina, South Africa, India, and China all experienced unusually cold summers in 1816. The average summer temperature in the northeastern United States was about 3°C to 6°C (5°F to 10°F) below average. There was snow in New England in every month of the year in 1816.

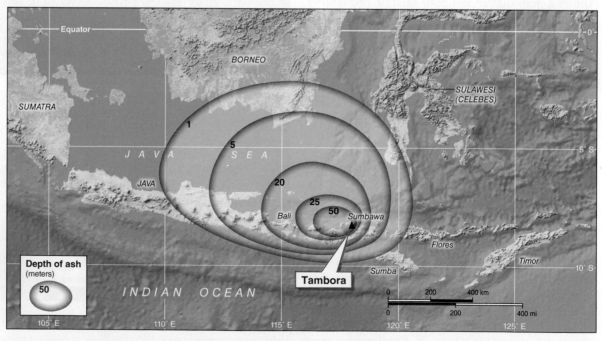

Story Map

The World's Deadliest Volcanoes

Available at
www.saplinglearning.com.

Figure 15.32 Geographic extent of the Tambora ashfall. The ash that settled on the ground was 1 m (3.3 ft) deep as far from the volcano as 800 km (500 mi) away in southern Borneo.

Unfortunately, Tambora's effects on humanity were not limited to unseasonable snowstorms. Tambora triggered crop failure and disease outbreaks and changed rainfall patterns as well.

Crop Failure

Crop failure was widespread, leading to hunger and starvation in New England, the United Kingdom, Germany, and much of the rest of Europe. There is almost no vintage 1816 wine from Europe because the grape harvests were destroyed by the low temperatures. New England experienced an unusual number of killing frosts and record low agricultural harvests during the summer of 1816.

Many European cities had already been shaken politically by the Napoleonic Wars, and the food shortages after the eruption of Tambora sparked social unrest, riots, arson, and looting. Great Britain even stopped collecting income taxes in 1816 in response to food shortages and a hungry and volatile populace.

Typhus Outbreak

Tambora is blamed for a typhus epidemic between 1816 and 1819 in Europe. Typhus is an infectious disease that causes high fever, headache, and a dark red rash. The epidemic is thought to have started in Ireland, spread to England, and then moved south into continental Europe. Some 65,000 people died of typhus during this period. People were vulnerable to this disease because of the poor nutrition resulting from crop losses caused by the eruption.

Indian Monsoon

Tambora's aerosols in the upper atmosphere are thought to have changed the Asian monsoon rainfall pattern (see Section 5.3), causing widespread crop failures, severe hunger, and, as a result, greater susceptibility to diseases. Cholera (a disease marked by severe gastrointestinal symptoms that can be fatal) outbreaks and famine occurred in 1816 in the Bengal region of eastern India because rainfall from the monsoon came late and heavy, causing severe flooding. The cholera outbreak spread into parts of Europe, China, and Russia as people traveled between regions and transmitted the disease.

The atmosphere, biosphere and people, lithosphere, and hydrosphere are all connected in ways that are sometimes difficult to see until an event like Tambora reveals the connections.

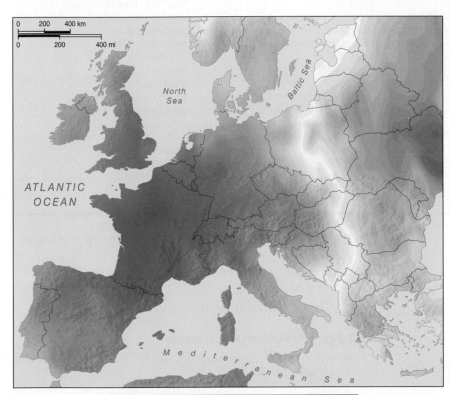

Figure 15.33 Europe's Year without a Summer. This reconstruction of summer temperatures (in degrees Celsius departure from today's average) for Europe in 1816 comes mostly from tree ring analysis. The cooling in the Northern Hemisphere lagged the Tambora eruption by a year because it took several months for the volcanic aerosols to circulate in the upper atmosphere and cool Earth's surface.

CHAPTER 15 Study Guide

Focus Points

15.1 About Volcanoes

- **Types of volcanoes:** There are four major types of volcanoes: shield volcanoes, stratovolcanoes, lava domes, and cinder cones.
- **Lava:** Lava, the most conspicuous product of volcanic activity, ranges from fluid mafic to thick felsic lava.
- **Pyroclasts:** Volcanoes eject materials ranging in size from fine ash to large blocks into the air.
- **Gases:** Gases, particularly carbon dioxide and water vapor, are a significant component of volcanic emissions.
- **Volcanic landforms:** Landforms resulting from volcanism include columnar jointing, large igneous provinces, and calderas.

15.2 Pele's Power: Volcanic Hazards

- **Shield volcanoes:** Shield volcanoes have gentle, effusive eruptions.
- **Stratovolcanoes:** Stratovolcanoes erupt both effusively and explosively. Explosive eruptions occur when gas in the magma chamber expands rapidly.
- **Volcanic geohazards:** Lahars and pyroclastic flows are the two greatest threats posed by stratovolcanoes.
- **Eruption prediction:** Scientists can sometimes predict a volcanic eruption by monitoring gas emissions and earthquake activity.

15.3 Tectonic Hazards: Faults and Earthquakes

- **Fault types:** Faults occur as normal faults, reverse faults, and strike-slip faults.
- **Fault indicators:** Fault scarps indicate normal and reverse faults. Offset features indicate strike-slip faults.
- **Earthquakes:** Earthquakes are caused when the crust suddenly breaks and releases built-up stress energy in the form of seismic waves.
- **Seismic belts:** Earthquakes occur mainly in seismic belts that coincide with plate boundaries.

15.4 Unstable Crust: Seismic Waves

- **Intensity and magnitude:** Earthquake intensity is determined by the amount of damage done to structures. Earthquake magnitude is determined by the degree of measured ground shaking.
- **Earthquake prediction:** Scientists cannot predict earthquakes.
- **Saving lives:** Building codes and retrofitting greatly strengthen buildings and save human lives.

15.5 Geographic Perspectives: The World's Deadliest Volcano

- **Tambora:** The 1815 eruption of Tambora was the strongest and deadliest volcanic eruption in recorded history.

Key Terms

active volcano, 453
aftershock, 468
angle of repose, 455
ash (volcanic), 453
caldera, 460
cinder cone, 454
columnar jointing, 459
effusive eruption, 462
epicenter, 468
explosive eruptions, 462
extinct volcano, 453
fault scarp, 467
felsic lava, 457
flank eruption, 453
focus, 467
foreshock, 468
geohazard, 452
intermediate lava, 456
joint, 459
lahar, 464
lapilli, 457
large igneous province (LIP), 460
lava, 453

lava bomb, 458
lava dome, 454
liquefaction, 473
mafic lava, 456
magma, 453
magma chamber, 453
Modified Mercalli Intensity (MMI) Scale, 472
moment magnitude scale, 474
normal fault, 467
pumice, 458
pyroclast, 453
pyroclastic flow, 464
reverse fault, 467
seismograph, 471
shield volcano, 453
stratovolcano, 453
strike-slip fault, 467
tsunami, 452
volcanic block, 458
volcanic crater, 453
Volcanic Explosivity Index (VEI), 463
volcanic vent, 453

Concept Review

The Human Sphere: Killer Waves

1. What is a tsunami? How are tsunamis generated? Why are tsunamis geohazards?

15.1 About Volcanoes

2. What are the four kinds of volcanoes? Which is the smallest volcano type, and which is the largest? Describe how each is built up.

3. What are the three types of lava? Explain how each behaves and what causes it to behave that way.

4. Give examples of the types of materials volcanoes produce. Briefly describe each type.

5. What is a joint? What is columnar jointing? In which kind of lava can it be found?

6. What is a large igneous province? Give three examples of where they can be found.

7. What is a caldera, and how does it form?

15.2 Pele's Power: Volcanic Hazards

8. What are the two types of volcanic eruptions, and what determines which type of eruption will occur?

9. Explain why volcanic gases come out of solution and what happens when they do.

10. Why does felsic magma/lava not release gas as easily as mafic magma/lava? How is this significant?

11. What is the VEI? With each whole number increase, how much more material is erupted into the atmosphere?

12. What are the two most deadly products of volcanic eruptions? Explain why each is so hazardous.

13. Can scientists predict volcanic eruptions? Explain.

15.3 Tectonic Hazards: Faults and Earthquakes

14. Describe the three types of faults. For each type, what are the direction of force and the type of movement?

15. Explain how an earthquake forms using the terms *stress* and *friction*. What is elastic rebound theory in this context? What is the *stick-slip* process?

16. Define and briefly explain the following terms: focus, epicenter, and seismic waves.

17. What is a foreshock? What is an aftershock? What causes them?

18. Describe the geographic pattern of earthquakes worldwide. Where do most earthquakes occur?

15.4 Unstable Crust: Seismic Waves

19. Compare a body wave to a surface wave. Where does each travel?

20. Compare a compressional wave with a shear wave. What kind of movement does each produce?

21. What scientific instrumentation is used to measure ground shaking?

22. Compare P waves, S waves, L waves, and R waves in terms of the sequence of their arrival after an earthquake and the strength of the ground shaking they cause.

23. What does the Modified Mercalli Intensity Scale indicate about an earthquake? What evidence does this scale use to rank earthquakes?

24. What information about an earthquake does the moment magnitude scale provide?

25. What is liquefaction? On what kind of ground does it occur?

26. Can scientists predict earthquakes?

27. In what ways can people reduce their vulnerability to earthquakes?

15.5 Geographic Perspectives: The World's Deadliest Volcano

28. What kind of volcano is Tambora? In the context of plate tectonics, explain how Tambora was formed.

29. Given the VEI ranking of Tambora's 1815 eruption, how much pyroclastic material did it eject into the atmosphere?

30. When was the Year without a Summer, and how does the term *volcanic winter* relate to it?

31. Outline the negative effects the Tambora eruption set in motion for various parts of the world.

Critical-Thinking Questions

1. Are you vulnerable to volcanic hazards where you live? If you are unsure, what kind of questions would you ask to find out?

2. Is there a risk of an earthquake occurring where you live? If you do not know, what kinds of questions could you ask to find out?

3. If a VEI 6 or greater eruption occurred today within 20 miles of where you live, what effects do you think it would have locally (where you live) and globally?

4. Do you think scientists will ever be able to predict accurately when a given region will be hit by an earthquake?

5. In Sri Lanka, many elephants ran to high ground minutes before the 2004 tsunami struck, even though their unknowing riders ordered them to stop. Similarly, there are many eyewitness accounts of animals such as horses and dogs acting strangely or panicking minutes before an earthquake strikes. What do you think animals may be sensing that humans and scientific instruments are not sensing? Do you think scientists should pursue further research into this area, or would it be a waste of money?

Test Yourself

Take this quiz to test your chapter knowledge.

1. **True or false?** Shield volcanoes produce effusive eruptions.

2. **True or false?** Lava viscosity is in large part related to the silica content of the lava.

3. **True or false?** Lava is one of the most deadly hazards of volcanoes.

4. **True or false?** Of the four types of waves, S waves travel fastest and are the first to arrive after an earthquake.

5. **Multiple choice:** Which of the following is not associated with explosive volcanic eruptions?

a. stratovolcano
b. felsic magma
c. caldera formation
d. large igneous province

6. **Multiple choice:** Which of the following types of seismic waves produces a rolling movement on the surface of Earth's crust?

a. P waves
b. S wave
c. L waves
d. R waves

7. **Multiple choice:** About how much more ground shaking does a magnitude 8 earthquake create than a magnitude 5 earthquake?

a. 100 times more
b. 1,000 times more
c. 10,000 times more
d. 100,000 times more

8. **Multiple choice:** Which of the following is not a type of pyroclast?

a. bombs
b. ash
c. lapilli
d. lava flows

9. **Fill in the blank.** A _____ is a slurry of mud created when a snow-capped volcano erupts.

10. **Fill in the blank.** The _____ is the point directly over an earthquake's focus.

Online Geographic Analysis

Currently Active Volcanoes

In this exercise we analyze volcanic recent activity.

Activity

Go to http://volcano.si.edu. This is the Smithsonian Institution's Global Volcanism Program. Click on the "Reports" tab at the top of the page and select the "Smithsonian/USGS Weekly Volcanic Activity Report" option from the pull-down menu. You will see a map showing volcanic activity around the world for the past week.

1. **In what geographic region is most of the volcanic activity occurring?**

2. **Given your answer to the previous question, in what tectonic environment is most of the volcanic activity occurring?**

Select any volcano on the map. Then click its name to pull up more information about it.

3. **What is the name of the volcano you selected? In what country is it located?**

4. **List the geographic coordinates of your volcano and its summit elevation.**

Select the "General Information" tab provided below the volcano's photo and read the information provided. Then click the "Photo Gallery" tab and scroll down through the photos and read their descriptions.

5. **What type of volcano did you choose?**

6. **What are the notable features of your volcano? For example, does it have a lava dome at the summit? Does it have a summit crater? Is the type of lava it produces discussed?**

Select the "Eruptive History" tab provided below the volcano's photo.

7. **When did your volcano last erupt? On the VEI scale, how large was that eruption?**

8. **What has been the largest eruption in your volcano's history?**

9. **If there were multiple years of large eruptions, what is the most recent date of your volcano's largest eruption?**

10. **Choose two other volcanoes and answer Questions 3–9 for them.**

Picture This. *Your Turn*

Name That Fault

These three photos show examples of the three fault types discussed in this chapter. Fill in the blank spaces above each photo with one of the following types of faults: normal fault, reverse fault, or strike-slip fault. You may use each more than once. *(Left. Courtesy of Rick Scott; center. © David K. Lynch; right. Russell L. Losco, Lanchester Soil Consultants, Inc.)*

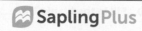

For animations, interactive maps, videos, and more, visit www.saplinglearning.com. Sapling Plus

The water that the Sun evaporates from the oceans precipitates back down on the continents. As the force of gravity propels water back to the oceans, the water erodes Earth's surface. Through time, even the highest mountains are reduced to grains of sand by the action of flowing water, wind, and coastal waves. Part IV explores the work of weathering, erosion, and sediment deposition in sculpting Earth's surface.

CHAPTER 16
Weathering and Mass Movement

Weathering reduces rocks to fragments. Gravity moves weathered material downslope, sometimes with deadly speed.

CHAPTER 17
Flowing Water: Fluvial Systems

Flowing water is the most important agent of erosion. Streams erode Earth's land surface and also build it up by depositing the sediments they transport.

CHAPTER 18
The Work of Ice: The Cryosphere and Glacial Landforms

Where they form, glaciers flow downslope and gouge deeply into Earth's surface, creating unique landforms.

CHAPTER 19
Water, Wind, and Time: Desert Landforms

Desert landforms are shaped by flowing water and wind.

CHAPTER 20
The Work of Waves: Coastal Landforms

Coastal beaches and rocky shores are shaped by wave energy.

(Chapter 16: MediaProduction/Getty Images; Chapter 17: travellight/Shutterstock; Chapter 18: Jon Arnold/Danita Delimont Stock Photography; Chapter 19: Bruce Gervais; Chapter 20: Trevor Cole/Getty Images)

Weathering and Mass Movement

Chapter Outline *and Learning Goals* ◎

16.1 Weathering Rocks

◎ Distinguish between physical weathering and chemical weathering and provide examples of each.

16.2 Dissolving Rocks: Karst Landforms

◎ Describe different kinds of karst landforms and explain how each forms.

16.3 Unstable Ground: Mass Movement

◎ Identify different types of mass movements and describe their behavior and causes.

16.4 Geographic Perspectives: Deadly Mass Movements

◎ Assess the threats to people from mass movements and explain how human vulnerability to those threats can be reduced.

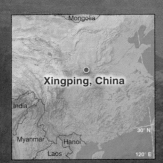

Dramatic limestone outcrops rise into the sky in this photo taken near the town of Xingping in central China. These unusual mountains were formed as limestone bedrock was gradually dissolved by rainwater in a process called karst weathering.

(MediaProduction/Getty Images)

To learn more about karst weathering processes, turn to Section 16.2.

THE HUMAN SPHERE Weathering Mount Rushmore

Figure 16.1 Mount Rushmore maintenance. Climbers descend the Mount Rushmore National Memorial and caulk minute cracks with a silicone sealant. The sealant prevents water from entering the cracks and freezing and widening the cracks through the process of *frost wedging.* *(Courtesy of NPS, Mount Rushmore National Memorial)*

The Mount Rushmore National Memorial is carved from a body of granite in the Black Hills of South Dakota. Like any other rock, the granite that makes up this monument is continually subject to the effects of weathering that break down rocks over time. Weathering acts only on the surfaces of rock. Because the rock monument's surface is sculpted, its rock is more susceptible to weathering as there is a greater surface area on which weathering can act. To protect the monument from the weathering process, the U.S. National Park Service closely monitors and maintains the structure **(Figure 16.1)**.

Scientists have developed a system that identifies "rock blocks" that provide key support for large portions of the monument. If these blocks move, other portions of the structure are more susceptible to movement as well. The National Park Service monitors these key rock blocks using laser-based instruments that can measure movement to within 0.0003 cm (0.00001 in). The instruments take measurements four times each day. In the last two decades since monitoring was put into use, it has revealed that the blocks move very little and the monument is structurally sound.

16.1 Weathering Rocks

◎ **Distinguish between physical weathering and chemical weathering and provide examples of each.**

The adage "Nothing lasts forever" is certainly true of Earth's physical landscapes. Even the mightiest and most massive mountains are ephemeral features in the great span of geologic time. In Part III,

we examined the work of Earth's geothermal energy in building Earth's surface and creating vertical relief. In Part IV, we explore how these uplifted landforms are torn down. Solar energy and gravity drive the wearing down of the crust's surface. The Sun creates differences in atmospheric pressure, and these differences drive the wind and ocean waves. Coastal waves (generated by the wind) relentlessly batter coastal areas, and winds in coastal areas and deserts also transport and deposit sediments. The Sun also evaporates water into the atmosphere. Winds transport the water vapor over land, where water precipitates and falls as rain and snow. Gravity sets the precipitated water in motion in streams and flowing glaciers that carve into Earth's land surface.

Tectonic forces act sporadically to lift mountains, but the forces set in motion by the Sun and gravity that reduce vertical relief are present nearly everywhere on land, and they never rest. **Weathering** is the process by which solid rock is dissolved and broken apart into smaller fragments. **Erosion** is the scouring and stripping away of rock fragments loosened by weathering. Streams and glaciers, for example, erode surfaces by scouring bedrock and sediment and dislodging loose material; coastal waves crash against rocks and break them apart; and wind in deserts carries sand and fine particles that act as sandpaper to abrade (wear away) rock surfaces.

Weathering breaks rocks apart or decomposes them through chemical reactions with water. All rocks weather: Unweathered (or "fresh") rock surfaces are either newly formed (such as happens when lava cools to basalt) or newly exposed (such as happens when a rock breaks). Some rocks, such as metamorphic rocks or crystalline intrusive igneous rocks, are very resistant to weathering. But given enough time, all rocks exposed at the surface weather and erode.

Sediment transport is the movement of rock fragments (sediments) that have been weathered and eroded. Wind and flowing streams and glaciers carry and move sediments. Everywhere, even in deserts where there is little water, streams are the most important means by which sediments are transported. Around the margin of ice sheets, in deserts, and in coastal areas, wind can play an important role in sediment transport.

Deposition occurs where sediments in transport are laid down (deposited). Deposition occurs primarily where water stops flowing because it no longer has kinetic energy to carry sediments. Lakes, stream floodplains, wetlands and coastal estuaries, and the ocean are the most common environments of deposition. In deserts, deposition occurs where the strength of the wind is reduced, such as in low-lying areas. Weathering and erosion and sediment transport and deposition are all processes of **denudation**—the lowering and wearing away of Earth's surface.

Physical Weathering

There are two main types of weathering: physical weathering and chemical weathering (which is discussed in the following section). **Physical weathering** breaks rocks down into smaller pieces,

or clasts—collectively called *debris*—without altering the chemical makeup of the rock. The processes of physical weathering have their greatest effect at high elevations and high latitudes.

Natural cracks in rock, called *joints*, provide surface areas on which physical weathering processes can act. Nearly all outcrops have joints. Tectonic stresses near geologic faults commonly fracture rocks and create jointing. As igneous rocks cool from magma and lava, joints in the rock develop during the cooling process. Sedimentary rocks usually have *horizontal joints* at their *bedding planes* (the surfaces where their layers meet), but sedimentary rocks also form *vertical jointing* as a result of stresses on the rock, as shown in **Figure 16.2**.

Pressure-release jointing occurs when overlying rocks and sediments, referred to as *overburden*, are removed from rocks that formed at great depths. After the overburden is removed, the enormous pressure associated with deep burial lessens, and the rock expands slightly, creating a network of joints. Pressure-release jointing is most common in intrusive igneous rocks and metamorphic rocks.

In the process of **exfoliation**, a type of pressure-release jointing, joints form parallel to the rock surface, creating sheetlike slabs of rock resembling the layers of an onion. These slabs peel off through *sheeting*, creating broken horizontal slabs of rock that can slip off the rock face. Sheeting slabs can be a few millimeters thick or several meters thick. Exfoliation sometimes rounds the edges of large granite plutons, creating *exfoliation domes*, such as Half Dome in Yosemite National

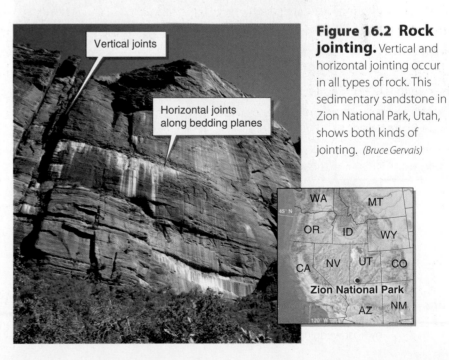

Vertical joints

Horizontal joints along bedding planes

Figure 16.2 Rock jointing. Vertical and horizontal jointing occur in all types of rock. This sedimentary sandstone in Zion National Park, Utah, shows both kinds of jointing. *(Bruce Gervais)*

Zion National Park

Park (see Figure 14.10) or Enchanted Rock in Texas, shown in **Figure 16.3**.

In places where temperatures routinely drop below freezing, frost wedging is the most important type of physical weathering. **Frost wedging** is the process by which water in a joint in rock freezes and expands, causing the opening to grow. When liquid water freezes to ice, it expands by almost 10%. If water freezes in a confined space, such as in a house water pipe or a joint in a rock, it applies considerable force as it expands against the material confining it. Ice formation can break water

Sheeting slabs

Figure 16.3 Exfoliation and sheeting. Exfoliation and sheeting have rounded the edges of Enchanted Rock near Fredericksburg, Texas, creating an exfoliation dome. *(Photo Researchers/Getty Images)*

Figure 16.4 Frost wedging. This granite boulder (about 2 m [6.5 ft] across) in the Sierra Nevada, in California, has been broken apart by frost wedging. *(Bruce Gervais)*

pipes, and it can break rocks. Over time, repeated freezing and thawing of water in rock joints can break and shatter rocks, as shown in **Figure 16.4**.

Frost wedging is not the only type of physical weathering. In the process of **salt weathering**, salt crystals grow in pore spaces within a rock and exert pressure as they grow. This pressure within the pore spaces dislodges individual mineral grains within the rock and weakens the rock. In order for salt crystals to grow, the rock must be continually wetted with salt water and then dried. Salt weathering occurs mostly in arid climates where evaporation is high and in coastal areas where rocks are continually coated with salt spray from waves. **Tafoni** are rounded pits or cavities on the surface of a rock that form through salt weathering. They are most common in coastal regions **(Figure 16.5)**.

Similarly, during *hydration*, water is added onto minerals in rock, without changing the minerals chemically. One example of hydration occurs as the mineral anhydrite forms gypsum ($CaSO_4 * 2H_2O$). When water is added onto minerals and becomes part of a rock's structure, the rock often weakens as the minerals expand and create stresses on the rock. Hydration mainly changes the texture, composition, and volume of minerals in rocks. Repeated hydration and *dehydration* (the removal of water from rock) cycles within a rock can lead to its disintegration without changing its chemical structure.

Biophysical weathering is weathering caused by any living organism such as plants and animals. Animals from small rodents to elephants can physically break rock down into smaller pieces. **Root wedging** occurs when plant roots that are seeking water grow into joints. Many city residents have seen sidewalks buckled by tree roots pushing up from below. The same thing happens in natural settings: As roots grow, they can force rocks apart **(Figure 16.6)**.

Chemical Weathering

Physical weathering breaks rock apart without altering its chemical structure. **Chemical weathering**, in contrast, changes the minerals in rock through chemical reactions involving water. Physical and chemical weathering both occur at the same time and affect each other's rates. For example, chemical weathering occurs mainly on the surfaces of rocks, and physical weathering assists in the process of chemical weathering by breaking rocks into smaller pieces and creating a greater surface area on which chemical weathering can act **(Figure 16.7)**. Chemical weathering is dominant where temperatures are above freezing and there is ample moisture. Accordingly, rocks in the lowland tropics are most susceptible to chemical weathering.

A number of chemical reactions cause chemical weathering of rock. When carbon dioxide dissolves in water, a carbonic acid (H_2CO_3) solution is produced. **Carbonation** is a chemical weathering process in which carbonate rock such as limestone is dissolved in a carbonic acid solution and carried away. In Section 16.2, we discuss the important process of carbonation in more detail in the context of limestone *karst* landscapes.

Hydrolysis occurs when water reacts with and combines with minerals in rocks to form new minerals. Typically, the new minerals are softer and more easily eroded than the original minerals. The mineral feldspar found in granite, for example, is reduced to clay and salts through hydrolysis. These are relatively weak minerals and are easily carried away by water.

In *oxidation*, oxygen atoms combine with the minerals in rocks and weaken them. Just as iron rusts, rock containing iron oxide compounds, such as hematite, rusts (oxidizes) into a softer, reddish-brown rock.

Biochemical weathering refers to chemical weathering resulting from the activities of living organisms. Many plant roots, bacteria, lichens, and fungi obtain nutrients by producing acids that break down minerals in rocks.

Unique Landforms Created by Weathering

Weathering seldom acts equally across the surface of a rock. The different shapes and materials of which rocks are made cause weathering to happen at different rates across the rock surface. This process is called *differential weathering*. Angular blocks, for example, experience the most weathering along their edges and corners, causing them to become rounded through time. This process, called **spheroidal weathering**, leads to the formation of rounded or spherical boulders and outcrops, as illustrated in **Figure 16.8** on the following page.

Dramatic and unusual landforms are sometimes created through weathering. Noteworthy examples include *monoliths, arches,* and *hoodoos.* A monolith is any single massive rock that rises above its surroundings. An arch is a natural bridge, and a hoodoo is a type of column **(Figure 16.9** on the following page**)**.

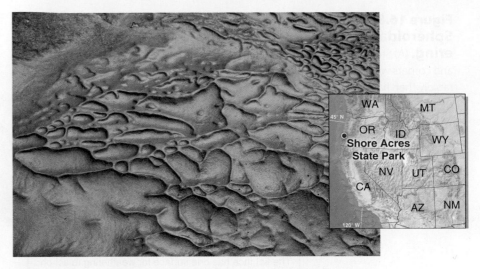

Figure 16.5 Tafoni. This sandstone in Shore Acres State Park, in southwestern Oregon, shows the peculiar honeycombed pattern of tafoni, the rounded surface cavities caused by salt weathering. This photo shows an area about 1 m (3.3 ft) across. *(Bruce Gervais)*

Figure 16.6 Root wedging. This western juniper (*Juniperus occidentalis*) near South Lake Tahoe, California, is growing in a joint in granite and pushing the rock apart. *(Bruce Gervais)*

Figure 16.7 Weathering and surface area. The more fragmented rock becomes through physical weathering, the greater its surface area on which chemical weathering takes place. (A) The height and width of this block are both 2 m. To determine the surface area of one side, we multiply the height by the width. Each side of the block is therefore 4 m^2 ($2 \times 2 = 4$ m^2). There are 6 sides on the block, so the single block's total surface area is 4 m^2 × 6 sides, or 24 m^2. (B) When the same block is broken into four blocks, its surface area greatly increases. Each new block's side is 1 m^2. There are 6 sides per block, so each block's total surface area is 6 m^2 (1 m^2 × 6 sides). Because there are eight 6 m^2 blocks (instead of just one 24 m^2 block), the total surface area of the original block doubles from 24 m^2 to 48 m^2 (6 m^2 × 8 blocks).

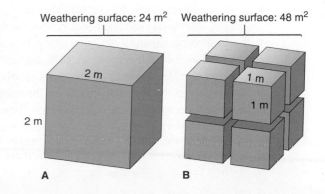

Weathering surface: 24 m^2 Weathering surface: 48 m^2

2 m

2 m

1 m

1 m

A

B

Figure 16.8
Spheroidal weathering.
(A) Sharp edges and corners weather more rapidly than flat surfaces. As a result, weathering rounds the edges of rocks over time. (B) Granite boulders in the Namib Desert, Namibia, have been rounded by weathering. *(© Ruedi Walther)*

An edge has two surfaces for weathering to act on.

A corner has three surfaces for weathering to act on.

The flat face has one surface for weathering to act on.

A

B

A

B

C

Figure 16.9 Landforms created by weathering.
(A) Devil's Tower in Wyoming is a resistant column of intrusive igneous rock that was once surrounded by sedimentary rock. The less resistant sedimentary rock has been weathered and eroded away, leaving this monolith. (B) A natural arch forms, typically in sandstone, when weak joints and bedding planes erode. With further weathering, all natural arches, including this one in Arches National Park in Utah, called Broken Arch, will eventually collapse. (C) Hoodoos are tall, columnar rock formations, usually capped with a layer of rock that is resistant to erosion. The resistant cap protects the softer rocks beneath it from erosion. These hoodoos are near Drumheller in Alberta, Canada. *(A. Joel Sartore/Getty Images; B. Tom Till/Alamy; C. Wayne Barrett & Anne MacKay/Getty Images)*

Weathering acts on weaker minerals within rocks more than on relatively resistant minerals. As a result, resistant minerals, such as quartz, persist through time, while weaker minerals break down. Many beaches are composed of resistant quartz grains that were once embedded within rocks, as the **Picture This** feature below discusses.

16.2 Dissolving Rocks: Karst Landforms

◎ **Describe different kinds of karst landforms and explain how each forms.**

In 1864, when Jules Verne's book *Journey to the Center of the Earth* was published, its readers were transported to enormous subterranean caves lit by bioluminescent bacteria living on the caves' ceilings and filled with a deep subterranean ocean, on the shores of which were giant mushrooms, petrified forests, and even living dinosaurs. Although these phenomena are not real, Verne's imagination no doubt was inspired at least in part by subterranean karst caves.

Karst Processes

Karst refers to an area dominated by the weathering of carbonate rocks, usually limestone. The term *karst* is the German form of the Slovenian term *kras*, which means "barren landscape." Recall from Section 16.1 that the process of carbonation dissolves limestone and other carbonate rocks.

As rain falls through the atmosphere, it absorbs a small amount of carbon dioxide from the air. Water also absorbs carbon dioxide as it flows through soil that is rich in organic substances. Once water has absorbed carbon dioxide, it forms a weak carbonic acid solution. This acidic water then dissolves the calcium carbonate in limestone (forming calcium bicarbonate) and carries it away. This process occurs most where water flows through joints in the rock (see Section 16.1).

In some cases, acids other than carbonic acid dissolve carbonate rocks. In the unusual case of

---

Picture This The Life of a Sand Grain

Many minerals in rocks are subject to chemical weathering, particularly oxidation and hydrolysis. However, quartz, a mineral found in many rocks, is extremely resistant to chemical weathering and may persist for millions of years as grains of quartz sand.

Many beaches of the world are made up in large part of quartz sand grains that were long ago weathered out of bedrock. An individual sand grain on the beach might have been weathered from granite high on a mountain range long ago. From there, it might have been transported to a beach, where the waves washed over it for millennia. This ancient beach sand could then have been buried and lithified to form new sedimentary rock, only to be uplifted by geologic forces into a different mountain range, with the original quartz grain in it. Weathering might have torn down this mountain range, too, eventually creating a new beach many millions of years later. This process may be repeated several times through hundreds of millions of years.

Each grain of sand on a beach is a product of Earth's deep geologic history. In this sense, when we walk across a beach, such as this one on the shoreline of Point Reyes in northern California, we also walk across a portion of Earth's ancient history.

Consider This

1. How does differential weathering result in a beach composed of quartz sand?
2. How can quartz grains be older than the bedrock from which they come?

(Bruce Gervais)

Carlsbad Caverns and Lechuguilla Cave in New Mexico, for example, sulfuric acid that forms from natural oil deposits is dissolving the limestone bedrock.

Areas with limestone bedrock are home to the most widespread types of karst topography. All carbonate rocks, including marble, dolomite, gypsum, and chalk, are, however, subject to dissolution. **Dissolution** is the general process in which a mineral completely dissolves in water. Carbonation, for example, is a form of dissolution.

The process of dissolution acts fastest in areas that are cool but not frozen. Cold water holds more carbon dioxide than warm water, making a stronger carbonic acid, which enhances dissolution. But when water is frozen, chemical reactions slow or stop. Some Arctic regions have carbonate rocks, but dissolution is greatly slowed there because water is frozen and chemically nonreactive for most of the year. Carbonate rocks susceptible to weathering by karst processes cover about 15% of Earth's land surface **(Figure 16.10)**.

A Riddled Surface: Karst Topography

A **sinkhole** (or *doline*) is a depression in Earth's surface that results from the dissolution of carbonate rock underground. In the United States, sinkholes are particularly common in Kentucky, Tennessee, Michigan, and Florida.

On August 12, 2013, a sinkhole opened up beneath a condo building near Orlando, Florida **(Figure 16.11A)**. Over a span of about 15 minutes, half of the building collapsed into the sinkhole which measured 18 m (60 ft) in diameter, giving the occupants barely enough time to safely evacuate. Those regions where karst processes prevail, such as Florida, are often referred to as having limestone *karst topography*.

Florida has the greatest number of sinkholes of any U.S. state. Florida insurance companies pay out about $400 million for damage done by the state's sinkholes each year. Florida also has the greatest number of **sinkhole lakes**: sinkholes that are filled with water **(Figure 16.11B)**.

Many limestone karst regions are riddled with steep-walled depressions called collapse sinkholes. A **collapse sinkhole** forms where the ceiling of a subterranean cave has collapsed. **Figure 16.12A** on page 494 illustrates the process of collapse sinkhole formation.

When the lowest level of a collapse sinkhole lies below the water table, a *cenote* forms. Most of the Yucatán Peninsula of Mexico is composed of limestone, and it has many karst features, including cenotes **(Figure 16.12B** on page 494**)**. Another spectacular example of a collapse sinkhole is explored in the **Picture This** feature on page 495.

There are few surface streams in karst regions because water flows underground through cave

Figure 16.10 Karst regions. This world map shows carbonate rocks exposed at the surface and subject to weathering by karst processes.

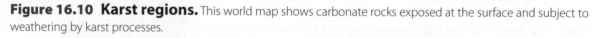

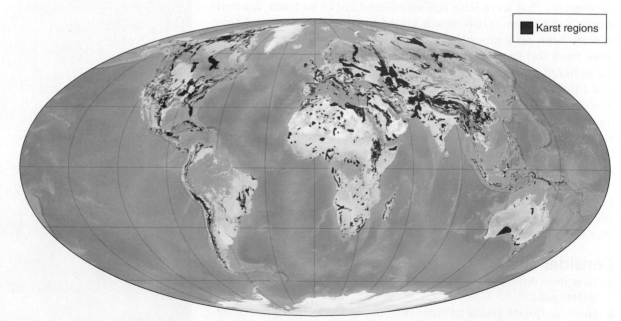

Karst regions

Figure 16.11 Sinkholes. (A) Sinkholes form as limestone at Earth's surface dissolves and is carried away by water. Sinkholes can give way quickly, destroying structures built on them (inset photo). When the lowest elevation of a sinkhole is lower than the water table, a sinkhole lake forms. (B) This Landsat satellite image of Florida is centered on Orlando. The dark areas are surface waters. Most of the lakes visible here are sinkhole lakes. Some of them are 16 km (10 mi) across. *(A. Red Huber/Orlando Sentinel/MCT via Getty Images; B. EROS Center, U.S. Geological Survey)*

Web Map
Sinkholes
Available at
www.saplinglearning.com

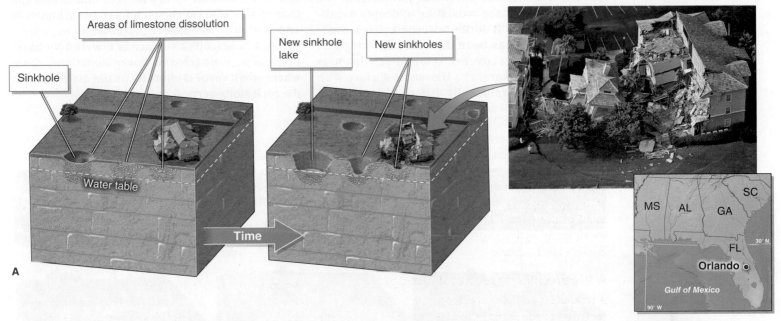

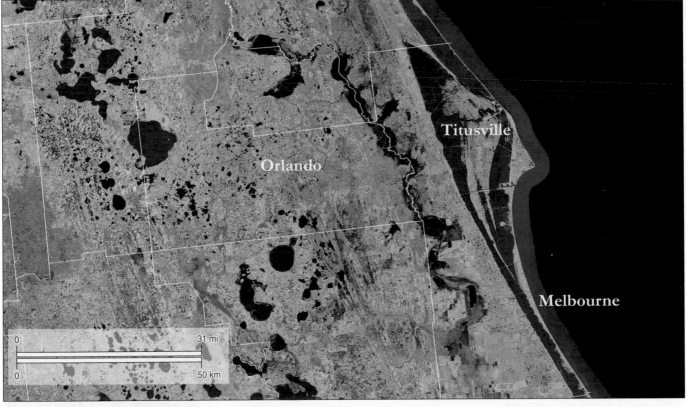

systems. The surface streams that are found in karst regions usually flow for only short stretches and then disappear into the ground. Such streams are called **disappearing streams (Figure 16.13)**.

Many other unique landforms are associated with limestone karst topography **(Figure 16.14** on page 496**)**. *Limestone pavement*, for example, is a type of bare surface consisting of deeply weathered limestone. It forms where exposed limestone bedrock has been dissolved by rainwater and an overlying cover of soil and vegetation is absent. *Cockpit karst* is a limestone surface with a topography dramatically different from that of limestone pavement. Cockpit karst consists of vegetated rounded hills formed by limestone weathering. In other cases, karst processes create *karst spires* rather than rounded hills. In Madagascar, a "stone forest" called the Tsingy seemingly grows from the ground. Sharp pinnacles and pointed spires of limestone form a labyrinthine landscape that covers several thousand square kilometers and is almost completely inaccessible to people. When a landscape's surface is lowered by karst processes, pinnacles of *tower karst* may form where weak vertical joints focus the dissolution of the rock along vertical planes.

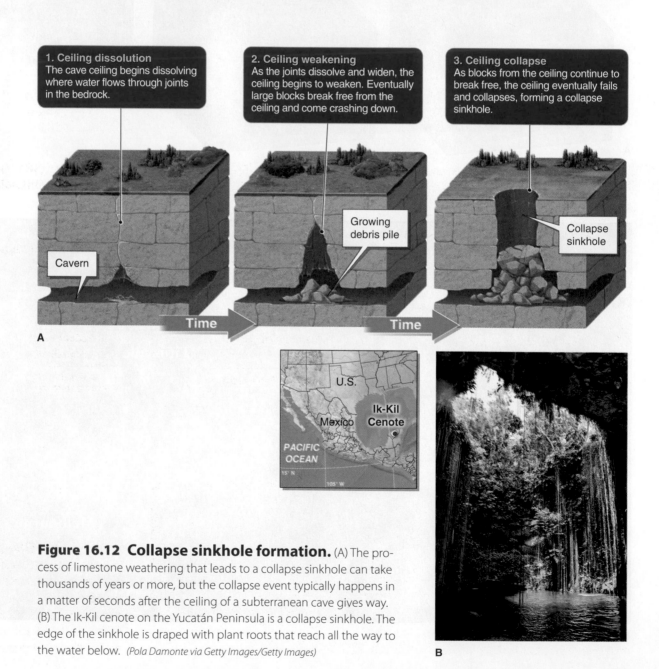

1. Ceiling dissolution
The cave ceiling begins dissolving where water flows through joints in the bedrock.

2. Ceiling weakening
As the joints dissolve and widen, the ceiling begins to weaken. Eventually large blocks break free from the ceiling and come crashing down.

3. Ceiling collapse
As blocks from the ceiling continue to break free, the ceiling eventually fails and collapses, forming a collapse sinkhole.

Cavern

Growing debris pile

Collapse sinkhole

Time

Time

A

U.S.

Mexico

Ik-Kil Cenote

PACIFIC OCEAN

B

Figure 16.12 Collapse sinkhole formation. (A) The process of limestone weathering that leads to a collapse sinkhole can take thousands of years or more, but the collapse event typically happens in a matter of seconds after the ceiling of a subterranean cave gives way. (B) The Ik-Kil cenote on the Yucatán Peninsula is a collapse sinkhole. The edge of the sinkhole is draped with plant roots that reach all the way to the water below. *(Pola Damonte via Getty Images/Getty Images)*

Picture This The Great Blue Hole, Belize

(Greg Johnston/Getty Images)

The Great Blue Hole in Belize is a submerged collapse sink-hole that formed during the most recent glacial period, when sea levels were about 85 m (280 ft) lower than they are today. As the continental ice sheets melted about 10,000 years ago (see Section 7.2), sea level rose, and the collapse sinkhole was inundated with seawater. Coral reefs now fringe the sinkhole and support rich marine biodiversity. The Great Blue Hole, which is about 300 m (984 ft) across and 125 m (410 ft) deep, is among the largest sinkholes on Earth.

Consider This

1. How did the Great Blue Hole form?
2. Why could this collapse sinkhole not have formed after sea level rose?

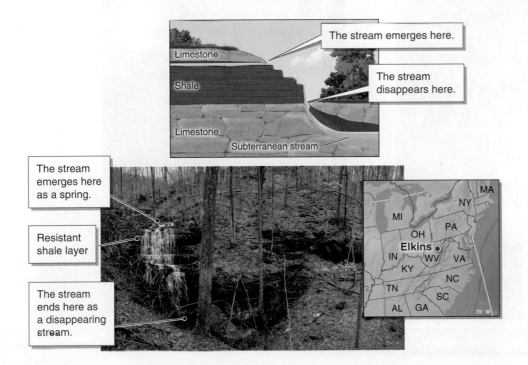

Video

Groundwater: Sources to Streams

Available at www.saplinglearning.com

Figure 16.13 Disappearing stream.
This small stream in West Virginia is spilling over a resistant shale cliff that is sandwiched between layers of limestone. Notice that the stream emerges from above the shale as a spring and flows back into a subterranean channel as a disappearing stream. (© James Van Gundy)

Figure 16.14 Limestone karst topography. (A) This limestone pavement near Malham, North Yorkshire, England, has weathered along relatively weak linear joints in the bedrock. (B) The Chocolate Hills, in Bohol Province in the Philippines, are a spectacular example of cockpit karst topography. (C) The Tsingy is a weathered limestone plateau in the Melaky region of western Madagascar. Vertical jointing in the limestone bedrock created the sharp karst spires. (D) Southern China has extensive exposures of limestone and the world's best examples of tower karst. This 400 m (1,312 ft) high karst tower is in Zhangjiajie National Forest Park, near Zhangjiajie City in northern Hunan Province, China. *(A. Adam Burton/robertharding/Getty Images; B. Per-Andre Hoffmann/Getty Images; C. Iñaki Caperochipi/Getty Images; D. Feng Wei Photography/Getty Images)*

A

B

C

D

A Hidden World: Subterranean Karst

Some of the most dramatic and enchanting environments in physical geography are limestone cave (*cavern*) systems. Limestone caves are formed over millions of years through the dissolution of limestone by rainwater **(Figure 16.15)**. Within the cave systems, remarkable cave formations often grow from the ceilings, floors, and walls.

In the open air of karst caves, water seeping from the ground surface outside enters through the cave ceiling and drips and flows across the ceiling and down the walls. Minerals such as calcium carbonate are slowly precipitated out of the flowing and dripping water. These minerals accumulate into cave formations called **speleothems**.

Dripstones are speleothems formed by precipitation of calcium carbonate by dripping water. *Flowstones* are sheetlike calcium carbonate deposits formed along the edges where water drops flow down before they drip. Dripstone and flowstone structures form extremely slowly and are commonly several hundred thousand years old. Among the most common dripstones are **stalactites**, formations that grow from the ceiling downward. Conversely, **stalagmites** grow from the cave floor upward as water drips from stalactites. A **limestone column** is a cylindrical dripstone that results when a stalactite joins with a stalagmite. Stalactites often form a hollow structure called a *soda straw* when they first start forming. Examples of speleothems are shown in **Figure 16.16** on the following page.

When the climate is wetter, more water infiltrates the ground from the surface and drips from speleothems, and calcium carbonate builds slightly faster than during dry periods. If, however, the climate becomes wet enough to raise the water table and flood the cave again, the speleothems will stop growing, and water will slowly dissolve them again. Similarly, if water stops flowing into the cave and the ceilings and walls become dry, speleothem growth will stop.

Figure 16.15 Limestone cave formation.

Groundwater dissolution
Limestone cave systems form as groundwater flows downward through joints in limestone bedrock and dissolves calcium carbonate in the rock. Subterranean caves slowly begin to form as limestone is removed by this dissolution process. Dissolution takes place above the water table where water is trickling through the joints and below the water table where the water flows.

Cave system formation
Over several million years, a labyrinth of caves forms in the limestone bedrock. Caves below the water table are flooded; caves above the water table are dry. Cave formations grow only where rainwater finds its way through the ground above and drips from the ceilings and walls of the cave. Water flows between caves through underground stream systems. Where the ceiling collapses, a collapse sinkhole forms, allowing explorers to enter the cave system.

Animation
Limestone Cave Formation
Available at
www.saplinglearning.com

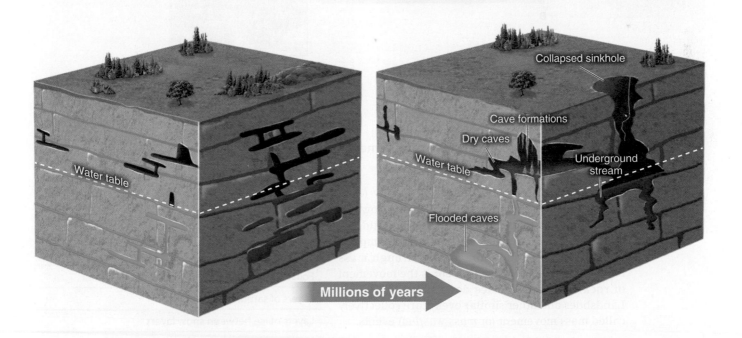

Figure 16.16 Dripstone development. (A) Dripstones form as water drips through the open air of a cave. As it does so, it deposits calcium carbonate, and the cave formations begin to grow. (B) The Frasassi Caves, near the town of Genga, Italy, contain excellent examples of dripstones. The photo on the right shows a polished cross-section of a stalactite and the hollow soda straw center that forms in the stalactite. *(Left. DEA/A. DE GREGORIO/Getty Images; right. © David Lynch)*

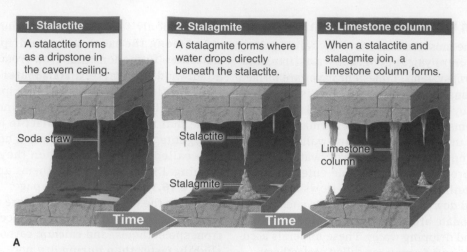

1. Stalactite
A stalactite forms as a dripstone in the cavern ceiling.

Soda straw

2. Stalagmite
A stalagmite forms where water drops directly beneath the stalactite.

Stalactite

Stalagmite

3. Limestone column
When a stalactite and stalagmite join, a limestone column forms.

Limestone column

Time   Time

A

Stalactite

Stalagmite

Limestone column

B

Most recent layers

Hollow "soda straw" center

Oldest layers

Romania
● **Frasassi Caves**   45° N
Italy
Greece
Turkey
Mediterranean Sea
15° E   30° E
Libya

16.3 Unstable Ground: Mass Movement

◎ **Identify different types of mass movements and describe their behavior and causes.**

In this part of *Living Physical Geography* (Part IV), we explore the eroding and tearing down of the landscapes that geothermal energy builds up through plate tectonics. One particularly important process of denudation occurs through mass movement events. **Mass movement** is the movement of rock, soil, snow, or ice downslope by gravity. Landslides and other similar events are collectively called mass movement (or *mass wasting*) events.

Why Mass Movement Occurs

A slope is *stable* when it is unlikely to move *(fail)* and *unstable* when it has moved (experienced

mass movement) in the past or is likely to do so soon. Several factors, summarized in **Table 16.1**, create weak layers within rocks, regolith (the loose Earth material that covers bedrock), or snow and potentially create unstable slopes.

Table 16.1   **Factors That Make Slopes Unstable**
Geologic faults
Jointing in rocks
Foliation planes in metamorphic rock
Layers of saturated clay or sand
Layers of ice between snow layers

Conversely, several factors increase slope stability. Friction and electrical charges hold soil particles together. Soil moisture and plant roots

1
1
1
1
1
1
1
1
1
1
1
1
1
1
1
1
1
1
1
1
1
1
1
1
1
1
1
1
1
1
1
1
1
1
1
1

Figure 16.17 Slope stability. (A) The angle of repose is a result of the relationship between gravitation and friction among particles, and it varies depending on the shape, size, and wetness of those particles. (B) If friction is equal to or greater than the downslope force, the material will not move. If friction is less than the downslope force, the material will move. In this example, the downslope force is increased by steepening of the slope, as might occur where a hillside is cut by a new road.

also anchor regolith and keep it from moving. Vegetated soil that is moist but not saturated is far more stable than dry and unvegetated soil. The surface tension of water and the roots of plants both help bind particles together. A material's angle of repose (the steepest angle a pile of sediments can form; see Section 15.1) is determined by the equilibrium between gravity pulling particles downward and friction holding particles in place. Larger particles have steeper angles of repose than smaller particles. The stability of a slope depends on the relationship between *friction*, which keeps the material in place, and the *downslope force* (or *gravitational force*), which causes the material to slip downhill **(Figure 16.17)**.

Mass movements can be caused by any factor that increases the downslope force or decreases the resistance force. **Table 16.2** summarizes the primary factors that change these forces and result in slope failure.

Types of Mass Movements

There are many different types of mass movements, and they range in duration from taking hundreds of years to only a few seconds to play out. In the material that follows, we explore mass movement types, starting with the slowest and most gradual and working up to the fastest and most dangerous, in the sequence shown in **Table 16.3**.

Table 16.3	Types of Mass Movements
TYPE OF MASS MOVEMENT	**RATE OF MOVEMENT**
Soil creep	Slow
Slumps	↕
Flows and landslides	
Avalanches	
Rockfall	Fast

Table 16.2	Factors That Cause Mass Movements
FACTOR	**EFFECT ON FRICTION AND DOWNSLOPE FORCE**
Earthquakes	Ground shaking can increase downslope force. Ground shaking can also separate particles of regolith and decrease the friction that holds them together.
Rivers and roadcuts	Steeper slopes are subject to greater downslope force. Rivers often undercut their banks and make them steeper. Roadcuts may make slopes steeper and more likely to fail.
Ground saturation	As soils become saturated, they become heavier, and the downslope force increases. Saturation also moves soil particles farther apart, reducing friction between them. Mass movement commonly follows storms that bring heavy, soaking rains. Broken water pipes on steep slopes can also lead to mass movement.
Weathering	Increased weathering weakens the integrity and strength of rocks and regolith.
Removal of vegetation	Removal of vegetation and its anchoring roots weakens the ground and decreases the resistance force.

Soil Creep

Soil creep is the imperceptible downslope movement of soil and regolith as their volume changes in seasonal expansion–contraction cycles. Clay particles in soil expand when wetted or warmed and then, as the season changes, contract when dried or cooled. In addition, wet soils that freeze in winter can expand by about 10%. When thawed in spring, they contract by the same amount. As they expand, soils move outward perpendicularly away from the sloped surface. As they contract, they settle downward vertically. These movements result in soil creep **(Figure 16.18)**.

These small movements add up to significant downslope migration of soils after many years. Soil creep is too slow to be seen in motion, but its effects are easy to see in a landscape **(Figure 16.19)**.

A special type of soil creep, called **solifluction**, occurs in northern and alpine tundra (see Section 9.4). In solifluction, freeze–thaw expansion–contraction cycles cause the soil to flow slowly downslope in overlapping sheets, as shown in **Figure 16.20**.

Another type of imperceptible soil movement is caused by the movements of livestock, particularly cattle, on the slopes of hills. This phenomenon is explored in the **Picture This** feature on the following page.

Slumps

A **slump** is a type of mass movement in which regolith detaches and slides downslope along a spoon-shaped plane, called a *failure surface*, and comes to rest more or less as a unit. Slumps are often called *rotational slides* because they follow the concave failure surface over which regolith moves. Together, the topmost point of detachment of the slump and the resulting cliff are called the *head scarp*. The base of the slump is the *toe*, as shown in **Figure 16.21**.

Figure 16.18 Soil creep. In thick regolith, soil creep is fastest near the soil surface and slows down nearer bedrock.

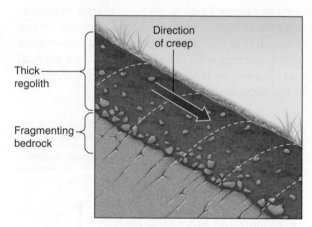

Figure 16.20 Solifluction in the Tian Shan Mountains, Kyrgyzstan.

Soil is slowly flowing downhill in sheets.

(© Marli Miller)

Figure 16.19 Soil creep and its effects. Structures and vegetation on hillsides sometimes sag downslope because of soil creep. *(Bruce Gervais)*

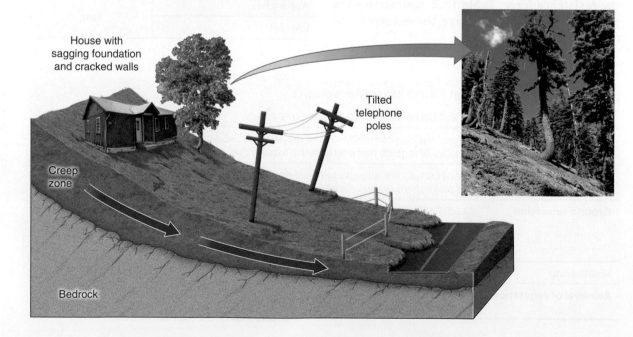

Picture This Cattle Terraces

These parallel ridges, called *cattle terraces* (or *livestock terraces*), are near Hollister, California. Cattle have created the terraces by walking along the hillside. There is nothing special about the soils where cattle terraces occur. The terraces form because the slope is too steep for the cattle to walk straight up or down, so they must instead follow a line of equal elevation, or a *contour*. Cattle terraces are common on slopes of intermediate steepness where cattle graze in large numbers. If a slope is too gentle, cattle will not follow the contours, and terraces will not develop. Likewise, if a slope is too steep, cattle will not graze on it for fear of falling. Cattle terraces are a form of mass movement similar to soil creep or solifluction.

Consider This

1. If you went to a new area and saw what looked like cattle terraces, what evidence would you look for to support the hypothesis that cattle made the terraces?

2. If you were in doubt about whether cattle made the hillside feature, what alternative explanation might you propose?

(Laura Alice Watt)

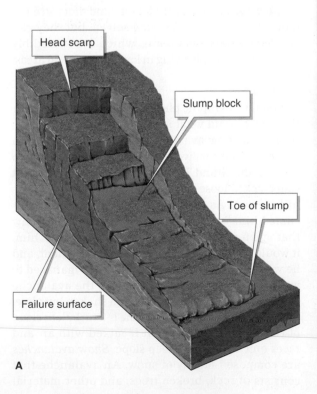

A

B

Figure 16.21 Slumps. (A) Slumps can be triggered by earthquakes or heavy rains. Their downhill rate of movement ranges from millimeters per day or less to several meters per minute. They range in size from a few meters to a few kilometers across. (B) Heavy spring rains led to this slump that occurred in Stone Canyon near Los Angeles. *(Tom McHugh/Science Source)*

Flows and Landslides

Flows and landslides are common occurrences in all mountainous regions. **Landslide** is a general term for the sudden and rapid movement of rock or debris down a steep slope. *Flow* is a general term that includes any landslides that have large amounts of water mixed into them.

There are several different kinds of flows. Their surface areas vary from a few square meters to tens of square kilometers. The rate of movement depends on the type of flow, the water content of the moving material, and the steepness of the slope. On gentle slopes, flows may move at the rate of a slow walk, but on steeper slopes, it may be impossible for a human to outrun them.

A **mudflow** is a fast-moving flow composed of fine-grained material mixed with water that forms mud. In a **debris flow**, a fast-flowing slurry of mud is mixed with large objects, such as rocks and vegetation. Volcanic eruptions sometimes produce debris flows called *lahars* (see Section 15.2). One of the most deadly debris flows in history, however, occurred in northern Venezuela not because of a volcanic eruption, but because of heavy rains **(Figure 16.22)**.

All landslides occur in mountainous terrain where slopes are steep (but not vertical). They can move at rates of several hundred kilometers per hour and come to rest within minutes of the initial movement beginning.

A **rock slide** is a landslide that consists predominantly of rocks and broken rock fragments, and a **debris slide** (or *debris avalanche*) consists of regolith and other material, such as soil and trees. Heavy rains and earthquakes often trigger the slope failure that gives rise to landslides. **Figure 16.23** details the techniques scientists are using to monitor ground movement on unstable slopes to warn people in harm's way before a slope fails so they can evacuate.

Landslides also occur on the seafloor where slopes are steep. Much of the coastline of the Hawaiian Islands, for example, is dominated by enormous head scarps formed by undersea debris slides that occurred before the islands were settled by people **(Figure 16.24)**. These undersea landslides are not dangerous to people by themselves, but they can create very large tsunamis, which are incredibly hazardous to people living in low-lying coastal areas.

Figure 16.22 Caraballeda debris flow. (A) In December 1999, after heavy rains, a debris flow devastated portions of Caraballeda, in the state of Vargas, Venezuela. Up to 30,000 people died, and 75,000 were displaced. The damage totaled some $3 billion. (B) Boulders the size of small houses were transported in the debris flow. *(U.S. Geological Survey, photo by Matthew C. Larsen)*

A

B

Avalanches

National Geographic photographer and adventurer Jimmy Chin did what few people have ever done: He rode a snow avalanche 300 m (984 ft) down a mountain and survived. While Chin was on a photo shoot in the Grand Tetons of Wyoming, a large, wet slab cracked loose around him as he began a downhill ski run. As the churning cloud of snow consumed him, he struggled to stay on top. He knew that when it came to rest, if the snow buried him, it would compact and entomb him like cement, and he would suffocate within minutes. He managed to work his way to the surface before the avalanche came to a rest. He was bruised and shaken but alive.

Broadly defined, an **avalanche** is a turbulent cloud of rock debris or snow that is mixed with air and races quickly down a steep slope. *Snow avalanches* are composed mostly of snow. An avalanche that consists of rock, broken trees, and other material

Laser measurements
Lasers located on unstable soil are beamed and reflected off an unmoving object nearby, such as an outcrop. Small changes in ground position are detected by the reflected laser signal, providing extremely precise measurements of even small movements.

GPS sensors
A network of sensors connected to the GPS (Global Positioning System) is installed on unstable slopes. Even the smallest movements of the soil are measurable using the GPS.

Borehole cables
Scientists drill deep boreholes through the ground and lower wire cables into the holes. If the ground begins slowly moving, the cables will be pinched and send a signal that the ground is shifting.

Figure 16.23 SCIENTIFIC INQUIRY: Can scientists predict dangerous landslides? The key to developing an early warning system for deadly landslides is monitoring ground movement long before the slope fails. Heavy rains are more likely to trigger a mass movement event on slopes that have been moving before the rains. Long-term monitoring stations allow scientists to develop a history of slope movement for a given location. Scientists issue evacuation warnings if the slope shows signs of moving, although accurate predictions of the timing of slope failure are not possible. *(Clockwise from top left. © 2014 Nicole Feidl and UNAVCO; © TUM, Forschungsgruppe alpEWAS; © TUM, Forschungsgruppe alpEWAS)*

A

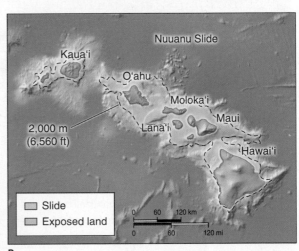

B

Figure 16.24 Undersea debris slides in the Hawaiian Islands. (A) The north coast of Moloka'i, shown here, has been shaped by large debris slides. Rising out of the ocean over 1,000 m (3,300 ft), these sea cliffs are the highest in the world. (B) The debris slides that shaped the sea cliffs continued down the steep submarine slopes surrounding the Hawaiian Islands. The olive areas in this map show the extent of these submarine debris slides. The 2,000 m (6,560 ft) depth contour is shown. The Nuuanu slide, shown here, is larger than even the largest island, Hawai'i.
(© Richard J. Anderson)

is called a *debris avalanche*. An avalanche is mixed with air and is therefore extremely turbulent. It gains such momentum that it is capable of knocking down trees and buildings in its path. Snow avalanches often happen in the same area repeatedly, forming **avalanche chutes** through which snow and debris avalanches regularly move (**Figure 16.25**).

Rockfall

Rockfall occurs when rocks tumble off a vertical or nearly vertical cliff face (**Figure 16.26**). As they fall, rocks are broken apart into smaller fragments and dislodge other rocks as well. Rockfall is particularly common along roadcuts. Near rocky cliffs, the asphalt surface of roads is often pitted by earlier rockfall events.

Rockfall caused by frost wedging on steep mountain faces often creates piles of rock. The pieces of angular broken rock that accumulate at the angle of repose at the base of a steep slope or cliff are called **talus** (or *scree*). Repeated rockfall events in the same location can carve notches called *rockfall chutes* into the bedrock. Rockfall chutes also function as avalanche chutes as they often channel frequent snow avalanches. Talus accumulates at the base of rockfall chutes in cone-shaped piles called **talus cones**. A **talus apron** forms where two or more talus cones merge at the foot of a cliff. **Figure 16.27** illustrates these landforms.

The agents of weathering and mass movement act simultaneously and continuously to reduce the vertical relief that Earth's internal geothermal energy has built up. **Figure 16.28** reviews these processes in the context of the Grand Canyon.

Figure 16.25 An avalanche chute. Avalanche chutes form where the topography of mountainous areas creates frequent snow avalanches. This snow avalanche is moving down an avalanche chute in the Pamirs, a mountainous region in Tajikistan. *(Medford Taylor/ Getty Images)*

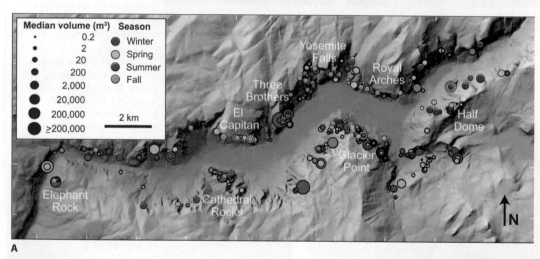

A

B

Figure 16.26 Rockfall in Yosemite Valley.

(A) This map shows the locations of rockfall events in Yosemite Valley over 150 years. The color indicates the season in which each rockfall event occurred, and the size of the colored circle indicates the amount of material that fell. (B) This photo shows a major rockfall event that occurred on October 11, 2010. (C) Large angular boulders dot the floor of Yosemite Valley. Many of these boulders, which originated from prehistoric rockfall events, traveled far into the valley before they came to rest. *(A. Greg Stock, Yosemite National Park, National Park Service; B. © Tom Evans, www. elcapreport.com; C. Steven M. Bumgardner, Yosemite National Park, National Park Service)*

Video

Yosemite Nature Notes: Rockfall

Available at www.saplinglearning.com

C

Figure 16.27 Talus. Talus has accumulated in several talus cones, which have merged into a talus apron at Moraine Lake in the Canadian Rockies in Banff National Park. Talus always settles at its steepest stable angle, or the angle of repose. *(Bruce Gervais)*

Figure 16.28 Carving the Grand Canyon.

(Gary Crabbe/AGE Fotostock)

Weathering

Differential weathering
These layers of sedimentary rock are composed of different materials and have different resistances to weathering, creating the stair-step pattern seen here.

Jointing
Weathering occurs fastest along weak areas or cracks in the rock, called joints.

Chemical weathering
Iron minerals (such as hematite) in the rock are oxidizing (rusting). This chemical weathering process weakens the rock.

Physical weathering
Frost wedging widens cracks in the sandstone, breaking the rock apart.

Mass movement

Rockfall
This boulder is perched precariously and will eventually fall in high wind, during an earthquake, or as weathering continues.

Rock slide
These lobes protruding into the river were formed by rock slides. They formed temporary dams until the river cut through them.

Talus and angle of repose
Frost wedging and other types of weathering break off blocks of rock that accumulate in a talus apron that settles at the angle of repose.

Erosion

Stream erosion
(see section 17.2)
Like a conveyor belt, the Colorado River carries material that slides into it downslope toward the ocean. This process deepens and widens the canyon through time.

Vegetation
These dark areas are covered by streamside plants. At the river's edge, plant roots anchor sandy soil and reduce erosion. Plant roots can also cause physical weathering as they break rocks into smaller fragments.

GEOGRAPHIC PERSPECTIVES
16.4 Deadly Mass Movements

◎ **Assess the threats to people from mass movements and explain how human vulnerability to those threats can be reduced.**

According to the USGS, mass movement events are among the world's most deadly natural disasters. Mass movements cause an estimated $2 billion to $4 billion in damage and an average of 50 deaths each year in the United States. Worldwide, mass movements cause significant loss of life **(Figure 16.29)**.

Causes of Deadly Mass Movements

As we have discussed in this chapter, the factors that cause most mass movement events are the undercutting of steep slopes, the removal of vegetation, earthquakes, and heavy and prolonged rainfall. Many mass movement disasters are preceded by deforestation and slope steepening through development, which weakens the slopes. An earthquake or saturation of the ground by heavy rainfall then triggers slope movement. Major mass movement events causing 1,000 deaths or more have occurred, on average, about every 4.5 years over the past 50 years **(Figure 16.30)**.

Less obvious hazards are created by landslides when rivers are blocked by debris. River canyons are frequently sites of slides because stream erosion cuts steep walls that become unstable through time. These slides often settle across rivers, damming them and creating lakes behind the dam debris. The 2008 Sichuan earthquake created 66 new lakes as landslide debris dammed the streams. More than 40 of these debris dams had to be carefully removed because scientists feared that the dams would suddenly break, allowing the impounded water to burst forth and submerge communities downstream.

Assessing the Risk

Many developing countries are particularly vulnerable to mass movement disasters. Factors that increase the possibility of a disaster include heavy rainfall brought by the summer monsoon (see Section 5.3) or by tropical cyclones (see Section 6.3), farming on steep slopes (see Section 11.1), *shantytown* (slum) settlements on steep hillsides, large populations, and poverty. In many large urbanized regions, such as Rio de Janeiro, Brazil, and Manila in the Philippines, human population growth and economic marginalization have driven uncontrolled settlement expansion onto steep and unstable slopes, where mass movements are more likely to occur.

Using satellite data and a GIS (see Section 1.4), scientists have mapped the areas of greatest mass movement susceptibility **(Figure 16.31)**. But as useful as they are for identifying areas of risk,

Story Map

Deadly Mass Movements

Available at
www.saplinglearning.com

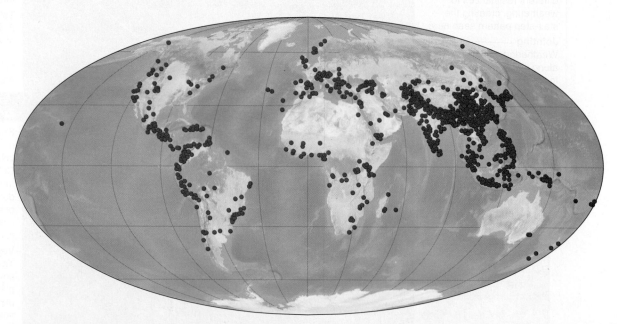

Figure 16.29 Map of deadly mass movements. This map shows the locations of 2,620 mass movement events that took place between 2004 and 2010. These mass movements resulted in about 32,300 human deaths. Landslides that were triggered by earthquakes are not included. If they were included, the number of people who lost their lives would be considerably higher.

A

YEAR	LOCATION	ESTIMATED FATALITIES	TRIGGER
1999	Vargas, Venezuela	30,000	Heavy rains
2005	Pakistan and India	25,500	Kashmir earthquake
1985	Colombia (Tolima)	23,000	Eruption of Nevado del Ruiz volcano
2008	China (Sichuan)	20,000	Sichuan earthquake
1970	Peru (Ancash)	18,000	Earthquake
1998	Honduras, Guatemala, Nicaragua, and El Salvador	10,000	Heavy rains from Hurricane Mitch

B

Landslide risk

Slight   Moderate   Severe

Figure 16.30 Deadly mass movements. (A) On August 14, 2017, a debris slide triggered by heavy rains killed 1,141 people and destroyed 3,000 homes in a mountainous area near Freetown, Sierra Leone. This photo shows rescue teams searching for victims the day after the slide. (B) This table lists recent mass movement disasters that took 10,000 lives or more. Note that in some cases, multiple mass movement events are grouped together under a single trigger event, such as the Sichuan earthquake in 2008 and Hurricane Mitch in 1998. *(Anadolu Agency/Getty Images)*

Video

New NASA Model Finds Landslide Threats

Available at www.saplinglearning.com

Figure 16.31 Global landslide risk map. This map of landslide risk, which includes all types of ground movement, was made using satellite data and a GIS. To make the map, scientists analyzed slope steepness (steeper slopes have a higher risk), land-use types (land with less vegetation has a higher risk), and soil types (coarse-grained soils have a higher risk). The black dots show areas where significant landslides have occurred.

such maps do not provide real-time warnings when shifts such as changes in weather elevate the risk of mass movements.

To address this problem, scientists have developed a risk-assessment program that is based on rainfall rates and amounts. It is available in real time to anyone with an Internet connection. The TRMM (Tropical Rainfall Measuring Mission), run by NASA and the Japan Aerospace Exploration Agency (JAXA), evaluates landslide risk by using satellites to measure rainfall amounts and rainfall intensity for all tropical areas. These data are used in conjunction with a mass movement risk map (see Figure 16.31) to provide real-time warnings **(Figure 16.32)**.

At a global scale, identifying broad regions most at risk of landslides is a relatively simple task. However, at a local geographic scale, identifying which specific slopes will move and which towns and villages are at risk is far more difficult. Although monitoring techniques such as those shown in Figure 16.23 can detect ground movements, there is not yet a warning system for mass movements analogous to the U.S. tornado warning system (see Section 6.2), which has proven very effective at giving people enough lead time to get out of harm's way. Scientists are trying to develop a similar system that gives residents warning of landslide hazards. Such a system could reduce our vulnerability to these natural disasters and save many human lives.

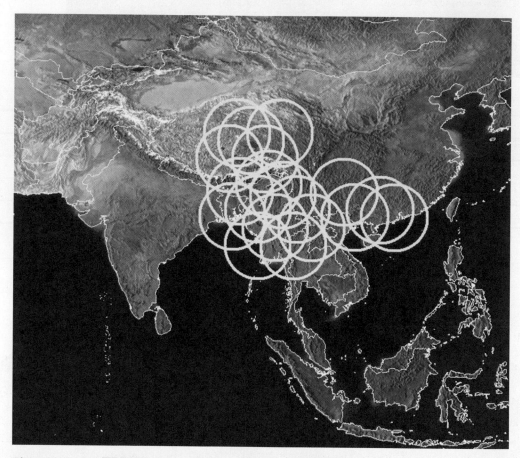

Figure 16.32 TRMM map of potential landslide areas. This map shows landslide risks for South Asia for January 8, 2018. This map is based on 3-day average rainfall intensities and rainfall totals. Areas with increased landslide potential are circled in yellow. Areas that are likely or very likely to experience landslides are colored orange. *(NASA)*

Chapter 16 Study Guide

Focus Points

16.1 Weathering Rocks

- **Physical weathering:** Rates of physical weathering are greatest at high elevations and high latitudes.
- **Frost wedging:** Frost wedging is the most important type of physical weathering where temperatures fall below freezing.
- **Chemical weathering:** Chemical weathering is dominant where temperatures remain above freezing and there is plenty of moisture.

16.2 Dissolving Rocks: Karst Landforms

- **Karst processes:** Carbonate rocks dissolve in naturally acidic rainwater (a form of chemical weathering), forming karst topography. Most karst topography is found in limestone bedrock.
- **Surface karst topography:** Limestone pavement, sinkholes, collapse sinkholes, disappearing streams, cockpit karst, and tower karst are among the most prominent surface features in karst topography.
- **Subterranean karst:** Subterranean landforms, such as caves and speleothems, are common in karst regions.

16.3 Unstable Ground: Mass Movement

- **Causes of mass movement:** When the downslope force exceeds the resistance force, a slope will fail, and mass movement will occur. Many factors, such as earthquakes, heavy rains, and roadcuts, can trigger mass movement events.
- **Mass movement speed:** Material moves downslope at speeds ranging from slow soil creep to rapid and deadly avalanches and rockfall.

16.4 Geographic Perspectives: Deadly Mass Movements

- **Human toll:** Many human fatalities caused by landslides occur because people live on steep and unstable slopes that have been weakened by heavy rains.

Key Terms

avalanche, 502
avalanche chute, 504
carbonation, 489
chemical weathering, 488
collapse sinkhole, 492
debris flow, 502
debris slide, 502
denudation, 487
deposition, 487
disappearing stream, 494
dissolution, 492
erosion, 487
exfoliation, 487
frost wedging, 488
karst, 491
landslide, 502
limestone column, 497
mass movement, 498
mudflow, 502
physical weathering, 487

rock slide, 502
rockfall, 504
root wedging, 488
salt weathering, 488
sediment transport, 487
sinkhole, 492
sinkhole lake, 492
slump, 500
soil creep, 500
solifluction, 500
speleothem, 497
spheroidal weathering, 489
stalactite, 497
stalagmite, 497
tafoni, 488
talus, 504
talus apron, 504
talus cone, 504
weathering, 487

Concept Review

The Human Sphere: Weathering Mount Rushmore

1. Why is weathering a concern for the people responsible for maintaining the Mount Rushmore National Memorial?

16.1 Weathering Rocks

2. Geothermal energy builds vertical relief. Where does the energy that reduces vertical relief come from?

3. Compare and contrast the processes of weathering and erosion. What does weathering do to rocks? What does erosion do to rocks and rock fragments?

4. What is denudation? What processes denude Earth's surface?

5. What is physical weathering? Geographically, where is it most common? Give examples of physical weathering processes.

6. Why are joints important to the process of weathering?

7. What is exfoliation? Where does it occur?

8. What is frost wedging? Where does it take place?

9. Besides frost wedging, which two other types of physical weathering break rocks apart? Briefly explain each of them.

10. Describe chemical weathering and list examples of chemical reactions that cause it.

11. Geographically, where is chemical weathering the dominant weathering process?

12. What is differential weathering? How does it relate to jointing and bedding planes?

13. List three examples of landforms created by differential weathering.

16.2 Dissolving Rocks: Karst Landforms

14. What is karst? What processes give rise to karst landforms?

15. In which type of bedrock is karst most commonly found?

16. What is a sinkhole, and how does one form?

17. What is a collapse sinkhole, and how does one form?

18. What is a sinkhole lake?

19. What is a karst cave? How do karst caves form?

20. Explain what a disappearing stream is and how it forms.

21. What is limestone pavement?

22. Explain how cockpit karst and tower karst form.

23. What are speleothems? Why do they form only when the water table drops below a karst cave?

24. What are dripstones and flowstones?

25. Compare and contrast stalactites, stalagmites, and limestone columns. How does each form?

16.3 Unstable Ground: Mass Movement

26. What is a mass movement event? What single factor do all kinds of mass movements share?

27. What are resistance and downslope forces in the context of mass movement events? Give examples of how each force can change.

28. What happens when the resistance force exceeds the downslope force? What happens when the downslope force exceeds the resistance force?

29. Review the different types of mass movements and the settings in which each occurs.

30. What are the differences between flows and landslides?

31. Define *avalanche*. Are all avalanches composed of snow?

32. What is talus? What is a talus cone? How do rockfall chutes relate to talus cones?

16.4 Geographic Perspectives: Deadly Mass Movements

33. What natural and anthropogenic factors increase the chance of a mass movement event?

34. Can scientists predict landslides with accuracy?

35. Explain how scientific tools have helped reduce human vulnerability to mass movement disasters.

Critical-Thinking Questions

1. Chemical weathering occurs faster as temperatures warm and rainfall increases. On a global scale, how might human activities affect the rate of chemical weathering?

2. In terms of geologic time, do you think that landforms resulting from differential weathering, such as natural arches and hoodoos, are short-lived or permanent features? Explain.

3. Are there subterranean caves near where you live? If you do not know, how can you find out?

4. The process of weathering has been going on since Earth first formed, and it never stops, yet many rocks are still "fresh" and relatively unweathered. Why haven't all of Earth's rocks been weathered away?

Test Yourself

Take this quiz to test your chapter knowledge.

1. True or false? Frost wedging is a type of physical weathering.

2. True or false? Chemical weathering occurs mostly at high latitudes.

3. True or false? Debris flows move slowly and are rarely fatal.

4. True or false? A snow avalanche is a type of mass movement.

5. True or false? Only limestone rock forms karst topography.

6. True or false? Dripstones and flowstones are two common types of speleothems.

7. Multiple choice: Which of the following increases the resistance force on a slope?

  a. a roadcut   c. vegetation growth
  b. an earthquake   d. heavy rains

8. Multiple choice: Which of the following requires vertical joints to form?

  a. a collapse sinkhole   c. tower karst
  b. limestone pavement   d. cockpit karst

9. Fill in the blank: Natural arches are the result of
_____.

10. Fill in the blank: A pitted surface caused by weathering in a coastal area is most likely _____.

Online Geographic Analysis

Landslide Areas

In this exercise we examine landslide activity in California.

Activity

In this online interactive exercise we will examine landslide areas that are under active watch by the USGS. Go to https://landslides.usgs.gov/monitoring/. Click the "Maps" link in the menu on the left. Then click the Monitoring Stations map to bring up the list of monitoring stations and select the "U.S. Highway 50, CA" link.

1. When did the landslides occur along Highway 50 between Placerville and South Lake Tahoe in California?

2. How often do the sensors sample data? How often are the data transmitted to the USGS computer?

3. At the bottom of the page there is a photo of a scientist monitoring the Cleveland Corral landslide. What instrument is being used?

4. Another photo at the bottom of the page shows the Cleveland Corral landslide. What time of year does this landslide typically move? Why does movement occur then?

5. Click the "Rainfall" link. Why is rainfall monitored for the slide area?

6. What is the cumulative total rainfall for the slide area?

7. Scroll down on the page until you see the "Groundwater Pressure" heading. Based on what you see in the graphs, has the groundwater pressure changed over the period shown?

8. Scroll down on the page until you see the "Slope Movement" heading. Examine the graphs. Describe slope movement for the time shown.

9. Scroll down to the "Geophone" heading. What is a geophone, and what does it record? Why does it record these variables?

10. Exit to the map page and briefly examine several others slides on the map. What is the single most important closely monitored variable that triggers landslides?

Picture This. *Your Turn*

Mass Movement

Fill in the blanks on the diagram, using each of the following terms only once.

1. Head scarp
2. Lahar
3. Landslide
4. Rock slide
5. Rockfall

6. Rockfall chute
7. Slump
8. Slump toe
9. Solifluction

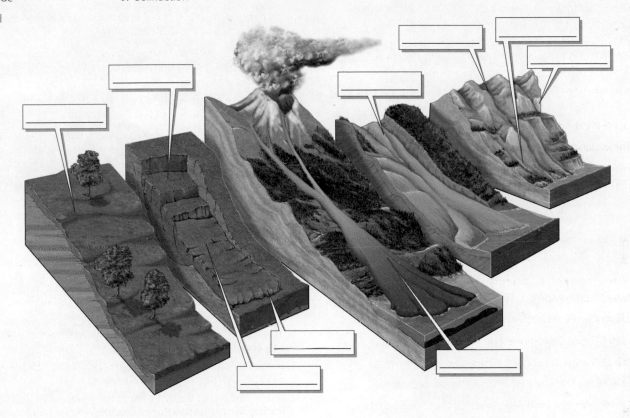

For animations, interactive maps, videos, and more, visit www.saplinglearning.com. **Sapling**Plus

Flowing Water: Fluvial Systems

Chapter Outline *and Learning Goals* ◎

This photo shows a sunrise over Horseshoe Bend on the Colorado River, located near Page, Arizona. Long ago at this location, the Colorado River meandered on an ancient flat floodplain. When the region was lifted by geologic forces, the Colorado River cut downward into the sandstone bedrock below, forming an *entrenched meander*.

(travellight/Shutterstock)

To learn about entrenched meanders, turn to Section 17.2.

THE HUMAN SPHERE

Living on Floodplains

Figure 17.1 The Fertile Crescent. The first permanent human settlements, the development of agriculture, and animal domestication began on the corridor of land extending from the lower Nile River to the Tigris and Euphrates rivers (shaded area).

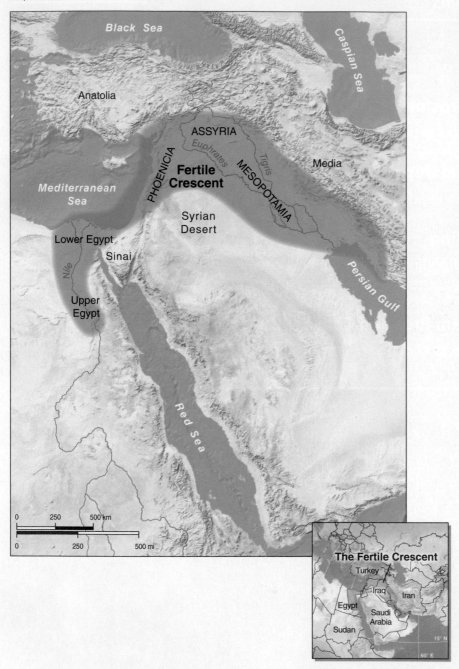

Video

Building Smart Down by the River

Available at www.saplinglearning.com

People have always lived near large rivers. Early subsistence agriculture and the first sedentary (nonmigratory) human societies began over 10,000 years ago on the large swath of land called the *Fertile Crescent* formed by the lower Nile River and the Tigris and Euphrates rivers in ancient Mesopotamia. Today, this area spans Iraq, western Iran, southern Turkey, Syria, Lebanon, Israel, and Egyp*t* **(Figure 17.1)**.

The land in the floodplains along most large rivers is flat and desirable for settlement. **Floodplains** are the areas of flat land near a stream that experience flooding on a regular basis. They are composed of deep layers of sediment, as we will discuss in more detail in Section 17.3. The soils deposited by occasional flooding, called *alluvial soils*, or entisols (see Section 11.1), are rich and fertile. Rivers also provide corridors for transportation inland from the ocean. Because of these benefits, many major cities have been established along rivers. Familiar examples include Buenos Aires, Cairo, Jakarta, London, Madrid, Paris, Portland (Oregon), Rome, Shanghai, Tokyo, and Washington, DC.

A problem arises, however, when rivers overflow their banks. Like all other physical systems, rivers are spatially and temporally dynamic. Their flat floodplains were built layer by layer as they spilled out of their channels many times in the past and deposited the sediments they carried.

Scientists anticipate that flooding on stream floodplains will worsen because of population growth and increased precipitation variability due to climate change. In the United States, roughly half of all disasters that the Red Cross responds to are flooding related. Research indicates that by 2030 the number of people affected by flooding along streams could increase from the current 20 million people to 50 million people.

17.1 Stream Patterns

◎ **Identify the various geographic patterns of stream systems.**

Water is one of the most important compounds on Earth. In the hydrologic cycle, water evaporates from the oceans, enters the atmosphere, condenses into clouds, and precipitates over land (see Section 4.1). The resulting surface runoff flows back to the oceans in streams. A **stream** is a channel in which water flows downhill by the force of gravity. The term *stream* includes flows from the smallest rivulet of water to the mightiest rivers on Earth. Streams are also called rivers and, depending on the geographic region, by many other names, including brooks, runs, creeks, washes, and wadis.

Drainage Basins

Streams that connect to each other form a drainage basin. A **drainage basin** (or *watershed*) is a geographic region drained by a single trunk stream and its tributaries. A **trunk stream** is a single large stream into which smaller tributaries merge. A **tributary** is a stream that joins with other streams to form a larger stream. **Figure 17.2** illustrates these concepts using the Amazon River.

Topography determines the boundaries of drainage basins. There is no set limit to their size. A single drainage basin can cover a small hillside or a portion of a continent (**Figure 17.3** on the following page).

A ridge or highland that separates drainage basins and defines their boundaries is called a **drainage divide**. A ridge or highland that separates two drainage systems, which have trunk streams that each empty into different ocean basins, is a **continental divide (Figure 17.4** on the following page).

Some drainage basins and the streams within them have no outlet to the oceans. In these cases, streams flow into a low-lying basin on land. Such regions are said to have an **internal drainage**

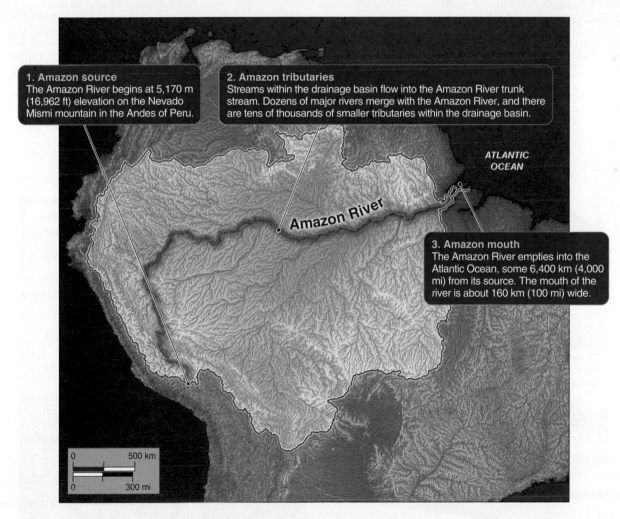

1. Amazon source
The Amazon River begins at 5,170 m (16,962 ft) elevation on the Nevado Mismi mountain in the Andes of Peru.

2. Amazon tributaries
Streams within the drainage basin flow into the Amazon River trunk stream. Dozens of major rivers merge with the Amazon River, and there are tens of thousands of smaller tributaries within the drainage basin.

ATLANTIC OCEAN

Amazon River

3. Amazon mouth
The Amazon River empties into the Atlantic Ocean, some 6,400 km (4,000 mi) from its source. The mouth of the river is about 160 km (100 mi) wide.

0   500 km
0   300 mi

Figure 17.2 Amazon River drainage basin.
Drainage basins are named after their trunk streams. The Amazon River drainage basin (the outlined light area), which encompasses 6.1 million km² (2.4 million mi²), is the largest drainage basin in the world. The "source" of a stream is where it begins, and the "mouth" is where the stream ends.
(NASA/Science Source)

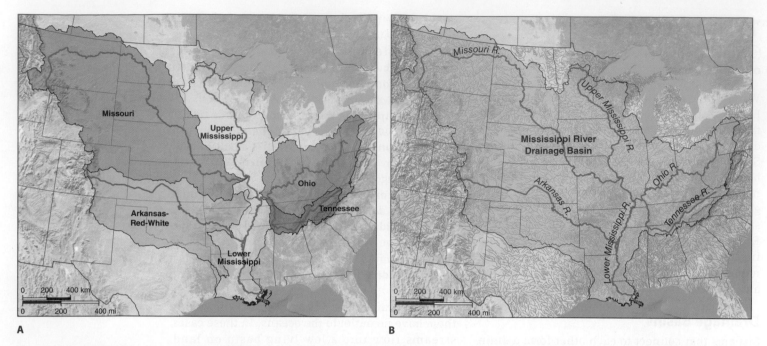

A

B

Figure 17.3 Mississippi River drainage basin and tributaries. (A) The Mississippi River drainage basin is composed of six major sub-basins, each of which is named after the trunk stream within the sub-basin. The Mississippi River drainage basin is divided into the Upper Mississippi and the Lower Mississippi rivers. (B) This map shows all the large streams within the Mississippi River drainage basin. The major tributaries feeding into the Mississippi River are shown and labeled.

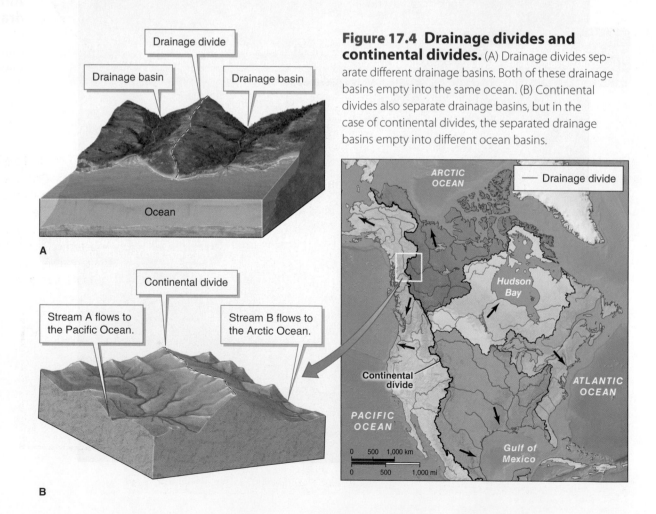

Figure 17.4 Drainage divides and continental divides. (A) Drainage divides separate different drainage basins. Both of these drainage basins empty into the same ocean. (B) Continental divides also separate drainage basins, but in the case of continental divides, the separated drainage basins empty into different ocean basins.

system. As **Figure 17.5** shows, every continent (except Antarctica) has internal drainage systems, and a large portion of Eurasia has internal drainage.

Internal drainage systems are located in arid regions where evaporation rates exceed precipitation. The water that collects in a large terrestrial basin can leave only through evaporation or infiltration into the ground. During evaporation, salts and other dissolved minerals remain behind in the standing water. For this reason, many internal drainage basins contain saline lakes **(Figure 17.6)**.

Trunk streams and their tributaries form different drainage patterns within drainage basins, depending on the topography and the type of surface over which they flow. Although there are many different kinds of drainage

Figure 17.5 Global drainage basins. Drainage basins for each of the world's oceans are grouped by color. Drainage basins emptying into the Pacific Ocean, for example, are shown in blue. Internal drainage basins are brown.

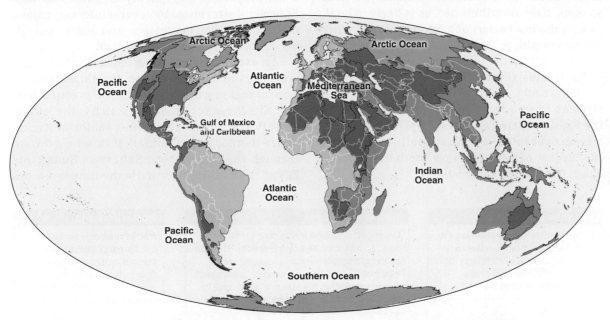

Web Map
Drainage Basins
Available at
www.saplinglearning.com

Figure 17.6 The Dead Sea. The water in the Dead Sea is nearly nine times as salty as ocean water. This high salinity makes the water very dense, allowing people to float on it easily. The sea's name reflects the fact that the water is too salty to support life, except for some salt-tolerant bacteria. *(Kord.com/Getty Images)*

patterns, six common ones are illustrated in **Figure 17.7**. The most common drainage pattern is **dendritic drainage** (from the Greek word *dendro*, which means tree or branches), in which the pattern of the streams resembles the branches of a tree.

Stream Order and Stream Permanence

The geographer Arthur Strahler (1918–2002) created a **stream order** system that ranks streams based on the number of tributaries flowing into them. The higher the stream order number, the larger the stream. The smallest streams, *first-order streams*, have no tributaries, as is likely to be the case in the **headwaters** of a drainage basin, where streams originate. A *second-order stream* is formed at the confluence (merging point) of two first-order streams. **(Figure 17.8)**.

About 80% of all streams are third-order streams or smaller. Very few streams are above the eighth order. The Mississippi River, the 15th largest river in the world, is a tenth-order stream; the largest-order stream on Earth, the Amazon River, is a twelfth-order stream.

Stream Permanence

Stream order is related to stream permanence. A stream that runs dry during part or most of the year is an **intermittent stream**. Many first-order streams are intermittent streams. In contrast, an **ephemeral stream** flows briefly, for hours or days, only after heavy rainstorms. **Permanent streams** flow year-round. Intermittent and permanent streams flow where and when the water table (see Section 11.3) is at or above the level of the ground.

Ephemeral streams are most common in arid regions, where thunderstorms bring sudden downpours. They are short-lived because their water evaporates and quickly infiltrates the soil. Ephemeral streams go by several informal names, including *arroyos*, *dry washes*, and *wadis*, depending on the area in which they occur.

An **exotic stream** is a permanent stream that originates in a humid region and flows through an arid region where there is little precipitation. The Nile River, for example—which is the longest river in the world—begins in equatorial Africa, where there is ample rainfall. It flows northward through the arid eastern Sahara in Sudan and Egypt as an exotic stream. By the time it reaches

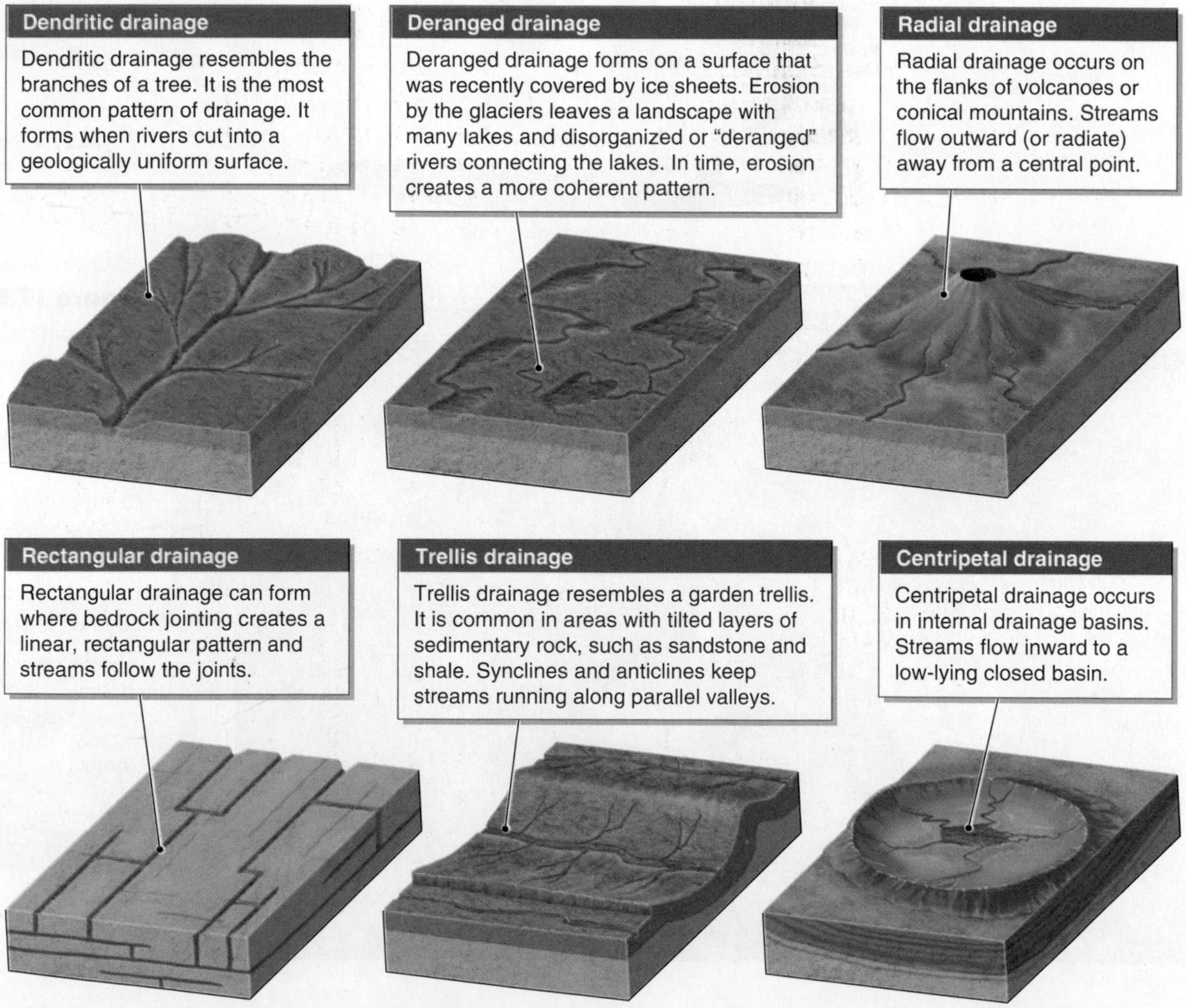

Figure 17.7 Natural stream drainage patterns.

Dendritic drainage

Dendritic drainage resembles the branches of a tree. It is the most common pattern of drainage. It forms when rivers cut into a geologically uniform surface.

Deranged drainage

Deranged drainage forms on a surface that was recently covered by ice sheets. Erosion by the glaciers leaves a landscape with many lakes and disorganized or "deranged" rivers connecting the lakes. In time, erosion creates a more coherent pattern.

Radial drainage

Radial drainage occurs on the flanks of volcanoes or conical mountains. Streams flow outward (or radiate) away from a central point.

Rectangular drainage

Rectangular drainage can form where bedrock jointing creates a linear, rectangular pattern and streams follow the joints.

Trellis drainage

Trellis drainage resembles a garden trellis. It is common in areas with tilted layers of sedimentary rock, such as sandstone and shale. Synclines and anticlines keep streams running along parallel valleys.

Centripetal drainage

Centripetal drainage occurs in internal drainage basins. Streams flow inward to a low-lying closed basin.

the Mediterranean Sea, its flow is much reduced because of evaporation, infiltration into the ground, and water diversions by people.

Anthropogenic Intermittent Streams

Diversions of water from streams by people have made many formerly permanent streams run low or run dry. When this happens, human conflict may ensue, particularly where a stream crosses state or international borders. In the Middle East, for example, Turkey diverts water from the Tigris and Euphrates rivers near their headwaters, causing political friction with downstream neighbors such as Iraq. Likewise, Ethiopia is constructing a $5 billion hydroelectric dam called the Grand Ethiopian Renaissance Dam, which will be the largest in Africa. The Grand Ethiopian Renaissance Dam will divert water from the Nile River and reduce water supplies for Egypt downstream.

The Colorado River and many of its tributaries in the southwestern United States provide water for 30 million people in seven western states. Water diversions are so extensive in this area that during dry periods, the Colorado River does not have enough water flowing in its channel to reach the Gulf of California in Mexico, where it historically emptied. The **Picture This** feature below highlights the Gila River, a tributary of the Colorado River, in this context.

Figure 17.8 Stream order. Stream order increases only when two streams of the same order join. In this example, two third-order streams, shown in yellow, merge to form a fourth-order stream, shown in red.

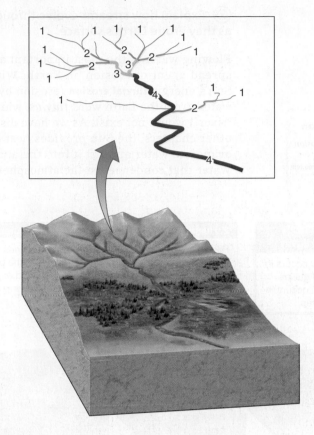

Picture This **The Gila Reborn**

(© Darryl Montgomery)

This muddy trickle of water is the brief rebirth of the Gila River. The Gila River's headwaters are in the relatively wet highlands of western New Mexico. As it flows out of these mountains and through the desert, several tributaries join it. In the past, the Gila flowed all the way to and joined the Colorado River near Yuma, Arizona. But water diversions now prevent the Gila from reaching the Colorado in all but the wettest years. Typically, the Gila runs dry downstream of Phoenix. In 2010, however, heavy rains in its headwaters allowed the Gila to fill its streambed downstream of Phoenix, if only for a little while. This was the first time it had done so in 5 years.

Consider This

1. Is the Gila River an exotic stream? Explain your answer.
2. Is the Gila River a permanent stream, an intermittent stream, or an ephemeral stream? Explain your answer.

17.2 The Work of Water: Fluvial Erosion

◎ **Explain how streams evolve through time as they erode Earth's surface.**

Flowing water is the most important and widespread agent of erosion on Earth. Without the Sun's energy, **fluvial erosion** (erosion by running water; from the Latin word *fluvius*, which means "river") would not exist. As we have discussed in other chapters, the Sun provides heat energy to evaporate water and lift it into the atmosphere. Water that condenses in the atmosphere falls to

Earth as precipitation, and Earth's gravitational field pulls that water downslope. The kinetic energy of the stream's flowing water works to *incise* (cut into; also called *downcut*) Earth's crust and transport rock fragments downslope toward the oceans.

Fluvial erosion begins as precipitation falls on the ground. Some of the water flows into and through the ground, where it contributes to soil moisture and groundwater. Where the ground surface is sloped, water also flows downslope in thin sheets called *sheet wash*.

As water collects in low-lying areas, it begins to form shallow channels. Newly forming stream channels are called *rills*. Rills develop into *gullies* as they deepen. The flat areas between stream channels, where sheet wash occurs, are *interfluves*. Stream channels are always lower than their adjacent interfluves.

As the amount of water and the speed of flow in a stream channel increase, the stream gains more kinetic energy to do the work of carrying soil particles and rock fragments downslope. **Figure 17.9** illustrates the process of **headward erosion**, by which a stream channel lengthens upslope through time, forming new tributaries by fluvial erosion. The spatial scale of headward erosion ranges from a few meters to hundreds of kilometers or more, and its time scale varies from weeks to thousands of years or longer, depending on the geographic setting, geology, and climate.

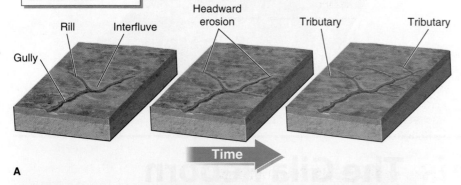

1. Rills, gullies, and interfluves
Stream flow occurs in the rills and gullies. Sheet wash occurs in the interfluves.

2. Headward erosion
Rills migrate upslope through headward erosion, which deepens and lengthens them.

3. New tributaries
New tributaries develop in the process of headward erosion.

The Volume of Water: Stream Discharge

Stream discharge is the volume of water flowing past a fixed point within a stream channel. It is expressed in cubic meters per second (m^3/s) or cubic feet per second (ft^3/s). Discharge can be visualized as the number of 1 m (or 1 ft) cubes of water passing beneath a bridge every second.

Stream discharge differs in urbanized paved areas and rural areas, as the *stream hydrograph* shown in **Figure 17.10** illustrates. Urbanization changes the permeability of the ground surface and therefore the discharge of streams. Paved surfaces are impermeable, and water flows from them quickly into nearby streams rather than infiltrating the ground. As a result, streams in urbanized areas experience sudden spikes in discharge during heavy precipitation events, and discharge drops quickly after storms.

Discharge is important because as more water flows through a stream channel, and as the water flows faster, the stream has more energy to erode Earth's surface. The discharge of a stream is influenced by climate, stream order, season, and surface permeability **(Table 17.1)**.

Figure 17.9 Headward erosion. (A) Headward erosion begins as rainwater collects in low-lying areas to form rills and gullies. (B) The process of headward erosion is visible in this photo of the Namib Desert in Namibia. Note the gemsbok (a type of antelope) in the top center for scale. *(National Geographic Creative/Alamy)*

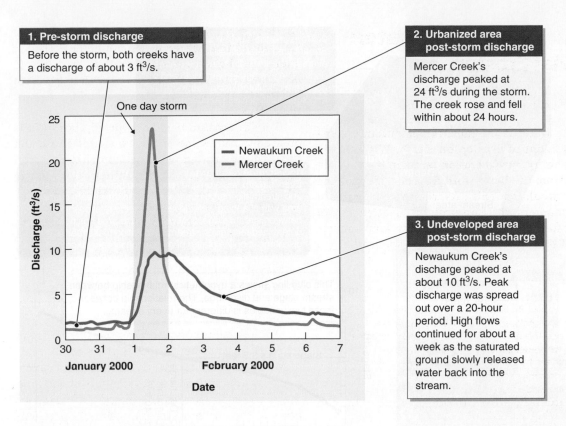

1. Pre-storm discharge

Before the storm, both creeks have a discharge of about 3 ft³/s.

2. Urbanized area post-storm discharge

Mercer Creek's discharge peaked at 24 ft³/s during the storm. The creek rose and fell within about 24 hours.

3. Undeveloped area post-storm discharge

Newaukum Creek's discharge peaked at about 10 ft³/s. Peak discharge was spread out over a 20-hour period. High flows continued for about a week as the saturated ground slowly released water back into the stream.

Figure 17.10 Urbanization affects stream discharge.

This stream hydrograph shows discharge in cubic feet per second for two streams of similar size in western Washington State from January 30 to February 7, 2000. On February 1, rainfall from a storm increased the discharge for both streams. Mercer Creek is in a paved, urbanized area near Bellevue, Washington. Its discharge peaked higher and faster than that of Newaukum Creek, which is in an undeveloped area. *(Source: USGS Fact Sheet 076-03)*

Table 17.1	Stream Discharge Factors

1. Climate: Streams in wet regions have greater discharge than streams in arid regions.

2. Stream order: First-order streams are often intermittent or have little discharge, even in the wettest of climates. Higher-order streams have more discharge.

3. Season: Peak flow periods depend on the timing of precipitation or snowmelt. Heavy rainfall can increase stream discharge to several hundred times its typical flow.

4. Surface permeability: Permeable soil and rock allow water to infiltrate the ground rather than flow over it in a stream channel. Greater permeability reduces surface stream discharge.

In calculations, stream discharge is symbolized by Q. Discharge is measured by multiplying the stream channel area (A) by the speed of water flow, or velocity (v): $Q = A \times v$. Stream channel area is calculated by multiplying channel width by channel depth. The calculation of stream discharge takes into account the variations in flow velocity, stream channel area, and river height that are unique to each stream **(Figure 17.11** on the following page**)**.

The Amazon River has the highest average discharge in the world, about 200,000 m³/s (7 million ft³/s). The Amazon's flow constitutes nearly 20% of the world's total discharge by rivers into the oceans. By comparison, the Mississippi River, the 15th largest river in the world, has an average discharge of about 16,800 m³/s (600,000 ft³/s).

Carrying Materials: Stream Load

The force of flowing water, called *hydraulic action*, frees rock fragments and sets them in motion. Depending on the stream discharge, the size of the fragments in transport ranges from tiny grains of sand and silt to large rocks.

As rock material moves in the stream bed, abrasion rounds and polishes the fragments. **Abrasion** is the process by which movement of one material wears away another material. The continual tumbling of rocks in a streambed abrades other rocks and the channel bedrock **(Figure 17.12** on the following page**)**. Heavy rocks in a stream channel move only during the highest flows, such as those that follow heavy rains or snowmelt.

The material that moves within a stream channel is the *stream load*. The size of the particles in the stream load is directly related to stream discharge and flow velocity. There are three kinds of stream loads: dissolved load, suspended load, and bed load. The **dissolved load** consists of soluble minerals that are carried in solution by a stream. The dissolved load is invisible. The **suspended load** is made up of small particles of clay and silt that

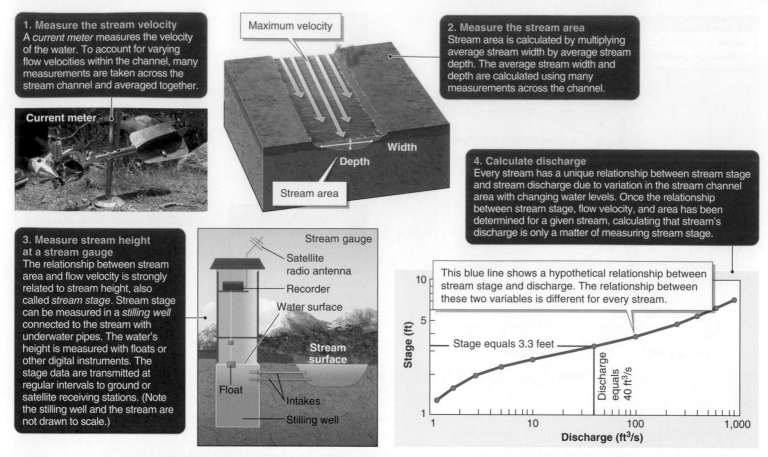

1. Measure the stream velocity
A *current meter* measures the velocity of the water. To account for varying flow velocities within the channel, many measurements are taken across the stream channel and averaged together.

Current meter

Maximum velocity

Width

Depth

Stream area

2. Measure the stream area
Stream area is calculated by multiplying average stream width by average stream depth. The average stream width and depth are calculated using many measurements across the channel.

4. Calculate discharge
Every stream has a unique relationship between stream stage and stream discharge due to variation in the stream channel area with changing water levels. Once the relationship between stream stage, flow velocity, and area has been determined for a given stream, calculating that stream's discharge is only a matter of measuring stream stage.

3. Measure stream height at a stream gauge
The relationship between stream area and flow velocity is strongly related to stream height, also called *stream stage*. Stream stage can be measured in a *stilling well* connected to the stream with underwater pipes. The water's height is measured with floats or other digital instruments. The stage data are transmitted at regular intervals to ground or satellite receiving stations. (Note the stilling well and the stream are not drawn to scale.)

Stream gauge

Satellite radio antenna

Recorder

Water surface

Stream surface

Float

Intakes

Stilling well

This blue line shows a hypothetical relationship between stream stage and discharge. The relationship between these two variables is different for every stream.

Stage equals 3.3 feet

Discharge equals 40 ft³/s

Stage (ft)

Discharge (ft³/s)

Figure 17.11 SCIENTIFIC INQUIRY: How is stream discharge calculated?   Calculating the discharge of streams is important for a number of applications, including water conservation, determination of water allocations among different regions and groups, and flood control. *(U.S. Geological Survey)*

Figure 17.12 Stream abrasion.

The greatest discharge in streams draining from the Himalayas occurs between May and September, with the onset of the summer snowmelt season and warm monsoon rains. Stream abrasion, which rounds the rocks in the stream, occurs mostly during these peak flows. This photo was taken in July near Dharamsala, India, in the foothills of the Himalayas. *(© Roopaushree)*

Rocks rounded by abrasion

Dharamsala

Pakistan

Nepal

India

Figure 17.13 Contrast in suspended loads. This photo shows the confluence of the Drava and Danube rivers near Osijek, Croatia. Arrows show the direction of flow. The Drava River has a comparatively low suspended load and thus looks clear compared with the murky Danube River. *(© Mario Romulić)*

are light enough to remain suspended in the flowing water. The suspended load gives some streams their muddy appearance, as shown in **Figure 17.13**.

A stream's **bed load** is the moving material in a stream channel—ranging from sand grains to gravel, cobbles, and large rocks—that is too heavy to become suspended in the current. Sand grains are moved along in a bouncing or hopping motion called **saltation**. Larger rocks are moved by a sliding, dragging, and tumbling motion called **traction**. Traction sufficient to move large rocks occurs only during peak flow periods. **Figure 17.14** illustrates the three types of stream loads.

Geomorphologists refer to the size of fragments a stream can transport as *stream competence*. Streams that can transport large fragments, such as boulders, have high stream competence. *Stream capacity* refers to the total amount of material (of all sizes) that a stream can transport. During periods of peak discharge, when the stream has more kinetic energy, stream capacity and competence increase.

Finding an Equilibrium: Grading and Stream Gradient

Streams are referred to as *graded* if sediment transport along the streambed is equal to the rate of sediments entering the stream along the length

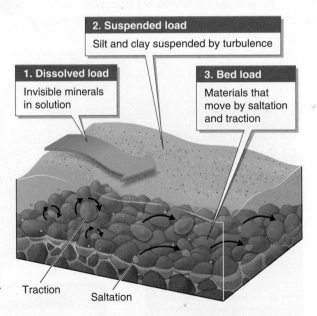

Figure 17.14 Three types of stream loads.

of the stream. In other words, in a graded stream, sediment is moving through the channel, and there is no net loss or gain of sediments. Along the length of a graded stream, there is a dynamic equilibrium (or balance) between erosion, transportation, and deposition of sediments within the stream channel. In theory, a fully graded stream has a smooth *longitudinal* (lengthwise) profile (see **Figure 17.15**). Few streams, however, have such smoothly graded profiles because of the different rock types, geologic activity, rock slides, and climates within any given drainage basin.

Stream gradient (or *stream slope*) refers to the steepness of a stream, or the drop in elevation of the stream channel in the downstream direction. High-gradient streams drop quickly in elevation, and low-gradient streams are almost flat. Stream gradient can be expressed as meters of elevation change per kilometer (or feet per mile) or as a percentage, where the vertical drop in a stream is divided by the horizontal distance of the section of

stream. For example, if a stream drops 100 m over a 1 km distance, its gradient is 100 m per kilometer. Calculated as a percentage, its gradient is 10% (100 m / 1,000 m × 100 = 10%). A stream that drops only 1 meter over a 1 km distance has a gradient of 0.1% (1 m / 1,000 m × 100 = 0.1%) (see Figure 17.15).

Base level is the lowest level a stream can reach—usually sea level. A stream can erode Earth's surface only where it is above base level. The higher a stream is above base level, the more energy it has to erode the land surface. At base level, a stream becomes standing water, loses its capacity and competence, and drops the sediments it is carrying. Figure 17.15 contrasts high-gradient and low-gradient sections of a stream in relationship to base level.

Figure 17.15 Stream gradient environments.

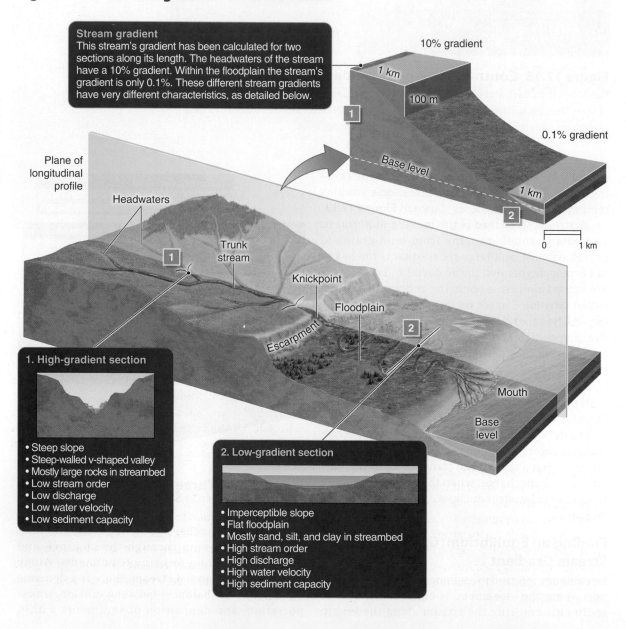

Stream gradient
This stream's gradient has been calculated for two sections along its length. The headwaters of the stream have a 10% gradient. Within the floodplain the stream's gradient is only 0.1%. These different stream gradients have very different characteristics, as detailed below.

10% gradient
1 km
100 m
1
0.1% gradient
Base level
1 km
2

0   1 km

Plane of longitudinal profile

Headwaters
1
Trunk stream
Knickpoint
Floodplain
Escarpment
2
Mouth
Base level

1. High-gradient section
• Steep slope
• Steep-walled v-shaped valley
• Mostly large rocks in streambed
• Low stream order
• Low discharge
• Low water velocity
• Low sediment capacity

2. Low-gradient section
• Imperceptible slope
• Flat floodplain
• Mostly sand, silt, and clay in streambed
• High stream order
• High discharge
• High water velocity
• High sediment capacity

Figure 17.16 Niagara Falls knickpoint. (A) A waterfall migrates upstream when the more resistant lip of rock that forms it is undercut and collapses into the plunge pool. (B) Niagara Falls, on the United States–Canada border, consists of three waterfalls. The largest is Horseshoe Falls in Canada; American Falls and Bridal Veil Falls are in the United States. The Niagara knickpoint has retreated upstream (south) by almost 11 km (7 mi) during the past 10,000 years. *(Oleksiy Maksymenko/Getty Images)*

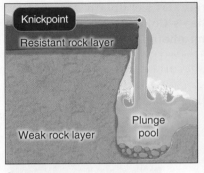
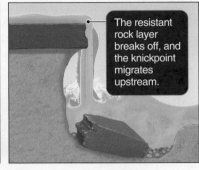
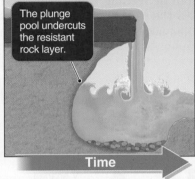

A

B

Cutting with Water: Waterfalls, Valleys, and Canyons

A **knickpoint** is a location where there is an abrupt increase in stream gradient over a short distance. Waterfalls may form at a knickpoint where a layer of rock that is resistant to erosion overlies a relatively weak and soft layer of rock. The resistant layer of rock creates an **escarpment**, a long cliff face or steep slope over which the water flows (see Figure 17.15). As the stream spills over the lip of the resistant rock layer, it may form a bowl called a **plunge pool** at the base of the waterfall. A plunge pool is created by abrasion caused by rocks circulating at the base of the waterfall. Eventually, a plunge pool can undercut the support of the resistant layer above, causing the rock above the pool to collapse. As a result, the knickpoint migrates upstream. This process is illustrated in **Figure 17.16**.

The rate of stream incision and the resulting topography depend on the stream's elevation above base level, stream discharge, and flow velocity, as well as the hardness and structure of the bedrock the stream cuts through. Stronger rocks such as metamorphic rocks are much more resistant to incision than weaker rocks such as sandstone. Most streams cut V-shaped valleys in their headwaters, but this is not always the case, as shown in **Figure 17.17** on the following page.

Lifting Streams: Stream Rejuvenation

Stream rejuvenation is a process in which a stream gains erosional energy as its base level is lowered relative to its drainage basin. Stream rejuvenation is usually a consequence of geologic uplift, but lowered global sea level can also cause stream rejuvenation.

The higher above its base level a stream lies, the more it cuts into the sediments and bedrock beneath it. Eventually, barring further geologic uplift, the stream returns to its original gradient through weathering, mass wasting, erosion, and deposition **(Figure 17.18)**.

As we will discuss in more detail in Section 17.3, a stream that flows across a flat flood-plain may form looping bends called **meanders**.

If geologic uplift creates a level plateau, rather than a sloped surface, stream meanders may be preserved as *entrenched meanders* as the stream erodes the plateau **(Figure 17.19)**. Entrenched meanders are found throughout the southwestern United States, where geologic uplift raised what was a large, flat floodplain and shallow sea to form today's Colorado Plateau (see the chapter-opening photo). The uplift took place

Figure 17.17 Types of valleys and canyons. *(Top. kavram/Getty Images; center. Matteo Colombo/Getty Images; bottom. Kerrick James/Getty Images)*

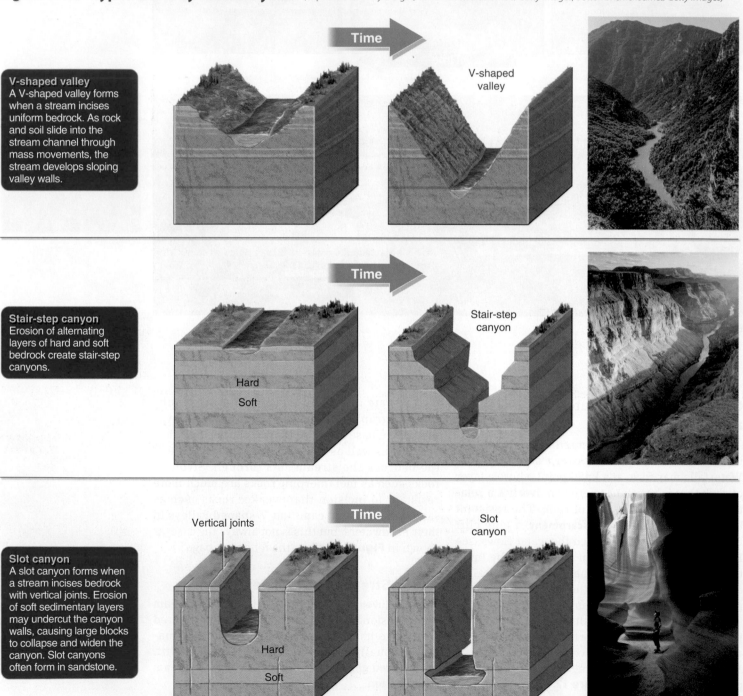

V-shaped valley
A V-shaped valley forms when a stream incises uniform bedrock. As rock and soil slide into the stream channel through mass movements, the stream develops sloping valley walls.

V-shaped valley

Stair-step canyon
Erosion of alternating layers of hard and soft bedrock create stair-step canyons.

Hard
Soft

Stair-step canyon

Slot canyon
A slot canyon forms when a stream incises bedrock with vertical joints. Erosion of soft sedimentary layers may undercut the canyon walls, causing large blocks to collapse and widen the canyon. Slot canyons often form in sandstone.

Vertical joints

Slot canyon

Hard

Soft

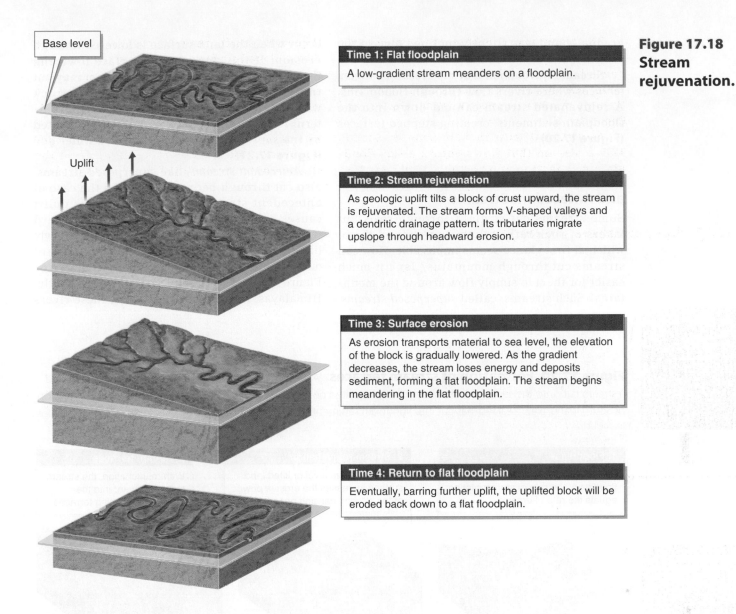

Base level

Uplift

**Figure 17.18
Stream
rejuvenation.**

Time 1: Flat floodplain

A low-gradient stream meanders on a floodplain.

Time 2: Stream rejuvenation

As geologic uplift tilts a block of crust upward, the stream is rejuvenated. The stream forms V-shaped valleys and a dendritic drainage pattern. Its tributaries migrate upslope through headward erosion.

Time 3: Surface erosion

As erosion transports material to sea level, the elevation of the block is gradually lowered. As the gradient decreases, the stream loses energy and deposits sediment, forming a flat floodplain. The stream begins meandering in the flat floodplain.

Time 4: Return to flat floodplain

Eventually, barring further uplift, the uplifted block will be eroded back down to a flat floodplain.

Figure 17.19 Entrenched meanders. (A) After geologic uplift, a flat floodplain may form entrenched meanders. (B) The San Juan River in Utah has superb examples of entrenched meanders. These meanders formed long ago on a flat floodplain and were preserved as they cut through sedimentary rocks beginning about 15 to 20 million years ago. This photo was taken at Goosenecks State Park in Utah; the park's name refers to the similarity between the meanders and the curve of a goose's neck. *(Wild Horizon/Getty Images)*

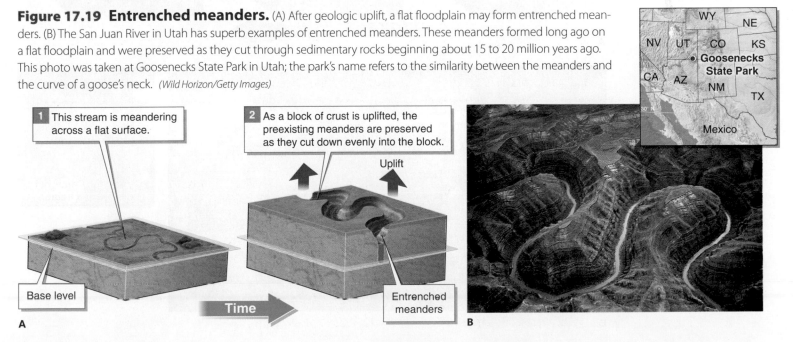

1 This stream is meandering across a flat surface.

2 As a block of crust is uplifted, the preexisting meanders are preserved as they cut down evenly into the block.

Uplift

Base level

Time

Entrenched meanders

WY

NE

NV   UT   CO   KS

Goosenecks
State Park

CA   AZ   NM   TX

Mexico

A   B

gradually and was slightly inclined toward the southwest.

Stream rejuvenation may also result in *stream terraces* where rivers flow through floodplains. A rejuvenated stream can cut down into the floodplain sediments, creating stepped terraces **(Figure 17.20)**.

Cutting through Mountains: Superposed and Antecedent Streams

Some streams cut straight through resistant layers of rock rather than flow around them, which at first glance seems impossible. How can streams cut through mountains? Isn't it much easier for them to simply flow around the mountains? Such streams, called *superposed streams*,

occur when the land surface is lowered through erosion. If the rate of land surface lowering through erosion is gradual, the stream may cut into the buried underlying ridge of bedrock while maintaining its original drainage pattern. The resulting cut through the ridge created by the *superposed stream* is called a *water gap* **(Figure 17.21)**.

Antecedent streams, like superposed streams, also cut through bedrock ridges. In the case of antecedent streams, however, geologic uplift causes a preexisting stream to cut downward through resistant ridges of rock as they are slowly lifted. Almost all of the rivers draining southward out of the Himalayas are antecedent streams (see Figure 13.30). The streams were there before the Himalayas. As the Himalayas rose, these rivers

Figure 17.20 Formation of stream terraces. (A) Stream terraces may form as a region is uplifted and a stream cuts into the sediments that have been deposited on a floodplain. (B) Cave Stream, in the Canterbury region of South Island, New Zealand, has cut into floodplain sediments, forming two terraces. *(G. R. 'Dick' Roberts © Natural Sciences Image Library.)*

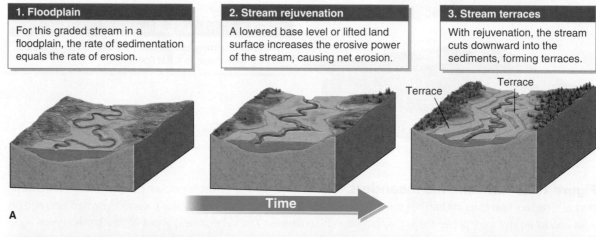

1. Floodplain

For this graded stream in a floodplain, the rate of sedimentation equals the rate of erosion.

2. Stream rejuvenation

A lowered base level or lifted land surface increases the erosive power of the stream, causing net erosion.

3. Stream terraces

With rejuvenation, the stream cuts downward into the sediments, forming terraces.

Time

A

B

Figure 17.21 Superposed stream water gap. (A) In this diagram, a stream cuts through underlying bedrock structures as the landscape surface is gradually lowered through erosion. (B) This satellite image shows the Potomac River at the borders of West Virginia, Virginia, and Maryland. The Potomac River is a superposed stream. It cut through the Blue Ridge Mountains just downstream of the town of Harpers Ferry, West Virginia. Beginning about 480 million years ago, erosion began to expose the underlying layers of igneous and metamorphic rocks, and the river maintained its course by cutting through them rather than flowing around them, creating the Potomac Water Gap. *(NASA Earth Observatory images by Jesse Allen, using Landsat data from the U.S. Geological Survey.)*

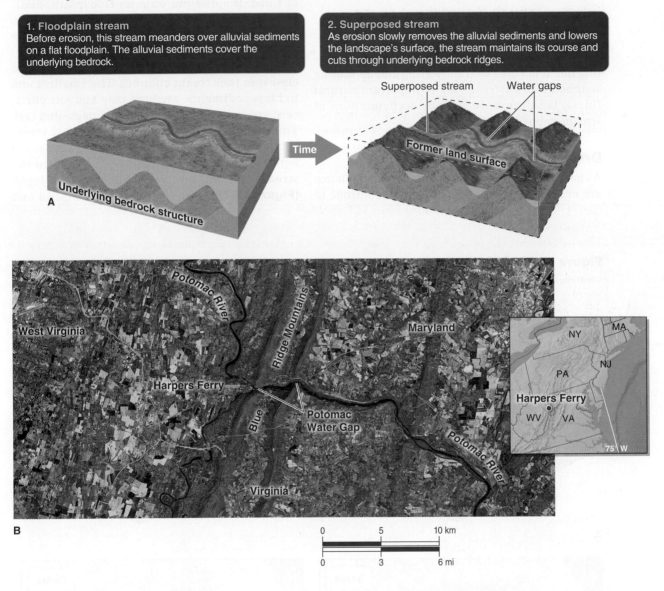

1. Floodplain stream
Before erosion, this stream meanders over alluvial sediments on a flat floodplain. The alluvial sediments cover the underlying bedrock.

2. Superposed stream
As erosion slowly removes the alluvial sediments and lowers the landscape's surface, the stream maintains its course and cuts through underlying bedrock ridges.

largely maintained their southward course and cut into the mountains.

Distinguishing a water gap caused by either superposition or antecedence is not possible with only a quick glance at the stream and its land surface. The tectonic and erosional history of a region must be known before it is possible to determine how the water gap formed. In some cases, streams with water gaps no longer flow in their former channels. When that occurs, a *wind gap* results.

Stream Capture: Diverting the Waters

Stream capture (or *stream piracy*) is the diversion of one stream into another as headward erosion merges the two streams. When a stream's tributaries migrate upslope through headward erosion and cut through the drainage divide of another drainage basin, streams within that drainage basin can be diverted into the new stream channel, as illustrated in **Figure 17.22**.

17.3 Building with Streams: Fluvial Deposition

◎ **Discuss landforms resulting from the deposition of sediment by streams.**

Streams shape Earth's surface not only by cutting into it but also by transporting and depositing sediments downstream. Many important natural fluvial landforms are built by accumulations of stream-deposited sediments.

Dropping the Load: Stream Sorting

As long as stream capacity and stream competence are maintained, the stream load will continue to

be transported. Where stream capacity or stream competence is reduced, material carried by the stream begins to settle out. Streams can deposit sediments anywhere along their length. Where a stream stops flowing at base level, it loses all kinetic energy, and all the material it carries, except its dissolved load, settles out by gravity.

As a stream overflows its banks during times of flood, it spills out onto its floodplain. Sediments deposited on a floodplain by a stream are referred to as **alluvium**. Alluvium is sorted by size as it is deposited by streams. The heaviest material, such as gravel and sand, settles out first and closest to the stream channel. The smallest and lightest sediments, such as clay and silt particles, are transported farthest and deposited last. Thus, along a floodplain with each flood event, sediments are sorted from largest to smallest both horizontally with distance away from the stream channel and vertically through depth **(Figure 17.23)**. The largest particles (sand and

Figure 17.22 Stream capture. (A) This model shows the process of headward erosion that leads to stream capture. The red lines indicate the area of greatest headward erosion where the point of stream capture will eventually occur. (B) Beginning about 3 mya (million years ago) the headwaters of the Niger River cut through the drainage divide of the adjacent Taoudeni drainage basin. As a result, the Niger River captured a significant portion of the Taoudeni Basin's streams.

A

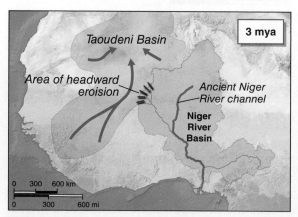

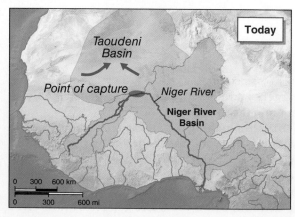

B

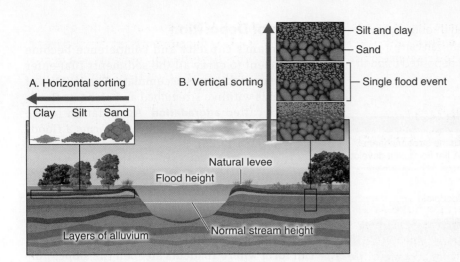

A. Horizontal sorting

B. Vertical sorting

Clay Silt Sand

Silt and clay

Sand

Single flood event

Natural levee

Flood height

Layers of alluvium

Normal stream height

Figure 17.23 Horizontal and vertical sediment sorting. This diagram shows a cross section of layers of alluvium deposited on a stream floodplain. During floods, sediments within each layer are sorted by size both horizontally (A) and vertically (B). With repeated flooding, horizontal sorting results in coarse sediments building natural levees near the stream channel.

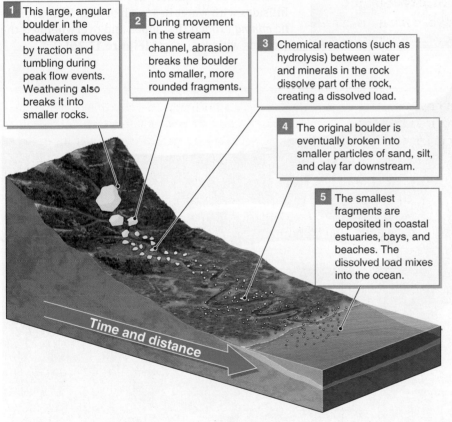

1 This large, angular boulder in the headwaters moves by traction and tumbling during peak flow events. Weathering also breaks it into smaller rocks.

2 During movement in the stream channel, abrasion breaks the boulder into smaller, more rounded fragments.

3 Chemical reactions (such as hydrolysis) between water and minerals in the rock dissolve part of the rock, creating a dissolved load.

4 The original boulder is eventually broken into smaller particles of sand, silt, and clay far downstream.

5 The smallest fragments are deposited in coastal estuaries, bays, and beaches. The dissolved load mixes into the ocean.

Time and distance

Figure 17.24 Sorting sediments along the length of a stream. Through abrasion and chemical weathering, rocks in a streambed are slowly broken down into smaller and smaller fragments. These fragments are sorted according to size along the stream's length.

gravel) accumulate near the stream channel in raised barriers that parallel the stream channel, called *natural levees*.

Sediment size sorting also happens along the longitudinal length of a stream, from the headwater tributaries down to the stream's mouth. Generally, the largest boulders are high in the headwaters because they are too heavy to be carried very far. They eventually break into smaller and lighter fragments that will move more easily.

Cobbles and pebbles will be moved farther downstream than boulders. Sand, silt, and clay settle out of the water still farther downstream.

To get a sense of how the sorting process works, imagine a white quartz boulder in a stream channel where all other rocks are dark. Because the boulder is white, we can easily see it and follow it from a high mountainous area and on downstream as it is broken into ever-smaller fragments over time **(Figure 17.24)**.

Figure 17.25 Aggradation. Floodplains are formed through aggradation when sediment enters a stream faster than the stream can carry it away. Floodplains are composed of deep layers of alluvium deposited through aggradation.

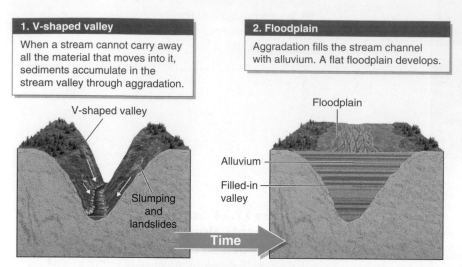

1. V-shaped valley

When a stream cannot carry away all the material that moves into it, sediments accumulate in the stream valley through aggradation.

V-shaped valley

Slumping and landslides

2. Floodplain

Aggradation fills the stream channel with alluvium. A flat floodplain develops.

Floodplain

Alluvium

Filled-in valley

Time

Places of Deposition

If a stream's capacity and competence become insufficient to carry all the sediments that enter the stream, sediments accumulate. The buildup of sediments within a streambed and along its floodplain is called **aggradation**. Aggradation in the low-gradient sections of a stream forms a floodplain (**Figure 17.25**).

Human activities can greatly accelerate the rate of stream aggradation. Sediment input into a stream channel increases where vegetation is removed by fire or human activity. Sediments from unpaved roads such as logging roads in steep terrain can wash into streams as well. Landslides can form where roadcuts steepen the slope, putting large volumes of material into a stream channel (see Section 16.3). During the Gold Rush in California in the middle 1800s, miners blasted the hillsides with water in search of gold. This *hydraulic mining* eroded the slopes and overwhelmed the streams with sediments (**Figure 17.26**).

Figure 17.26 Hydraulic mining during the Gold Rush. Miners used water cannons called *water monitors* to loosen sediments, and the gold contained in them, from the foothills of the Sierra Nevada. Mining put more sediment into the streams than the streams could carry, resulting in aggradation in the stream channels. The practice was banned in 1884 because the sediments were making the streams (and even the San Francisco Bay far downstream) too shallow for boats to navigate. This photo shows mining operations in the 1860s in what is today Malakoff Diggins State Historic Park (SHP) in California. *(U.S. Geological Survey/photo by Carlton Watkins)*

Aggradation also occurs where streams flow out of glaciers. Glaciers flow downslope and grind the bedrock beneath them into rock fragments (see Section 18.2). A glacier's outlet stream can move this material only during peak flow periods in late summer or during storms that bring heavy rainfall. At most times, a stream that flows out of a glacier occupies only a small portion of the streambed, flowing around elongated islands of gravel and sand. Streams that form intertwining channels around sediments in the streambed are called **braided streams (Figure 17.27)**. Although most braided streams originate from glaciers, they can form in other settings, such as where mining activities cause large quantities of sediment to enter a stream.

An ephemeral desert stream may form an alluvial fan where it exits a steep canyon and spills out onto a flat plain. An **alluvial fan** is a gently sloping fan-shaped accumulation of sediment that is deposited at the base of a mountain by an ephemeral stream in arid regions **(Figure 17.28)**. After heavy rains, as a stream exits a steepwalled V-shaped canyon, its velocity slows, and its capacity and competence decrease. As a result,

Figure 17.27 A braided stream. This braided stream originates from Franz Josef Glacier in New Zealand. *(David Wall/Alamy)*

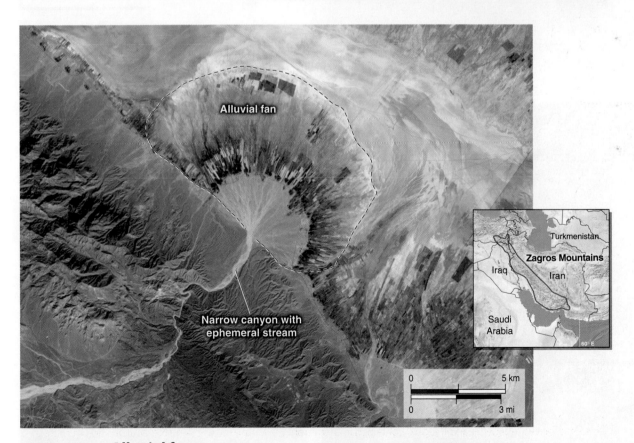

Figure 17.28 Alluvial fan. This satellite image shows an alluvial fan (outlined with a dotted line) in the Zagros Mountains of southern Iran. The green crops seen here get their water from pumped groundwater, not surface streams. *(NASA image created by Jesse Allen, using data from NASA/GSFC/METI/ERSDAC/JAROS, and the U.S./Japan ASTER Science Team.)*

Animation

Stream Meanders

Available at
www.saplinglearning.com

▶

Figure 17.29 Stream meanders. (A) A meander develops on a flat floodplain as the thalweg cuts into the outside edge of the river bank. This process takes decades or longer. (B) This small stream near Cairns, Australia, developed its meanders and meander scars in the process outlined in Part A of this figure. (C) Stream meanders are most common in low-gradient streams. But high-gradient streams, such as this small creek in the central Sierra Nevada in California, also develop meanders where the rivers flow through flat meadows. *(B. © G.R. "Dick" Roberts/Natural Sciences Image Library; C. Bruce Gervais)*

1. Before meanders

A straight stream channel on a flat floodplain will develop meanders when the fastest flow, called the *thalweg*, increases erosion on one bank.

2. Meanders

A cut bank forms on the outside bend of a meander where flow velocity is highest. On the inside of a meander, where flow velocity is lowest, sediments are deposited to form a point bar.

3. Meander neck

The meander loop deepens as erosion on the cut banks continues. As the meander deepens a meander neck forms.

4. Meander cutoff

At the meander neck, a cutoff forms as the two meanders join. The abandoned stream channel creates an oxbow lake if it is filled with water, or a meander scar if it becomes dry.

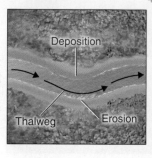
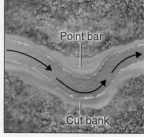
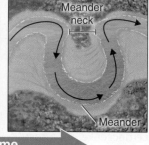
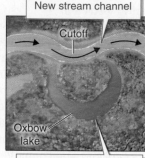

A

Time

B

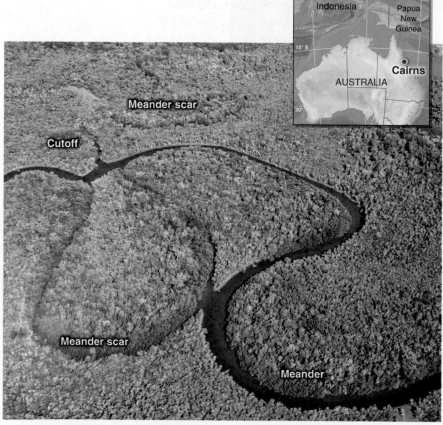

C

any sediment the stream was carrying drops out. During the brief time the ephemeral stream flows (hours or days), it occupies only a small portion of the fan at any one time. Over many centuries or longer, the stream migrates back and forth across the fan, depositing new sediments across the fan and building it up.

Stream Meanders and Floodplains

In steep-walled valleys and canyons, streams are mostly confined to their channels and cannot cause extensive flooding. On floodplains, however, natural streams are spatially dynamic and routinely leave their channels and flood the surrounding area. Through time, stream channels shift across the floodplain through the process of meandering. Recall from Section 17.2 that a meander is a looping bend in a stream channel that forms on a floodplain. Meanders develop when the fastest-flowing portion of the stream, called the **thalweg**, erodes one bank more than the other. That bank, called a **cut bank**, where erosion exceeds deposition, becomes the outside edge of the meander. On the opposite bank, where the flow is slower, deposition exceeds erosion, and the accumulation of sediments forms a **point bar**.

As cut banks and point bars develop, through time new meanders form and old meanders are abandoned. As meander loops converge, they form *meander necks*. These are places where a meander is cut off and a portion of the stream channel is abandoned. When this happens, a water-filled **oxbow lake** forms. Where an oxbow lake dries up, a *meander scar* results. **Figure 17.29** illustrates these meander features.

Tributaries entering a floodplain sometimes flow parallel to the meandering trunk stream because they cannot flow over the higher floodplain and natural levees bordering the trunk stream channel. As a result, permanently flooded areas called *wetlands* (also called *swamps, backswamps,* or *bayous*) and *yazoo streams* form **(Figure 17.30)**.

The End of the Line: Base Level

All streams stop flowing at base level. All streams eventually find the ocean, with the exception of those in internal drainage systems (see Section 17.1).

Along the way to the ocean, natural dams (such as beaver dams or landslide debris) may form a *temporary base level* (also called a *local base level*) as a pond or lake develops behind the dam. Over time, fluvial erosion will eventually cut through the obstruction, and the stream will flow freely again.

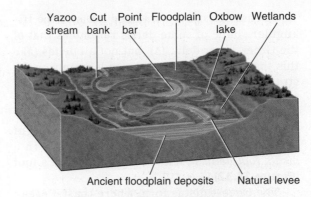

Figure 17.30 Floodplain features.

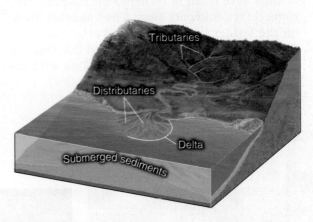

Figure 17.31 Features of a stream and its delta. Distributaries form a branching network of channels in a delta. The sediments visible on the surface of a delta are only a small fraction of the total sediment accumulations that make up the delta. Most of the sediments in a delta are hidden beneath the surface of the water.

Streams are also dammed by people. Damming creates an artificial lake called a **reservoir**. Stream capacity and stream competence are zero in lakes and reservoirs, and as a result, suspended sediments settle out.

A **delta** is an accumulation of sediments that forms where a stream reaches base level. Because of the amount of material deposited in a delta, the stream channel is likely to become blocked, forcing the stream into a new channel around the sediments. (This is the same process that occurs in a braided stream.) In a delta, the trunk stream can shift and form a smaller branching network of streams called **distributaries** (or *distributary channels*). As illustrated in **Figure 17.31**, tributaries occur at the headwaters of a drainage basin, and distributaries occur in the delta.

Stream deltas are so named because the tri-angular shape of some deltas resembles that of the Greek letter delta (Δ), but not all deltas take this shape. A delta's shape depends mainly on the stream's suspended load and on the influence of currents at the stream's mouth. An *arcuate delta* is triangular and may have a braided or branch-ing pattern of distributaries. As you might expect from its name, the distributaries in a *bird's foot delta* resemble toes on a bird's foot **(Figure 17.32)**.

The largest deltas form where coastal ocean currents are at a minimum and major streams carry a heavy load of suspended sediments. Where coastal currents are strong, in contrast, river mouths have small or no deltas because the cur-rents carry sediments away. The Columbia River on the border of Oregon and Washington, the Congo River in western Africa, and the Amazon River are examples of major streams with no coastal deltas.

In many deltas, the land elevation drops due to natural **subsidence**—the lowering of land elevation through the compaction of sediments. Deltas keep pace with natural subsidence as their streams continually deposit new sediments that build up the land. As long as sediments are renewed by flu-vial deposition, there is an equilibrium between subsidence and sedimentation.

When artificial levees are built around a stream to control water flow, however, this equi-librium is disrupted. With levees in place, the river can no longer flood and form new meanders. The river's sediments are carried through and past the delta to the seaward edge and are deposited offshore rather than in the delta. The interior delta becomes sediment starved, and as the land ele-vation sinks through natural subsidence, the sea gradually encroaches on the land. The **Picture This** feature on the following page highlights the effects of this process in New Orleans, which is built on the Mississippi River delta.

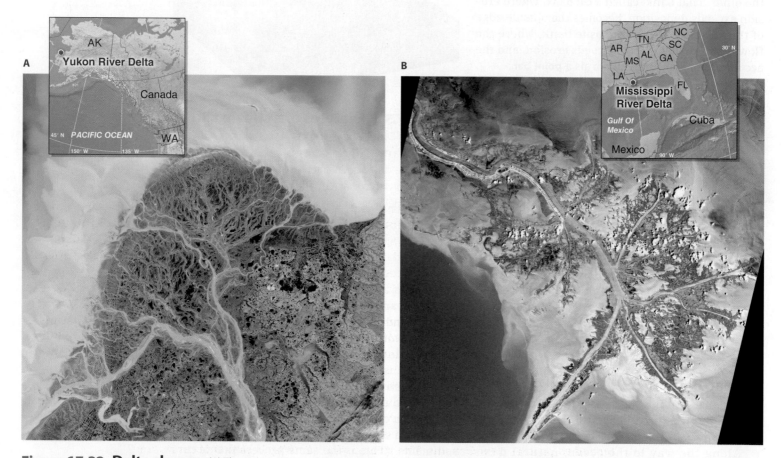

Figure 17.32 Delta shapes. (A) The Yukon River originates in British Columbia, Canada, and empties into the Bering Sea in Alaska. Its delta has the classic arcuate shape with branching networks of distributaries. The swirling brown color of the sea is suspended sediments in the water that have been transported by the Yukon River. The sediments will eventually settle to the seafloor and add to the delta. (B) The Mississippi River forms a bird's foot delta (a portion of which is shown here). Suspended sediments transported by the river are visible in the ocean. *(A. NASA Earth Observatory image created by Jesse Allen and Robert Simmon, using Landsat data provided by the United States Geological Survey; B. NASA/GSFC/METI/ERSDAC/JAROS, and U.S./Japan ASTER Science Team)*

Picture This **The Sinking City**

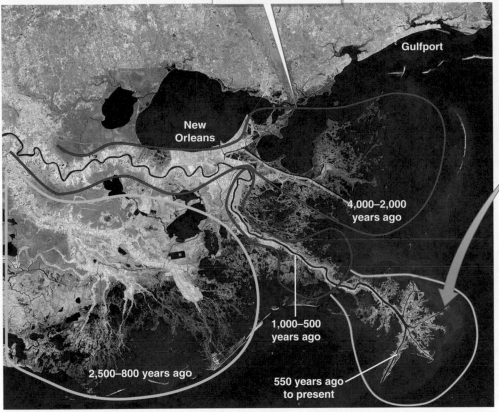

Locations of former Mississippi river mouths

Gulfport

New Orleans

4,000–2,000 years ago

1,000–500 years ago

2,500–800 years ago

550 years ago to present

(EROS Center, U.S. Geological Survey)

(Jupiterimages/Getty Images)

NC
SC
TN
AR
MS AL GA
LA
FL
Mississippi
River Delta
Gulf of Mexico   Cuba
Mexico
30° N
90° W

New Orleans is a sinking city. The past equilibrium between sedimentation and subsidence in the Mississippi River delta was disrupted when artificial levees were built around the river, mostly by the U.S. Army Corps of Engineers in the 1930s and 1940s. The levees were built to protect the city from flooding by the Mississippi River. Before the levees were built, the mouth and distributaries of the Mississippi River shifted every few hundred years to a new location, building the delta with fresh new sediments. In this satellite image, the former positions of the river mouth are outlined in different colors, and their ages are given. Once the levees and upstream dams were in place, the trunk stream channel was stabilized to protect New Orleans. The river could no longer shift, flood, and build.

Some parts of New Orleans are now 2 to 6 m (6.5 to 20 ft) below sea level, and the nearby coastal wetlands are riddled with holes and canals that were not present historically (inset). In many regions, the wetlands cannot grow fast enough to keep pace with subsidence. Because of the loss of coastal wetlands as well as rising sea level (see Section 18.5), New Orleans is increasingly vulnerable to flooding due to hurricane storm surges.

Consider This

1. How did the Mississippi River's meandering process change after levees and upstream dams were put in place?
2. Why is New Orleans sinking?

17.4 Rising Waters: Stream Flooding

◎ Describe two modes of stream flooding and explain how people reduce their vulnerability to flooding.

A **flood** is inundation by water in a region not normally covered by water. When the volume of water moving through a stream exceeds the capacity of the stream channel, flooding occurs. Most low-gradient stream areas are naturally prone to flooding during periods of peak discharge. This is particularly true of streams in urbanized areas, where impermeable surfaces create rapid inflows of water into the stream channel (see Figure 17.10). There are two main types of flooding: flash floods and seasonal floods.

Flash Floods

Sudden, intense rainfall or the collapse of a dam may cause a **flash flood**. These floods are very dangerous because they often occur with little or no warning. They are most common in arid regions where rains come as sudden thunderstorm showers. Dry streambeds in deserts can quickly become torrents of raging water following a thunderstorm. Hikers in slot canyons (see Figure 17.17) are particularly vulnerable to flash floods during summer thunderstorm activity; these canyons can fill with water within minutes, and it is nearly impossible to get out of a slot canyon quickly.

The worst flash flood disaster in the United States, called the Johnstown Flood, was human-caused. At about 3 p.m. on May 31, 1889, the South Fork Dam on the Conemaugh River in Pennsylvania collapsed. The earthen dam, which was poorly constructed, failed when torrential rains saturated and weakened it. Twenty million tons of water formed a surge of water 20 m (60 ft) high that raced down the valley leading to Johnstown. In only 10 minutes, the water destroyed much of the town of 30,000 people. More than 2,000 people died, and $17 million in damage resulted.

Seasonal Floods

In contrast to flash floods, **seasonal floods** are predictable floods that occur with seasonally heavy rain or snowmelt. Seasonal floods follow prolonged rains that saturate the soil and inhibit more water from infiltrating the ground. If rains continue, water must flow over land rather than into the ground. As a result, water spills out of the stream channel and inundates the surrounding floodplain. Seasonal flooding poses less of a threat to human life than flash flooding because often the water rises slowly and predictably, and there is ample time to evacuate.

Seasonal flooding is common throughout South Asia because of its heavy monsoon rains (see Section 5.3) and tropical cyclones (see Section 6.3). South Asia often floods when the rainy monsoon season commences. The monsoon rain in 2017 was particularly heavy and resulted in extensive flooding throughout northern India and Bangladesh **(Figure 17.33).**

Figure 17.33 South Asia monsoon flooding. Heavy monsoon rains in 2017 were measured between August 10 and 16 using satellite-based sensors. The highest rainfall is indicated with purple on this map. Over 100 cm (40 in) of rain fell in parts of Bangladesh over the 7-day period. The inset photo shows the city of Dhaka in Bangladesh paralyzed by the floodwaters.   *(A. NASA/JAXA, Hal Pierce; B. NurPhoto/Getty Images)*

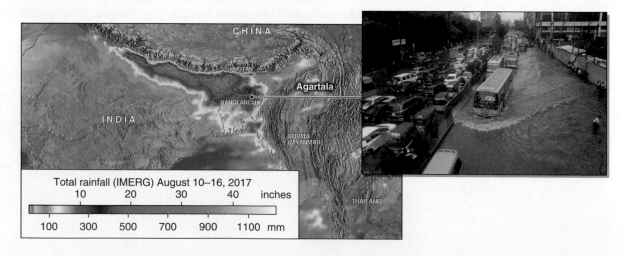

Total rainfall (IMERG) August 10–16, 2017

Controlling the Waters

The most effective tools in reducing the risk of flooding on floodplains are dams, levees, and stream bypasses. Engineers design these flood control structures to handle the worst-case flood events. As **Figure 17.34** discusses, each of these tools has advantages and disadvantages.

There are two ways to summarize the risk of flooding for a given location: the annual probability and the recurrence interval. The *annual probability* of flooding specifies the chances of a given discharge amount in a locality for any given year. For example, for a river that normally flows at 10,000 ft³/s, the annual probability of a 50,000 ft³/s flow might be calculated to be 1%. Such calculations are based on historical climate and stream flow data. A 1% flooding probability means that in any given year, there is a 1% chance of a 50,000 ft³/s discharge event. With each passing year, the probability increases arithmetically, such that after 100 years, there is a 100% chance of a 50,000 ft³/s discharge event.

The *recurrence interval* states how many years, on average, will pass between high-discharge events. In the example just cited, the recurrence interval for a 50,000 ft³/s discharge event is once every 100 years. Such floods are called *100-year floods*. A *500-year flood* is a discharge event that happens once every 500 years, on average, and has an annual probability of 0.2%. These events can be expected, on average, every 500 years. Because recurrence intervals rely on averages, if a 500-year flood occurs, it does not mean that another 500 years will pass before such a flood happens again. Another 500-year flood could occur at any time. Furthermore, climate change is resulting in increased rainfall variability nearly everywhere, in some cases greatly increasing the probability of extreme flooding.

GEOGRAPHIC PERSPECTIVES
17.5 **Dam Pros and Cons**

◎ **Assess the benefits and drawbacks of dams.**

According to the U.S. Army Corps of Engineers National Dams Inventory, there are more than 85,000 public and private dams in the United States. Most of them are small embankment dams (or earthen dams) that impound small reservoirs. **Figure 17.35** shows the history of dam building in the United States. Canada has about 10,000 dams, and worldwide there are about 850,000 dams.

Figure 17.34 **Flood management structures.** *(Left. Akira Kaede/Getty Images; center. Karl Johaentges/LOOK-foto/Getty Images; right. Matthew D White/Getty Images)*

Dam	Levee	Stream bypass
What is it? An earthen or concrete structure that blocks a stream and impounds water, forming a reservoir.	**What is it?** Earthen walls running parallel to a stream to keep water in its channel.	**What is it?** Outlet gates called *weirs* open at times of high flow to divert water out of the stream channel into a separate flood channel called a *stream bypass*.
Advantages 1. Reduces peak flow by capturing water in the reservoir for later use, such as for agriculture or city use. 2. Generates mostly carbon-free hydroelectricity.	**Advantages** 1. Effectively controls flooding if built to withstand large flood events.	**Advantages** 1. Effective means of reducing discharge within the main stream channel. 2. Periodically flooded land improves habitat for many organisms and renews alluvial soils for agriculture.
Disadvantages 1. Floods communities and reduces riparian wildlife habitat. 2. Sedimentation reduces the reservoir capacity. 3. Dams sometimes fail catastrophically.	**Disadvantages** 1. May fail when built poorly or not properly maintained. 2. Levees built too close to the stream channel constrict the stream and reduce riparian habitat. 3. Deltas subside when levees prevent sediment deposition onto the floodplain, as in the case of New Orleans.	**Disadvantages** 1. Requires large amounts of undeveloped land that can be periodically flooded.

Figure 17.35 **U.S. dams and reservoirs built since 1850.** These four maps show the history of dam building in the United States. Since 1850, the number of dams has grown significantly, reflecting the importance of these structures to human society. Most of the building of dams took place between 1950 and 2000; dam building has since decreased. *(Courtesy of James Syvitski, University of Colorado at Boulder)*

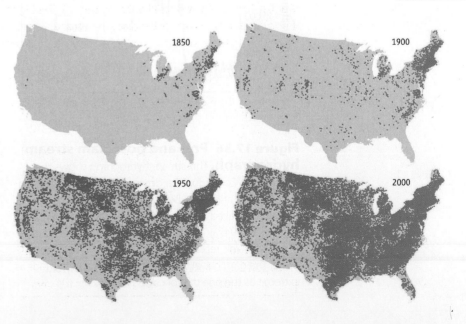

1850

1900

1950

2000

The Pros of Dams

Dams are extremely important structures in most countries and provide many important benefits to society. But there are drawbacks to dams as well. In this section, we weigh the pros and cons of large dam projects.

Water Storage

Water is life, and dams provide water. A dam allows communities to store water in a reservoir, making it available at peak demand periods and in dry seasons. In the United States, dams have played an essential role in the development of arid Western states, and they currently play a crucial role in the development of many growing economies around the world. Most of the large cities of the arid western United States, including Phoenix, Las Vegas, Tucson, and Los Angeles, could not have developed to their current sizes without water from reservoirs. The agricultural heartland of California, the Central Valley, would not exist without the extensive series of dams that catch spring snowmelt from the Sierra Nevada and save it for the parched summer months.

Flood Control

Dams reduce flood risk by reducing the peak flows of streams during high-precipitation events. The stream hydrograph in **Figure 17.36** shows the changes in stream hydrology that resulted from the installation of a dam.

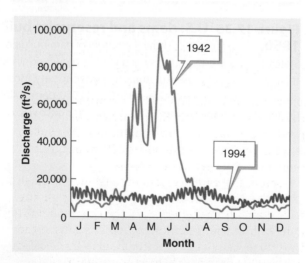

Figure 17.36 Pre- and post-dam stream hydrograph. This stream hydrograph shows the flows in the Colorado River downstream of Glen Canyon Dam in 1942 and 1994. Before the dam was built, stream discharge peaked at over 90,000 ft³/s with the late spring snowmelt (1942 line), creating a major flood event. The 1994 levels represent typical flows with the dam in place. The dam has eliminated peak discharge periods as extreme as the one that occurred in 1942.

Figure 17.37 Hydroelectricity generation. Water rushing from a reservoir spins large turbines that generate electricity in hydroelectric plants.

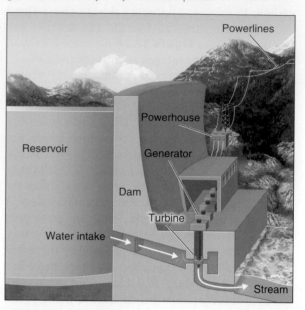

Hydroelectricity

Carbon-free energy sources, such as wind power, solar power, and hydroelectricity, are essential to addressing the problem of climate change (see Section 7.5). Many large dams generate carbon-free hydroelectricity by using the kinetic energy of the water rushing out of the base of the dam to spin steel turbines that produce electrical energy **(Figure 17.37)**.

Only about 6% of the energy demand in the United States is met by hydroelectricity. In contrast, about 96% of Norway's electricity is produced by hydroelectric power plants. Iceland produces 70% of its electricity from streams (and the rest comes from geothermal energy). New Zealand, Austria, Switzerland, and Canada each produce about 60% of their electricity from streams. Roughly half of the world's renewable energy comes from hydroelectricity.

Transportation and Recreation

Because of dams, many former rivers are now series of long, narrow lakes that allow large vessels to safely navigate. The upper Mississippi River, from its headwaters to St. Louis, Missouri, has 43 dams and locks. A *lock* is an enclosure that raises and lowers boats between stretches of water with different elevations. These structures slow stream flow and create lakes, providing a transportation corridor essential to the region's economy.

Large reservoirs created by dams provide recreational opportunities for millions of people

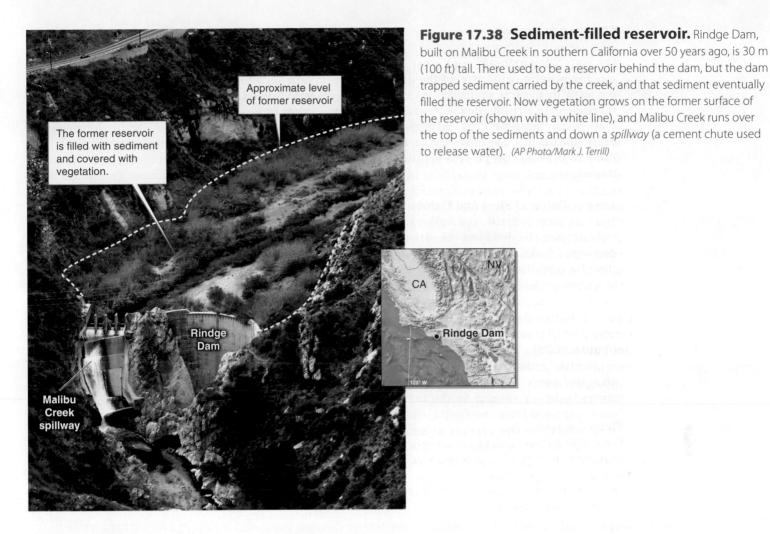

Figure 17.38 Sediment-filled reservoir. Rindge Dam, built on Malibu Creek in southern California over 50 years ago, is 30 m (100 ft) tall. There used to be a reservoir behind the dam, but the dam trapped sediment carried by the creek, and that sediment eventually filled the reservoir. Now vegetation grows on the former surface of the reservoir (shown with a white line), and Malibu Creek runs over the top of the sediments and down a *spillway* (a cement chute used to release water). *(AP Photo/Mark J. Terrill)*

each year. Activities include water skiing, boating, fishing, swimming, and camping. Recreation on reservoirs is an important part of many local economies.

The Cons of Dams

Reservoirs behind dams provide water for thirsty cities and crops, but they may also destroy stream habitat or preexisting ecosystems and human communities. Other drawbacks of dams are described in the following sections.

Reservoir Sedimentation

Reservoirs behind dams fill up with sediment because they form a temporary base level. The sediments that accumulate behind dams would have, in the absence of the dams, been deposited along coastal regions to help build and maintain coastal beaches. Without these sediments, the beaches erode (see Section 20.3). Furthermore, as a reservoir fills with sediments, the reservoir's capacity to hold water decreases and sometimes disappears entirely, so that the useful life of the reservoir is limited, as shown in **Figure 17.38.**

How quickly a reservoir fills with sediments depends on how much sediment is transported in the stream and deposited each year.

The Aswan High Dam on the Nile River in Egypt was completed in 1970, creating artificial Lake Nasser. The Nile River carries a heavy suspended load as it runs through the desert of Sudan and Egypt. The river deposits 120 metric tons of sediment each year in Lake Nasser.

Before the emplacement of the dam, the clays and silts the river carried settled out on the floodplain along the length of the river and built it up with nutrient-rich alluvium. The soils of the Nile floodplain had been sustainably farmed for over 5,000 years using these continually replenishing rich alluvial soils. A widespread and economically important brick-making industry relied on the sediments as well. Since the dam was built, because flooding has not occurred and new sediments have not been added, synthetic (artificial) fertilizers have been needed to maintain agricultural output along the Nile floodplain, and the brick-making industry in many areas is gone.

Clearwater Erosion

Because sediments are trapped within reservoirs, the water downstream runs clear. Because the moving water downstream of the dam is no longer working to carry sediments, its kinetic energy increases. As a result, downstream of dams, fluvial erosion increases. The increased streambank erosion leads to a loss of vegetation and the wildlife that uses it. This process is called *clearwater erosion*.

Loss of Cultural Sites and Habitat

When a dam is built, the valleys upstream of the dam are flooded, and the stream ecology is destroyed. Lake Powell, the reservoir created by Glen Canyon Dam in Arizona, flooded extensive stretches of the lower Colorado River valley. Slot canyons, natural arches, wildlife habitat, and many active Native American communities and their archaeological sites were lost beneath the reservoir **(Figure 17.39)**.

Similarly, the Columbia River watershed has 46 major dams. These dams have been built in rivers in which salmon by the tens of millions once migrated from the Pacific Ocean to spawn (reproduce) in the gravels of small tributary streams. The dams now block the migration of salmon. "Fish ladders" and fish hatcheries have been installed on some dam sites. Fish ladders are intended to allow salmon to bypass the structures, but they have limited success. Fish hatcheries nurture domesticated (raised in captivity) salmon, which are genetically less diverse than wild salmon. As a result, and in conjunction with other factors such as commercial fishing, salmon populations throughout the Pacific Northwest are now a fraction of their former numbers.

In response to these problems, some "deadbeat dams" (those that are no longer useful) are being removed. For example, the Elwha Dam, on the Elwha River in Olympic National Park, Washington State, was removed in 2012, and removal of the Glines Canyon Dam on the same river was completed in 2014 **(Figure 17.40)**. California's San Clemente Dam was removed in 2015, allowing protected steelhead trout (*Oncorhynchus mykiss*) to again migrate up the Carmel River to spawn. There is a growing list of other deadbeat dams in California, Oregon, and Washington that could soon come down.

Flooded Communities and Farmland

Reservoirs created by dams can destroy communities and productive farmland. The Three Gorges Dam on the Yangtze River in China is the largest hydroelectric power station in the world. For many, the dam is a technological marvel and represents an important move away

Figure 17.39 Glen Canyon Dam and Lake Powell. (A) Glen Canyon Dam was completed in 1966. At 216 m (710 ft), it is the fourth tallest dam in the United States. It created Lake Powell reservoir. (B) Hundreds of archaeological sites, such as this submerged Puebloan structure at Doll Ruin in Moqui Canyon, now lie hidden beneath Lake Powell. *(A. Matthew Micah Wright/Getty Images; B. Photo courtesy of the NPS Submerged Resources Center)*

A

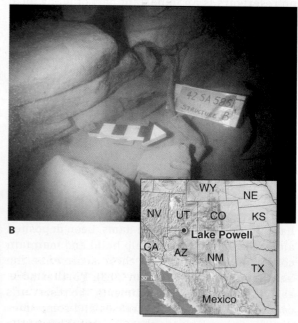

B

Figure 17.40 Glines Canyon Dam removal. The Glines Canyon Dam, built in 1927, was 64 m (210 ft) high. When this "deadbeat dam" was removed in 2014, it was the largest dam demolition project in U.S. history. The removal of the dam gave an estimated half million salmon access to ancestral spawning grounds within the Elwha River watershed for the first time in 80 years. This photo shows the dam partially removed. *(Keith Thorpe/Peninsula Daily News via AP)*

from reliance on energy from coal. To its critics, however, this dam has become symbolic of environmental sacrifice in the name of economic progress. The dam and reservoir forced the relocation of 1.5 million people, many of whom had lived in the area for generations. Some 1,300 archeological sites and 8,000 years of human history were erased when the waters rose. This process continues in China, which has some 80 major dams planned, under construction, or completed along major rivers in its western reaches. (See Section 13.5 to learn about China's planned water "megaprojects.")

Climate Change

Reservoirs, particularly reservoirs located in the tropics, produce methane gas through anaerobic decomposition of underwater organic material. Methane is about 20 times more potent as a greenhouse gas than carbon dioxide. Large reservoirs in the tropics can create more greenhouse gases, in the form of methane, than coal-burning power plants that emit carbon dioxide. Midlatitude reservoirs also produce methane, but they produce less than tropical reservoirs.

Taking the Good with the Bad

Although dams provide many societal benefits, these benefits come at a cost. In some cases, the cost is relatively slight, while in others it is so steep that maintaining a dam no longer makes sense. Massive new hydroelectric projects in developed countries such as the United States are unlikely because the environmental and societal costs far outweigh the benefits. In developing countries, such as China, Brazil, and India, however, new large dam projects are on the upswing to meet the power, water, and flood control needs of growing populations and growing economies.

Chapter 17 **Study Guide**

Focus Points

The Human Sphere: Living on Floodplains

Human civilization: Human civilization began on floodplains, and human populations today are increasingly vulnerable to flooding on floodplains.

17.1 Stream Patterns

- **Drainage basins:** A drainage basin is composed of tributary streams and a trunk stream. It is delineated by topographic relief.
- **Drainage and continental divides:** Drainage divides separate different drainage basins. Continental divides separate drainage basins that empty into different oceans.
- **Drainage patterns:** Stream tributaries form different geographic patterns. The most common pattern is dendritic drainage.
- **Stream order:** Most streams are first-, second-, or third-order streams.

17.2 The Work of Water: Fluvial Erosion

- **Fluvial erosion:** Fluvial erosion is the most important and widespread agent of erosion on Earth.
- **Stream discharge:** Stream discharge is determined by climate, stream order, season, and ground permeability. High-discharge streams have more energy to transport sediments than low-discharge streams.
- **Stream rejuvenation:** Entrenched meanders, stream terraces, and antecedent streams result when a stream is uplifted.

17.3 Building with Streams: Fluvial Deposition

- **Stream sorting:** During flooding, streams drop large sediments first and the lightest materials last, resulting in sediments sorted both vertically and horizontally away from the stream channel.
- **Aggradation landforms:** Floodplains and braided streams form through aggradation.
- **Meanders:** Stream channels on floodplains are dynamic and form meanders and meander cutoffs.
- **Flooding:** When streams overflow their channels and cover the floodplain with a sheet of standing water, they add sediments to the floodplain.
- **Delta subsidence:** The input of new sediments with flooding in a delta keeps pace with the natural rate of subsidence through sediment compaction. Artificial levees prevent flooding and disrupt this equilibrium.

17.4 Rising Waters: Stream Flooding

- **Types of flooding:** Flash floods and seasonal floods are the two modes of stream flooding.
- **Flood mitigation:** Dams, levees, and stream bypasses help reduce flooding events.

17.5 Geographic Perspectives: Dam Pros and Cons

- **Dam pros:** Dams provide many benefits to human society, including flood control, water availability, and hydroelectricity generation.
- **Dam cons:** Dams create many problems, including destruction of habitat, farmland, and human communities and beach erosion resulting from sediment starvation.

Key Terms

abrasion, 521
aggradation, 532
alluvial fan, 533
alluvium, 530
base level, 524
bed load, 523
braided stream, 533
continental divide, 515
cut bank, 535
delta, 535
dendritic drainage, 518
dissolved load, 521
distributary, 535
drainage basin, 515
drainage divide, 515
ephemeral stream, 518
escarpment, 525
exotic stream, 518
flash flood, 538
flood, 538
floodplain, 514
fluvial erosion, 520
headward erosion, 520
headwaters, 518
intermittent stream, 518

internal drainage, 515
knickpoint, 525
meander, 526
oxbow lake, 535
permanent stream, 518
plunge pool, 525
point bar, 535
reservoir, 535
saltation, 523
seasonal flood, 538
stream, 515
stream capture, 529
stream discharge, 520
stream order, 518
stream rejuvenation, 525
subsidence, 536
suspended load, 521
thalweg, 535
traction, 523
tributary, 515
trunk stream, 515

Concept Review

The Human Sphere: Living on Floodplains

1. What are the benefits of locating a large city near a river on a floodplain? What problems arise for cities on floodplains?

17.1 Stream Patterns

2. What are tributaries and trunk streams? How do these phenomena fit into drainage basins?

3. What is the difference between a drainage divide and a continental divide?

4. What is an internal drainage basin? Why are lakes found in internal drainage basins typically saline?

5. Compare the different types of drainage patterns. Which is the most common on Earth's surface?

6. What happens to stream order when a first-order stream merges with a second-order stream? What happens when two second-order streams merge? What is the stream order of the Amazon River?

7. Compare and contrast intermittent, permanent, ephemeral, and exotic streams.

17.2 The Work of Water: Fluvial Erosion

8. What is fluvial erosion? Why is it the most important agent of erosion?

9. What is headward erosion? How do rills and gullies relate to headward erosion?

10. What is stream incision? Where does it occur?

11. What is stream discharge? What four main factors determine stream discharge?

12. What is a stream hydrograph? Compare natural stream discharge with discharge of streams in urbanized areas.

13. What is abrasion? How does it relate to fluvial erosion?

14. What is a stream load? List three types of stream loads.

15. Compare stream capacity with stream competence. What factors increase each?

16. In which situations do stream capacity and stream competence decrease? What happens to the stream load when they decrease?

17. What does it mean to say that a stream is *graded*? Compare a graded stream with one that is not graded.

18. What is base level?

19. Why do some streams have V-shaped valley walls, while others form stair-step canyons or slot canyons?

20. What is a knickpoint? Why do knickpoints migrate upstream?

21. What is stream rejuvenation? What rejuvenates a stream?

22. How do entrenched meanders and stream terraces form?

23. What is a water gap? Compare superposed streams with antecedent streams.

24. What is stream capture? How does one stream capture another?

17.3 Building with Streams: Fluvial Deposition

25. What does it mean to say that streams "sort" their load? Why is the load sorted? How is it sorted?

26. What is aggradation? Where does it occur? How does it relate to stream capacity and stream competence?

27. What are braided streams? Where do they form?

28. What are alluvial fans? Where do they form?

29. What is a stream meander? What is a thalweg, and how does it relate to stream meander development? How are point

bars and cut banks related to the development and migration of meanders?

30. Where do natural levees form? How do they form?

31. Identify several floodplain landforms and describe how they form.

32. Why do deltas form at base level? What are the two main delta shapes?

33. What are distributaries? Where are they found? How do they form?

34. What is subsidence? Why do deltas experience subsidence? Explain how a delta maintains equilibrium between sedimentation and subsidence.

35. How do artificial levees disrupt the equilibrium between sedimentation and subsidence?

17.4 Rising Waters: Stream Flooding

36. Why do floods happen? What are the two kinds of flooding?

37. What three major approaches are used to control floods? Explain how each works and the advantages and disadvantages of each.

17.5 Geographic Perspectives: Dam Pros and Cons

38. How do dams provide water?

39. How do dams generate hydroelectricity?

40. Why do sediments settle out in reservoirs?

41. How do dams result in habitat loss? Relate your answer to the salmon runs of the Pacific Northwest.

42. How do dams relate to climate change?

Critical-Thinking Questions

1. Fluvial erosion, which never ceases, cuts into mountains and reduces their height. Why haven't streams worn all of Earth's mountains flat?

2. Rocks are much harder than water. How does water carve valleys into mountains composed of solid bedrock?

3. Some stream channels naturally follow fault lines. Earthquakes make rocks there weak and cause them to easily erode. How does this fact relate to problems with some dams on streams?

4. Do you live in an area where floods occur? If you do not know, what kinds of information would you need to find out in order to answer this question?

5. Which argument do you think is stronger: the argument for dam removal or the argument for dam building? Does your answer depend on the circumstance? Explain.

Test Yourself

Take this quiz to test your chapter knowledge.

1. **True or false:** Every drainage basin has a trunk stream.

2. **True or false:** The merging of two second-order streams creates a fourth-order stream.

3. **True or false:** Flash floods are typically more dangerous than seasonal floods.

4. **True or false:** Exotic streams flow through arid regions and dry up before reaching the ocean.

5. **Multiple choice:** Which of the following features is not associated with stream meanders?

   a. point bar
   b. oxbow lake
   c. knickpoint
   d. alluvium

6. **Multiple choice:** Waterfalls migrate upstream through the process of _____.

   a. headward erosion
   b. subsidence
   c. stream meandering
   d. stream aggradation

7. **Multiple choice:** Internal drainage basins are most likely to have which of the following?

   a. escarpments
   b. permanent streams
   c. plunge pools
   d. saline lakes

8. **Multiple choice:** Stream terraces are created when which of the following occurs?

   a. flooding
   b. geologic uplift
   c. the building of a new dam
   d. the merging of tributary streams

9. **Fill in the blank:** At _____, a stream stops flowing and becomes standing water.

10. **Fill in the blank:** A(n) _____ separates two different drainage basins that flow into different oceans.

Online Geographic Analysis

Stream Flow in the United States

In this exercise, we will examine stream flow data for streams monitored by the USGS.

Activity

Go to https://waterwatch.usgs.gov. Click the "Current Streamflow" link on the menu on the left.

1. **Each dot on this U.S. map represents a stream gauge location. Describe the geographic pattern of stream gauges you see on the map.**

2. **What do the stream gauge location dot colors indicate? Do you see a pattern to the red- or blue-coded locations?**

Now click the state where you currently live. From there, click the stream flow gauge nearest you. Using the pop-up screen, answer the following questions.

3. **What is the name of the stream on which your stream gauge is located?**

4. **What is the current discharge of this stream (in cubic feet per second [cfs])?**

5. **What is the current stage of this stream?**

6. **What is the length of the stream gauge record for the gauge you chose?**

7. **What percentage of normal is your stream currently flowing?**

Go back out to the U.S. map and click Texas. From the Texas map, select a stream gauge from the Houston area in the southeast. (You can hover over the gauge locations to see where they are.)

8. **Answer Questions 3 through 7 for this gauge in Texas.**

Next click the "Hydrograph" tab. You will be directed to another page. Enter March 1, 2017 in the "Begin date" field. Enter December 15 in the "End date" field. Click the "Go" button.

9. **Hurricane Harvey struck the Houston area in late August in 2017 and caused widespread flooding. What was the peak discharge, in cubic feet per second, for the stream gauge you selected during Hurricane Harvey?**

10. **Scroll down to the gauge height (also called the *stage height*) graph. What was the peak gauge height, in feet?**

Picture This. *Your Turn*

Identification of Stream Features

Fill in the blanks on the diagram, using each of the terms below only once.

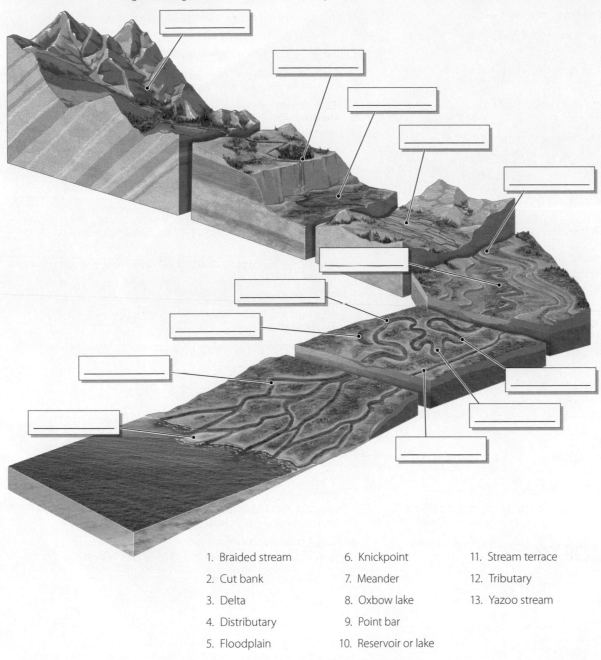

1. Braided stream
2. Cut bank
3. Delta
4. Distributary
5. Floodplain

6. Knickpoint
7. Meander
8. Oxbow lake
9. Point bar
10. Reservoir or lake

11. Stream terrace
12. Tributary
13. Yazoo stream

For animations, interactive maps, videos, and more, visit www.saplinglearning.com.

The Work of Ice: The Cryosphere and Glacial Landforms

Chapter Outline *and Learning Goals* ◎

18.1 About Glaciers

◎ Explain how glaciers form and move and describe different glacier types and their geographic settings.

18.2 Carving by Ice: Glacial Erosion

◎ Explain how glaciers cut into rocks and identify landforms caused by glacial erosion.

18.3 Building by Ice: Glacial Deposition

◎ Identify landforms created from glacial sediments and explain how they formed.

18.4 Frozen Ground: Periglacial Environments

◎ Identify features unique to areas with permanently frozen soils and explain environmental changes taking place in those areas.

18.5 Geographic Perspectives: Polar Ice Sheets and Rising Seas

◎ Assess the role of polar ice sheets in current and future sea-level changes and explain why sea-level rise presents problems for human society.

In this long-exposure photo, the zigzagging orange lines are the headlights of cars driving on the *Trollstigen* (which means "troll's path"), a steep and winding road in northern Norway. The road climbs up the side of a U-shaped valley that was carved by a glacier over 15,000 years ago. Natural climate change long ago melted the ice that made this valley.

(Jon Arnold/Danita Delimont Stock Photography)

To learn more about glacial landforms like this one, turn to Section 18.3.

THE HUMAN SPHERE The Mammoth Hunters

Figure 18.1 Mammoth remains. Preserved mammoth tusks, such as these found in Siberia, provide important information for scientists studying these ancient animals. Mammoth "hunters" (shown here) sell their prized ivory tusks. Tusks of this size may sell for as much as $100,000 each. An estimated 50 tons of mammoth bones are found each year in Russia alone. *(Francis Latreille/Reuters)*

For 5 million years, the woolly mammoth (*Mammuthus primigenius*)—a close relative of the modern African elephant—roamed across much of the Northern Hemisphere. Their numbers were in the millions. By about 9,600 years ago, however, all mainland mammoth populations in North America and Eurasia were gone. The last remaining mammoth holdout was Wrangel Island, in the northeastern corner of Siberia. Wrangel Island's mammoth population disappeared about 4,000 years ago, right after humans found the island. Human hunting was the main cause of the extinction of the mammoth and

many other large animals, called *megafauna*, that also lived at that time in many parts of the world.

The remains of some mammoths have been preserved in the frozen northern soils for millennia. Now those soils are thawing as the **cryosphere**—the frozen portion of the hydrosphere—is changing in response to anthropogenic climate change. High latitudes are warming at twice the rate of the midlatitudes because of the ice–albedo positive feedback (see Section 7.1). As the frozen soils thaw, the mammoth remains become exposed, allowing scientists and collectors to find them **(Figure 18.1)**.

18.1 About Glaciers

◎ **Explain how glaciers form and move and describe different glacier types and their geographic settings.**

As we saw in Chapter 7, "The Changing Climate," natural climate change is always occurring. If we could somehow step foot on Earth 20,000 years ago, we would see a very different world. Earth was in a glacial period. Huge ice sheets covered about 30% of Earth's land surface in the Northern Hemisphere, compared to today's land ice coverage of 10%. Most of the locations where northern cities—such as Seattle, Chicago, Toronto, Montreal, and New York City—now sit were covered by immense ice sheets. The world was colder, windier, and dustier. Growing crops the way we do today and feeding the current world population would have simply been impossible at that time.

Evidence of these glacial conditions takes the form of sediments composed of unsorted rock fragments called till. **Till** (discussed more in Section 18.3) is any debris deposited by a glacier without the influence of running water. Such sediments are commonplace throughout northern Eurasia and northern North America. Glacial till ranges in size from tiny silt particles to massive boulders **(Figure 18.2A)**. In some places, large boulders seemed to have been placed randomly across the countryside, as if by giants. Many of them do not match the local bedrock, indicating that they came from far away **(Figure 18.2B)**.

Originally, scientists thought water in streams created these unsorted glacial sediments and transported these large boulders. But such massive boulders would be too large for water to transport. Furthermore, the sediments that streams deposit are sorted by size (see Section 17.3). Why are these sediments unsorted? In 1837, the Swiss-American geologist Louis Agassiz was the first to propose that these sediments and boulders were transported by now-melted glaciers. His colleagues intensely criticized his "ice age theory." Within a decade or so, however, it was embraced because the physical evidence overwhelmingly supported it.

Large boulders transported long distances by a glacier are called **glacial erratics**. Originally, glacial erratics and glacial till were attributed to the great biblical flood. It was thought that the boulders "drifted" into place, and they were therefore called *drift*. This term has persisted: Today, all sediments relating to the activity of glaciers and the streams flowing out of glaciers are collectively called *drift*.

A

B

Figure 18.2 Glacial sediments. (A) A geography student points out unsorted glacial till deposits that are exposed on a cliff face near Criccieth, Wales. (B) This isolated boulder on the coast of southern Sweden is locally called "Glumsten." The boulder served as a guidepost for sailors for hundreds of years, until a lighthouse was built. Note the nearby picnic table for scale.   *(A. redsnapper/Alamy; B. © Lars Wikander)*

What Is a Glacier?

A **glacier** is a large mass of ice that is formed from the accumulation of snow and flows slowly downslope. Glaciers form slowly, one snowflake at a time. Eventually, if snowfall accumulates over many years and is compressed by gravity, it turns into **firn**. Firn, which is also called *névé*, is snow that has been compressed into granular ice and has an air content of roughly 50% of the total volume. With further time and compression, firn turns into glacial ice. **Glacial ice** is ice that has an air content of less than 20% of the total volume. Because of its lack of trapped air, glacial ice absorbs longwave orange and red light and scatters and transmits blue light. This gives glacial ice its brilliant blue color. Glacial ice usually forms visible annual layers, like the rings of a tree. Seasonal properties of snow and air temperature cause these layers. Darker layers are ice with lower air content that was formed during the winter **(Figure 18.3)**.

Glaciers do not form in environments where snow melts away in summer or where there is insufficient snowfall. Glacial ice can develop only where snowfall exceeds snow loss through melting and sublimation (the process in which ice transforms directly into water vapor without first melting into a liquid). In such places, snowfall accumulates year after year. In addition, the slope where snow is accumulating cannot be too steep. If it is, the snow will slip downslope in avalanches rather than accumulate and form a glacier. The time it takes for glacial ice to form varies from one to several decades, or more, depending on snowfall rates. Heavy snowfall produces glacial ice more quickly.

Flowing Ice

Glaciers move downslope through the processes of *basal sliding* and *plastic deformation*. Basal sliding causes a piece of ice to slide down the sloped hood of a car, for example, as liquid water lubricates the base of the ice. The same phenomenon occurs in glaciers where the bedrock is smooth and liquid water is present at the base of the glacier.

Ice is a solid. From the surface of the glacier down to 60 m (200 ft), an area referred to as the *brittle zone*, the ice is brittle and easily breaks. Cracks called **crevasses** form in the brittle zone when the glacier flows over an upward

1. Fresh snow
New snowflakes have a six-armed structure that keeps them separated initially. The air content of fresh snow is about 90%.

2. Firn (or névé)
As snowfall accumulates, gravity compresses the snow, and the snowflakes fit together more tightly. Compression and partial melting squeeze air out of the snow, forming a granular ice called *firn*. Firn's air content is roughly 50% by volume.

3. Glacial ice
With increasing compression caused by the growing weight of the snowpack above, through time most of the air bubbles are squeezed out of the firn. Glacial ice has an air content of 20% or less by volume.

A

B

Figure 18.3 Glacial ice. (A) Glacial ice forms as snowfall accumulates. (B) Grey Glacier in Torres del Paine National Park in southern Chile displays the intense blue color and layering typical of glacial ice. *(Michele Falzone/Alamy)*

protrusion in the bedrock beneath **(Figure 18.4)**. Deeper than 60 m is the *plastic deformation zone,* where ice is not brittle but instead behaves plastically and can slowly deform and flow. We discussed plastic deformation in Section 12.3 in the context of the solid rock of Earth's mantle, which slowly deforms and flows. Glaciers flow because the ice crystals deep within a glacier are not strong enough to support the overlying weight of the glacier. Internal shearing within ice crystals allows gravity to deform the ice, causing it to flow downslope.

A glacier flows downslope at an average rate of a few centimeters per day. It does not always move steadily downslope, however. Its speed varies depending on several factors, including

Figure 18.4 Glacier crevasses and plastic deformation. (A) Crevasses form in the brittle zone where ice flows over protrusions in the bedrock. The transition between the brittle zone above and the plastic deformation zone below is called the *transition zone.* (B) These crevasses have formed on Belcher Glacier as it flows toward the sea on Devon Island, Nunavut, Canada. The blue arrows show the direction of ice movement. *(© Brad Danielson, University of Alberta)*

Animation

Glacier Crevasses and Plastic Deformation

Available at
www.saplinglearning.com

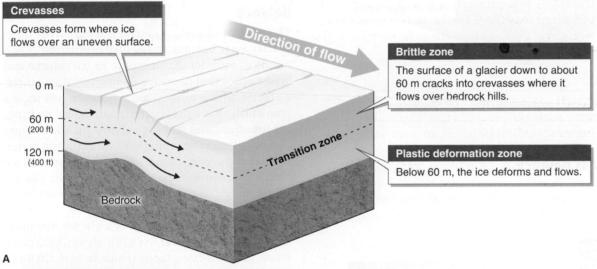

Crevasses
Crevasses form where ice flows over an uneven surface.

Direction of flow

Brittle zone
The surface of a glacier down to about 60 m cracks into crevasses where it flows over bedrock hills.

0 m

60 m
(200 ft)

120 m
(400 ft)

Transition zone

Plastic deformation zone
Below 60 m, the ice deforms and flows.

Bedrock

A

Devon Island
ice cap

Crevasses

AK   Canada

**Hubbard
Glacier**

*Gulf of
Alaska*

150° W   165° W

B

Figure 18.5 Glacier flow rates. Ice flow in a glacier is fastest at its surface and center.

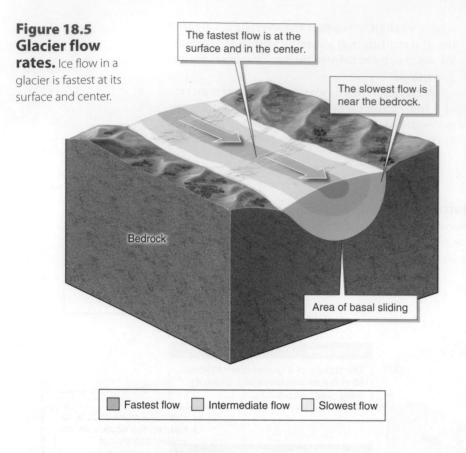

The fastest flow is at the surface and in the center.

The slowest flow is near the bedrock.

Bedrock

Area of basal sliding

■ Fastest flow   ■ Intermediate flow   □ Slowest flow

Figure 18.6 Glacier mass balance. A glacier grows when snowfall in the accumulation zone exceeds the loss of ice in the ablation zone. It shrinks when snowfall in the accumulation zone is less than ice loss in the ablation zone. The equilibrium line represents the elevation at which annual ice gain and ice loss are equal.

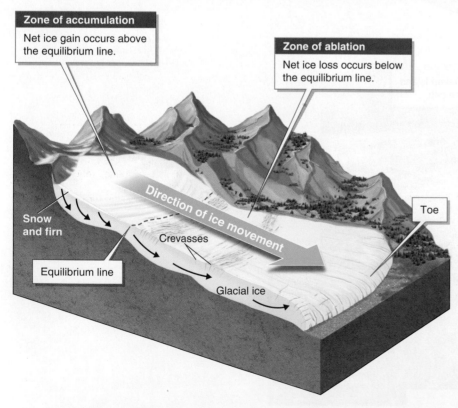

Zone of accumulation
Net ice gain occurs above the equilibrium line.

Zone of ablation
Net ice loss occurs below the equilibrium line.

Snow and firn

Direction of ice movement

Crevasses

Toe

Equilibrium line

Glacial ice

air temperature, the roughness and steepness of the topography over which it is flowing, and the amount of water lubricating its base. During a *glacial surge*, the speed of a glacier accelerates for a period of a few days to a month or more. Glacial surge speeds can be on the order of 30 m (100 ft) in a single day or even faster. Glacial surges typically occur in the warm summer months when portions of the glacier melt.

Like water in a stream, ice within a glacier flows at different speeds. Friction with the bedrock slows down the ice at the margins (edges). The areas of fastest flow are farthest from the bedrock—namely, near the surface and central portions of the glacier, as illustrated in **Figure 18.5**.

Inputs and Outputs: Glacier Mass Balance

The volume of ice in a glacier changes over time. **Glacier mass balance** is the difference between inputs to a glacier that increase its ice volume and losses of ice that decrease its volume. As a whole, a glacier with a neutral mass balance has inputs (snowfall) that equal outputs caused by melting, sublimation, and iceberg *calving* (in which portions of glaciers breaking off into standing water, such as the ocean). When snowfall is less than ice loss, a *negative mass balance* exists for the glacier. When snowfall exceeds ice loss, a glacier has a *positive mass balance*.

Along most of the length of a glacier, the mass balance is not neutral. At high elevations, more snow falls on the glacier than is lost through melting and sublimation. The area of a glacier where ice gain from snowfall exceeds ice loss is the **accumulation zone**. Below, in the **ablation zone**, ice loss from melting and sublimation exceeds ice accumulation. At the **glacier toe**, the lowest end of a glacier, there is only ice loss. The transition between the accumulation zone and the ablation zone is the **equilibrium line** (or *firn line*), the elevation at which a glacier's ice accumulation and ice loss are equal over a period of a year. Firn forms only above the equilibrium line **(Figure 18.6)**.

Glacial advance occurs when a glacier has a positive mass balance and the glacier toe advances forward and downslope. When a glacier has a negative mass balance, the glacier toe moves upslope in the process of **glacial retreat**. The toe of a glacier continually advances and retreats in response to changes in glacier mass balance. As shown in **Figure 18.7**, however, the ice within a glacier is always flowing downslope through the glacier, even when the glacier toe is retreating upslope.

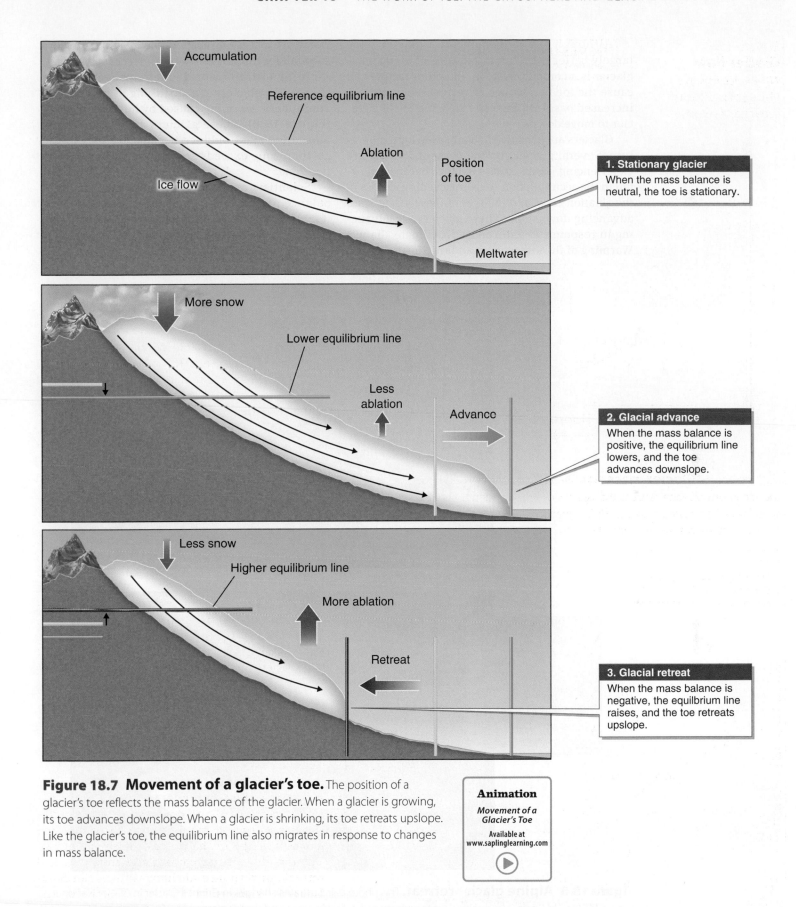

Figure 18.7 Movement of a glacier's toe. The position of a glacier's toe reflects the mass balance of the glacier. When a glacier is growing, its toe advances downslope. When a glacier is shrinking, its toe retreats upslope. Like the glacier's toe, the equilibrium line also migrates in response to changes in mass balance.

Animation

Movement of a Glacier's Toe

Available at
www.saplinglearning.com

Although the movement of a glacier's toe largely reflects the mass balance of a glacier, a glacier is a complex system, and other factors can cause the toe to advance or retreat. For example, increased basal sliding can cause the toe of a glacier to move downslope.

Glaciers are sensitive to changes in air temperature. Several glaciers on the coast of Norway are experiencing positive mass balance because snowfall in the accumulation zone is exceeding ice loss in the ablation zone, and the toes of these glaciers are advancing downslope. These glaciers are advancing in response to anthropogenic climate change. Warming of the atmosphere and the North Atlantic Ocean are increasing evaporation, resulting in more snowfall for the glaciers in parts of coastal Norway.

In mountains worldwide, however, 90% of the world's glaciers are shrinking in response to atmospheric warming caused by human activities (**Figure 18.8**). These glaciers are experiencing a negative mass balance because ice loss in the ablation zone exceeds snowfall in the accumulation zone.

Scientists are carefully measuring the mass balance of glaciers around the world and monitoring them as Earth's climate changes (see Section 7.5). **Figure 18.9** explores how scientists monitor glaciers.

Figure 18.8 Alpine glacier retreat. This photo pair illustrates changes in Grinnell Glacier in Glacier National Park in Montana's Rocky Mountains. The top photo was taken in 1940, and the bottom photo from the same location was taken in 2014. During this time period, the glacier lost 45% of its ice. At this rate, Grinnell Glacier will be gone by 2030. *(Top. Courtesy GNP Archives; bottom. Lisa McKeon, USGS)*

Photographic evidence
Historical photos provide important quantitative information about glacier mass balance over time. These photos show that the South Cascade Glacier in Washington State has retreated about 1.6 km (1 mi) upslope since 1958.

Field measurements
One means of measuring glacier mass balance is to place a grid of "ablation stakes" on the surface of a glacier. By measuring snow and ice accumulation or loss at each stake, scientists can measure the glacier's mass balance. Scientists compare temperature and precipitation data with glacier mass balance data to see how the glacier is responding to these climate variables. Here, two researchers from the USGS are setting a measurement stake on the South Cascade Glacier.

A   B

Satellite remote sensing
Satellites can be used to detect changes in glacier mass balance in many different ways, including sensing subtle changes in gravity around a glacier and detecting changes in the surface elevation of the glacier and in the position of its toe. In this 2016 false-color Landsat image of the Columbia Glacier in the Chugach Mountains of southeastern Alaska, snow and glacial ice are blue, healthy vegetation is green, bare ground is brown, and water is black. The position of the glacier's terminus is shown here for each decade since mid-1980s, when satellites began imaging the glacier. Since 1986 the Columbia Glacier has lost half its mass, and its terminus has retreated about 20 km (12 mi) inland (northward). If it continues, it will be gone by 2050.

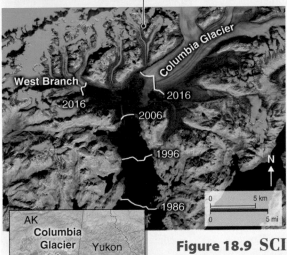

C

Graphed research data
The mass balance research data for three benchmark glaciers—South Cascade Glacier in Washington and Wolverine and Gulkana glaciers in Alaska—are graphed here for the period 1958 to 2013. The y-axis shows *meters of water equivalent (MWEQ)*: the depth of liquid water that has been lost from the glacier. In Alaska, 99% of the glaciers are shrinking. This pattern of glacial retreat is being observed worldwide as a result of climate change.

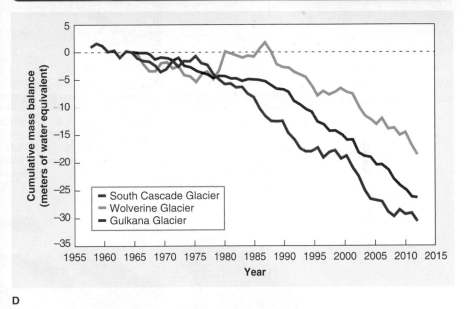

D

Figure 18.9 SCIENTIFIC INQUIRY: How do scientists monitor glaciers? Since the 1950s, the U.S. Geological Survey (USGS) has been monitoring several "benchmark" glaciers. Three of them are featured in this figure. Understanding changes in glacier mass balance is essential for understanding the cryosphere's response to atmospheric warming. When glaciers melt, the loss of ice and snow allows the ground to absorb more sunlight and warm faster, creating an ice–albedo positive feedback (see Section 7.1). Melting glaciers also contribute to sea-level rise that threatens coastal populations. (*A. Photo by L. Wernstedt, U.S. Forest Service; B. Photo by U.S. Geological Survey; C. NASA images by Jesse Allen and Robert Simmon, using Landsat 4, 5, 7, and 8 data from the USGS Global Visualization Viewer; D. U.S. Environmental Protection Agency*)

Two Types of Glaciers

Today, glaciers are on every continent except Australia. They are found near the equator at high elevations and in the Arctic and Antarctica at sea level **(Figure 18.10A)**. There are two basic types of glaciers: alpine glaciers and ice sheets. **Alpine glaciers** (also known as mountain glaciers) are glaciers that are found in mountainous areas. Alpine glaciers are divided into four types: A **cirque glacier** (pronounced SIRK) is a glacier that forms at the head of a valley, a **valley glacier** is a glacier that occupies a mountain valley, a **piedmont glacier** is a lobe of ice that forms as a valley glacier flows onto a flat plain, and an **ice cap** is a dome of ice that sits over a high mountain region and has an extent of 50,000 km² (19,000 m²) or less **(Figure 18.10B)**.

In contrast to alpine glaciers, **ice sheets** are domed sheets of ice that are larger than 50,000 km² and can cover a significant portion of a continent. Only Greenland and Antarctica have ice sheets,

Figure 18.10 World map of alpine glaciers and types.
(A) There are roughly 198,000 alpine glaciers in the world. Most alpine glaciers are found in the Himalayas and on the Tibetan Plateau. Western North America has the second highest number of glaciers. (B) Cirque glaciers, valley glaciers, piedmont glaciers, and ice caps are the primary types of glaciers in mountainous areas. The cirque is the smallest alpine glacier type, and an ice cap is the most geographically extensive. *(USGS)*

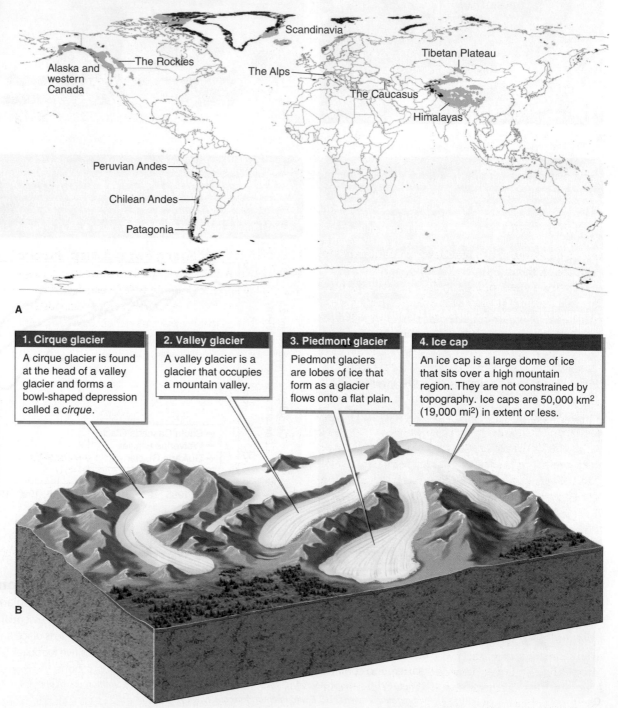

1. Cirque glacier
A cirque glacier is found at the head of a valley glacier and forms a bowl-shaped depression called a *cirque*.

2. Valley glacier
A valley glacier is a glacier that occupies a mountain valley.

3. Piedmont glacier
Piedmont glaciers are lobes of ice that form as a glacier flows onto a flat plain.

4. Ice cap
An ice cap is a large dome of ice that sits over a high mountain region. They are not constrained by topography. Ice caps are 50,000 km² (19,000 mi²) in extent or less.

Figure 18.11 Features of ice sheets. (A) As the contour lines show, the Greenland ice sheet reaches over 3,000 m (9,850 ft) above sea level in elevation, and the Antarctic ice sheet reaches over 4,000 m (13,125 ft) above sea level. Note that the two ice sheets are not shown at their relative sizes. (B) As an ice sheet or ice cap flows outward and downslope, several different types of glaciers and glacial features are formed.

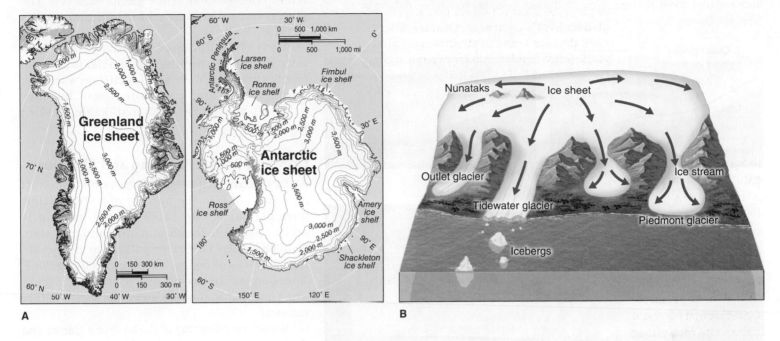

and together these two ice sheets contain over 99% of all ice on Earth **(Figure 18.11A)**.

As an ice sheet or ice cap flows slowly outward from its center, it may surround a high mountain peak, forming a **nunatak** (pronounced nun-uh-TAK): an area of bedrock that protrudes above the glacier **(Figure 18.11B)**. Lobes of ice that spill from the margins of the ice sheet form outlet glaciers and piedmont glaciers. An **outlet glacier** is a glacier that flows out of an ice sheet or ice cap through a constricted valley, often directly into the ocean. *Ice streams* are areas where the speed of flow of an outlet glacier is accelerated, as when ice travels over steep terrain or through a constricted valley. Outlet glaciers and valley glaciers that reach sea level and calf icebergs are called **tidewater glaciers**. An **iceberg** is a large block of ice that breaks from the toe of a tidewater glacier (or an *ice shelf*, discussed below) and floats in open water (see Figure 18.11B).

Most tidewater glaciers are located above 55 degrees north and south latitude. The lowest-latitude tidewater glacier is San Rafael Glacier, found at 46.5 degrees south latitude in Chile. Greenland has the most tidewater glaciers. The largest of them, Jakobshavn Glacier, produces 10% of all icebergs originating from Greenland. The busy shipping lanes between Europe and North America run through the *iceberg fields* in the North Atlantic Ocean **(Figure 18.12)**.

Figure 18.12 Iceberg hazard map. Shipping lanes between Europe and North America pass through iceberg fields in the North Atlantic Ocean. The most hazardous area is outlined in yellow. Many scientists suspect that an iceberg from the Jakobshavn Glacier sank the *RMS Titanic* in April 1912, a tragic accident in which more than 1,500 people died. *(U.S. Coast Guard Navigation Center)*

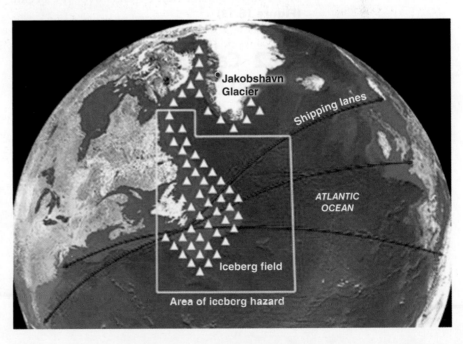

Figure 18.13 Ice shelves. (A) Ice shelves form as ice sheets or outlet glaciers flow offshore and float on the water. When an ice shelf breaks, it calves large, table-like icebergs. (B) About 90% of an iceberg is submerged and hidden beneath the water. This iceberg broke free from an ice shelf in Antarctica in 2002. It is about 220 m (720 ft) thick and rises about 20 m (65 ft) above the sea. It was photographed drifting in the Southern Ocean. *(Mlenny Photography/Getty Images)*

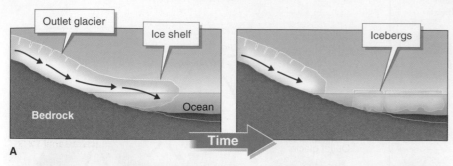

A

B

Enormous icebergs are produced where ice sheets flow into the sea, an area called an ice shelf. An **ice shelf** is the portion of an ice sheet or an outlet glacier that extends over the ocean **(Figure 18.13)**.

18.2 Carving by Ice: Glacial Erosion

◎ **Explain how glaciers cut into rocks and identify landforms caused by glacial erosion.**

Erosion by glaciers is less geographically widespread than erosion by streams. Where glaciers occur, however, they exert a powerful erosive force and are capable of grinding down and wearing away the hardest of bedrock. Even in the tropics, at high elevations, glacial erosion is a major factor that shapes landscapes.

As mountain ranges are uplifted through tectonic forces, they enter colder reaches of the troposphere where glaciers will form. As the glaciers form and begin flowing, they grind away at the mountains; the higher mountains are lifted, the more they are eroded by glaciers. Scientists refer to this process as the *glacial buzz saw*. The glacial buzz saw limits how high mountains can rise.

Many of Earth's most spectacular landforms were carved during the last glacial maximum (see Section 7.2). With natural climate warming, those glaciers long ago melted away, but they have left behind their unmistakable mark of erosion.

Grinding Rocks: Plucking and Abrasion

At first glance, it seems impossible that soft ice could cut into the hardest of bedrock. How does glacial ice do this? Glacial erosion is accomplished through plucking and abrasion. **Plucking** involves a glacier pulling up and breaking off pieces of bedrock as it moves downslope. Abrasion, as discussed in Section 17.2, occurs when one material wears away (abrades) another material.

Water accumulating at the base of a glacier and then refreezing around protrusions in the bedrock allows the glacier to pull loose, or pluck, fragments of bedrock. The ice grabs the bedrock and pulls it up, much as you would pull a staple out of paper with a staple remover. Fragments of bedrock become embedded in the base of the glacier as a result of glacial plucking.

As a glacier flows downslope, the rock fragments that have been plucked and embedded in its base grind against, or abrade, the bedrock much like sandpaper sanding wood. Grooves are gouged into the surface of bedrock by dragged rocks and boulders. These grooves are called **glacial striations**. In addition, a smoothed surface called **glacial polish** forms on the bedrock over which the glacier flows **(Figure 18.14A)** as fine grains of sand, clay, and silt sandwiched between the moving ice and the bedrock abrade and polish the surface of the bedrock. This process pulverizes the sand and clay fragments into silt, the smallest of all rock particles. This very fine powder of silt is called *glacial flour*. When glacial flour is suspended in water, it can color the water bright turquoise or cyan **(Figure 18.14B)**.

In North America and Europe, many people live in glacially formed landscapes but aren't aware of it. New York's Central Park is a particularly interesting landscape in this regard. See the **Picture This** feature on the following page to learn more about the glacial history of Central Park.

Figure 18.14 Glacial striations and glacial flour. (A) This sandstone outcrop in the Wasatch Mountains in Utah shows prominent striations and polish. The glacial striations are visible as parallel grooves, and the glacial polish has produced the smoothed surface. The scale of the photo is about 1 m (3.3 ft) across. (B) Moraine Lake, Banff National Park, Canada, is world famous for its brilliant blue water. Its color comes from the uniformly small particles of glacial flour suspended in the water, reflecting mostly blue wavelengths of sunlight.   *(A. Robert_Ford/Getty Images; B. Bruce Gervais)*

Picture This Landscaping with Glaciers

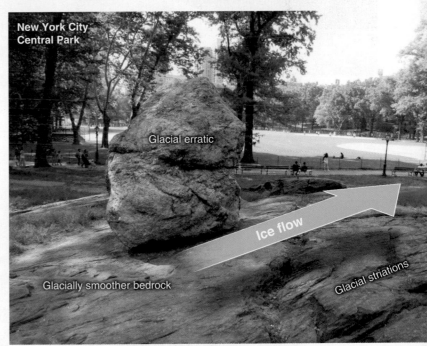

(Bruce Gervais)

Some 42 million people visit New York City's Central Park each year, but few of them know that ice built the park as much as humans did. Twenty thousand years ago, the Laurentide ice sheet covered what is now New York City to a depth of about 300 m (1,000 ft). The ice sheet flowed from north to south and carved into the extremely hard Manhattan schist metamorphic bedrock. Plucked rocks and boulders were incorporated into the base of the ice and dragged across the bedrock, gouging deeply into it. Although the ice melted 12,000 years ago, its handiwork can still be seen throughout the park.

All of the exposures of Manhattan schist bedrock throughout the park (including the one pictured here) have been rounded and smoothed by the ice. Dozens of the park's bedrock exposures have a unique submarine-like shape. These landforms, called *roches moutonnées* (see Figure 18.22), are formed only beneath glaciers. Hundreds of huge glacial erratics are found throughout the park as well. The park's landscape designers intentionally left many of the largest ones exactly where the ice deposited them 12,000 years ago. The glacial erratic shown here is not composed of local Manhattan schist but matches the bedrock found hundreds of miles to the north, in Canada.

Consider This

1. Why doesn't the glacial erratic rock in the photo match the local Manhattan schist bedrock?
2. In what direction would you expect the glacial striations throughout Central Park to be oriented? Why?

Figure 18.15 Rock debris on glaciers.

Valley glaciers typically accumulate a heavy load of rock debris that tumbles onto the glaciers' surfaces as they flow through steep-walled valleys. This photo shows the Kaskawulsh Glacier in Kluane National Park, Yukon, Canada. All of the dark stripes on it are composed of rock debris ranging in size from small particles of sand to large boulders. Arrows show the direction of ice flow. *(DEA/F. BARBAGALLO/Getty Images)*

A

B

Figure 18.16 Supraglacial and subglacial streams. (A) This supraglacial stream is flowing on top of the Greenland ice sheet. (B) This subglacial stream is an outlet for the Glacier du Mont Miné in the Alps of Switzerland. *(A. The Asahi Shimbun/Getty Images; B. Scott Montross, Montana State University)*

Transporting Rocks: Ice and Glacial Streams

In addition to rock fragments they have picked up through plucking, most glaciers carry rock debris that has fallen onto their surface through mass movement of overlying slopes. Any rocks on or within the glacier are transported downslope with the flowing ice **(Figure 18.15)**. Glaciers are like conveyor belts that transport ice and rock debris downslope. Rock material can be transported all the way to the toe of the glacier, where it is deposited in piles of unsorted sediments called *moraines* (see Section 18.4).

In summer, glaciers produce streams of meltwater that flow over the surface of the glaciers **(Figure 18.16)**. These are called *supraglacial streams*. There are also streams in tunnels inside the glacier and beneath the glacier, between its base and bedrock. These are called *subglacial streams*. Subglacial streams transport glacial sediments downslope. Much like streams that form a delta, subglacial streams exiting a glacier (called *outlet streams*) form a flat **outwash plain** as the sediments they carry accumulate through aggradation. Because of their heavy sediment loads, streams on outwash plains are typically braided (see Figure 17.27). Outwash plains are examples of *glaciofluvial* landforms, meaning that flowing ice and flowing water combine to create them.

A proglacial lake may form at the toe of a glacier where depressions have been excavated by ice or where sediments dam an outlet stream **(Figure 18.17)**. Proglacial lakes are becoming more common, particularly in the Himalayas, as valley glaciers and outlet glaciers retreat upslope in response to the atmospheric warming of the past century.

Erosional Landforms

Glacial geomorphology is the study of landforms created by glaciers. Understanding how glaciers influence a landscape is important for understanding and describing the development of a region's topography, as well as its climate and biogeographic history. In the remainder of this section, we explore landforms created by glacial erosion, starting with those that form in high-elevation settings and concluding with those found at sea level.

Cirques and Tarns

Cirques and tarns are prominent landscape features often found at the highest elevations in mountain ranges. A **cirque** is a bowl-shaped valley with steep rock walls, called *headwalls*, formed by a cirque glacier. As cirque glaciers flow downslope, they often carve depressions in the bedrock just downslope of the cirque. Tarns are often found just downslope of cirques. A **tarn** is a mountain lake that forms within or just below a cirque as meltwater draining from a glacier or snow fills a depression created by the glacier **(Figure 18.18)**.

Figure 18.17 Proglacial lake. Bear Glacier in the Kenai Fjords Park in Alaska terminates in a proglacial lake filled with icebergs. The Pacific Ocean is in the foreground. *(Flyver/Alamy)*

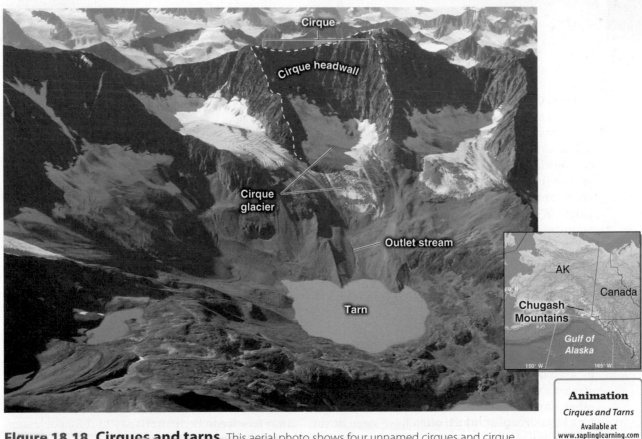

Figure 18.18 Cirques and tarns. This aerial photo shows four unnamed cirques and cirque glaciers and three tarns in the Chugach Mountains of southeastern Alaska. This landscape has been significantly shaped by glacial erosion. *(U.S. Geological Survey, photo by Bruce Molina)*

Animation

Cirques and Tarns
Available at
www.saplinglearning.com

Figure 18.19 Arêtes, cols, and horns. (A) Hikers pass through a col while following along the sharp spine of an arête (traced with a dotted line) in the Alps. (B) Shivling, which stands at 6,543 m (21,466 ft), is one among hundreds of horns in the Himalayas. *(A. Arterra/Getty Images; B. Travel Ink/Getty Images)*

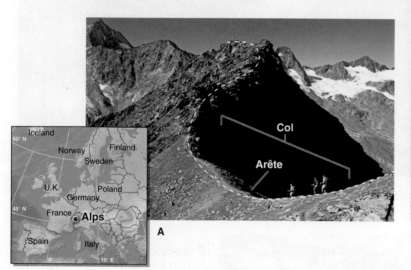

Arêtes, Cols, and Horns

Where valley glaciers erode two opposing sides of a mountain, an elongated and sharp, steep-sided ridge, called an **arête** (pronounced ar-eh-TAY), can form. The lowest area along the arête is a **col** (or *pass*) **(Figure 18.19A)**. Occasionally cirques erode into the base of a mountain from three sides. When this happens, a pyramid-shaped mountain peak called a **horn** forms. The cirque glaciers do not erode near the top of the horn; instead, the glaciers cut into the base of the mountain from three sides, undercutting the headwalls and causing the sides to steepen through rockfall **(Figure 18.19B)**.

Glacial Valleys and Paternoster Lakes

Recall from Section 17.2 that a stream's energy is focused on its narrow stream channel. As a result, streams cut V-shaped valleys. Glaciers, on the other hand, carve broad U-shaped **glacial valleys** (or *glacial troughs*), which often have steep or vertical valley walls. Because the base of a glacier is mostly flat, plucking and abrasion carve a flat valley. Where the sides of the valley are undercut by

the glacier, rockfall often causes steep valley walls to develop.

Like trunk streams and tributary streams (see Section 17.1), *trunk glaciers* have a larger ice volume than *tributary glaciers*. As a result, they have more erosive power and cut more deeply into the bedrock. The surface heights of trunk and tributary glaciers, however, are equal. After the glaciers have melted, a *hanging valley* is left where a tributary glacial valley feeds into a larger glacial valley with a deeper valley floor, as shown in **Figure 18.20**.

Relatively resistant portions of bedrock and ridges of debris called *recessional moraines* (which we will discuss further in Section 18.4) sometimes form obstructions to stream flow through a glacial valley. A series of small lakes may form as these obstructions dam the stream flowing through the glacial valley. Lakes that form in this way are **paternoster lakes** (from Latin, meaning "our father," in reference to their similarity to religious rosary beads) **(Figure 18.21)**.

Also found on the floors of glacial valleys are protruding outcrops of bedrock called

A

B

Figure 18.20 Glacial valley and hanging valley. (A) Yosemite Valley, in the Sierra Nevada of California, is a U-shaped valley formed by glaciers. (B) Bridalveil Creek flows through a U-shaped hanging valley and then plunges over the vertical edge of Yosemite Valley to form Bridalveil Fall. *(A. Alice Cahill/Getty Images; B. David Gomez/Getty Images)*

Figure 18.21 Paternoster lakes. Glacier National Park in Montana has many glacial valleys with paternoster lakes. This photo was taken from the vantage point of Grinnell Glacier, overlooking Grinnell and Josephine lakes (in the foreground). Note the U-shaped valley. *(Design Pics Inc/Alamy)*

Figure 18.22 Roche moutonnée. A roche moutonnée is formed where glacial ice flows over a rock outcrop, abrades the upstream side of the outcrop, and plucks the downstream side, creating an asymmetrical hill. The blunt end faces the direction of the glacier's movement. This roche moutonnée, with glacial striations, is one of hundreds found throughout Central Park in New York City. (*Bruce Gervais*)

roches moutonnées (derived from the French *mouton*, for their resemblance to sheep). A **roche moutonnée** is an elongated and asymmetrical ridge of glacially carved bedrock **(Figure 18.22)**.

Drowned Glacial Valleys: Fjords

Many high-latitude coastlines, such as those of southern Chile, southern Alaska, Norway, Iceland, and Greenland, look remarkably similar because they were formed in the same way and are dominated by fjords. **Fjords** are long U-shaped glacial valleys that have been flooded by the sea. Global sea level was drawn down by some 85 m (280 ft) during the last glacial maximum, when ice sheets covered much of North America and Eurasia (see Section 7.2). Outlet glaciers flowing from these ice sheets, as well as valley glaciers, eroded deep glacial valleys that reached below the present-day sea level. By roughly 10,000 years ago, most of this ice had melted. Sea level rose 85 m, and the glacial valleys flooded, creating fjords **(Figure 18.23)**.

Climate and Glacial Landforms

Latitude and elevation are important factors that determine whether glaciers will develop in a region and give rise to landforms like those we have explored in this section. Ultimately, climate controls the growth of glaciers in any given region. Like fluvial landforms (described in Chapter 17, "Flowing Water: Fluvial Systems"), glacial landforms develop and evolve through time as climate changes **(Figure 18.24)**.

Figure 18.23 Fjords. Preikestolen (Pulpit Rock) (note the people on the rock for scale) overlooks Lysefjordin, Norway. The vertical drop from Preikestolen to the fjord is 604 m (1,982 ft). The inset satellite image shows the many deep fjords of coastal Norway. (*Above. Anders Blomqvist/Getty Images; inset. Jeff Schmaltz, MODIS Rapid Response Team, NASA/GSFC*)

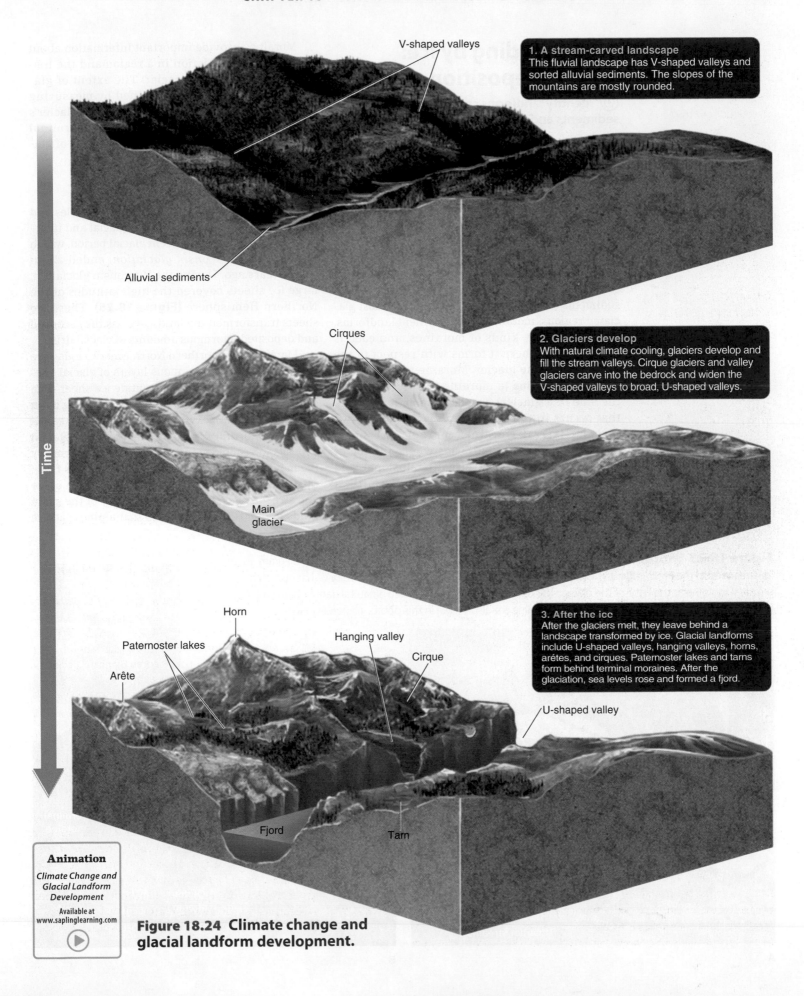

1. A stream-carved landscape
This fluvial landscape has V-shaped valleys and sorted alluvial sediments. The slopes of the mountains are mostly rounded.

V-shaped valleys

Alluvial sediments

2. Glaciers develop
With natural climate cooling, glaciers develop and fill the stream valleys. Cirque glaciers and valley glaciers carve into the bedrock and widen the V-shaped valleys to broad, U-shaped valleys.

Cirques

Main glacier

3. After the ice
After the glaciers melt, they leave behind a landscape transformed by ice. Glacial landforms include U-shaped valleys, hanging valleys, horns, arétes, and cirques. Paternoster lakes and tarns form behind terminal moraines. After the glaciation, sea levels rose and formed a fjord.

Horn

Hanging valley

Paternoster lakes

Cirque

Arête

U-shaped valley

Fjord

Tarn

Time

Animation

Climate Change and Glacial Landform Development

Available at
www.saplinglearning.com

Figure 18.24 Climate change and glacial landform development.

18.3 Building by Ice: Glacial Deposition

◎ **Identify landforms created from glacial sediments and explain how they formed.**

Glaciers and streams have many aspects in common. Both are composed of water. Both flow downslope. Both carry sediments. And both deposit sediments where they stop flowing. This section examines landforms made by glacial sediments.

Deposits by Alpine Glaciers

Alpine glaciers transport material of all sizes and pile it into jumbled, unsorted moraines, which are composed of till. A **moraine** is a heap of unsorted sediments (till) deposited by a glacier. In most glaciated regions, moraines are prominent landforms. There are many kinds of moraines, and each is identified by where it forms with respect to the movement of the glacier. Moraine types include the **lateral moraine** (a moraine that occurs along the side of a glacier), **medial moraine** (a moraine that occurs in the center of a glacier where two or more lateral moraines merge), **recessional moraine** (a moraine that forms where the toe of a glacier pauses as it is retreating upslope), and **terminal moraine** (a moraine that marks the farthest advance of the glacier). These moraine types are shown in **Figure 18.25**.

Moraines provide important information about the history of glaciation in a region and the history of an individual glacier. The extent of glacial recession can be ascertained by measuring the distance of the glacier's toe from the glacier's terminal moraine. When the age of the terminal moraine is known, the average annual rate of glacier recession can also be determined.

Deposits by Ice Sheets

We saw in Section 7.2 that Milankovitch cycles and the resulting orbital forcing cause glacial and interglacial cycles. The most recent glacial period, which is called the *Wisconsin glaciation*, ended about 12,000 years ago. During the Wisconsin glaciation, large ice sheets covered the high latitudes of the Northern Hemisphere **(Figure 18.26)**. These ice sheets transformed the landscapes as they scoured and deposited enormous amounts of glacial drift.

Many areas of northern North America today are covered by nearly continuous layers of glacial sediments deposited by the Laurentide ice sheet. The sediments were deposited unevenly, resulting in an undulating, hummocky, and mounded surface called a *ground moraine*. In some places, these glacial deposits were molded into identifiable landforms by the moving ice sheet or by meltwater streams flowing beneath the glacier, or by a combination of ice and streams. Ice sheets produce many of the same kinds of depositional features that alpine glaciers

Figure 18.25 Moraines. (A) The many different types of moraines are illustrated in this diagram. (B) An outlet stream from the Piedras Blancas Glacier in Patagonia, in southern Argentina, has cut a V-shaped notch into the terminal moraine at the base of the glacier. The terminal moraine has formed a natural dam behind which a tarn has formed. A lateral moraine and a recessional moraine are also visible in this photo. *(© Michael Schwab)*

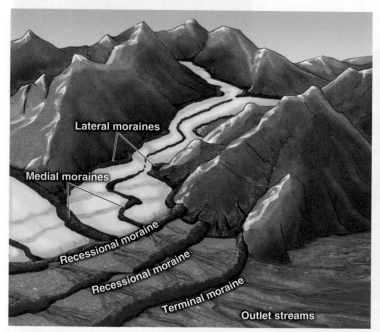

A

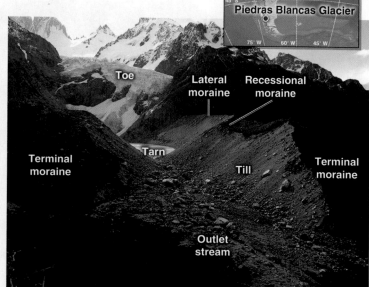

B

create, such as terminal and recessional moraines. They also create landforms that are unique to ice sheets. **Drumlins** (Irish Gaelic for "hills"), for example, are elongated hills composed of till that a moving ice sheet deposits **(Figure 18.27)**.

Another common landform deposited by ice sheets is an **esker** (meaning "ridge" in Irish Gaelic), a long ridge of sorted sand and gravel deposited by a subglacial stream. Eskers may run continuously for tens of kilometers in length parallel to the direction of ice sheet movement **(Figure 18.28** on the following page**)**.

Many areas across North America consist of rounded hills called *kames*, which are mounded accumulations of glaciofluvial sediments. Although the formation of kames is not well understood, scientists suspect that supraglacial streams and wind transported and deposited sediments on top of the melting ice sheet. After the ice melted, the sediments settled in mounds composed of sand and gravel.

Subglacial streams flowing beneath the Laurentide ice sheet deposited sediments on flat outwash plains where they exited the glacier. Large stagnant blocks of ice that broke from the retreating ice sheet were buried in these sediments. When the blocks eventually melted, they formed depressions called *kettle holes*. When kames and kettle holes form in the same area, the result is **kame-and-kettle topography**, a landscape dominated by irregular mounds and shallow depressions or lakes. **Figure 18.29** on the following page illustrates the formation of kame-and-kettle topography.

Figure 18.26 Ice sheets of the Wisconsin glaciation. Shown here at their maximum extent 20,000 years ago, the Laurentide ice sheet and the Cordilleran ice sheet covered almost all of Canada and the northernmost United States. Northern Eurasia was covered by the Scandinavian ice sheet and the Siberian ice sheet. These ice sheets were up to 3 km (2 mi) thick. Because sea level was lower during the Wisconsin glaciation, North America and Eurasia were connected by a corridor of dry land.

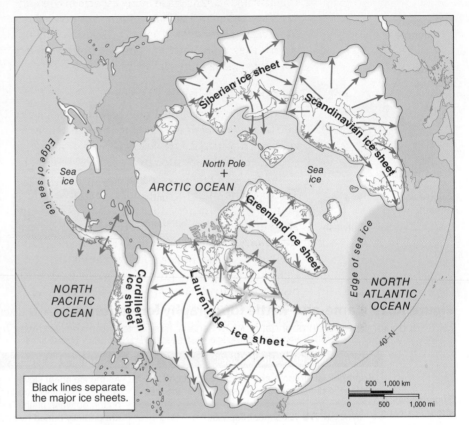

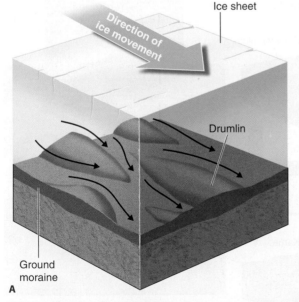

Figure 18.27 Drumlins. (A) Drumlins are hills composed mostly of clays and are no higher than 50 m (160 ft) and a few hundred meters in length. They are formed as an ice sheet advances. The tapering end of a drumlin points in the direction the ice was advancing. (B) *Drumlin fields*, such as this one in Dane County in southern Wisconsin, cover portions of southern Canada and the northern United States. Contour plowing and cropping, following the natural shape of the drumlin, reduce erosion. The ice sheet movement direction was toward the viewer. *(Kevin Horan/Getty Images)*

Figure 18.28 Eskers. (A) Subglacial streams flow in tunnels beneath an ice sheet, depositing sand and gravel in their channels. After the ice sheet melts, these glacial sediments create eskers. Eskers are up to 50 m (160 ft) in height and kilometers in length. (B) This esker is located near Dahlen, North Dakota. *(All Canada Photos/Alamy)*

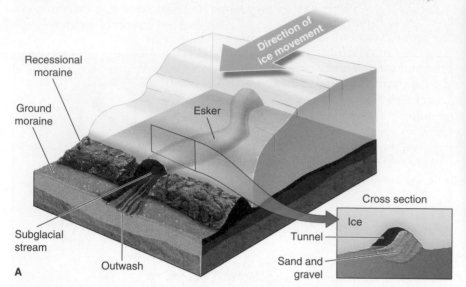

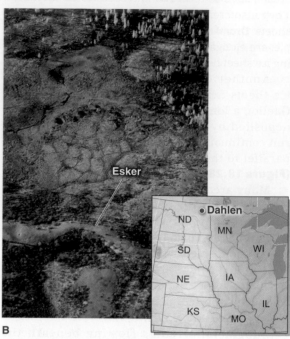

Figure 18.29 Kame-and-kettle topography formation. (A) Kame-and-kettle topography forms as an ice sheet melts. (B) Kame-and-kettle topography is found throughout much of Canada. This photo is from the Northwest Territories. *(Thomas and Pat Leeson/Science Source)*

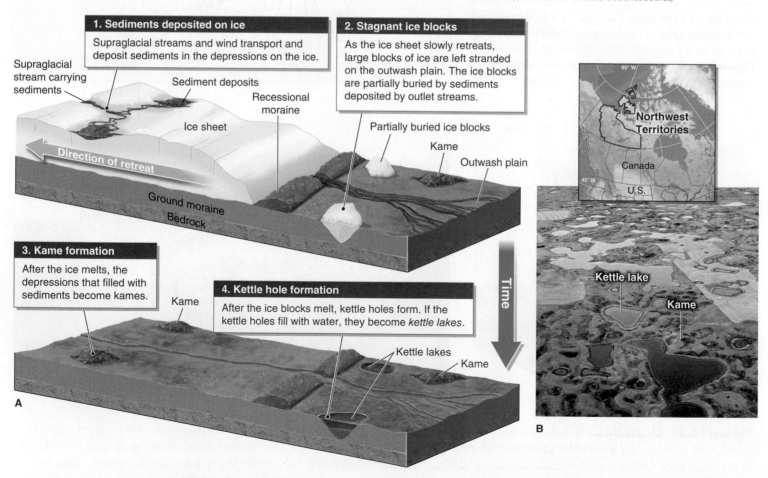

In addition to creating kame-and-kettle topography, the retreat of the Laurentide ice sheet left a series of recessional moraines across much of the upper Midwest of the United States and the province of Ontario in Canada. From the ground, these recessional moraines can be difficult to see because they are very large and flat. **Figure 18.30** maps their extent.

Glacial Dust: Loess

About 10% of the surface of the continents, mostly at midlatitudes, is covered by loess deposits. **Loess** (pronounced LUHSS) is made up of wind-deposited dust that often originates as glacial flour deposited on glacial outwash plains. Electrical charges on these tiny silt and clay particles cause the fine dust to stick together where it settles, forming large accumulations of loess. Deposits of loess are typically unstratified (lacking horizontal layers).

Many loess deposits were formed by processes that are no longer active. The Pleistocene ice sheets and mountain glaciers in North America and Scandinavia formed loess accumulations when summer meltwater deposited silt and clay onto the outwash plain. Cold winters reduced the stream flow, exposing the sediments to strong katabatic winds (see Section 5.3). These winds picked up and transported the fine sediments and deposited them in loess accumulations.

About 30% of the United States is covered by loess deposits, and eastern Europe has extensive

Figure 18.30 North American recessional moraines. The many recessional moraines of the Great Lakes region are shown here in gray. The gray moraines mapped in the darker blue area were formed by the Laurentide ice sheet some 15,000 years ago. Moraines in southern Illinois, Indiana, and Ohio (in darker shades of blue) are from the previous glacial period, called the *Illinoian glaciation*, that occurred some 150,000 years ago. Southwestern Wisconsin does not have glacial drift (or glaciofluvial deposits) and is called the "driftless area." Many scientists think the glaciofluvial deposits in the driftless area were washed away long ago in a large flood event caused by the sudden draining of one of the ancient lakes in the area.

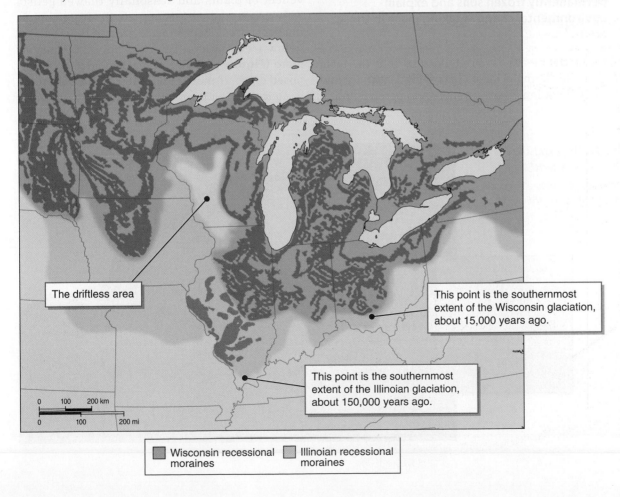

The driftless area

This point is the southernmost extent of the Wisconsin glaciation, about 15,000 years ago.

This point is the southernmost extent of the Illinoian glaciation, about 150,000 years ago.

0   100   200 km
0   100   200 mi

Wisconsin recessional moraines   Illinoian recessional moraines

**Figure 18.31
Global loess
deposits.** Loess
deposits are found mostly
at midlatitudes. The inset
photo shows a roadcut
through the Loess Hills in
the Missouri River valley
in western Iowa. Nesting
bank swallows excavated
the holes. (*Riparia riparia*).
(*© Lee Rentz/NHPA/Photoshot*)

Loess deposits

loess deposits, as shown in **Figure 18.31.** Not all
loess is formed by glacial processes. In some areas,
such as central China, loess was formed by windblown dust originating in deserts rather than glacial outwash plains.

18.4 Frozen Ground:
Periglacial Environments

◎ **Identify features unique to areas with
permanently frozen soils and explain
environmental changes taking place in those
areas.**

Periglacial means "around glaciers," but the term
refers to all unglaciated areas at high latitudes and
high elevations that are subject to persistent and

intense freezing. Like glaciers, periglacial environments are some of the most rapidly changing
physical systems on Earth as a result of climate
change. As they change, they are also capable of
creating significant climate feedbacks that could
further change the global climate system.

Permafrost

In many regions of Alaska, northern Canada,
and Siberia, the ground just beneath the surface
veneer of plants and seasonally thawed gelisol
soils (see Section 11.1) is perpetually frozen, forming permafrost. **Permafrost** is ground that remains
below freezing continuously for two years or
more (**Figure 18.32**). Not all permafrost is composed of ice. If there is no water in the soil, there
is no ice—only unconsolidated rock fragments at

Figure 18.32 Permafrost. This exposure of
permafrost is near Cherskii, Russia, in eastern Siberia.
Note the person in the foreground for scale. (*© Katey M.
Walter Anthony, University of Alaska Fairbanks (under NSF #0099113)*)

Surface veneer
of vegetation

Permafrost

a temperature below freezing. Loose frozen sand, for example, is permafrost.

Although permafrost, by definition, remains frozen year-round, in most regions the topmost portion of the ground—called the **active layer**—thaws each summer and refreezes in fall. The depth of the active layer decreases farther north and at higher elevations in mountainous areas. When the active layer thaws, ice turns to liquid water, forming lakes and shallow wetlands called *bogs* **(Figure 18.33A)**. The boreal forest and northern tundra biomes (see Section 9.4) are found in areas with permafrost.

About 25% of the soils in the Northern Hemisphere are permafrost. At high latitudes, permafrost is typically unbroken, or *continuous*. Farther south, permafrost may be *isolated, sporadic,* or *discontinuous* **(Figure 18.33B)**.

Most permafrost is found at 60 degrees north latitude and higher, but it can be found at all latitudes at high elevations. Near the equator, permafrost is found at elevations higher than 5,000 m (16,400 ft). Mount Kilimanjaro in Kenya, for example, located at 3 degrees south latitude, has discontinuous permafrost soil. The thickest permafrost is found in northeastern Russia. The permafrost there is 1,650 m (5,445 ft) thick and over a half million years old.

The Southern Hemisphere has relatively little permafrost due to its lack of land at high latitudes. Only 0.3% of Antarctica's land is not covered by the Antarctic ice sheet, and all of that exposed land is permafrost. The Patagonian ice fields in South America and the highlands of the Southern Alps in New Zealand have discontinuous permafrost.

As discussed in Section 7.4, the cryosphere is changing rapidly in response to warming of the atmosphere. Changes are rapidly occurring in periglacial environments, particularly in the active layer, as ground temperatures increase (see the **Picture This** feature on page 574).

Periglacial Features

Several landforms and phenomena are found only in periglacial environments. Two periglacial landforms are pingos and patterned ground. A *pingo* is a hill with a core of ice **(Figure 18.34A** on page 575**)**. A pingo forms as liquid water from below is forced up through a layer of permafrost. As the water pushes through the permafrost, it forms a mound. Pingos grow at a rate of a few centimeters per year. *Patterned ground* is formed as freeze–thaw cycles slowly wedge soil apart and, over time, develop polygon shapes **(Figure 18.34B** on page 575**)**.

Figure 18.33 Periglacial features. (A) Continuous permafrost occurs where the permafrost is 100 m (330 ft) or more thick. Permafrost that is thinner and interrupted by unfrozen ground, called *talik*, is discontinuous permafrost. (B) Permafrost is widely distributed throughout the Northern Hemisphere. This map shows discontinuous and continuous permafrost areas as well as areas with sporadic and isolated permafrost.

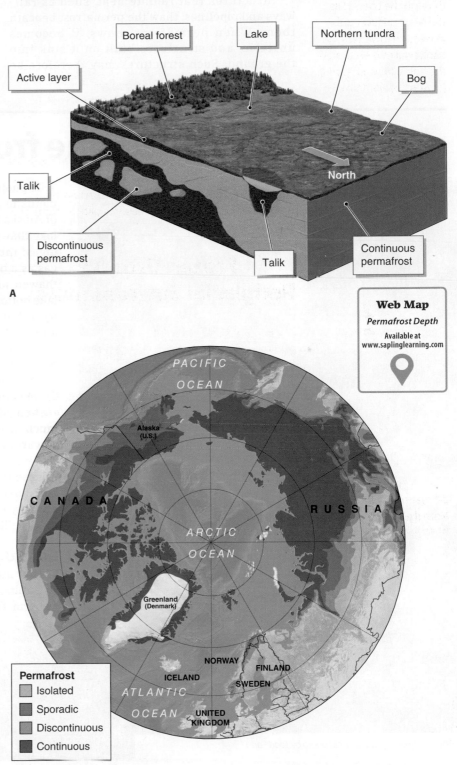

Web Map

Permafrost Depth

Available at
www.saplinglearning.com

Trees growing on the active layer above permafrost are shallow-rooted and unstable because their roots cannot penetrate the permafrost just below. When the active layer deepens during particularly warm summers, the trees may become tilted, creating a *drunken forest* **(Figure 18.35)**.

Structures that radiate heat, such as railways and pipelines, thaw the permafrost beneath them. When permafrost thaws, it becomes unstable, and structures built on it sink into the ground. Such structures may therefore be built elevated above the ground to avoid these problems.

The 1,300 km (800 mi) Trans-Alaska Pipeline moves heated oil from the North Slope of Alaska all the way to the Pacific coast. Where it passes over permafrost, the pipeline has been elevated about 2 m (6.5 ft) above the ground to prevent melting of the permafrost, which would cause the pipe to rupture (see Figure 9.26). China's Qinghai–Tibet railway is also engineered for permafrost conditions **(Figure 18.36** on page 576**)**.

Picture This Methane from Permafrost

(University of Alaska Fairbanks photo by Todd Paris)

AK
● **Fairbanks**

Gulf of Alaska

Canada

45° N *PACIFIC OCEAN*

150° W   135° W

This photo shows researcher Katey Walter Anthony lighting methane gas emerging from a frozen lake on the University of Alaska Fairbanks campus. As the permafrost beneath the lake thaws, anaerobic microorganisms (called *methanogens*) digest the organic carbon in the soils, producing methane gas as a by-product. Carbon dioxide is also released as the thawed soils decompose. The methane bubbles up from the lake bottom and is trapped beneath the frozen lake's cover of ice. If the ice is drilled through and the gas ignited, it forms a fireball. This process of permafrost thawing and *methanogenesis* (methane generation) is happening throughout the Northern Hemisphere's permafrost soils.

Two observations are becoming increasingly concerning for climate scientists. First, there are some 1.7 trillion metric tons of carbon stored in northern permafrost—twice as much as is currently in the atmosphere. Second, the Arctic is warming about twice as fast as the global average, and this warming has the potential to thaw the permafrost and release the stored carbon in the form of methane and carbon dioxide. Scientists are concerned that permafrost thawing may create a positive feedback loop (see Section 1.3) that could accelerate the warming trend already under way.

Consider This

1. How are Arctic lakes becoming sources of methane?
2. Outline the positive feedback cycle that could result from permafrost methane production.

Figure 18.34 Pingo and patterned ground. (A) Pingos can reach 50 m (160 ft) in height and range from the size of a car to the size of a hill, like the one shown here in the Mackenzie River delta in Canada. (B) This patterned ground is in the Farnell Valley, Antarctica. *(A. Jason Pineau/ Getty Images; B. Maria Stenzel/National Geographic Creative)*

Figure 18.35 Drunken forest. This drunken forest is near Fairbanks, Alaska. *(Tingjun Zhang, College of Earth and Environmental Science, Lanzhou University, China)*

Figure 18.36 Engineering for permafrost. The Qinghai–Tibet railway, completed in 2006, has the highest elevation of any railway in the world. It is 1,956 km (1,215 mi) long and connects Xining, Qinghai Province, to Lhasa, Tibet Autonomous Region, China. Much of it is built on permafrost. If ice thaws beneath the tracks, the tracks will move, and such movements could derail a high-speed train. To prevent this, the tracks are raised above the permafrost, and in places vertical pipes circulate liquid nitrogen beneath the pilings to keep the soils around them frozen. *(View Stock/Getty Images)*

GEOGRAPHIC PERSPECTIVES
18.5 Polar Ice Sheets and Rising Seas

◎ **Assess the role of polar ice sheets in current and future sea-level changes and explain why sea-level rise presents problems for human society.**

One of the most dire aspects of anthropogenic climate change is sea-level rise. Many hundreds of millions of people live at or near sea level, and rising seas threaten to displace them and create significant geopolitical strife in the process.

Warming of the atmosphere is quickly changing Earth's cryosphere. The cryosphere, as we saw in Chapter 7, is fundamentally significant for two reasons: Ice–albedo positive feedbacks can enhance the warming trend that is already under way, and melting glaciers drain water into the oceans and cause sea level to rise. Low-lying coastal and island populations are vulnerable to changes in sea level.

Alpine glaciers collectively contain enough water to raise sea level by 0.45 m (1.48 ft) if they were to melt entirely. But this amount of water pales in comparison to that contained in the polar ice sheets, as **Table 18.1** on the following page shows. Together, the two ice sheets have enough water in them to cause 80 m (262 ft) of sea-level rise.

Table 18.1   Global Ice Volume

ICE LOCATION	APPROXIMATE PERCENTAGE OF CRYOSPHERE	VOLUME (KM³)	POTENTIAL SEA-LEVEL RISE (M)	POTENTIAL SEA-LEVEL RISE (FT)
Antarctica	91%	29,528,300	73.32	240.54
Greenland	8%	2,620,000	6.55	21.49
Alpine glaciers and all other ice	1%	180,000	0.45	1.48
Total	100%	32,328,300	80.32	263.51

Data from http://pubs.usgs.gov/fs/fs2-00/.

Given the amount of ice contained in the ice sheets of Greenland and Antarctica and the rise in global sea level they are capable of causing, scientists closely monitor them for signs of change. Even small changes have a major influence on global sea level. Unfortunately, Earth's two ice sheets are losing ice because of anthropogenic climate change.

The Greenland Ice Sheet

Video

Greenland Ice Loss 2002–2016

Available at www.saplinglearning.com

Greenland is shedding more ice into the ocean through melting and calving where outlet glaciers reach the ocean than it is accruing through snowfall. In other words, the Greenland ice sheet has a negative mass balance. Data from NASA's *GRACE* satellite indicate that Greenland is losing 286 Gt/yr (gigatons per year) of ice. This is equivalent to 286 km³ (68 mi³) of ice each year. The Greenland ice sheet mass is decreasing for two reasons: Its surface is melting and its outlet glaciers flowing to the ocean are speeding up and calving more ice.

Greenland's Melting Surface

Greenland's albedo has decreased in some areas by as much as 15% relative to the 2000–2009 average. There are three reasons for this decrease in albedo. First, warmer summer temperatures on the ice sheet are partially melting snow. When snow refreezes, it clumps together, forming rounded crystals of ice rather than angular ice crystals of snow. These rounded clumps are less reflective and absorb more sunlight than unmelted snow. Second, natural pink bacterial blooms on the ice surface (called "watermelon ice") reduce the albedo as well. Finally, and perhaps most importantly, dark soot from industrial emissions and fires from across the Northern Hemisphere settles on and darkens the ice surface (**Figure 18.37**).

Figure 18.37 Greenland darkening.

(A) This map shows changes in Greenland's albedo relative to the 2000–2009 average. Greenland's albedo has decreased by as much as 15% along the exterior margins (purple areas). (B) From the air, the dirty ice surface is clearly visible (top photo taken near Qaanaaq, Greenland, population 656). The dark soot (and bacteria) covering the ice heats up more than the surrounding ice because it has a lower albedo. As a result, it melts down into the ice, forming pools of water filled with black sludge called *cryoconite* (bottom photo). The larger cryoconite pools shown here are about 20 cm (8 in) in diameter. *(A. Arctic Program/NOAA; B: Top. Kyodo News/Getty Images; bottom. The Asahi Shimbun/Getty Images)*

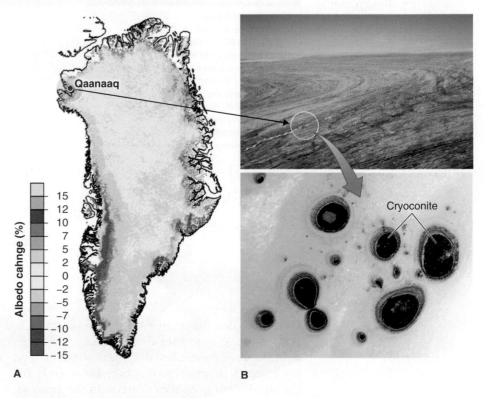

Greenland's Outlet Glaciers

The Greenland ice sheet is also losing mass because outlet glaciers have more than doubled their rate of flow to the ocean during the past few decades. Scientists are not sure why this is happening but suspect that supraglacial lakes and moulins that form during the summer snowmelt could play an important role **(Figure 18.38)**. A **moulin**, a vertical shaft in a glacier through which meltwater flows, may allow meltwater from a supraglacial lake to reach the base of the glacier.

The Antarctic Ice Sheet

Until recently, most scientists assumed that it would take centuries for Antarctica's climate and ice to begin responding to the warming of the atmosphere. But changes to ice in Antarctica are happening as fast as or even faster than the changes occurring in Greenland.

Antarctica's Ice Shelves

Outlet glaciers that feed into the sea from the main ice sheets of Antarctica form floating ice shelves that are 350 m (1,150 ft) thick and thicker. The ice shelves are weakened and break apart as they are thinned by warm ocean water circulating beneath. Summer meltwater lakes also form on their surface and bore holes through the ice as moulins form, further weakening the shelves.

These ancient ice shelves appear to be sensitive indicators of climate change; several of them have broken apart in the past several years. The Antarctic Peninsula has (or had) several ice shelves collectively named the Larsen ice shelf. In a series of dramatic events between 1989 and 1995, the 4,000-year-old Larsen A ice shelf completely disintegrated. Scientists were surprised to see that almost the entire Larsen B shelf broke apart in just a few months in 2002. Larsen B was 10,000 years old. Then, in early 2018 and equally alarming, about 10% of the Larsen C ice broke away, calving one of the largest icebergs ever recorded. The immense iceberg was 350 m (1,150 ft) thick and 5,800 km² (2,200 mi²) in extent—about the size of the state of Delaware **(Figure 18.39)**.

Video

Greenland's Thinning Ice

Available at www.saplinglearning.com

Figure 18.38 Influences on the flow of a Greenland outlet glacier. *(1. The Asahi Shimbun/Getty Images; 2. © Konrad Steffen)*

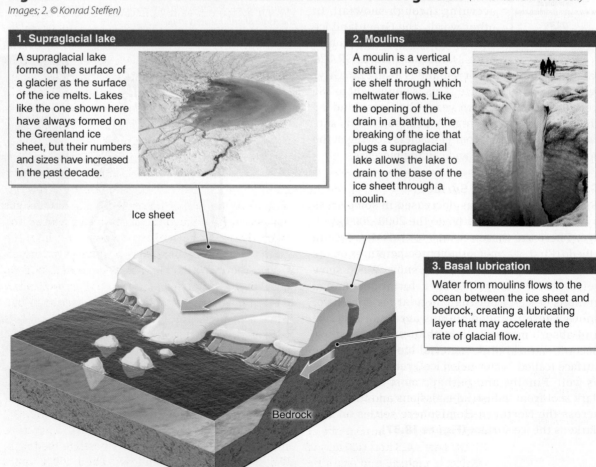

1. Supraglacial lake

A supraglacial lake forms on the surface of a glacier as the surface of the ice melts. Lakes like the one shown here have always formed on the Greenland ice sheet, but their numbers and sizes have increased in the past decade.

2. Moulins

A moulin is a vertical shaft in an ice sheet or ice shelf through which meltwater flows. Like the opening of the drain in a bathtub, the breaking of the ice that plugs a supraglacial lake allows the lake to drain to the base of the ice sheet through a moulin.

3. Basal lubrication

Water from moulins flows to the ocean between the ice sheet and bedrock, creating a lubricating layer that may accelerate the rate of glacial flow.

Ice sheet

Bedrock

Figure 18.39 Antarctic ice shelves and their disintegration. (A) Ten large ice shelves flank the West and East Antarctic ice sheets. (B) The extent of the former Larsen A and B ice shelves is shown here. Following in the pattern of the Larsen A and Larsen B ice shelves, the process of disintegration appears to have begun for the Larsen C ice shelf with the large portion that broke off in 2018.

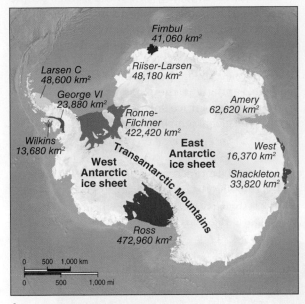

A

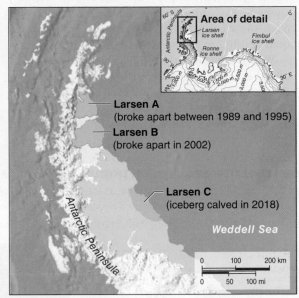

B

Antarctica's Outlet Glaciers

Because ice shelves are already in the ocean, they do not cause a significant change in sea level when they break apart. However, ice shelf disintegration does have an impact: Once an ice shelf is gone, the outlet glaciers behind it flow to the sea faster because they are no longer slowed by the buttressing effects of the ice shelf. The glaciers that feed the Larsen A and Larsen B ice shelves, for example, began flowing 300% faster into the ocean after the ice shelves were gone.

Furthermore, the rate of flow from several outlet glaciers on the West Antarctic ice sheet, including the Smith, Pope, and Kohler glaciers, has increased by as much as six times since monitoring of them began in 1996. The flow of other outlet glaciers in West Antarctica is also accelerating. The base of the ice sheet and these outlet glaciers lies below sea level. Scientists suspect that increasingly warmer ocean water is circulating beneath and melting the undersides of the glaciers and allowing them to become unstuck from the bedrock, accelerating their rate of flow.

Video

Antarctic Ice Loss 2002–2016

Available at www.saplinglearning.com

NASA's *GRACE* satellite shows that Antarctica as a whole is losing 127 Gt/yr of ice, or 127 km³ (30 mi³). Together, Greenland and Antarctica are now shedding ice at an annual rate of 413 Gt. That's 413 km³ (100 mi³) of ice that is melting and draining into the oceans each year.

Sea Level on the Rise

Much as ice cubes dropped into a glass of water raise the water level in the glass, as outlet glaciers flow into the ocean, they displace seawater and raise sea level. The faster these outlet glaciers flow into the ocean, the more sea-level rise they cause.

Estimates of how much sea level will rise in the coming decades vary widely. According to the IPCC (Intergovernmental Panel on Climate Change) *Fifth Assessment Report* (see Section 7.4), by the year 2100 sea level will likely be 1 to 2 m (3.3 to 6.6 ft) higher than it is today. There are many factors that complicate these estimates, including the rate of increase in greenhouse gas emissions, how much the ocean water warms and rises through thermal expansion, and how the giant ice sheets respond to atmospheric warming.

Predicting what the polar ice sheets will do is difficult because their behavior is nonlinear, meaning that the rate at which they change can change. For example, the surface of the ice sheets will begin melting if ice-sheet thinning causes the elevation of the ice sheets to decrease, leaving the sheets exposed to warmer air. Surface melting would lower their elevation even more, triggering a positive feedback. Estimates based on computer models indicate that warming of about 1.6°C (2.9°F) above the global preindustrial temperature could be enough to trigger this positive feedback. If that happened, it would take about 2,000 years to melt both ice sheets entirely.

Between 1900 and 2000, sea level rose by about 23 cm (9 in), at an average rate of about 2 mm (0.08 in) per year. This rate has recently accelerated, and now sea level is rising at 3.2 mm (0.13 in) per year **(Figure 18.40A)**. By the end of this century, however, the annual rate of sea-level rise could reach some 6 mm (0.24 in) per year, and sea level could be 1 to 2 m (3.3 to 6.6 ft) higher than it is today.

Worldwide, more than 100 million people live in coastal areas at elevations within 1 m (3.3 ft) of the high-tide line **(Figure 18.40B)**. In the United States, 3.7 million people live within 1 m of the high-tide line. Worldwide, 634 million people live within

10 m (33 ft) of sea level. Sea level isn't going to rise 10 m in the foreseeable future, but a 1-meter rise in sea level would have major destabilizing social and economic effects on these coastal populations.

In addition, currently 50% of the world's people live in cities. By 2100, the world's urban population will increase to about 85%. Many of the world's largest cities are coastal, and the number of people vulnerable to sea-level rise will certainly increase.

Given their sheer size and their potential to raise sea level, and given their potential sensitivity to warming, the polar ice sheets will be closely monitored as the global temperature continues to climb.

Figure 18.40 Global sea-level rise and coastal populations. (A) This graph shows the rate of sea-level rise since 1990, as measured by satellites. (B) Low-lying populated areas are the first to face the challenges of sea-level rise. The Maldives, for example, is a nation built on coral atolls that are only a few feet above high tide. The capital city of Malé, population 133,412, is shown here. Malé is one of the most densely populated cities in the world. As sea level continues rising, inhabiting low islands like this will become impossible. *(Levente Bodo/Alamy)*

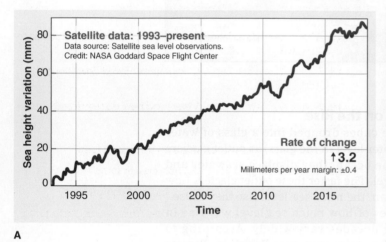

A

B

Chapter 18 **Study Guide**

Focus Points

18.1 About Glaciers

- **Glacier formation:** Glaciers form through snow accumulation. Snow is compressed by gravity to form granular snow, then firn, and then glacial ice.
- **Glacier movement:** Glaciers move downslope at an average rate of a few centimeters per day through basal sliding and plastic deformation. At times, they move faster in glacial surges.
- **Toe position:** The position of a glacier's toe largely reflects the mass balance of the glacier.
- **Glacial retreat:** Most alpine glaciers are retreating as a result of atmospheric warming.
- **Types of glaciers:** Glaciers can be categorized as alpine glaciers or ice sheets.

18.2 Carving by Ice: Glacial Erosion

- **Glacial erosion:** Glaciers are important agents of erosion. They cut into bedrock by plucking it and abrading it.
- **Glacial transport:** Glaciers transport rock fragments downslope by means of ice flow and subglacial streams.
- **Glacial landforms:** Glaciers make many unique landforms, including cirques, tarns, horns, and U-shaped valleys.

18.3 Building by Ice: Glacial Deposition

- **Glaciers and streams:** Glaciers, like streams, deposit sediments where they stop flowing.
- **Moraines:** Glaciers deposit till to form moraines.
- **Ice sheet deposits:** Much of northern North America is covered by glacial sediments deposited by ice sheets of the Wisconsin glaciation.
- **Loess:** Glacial sediments on outwash plains are an important source of loess deposits.

18.4 Frozen Ground: Periglacial Environments

- **Permafrost:** Permafrost covers about 25% of the Northern Hemisphere.
- **Release of carbon:** Permafrost is thawing in response to climate change, releasing methane and carbon dioxide into the atmosphere.

18.5 Geographic Perspectives: Polar Ice Sheets and Rising Seas

- **Distribution of ice:** About 99% of the cryosphere is found in the ice sheets of Greenland and Antarctica.
- **Ice loss:** Greenland and Antarctica are losing about 286 billion tons of ice each year as a result of melting and the increased rate of outlet glacier flow to the sea.

Key Terms

Concept Review

The Human Sphere: The Mammoth Hunters

1. About when did the mainland mammoth populations in North America and Eurasia disappear? Why did the mammoth go extinct?

2. Why do people seek the remains of mammoths preserved in frozen ground?

18.1 About Glaciers

3. What is glacial drift? Give examples of drift.

4. What is a glacier? Where are glaciers found?

5. What are the steps in glacier formation? What two requirements must be met before a glacier can begin to form?

6. By what two processes do glaciers move?

7. What is glacier mass balance? Compare and contrast the accumulation zone, the ablation zone, and the equilibrium line.

8. What is a glacier's toe? Why does it move upslope and downslope?

9. What does it mean to say that a glacier is *in retreat*? What causes a glacier to retreat?

10. Compare and contrast the geographic settings of an alpine glacier and an ice sheet. How do their sizes compare?

11. What are the four types of alpine glaciers? Describe where each occurs.

12. What are the names of the two ice sheets? Which is larger?

13. What is an iceberg? How does one form?

18.2 Carving by Ice: Glacial Erosion

14. What effect does glacial erosion have on mountain height?

15. What are plucking and abrasion? How do these processes allow glacial ice to carve into hard bedrock?

16. What are glacial striations and glacial polish?

17. What is glacial flour? What effect does it have on the appearance of lake water?

18. By what two means are rock fragments moved downslope by glaciers?

19. What is an outwash plain? Where and how do they form?

20. What is a glaciofluvial process?

21. What is a proglacial lake? Where do these lakes form?

22. Describe a cirque and explain how cirques form.

23. What are arêtes, horns, and cols, and how do they form?

24. Describe the shape of a glacial valley. What kind of glacier forms glacial valleys?

25. Compare a trunk glacier with a tributary glacier. Describe how a hanging valley forms from these glaciers.

26. What are paternoster lakes, and how do they form?

27. What is a fjord? Why are there no fjords at middle and low latitudes?

18.3 Building by Ice: Glacial Deposition

28. Why are stream deposits sorted but glacial deposits unsorted?

29. What is till? What is the relationship of till to moraines?

30. Differentiate between lateral, medial, terminal, and recessional moraines.

31. Describe the formation of drumlins, eskers, and kettle holes in the context of ice sheet deposits.

32. What is loess, and how does it form?

18.4 Frozen Ground: Periglacial Environments

33. What does *periglacial* mean? Are there glaciers in periglacial environments?

34. What is permafrost, and how is it defined? Where does it occur?

35. Why are many Arctic lakes emitting methane?

36. What are pingos? What is patterned ground? Explain how they form.

37. Explain why drunken forests occur. Why might rail lines and pipelines sink into permafrost? How do engineers prevent sinking from happening?

18.5 Geographic Perspectives: Polar Ice Sheets and Rising Seas

38. On what continent is most of the cryosphere's ice stored? As a percentage, how much of the cryosphere is on that continent?

39. Describe Greenland's loss of ice in terms of changing ice sheet albedo and outlet glacier flow rates.

40. What is a supraglacial lake? How is it related to the speed of outflow glaciers on Greenland?

41. What happens to the speed of an outlet glacier when an ice shelf that supports it breaks apart?

42. What large ice shelves have recently broken apart? What ice shelf appears to be in the process of breaking apart?

43. How much has sea level risen in the past century? How much could sea level rise by the end of this century?

44. How many people worldwide live in coastal areas within 1 m of sea level?

Critical-Thinking Questions

1. If you are hiking outdoors and see a large, isolated boulder, what information would you need to determine if the boulder is a glacial erratic?

2. Why are the toes of some glaciers advancing rather than retreating upslope, even though the atmospheric temperature over those glaciers is increasing?

3. How would a 1 m rise in sea level affect you where you live? What information should you find to answer this question? How would you find that information?

4. What does it mean to say that ice sheets have "nonlinear" behavior in response to warming? Explain how nonlinear behavior works and why its effects are important.

Test Yourself

Take this quiz to test your chapter knowledge.

1. **True or false?** Glacier mass balance refers to the balance between ice gain and ice loss on a glacier.

2. **True or false?** For most glaciers, the zone of equilibrium is moving upslope with global warming.

3. **True or false?** Till is deposited by streams.

4. **True or false?** Melting of an ice shelf in itself does not cause significant sea-level rise.

5. **Multiple choice:** Which of the following is found below a cirque?
   a. tarn   c. col
   b. horn   d. arête

6. **Multiple choice:** Which of the following indicates the maximum downslope position of a glacier in the past?
   a. lateral moraine   c. terminal moraine
   b. medial moraine   d. recessional moraine

7. **Multiple choice:** Which of the following is produced by a subglacial stream?
   a. kettle hole   c. ground moraine
   b. esker   d. loess

8. **Multiple choice:** Which of the following has the largest volume of ice?
   a. all alpine glaciers combined   c. Antarctica
   b. Greenland   d. Iceland

9. **Fill in the blank:** A(n) _____ forms U-shaped valleys in mountains.

10. **Fill in the blank:** A(n) _____ is a drowned U-shaped valley.

Online Geographic Analysis

Ice Loss in Greenland and Antarctica

In this exercise we examine ice loss at the poles.

Activity

Go to https://climate.nasa.gov/vital-signs/land-ice/. This NASA page shows data of the ice mass balance for the Antarctic ice sheet (top graph) and the Greenland ice sheet (bottom graph).

1. **When did the measurements of the Antarctic ice sheet begin?**

2. **What is the current mass balance of the Antarctic ice sheet (given as the "rate of change")?**

3. **The data graphed are cumulative changes (changes added up over time) in the ice sheet. Hover your cursor over the most recent cumulative measurement on the graph. What is the value?**

4. **Find the lowest cumulative value on the graph. What is it, and when did it occur?**

5. **What is the highest value, and when did it occur?**

6. **Scroll down to the Greenland ice sheet graph and answer the Questions 1 through 5 for the Greenland data.**

Go to https://grace.jpl.nasa.gov. In the search box in the upper right type in "Prolific Earth Gravity." Click on the "Prolific Earth Gravity Satellites End Science Mission" link. Read the information about the satellites and answer the following questions.

7. **The ice sheet mass balance data are generated by the GRACE twin satellites. What does the acronym GRACE stand for?**

8. **Explain briefly how GRACE measures ice volume in the ice sheets.**

9. **What is GPS? (See Chapter 1, "The Geographer's Toolkit.") Why does GRACE rely on GPS for its measurements?**

10. **What other physical phenomena on Earth does GRACE measure?**

Picture This. *Your Turn*

Identifying Glacial Landforms

Fill in the blanks on the diagram, using each of the terms below only once.

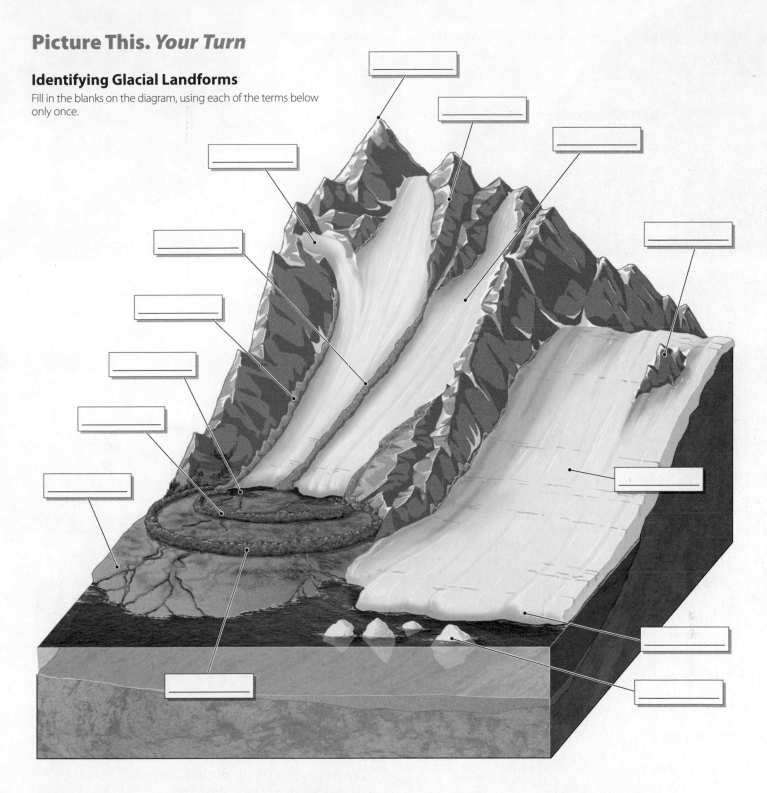

1. Arête
2. Cirque
3. Horn
4. Ice sheet
5. Ice shelf

6. Iceberg
7. Lateral moraine
8. Medial moraine
9. Nunatak
10. Outwash plain

11. Proglacial lake
12. Recessional moraines
13. Terminal moraine
14. Valley glacier

For animations, interactive maps, videos, and more, visit www.saplinglearning.com.   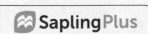 Sapling Plus

Water, Wind, and Time: Desert Landforms

Chapter Outline *and Learning Goals* ◎

19.1 Desert Landforms and Processes

◎ List important examples of desert landforms and describe the processes that form them.

19.2 Desert Landscapes

◎ Explain how various desert landscapes develop.

19.3 Geographic Perspectives: Shrinking Desert Lakes

◎ Assess the effects of water diversions on desert lakes.

These deeply weathered granite boulders are part of the Alabama Hills in eastern California. In the background, the Sierra Nevada's Mount Whitney, the highest point in the lower 48 states, and other peaks tower into the clouds. The Alabama Hills are in the Mojave Desert, a desert formed mostly by the rain shadow effect of the Sierras.

(Bruce Gervais)

To learn more about how deserts form, go to Section 19.1.

THE HUMAN SPHERE Flooding in the Driest Place on Earth

The Atacama Desert in northern Chile is a thin 1,000 km (600 mi) strip of coastal plain known as the driest place on Earth **(Figure 19.1)**. In the heart of the Atacama, no rainfall has ever been recorded, there is no water, and there are no visible signs of life. Nothing grows in the soil, and nothing burrows in the ground or flies in the air. NASA tests rover vehicles in the Atacama because the inhospitable conditions there are comparable to those of Mars.

A "perfect storm" of factors together create the driest place on Earth. First, the Atacama is located in the subtropical high-pressure zone (see Section 5.2). Second, this extremely arid desert lies in the rain shadow created by the Andes (see Section 4.4). Third, the water from the cold Peru (Humboldt) Current circulating up from Antarctica is slow to evaporate, leaving little moisture in the air.

Despite all this, more than a million people live in the Atacama Desert. Antofagasta, a port city in northern Chile, has a population of about 360,000 and is the fourth largest city in Chile.

The city receives a miniscule 4 mm (0.16 in) of rain each year, on average. It has gone as long as 4 years without measurable rainfall. Like other cities in most desert regions, however, Antofagasta is built on a surface created by flowing water. The Atacama is situated at the foot of the Andes, one of the highest mountain ranges in the world. The city gets its water from streams draining out of the Andes, and Antofagasta also has plans to build a desalination plant to filter water from the ocean.

Paradoxically, the greatest threat to Antofagasta is not too little water but too much water. On rare occasions during strong El Niño years, flooding rains fill the normally dry canyons in the Andes foothills to the east of the city with raging torrents of water. The bare hills have little to no vegetation to anchor soils, and the canyons turn into raging slurries of mud that race down onto the alluvial fans on which the city is built. Such flooding has happened eight times in the city's history. The most recent flood event occurred in 2015, when as many as 150 people died.

Figure 19.1 Atacama Desert.

The Atacama Desert is sandwiched between the Andes and the Pacific Ocean in northern Chile. The photo on the right is of the Valley of the Moon in the heart of the Atacama. *(Giulio Ercolani/ Alamy)*

19.1 Desert Landforms and Processes

◎ List important examples of desert landforms and describe the processes that form them.

We usually think of deserts as hot, sandy places. For many desert regions, that image is accurate. But some deserts are cold and rocky, and the polar deserts are covered by ice sheets. What exactly is a desert? All deserts have severe moisture deficits for most or all of the year (see Section 9.4). Generally speaking, **deserts** occur where less than 25 cm (10 in) of precipitation falls annually. More humid regions can also be deserts if evapotranspiration rates are high and moisture loss exceeds precipitation.

Three main factors contribute to an area's aridity: (1) the subtropical high-pressure zone found between about 20 and 30 degrees latitude north and south (see Section 5.2); (2) rain shadows on the leeward sides of major mountain ranges (see Section 4.4); and (3) inland locations at high latitudes or high elevations where the air is cold and its water vapor content is low. Between 20 and 30 degrees latitude, cold ocean currents may also contribute to aridity, as is the case for Antofagasta. The most geographically widespread cause of aridity is the influence of the subtropical high **(Figure 19.2)**.

Because of their aridity, all deserts have sparse plant cover and low biomass. As a result, no other environment on Earth is as exposed to the erosive effects of flowing water and wind, except perhaps coastal areas, as we will see in Chapter 20, "The Work of Waves: Coastal Landforms." We explored life in the desert biome in Section 9.4. In this chapter, we focus on the processes of desert geomorphology.

Weathering in the Desert

Physical weathering (see Section 16.1) happens more quickly than chemical weathering in deserts. Chemical weathering requires the presence

Figure 19.2 World map of deserts. The Köppen climate system (see Section 9.1 and Appendix V, "The Köppen Climate Types") classifies arid and semiarid regions as BWh and BWk climates. Arid climates are found on all continents. Deserts of 500,000 km² (193,000 mi²) or greater, excluding those of Antarctica and the Arctic, are included in this table. The table also lists the main cause of the aridity in each desert.

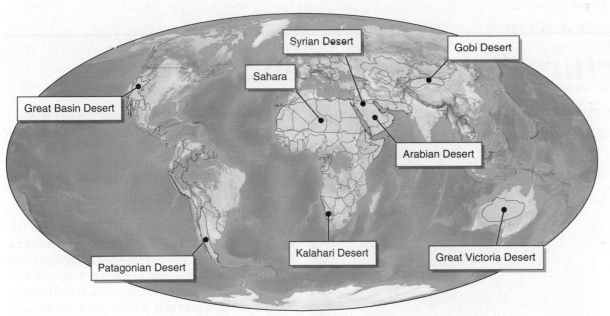

Web Map

Deserts of the World

Available at
www.saplinglearning.com

	AREA IN KM²	AREA IN MI²	MAIN CAUSE OF ARIDITY
Subtropical hot deserts			
Sahara (Africa)	9,100,000	3,500,000	Subtropical high
Arabian Desert (Middle East)	2,600,000	1,000,000	Subtropical high
Great Victoria Desert (Australia)	647,000	250,000	Subtropical high
Kalahari Desert (Africa)	570,000	220,000	Subtropical high
Syrian Desert (Middle East)	500,000	193,000	Subtropical high
Midlatitude cold deserts			
Gobi Desert (Asia)	1,300,000	500,000	Cold interior location
Patagonian Desert (South America)	670,000	260,000	Andes rain shadow
Great Basin Desert (North America)	500,000	193,000	Sierra Nevada rain shadow

of water, and water is only intermittently present in deserts. Chemical weathering therefore occurs slowly in deserts.

Chemical weathering is nonetheless important in deserts. When water from rainfall or dew enters a rock, it gradually dissolves minerals within the rock. In sandstone, for example, calcite is often the substance that holds quartz grains together. Quartz resists chemical weathering, but calcite does not. When calcite is dissolved through chemical weathering, the rock disintegrates and crumbles, leaving loose quartz grains. These grains can then be carried away by wind and water and deposited to form quartz sand dunes.

The exposed surfaces of many desert rocks are covered with **rock varnish**: a thin, weathered, and darkened surface formed by the activities of bacteria and the accumulation of windblown clay particles. Rock varnish is formed by a type of biochemical weathering that occurs on rock surfaces almost exclusively in arid and semiarid climates. In more humid climates, rain washes away the clay particles too fast for rock varnish to form.

Scratching the surface of a rock covered in rock varnish removes the thin patina of rock varnish and reveals the lighter-colored rock beneath. People around the world have drawn on rock varnish for thousands of years. Many prehistoric rock carvings, called **petroglyphs**, were created by scratching away rock varnish from a rock's surface. These images record a history of animals and people that goes back thousands of years, as discussed in the **Picture This** feature below.

The study of ancient environments such as those depicted in petroglyphs represents a line of scientific inquiry called *paleoecology*: the study of ancient ecosystems. **Figure 19.3** explores the environmental history of the Sahara further from a paleoecological standpoint.

Picture This **The Green Sahara**

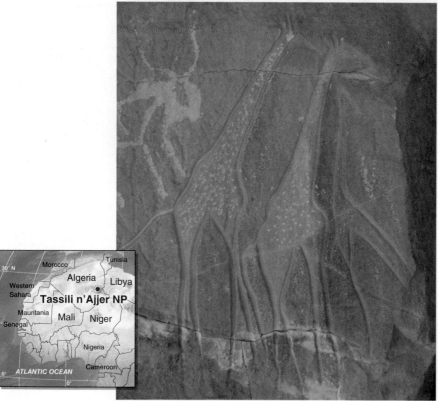

(Werner Forman/Getty Images)

This petroglyph of giraffes in Tassili n'Ajjer National Park, in southeastern Algeria, was made between 11,500 and 8,000 years ago. During the time when this petroglyph was made, the Sahara was wetter. Large animals such as giraffes, crocodiles, elephants, hippopotamuses, and aurochs (a now-extinct wild ancestor of cattle) lived throughout the region. This period is called the "Green Sahara" because most of northern Africa consisted of savanna woodland and seasonal tropical forest. Natural climate change has transformed northern Africa into a desert, and these wild animals now live no closer than about 1,000 km (620 mi) to the south. The petroglyphs found throughout the Sahara are a testament to how much northern Africa has changed in the past 8,000 years.

Consider This

1. What was the Green Sahara? When did it occur?
2. The meanings of most petroglyphs in the Sahara are unknown. Why do you think people made them?

Figure 19.3 SCIENTIFIC INQUIRY: Why did northern Africa become wetter, and how were people affected? Understanding Earth's natural history and how people have been affected by Earth's changing environments is one of the central themes in physical geography. Why was the Sahara wetter some 9,000 years ago? And how did people respond to this episode of natural climate change? *(© Mike Hettwer)*

Why did northern Africa become wetter?

Changes in Earth–Sun orbital geometry caused increased summer insolation in the Northern Hemisphere (see Section 7.2). Summer insolation values were about 7% higher 10,000 years ago than they are today. Warmer summers in northern Africa created strong onshore flow of moist air and summer monsoon rains (see Section 5.3). Climate was wetter as a result.

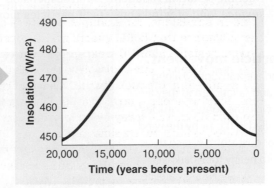

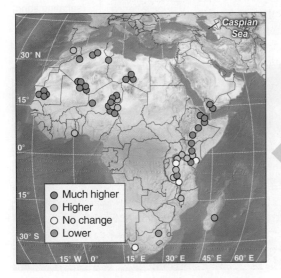

What is the evidence for a wetter climate?

About 9,000 years ago, lake levels in northern Africa were at their highest. This map compares the levels of African lakes 9,000 years ago with present-day lake levels. Some of those ancient lakes were large, and two were comparable in size to the Caspian Sea today (called out on the map for comparison). These lakes left behind beaches and fossils. The ages of the highest beaches and, therefore, when lake levels were highest has been determined by radiocarbon analysis or organic material such as shells (see Section 12.2). Other evidence comes from sediment cores (see Section 7.2) from various lakes in northern Africa as well as from the Mediterranean Sea and the Indian and Atlantic oceans.

How did people respond to northern Africa's wetter climate?

Petroglyphs in northern Africa show that the region was populated by nomadic hunting groups at the time the region was wetter. These groups began establishing semipermanent settlements and herding cattle, sheep, and goats. Bones, teeth, and artifacts have been preserved at human burial sites. These remains tell us about the diet, health, and age of the people who left them. They also reveal where people moved. Beginning about 6,000 years ago, human settlements in the interior of northern Africa gradually disappeared as surface water became unavailable and the region turned to desert. The Early Dynastic Period of Egypt began about 5,000 years ago. Research suggests that many groups dispersed from the interior Sahara to the Nile River in response to these changes.

Sculpting with Wind: Eolian Processes

Eolian (or *aeolian*; pronounced EE-ol-ee-un) means "of or relating to wind." Just as sediments are transported in a stream, wind moves particles through traction, saltation (bouncing), and suspension. The *surface load* of wind consists of particles moving by traction, rolling, and saltation. The *suspended load* consists of fine sediments of silt and clay that are suspended in the atmosphere. These particles can travel distances ranging from a few meters to thousands of kilometers as *dust storms* (also called *haboobs*) **(Figure 19.4)**.

Figure 19.4 Eolian particle movement.

(A) Sediments moved by wind are transported along the surface through traction and saltation, and they are suspended in the air as dust. Only the lightest particles of clay and silt remain suspended in the air by the wind. (B) On March 25, 2011, a major dust storm was so thick that it plunged Kuwait City into nearly total darkness during daylight hours. (C) The next day, NASA's satellite *Terra* captured the same dust storm as it moved south across the Arabian Peninsula. *(B. © Surya Murali; C. NASA images courtesy Jeff Schmaltz, MODIS Rapid Response Team at NASA GSFC)*

Web Map

Desert Storms

Available at
www.saplinglearning.com

A

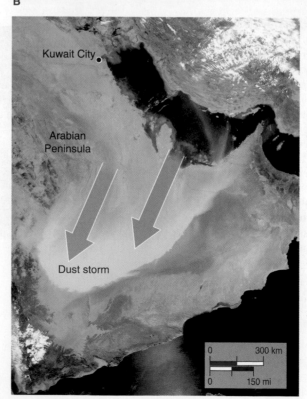

B

C

Figure 19.5 A ventifact. This sandstone ventifact, about 1.5 m (5 ft) high, is on Kangaroo Island on the south coast of Australia. Its unusual shape is the result of erosion by blowing sand. *(Gregory Dimijian, MD/Science Source)*

Dust storms can travel rapidly. The March 2011 dust storm shown in Figure 19.4 moved about 50 km/h (31 mph) as it crossed the Arabian Peninsula.

Saltating sand grains hop over the ground within the first meter or so of the ground surface, abrading and sandblasting rock surfaces and creating ventifacts. A **ventifact** is a wind-sculpted desert rock shaped by abrasion by saltating sand **(Figure 19.5)**.

If enough fine-grained sediment is transported away from a region by the wind, the land surface is lowered through the process of deflation. **Deflation** is the removal of sediments and lowering of the land surface by wind erosion.

Where the wind's speed increases as it flows around an obstruction, such as a large boulder, a blowout depression may occur. A *blowout depression* is a bowl-shaped depression formed in sand where wind flow becomes turbulent. Blowout depressions often form around pedestal rocks. A **pedestal rock** is a remnant of a layer of erosion-resistant *cap rock* supported by a slender column of less resistant rock **(Figure 19.6)**.

Yardangs are a type of desert ventifact with elongated ridges formed by deflation and relentless abrasion by saltating sand grains. Yardangs form where persistent winds move sand in the

Figure 19.6 Pedestal rock. This large rock balances on a narrow pedestal in the Sahara in remote eastern Libya. The cap rock is more resistant to weathering and erosion than the rock forming the pedestal beneath it. As abrasion continues to thin the pedestal, the rock will eventually come crashing down. Note that a blowout depression has formed at the base of the pedestal. (The dotted line traces the lip of the blowout depression.) *(© David Parker/Science Source)*

Figure 19.7 Yardangs. Yardangs are formed through deflation and abrasion where winds blow persistently from the same direction. These yardangs are in the White Desert in western Egypt. Note the pickup truck for scale in the lower left. *(Raimund Linke/Getty Images)*

Picture This Sahara Dust and Climate

One recurrent theme in physical geography is that physical systems are interconnected. A good example is the dust plumes that move from the Sahara westward on the trade winds. This image, developed from satellite data, shows a dust plume that stretched all the way from northern Africa to Florida on July 21, 2012. The darker the orange color, the higher the concentration of suspended material in the air.

Satellite measurements show that each year an average of 182 million tons of dust is blown off the Sahara, and about 27 million tons of this settles out on the Amazon rainforest. The dust from northern Africa contains iron and phosphorus. Iron and phosphorus promote the growth of phytoplankton (see Section 10.3) in the oceans. Therefore, the dust stimulates the growth of phytoplankton as it settles on the ocean. Similarly, the soils of the Amazon rainforest are nutrient poor (see Section 9.2), and the dust that settles there stimulates plant growth as well. Together, the marine phytoplankton and Amazon rainforest vegetation represent significant carbon sinks (see Section 7.3) and help slow the rate of carbon dioxide accumulation in the atmosphere.

Climate models do not yet agree on whether precipitation in northern Africa will increase or decrease in the coming decades as the global temperature continues to climb. If climate change brings wetter conditions to the Sahara, less dust will be suspended in the air as vegetation grows and anchors the soils. If climate change brings drier conditions,

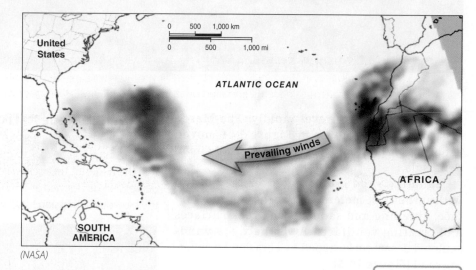

(NASA)

more dust will be suspended as vegetation is diminished.

Consider This

1. Dust has a high albedo and reflects sunlight. In the context of Earth's albedo, what will happen to the atmosphere's overall albedo if more dust from the Sahara is suspended? How would this affect Earth's radiation equilibrium temperature? (See Section 3.3.)
2. In the context of carbon drawdown out of the atmosphere, how could more dust in the atmosphere draw carbon dioxide out of the atmosphere and into the oceans?

Video

Satellite Tracks Saharan Dust to Amazon in 3-D

Available at www.saplinglearning.com

same direction for most of the year. They are generally two to three times longer than they are wide, and they can be up to hundreds of meters high and several kilometers long **(Figure 19.7)**.

Suspended material does not just end up on land surfaces. Dust may be held aloft for weeks and transported over the oceans, eventually setting and slowly accumulating as sediment on the seafloor (see Section 14.3). In some cases windblown dust from deserts plays important roles in Earth's physical systems, as the **Picture This** feature on the facing page explains.

Shifting Sand: Sand Dunes

A **sand dune** is a hill or ridge formed by the accumulation of windblown sand. Only about 10% of deserts are covered by sand dunes, and not all sand dunes are found in deserts. Where desert sand dunes do occur, they are prominent landforms. *Active* sand dunes migrate across the desert floor. In contrast, *stabilized* sand dunes are anchored by vegetation and are unable to move.

Individual sand grains migrate up the windward side of a sand dune, up and over the *dune crest*, and then slide down the **slip face** on the leeward side of the dune **(Figure 19.8)**. On the slip face, sand grains settle out at the angle of repose—the steepest angle at which loose sediments can settle (see Figure 16.17).

Sand dunes take many different physical configurations, depending on the physical setting in which they form. **Barchan dunes** (pronounced BAR-can) are crescent-shaped sand dunes with the crescent tips pointing downwind. **Star dunes** are star-shaped sand dunes that form as the prevailing wind changes direction through the year. **Transverse dunes** are long, narrow sand dunes with dune crests that run perpendicular (or transverse) to the prevailing wind. **Longitudinal dunes** have linear dune crests that are oriented parallel to the wind. Finally, **parabolic dunes** resemble barchan dunes, except their crescent tips are anchored by vegetation and point upwind.

Figure 19.8 Active sand dunes. Individual sand grains are moved in the direction of the prevailing wind. The dune as a whole migrates in the direction of the prevailing wind as well. When structures are built in the path of migrating dunes, those structures become engulfed by sand. *(Left. John Beatty/Getty Images; right. Don Smith/Getty Images)*

1. Windward side
Through saltation and traction, sand grains move up the windward side of the dune and over the crest. Sand ripples form on the windward side.

2. Leeward side
The slip face on the leeward side is protected from the wind. Sand grains move over the crest and slide down the slip face, settling out at the angle of repose.

3. Dune migration
As a whole, dunes slowly migrate in the direction of the prevailing wind. It is nearly impossible to stop sand dune migration. This house in Kolmanslop in the Namib Desert, is being buried by a migrating sand dune.

Windward side
Dune crest
Leeward side
Prevailing wind
Slip face
Prevailing wind
Slip face

Sudan
Nigeria
Ethiopia
ATLANTIC OCEAN
DRC
Tanzania
Angola
Namib Desert
Namibia
South Africa

Three factors determine what kind of sand dunes form: (1) the amount of sand available, (2) the prevailing wind direction, and (3) and the presence of vegetation **(Figure 19.9)**.

Singing Sand

About 30 sand dune locations around the world have been observed to periodically "sing" or "boom." As grains of sand slide down the slip face, each grain vibrates when it encounters other grains. Billions of particles of sand vibrating at the same frequency can create an audible sound. The pitch of the sound depends on the size of the sand grains, their silica content, and the depth of the sand dune.

Desertification and Stabilizing Dunes

Desertification is occurring in many arid regions where growing human populations put increasing pressure on already fragile semiarid environments. **Desertification** is the transformation of fertile land to desert, usually through a combination of overgrazing of livestock and drought. Arid western China, for example, is facing a growing

problem with desertification. As human activities there convert formerly vegetated regions to desert, dust storms are increasingly developing and sweeping eastward into populated areas.

One means of addressing desertification is to attempt to stabilize the sand so that vegetation can be reestablished. Plant roots anchor the sand and stop its movement. But successfully establishing vegetation on a shifting sand dune is extremely difficult. Where sand dunes are threatening human settlements, fences and netting may be used to slow the movement of the sand so that vegetation can be planted **(Figure 19.10)**.

Sculpting with Water: Fluvial Processes

Deserts are by definition arid places, so it may seem odd that the most important agent of erosion in most desert areas is not wind but flowing water. Deserts are particularly vulnerable to erosion by water because they have little vegetation. Plant roots anchor soil, and plants lessen the erosive effects of raindrops, sheet wash, and stream flow on the desert surface.

Figure 19.9 Kinds of sand dunes. Barchan and star dunes form where sand does not completely cover the desert floor. Where sand completely covers the desert floor, transverse and longitudinal dunes form. The presence of vegetation often results in parabolic dunes.

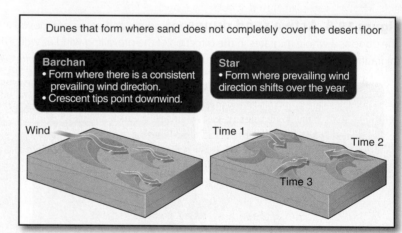

Dunes that form where sand does not completely cover the desert floor

Barchan
• Form where there is a consistent prevailing wind direction.
• Crescent tips point downwind.

Wind

Star
• Form where prevailing wind direction shifts over the year.

Time 1
Time 2
Time 3

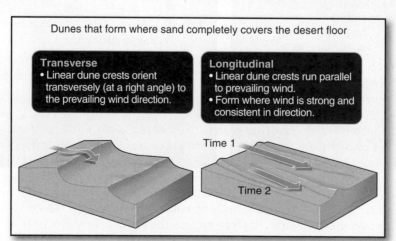

Dunes that form where sand completely covers the desert floor

Transverse
• Linear dune crests orient transversely (at a right angle) to the prevailing wind direction.

Longitudinal
• Linear dune crests run parallel to prevailing wind.
• Form where wind is strong and consistent in direction.

Time 1
Time 2

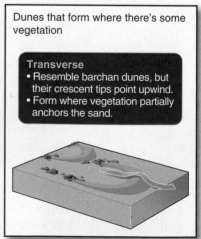

Dunes that form where there's some vegetation

Transverse
• Resemble barchan dunes, but their crescent tips point upwind.
• Form where vegetation partially anchors the sand.

Fluvial Erosion

Many deserts experience summer thunderstorms and sudden flooding rains. The force of impact from raindrops dislodges particles. As water flows over the bare ground surface, water picks up these particles and transports them downslope. As tributaries merge, discharge increases, and streams gain erosive power. Ephemeral desert streams (streams that flow only after heavy downpours) are normally dry, but they run muddy brown from the suspended sediments they carry after thunderstorms.

Each time rainfall fills the channel of an ephemeral stream with water, sediments move downstream in the stream channel. Small particles of silt, clay, and sand are transported farthest, but under high-discharge conditions, even large and heavy rocks will move. Once the rains cease, water infiltrates the ground, and the channel dries up again. The channel sediments do not move until the next time heavy rainfall fills the channel with water. These ephemeral streams go by many names, depending on the geographic locale. They are called *arroyos*, *dry washes*, or *coulees* in the United States. In the Middle East and northern Africa they are called *wadis*, in South Africa *dongas*, and in India *nullahs*.

Figure 19.10 Sand dune stabilization. These sand dunes in the Tengger Desert in Inner Mongolia of China are being slowed but not stopped with straw netting. The sand movement slows as it catches in the straw. *(Bruce Dale/Getty Images)*

Fluvial Deposition

Where streams exit steep canyons, the stream gradient flattens out, stream velocity is reduced, and the kinetic energy of the stream drops to almost nothing. As a result, the sediments carried by the stream are deposited, forming alluvial fans (see Section 17.3). Where alluvial fans emerge from adjacent valleys, they often coalesce to form a broad, sloping plain called a **bajada** (pronounced ba-HAH-da) **(Figure 19.11)**.

Where ephemeral desert streams terminate in an internal drainage basin (see Section 17.1), they form a temporary shallow lake called a **playa lake**. Standing water in the playa lake typically evaporates within days to weeks. As the water evaporates, the dissolved minerals in the water precipitate out of the water, forming evaporite deposits (see Section 14.3). These deposits form a flat, dry lake bed called a **playa**. Playas are among the flattest and smoothest of all natural landforms on Earth. The world's fastest cars have set land speed records on the Bonneville Salt Flats, a playa in Utah. The **Picture This** feature on page 596 explores an unusual phenomenon that occurs in playas.

Some internal drainage basins in deserts develop saline lakes. A **saline lake** is a permanent body of salty water occupying a playa. Evaporation

Figure 19.11 Bajada. Alluvial fans have coalesced to form a bajada in the Toiyabe Range in central Nevada. The gradual slope of the bajada can be seen in relationship to the level road. *(Larry Workman)*

rates are high in these desert lakes, and as a result, dissolved minerals such as salt become concentrated in the water. Utah's Great Salt Lake and the Dead Sea in the Middle East (see Figure 17.6) are saline lakes. Saline lakes are far saltier than the ocean. The Geographic Perspectives at the end of this chapter explores the environmental effects of human activities on saline lakes.

Picture This Racetrack Rocks

(Dennis Flaherty/Getty Images)

People have long known that rocks up to 300 kg (700 lb) in Death Valley's Racetrack Playa move across the playa floor, but it was unclear how the rocks moved. No one has ever actually witnessed the rocks in the act of moving, but their long trails provide indisputable evidence that they do move. This phenomenon has also been observed in nearby Bonnie Claire Playa in Nevada. (The rock shown here is about 15 cm [6 in] in diameter.)

In an attempt to capture a rock moving, time-lapse photography has been used, but without success.

Two separate rocks may start off with parallel trails and then diverge. The sizes of the rocks do not appear to influence their rate or direction of movement. Smooth rocks tend to travel in a less straight path than sharp-edged rocks.

Over the years, many hypotheses have been proposed to explain these rock movements. All involve a combination of strong wind, a slippery wetted-clay surface, and ice. Recently, researchers from Scripps Institution of Oceanography found that during winter, sheets of water on the playa floor freeze into thin panes of ice. The wind then breaks the ice into large pieces and moves them and the rocks embedded in them across the playa floor. Winds as slow as 16 km/h (10 mph) appear to be sufficient to move the ice and the rocks. Conditions necessary for rock movement are rare, and most rocks move at a rate of less than 6 m (20 ft) per minute for an hour or less once every 10 years.

Consider This

1. Why is gravity not a factor in the movement of rocks on Racetrack Playa?
2. Do you think the hypothesis presented here adequately explains how 300 kg rocks move? Explain.

19.2 Desert Landscapes

◎ Explain how various desert landscapes develop.

To many, deserts are among Earth's most beautiful and dramatic landscapes. In the desert, there is a contrast between the harsh, barren, and inhospitable nature of the landscape and its endless vistas, intriguing physical landforms, and sense of quiet timelessness. The most prominent desert landscapes include regs, ergs, mesa-and-butte terrain, inselbergs, basin-and-range topography, and badlands. Each of these landscapes bears the signatures of water, wind, and time.

Regs: Stony Plains

A **reg** is a flat, stony plain. All deserts have regs, but they go by different names, depending on the region. For example, regs are called *hamada* in the Sahara and *gibber plains* in Australia. Regs everywhere are formed by either of two different processes. The first process is *eolian sorting*, in which small particles of silt and clay are blown farthest, while heavier particles of sand travel shorter distances. As similarly sized particles accumulate in the same area, sand dunes composed of quartz sand grains, to particles of clay, to loess deposits composed of fine silt begin to form prominent desert landforms. As deflation carries these lighter particles away, an accumulation of rocks, called a **lag deposit (Figure 19.12)**, may develop.

The second process that forms regs produces a flat, rocky surface composed of tightly interlocked stones called **desert pavement.** Desert pavement resembles the surface of a cobblestone street. Beneath desert pavement are aridisol soils composed of clay and silt. These soils often contain few or no stones. If deflation were the process that forms desert pavement, one would expect to find stones within the soil that would later be brought to the surface as the lighter soil was blown away. This is not the case, and recent research indicates that, instead, desert pavement may form as desert soils are built up on bare bedrock. **Figure 19.13** illustrates how desert pavement forms.

The top sides of the rocks in regs are often coated with rock varnish. Overturning or removing the rocks reveals a lighter surface. In some areas of southern Peru, the desert's surface was scraped and overturned by people of the ancient Nazca civilization to draw huge geoglyphs on the desert floor **(Figure 19.14)**.

Figure 19.12 Lag deposit.

(A) A lag deposit forms when gravel and rocks are concentrated at the ground surface after smaller and lighter particles have blown away. (B) This lag deposit is southwest of Tamanrasset in southern Algeria. *(Taghit/Getty Images)*

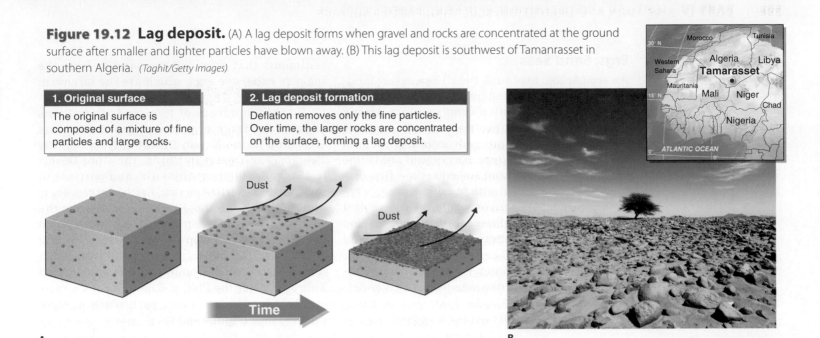

1. Original surface

The original surface is composed of a mixture of fine particles and large rocks.

2. Lag deposit formation

Deflation removes only the fine particles. Over time, the larger rocks are concentrated on the surface, forming a lag deposit.

Dust

Dust

Time

A

B

Figure 19.13 Desert pavement.

(A) Desert pavement forms as desert soils are built up when clays and silts accumulate beneath rocks. (B) This reg in the desert of western Namibia is composed of desert pavement. The scale bar in the photo represents the spatial dimensions in the foreground. *(Jean-Luc MANAUD/Getty Images)*

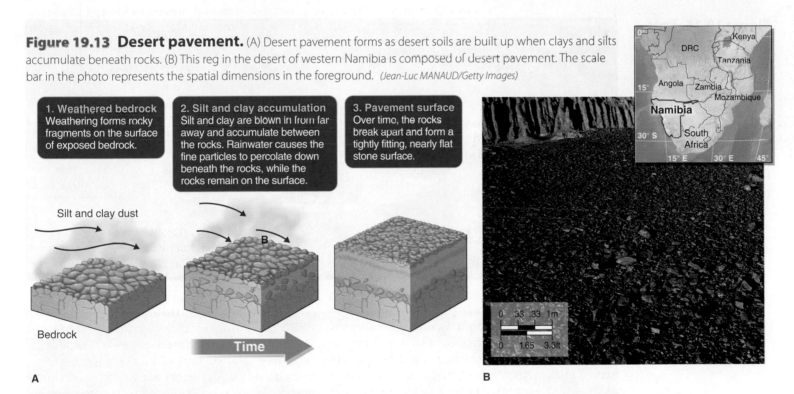

1. Weathered bedrock
Weathering forms rocky fragments on the surface of exposed bedrock.

2. Silt and clay accumulation
Silt and clay are blown in from far away and accumulate between the rocks. Rainwater causes the fine particles to percolate down beneath the rocks, while the rocks remain on the surface.

3. Pavement surface
Over time, the rocks break apart and form a tightly fitting, nearly flat stone surface.

Silt and clay dust

Bedrock

Time

A

B

North

Figure 19.14 Nazca geoglyphs.

This aerial photograph shows a drawing of a monkey (about 100 m [330 ft] across) carved into a reg in southern Peru. About two dozen strange figures and odd geometric shapes and patterns were etched into the Peruvian desert floor by the now-vanished Nazca civilization about 1,500 to 2,000 years ago. Although there are many hypotheses about why these works were created, their purpose remains unknown. *(Aaron Oberlander/Getty Images)*

Ergs: Sand Seas

An **erg** (Arabic for "dune field") is a desert landscape that is dominated by actively moving sand dunes. Ergs, also called *sand seas*, are among Earth's most dramatic landscapes. They are devoid of vegetation and, by definition, are at least 125 km² (50 mi²) in extent. Any type of sand dune may form in an erg, and some ergs are flat, featureless sheets of sand with few dunes.

Ergs are most common where the subtropical high creates arid conditions. Ergs cannot develop in humid climates where vegetation grows and stabilizes the sand. The source of sand for ergs varies, but it is often derived from ancient glacial outwash plains, dry lake beds, or river deltas. In much of the Middle East, such as Saudi Arabia, geologic uplift exposed ancient marine sediments that provided the sand to form the region's extensive ergs, which are the largest in the world.

It takes hundreds of thousands to millions of years for large ergs to develop. Large ergs are found in the Namib Desert of Namibia, the Taklamakan Desert in China, the Gobi Desert in Mongolia, interior Australia, and portions of southwestern North America. Earth's largest erg is *Rub' al Khali* (or the Empty Quarter) on the Arabian Peninsula (see page 142). It is estimated to have been active for over a million years. Covering 583,000 km² (225,000 mi²), it stretches from Saudi Arabia into Yemen, Oman, and the United Arab Emirates. After the Rub' al Khali, Earth's largest ergs are found in the Sahara, particularly in Algeria and Libya **(Figure 19.15)**.

Figure 19.15 Ergs.
This Land Rover navigates star dunes in the Erg Chebbi in Morocco. The dunes here reach up to 150 m (500 ft) in height. (B) The Sahara has the greatest concentration of ergs anywhere. Only the larger ergs are mapped here. *(funkyfood London - Paul Williams/Alamy)*

Erg Chebbi, Morocco

A

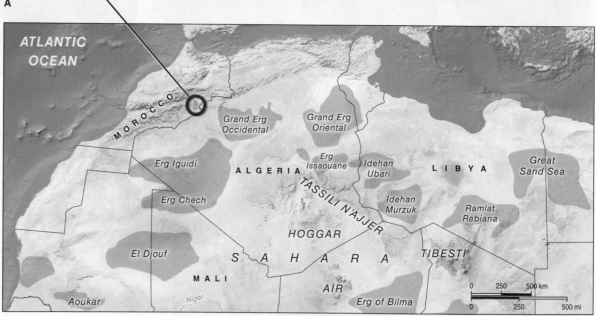

B

Mesa-and-Butte Terrain

Like sand seas, mesas and buttes come to mind for many when they picture a desert landscape. *Mesa-and-butte terrain*, however, is determined more by the composition and structure of the bedrock than by climate. Mesa-and-butte terrain occurs in humid climates, but it is easiest to see in arid desert climates, where it is not obscured by vegetation. The tepuis (large, flat mountains) in Venezuela and western Guyana, for example, form mesa-and-butte terrain that is covered by tropical rainforest.

A **mesa** is flat-topped elevated area with one or more cliff-face sides and is wider than it is tall. A **butte** is a medium-sized flat-topped hill with cliff-face sides and is taller than it is wide. A **chimney** is a slender spire of rock with a flat top. Most mesas, buttes, and chimneys occur where a plateau has been uplifted, weathered, and eroded by flowing water. Chimneys are far less common than mesas and buttes.

Although mesa-and-butte terrain can form on lava flows without geologic uplift, it is most common in sedimentary rocks with horizontal layers that have been uplifted. Examples of mesa-and-butte terrain are found throughout the southwestern United States, particularly in Monument Valley on the Colorado Plateau. **Figure 19.16** illustrates how mesa-and-butte terrain develops.

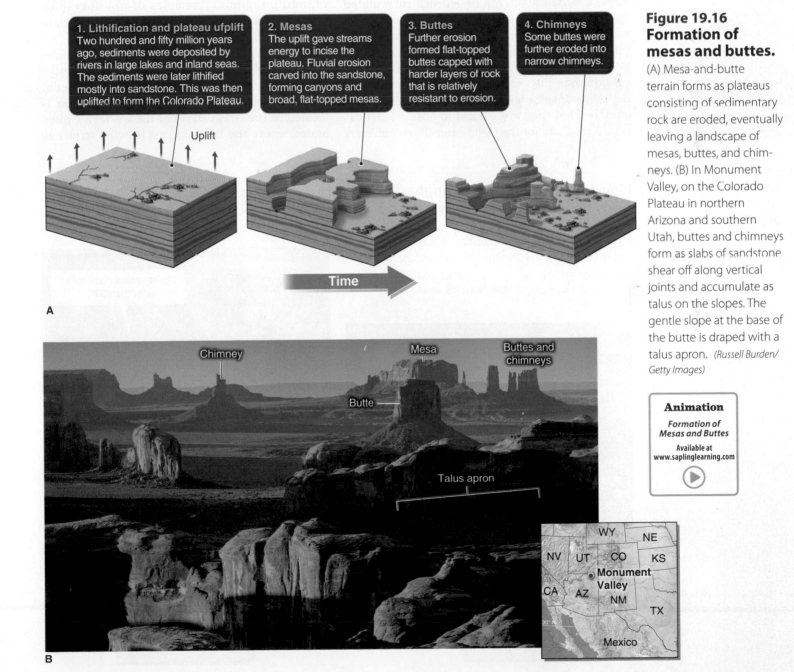

1. Lithification and plateau ufplift
Two hundred and fifty million years ago, sediments were deposited by rivers in large lakes and inland seas. The sediments were later lithified mostly into sandstone. This was then uplifted to form the Colorado Plateau.

2. Mesas
The uplift gave streams energy to incise the plateau. Fluvial erosion carved into the sandstone, forming canyons and broad, flat-topped mesas.

3. Buttes
Further erosion formed flat-topped buttes capped with harder layers of rock that is relatively resistant to erosion.

4. Chimneys
Some buttes were further eroded into narrow chimneys.

Uplift

Time

A

Chimney · Mesa · Buttes and chimneys · Butte · Talus apron

B

Figure 19.16 Formation of mesas and buttes.

(A) Mesa-and-butte terrain forms as plateaus consisting of sedimentary rock are eroded, eventually leaving a landscape of mesas, buttes, and chimneys. (B) In Monument Valley, on the Colorado Plateau in northern Arizona and southern Utah, buttes and chimneys form as slabs of sandstone shear off along vertical joints and accumulate as talus on the slopes. The gentle slope at the base of the butte is draped with a talus apron. *(Russell Burden/Getty Images)*

Animation
Formation of Mesas and Buttes
Available at
www.saplinglearning.com
▶

The surface of a flat-topped mesa has little or no *dip* (or slope). Dipping occurs where sedimentary rock has been folded or tilted through faulting (see Section 13.4). Where sedimentary rock at the ground surface dips gently, a *cuesta* forms **(Figure 19.17A)**. *Hogbacks* form where sedimentary rock at the ground surface dips steeply **(Figure 19.17B)**. And like mesas and buttes, cuestas and hogbacks are not restricted to desert environments; they are often associated with deserts, however, because they are easiest to see where there is little vegetation.

Inselbergs: Island Mountains

An **inselberg** (meaning "island mountain") is a large, weathered outcrop of bedrock surrounded by a flat plain. Inselbergs are exposed at the land surface through the process of *exhumation* (the removal of overlying sediments and rock). Inselbergs form because they are more resistant to erosion than the sediments and rocks surrounding them. Inselbergs are not restricted to desert climates, but they are easiest to observe in deserts. Kata Tjuta (or the Olgas) and Uluru (or Ayers

Rock) in central Australia are two dramatic examples of desert inselbergs formed from sedimentary rocks **(Figure 19.18)**.

Bornhardts are inselbergs that take on a more rounded form and are usually derived from erosion-resistant crystalline rocks such as granite or gneiss. As the land surface is lowered through erosion, a crystalline rock body, such as a pluton, batholith, or laccolith (see Section 14.2), is exposed and rounded through weathering **(Figure 19.19)**.

Basin-and-Range Topography

A *physiographic province* is a region with a unique topography that distinguishes it from the surrounding regions. Examples of physiographic provinces in the United States include the Atlantic Coastal Plain, the Cascade Range, and the Colorado Plateau. Canada's physiographic provinces include the St. Lawrence Lowlands in southern Quebec and the Interior Plains in the central territories. The **Basin-and-Range Province** is located in the Great Basin of the western United States, where the Earth's crust is being stretched

Figure 19.17 Cuestas and hogbacks. (A) A cuesta, such as Book Cliffs near Grand Junction, Colorado, is formed where sedimentary rock layers dip less than 30 degrees from horizontal. The dotted line traces the tilt of the sediments on this photo. (B) A hogback is composed of erosion-resistant sedimentary rock layers with a dip of 30 degrees or more from horizontal. This hogback is the San Rafael Reef, a steeply tilted (shown with dotted line) layer of sandstone and limestone in eastern Utah. *(A. Jim Wark/airphotona.com; B. FRANCOIS GOHIER/Science Source)*

Cuesta
Dip of sedimentary rocks is less than 30°.

< 30°

WY
NE
Book Cliffs
NV UT CO KS
CA AZ NM
TX
Mexico

A

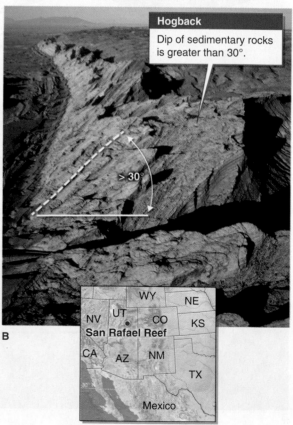

Hogback
Dip of sedimentary rocks is greater than 30°.

> 30°

WY
NE
NV UT CO KS
San Rafael Reef
CA AZ NM
TX
Mexico

B

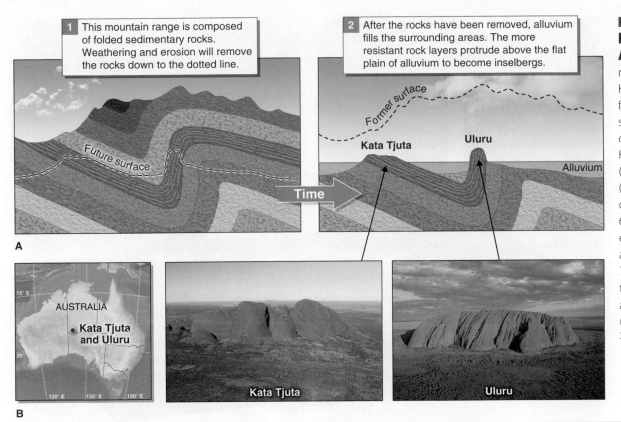

1 │ This mountain range is composed of folded sedimentary rocks. Weathering and erosion will remove the rocks down to the dotted line.

2 │ After the rocks have been removed, alluvium fills the surrounding areas. The more resistant rock layers protrude above the flat plain of alluvium to become inselbergs.

Former surface

Future surface

Kata Tjuta   Uluru

Alluvium

Time

A

AUSTRALIA

● **Kata Tjuta and Uluru**

15° S

30°

120° E   135° E   150° E

Kata Tjuta

Uluru

B

Figure 19.18 Inselbergs, Australia. (A) Over many millions of years, Kata Tjuta and Uluru were formed through the erosion of sandstone layers of differing hardness. (B) Kata Tjuta (left) and Uluru (right) are about 30 km (18 mi) apart. They are composed of the same 600-million-year-old erosion-resistant *arkose*, a type of sandstone. *Kata Tjuta* is the taller of the two, rising 546 m (1,791 ft) above the desert floor. *(B: Left. © Nigel Millett; right. SPL/Science Source)*

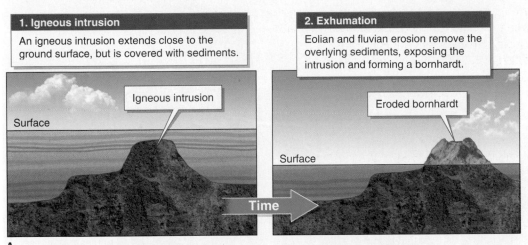

1. Igneous intrusion

An igneous intrusion extends close to the ground surface, but is covered with sediments.

2. Exhumation

Eolian and fluvian erosion remove the overlying sediments, exposing the intrusion and forming a bornhardt.

Surface

Igneous intrusion

Eroded bornhardt

Surface

Time

A

Figure 19.19 Bornhardt, Namibia. (A) Bornhardts are formed by exhumation of an igneous intrusion through erosion. Once the igneous intrusion is exposed, erosion begins to wear it down and round it. (B) Spitzkoppe, in Namibia, is a bornhardt composed of 120-million-year-old granite. Spitzkoppe's tallest peak rises 700 m (2,300 ft) above the desert floor and is 1,784 m (5,853 ft) above sea level. *(David du Plessis/Getty Images)*

Sudan

Nigeria   Ethiopia

0°

ATLANTIC OCEAN   DRC   Tanzania

15° S   Angola Zambia

Namibia

● **Spitzkoppe**

30° S   South Africa

0°   15° E   30° E   45° E

B

Figure 19.20 Basin-and-Range Province. (A) The Basin-and-Range Province is found in the Great Basin, where the North American crust is being stretched, creating a series of tilted fault blocks. The leading edges of these tilted blocks form parallel ranges (called horsts) surrounded by flat, alluvium-filled basins (called grabens). (B) This digital elevation map shows the geographic extent of the Great Basin. *(Courtesy Andrew Birrell)*

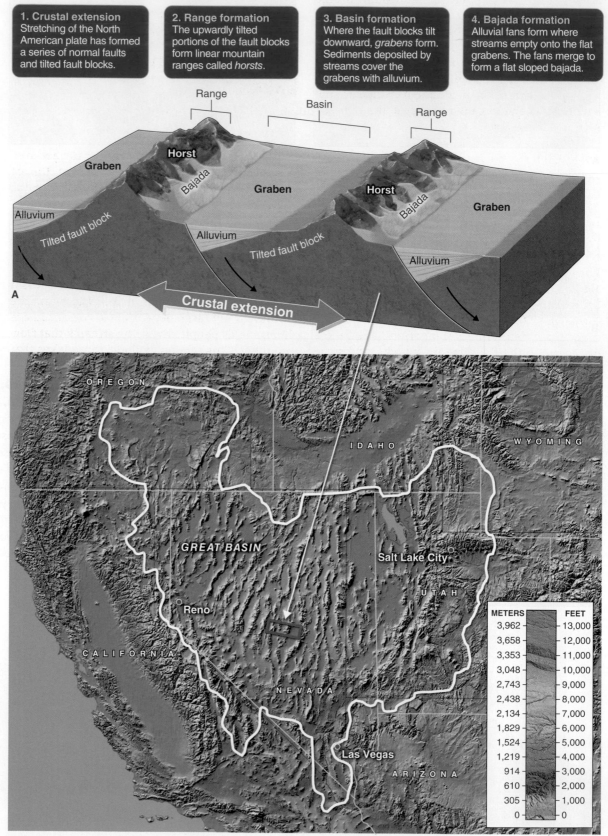

1. Crustal extension
Stretching of the North American plate has formed a series of normal faults and tilted fault blocks.

2. Range formation
The upwardly tilted portions of the fault blocks form linear mountain ranges called *horsts*.

3. Basin formation
Where the fault blocks tilt downward, *grabens* form. Sediments deposited by streams cover the grabens with alluvium.

4. Bajada formation
Alluvial fans form where streams empty onto the flat grabens. The fans merge to form a flat sloped bajada.

METERS	FEET
3,962	13,000
3,658	12,000
3,353	11,000
3,048	10,000
2,743	9,000
2,438	8,000
2,134	7,000
1,829	6,000
1,524	5,000
1,219	4,000
914	3,000
610	2,000
305	1,000
0	0

through rifting (see Section 13.3), forming a large internal drainage basin containing a series of tilted fault blocks called *horsts* and a series of valleys called *grabens*. **Figure 19.20** illustrates the process that has created the Basin-and-Range Province.

Basin-and-range topography is not restricted to deserts. It is included in this section, however, because it is a prominent desert landscape in western North America. Other comparable regions of widespread extension of continental crust are the Baikal Rift Zone in central Eurasia and the East African Rift System.

Badlands

Badlands are landscapes riddled by networks of gullies (dry stream channels). Badlands occur in arid to semiarid environments where slopes are largely unvegetated, and weak sedimentary rocks, such as shale, mudstone, and claystone, are eroded by infrequent but intense rainfall. They have loose, dry, clay-rich soils. Badlands are "bad" because they cannot be farmed and because they are difficult to cross. They range in size from a few grouped hillsides to thousands of square kilometers. Badlands often have fossils preserved in their sedimentary rocks, and some of them are brilliantly colored due to the presence of colorful minerals, such as hematite. Oregon's John Day Fossil Beds National Monument

and Arizona's Painted Desert are two examples of colorful badlands. **Figure 19.21** shows the spectacular badlands in the Qilian Mountains in China.

GEOGRAPHIC PERSPECTIVES
19.3 Shrinking Desert Lakes

◎ **Assess the effects of water diversions on desert lakes.**

Some of the world's largest lakes (in surface area but not in volume) are found in deserts. Desert lakes are often located in internal drainage basins and are saline (see Section 17.1). Like the mass balance of a glacier, the hydrologic balance of a desert saline lake may fluctuate a bit, but overall there is an equilibrium between water inputs and water outputs in such systems. A lake's level rises as water flows into it and drops as water leaves it by evaporation and infiltration of water into the ground. Many large desert lakes worldwide are shrinking as people divert the streams that flow into them. This section explores three lakes that have been diminished as a result of water diversions: the Aral Sea in Eurasia and Owens and Mono lakes in California.

Story Map

Shrinking Desert Lakes

Available at
www.saplinglearning.com

Figure 19.21 Badlands. The badlands of the Qilian Mountains in China are composed of siltstone, mudstone, and shale. The high levels of minerals like iron and manganese in the rocks give them their vibrant hues. *(VCG/Getty Images)*

Figure 19.22 Aral Sea map. (A) This map shows the Aral Sea as it looked before water diversions. The Amu Darya and Syr Darya rivers, originating in the Tian Shan mountain range to the southeast, once flowed into the Aral Sea basin but are now diverted for agriculture. (B) Cotton fields abound in the Aral Sea basin. Cotton plants need a lot of water, and standing irrigation water evaporates rapidly in the desert sunlight. *(Martin Barlow/AGE Fotostock)*

A

B

Figure 19.23 Aral Sea, 1960 and 2018. This satellite image compares the Aral Sea between April 2018 (green areas) and 1960 (traced outline). The original lake bed was about 400 km (250 mi) long, roughly the size of Wisconsin. Hundreds of rusting boats (inset) are now stranded far from the water's edge; they are a stark reminder of the fishery that once sustained local economies and people. *(Left. Kelly Cheng Travel Photography/Getty Images; right. NASA Earth Observatory image by Jesse Allen, using Terra MODIS data from the Land Atmosphere Near real-time Capability for EOS (LANCE) and Landsat data from the U.S. Geological Survey)*

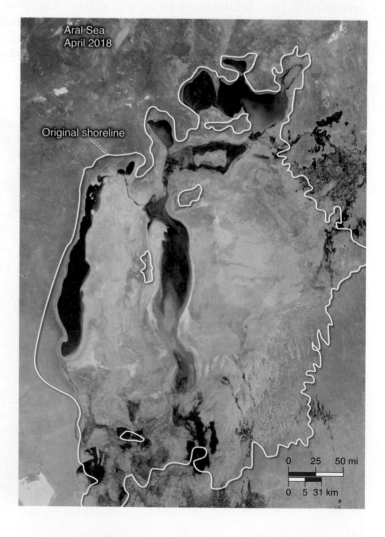

Asia's Aral Sea

The Aral Sea, straddling Kazakhstan and Uzbekistan in western Asia, was once the fourth largest lake in the world in surface area. In the 1960s, the rivers flowing into the Aral Sea, shown in **Figure 19.22**, were diverted by large irrigation canals as part of a massive Soviet project to grow cotton.

From the standpoint of producing cotton for export to global markets, the program has been a success. But without the input of water from the rivers, the Aral Sea has shrunk over time. It is now less than 10% of its original size and has become a series of shallow, briny lakes that are ecologically dead **(Figure 19.23)**. The loss of the Aral Sea brought an end to the lucrative inland fishery that once existed there. At the peak of the harvest, more than 50,000 tons of fish were caught in the Aral Sea each year.

Even before the water diversions, the Amu Darya and Syr Darya carried agricultural fertilizers and sewage. As the streams emptied into the Aral Sea, this waste accumulated in the lake's sediments. After the lake bed was exposed and dried, winds stirred up toxic dust storms and poisoned local communities. Rates of infant mortality, cancers,

and respiratory disease among local people skyrocketed. The area has the highest rate of esophageal cancer in the world. Furthermore, without the moderating effects of the lake water, the local climate has become more extreme: Winters are colder, and summers are windier and hotter.

The Great Basin Lakes

Because the Great Basin of the western United States is an internal drainage basin, no water within it flows to the ocean. Instead, streams drain to low spots within the basin, and then the water evaporates, leaving behind a saline lake. When water from streams that feed into these lakes is diverted, the lakes shrink. Many lakes in the Great Basin have experienced a fate similar to that of the Aral Sea. Utah's Great Salt Lake, Nevada's Pyramid Lake and Walker Lake, and California's Owens Lake and Mono Lake have all followed the same pattern.

California's Owens Lake

Owens Lake was a 500 km² (200 mi²) body of water that formed where the Owens River terminated in the lower Owens Valley **(Figure 19.24)**. In the

Figure 19.24 Owens Lake. (A) This map is of Mono Lake, Owens Valley, and the Los Angeles Aqueduct system. The inset photo shows the Los Angeles Aqueduct pipeline where it passes through Owens Valley. (B) An undated photograph of Owens Lake before water diversions is shown on the top. On the bottom, the playa lake bed is shown as it appears today. *(A. © Jenna Cavelle; B: Top. County of Inyo, Eastern California Museum; bottom. Lieuwe Hofstra)*

A

Owens Lake before water diversions

Owens Lake today

B

Figure 19.25 Mono Lake. (A) By the early 1980s, around 30% of Mono Lake's volume had been lost due to water diversions to Los Angeles. This graph is of Mono Lake levels from 1850 to 2017. (B) Mono Lake is known for its calcium carbonate *tufa* towers. Like cave formations, the tufa formations develop only while submerged. As the lake level lowered, many tufa formations became exposed, and their growth stopped. The tufa formations create a strange, otherworldly landscape that draws tourists from around the world to the lake. *(Danita Delimont/Getty Images)*

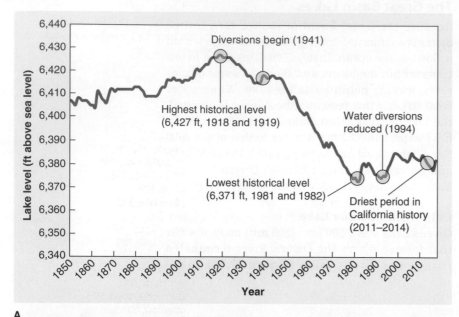

A

B

early 1900s, the city of Los Angeles needed water for its growing population and surreptitiously bought up land and water rights in the Owens Valley, some 266 km (165 mi) to the northeast of the city. Farmers who were not inclined to sell were coerced, sometimes under threat of violence, to do so. Beginning in 1913, the lower Owens River was diverted to Los Angeles through a system of pipes called the Los Angeles Aqueduct, and by 1926, Owens Lake was gone.

Today, Owens Valley has the dubious distinction of being the dustiest place in North America. The dust in Owens Valley's dust storms is about 30% salt and includes the toxic elements chromium, cadmium, and arsenic. As in the case of the Aral Sea, the toxicity of the local sediments arose chiefly from agricultural chemicals that rivers carried into the lake and deposited on the lake bed.

Saving Mono Lake

Mono Lake (pronounced MOH-no), which is somewhere between 1 million and 3 million years old, may be the oldest lake in North America. Several streams flow from the high Sierra Nevada to the west of the lake and empty into the Mono Basin to form Mono Lake. Located only 200 km (125 mi) north of Owens Lake, Mono Lake very nearly shared the fate of Owens Lake.

During the grab for Owens Lake water rights, the city of Los Angeles also acquired land in the Mono Basin and began pumping its water south. As stream inflow was reduced, the level of Mono Lake began dropping. The level dropped about 15 m (50 ft) between 1940 and 1980 **(Figure 19.25)**.

In the late 1970s, students and faculty from the University of California, Davis, started a "Save Mono Lake" campaign, and eventually, the issue of the lake's fate was brought to court. In 1994, legislation was passed to reduce water diversions and restore water to the Mono Basin. Since that time, the lake level has become stable, but it has not returned to pre-1941 levels. However, between 2011 and 2014 California experienced its driest period in its history. During that time, the lake level barely dropped, indicating that efforts to maintain Mono Lake's water level have so far been successful.

Desert lakes are sensitive systems that are easily lost. Both the Aral Sea and Owens Lake have been lost due to water diversions. As the case of Mono Lake shows, reversing the loss of these lakes is possible with an organized and sustained effort.

Chapter 19 Study Guide

Focus Points

19.1 Desert Landforms and Processes

- **Weathering:** Physical weathering is more prominent than chemical weathering in deserts.
- **Eolian erosion:** Erosion by wind is a significant factor in desert environments.
- **Sand dunes:** Sand dune type is determined mostly by the amount of sand and the direction of the prevailing wind.
- **Fluvial erosion:** Fluvial erosion is more important than eolian erosion in deserts. Fluvial erosion occurs mostly after heavy rainfall, when the channels of ephemeral streams are filled.
- **Fluvial deposition:** Streams deposit sediments on alluvial fans and playas.

19.2 Desert Landscapes

- **Desert landscapes:** Landscapes found in deserts include regs, ergs, mesa-and-butte terrain, inselbergs, basin-and-range topography, and badlands.

19.3 Geographic Perspectives: Shrinking Desert Lakes

- **Human diversions:** Water diversions by people have caused desert lakes to shrink.

- **Hydrologic balance:** The levels of desert lakes in internal drainage basins represent a balance between water flow into the lakes and water losses by evaporation and infiltration.
- **Shrinking lakes:** The Aral Sea and the lakes of the Great Basin are examples of desert lakes that have shrunk or are vulnerable to shrinking due to water diversions.

Key Terms

badlands, 603
bajada, 595
barchan dune, 593
Basin-and-Range Province, 600
butte, 599
chimney, 599
deflation, 591
desert, 586
desert pavement, 596
desertification, 594
eolian, 590
erg, 598
inselberg, 600
lag deposit, 596
longitudinal dune, 593

mesa, 599
parabolic dune, 593
pedestal rock, 591
petroglyph, 588
playa, 595
playa lake, 595
reg, 596
rock varnish, 588
saline lake, 595
sand dune, 593
slip face, 593
star dune, 593
transverse dune, 593
ventifact, 591
yardang, 591

Concept Review

The Human Sphere: Flooding in the Driest Place on Earth

1. Why does Antofagasta experience flooding when it is in one of the driest places on Earth?

19.1 Desert Landforms and Processes

2. What single physical characteristic defines all deserts?

3. What three major physical factors create deserts?

4. Which is more important in deserts: physical or chemical weathering? Why?

5. What is rock varnish? How does it form?

6. What is a petroglyph? On what surface were petroglyphs often made?

7. What was the Green Sahara? When did it exist? What caused it? What is the evidence for it? How were people affected by it?

8. What is eolian erosion? What similarities do eolian erosion and fluvial erosion share?

9. What is a ventifact, and how does it form? What are pedestal rocks and yardangs?

10. Describe the main physical aspects of a sand dune and explain how a sand dune migrates.

11. What are the major types of sand dunes?

12. Why do some sand dunes "sing"?

13. What is sand dune stabilization? How does it relate to reversing desertification?

14. Deserts are dry places. Why, then, is water the most important agent of erosion in deserts?

15. Give examples of local geographic names for ephemeral desert streams.

16. Where and how do bajadas form?

17. What is a playa lake? A playa? A saline lake? Explain how each forms.

19.2 Desert Landscapes

18. What is a reg? Compare and contrast the formation of lag deposits with the formation of desert pavement.

19. What is an erg? Where, geographically, are ergs most common?

20. What is mesa-and-butte terrain? Explain how it develops.

21. What are inselbergs? Explain how they develop.

22. What is basin-and-range topography? Explain how it develops.

23. What are badlands?

19.3 Geographic Perspectives: Shrinking Desert Lakes

24. Why do the levels of lakes in internal drainage basins fluctuate easily?

25. How are people causing large desert lakes worldwide to shrink?

26. Where was the Aral Sea? What happened to it?

27. Why are there now toxic dust storms in the Aral Sea Basin?

28. What do desert lakes in the Great Basin in the western United States have in common with the Aral Sea?

29. What happened to Owens Lake that nearly happened to Mono Lake? Why are there toxic dust storms in Owens Valley today?

30. Is Mono Lake's situation getting worse, stable, or improving? Explain.

Critical-Thinking Questions

1. Scraping the desert floor to make geoglyphs was no doubt a lot of work for the Nazca people. What purpose do you think these features served for the people who made them?

2. Regs composed of lag deposits and regs composed of desert pavement look alike on the surface. If you were in a desert and standing on a reg, what evidence would you look for to determine which process formed that particular reg?

3. What would happen to the great Saharan and Arabian ergs if the monsoon were to return to the interior desert, bringing summer rainfall?

4. In order to restore Owens Lake, what steps would need to be taken? Do you side with leaving things as is or restoring the lake? In your opinion, which option contributes more to the greater public good?

Test Yourself

Take this quiz to test your chapter knowledge.

1. True or false? All deserts are arid.

2. True or false? Water is the most important agent of erosion in deserts.

3. True or false? Desertification means the process of turning to desert.

4. True or false? Barchan dunes form where the prevailing wind shifts direction throughout the year.

5. Multiple choice: Which of the following result from fluvial deposition?
   a. bajadas
   b. ergs
   c. inselbergs
   d. sand dunes

6. Multiple choice: Which of the following desert landscapes is most likely to have been produced on an uplifted plateau?
   a. basin-and-range topography
   b. regs
   c. mesa-and-butte terrain
   d. inselbergs

7. Multiple choice: Rifting results in which of the following landscapes?
   a. badlands
   b. mesa-and-butte terrain
   c. basin-and-range topography
   d. ergs

8. Multiple choice: Which of the following is not a factor that determines the type of sand dunes that will form?
   a. amount of sand
   b. type of sand
   c. wind direction
   d. wind strength

9. Fill in the blank: A(n) _____ is a bowl-shaped depression formed in sand where the wind velocity increases.

10. Fill in the blank: The _____ is the leeward side of a sand dune where sand settles at the angle of repose.

Online Geographic Analysis

World of Change: Shrinking Aral Sea

In this exercise we will explore the shrinking Aral Sea.

Activity

Go to https://earthobservatory.nasa.gov. This is NASA's Earth Observatory web page. In the upper-left search box, type in "Aral Sea" and select the "World of Change: Shrinking Aral Sea" link. On the featured web page, satellite imagery of the Aral Sea is shown for every year since the year 2000. The lake level in 1960 is traced on each of the satellite images. Inspect the full sequence of satellite images, read the information provided below the imagery, and answer the following questions.

1. When did water diversions from the Aral Sea begin?

2. What do the green areas represent?

3. What do the white areas represent?

4. In which year are the Aral's water levels lowest?

5. Which year are water levels highest?

6. Using the scale bar in the lower-right corner, measure the distance between the 1960 lake level and the year in which the water levels are lowest from east (right) to west (left) at the maximum distance. What is the approximate distance you measured?

7. Now measure the same distance in the photo that you determined lake levels were highest. What is the approximate distance you measured?

8. The reading mentions the Kok-Aral Dike and Kok-Aral Dam. Why were these structures built?

9. When were the Kok-Aral Dike and Kok-Aral Dam completed?

10. Did the Kok-Aral Dike and Kok-Aral Dam have their intended effect? Explain.

Picture This. *Your Turn*

Desert Landform Identification

Fill in the blanks on the diagram, using each of the terms below only once.

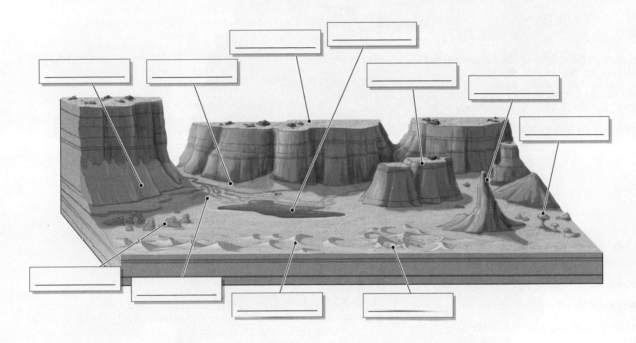

1. Bajada
2. Barchan dunes
3. Butte
4. Chimney

5. Mesa
6. Pedestal rock
7. Playa lake
8. Star dunes

9. Talus apron
10. Yardang
11. Wash

For animations, interactive maps, videos, and more, visit www.saplinglearning.com.    SaplingPlus

The Work of Waves: Coastal Landforms

Chapter Outline *and Learning Goals* ◎

20.1 Coastal Processes: Tides, Waves, and Longshore Currents

◎ Describe the role of tides and wave energy in coastal landform development.

20.2 Coastal Landforms: Beaches and Rocky Coasts

◎ Describe landforms on beaches and rocky coasts and explain how they form.

20.3 Geographic Perspectives: The Sisyphus Stone of Beach Nourishment

◎ Evaluate the practice of restoring eroded beaches through artificial replenishment of sand.

Columnar rock formations like the one in this photo are called *sea stacks*. The sea stack pictured here, locally named Dun Briste, which means "broken fort," is near Ballycastle, Northern Ireland. Its horizontal layers of sedimentary rock represent 350 million years of Earth history.

(Trevor Cole/Getty Images)

To learn about how landforms such as this form along rocky shorelines, turn to Section 20.2.

THE HUMAN SPHERE Mavericks

Figure 20.1 Riding wave energy. A surfer rides a large wave at the Mavericks Invitational in northern California. *(Ezra Shaw/Getty Images)*

For most of the year, the world's professional-class surfers are scattered around the globe, working to hone their surfing skills. Once a year, between January and March, 24 preselected surfers may get the phone call telling them that the Mavericks Invitational surf competition in Half Moon Bay in northern California is about to begin. They will have 24 hours to travel to Half Moon Bay, where they will surf some of Earth's largest and most powerful waves. The waves at Mavericks routinely reach heights of 10 m (33 ft), and waves as high as a six-story building, 24 m (80 ft), have been witnessed there **(Figure 20.1)**. These high coastal waves are the result of *swell*s, waves that are generated in the open ocean. As the swells approach the coast, they encounter a gradually sloping seafloor that channels their energy into a focused area and creates legendary surfing conditions.

Because waves are generated by wind, and wind is generated by solar heating and air pressure differences across Earth's surface, the erosional energy of coastal waves is ultimately derived from the Sun. The waves of the ocean possess a considerable amount of energy, and they never rest. This chapter explores the ways in which wave energy shapes the coast and the landforms that result.

20.1 Coastal Processes: Tides, Waves, and Longshore Currents

◎ **Describe the role of tides and wave energy in coastal landform development.**

We begin our exploration of coastal landforms by examining the daily sea-level changes caused by tides. Tides are important to the study of coastal processes because they allow the energy of waves to influence a greater vertical range of coastline than would be possible without tides.

Tidal Rhythms

If you have ever visited the ocean, you may have noticed the effects of tides. **Tides** are the rise and fall of sea level caused by the gravitational effects of the Moon and Sun. In this book, we use the term *tide* to refer to the twice-daily changes of sea level in coastal areas.

Tides are caused by the *tide-generating force* created by the gravitational pull of the Moon and Sun and by decreased gravitational pull on the side of Earth facing away from the Moon. These forces create two bulges of seawater on Earth, one on the side facing the Moon and one on the side opposite of the Moon. The tidal bulge beneath the Moon (called the *lunar tide*) is larger than the one on the opposite side of the planet **(Figure 20.2)**.

The Sun also exerts a gravitational tug on Earth and generates tides. The tide-generating force of the Moon is greater than that of the Sun because the Sun is nearly 400 times farther away from Earth than the Moon. Lunar tides are roughly twice as high as *solar tides* (tides caused by the Sun). Sometimes, tides are unusually high or low. When Earth, the Moon, and the Sun are aligned along the same axis, the tide-generating forces of the Moon and the Sun act together, and high tides called *spring tides* result. When the Moon is positioned at a right angle between Earth and the Sun, the Moon's pull acts perpendicularly to the Sun's, creating a lower *neap tide* **(Figure 20.3** on the following page**)**. Other factors, such as windiness and how close the Moon is to Earth, also influence tide levels.

Because of Earth's rotation on its axis and the revolution of the Moon around Earth, the daily rhythm of the tides runs on a schedule of approximately 24 hours and 50 minutes. This means that for 6 hours and 13 minutes, the tide is coming in as a *flood tide*, and for the next 6 hours and 13 minutes the tide is going out as an *ebb tide*. This pattern

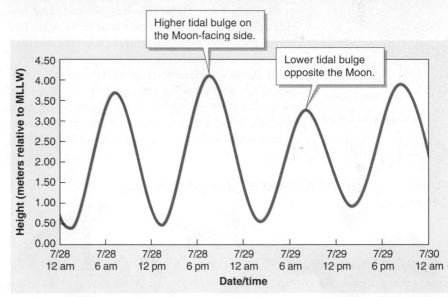

Side facing the Moon
The Moon's gravitational pull causes the oceans to mound up on the side facing the moon, creating a high tide. The Moon-facing side has the highest tide.

Side opposite the Moon
There is less gravitational pull by the Moon on the opposite side of Earth because it is farthest from the Moon. As a result, water bulges out and creates a high tide on the opposing side of Earth.

A

Higher tidal bulge on the Moon-facing side.

Lower tidal bulge opposite the Moon.

B

Figure 20.2 Lunar tides. (A) Earth's two tidal bulges are greatly exaggerated here to make them easier to see. The tidal bulge nearest the Moon is the larger of the two. Every location on Earth passes through a tidal bulge approximately every 12 hours. (B) This tide graph shows the twice-daily tidal rhythm (in meters) in relationship to the mean level of low water (MLLW) near Kodiak Island, Alaska, from July 28 to July 30, 2013.

Figure 20.3 Spring and neap tides.

The highest tides, called spring tides, occur when Earth, the Sun, and the Moon are aligned along the same axis (shown as a dotted line). In this position, the gravitational pull of the Moon and Sun are in the same direction, creating above-average tides. Spring tides occur during a full moon or during a new moon. The lowest tides, called neap tides, occur when the Moon is in half-phase and at a right angle to the line between Earth and the Sun.

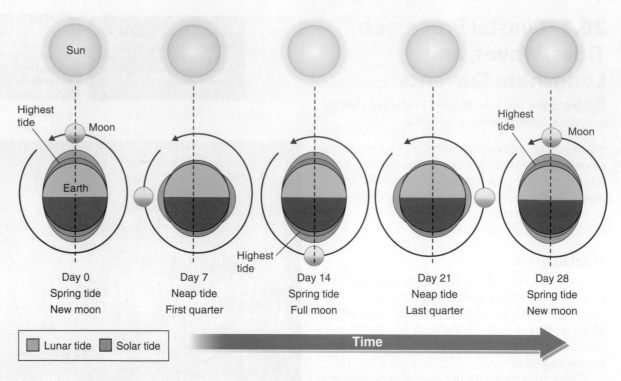

repeats twice daily on most coastlines. *High tide* occurs when the flood tide has peaked, and *low tide* occurs when the ebb tide is at its lowest. The shore area that is exposed during low tide but covered during high tide is called the *intertidal zone*.

If Earth's surface were perfectly smooth, without wind, and covered only with water, the *tidal range*—the difference between low tide and high tide levels—would be the same for any given latitude at all locations. However, differences in bathymetry,

Web Map

Tidal Ranges

Available at
www.saplinglearning.com

windiness, and the positions and shapes of coastlines affect the height of tides and create uneven tidal ranges across the oceans **(Figure 20.4)**.

The average tidal range on Earth is 100 cm (36 in). But in some regions, tides are barely noticeable, while in others, their effects dominate the coastline. The world's greatest tidal range occurs in the Bay of Fundy in eastern Canada **(Figure 20.5)**, where the maximum tidal range is 16.8 m (55 ft).

Ocean Waves

Ocean waves are caused by wind. As wind blows the surface of a body of water, molecules of air drag against molecules of water. Some of the kinetic energy from the moving air is transferred to the water. The wind energy absorbed by the water causes both surface ocean currents and waves. If there were no wind, ocean currents would be eliminated, and the oceans would be glassy smooth. Only infrequent wave-generating forces, such as earthquakes (see The Human Sphere in Chapter 15, "Geohazards: Volcanoes and Earthquakes") and coastal landslides, would make waves.

In the open ocean, wave height depends on the speed of the wind, the duration of the wind, and the **fetch**: the distance of open water over which the wind blows. Slow winds, blowing for a short time over a small fetch, such as an inland lake, produce small waves. Fast and sustained winds blowing over the open ocean generate large waves called *swells*. Large swells are generated in the open ocean; they can have wave heights of 2 to 10 m (6 to 33 ft) or more and *wavelengths* (the distance between wave crests) of 40 to 500 m (130 to 1,600 ft) or more.

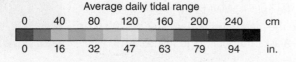

Average daily tidal range

0	40	80	120	160	200	240	cm
0	16	32	47	63	79	94	in.

Figure 20.4 World map of daily tidal ranges. This map shows average daily tidal ranges worldwide. Tidal ranges vary due to geometry of the coastline and the seafloor and prevailing wind speed and direction.

More visible waves with varying heights and wavelengths are superimposed on top of the swells. As these smaller waves move across the surface of the open ocean, they cause the water to appear to be flowing, but it is not. When the wind blows a flag, waves pass down the length of the flag as it ripples, but the fabric itself does not move down the length of the flag with the waves. Similarly, a floating object on the ocean's surface, such as a bottle, moves in a circular pattern as the waves pass beneath it. The bottle does not move forward horizontally with the waves. Individual water molecules, like the bottle, trace a circular path. Waves that cause this circular movement are called **waves of oscillation.**

As waves approach the shore, they become organized in long ridges, and the rotating circles in the waves of oscillation organize to form long cylinders. The radius of the topmost cylinder at the surface is equal to the *wave height*. Each wave is composed of a series of cylinders that are stacked vertically. With increasing depth, the diameter of the cylinders decreases. The movement of water ends at the *wave base*, at a depth equal to about three-quarters of the wavelength **(Figure 20.6)**.

As a wave approaches the coastline, the wave base comes in contact with the seafloor. As the wave continues into shallower water, friction with the seafloor slows the base more than the top and,

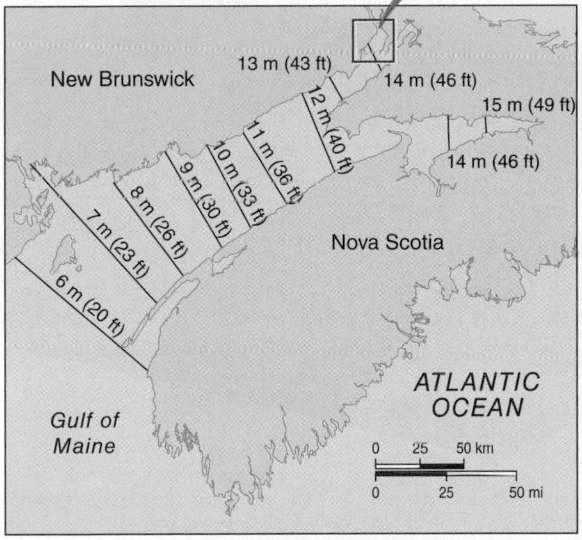

Figure 20.5 **The world's greatest tidal range.**

The unusually high tides in the Bay of Fundy are caused by the shape of the coastline. As water enters the bay, it is constricted and piles up in the narrow funnel of the bay. The inset photos show the same location at Hopewell Cape in the Bay of Fundy, in New Brunswick, at low tide (top) and at high tide (bottom). Note the people in the top photo for scale. *(Top. Danita Delimont/Getty Images; bottom. Edward Kinsman/Getty Images)*

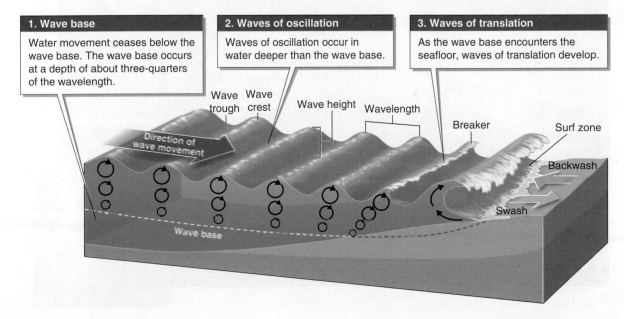

Figure 20.6 Wave features. Wave motion ceases at the wave base (shown here as a dotted line). In shallow water, the wave base encounters the seafloor, creating waves of translation that form breakers.

1. Wave base
Water movement ceases below the wave base. The wave base occurs at a depth of about three-quarters of the wavelength.

2. Waves of oscillation
Waves of oscillation occur in water deeper than the wave base.

3. Waves of translation
As the wave base encounters the seafloor, waves of translation develop.

as a result, the wave crest grows higher. The wave crest also travels faster than the wave base, causing the wave to shear in an elliptical motion. In this type of wave, called a **wave of translation**, water (and any object in it) moves forward horizontally in the direction of wave movement. Eventually, the wave crest collapses, forming a *breaker* (see Figure 20.6).

On a beach, breakers form in the *surf zone*. As breakers collapse on themselves in their forward momentum, they create a **swash**: a rush of water up a beach. The slope of the beach will limit how far the swash can travel up the beach. The water flows

back down the beach as **backwash** and returns to the ocean (see Figure 20.6).

As waves approach the shoreline, they bend so that their orientation is roughly parallel to the shoreline. This process is called **wave refraction**. Waves bend (refract) around the contours of the coastline because of the underwater topography. As a wave approaches the shore, the side of the wave closest to the shore encounters the seafloor first and slows more than the portion farther away, in deeper water. Even where the coast is uneven, with rocky prominences called **headlands**

Figure 20.7 Wave refraction. (A) As a wave approaches the shoreline at an angle, it wraps around the contours of the coastline. (B) Waves refract around a headland near the town of Bulli in New South Wales, Australia. *(Andrew McInnes/Alamy)*

1. Waves approaching shore
As waves approach the shoreline at an angle, the wave bases nearest to shore encounter shallow water first and slow down due to friction with the seafloor.

2. Wave refraction
The same wave farther from shoreline, in deeper water, continues traveling at the same speed, causing the wave to be aligned parallel to the shoreline. As a result, the wave refracts around the contours of the coastline.

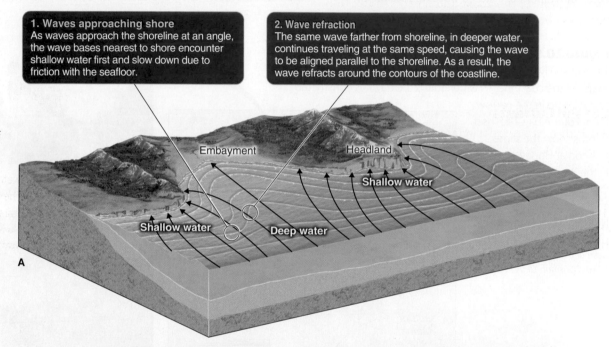

and concave sandy beaches called **embayments**, waves refract around the headlands and maintain a mostly parallel approach to the shoreline **(Figure 20.7)**.

Sometimes enormous waves called *rogue waves* develop in the open ocean. Scientists are unsure exactly how rogue waves form, but they appear to be mostly the result of *phase alignment* among waves of oscillation. This occurs when two or more wave crests merge into a single wave with a greatly amplified wave crest height **(Figure 20.8)**.

Rogue waves can pose serious danger, possibly damaging or even sinking vessels out at sea. When they form near the coast and move ashore, they become *sneaker waves* or *sleeper waves*. Coastal sneaker waves are extremely dangerous; many people walking along the beach or on ocean jetties that extend out to sea have been surprised and drowned by these waves.

Coastal Currents

Although wave refraction causes waves to approach a beach nearly parallel to the shoreline, most waves wash ashore at a slight angle. This angled approach of waves creates a **longshore current**, which flows parallel to the beach in the direction of wave movement. **Longshore drift** is the movement of sediment down the length of the beach in the direction of wave movement **(Figure 20.9)**. Longshore drift transports sand,

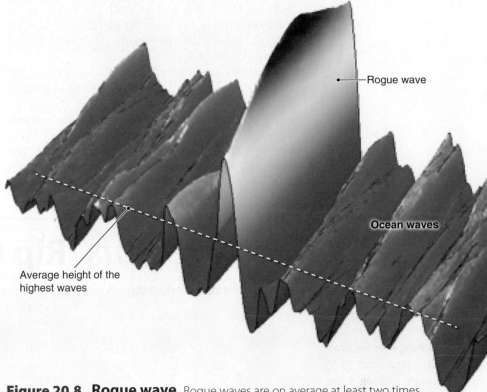

Figure 20.8 Rogue wave. Rogue waves are on average at least two times taller than the highest waves (shown with a dotted line). They are exceedingly rare, and their existence has only been confirmed in the past few years through computer modeling. This image shows a computer simulation of a rogue wave. *(Birkholz, S. et al. Ocean rogue waves and their phase space dynamics in the limit of a linear interference model. Sci. Rep. 6, 35207; doi: 10.1038/srep35207 (2016). Courtesy Günter Steinmeyer)*

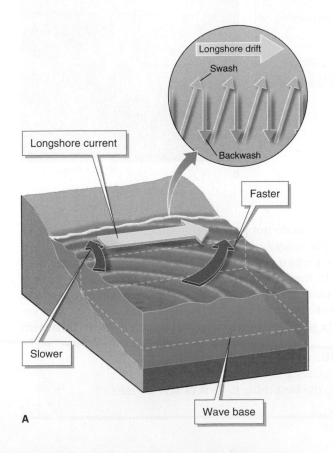

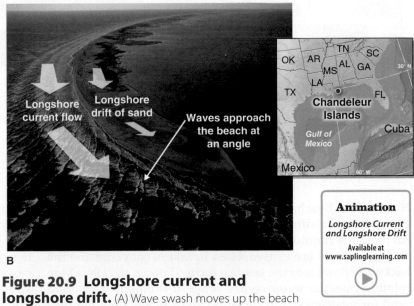

Figure 20.9 Longshore current and longshore drift. (A) Wave swash moves up the beach at an angle, but gravity pulls the backwash straight down the slope of the beach. As a result, sand follows a zigzag path down the beach in the process of longshore drift. (B) Along the Chandeleur Islands, off the coast of Louisiana, longshore current and longshore drift result as incoming waves approach the beach at an angle. The resulting movement of sand may gradually lengthen the island. *(Annie Griffiths/Getty Images)*

both on the beach and beneath the water in the surf zone. Only the topmost meter or so of sand actively moves. Deeper layers move only during strong storms, if at all.

Beaches are generally safe places to enjoy, but dangers sometimes lurk. In addition to extremely rare sneaker waves washing ashore, dangerous rip currents can form on beaches. **Rip currents** are short-lived jets of water that form as backwash from breakers flows back into the sea. The **Picture This** feature below explores them further.

20.2 Coastal Landforms: Beaches and Rocky Coasts

◎ **Describe landforms on beaches and rocky coasts and explain how they form.**

Why are some coastlines flat with miles of sandy beach, while others are composed of steep and rocky cliffs? The tectonic setting determines what kind of landforms develop in any given stretch of coast. Most importantly, plate movements can lift or

Picture This **Rip Currents**

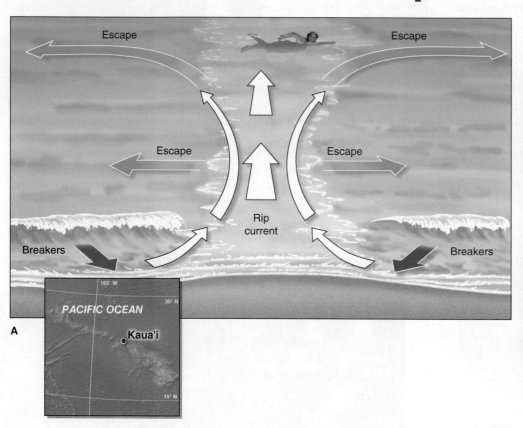

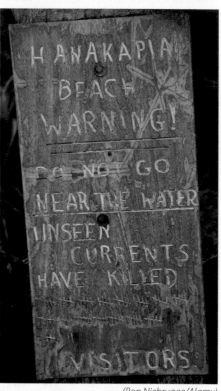

B   *(Ron Niebrugge/Alamy)*

Hanakapi'ai Beach, on the island of Kaua'i, is marked with a sign warning visitors to avoid the water due to rip currents. An ominous running tally of fatalities is marked on the sign.

Rip currents are caused when breakers converge and the backwash flows into the sea at a focused point, usually where relatively deeper water occurs between submerged *shoals* of sand (or *sandbanks*). Rip currents can form any time, but they are most likely to form when wave heights are at a maximum. They range in width from a few meters to hundreds of meters. Some rip currents disappear at the breakers, but others may continue for hundreds of meters beyond the surf zone.

Rip currents do not pull people under water. Instead, they carry people away from shore. Even the strongest swimmers may drown when they become too exhausted to swim back to safety. In the United States, about 150 people drown in rip currents each year. If you are caught in a rip current, the best strategy is to swim parallel to the beach to get out of the current rather than try to fight it by swimming toward the beach.

Consider This

1. What causes a rip current?
2. What is the best thing to do if caught in one?

lower Earth's crust and, in so doing, determine the geomorphic features of the coastline. At active continental margins near convergent plate boundaries (see Section 13.2), the converging plates compress and raise the crust. Similarly, the weight of sediments or ice sheets (see Section 18.1) can depress the crust; if this weight is removed, the crust rebounds upward through isostasy. **Isostasy** is the gravitational equilibrium between the lithosphere and the support of the asthenosphere below.

When a coastline lifts to higher elevation, an emergent coast forms. At an **emergent coast**, either sea level is dropping or the land is rising. Emergent coasts are dominated by erosional landforms such as steep cliffs and rocky shorelines. The west coasts of North and South America, for example, are tectonically active emergent coasts with rocky shorelines. The emergent coast of eastern Canada is rising because the weight of the Laurentide ice sheet is gone.

Most plate boundaries do not coincide with continental margins. Instead, a single lithospheric plate holds both a continent and a *coastal plain* that extends beneath the sea as a flat, gently sloping continental shelf (see Section 10.1). These passive continental margins are often **submergent coasts**, where either sea level is rising or the land surface is subsiding. Submergent coasts are dominated by depositional landforms such as beaches and coastal wetlands. The southeastern United States and Gulf Coast, for example, have gentle slopes with sandy beaches and coastal wetlands.

Coastal sea level can also change without the raising or the lowering of the lithosphere. About 15,000 years ago, during the most recent glacial period, sea level was 85 m (280 ft) lower than it is today because so much water was frozen on land in ice sheets (see Section 7.2). When the glacial period ended, the meltwater from these ice sheets raised sea level by 85 m. Change in global sea level as a result of change in the amount of water in the oceans is called **eustasy**.

Either way, an emergent coast occurs where sea level falls (through eustasy) or where the coast is lifted (through isostasy). Likewise, a submergent coast occurs where sea level rises (eustasy) or where the coast sinks (isostasy).

Other important factors that contribute to the shape of a coastline are the amount of wave energy reaching it, the presence of living structures such as coral reefs, the tidal range, and the amount and type of sediment available. **Table 20.1** groups coasts into five main categories. In this section, we discuss two of the most common types of coasts: *beaches* and *rocky shorelines*.

Table 20.1 **Types of Coasts**

TYPE OF COAST	DESCRIPTION/SETTING	TEXT REFERENCE
1. Depositional coasts: Beaches and deltas		
Beach	Sand-dominated coast	Section 20.2
Barrier island	Sand-dominated coast, with a water body separating the island from the mainland	Section 20.2
Delta	Formed at river mouths	Section 17.3
2. Erosional coasts: Rocky shorelines		
Bedrock	Found anywhere bedrock is exposed to coastal wave energy	Section 20.2
Limestone karst	Various settings; common in Caribbean islands and in Croatia	Section 16.2
Lava flow	Various settings; found on Hawai'i	Section 13.4
3. Organic coasts		
Coral reef	Dominated by living reefs; restricted to the tropics and subtropics	Section 10.2
Mangrove	Dominated by mangroves; restricted to the tropics and subtropics	Section 10.2
Estuary	Formed where freshwater meets salt water at river mouths	Section 10.2
Anthropogenic	Includes any coast that has undergone significant modification by human activities	Chapter 10, "Ocean Ecosystems"
4. Flooded coasts		
Ria	Drowned V-shaped river valley; found at the mouths of rivers and where estuaries form	Section 10.2
Fjord	Drowned U-shaped glacial valley at high latitudes	Section 18.2
5. Ice coasts		
Glaciers and ice shelves	Generally found at latitudes higher than 55 degrees	Chapter 18, "The Work of Ice: The Cryosphere and Glacial Landforms"
Glacial sediments	Found where continental ice sheets deposited glacial sediments; common in northeastern Canada and northern British Isles	Section 18.3

Beaches

A beach consists of an *offshore zone*, a *nearshore zone*, a *foreshore zone* (or intertidal zone), and a *backshore zone*. A raised ridge of sand, called a *berm*, separates the foreshore and the backshore zones **(Figure 20.10)**. Waves that break on the gentle slope of a beach have less energy than those that break on a rocky coast. As waves move toward the beach, the base of each wave makes contact with the seafloor in the nearshore zone. As a result, the wave's energy is reduced as it works to transport sediments on the seafloor before encountering the foreshore zone.

The composition of beach sediments varies, depending on the source of the sediments. Tan and beige beaches of quartz are the most common because quartz is particularly resistant to weathering and erosion and persists for a long time (see Section 16.1). Not all beaches are composed of quartz sand, however. Weathering and erosion of coral reefs results in brilliant white beaches consisting of broken fragments of coral **(Figure 20.11A)**. Many volcanically active areas have black beaches made of basalt particles **(Figure 20.11B)**.

Not all beaches are composed of sand. Where weathering and mass movement of nearby cliff faces are occurring, larger particles, such as gravel or cobbles, form a **shingle beach (Figure 20.11C)**.

Beach Landforms

Beaches are like rivers of sediment. Longshore drift transports sediments down the length of the beach. Surface sediment, usually sand, is continually entering the beach from the "upstream" end, and it is continually leaving the beach on the "downstream" end.

The amount of sand available to a beach is that beach's *sediment budget* (the difference between sand inputs and sand outputs). When the sediment budget is in equilibrium, sand inputs and outputs are equal, and the beach neither grows nor shrinks. When the sediment budget is positive, the beach accrues sand and grows, and when it is negative, the beach loses sand and is diminished. Typically, the sediment budget fluctuates seasonally. For example, sand on beaches often accumulates in summer. Then in winter the powerful waves from storms bring net erosion to the beach as they move the sand just offshore. Although it may fluctuate seasonally, over the course of a year, the sediment budget of most beaches is in equilibrium. If it isn't, the beach either diminishes or grows in size.

Figure 20.12 on page 622 shows four beach landforms commonly created by the movement of sand:

- *Coastal sand dunes* form where prevailing winds blow sand inland. Vegetation traps the sand, allowing it to accumulate into dune formations. Through time, many coastal sand dunes become stabilized by vegetation through the process of ecological succession (see Section 8.5).

- A **sandspit** is an elongated dry bar of sand that extends from a beach out into the water, usually parallel to the shore.

- A **baymouth bar** is an unbroken sandspit that forms across the mouth of a river or an estuary. A baymouth bars creates an enclosed brackish-water lagoon behind it and is formed by longshore drift.

- A **tombolo** forms when a sandspit connects an island to the mainland. Sandspits, baymouth bars, and tombolos are all formed by longshore drift.

Another common type of depositional coastal landform is a **barrier island**: an offshore sandbar that

Figure 20.10 Beach zones.

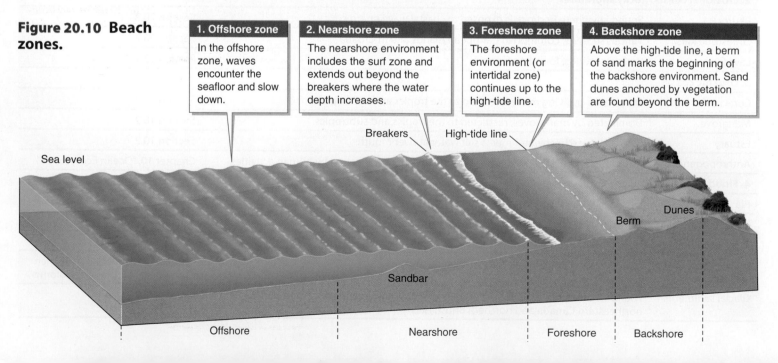

1. Offshore zone	2. Nearshore zone	3. Foreshore zone	4. Backshore zone
In the offshore zone, waves encounter the seafloor and slow down.	The nearshore environment includes the surf zone and extends out beyond the breakers where the water depth increases.	The foreshore environment (or intertidal zone) continues up to the high-tide line.	Above the high-tide line, a berm of sand marks the beginning of the backshore environment. Sand dunes anchored by vegetation are found beyond the berm.

Breakers   High-tide line

Sea level

Dunes

Berm

Sandbar

Offshore   Nearshore   Foreshore   Backshore

runs parallel to the coast. Worldwide, 2,149 barrier islands are recognized. These islands total more than 20,000 km (12,400 mi) in length, representing some 10% of all shorelines on Earth. Antarctica is the only continent without them. Barrier islands range in size from a few meters wide and less than a kilometer in length to several hundred meters wide and hundreds of kilometers in length. Barrier islands are separated from the mainland by sounds, bays, estuaries, and lagoons (see Section 10.1).

Barrier-island formation requires an uninterrupted supply of sand; tectonically stable, flat, and shallow coastal topography; and sufficient wave energy for longshore transport of sand. **Figure 20.13** on page 623 provides several examples of barrier islands.

A

B

C

Figure 20.11 Three kinds of beaches.
(A) The coral reefs surrounding the Galápagos Islands create white beaches composed of broken fragments of the reefs. (B) Iceland's beaches are dark gray or black because they are composed of sand derived from basalt, the material from which the island is made. This photo shows a black-sand beach near the village of Vík. (C) Beaches that are composed of gravel, pebbles, cobbles, or some combination of these materials are called shingle beaches. The particles found on shingle beaches are always smoothed and rounded by wave action. This shingle beach with cobbles is in Pembrokeshire, Wales.
(A. Ralph Lee Hopkins/Getty Images; B. Philippe Turpin/Getty Images; C. SPL/Science Source)

**Figure 20.12
Beach landforms.**
(Top left. © Ben Visbeek; top right.
© Emich Szabolcs, www.emich
.hu; bottom left. robertharding/
Alamy; bottom right. David Wall/
Alamy)

1. Coastal dunes

These coastal dunes in the Netherlands have been stabilized by vegetation.

2. Sandspit

This sandspit in Croatia extends into the water.

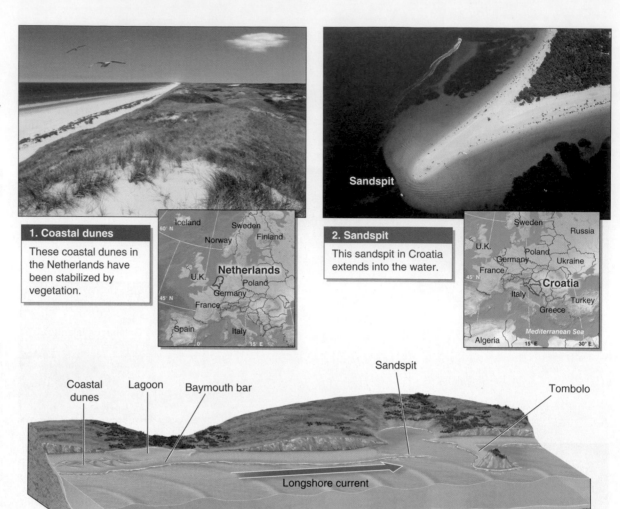

3. Baymouth bar

This baymouth bar encloses a lagoon in Dalyan, Turkey.

4. Tombolo

Mount Maunganui and the beach that connects it to North Island, New Zealand, form a tombolo.

A

Outer Banks barrier islands

Pamlico Sound

Cape Hatteras

0   10   20 mi
0   5   10 km

NY MA
MI
OH PA
IN
WV VA
KY
NC
TN SC Outer Banks
AL GA
75° W

B

PA
MD
WV Ocean City
VA
75° W

Figure 20.13 Barrier islands.

(A) Some barrier islands are located far from the coast. This satellite image shows the barrier islands that make up the Outer Banks of North Carolina. Pamlico Sound is a lagoon created by these barrier islands. Slatted "sand fences" (top photo) are used to help anchor sand in the backshore area. The Cape Hatteras Lighthouse (bottom photo) began operating in 1802 to warn ships of the ever-shifting shoals in the area. In 1999 it was moved inland 884 m (2,900 ft) due to beach erosion. (B) Ocean City, Maryland, is built on a barrier island. *(A: Left. NASA; right top. catnap72/Getty Images; right bottom. wbritten/Getty Images; B. Chris Parypa/Alamy)*

Barrier islands are particularly important to coastal communities because they absorb a significant amount of storm surge and wave energy from tropical cyclones (see Section 6.3). In so doing, they protect the mainland shoreline from the full force of storms.

Human Modification of Beaches

Beaches can lose sand or gain sand overnight during powerful storms that pound them with wave energy. Sand not only drifts "downstream" to other coastal areas but may also be removed from the coastal system if it is carried to deeper waters. In this situation, the beach's sediment budget is temporarily negative.

Over time, longshore drift returns sand to the beach, assuming that there is an adequate supply of sand feeding into the system from upstream. The beach's sediment budget will remain negative, however, if there is an inadequate supply of sand. Recall from Section 17.3 that sediments are transported to the coast by streams. Streams flowing into artificial reservoirs drop their sediments in the inland reservoir rather than on the coast. As a result, reservoirs become sediment traps that reduce the volume of sand flowing down a coast **(Figure 20.14)**.

When streams do not supply a steady input of sediment, beaches become sediment starved, and their sand is lost to erosion. By some estimates, as much as 90% of the beaches in the United States are experiencing net erosion (see the Geographic Perspectives at the end of the chapter). These losses of sand present problems because beaches

protect coastal development from the impact of storm energy, support local coastal economies, and provide habitat for many organisms, including shorebirds and nesting sea turtles. According to the Federal Emergency Management Agency (FEMA), beach erosion due to sea-level rise (see Section 18.5) and inland sediment traps is likely to destroy 25% of houses within 150 m (500 ft) of the coast in the United States in the next 60 years.

Given the extent of the problem of sand loss and the economic importance of beaches, scientists have developed several means of studying sediment flow on beaches to better manage beaches and reduce erosion **(Figure 20.15)**.

One method commonly employed to slow the erosion of sand is the use of groins. A **groin** is a linear structure made of concrete or stone that extends from a beach into the water **(Figure 20.16** on page 626). Sand moving down the beach encounters the groin and accumulates against it. A series of groins in a row is called a *groin field*. Groin fields slow, but do not stop, the loss of sand from a beach.

One problem with the use of groins is that they are hard structures that mar the appearance and character of the beach. They also reduce the amount of sand transported downstream to other beaches, which, in turn, become sediment starved and subsequently require the construction of more groins. This outcome can result in legal conflicts between beach landholders.

Like groins, **jetties** are design to slow the flow of sand on a beach. Jetties are artificial walls placed at the mouths of harbors, usually in pairs,

Figure 20.14 Beach sediment starvation.

Coastal beaches are fed sand by streams draining into the ocean. Reservoirs behind dams trap sediment upstream before it reaches the beach, potentially diminishing the sand supply for coastal beaches.

Animation

Beach Sediment Starvation

Available at
www.saplinglearning.com

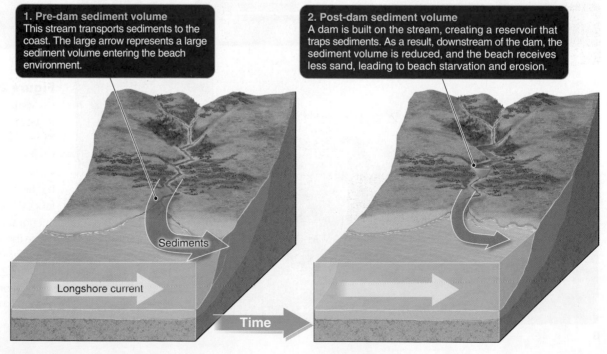

1. Pre-dam sediment volume
This stream transports sediments to the coast. The large arrow represents a large sediment volume entering the beach environment.

2. Post-dam sediment volume
A dam is built on the stream, creating a reservoir that traps sediments. As a result, downstream of the dam, the sediment volume is reduced, and the beach receives less sand, leading to beach starvation and erosion.

Sediments

Longshore current

Time

Nearshore instrumentation
Long metal pipes are driven deep into the sediments just offshore. Scientific instruments will be secured to the anchored pipes. The instruments will measure many variables, including water movement and sediment transport.

Radar
A radar system mounted on the berm measures the speed and direction of the longshore current.

Photography
A camera is mounted on the railing of Cape Hatteras lighthouse to provide additional information about longshore current and wave behavior.

Dye tracer
A green dye is dropped into the ocean to trace water movement in the nearshore environment. Movement of the dye is tracked from the air.

Truck-mounted LIDAR and radar
This truck carries LIDAR and radar equipment that is used to map changing beach topography and bathymetry.

Figure 20.15
SCIENTIFIC INQUIRY: How do scientists study beach erosion?

Understanding sediment flow within a beach system is central to efforts to reduce erosion and preserve the beach. This series of photos documents the USGS project that is monitoring Cape Hatteras Beach in North Carolina. *(U.S. Geological Survey, Coastal and Marine Geology Program)*

Video

The Importance of Accurate Coastal Elevation and Shoreline Data

Available at
www.saplinglearning.com

Figure 20.16 Groins and jetties. (A) Artificial means of managing sand movement include groins and jetties. (B) A groin field slows the rate of sand loss, and jetties keep the harbor clear of sediments, near Arles, France. *(Ausloos Henry/AGE Fotostock)*

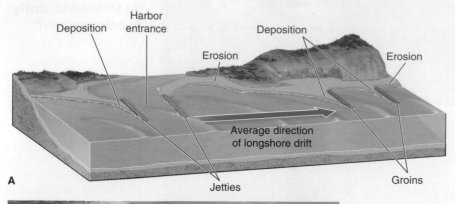

to prevent sand from closing the harbor entrance (see Figure 20.16A). Without jetties, frequent and costly dredging is required to keep a harbor deep and navigable by boats.

Seawalls are artificial hard structures designed to protect backshore environments from wave erosion during large storms. They are normally concrete or metal walls built parallel to the beach above the intertidal zone. Although they effectively reduce backshore erosion, they can increase the loss of beach sand by reflecting the energy of storm waves, which therefore erode more sand from the beach. They also prevent natural replenishment of backshore dunes with beach sand **(Figure 20.17)**.

Rocky Coasts

Rocky coasts often have deeper water leading up to the coastline than beaches do. As a result, wave energy is not expended on the seafloor, as it is in beach environments. Instead, the waves expend the full force of their energy against the rocks. In these environments, waves are a powerful and relentless force of erosion.

Water weighs 1 kg per liter (8.3 lb per gallon) and can be compressed very little. The weight and force of crashing waves pry rocks apart as air and water are forced into even the tiniest joints. There are many other landforms common to rocky coasts. **Sea stacks** are vertical towers of rock separated from but near a rocky coastal cliff face. **Wave-cut benches** are flat coastal platforms that result from wave erosion and exposed at low tide, and **wave-cut notches** are notches that form at the base of a coastal cliff as a result of weathering and erosion. These landforms are illustrated in **Figure 20.18**.

Figure 20.17 A seawall in Honolulu, Hawai'i. This seawall has stabilized the position of the coastline but at the expense of Waikiki Beach in Hawai'i. Although seawalls may effectively check the loss of coastal land, they accelerate the erosion of sand and often create unwelcoming environments. *(Carrie Thompson/Alamy)*

Arch and sea stack

Arches are formed by differential erosion of a coastal cliff. When the arch collapses, a sea stack will form. Sea stacks are steep or vertical towers of rock separate from, but close to, a rocky coast. The coast pictured here is near Étretat, in northern France.

Headland and embayment

Headlands occur where resistant rock outcrops protrude into the sea. Embayments are concave shorelines where beaches form. This coastline is in County Kerry, Ireland.

Figure 20.18 Landforms of rocky coasts.

(Clockwise from top left. Christian Goupi/Getty Images; Gareth McCormack/Getty Images; Christopher Biggs/Getty Images; Skyscan Photolibrary/Alamy; Reinhard Dirscherl/AGE Fotostock)

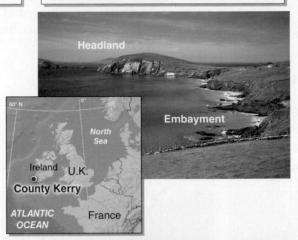

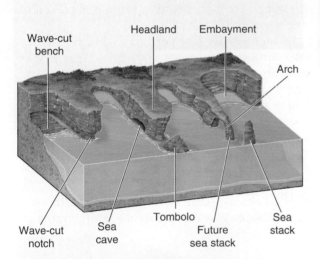

Sea cave

Sea caves, such as this one in Cabo San Lucas in Mexico, form where waves erode weaker areas of bedrock along joints.

Wave-cut notch

Weathering and erosion have formed a wave-cut notch at the base of this limestone island in the Republic of Palau.

Wave-cut bench

Wave-cut benches are flat coastal platforms created by erosion from waves and exposed at low tide. This wave-cut bench is near Bridgend in Southern Wales.

On rocky shores, marine terraces indicate the presence of an emergent coastline. A **marine terrace** is a wave-cut bench that has been elevated above sea level **(Figure 20.19)**. Either isostatic or eustatic changes can create a marine terrace.

Where a steep coastal cliff is composed of weak rock such as sandstone, the cliff may be undercut and break off in large pieces. Landslides may even develop, resulting in a situation called *cliff retreat*, in which a cliff face retreats inland. The coastal cliffs in Big Sur, California, shown in **Figure 20.20**, provide an example of cliff retreat.

Figure 20.19 Marine terrace. (A) This diagram illustrates the formation of a marine terrace by geologic uplift. Its formation is an example of isostasy because the crust has been uplifted. (B) This photo shows a marine terrace on an emergent coastline in northern Scotland near the town of Durness. The flat marine terrace is a wave-cut bench that was exposed after the land was uplifted. *(B. © Ian Gordon)*

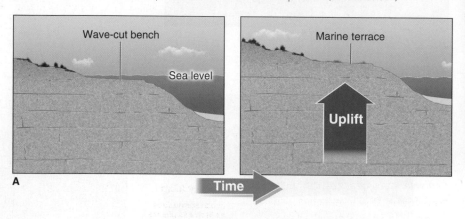

Figure 20.20 Cliff retreat. The cliffs of Big Sur in California have retreated at an average rate of 18 cm (7 in) per year over the past 50 years. Highway 1 is frequently closed as a result of landslides. In this March 2011 photo, a landslide followed several days of heavy rain that saturated and weakened the steep cliff. *(AP Photo/Monterey Herald, Orville Myers)*

GEOGRAPHIC PERSPECTIVES

20.3 The Sisyphus Stone of Beach Nourishment

◎ Evaluate the practice of restoring eroded beaches through artificial replenishment of sand.

Superstorm Sandy made landfall in New Jersey on October 22, 2012. About 94% of New Jersey's beaches and backshore dunes, as well as the structures on them, were damaged by the storm surge and waves that Sandy generated. Many of the beaches affected by the storm lost sand. Some 60% of New Jersey's beaches were narrowed from 15 to 30 m (50 to 100 ft). More recently, in 2017 Hurricane Irma caused similar catastrophic beach erosion in Florida (**Figure 20.21**).

What Is Beach Nourishment?

Beach erosion may occur in a single major storm event, but as we saw in Section 20.2, it may also be a gradual process in which sand is slowly lost over many years. There are three primary options available to communities that want to address the problem of a negative sediment budget and resulting beach erosion: (1) relocation of development away from the beachfront, (2) construction of hard structures such as the groins and seawalls discussed in Section 20.2, and (3) **beach nourishment** (or *beach replenishment*): the artificial replenishment of a beach with sand.

Beach nourishment can be used after a single large storm, such as Sandy, has significantly eroded a beach. More often, however, beach nourishment is a routine maintenance process used to restore sand on a beach. The most common way to obtain sand is through offshore dredging. A large ship vacuums up sand from the seafloor and pumps it onshore through large metal pipes. Bulldozers then push the sand into place (**Figure 20.22**). Sand that is dredged off the coast and added to a beach is considered *artificial sand*.

Because beach nourishment does not address the underlying cause of sand loss, the average lifetime for an artificial beach is 5 years. Sometimes the sand lasts less than 1 year, and sometimes it persists for a decade. On average, Virginia Beach, Virginia, has been nourished with sand every 1.3 years since 1951 (**Figure 20.23** on the following page).

Is Beach Nourishment Worth the Expense?

As we have seen in this chapter, beaches are spatially dynamic landforms. Human activities have changed the dynamics of sand movement along

Figure 20.21 Hurricane Irma beach erosion. When Hurricane Irma made landfall in Florida on September 11, 2017, the storm removed many beaches. This section of South Ponte Vedra Beach in Jacksonville, Florida, once had an extensive beach. Erosion from the storm removed the beach and left these homes on the verge of collapse. *(Paul Hennessy/Alamy)*

Story Map
Beach Formations and Storm Resilience
Available at
www.saplinglearning.com

Figure 20.22 Beach nourishment. On this beach in Barcelona, Spain, sand is being siphoned off the seafloor by a ship and pumped to shore in a slurry of water. On the beach, bulldozers will shape the sand. *(imageBROKER/Alamy)*

Figure 20.23 Beach nourishment in Virginia. This photo shows the difference between a recently nourished portion of the beach and a portion that has not recently been replenished with sand. *(Photo by USACE)*

coastlines, such that negative sediment budgets are increasingly common for many beaches worldwide. Beach nourishment resets the clock on beach erosion by only a few years; it does not address the underlying problem.

Proponents of beach nourishment argue that just as the nation's roads and bridges must be maintained, so must the nation's beaches. Beaches are a public asset and an important part of the national economy. The U.S. Army Corps of Engineers oversees beach nourishment projects paid for by the federal government. The Corps sees beach nourishment as beach maintenance for the common good: The beaches have sand, and the hotels have tourists. The livelihoods of many people depend on vacationers attracted to beaches. The Corps estimates that it has saved $443 million in storm-related damages to hotels in Virginia Beach alone during the past decade.

Opponents of the practice claim that beach nourishment is an exercise in futility, a Sisyphean task. In Greek mythology, King Sisyphus was made to roll a huge stone up a hill repeatedly for eternity as punishment for offending the gods. Just as he was nearly at the top of the hill, the stone would slip and roll back down again, and he would then have to start over. By some estimates, the cost of maintaining beaches in the United States with beach nourishment will be about $6 billion per year by 2020. By the end of the century, the total costs could approach $90 billion. The task of beach nourishment will continue to become more challenging as sea levels continue to rise. Given its costs, and given that it is only a temporary remedy, is beach nourishment a viable long-term strategy for addressing the problem of beach erosion?

At present, the answer appears to be yes. According to the Atlantic States Marine Fisheries Commission, the practice will increase in the coming decades. By 2050, southeastern Florida will have completed 100 beach nourishment projects and moved a total of 76 million cubic meters (100 million cubic yards) of sand. Many major tourist beaches around the world are experiencing erosion, including beaches in Australia, Hawai'i, Hong Kong, and Cancún, Mexico. All of them are relying on beach nourishment to maintain their beaches and keep the tourists coming.

Chapter 20 Study Guide

Focus Points

20.1 Coastal Processes: Tides, Waves, and Longshore Currents

- **Tides:** Tides increase the vertical range of wave erosion in coastal areas.
- **Waves:** Ocean waves are generated by wind. As they approach the shore, they slow down, grow higher, and break on the shoreline.
- **Sediment transport:** Wave energy creates longshore currents that move sediments along a coastline.

20.2 Coastal Landforms: Beaches and Rocky Coasts

- **Emergent coasts:** Most emergent coasts have erosional landforms such as steep cliffs and rocky shorelines.
- **Submergent coasts:** Most submergent coasts have depositional landforms such as beaches and wetlands.
- **Coast types:** Beaches and rocky coasts are two common coast types.
- **Beach landforms:** Beach landforms include coastal sand dunes, baymouth bars, sandspits, tombolos, and barrier islands.
- **Sand loss from beaches:** Artificial reservoirs trap sediment transported by streams and prevent it from reaching coastal beaches downstream, starving the beaches of sand.
- **Solutions to sand loss:** Groins, seawalls, and beach nourishment address sand loss on beaches.
- **Erosion of rocky coasts:** The force and weight of crashing waves break rock apart, forming various rocky landforms. These landforms include sea stacks and arches, headlands and embayments, wave-cut notches, and wave-cut benches.

20.3 Geographic Perspectives: The Sisyphus Stone of Beach Nourishment

- **Beach erosion:** Beach erosion is a widespread problem that threatens coastal development and economies.
- **Beach nourishment:** Beach nourishment is a common response to beach erosion that has drawbacks.

Key Terms

backwash, 616
barrier island, 620
baymouth bar, 620
beach nourishment, 629
embayment, 617
emergent coast, 619
eustasy, 619
fetch, 614
groin, 624
headland, 616
isostasy, 619
jetty, 624
longshore current, 617
longshore drift, 617
marine terrace, 628

rip current, 618
sandspit, 620
sea stack, 626
seawall, 626
shingle beach, 620
submergent coast, 619
swash, 616
tide, 613
tombolo, 620
wave of oscillation, 615
wave of translation, 616
wave refraction, 616
wave-cut bench, 626
wave-cut notch, 626

Concept Review

The Human Sphere: Mavericks

1. How do waves derive their energy from the Sun?

20.1 Coastal Processes: Tides, Waves, and Longshore Currents

2. What are tides? Why do they occur? Compare and contrast flood tides and ebb tides.

3. Compare and contrast neap tides and spring tides. Explain why each occurs.

4. How do waves form on the ocean's surface?

5. Compare and contrast a wave of oscillation and a wave of translation. Which one moves floating objects and water forward?

6. What factors contribute to wave height?

7. Why do waves approach nearly parallel to the coast?

8. What are longshore currents and longshore drift, and what causes them?

20.2 Coastal Landforms: Beaches and Rocky Coasts

9. What are emergent coastlines and submergent coastlines? What causes each? What landforms are most common in each?

10. What are eustasy and isostasy? What causes each?

11. What are the five major categories of coastlines?

12. Why are some beaches white and others black?

13. What is a shingle beach?

14. What is a barrier island?

15. How do inland dams and their reservoirs affect sediment transport to beaches?

16. How do scientists monitor movement of sand on a beach? What instruments do they use?

17. What physical structures are constructed on beaches to slow the loss of sand?

18. What landforms are found along rocky coastlines?

19. What is a marine terrace? In what two ways do marine terraces form?

20. What is cliff retreat? What causes it?

20.3 Geographic Perspectives: The Sisyphus Stone of Beach Nourishment

21. Why is beach erosion a problem?

22. What is beach nourishment?

23. What underlying problems cause beach erosion around the world?

24. How well does beach nourishment address the problem of beach erosion? List the pros and cons of this approach.

Critical-Thinking Questions

1. A recent study reports that ocean wave height is increasing worldwide. What effect, if any, do you think this increase will have on coastal erosion? Explain.

2. Think about the coast nearest to where you live or the last coast you visited. Is it an emergent coast or a submergent coast? What information do you need to answer this question?

3. When was the last time you went to a beach? Do you think that beach has been nourished? How would you find out?

4. Critics argue that because places such as Virginia Beach benefit from the tourism-related revenue, they should pay the entire bill for the cost of beach nourishment. The federal beach nourishment program has been criticized as "welfare for the rich." Critics believe that taxpayers from outside the state where a beach is located, who will probably never visit that beach or benefit from it, should not pay for its maintenance. What is your opinion on this issue? How do you think beach erosion should be addressed?

Test Yourself

Take this quiz to test your chapter knowledge.

1. **True or false?** The lunar tide is higher than the solar tide.

2. **True or false?** Waves on the ocean form because of ocean currents.

3. **True or false?** Reservoirs supply sediments to coastal areas and help beaches grow.

4. **True or false?** Marine terraces form only on emergent coasts.

5. **True or false?** Seawalls are designed to protect sand on beaches.

6. **Multiple choice:** Which of the following is built to keep harbor entrances free of sediments?

    a. groin
    b. jetty
    c. sea stack
    d. tombolo

7. **Multiple choice:** Which of the following is not a landform that results from deposition?

    a. tombolo
    b. sandspit
    c. baymouth bar
    d. headland

8. **Multiple choice:** Complete this sentence: Below the wave base, _____.

    a. wave motion ceases
    b. waves of oscillation begin
    c. baymouth bars form
    d. wave refraction occurs

9. **Fill in the blank:** If caught in a _____, you should swim parallel to shore until you are out of it.

10. **Fill in the blank:** Waves wrap around coastal contours because of _____.

Online Geographic Analysis

NOAA Tide Data

In this online interactive exercise, we will examine tide data.

Activity

Go to https://tidesandcurrents.noaa.gov. This NOAA web page provides real-time tide data for stations in the United States. First, examine Figure 20.4 and get a general sense of which U.S. region has the highest tidal range. Then, on the NOAA web page, click that state. Pan to the coastal area where tidal stations are and select one of them. In the pop-up window, click the "Home Page" link and answer the following questions.

1. **What state is your tidal station in? What is its name?**

2. **When was your tidal station established?**

3. **When is the next high tide? When is the next low tide?**

4. **How high and low will the tides be?**

5. **Units are given in feet. What are these data of tide heights in relationship to?**

6. **What is the mean tidal range at this station?**

7. **What is the diurnal tidal range? What is the difference, do you think, between mean and diurnal tide levels?**

8. **Scroll down to the chart showing the tidal levels. What is the current water level for your station?**

9. **Looking at the graph, are all high tides equal? If not, why aren't they? (See Figure 20.2 for help.)**

10. **Go back to Figure 20.4. Find another state with the lowest tide levels you see on the map and repeat Questions 1 through 9.**

Picture This. *Your Turn*

Coastal Landforms

Fill in the blanks on the diagram, using each of the terms only once.

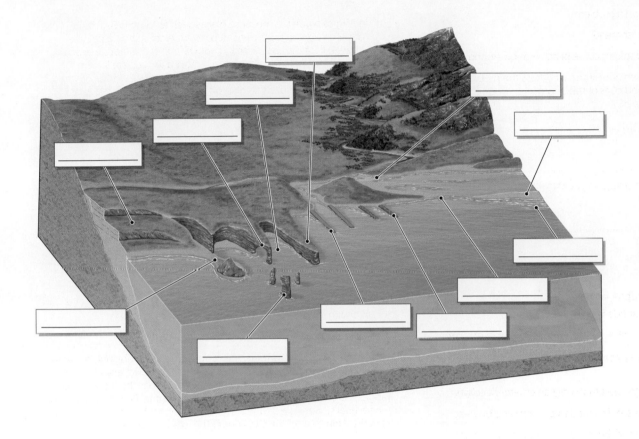

1. Arch
2. Barrier island
3. Baymouth bar
4. Embayment
5. Groin
6. Headland
7. Jetty
8. Marine terrace
9. Sandspit
10. Sea stack
11. Surf
12. Tombolo

For animations, interactive maps, videos, and more, visit www.saplinglearning.com.   **Sapling**Plus

Appendix I Unit Conversions

Unit Conversions—Metric to U.S. Customary

METRIC	MULTIPLIER	U.S. CUSTOMARY
Length		
millimeters (mm)	0.0394	inches (in)
meters (m)	3.28	feet (ft)
centimeters (cm)	0.394	inches (in)
kilometers (km)	0.621	miles (mi)
nautical mile	1.15	statute mile
Area		
square centimeters (cm^2)	0.155	square inches (in^2)
square meters (m^2)	10.8	square feet (ft^2)
square meters (m^2)	1.12	square yards (yd^2)
square kilometers (km^2)	0.386	square miles (mi^2)
hectares	2.47	acres
Volume		
cubic centimeters (cm^3)	0.061	cubic inches (in^3)
cubic meters (m^3)	35.3	cubic feet (ft^3)
cubic meters (m^3)	1.31	cubic yards (yd^3)
milliliters (ml)	0.0338	fluid ounces (fl oz)
liters (L)	1.06	quarts (qt)
liters (L)	0.264	gallons (gal)
cubic meters (m^3)	0.0008	acre-feet (acre-ft)
Mass		
grams (g)	0.0353	ounces (oz)
kilograms (kg)	2.2	pounds (lb)
kilograms (kg)	0.0011	tons (2,000 lb)
metric tons (tonne) (t)	1.1	short tons (tn)
Velocity		
meters per second (m/s)	3.28	feet per second (ft/s)
kilometers per hour (km/h)	0.62	miles per hour (mph)
knots (kt) (nautical mph)	1.15	miles per hour (mph)
Temperature		
degrees Celsius (°C)	1.8 then add 32	degrees Fahrenheit (°F)
Celsius degree (°C)	1.8	Fahrenheit degree (°F)

Unit Conversions—U.S. Customary to Metric

U.S. CUSTOMARY	MULTIPLIER	METRIC
Length		
inches (in)	25.4	millimeters (mm)
inches (in)	2.54	centimeters (cm)
feet (ft)	0.305	meters (m)
miles (mi)	1.61	kilometers (km)
statute miles	0.8684	nautical miles
Area		
square inches (in^2)	6.45	square centimeters (cm^2)
square feet (ft^2)	0.0929	square meters (m^2)
square yards (yd^2)	0.836	square meters (m^2)
square miles (mi^2)	2.59	square kilometers (km^2)
acres	0.405	hectares
Volume		
cubic inches (in^3)	16.4	cubic centimeters (cm^3)
cubic feet (ft^3)	0.0283	cubic meters (m^3)
cubic yards (yd^3)	0.765	cubic meters (m^3)
fluid ounces (fl oz)	29.6	milliliters (ml)
quarts (qt)	0.946	liters (L)
gallons (gal)	3.79	liters (L)
acre-feet (acre-ft)	1,233.5	cubic meters (m^3)
Mass		
ounces (oz)	28.4	grams (g)
pounds (lb)	0.454	kilograms (kg)
tons (2,000 lb)	907.0	kilograms (kg)
short tons (tn)	0.907	metric tons (tonne) (t)
Velocity		
feet per second (ft/s)	0.3049	meters per second (m/s)
miles per hour (mph)	1.6	kilometers per hour (km/h)
miles per hour (mph)	0.8684	knots (kt) (nautical mph)
Temperature		
degrees Fahrenheit (°F)	0.556 after subtracting 32	degrees Celsius (°C)
Fahrenheit degree (°F)	0.556	Celsius degree (°C)

Appendix II Topographic Maps

Topographic maps, or "topo" (pronounced TOE-poe) maps, are often used by geographers when details of Earth's surface are needed. The U.S. Geological Survey (USGS) and other government agencies oversee The National Map, a program that makes topographic maps for the United States (available at http://nationalmap.gov).

Aerial photography and satellite imagery provide much of the data from which topographic maps are made. USGS topographic maps range in scale from 1:24,000 to 1:1,000,000 (see Section 1.4).

The scale 1:24,000 is the most commonly used. Such maps represent 7.5 minutes of the geographic grid; they are therefore often referred to as 7.5-minute topos.

USGS topographic maps provide information about surface relief using contour lines. Human development, agriculture, waterways, glaciers, and vegetation are also shown on topographic maps. **Figure A.1** shows a sample of a USGS topographic map of Crater Lake in Oregon and provides some of the symbols commonly used on topo maps.

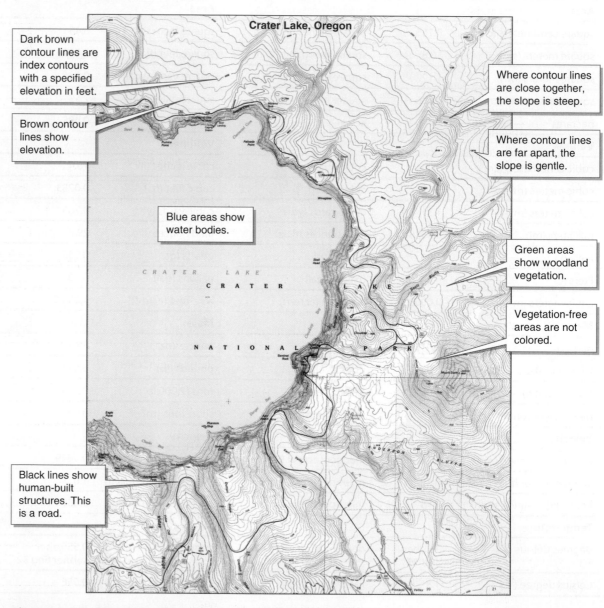

Crater Lake, Oregon

Dark brown contour lines are index contours with a specified elevation in feet.

Brown contour lines show elevation.

Where contour lines are close together, the slope is steep.

Where contour lines are far apart, the slope is gentle.

Blue areas show water bodies.

Green areas show woodland vegetation.

Vegetation-free areas are not colored.

Black lines show human-built structures. This is a road.

Figure A.1A Topographic map. *(The National Map/USGS)*

Figure A.1B Topographic symbols. *(The National Map/USGS)*

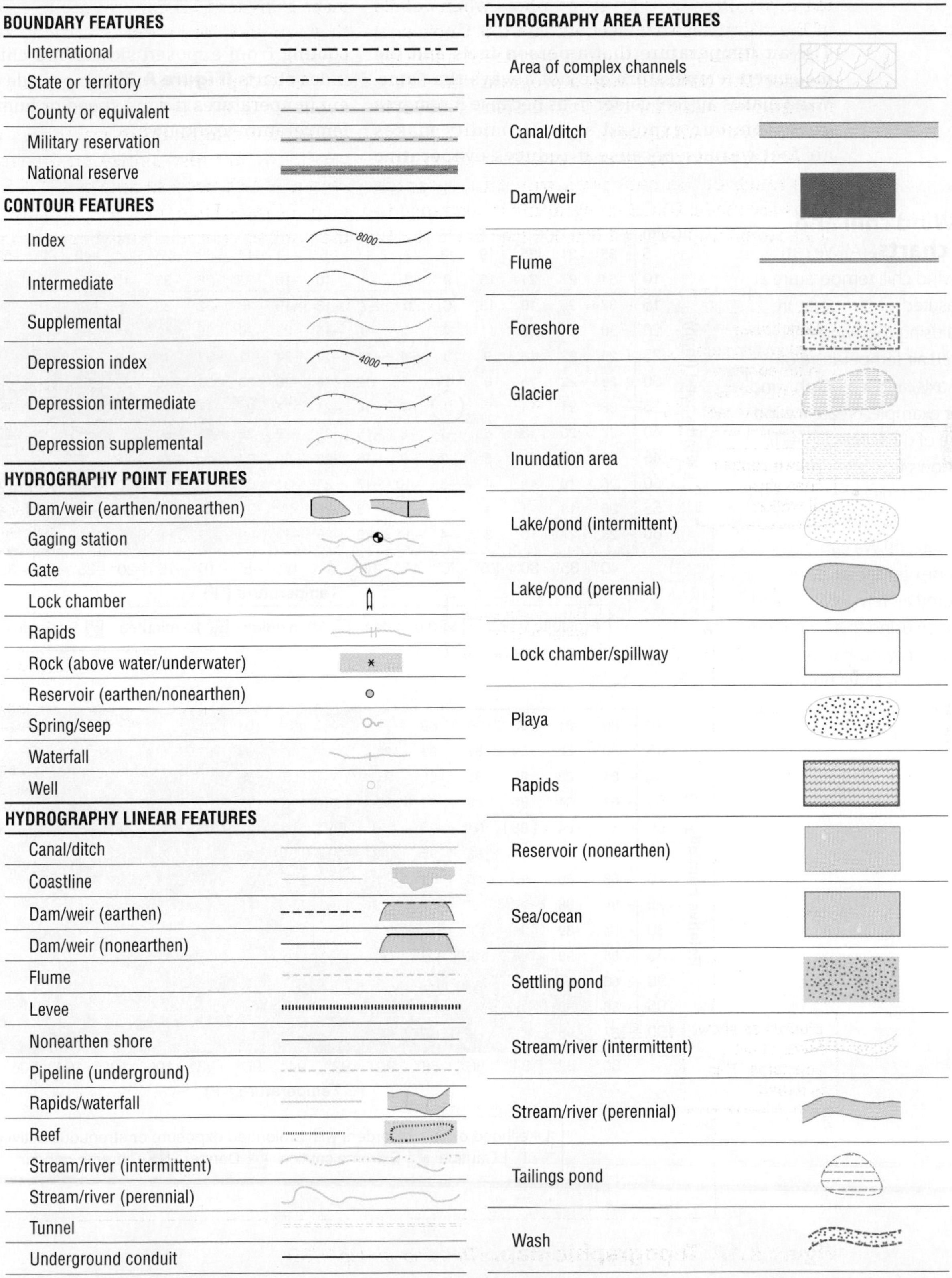

BOUNDARY FEATURES

International	— ·· — ·· — ·· —
State or territory	— — — — — —
County or equivalent	— — — — — —
Military reservation	— — — — — —
National reserve	▬ ▬ ▬ ▬ ▬

CONTOUR FEATURES

Index	8000
Intermediate	
Supplemental	
Depression index	4000
Depression intermediate	
Depression supplemental	

HYDROGRAPHY POINT FEATURES

Dam/weir (earthen/nonearthen)	
Gaging station	
Gate	
Lock chamber	
Rapids	
Rock (above water/underwater)	✳
Reservoir (earthen/nonearthen)	
Spring/seep	
Waterfall	
Well	

HYDROGRAPHY LINEAR FEATURES

Canal/ditch	
Coastline	
Dam/weir (earthen)	— — — —
Dam/weir (nonearthen)	
Flume	
Levee	
Nonearthen shore	
Pipeline (underground)	
Rapids/waterfall	
Reef	
Stream/river (intermittent)	
Stream/river (perennial)	
Tunnel	
Underground conduit	

HYDROGRAPHY AREA FEATURES

Area of complex channels	
Canal/ditch	
Dam/weir	
Flume	
Foreshore	
Glacier	
Inundation area	
Lake/pond (intermittent)	
Lake/pond (perennial)	
Lock chamber/spillway	
Playa	
Rapids	
Reservoir (nonearthen)	
Sea/ocean	
Settling pond	
Stream/river (intermittent)	
Stream/river (perennial)	
Tailings pond	
Wash	

Appendix III Wind Chill and Heat-Index Charts

The air temperature that a person feels and the measured temperature are not always the same. Wind makes air feel colder to us because it removes heat from our exposed skin. Humidity makes air feel warmer because it reduces evaporative cooling from exposed skin. Wind chill and heat-index charts **(Figure A.2)** can provide these apparent temperatures if wind speed or humidity and air temperature are known.

Figure A.2 Wind chill and heat-index charts. (A) We can determine the wind chill temperature at a particular measured air temperature and wind speed using this wind chill chart. A measured air temperature of 20°F (horizontal axis) and a 35 mph wind (vertical axis), for example, create a wind chill temperature of 0°F (circled in red). The chart also shows frostbite times for exposed skin. In the darkest region, at the bottom right, frostbite occurs within 5 minutes of exposure. (B) We can determine the heat-index temperature at a particular measured air temperature and relative humidity by using this heat-index chart. A measured air temperature of 84°F (horizontal axis) and a relative humidity of 60% (vertical axis) create a heat-index temperature of 88°F (circled in red).

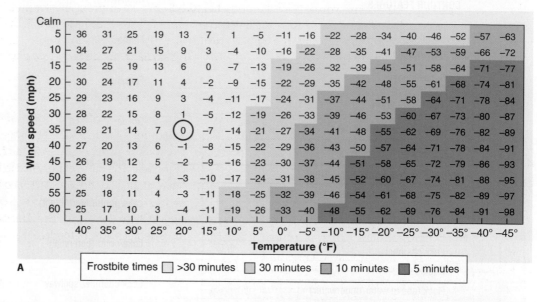

A

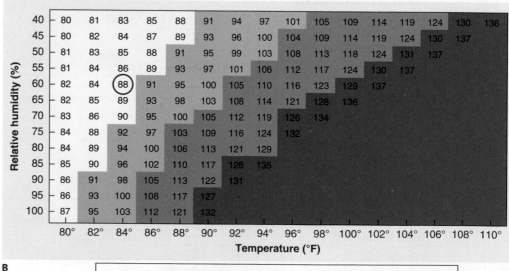

B

Appendix IV Analemma Chart

You can find the solar declination (the latitude of the subsolar point) for any day of the year by using an analemma chart **(Figure A.3)**. To use this type of chart, first find the date of interest and then read the vertical axis to the left to find the latitude of the subsolar point. For example, on October 20 the subsolar point is 10 degrees south latitude.

The figure-eight shape of the analemma reflects the actual path that the noontime Sun traces out over the course of a year. In other words, if we were to take a photo of the Sun each day at noon for a year and stitch together all 365 photos, the Sun would trace the shape of the analemma.

The figure-eight shape is caused by changes in the speed of Earth's orbit. As Earth revolves around the Sun, it moves fastest betweeen December and January, when Earth is closest to the Sun (the perihelion), and slowest between June and July, when Earth is farthest from the Sun (the aphelion). The *equation of time* at the top of the graph represents the difference between *mean solar time* (the time a local clock reads) and *observed solar time* (the time when the Sun crosses precisely overhead each day). On October 20, for example, the Sun is *fast* and crosses directly overhead 16 minutes ahead of local noon. On February 10, the Sun is *slow* by 14 minutes.

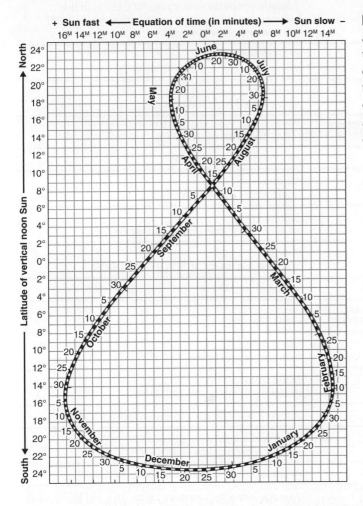

Figure A.3 Analemma chart. The analemma chart indicates the latitude where the Sun's rays are directly overhead for each day of the year. The dates of the year are listed in the figure eight, and the solar declination is listed on the vertical axis to the left. For example, on August 3, the Sun's rays shine directly overhead at 18 degrees north latitude.

Appendix V The Köppen Climate Classification System

As we learned in Section 9.1, the Köppen climate classification system recognizes six basic climate types: A, B, C, D, E, and H. Except for B, each type is identified by temperature characteristics; B climates are identified by aridity. Each of the major climate types is subdivided further based on temperature and precipitation characteristics. The H climate designation is reserved for elevations above 1,500 m (4,920 ft). **Table A.1** provides the specific characteristics for each climate type. For a map of these climate types and the biomes associated with them, see Figure 9.3.

Table A.1	Köppen Climate Types
FIRST LETTER	**CHARACTERISTICS**
A	• Tropical climates • Average monthly temperature above 18°C (64°F)
B	• Dry climates • Average annual precipitation below 76 cm (30 in)
C	• Mild midlatitude climates • The warmest month is above 10°C (50°F); the coldest month is between −3°C and 18°C (26°F and 64°F)
D	• Severe midlatitude climates • Average temperature is above 10°C (50°F) in the warmest month; average temperature is below 0°C (32°F) in the coldest month
E	• Polar climates • Warmest month is below 10°C (50°F)
H	• Highlands • Variable cold highland climates
SECOND LETTER	**CHARACTERISTICS**
f	• Precipitation in all months
m	• Monsoon • Short winter dry season
s	• Summer dry • At least 70% of precipitation in winter
t	• Tundra • Warmest month 0°C to 10°C (32°F to 50°F)
w	• Winter drought
THIRD LETTER	**CHARACTERISTICS**
a	• Hot summer • Warmest month above 22°C (72°F)
b	• Mild summer • Warmest month below 22°C (72°F)
c	• Above 10°C (50°F) for 1 to 4 months
d	• Bitter cold winter • Coldest month below −38°C (−36°F)
h	• Average yearly temperature above 18°C (64°F)
k	• Average yearly temperature below 18°C (64°F)

Glossary

A

ablation zone The area on a glacier where ice loss from melting and sublimation exceeds ice accumulation. (p. 554)

abrasion The process by which movement of one material wears away another material. (p. 521)

absolute age An age that is specified in years before the present. (p. 379)

absorption The ability of an object or a material to assimilate electromagnetic energy and convert it to another form of energy, usually thermal infrared energy. (p. 69)

abyssal plain A flat plain on the ocean floor between 4,000 and 6,000 m (13,000 and 20,000 ft) below sea level. (p. 307)

accreted terrane A mass of crust transported by plate movement and fused onto the margin of a continent. (p. 409)

accretionary prism A mass of folded sediments and rock on the sea-floor formed where oceanic lithosphere is subducting. (p. 405)

accumulation zone The area of a glacier where ice gain from snowfall exceeds ice loss. (p. 554)

acid rain (or *acid deposition*) Any form of precipitation (including snow) that has a pH lower than 5.0; also any dry acidic particles that precipitate. (p. 51)

active continental margin A margin of a continent that follows a plate boundary, typically characterized by a deep-sea trench and a narrow continental shelf. (p. 399)

active layer The top layer of permafrost that thaws each summer and refreezes in fall. (p. 573)

active volcano A volcano that has erupted during the past 10,000 years and is likely to erupt again. (p. 453)

adiabatic cooling The cooling of an air parcel through expansion. (p. 111)

adiabatic warming The warming of an air parcel through compression. (p. 111)

advection The horizontal movement of a property of the atmosphere, such as heat, humidity, or pollution. (p. 66)

advection fog Fog that results from moist air moving over a cold surface, such as a lake or a cold ocean current, which lowers air temperature to the dew point. (p. 117)

aerosols (or *particulate matter*) Microscopic solid or liquid particles suspended in the atmosphere. (p. 42)

aerovane A combination of an anemometer and a wind vane that measures wind speed and direction. (p. 135)

aftershock A small earthquake that follows the main earthquake. (p. 468)

aggradation The buildup of sediments in a streambed and along the stream's floodplain. (p. 532)

air mass A large region of air that is uniform in temperature and humidity. (p. 165)

air parcel A body of air of uniform humidity and temperature. (p. 107)

air pollutant Any airborne substance that occurs in concentrations high enough to endanger human health. (p. 49)

air pressure The force exerted by molecules of air against a surface. (p. 43)

albedo The reflectivity of a surface, given as the percentage of incoming radiation that the surface reflects. (p. 69)

alluvial fan A gently sloping fan-shaped accumulation of sediment that is deposited at the base of a mountain by an ephemeral stream in arid regions. (p. 533)

alluvium Sediments deposited on a floodplain by a stream. (p. 530)

alpine glacier Any glacier in a mountainous area. (p. 558)

alpine tree line (or *timberline*) The upper limit of trees in mountains, defined by low temperatures. (p. 292)

alpine tundra The cold, high-elevation treeless biome with herbaceous perennials. (p. 290)

analemma chart A chart used to find the solar declination (the latitude of the subsolar point) for any day of the year. (pp. 78, A-5)

anemometer (pronounced an-eh-MOM-eter) An instrument used to measure wind speed. (p. 135)

angle of repose The steepest angle at which loose sediments can settle. (p. 455)

Antarctic Circle The 66.5-degree south parallel. (p. 79)

Anthropocene (pronounced an-THROP-a-seen) The age of human transformation of Earth's physical systems; generally seen as starting in 1800. (p. 372)

anthropogenic Created or influenced by people. (pp. 4, 42)

anticline A fold in the crust with an archlike ridge. (p. 413)

anticyclone A meteorological system in which air flows away from a high-pressure region, creating clockwise circulation in the Northern Hemisphere and counterclockwise circulation in the Southern Hemisphere. (p. 136)

apparent temperature The temperature perceived by people as a result of low air temperatures coupled with wind or high air temperatures coupled with atmospheric humidity; apparent temperatures can be found using the *wind chill* and *heat index* charts (see Appendix III: Wind Chill and Heat-Index Charts). (p. 104)

aquiclude A sediment or rock layer that lacks pores and cannot contain water. (p. 350)

aquifer A sediment or rock layer with pores that contain water. (p. 350)

Arctic amplification The tendency of high-latitude regions to warm faster than the rest of the planet due to the ice–albedo positive feedback. (p. 213)

Arctic Circle The 66.5-degree north parallel. (p. 79)

arête (pronounced ar-eh-TAY) A steep-sided, sharp ridge formed where glaciers erode opposite sides of a mountain. (p. 564)

artesian well A well that has been drilled through an aquiclude into a confined aquifer below that may gush water. (p. 354)

ash (volcanic) Fine volcanic powder consisting of pulverized rock particles and solidified droplets of lava. (p. 453)

asthenosphere The layer of the mantle, which deforms and flows, found between about 100 and 200 km (62 and 124 mi) in depth. (p. 384)

atmosphere The layer of gases surrounding Earth. (pp. 8, 40)

atoll A ring of coral reefs with an interior lagoon, formed around a sinking volcano. (p. 312)

aurora australis (southern lights) Display of light in the southern hemisphere caused by energized molecules in the ionosphere. (p. 49)

aurora borealis (northern lights) Display of light in the northern hemisphere caused by energized molecules in the ionosphere. (p. 49)

avalanche A turbulent cloud of rock debris or snow that is mixed with air and races quickly down a steep slope. (p. 502)

avalanche chute A notch in mountain slopes formed by repeated snow and debris avalanches. (p. 504)

B

backwash The flow of water from breakers down the slope of a beach back toward the ocean. (p. 616)

badlands Topography that occurs in arid to semiarid environments where slopes are largely unvegetated and weak sedimentary rocks, such as shale, mudstone, and claystone, are eroded by infrequent but intense rainfall. (p. 603)

bajada (pronounced ba-HAH-da) A broad, sloping plain formed from merged alluvial fans. (p. 595)

barchan dune (pronounced BAR-can) A crescent-shaped sand dune with the crescent tips pointing downwind. (p. 593)

barometer An instrument used to measure air pressure. (p. 131)

barrier island An offshore sandbar that runs parallel to the coast. (p. 620)

barrier reef A coral reef that runs parallel to the shoreline and forms a deep-water lagoon behind it. (p. 312)

basalt A dark, heavy, fine-grained volcanic rock that constitutes oceanic crust. (p. 385)

base level The lowest level a stream can reach, usually sea level; a stream becomes standing water at base level. (p. 524)

Basin-and-Range Province A landscape in the southwestern United States where the crust is being stretched apart through rifting, forming a large internal drainage basin with a series of tilted fault blocks. (p. 600)

batholith A body of intrusive igneous rock hundreds of kilometers in extent formed by the movement and fusion of numerous plutons. (p. 431)

bathymetry Measurement of depth and topography beneath the surface of a body of water. (p. 306)

baymouth bar A continuous sandspit formed by longshore drift that creates a lagoon of brackish water behind it. (p. 620)

beach nourishment The artificial replenishment of a beach with sand. (p. 629)

bed load Material in a stream channel such as sand, gravel, and rocks that is too heavy to become suspended in the current. (p. 523)

bedrock Rock that is structurally part of and connected to Earth's crust. (p. 337)

biodiversity The number of living species in a specified region. (p. 231)

biogeographic realm A continental-scale region that contains genetically similar groups of plants and animals. (p. 247)

biogeography The study of the geography of life and how it changes through space and time. (p. 231)

bioluminescence The production of light through chemical means by living organisms. (p. 321)

biomass The dry weight of living material in a given area. (p. 254)

biome An extensive geographic region with relatively uniform vegetation structure. (p. 267)

biosphere All life on Earth. (p. 8)

boreal forest (or *taiga*) The cold coniferous forest biome found in North America and Eurasia, with vegetation made up of needle-leaved evergreen trees and an understory of mosses, lichens, and herbaceous plants. (p. 286)

braided stream A stream that forms intertwining channels around sediments in the streambed. (p. 533)

butte A medium-sized flat-topped hill with cliff-face sides that is taller than it is wide. (p. 599)

buttress roots Roots that stabilize tall tropical rainforest trees with a camera tripod–like structure. (p. 270)

bycatch Unwanted organisms caught by industrial fishing methods, which are usually thrown back to sea dead. (p. 326)

C

caldera A large depression that forms when a volcano's magma chamber empties and collapses after the volcano erupts. (p. 460)

capillary fringe The region of transition between the zone of aeration and the zone of saturation. (p. 352)

carbon cycle The movement of carbon through Earth's physical systems. (p. 208)

Carbon dioxide (CO$_2$) A colorless, odorless gas that is an air pollutant produced by the burning of fossil fuels and animal respiration. (p. 49)

carbon footprint The amount of greenhouse gases (particularly carbon dioxide) an activity generates. (p. 222)

carbon monoxide (CO) A toxic odorless and invisible gas. (p. 50)

carbonation A chemical weathering process in which carbonate rock such as limestone is dissolved in a carbonic acid solution (H$_2$CO$_3$) and carried away. (p. 489)

cartography The science and art of map making. (p. 18)

cementation A process in which minerals fill the spaces between loose particles and bind them together to form sedimentary rock. (p. 428)

Cenozoic era The past 66 million years of Earth history, marked by a persistent global cooling trend that began about 55 million years ago. (p. 202)

CFCs (chlorofluorocarbons) A class of ozone-degrading compounds used mainly as refrigerants, aerosol propellants, and fire retardants. (p. 56)

chemical energy Energy in a substance that can be released through a chemical reaction. (p. 8)

chemical sedimentary rock Sedimentary rock that forms as dissolved minerals precipitate out and water evaporates. (p. 434)

chemical weathering The process by which minerals in a rock change through chemical reactions involving water. (p. 488)

chimney A slender spire of rock with a flat top. (p. 599)

chinook wind A warm and dry local downslope wind that forms on the leeward side of the Rocky Mountains. (p. 148)

cinder cone A small, cone-shaped volcano consisting of pyroclasts that settle at the angle of repose. (p. 454)

circle of illumination The line separating night from day, where sunrise and sunset are occurring. (p. 75)

cirque (pronounced SIRK) A bowl-shaped depression with steep headwalls formed by a cirque glacier. (p. 563)

cirque glacier A glacier that forms at the head of a valley. (p. 558)

cirrus A high cloud with a feathery appearance that is composed of ice crystals. (p. 116)

clastic sedimentary rock Sedimentary rock composed of broken pieces of other rocks. (p. 434)

climate The long-term average of weather and the average frequency of extreme weather events. (pp. 11, 199)

climate forcing factor A force that can change climate and is unaffected by the climate system. (p. 200)

cloud An aggregation of microscopic water droplets and ice crystals suspended in the air. (p. 99)

cloud droplets Microscopic drops of liquid water found in clouds. (p. 42)

cloud seeding The introduction of artificial substances, such as silver iodide, to modify or enhance precipitation from clouds. (p. 122)

coal An organic rock formed from the remains of terrestrial wetland forests that is widely used today as a fuel source. (p. 434)

col A low area or pass over a ridge formed by two cirque glaciers. (p. 564)

cold front A region where cold air advances on relatively warm air; sometimes associated with severe weather. (p. 183)

collapse sinkhole A sinkhole formed where the ceiling of a cavern has collapsed. (p. 492)

collision Convergence of the continental crust of two different plates. (p. 406)

collision and coalescence The process by which cloud droplets merge to form raindrops. (p. 118)

colonization The successful establishment of a population in a new geographic region. (p. 246)

columnar jointing A geometric pattern of angular columns that forms from joints in basaltic lava during cooling. (p. 459)

community The populations of organisms interacting in a geographic area. (p. 256)

competition An interaction between organisms that require the same resources. (p. 245)

condensation A change in the state of water from gas to liquid. (p. 99)

condensation nucleus A small particle in the atmosphere, about 0.2 μm in diameter, on which water vapor condenses. (p. 118)

conduction The process by which energy is transferred through a substance or between objects that are in direct contact. (p. 66)

cone of depression The cone-shaped lowering of the water table resulting from groundwater overdraft. (p. 356)

conformal map projection A map projection that preserves the true shapes of continents at the expense of their true areas. (p. 20)

consumer An organism that cannot produce its own food through photosynthesis. (p. 253)

contaminant plume The cloud of pollution that migrates through an aquifer away from its source. (p. 360)

continental arc A long chain of volcanoes formed on the margin of a continent above a subducting plate. (p. 404)

continental crust The crust that makes up the continents, composed mainly of granite. (p. 385)

continental divide A ridge or highland that separates drainage systems that empty into different ocean basins. (p. 515)

continental drift A theory proposed by Alfred Wegener which states that continents move slowly across Earth's surface. (p. 395)

continental effect The increase in seasonality with distance from the oceans. (p. 83)

continental shelf The shallow, sloping seafloor near continental margins. (p. 306)

contour lines Lines of equal elevation in relationship to sea level on a topographic map. (p. 23)

convection The transfer of heat through vertical movement of mass within a fluid (liquid or gas). (p. 66)

convective uplift The rising of an air parcel that is warmer and less dense than the surrounding air. (p. 113)

convergent evolution The evolutionary process in which two or more unrelated organisms that experience similar environmental conditions evolve similar adaptations. (p. 238)

convergent plate boundary A region where two lithospheric plates move toward each other. (p. 402)

convergent uplift The rising of air as a result of converging airflow. (p. 115)

coral bleaching The loss of coloration in corals caused by the absence of their mutualistic algae; occurs when corals have been stressed or have died. (p. 312)

Coriolis effect (or *Coriolis force*) The perceived deflection of moving objects in relationship to Earth's surface. (p. 134)

crevasse A crack that develops in the top 60 m of a glacier. (p. 552)

crust The rigid outermost portion of Earth's surface. (pp. 9, 375)

cryosphere The frozen portion of the hydrosphere. (pp. 9, 550)

crystallization The process in which atoms or molecules come together in an orderly patterned structure. (p. 425)

cumulonimbus A cloud that extends high into the atmosphere and is capable of strong vertical development and of producing severe weather. (p. 116)

cumulus A dome-shaped, bunched cloud, with a flat base and billowy upper portions. (p. 116)

cut bank The outside edge of a meander, where erosion exceeds deposition. (p. 535)

cyanobacteria Photosynthetic bacteria that were among the first forms of life to evolve on Earth about 3.5 billion years ago. (p. 376)

cyclone A meteorological system in which air flows toward a low-pressure region, creating counterclockwise circulation in the Northern Hemisphere and clockwise circulation in the Southern Hemisphere. (p. 136)

D

debris flow A fast-flowing slurry of mud mixed with large objects, such as rocks and vegetation. (p. 502)

debris slide A landslide that consists of a mixture of rocks, soil, and vegetation. (p. 502)

December solstice A seasonal marker that occurs when the subsolar point is at 23.5 degrees south, on about December 21. (p. 77)

decomposer An organism that breaks down organic material into simple compounds. (p. 253)

decompression melting The melting of hot mantle material into magma as a result of changes in pressure, which lower the melting point of the rock; occurs as mantle material rises to a shallow depth in the lithosphere. (p. 402)

deep-sea trench A long narrow valley on the seafloor; deep-sea trenches are the deepest parts of the oceans. (p. 307)

deflation The removal of sediments and lowering of the land surface by wind erosion. (p. 591)

delta An accumulation of sediment that forms where a stream reaches base level. (p. 535)

dendritic drainage A pattern of streams in a drainage basin that resembles the branching of a tree. (p. 518)

denudation The lowering and wearing away of Earth's surface. (p. 487)

deposition The laying down (depositing) of sediments that are being transported by flowing water or ice. (pp. 13, 487)

desalination The removal of dissolved minerals from seawater. (p. 346)

desert A region that receives less than 25 cm (10 in) of precipitation annually where drought-adapted vegetation predominates. (pp. 294, 586)

desert pavement A reg surface composed of tightly interlocked rocks that resembles a cobblestone street. (p. 596)

desertification The transformation of fertile land to desert, usually through a combination of overgrazing of livestock and drought. (p. 594)

dew point (or *dew-point temperature*) The temperature at which air becomes saturated. (p. 108)

dew-point depression The difference between the air temperature and the dew point. (p. 108)

Diablo winds Winds that originate in the Great Basin and are heated adiabatically as they descend to near sea level in northern California; often associated with major wildfires. (p. 149)

digital elevation model (DEM) A three-dimensional representation of land surface or underwater topography. (p. 24)

dike A sheetlike vertical igneous rock formation. (p. 431)

disappearing stream A stream that leaves the ground surface and flows into subterranean channels. (p. 494)

dispersal The movement of an organism away from where it originated. (p. 239)

dissolution The general process in which a mineral is completely dissolved in water; carbonation of limestone rock is a form of dissolution. (p. 492)

dissolved load Soluble minerals carried in solution by a stream. (p. 521)

distributary One of a branching network of streams in a delta. (p. 535)

divergent evolution The evolutionary process by which individuals in one reproductively isolated population evolve adaptations different from those of closely related individuals in another population. (p. 239)

divergent plate boundary A region where two lithospheric plates move apart. (p. 402)

doldrums A low-wind region near the equator, associated with the ITCZ. (p. 139)

Doppler radar An active remote sensing technology that uses microwave energy to measure the velocity and direction of movement of particles of precipitation within a cloud. (p. 26)

drainage basin A geographic region drained by a single trunk stream and the smaller tributaries that flow into the trunk stream. (p. 515)

drainage divide A ridge or highland that separates drainage basins and defines the boundaries of the basins. (p. 515)

drift net fishing An industrial fishing method in which large nets are suspended in the upper reaches of the ocean. (p. 326)

dropsonde An instrument package dropped from an airplane to record weather data such as wind speed, pressure, humidity, and air temperature. (p. 179)

drought A prolonged period of water shortage. (p. 349)

drumlin An elongated hill formed by a moving ice sheet. (p. 569)

dry adiabatic rate The rate of temperature change in an unsaturated parcel of air; 10°C/1,000 m (5.5°F/1,000 ft). (p. 111)

dynamic air pressure Air pressure caused by air movement. (p. 133)

E

earthquake A sudden shaking of the ground caused by movements of Earth's crust. (p. 382)

ecological disturbance A sudden event that disrupts an ecosystem. (p. 248)

ecological succession The step-by-step series of changes in an ecosystem that follows a disturbance. (p. 248)

ecology The study of the interactions between organisms and their environment. (p. 231)

ecosystem The living organisms within a community and the non-living components of the environment in which the organisms live. (p. 231)

effusive eruption A nonexplosive eruption that produces mostly lava. (p. 462)

El Niño A periodic change in the state of Earth's climate caused by the slackening and temporary reversal of the Pacific equatorial trade winds and increased sea surface temperatures off coastal Peru. (p. 153)

electromagnetic spectrum (EMS) The full range of wavelengths of radiant energy. (p. 67)

elevation The vertical distance of a land surface above or below mean sea level. (p. 23)

eluviation The process by which rainwater carries soil particles downward. (p. 337)

embayment A concave sandy beach between headlands. (p. 617)

emergent coast A coast where sea level is dropping or the land is rising. (p. 619)

endemic Restricted to one geographic area. (p. 240)

energy The capacity to do work on or to change the state of matter. (p. 7)

enhanced Fujita scale (EF scale) A system used to rank tornado strength based on damage done to the landscape. (p. 172)

environmental lapse rate The rate of temperature change with change in altitude in the troposphere. The average environmental lapse rate is 6.5°C per 1,000 m (3.6°F per 1,000 ft). (p. 45)

eolian (or *aeolian*) (pronounced ee-OL-ee-un) Of or relating to wind. (p. 590)

ephemeral stream A stream that flows briefly after heavy rainstorms. (p. 518)

epicenter The location on the ground's surface immediately above the focus of an earthquake, where earthquake intensity is usually greatest. (p. 468)

epipelagic zone The sunlit surface of the ocean down to 200 m (650 ft) below sea level. (p. 308)

epiphyte A plant that grows on the surface of another plant but does not take nutrients from that plant. (p. 270)

equal-area map projection A map projection that preserves the true area of continents at the expense of their true shape. (p. 20)

equator The line of latitude that divides Earth into two equal halves. The equator is exactly perpendicular to Earth's axis of rotation. (p. 14)

equilibrium line (or *snowline*) The elevation at which a glacier's ice accumulation and ice loss are equal over the period of a year. (p. 554)

erg A desert landscape dominated by actively moving sand dunes. (p. 598)

erosion Scouring and stripping away of rock fragments loosened by weathering through flowing water, ice, or wind. (pp. 13, 487)

escarpment A long cliff face or steep slope. (p. 525)

esker A long ridge of sorted sand and gravel deposited by a subglacial stream. (p. 569)

estuary A brackish-water ecosystem found at the mouth of a river that is influenced by tides. (p. 317)

eustasy A change in global sea level as a result of change in the amount of water in the oceans. (p. 619)

evaporation The change in the state of water from liquid to gas. (p. 99)

evaporite A deposit of one or more minerals resulting from the repeated evaporation of water from a basin. (p. 438)

evapotranspiration The combined processes of evaporation and transpiration. (p. 100)

evolution The process of genetically driven change in a population caused by selection pressures in the environment. (p. 237)

exfoliation A physical weathering process in which joints form parallel to a rock surface, creating sheetlike slabs of rock. (p. 487)

exhumation The removal of overlying rock and sediment to expose deeper rocks at the surface. (p. 431)

exotic stream A permanent stream that originates in a humid region and flows through an arid region. (p. 518)

explosive eruption A violent volcanic eruption produced by a stratovolcano in which pyroclasts are ejected high into the atmosphere. (p. 462)

extinct volcano A volcano that has not erupted for tens of thousands of years and can never erupt again. (p. 453)

extinct Refers to the permanent and global loss of a species. (p. 233)

extrusive igneous rock (or *volcanic rock*) Rock that has cooled from lava on the crust's surface. (p. 430)

eye The central area of lowest pressure in a tropical cyclone. (p. 176)

eyewall The area of fastest winds and heaviest precipitation surrounding the eye of a tropical cyclone. (p. 176)

F

fault A fracture in the crust where movement and earthquakes occur. (p. 413)

fault scarp A cliff face resulting from the vertical movement of a reverse or normal fault. (p. 467)

felsic lava Thick lava that resists flowing and has a temperature range from 650°C to 800°C (1,200°F to 1,500°F) and a silica content of 70% or more; it builds stratovolcanoes and lava domes. (p. 457)

fetch The distance of open water over which wind blows. (p. 614)

firn Snow that has been compressed into granular ice; has an air content of roughly 50% of the total volume. (p. 552)

fishery A region where fish are caught for human consumption. (p. 324)

fjord A U-shaped coastal glacial valley flooded by the sea. (p. 566)

flank eruption An eruption of lava out the side of a volcano rather than from the central vent. (p. 453)

flash flood A flood that occurs with sudden, intense rainfall or dam collapse, often with little or no warning. (p. 538)

flood Inundation by water in a region not normally covered by water, which results when stream discharge exceeds stream channel capacity. (p. 538)

floodplain An area of flat land near a stream that experiences flooding on a regular basis. (p. 514)

fluvial erosion Erosion by running water. (p. 520)

flux melting The process by which subducted water lowers the melting point of mantle material and causes it to melt. (p. 404)

focus (or *hypocenter*) The location where the initial movement occurs along a fault during an earthquake. (p. 467)

foehn wind (pronounced FEH-rn) A downslope wind that forms on the leeward size of the European Alps. (p. 148)

fog A cloud at or near ground level that reduces visibility to less than 1 km (0.62 mi). (p. 117)

fold A wrinkle in the crust resulting from deformation caused by geologic stress. (p. 413)

food web The ecological interconnections among organisms occupying different trophic levels. (p. 254)

foreshock A small earthquake that precedes a larger, main earthquake. (p. 468)

fossil The remains or the impression of an organism preserved in sedimentary rock. (p. 438)

fossil fuels Ancient remains of plants preserved in the lithosphere in the form of coal, oil, and natural gas. (p. 49)

fossil groundwater Water that entered an aquifer long ago and is no longer being replenished. (p. 357)

fracking (or *hydraulic fracturing*) The procedure by which water is pumped under high pressure into a shale formation to extract natural gas and oil. (p. 442)

friction layer The layer of the atmosphere where wind is slowed by friction with Earth's surface; extends about 1 km (3,280 ft) above the surface. (p. 134)

fringing reef A coral reef that forms near and parallel to a coastline. (p. 312)

frontal uplift The rising of warm air masses where they meet relatively cold air masses. (p. 115)

frost wedging A physical weathering process in which water trapped in an opening in a rock freezes and expands, causing the opening to grow. (p. 488)

fulgurite A glassy hollow tube formed where lightning strikes sand. (p. 169)

G

geoengineering The deliberate, global-scale modification of Earth's environments to improve living conditions for humans. (p. 122)

geographic grid A coordinate system that uses latitude and longitude to identify locations on Earth's surface. (p. 14)

geographic information system (GIS) A system that uses computers to capture, store, analyze, and display spatial data. (p. 26)

geography The study of the spatial relationships among Earth's physical and cultural features and how they develop and change through time. (p. 4)

geohazard A hazard posed to people by the physical Earth. (p. 452)

general circulation model (GCM) Mathematical simulation of the behavior of the atmosphere, oceans, and biosphere; used to create long-term climate projections. (p. 218)

geostrophic wind A high-altitude wind that experiences strong Coriolis force deflection and moves along a path parallel to the pressure gradient rather than across it. (p. 135)

geothermal energy Heat from Earth's interior. (p. 8)

glacial advance The forward and downslope movement of a glacier's toe; occurs when a glacier has a positive mass balance. (p. 554)

glacial erratic A large boulder transported a long distance by a glacier. (p. 551)

glacial ice Ice with an air content of less than 20%. (p. 552)

glacial period A cold interval during the Quaternary ice age. (p. 203)

glacial polish A smoothed bedrock surface resulting from glacial abrasion. (p. 560)

glacial retreat The upslope movement of the toe of a glacier; occurs when a glacier has a negative mass balance. (p. 554)

glacial striation A groove gouged into the surface of bedrock by glacial abrasion. (p. 560)

glacial valley (or *glacial trough*) A U-shaped valley carved by a glacier. (p. 564)

glacier A large mass of ice that is formed from the accumulation of snow and flows slowly downslope. (p. 552)

glacier mass balance The difference between inputs to a glacier that increase its ice volume and losses of ice that decrease its ice volume. (p. 554)

glacier toe The lowest end of a glacier. (p. 554)

global heat engine A system that involves movement of heat from low to high latitudes and low to high altitudes as a result of heating differences. (p. 74)

Global Positioning System (GPS) A global navigation system that uses satellites and ground-based receivers to determine the geographic coordinates of any location. (p. 18)

Gondwana The landmass that resulted when Pangaea split about 200 million years ago; it consisted of modern-day South America, Australia, Africa, India, and Antarctica. (p. 395)

granite A silica-rich rock composed of coarse grains. (p. 385)

great circle A continuous line that bisects the globe into two equal halves, such as the equator; it is the shortest distance between two points on Earth. (p. 21)

Great Pacific Garbage Patch The region of concentrated plastic litter formed by the North Pacific Gyre. (p. 327)

green economy A sustainable economic system that has a small environmental impact and is based on carbon-free energy sources. (p. 221)

greenhouse effect The process by which a planet's atmosphere is warmed as greenhouse gases (such as water vapor, carbon dioxide, and methane) and clouds absorb and counterradiate heat. (p. 72)

greenhouse gas A gas that can absorb and emit thermal infrared energy. (p. 41)

groin A linear structure of concrete or stone that extends from a beach into the water, designed to slow the erosion of sand. (p. 624)

ground-level ozone Ozone that forms near the land surface and is a secondary anthropogenic air pollutant. (p. 50)

groundwater Water found beneath Earth's surface in sediments and rocks. (p. 342)

groundwater discharge The movement of water out of an aquifer. (p. 352)

groundwater mining The process of extracting groundwater where there is little to no groundwater recharge. (p. 357)

groundwater overdraft Withdrawal of water from an aquifer faster than the aquifer is recharged at the site of a well. (p. 356)

groundwater recharge The movement of water into an aquifer. (p. 352)

groundwater remediation The process of cleaning a contaminated aquifer. (p. 360)

gyre A large, circular ocean current. (p. 311)

H

habitat The physical environment in which an organism lives. (p. 240)

habitat fragmentation Division and reduction of natural habitat into smaller pieces by human activity. (p. 296)

hail Hard, rounded pellets of ice that precipitate from cumulonimbus clouds with strong vertical airflow. (p. 120)

headland A rocky prominence of coastal land. (p. 616)

headward erosion The process by which a stream channel migrates upslope by forming new rills through fluvial erosion. (p. 520)

headwaters The region where a stream originates. (p. 518)

heat Energy transferred between materials or systems due to their temperature differences. (p. 65)

herbivore An organism that eats only plants. (p. 253)

Holocene epoch The current interglacial period of warm and stable climate; the most recent 10,000 years of Earth history. (p. 203)

horn (glacial) A steep, pyramid-shaped mountain peak formed by glaciers. (p. 564)

horse latitudes The low-wind regions centered on 30 degrees north and south. (p. 139)

hot spot A volcanically active and isolated location on Earth's surface. (p. 410)

humidity The amount of water vapor in the atmosphere. (p. 104)

humus (pronounced HYOU-mus) Organic material that makes up the topmost layers of soil (the O and A horizons). (p. 340)

hurricane The North American and Central American name for a tropical cyclone with sustained winds of 119 km/h (74 mph) or greater. (p. 176)

hurricane warning An alert that is issued by the National Weather Service 36 hours in advance for areas where tropical storm–force winds are imminent. (p. 182)

hurricane watch An alert that is issued by the National Weather Service 48 hours in advance for areas where tropical storm–force winds are possible. (p. 180)

hydrogen bond A bond between water molecules that results from the attraction between one water molecule's positive end and another's negative end. (p. 100)

hydrologic cycle The circulation of water within the atmosphere, biosphere, lithosphere, and hydrosphere. (p. 99)

hydrosphere All of Earth's water in its three phases: solid, liquid, and gas. (p. 9)

hydrothermal vent community A unique ecosystem found at volcanic hot springs that emit mineral-rich water onto the seafloor. (p. 324)

hygrometer An instrument used to measure humidity. (p. 104)

I

ice-albedo positive feedback Destabilizing positive feedback in the climate system in which the melting of ice and snow expose bare ground and ice-free water, and these surfaces absorb more solar energy, causing more warming. (p. 200)

iceberg A large block of ice that breaks from the toe of a glacier or an ice shelf and floats in the ocean or a lake. (p. 559)

ice cap A dome of ice that sits over a high mountain region and has an extent of 50,000 km^2 (19,300 m^2) or less. (p. 558)

ice-crystal process (or *Bergeron process*) The process by which ice crystals grow within a cloud to form snow. (p. 118)

ice sheet A flat sheet of ice that has an extent of 50,000 km^2 (19,300 mi^2) or more. (p. 558)

ice shelf The portion of an ice sheet or an outlet glacier that extends over the ocean. (p. 560)

igneous rock Rock that has cooled from magma or lava. (p. 427)

illuviation The process by which particles are deposited after being transported downward through the soil by rainwater. (p. 337)

infiltration The process by which water seeps into the ground through the force of gravity. (p. 350)

infrared radiation (IR) Electromagnetic radiation that has wavelengths between 0.75 and 1,000 μm; longer than visible radiation. (p. 68)

inner core The innermost layer of Earth, composed of solid iron and nickel. (p. 383)

inselberg A large, weathered outcrop of bedrock surrounded by a flat alluvium-filled plain. (p. 600)

insolation (or *incoming solar radiation*) Solar radiation that reaches Earth. (p. 68)

interglacial period A warm interval that occurs between glacial periods. (p. 203)

Intergovernmental Panel on Climate Change (IPCC) The world's leading governing body on climate change; a politically neutral body operating as part of the United Nations. The IPCC's mission is to assess the world's present and future climate and the ramifications of climate change for human societies. (p. 218)

intermediate lava Lava that is intermediate in characteristics between mafic lava and felsic lava; it has a temperature of about 800°C to 1,000°C (1,500°F to 1,800°F), a silica content between 50% and 70%, and medium viscosity. (p. 456)

intermittent stream A stream that runs dry during part or most of the year. (p. 518)

internal drainage A drainage pattern in which streams terminate in a low-lying basin on land. (p. 515)

intrusive igneous rock Rock that has cooled from magma below the crust's surface. (p. 430)

ionosphere A region of the upper mesosphere and the thermosphere between about 80 and 500 km (50 and 310 mi) above Earth's surface where gases are ionized by solar energy. (p. 48)

island arc A chain of islands formed where oceanic lithosphere of one plate subducts beneath oceanic lithosphere of another plate. (p. 405)

isobar A line drawn on a map connecting points of equal pressure. Isobars are quantitative representations of the changing molecular density of the air over a geographic region. (p. 134)

isostasy The gravitational equilibrium between the lithosphere and the support of the asthenosphere below. (p. 619)

ITCZ (short for *intertropical convergence zone*; or *equatorial trough*) The discontinuous band of thermal low pressure and thunderstorms that encircles Earth in the tropics. (p. 139)

J

jetty An artificial wall placed at the mouth of a harbor to prevent sand from closing the harbor entrance; usually placed in pairs. (p. 624)

joint A crack or weak plane in rock. (p. 459)

June solstice The seasonal marker that occurs when the subsolar point is 23.5 degrees north latitude, about June 21. (p. 78)

K

kame-and-kettle topography A glaciofluvial landscape created by an ice sheet and dominated by irregular mounds and shallow depressions or lakes. (p. 569)

Karman line The altitude below which 99.99% of the atmosphere's mass occurs; a widely accepted definition as the "top" of the atmosphere; occurs at 100 km (62 mi) altitude above Earth's surface. (p. 41)

karst An area dominated by the weathering of carbonate rocks, usually limestone. (p. 491)

katabatic wind (or *gravity wind*) Wind that forms mainly over ice sheets or glaciers when intensely cold, dense, and heavy air spills downslope by the force of gravity. (p. 149)

Keeling curve A graph that shows the change in atmospheric CO_2 concentrations since 1958. (p. 212)

kelp forest A coastal marine ecosystem dominated by kelp, found where ocean water is colder than 20°C (68°F). (p. 318)

keystone species A species whose effects support many other species within an ecosystem. (p. 244)

knickpoint A location where there is an abrupt increase in stream gradient over a short distance. (p. 525)

Köppen climate classification system (pronounced KUHR-pen) A system used to classify climate types based on temperature and precipitation. (pp. 269, A-6)

K-selected species Species that are generally long-lived and produce few offspring during their lifetime. (p. 252)

L

La Niña A phenomenon during which the low pressure over Indonesia is deeper than normal, and the high pressure near western South America is higher than normal. (p. 154)

laccolith A shallow, dome-shaped igneous rock body. (p. 431)

lag deposit A reg surface that results from the removal of smaller particles, such as clay and silt, by deflation. (p. 596)

lagoon A fully or partly enclosed stretch of salt water formed by a coral reef or sand spit. (p. 312)

lahar A thick slurry of mud, ash, water, and other debris that flows rapidly down a snow-capped stratovolcano when it erupts. (p. 464)

lake-effect snow Heavy snowfall that results as cold air moves over large, relatively warm bodies of water, such as the Great Lakes. (p. 187)

land breeze A local offshore breeze created by heating and cooling differences between water and land. (p. 148)

landslide A sudden rapid movement of rock or debris down a steep slope. (p. 502)

lapilli (pronounced la-PILL-eye) Marble- to golf ball-sized cooled fragments of lava. (p. 457)

large igneous province (LIP) An accumulation of flood basalts that covers an extensive geographic area. (p. 460)

large-scale perspective A geographic scale that pertains to a geographically restricted area and makes geographic features large to show more detail. (p. 5)

latent heat Energy that is absorbed or released during a change in the state of a substance, such as during evaporation or condensation of water. (p. 102)

lateral moraine A moraine that occurs along the side of a glacier. (p. 568)

latitude The angular distance as measured from Earth's center to a point north or south of the equator. (p. 14)

Laurasia The landmass that resulted when Pangaea split about 200 million years ago; it consisted of modern-day North America, Greenland, and Eurasia. (p. 395)

Laurentide ice sheet The large ice sheet that covered much of North America 20,000 years ago. (p. 207)

lava Hot molten rock that spills onto the surface of Earth's crust. (pp. 385, 483)

lava bomb A streamlined fragment of lava ejected from a volcano that cooled and hardened as it moved through the air. (p. 458)

lava dome (or *plug dome* or *volcanic dome*) A dome-shaped volcano of solidified lava; may be tens to hundreds of meters high and as wide at the base. (p. 454)

leaching The process by which rainwater carries dissolved nutrients downward. (p. 337)

liana A woody climbing vine. (p. 270)

lifting condensation level (LCL) The altitude at which an air parcel becomes saturated. (p. 111)

light gap A lighted gap in a forest that forms where a large tree has fallen. (p. 270)

lightning An electrical discharge produced by a thunderstorm. (p. 169)

limestone A chemical sedimentary rock or an organic sedimentary rock composed of at least 50% calcite; 22% of all sedimentary rocks are limestone. (p. 434)

limestone column A cylindrical speleothem that results when a stalactite joins with a stalagmite. (p. 497)

limiting factor A factor that prevents an organism from reaching its reproductive or geographic potential. (p. 240)

liquefaction The transformation of solid sediments into an unstable slurry as a result of ground shaking during an earthquake. (p. 473)

lithification The formation of sedimentary rock through compaction and cementation of loose sediments. (p. 428)

lithosphere The layer of Earth that consists of the rigid crust and the rigid lithospheric mantle beneath it and extends to a depth of about 100 km (62 mi) below the surface on average. (pp. 9, 385)

Little Ice Age (LIA) The natural cooling period from about 1350 to 1850 CE; felt mostly in the Northern Hemisphere. (p. 204)

loam Soil that consists of approximately 40% sand, 40% silt, and 20% clay and has a high organic content. (p. 339)

loess (pronounced lehss) Wind-deposited silt and clay sediments that originate mostly from glacial outwash plains. (p. 571)

longitude The angular distance as measured from Earth's center to a point east or west of the prime meridian. (p. 14)

longitudinal dunes Sand dunes with linear dune crests that are oriented parallel to the wind; they form where sand completely covers the desert floor. (p. 593)

longline fishing An industrial fishing method that employs thousands of baited hooks on lines up to 80 km (50 mi) in length. (p. 326)

longshore current A current that flows parallel to the beach, in the direction of wave movement. (p. 617)

longshore drift The process by which sediment on a beach is moved down the length of the beach in the direction of wave movement. (p. 617)

longwave radiation (LWR) Radiation with wavelengths longer than 4 μm. (p. 68)

lower mantle The layer of heated and slowly deforming solid rock between the base of the asthenosphere and the outer core. (p. 384)

M

mafic lava Lava that has low viscosity and easily flows. It has a temperature of 1,000°C to 1,200°C (1,800°F to 2,200°F) and a silica content of 50% or less; it builds shield volcanoes. (p. 456)

magma Melted rock that is below the surface of Earth's crust. (pp. 385, 453)

magma chamber The reservoir of magma beneath a volcano. (p. 453)

magnetosphere The outer edge of the magnetic field that surrounds Earth and shields it from the solar wind. (p. 387)

mangrove forest A coastal marine ecosystem dominated by saltwater-tolerant shrubs and trees, found in the tropics and subtropics. (p. 314)

mantle drag The dragging force between the asthenosphere and the overlying lithosphere. (p. 400)

mantle plume A mostly stationary column of hot rock that extends from deep in the mantle up to the base of the lithosphere. (p. 410)

map A flat two-dimensional representation of Earth's surface. (p. 5)

map scale A means of specifying the degree to which the real world has been reduced on a map. (p. 22)

March equinox The seasonal marker that occurs when the subsolar point is over the equator about March 20. (p. 78)

marine terrace A wave-cut bench that has been elevated above sea level. (p. 628)

mass movement (or *mass wasting*) Downslope movement of rock, soil, snow, or ice caused by gravity. (p. 498)

matter Any material that occupies space and possesses mass. (p. 7)

meander A looping bend in a stream channel on a floodplain. (p. 526)

medial moraine A moraine that occurs in the center of a glacier where two or more lateral moraines merge. (p. 568)

Medieval Warm Period (MWP) The naturally warm period from about 950 to 1250 CE; felt mostly in the Northern Hemisphere. (p. 205)

Mediterranean biome The biome characterized by hot, dry summers and winter rainfall, with vegetation adapted to drought and fire. (p. 282)

meridian A line on the globe that runs from the North Pole to the South Pole and connects points of the same longitude. (p. 16)

mesa A flat-topped elevated area with one or more cliff-face sides that is wider than it is tall. (p. 599)

mesocyclone The rotating cylindrical updraft within a supercell thunderstorm. (p. 168)

mesosphere The layer of the atmosphere between 50 and 80 km (30 and 50 mi) above Earth's surface. (p. 44)

metamorphic rock Rock formed by heat and pressure applied to preexisting rock. (p. 427)

midlatitude cyclone (or *extratropical cyclone*) A large cyclonic storm at midlatitudes. (p. 182)

mid-ocean ridge A submarine mountain range. (p. 307)

migration Seasonal movement of populations from one place to another, usually for feeding or breeding. (p. 234)

Milankovitch cycles Small changes in Earth–Sun orbital geometry that resulted in Quaternary glacial–interglacial cycles. (p. 204)

millibar (mb) A measure of atmospheric pressure; average sea level, pressure is 1013.25 mb. (p. 105)

mineral A naturally occurring, crystalline, solid chemical element or compound with a uniform chemical composition. (p. 425)

Modified Mercalli Intensity (MMI) Scale An earthquake ranking system based on the damage done to structures. (p. 472)

Moho The boundary that separates Earth's crust from the lithospheric mantle, which lies about 35 km (22 mi) below the surface on average. (p. 385)

moist adiabatic rate The rate of cooling in a saturated air parcel; usually about 6°C/1,000 m (3.3°F/1,000 ft). (p. 112)

moment magnitude scale An earthquake ranking system based on the amount of ground movement produced. (p. 474)

monsoon A seasonal reversal of winds, characterized by moist summer onshore airflow and dry winter offshore airflow. (p. 130)

montane forest The forest biome composed of needle-leaved trees in the Northern Hemisphere and broad-leaved trees in the Southern Hemisphere; found on windward sides of mountains with abundant precipitation. (p. 288)

moraine A heap of unsorted sediments (called till) deposited by a glacier. (p. 568)

moulin (pronounced moo-LAN) A vertical shaft in a glacier through which meltwater flows. (p. 578)

mountain breeze A local downslope breeze produced by heating and cooling differences in mountainous areas. (p. 147)

mudflow A fast-moving flow composed mostly of mud. (p. 502)

multicell thunderstorm A type of thunderstorm that forms under conditions of moderate wind shear, is organized in squall lines or clusters, and often produces severe weather. (p. 167)

mutualism A relationship between two species from which both species benefit. (p. 245)

N

negative feedback A process in which interacting parts in a system stabilize the system. (p. 9)

niche The resources and environmental conditions that a species requires. (p. 239)

nimbostratus Rain-producing low-level sheets of clouds. (p. 116)

nitric oxide (NO) A nontoxic and harmless gas that can oxidize to form nitrogen dioxide (NO_2), an important anthropogenic air pollutant. (p. 50)

nitrogen dioxide (NO_2) Toxic reddish-brown gas produced mainly by vehicle tailpipe emissions. (p. 50)

non-native (or *exotic*) A species that has been brought outside its original geographic range by people. (p. 230)

nor'easter A type of midlatitude cyclone that brings blizzard-like conditions to the Mid-Atlantic states and New England. (p. 187)

normal fault The result of tensional force as two fault blocks move apart, causing one fault block to slip downward in relationship to the other fault block. (p. 467)

northern tree line The northernmost limit of the boreal forest. (p. 290)

northern tundra The cold, treeless high-latitude biome with vegetation consisting of shrubs and herbaceous perennials; found north of the boreal forest throughout northern Eurasia and North America (p. 290)

nuisance flooding Extreme high tides that inundate low-lying coastal areas (p. 198)

nunatak (pronounced nun-uh-TAK) Bedrock that protrudes above a glacier. (p. 559)

O

ocean conveyor belt The global system of surface and deep-ocean currents that transfers heat toward the poles. (p. 207)

oceanic crust The crust beneath the oceans, composed mainly of basalt. (p. 385)

offshore wind A coastal wind that flows from land to sea. (p. 136)

omnivore An organism that eats both plants and animals. (p. 253)

onshore wind A coastal wind that flows from sea to land. (p. 136)

organic sedimentary rock Sedimentary rock composed mostly of organic material and derived from ancient organisms. (p. 434)

orogenesis The process of mountain building. (p. 415)

orogenic belt A linear mountain range. (p. 415)

orographic uplift The rising of air over mountains. (p. 113)

outcrop An exposed area of bedrock. (p. 427)

outer core The second innermost layer of Earth, composed of a liquid alloy of iron and nickel, which generates Earth's magnetic field. (p. 384)

outlet glacier A glacier that flows out of an ice sheet or ice cap through a constricted valley, usually into the ocean. (p. 559)

outwash plain A flat area of sediments deposited by glacial outlet streams. (p. 562)

oxbow lake A water-filled abandoned channel that results when a meander is cut off from the stream channel. (p. 535)

ozone (O_3) A molecule that is a pollutant in the lower atmosphere but blocks harmful solar UV radiation in the stratosphere. (p. 50)

ozonosphere (pronounced oh-ZO-no-sphere) The region of the stratosphere in which high concentrations of ozone molecules block ultraviolet radiation. (p. 47)

P

Pacific Ring of Fire The zone of volcanically active mountain chains resulting from subduction on the margins of the Pacific Ocean. (p. 406)

paleoclimate An ancient climate. (p. 208)

Pangaea The supercontinent formed about 300 million years ago by the fusion of all continents. (p. 394)

parabolic dunes Crescent-shaped sand dunes; their crescent tips are anchored by vegetation and point upwind. (p. 593)

parallel A line that forms a circle on the globe by connecting points of the same latitude. (p. 14)

particulate matter (PM) Tiny liquid and solid particles (aerosols) suspended in the atmosphere. (p. 52)

passive continental margin A margin of a continent that does not coincide with a plate boundary and is typically characterized by a broad, sloping continental shelf. (p. 399)

paternoster lake One of a series of small lakes that form behind glacial steps in a glacial valley. (p. 564)

peat Brownish-black, heavy soil found in wetlands, formed from the partially decomposed remains of plants. (p. 435)

pedestal rock A remnant of a layer of erosion-resistant rock supported by a slender column of less resistant rock. (p. 591)

pedogenesis The process of soil formation. (p. 337)

pelagic Of or referring to the open sea. (p. 321)

perched water table A localized water table that lies above the regional water table. (p. 352)

percolation The process by which rainwater moves through the soil through narrow, meandering channels. (p. 350)

periglacial Of or referring to unglaciated areas at high latitudes and high elevations subject to persistent and intense freezing. (p. 572)

permafrost Ground that remains below freezing continuously for 2 years or more. (p. 572)

permanent stream A stream that flows all year. (p. 518)

permeability The ease with which water can flow through soil, sediments, or rocks. (p. 350)

petroglyph A prehistoric rock carving often made by scratching away rock varnish from a rock's surface. (p. 588)

photochemical smog Air pollution formed by the action of sunlight on tailpipe emissions. (p. 50)

photosynthesis The process by which plants, algae, and some bacteria convert the radiant energy of sunlight to chemical energy. (p. 8)

photovoltaic cell (PV cell) A semiconductor that converts sunlight directly into electricity, using silicon cells. (p. 87)

physical geography The study of Earth's living and nonliving physical systems and how they change naturally through space and time or are changed by human activity. (p. 4)

physical weathering Breakdown of rocks into smaller pieces, or clasts, without alteration of the chemical makeup of the rock. (p. 487)

phytoplankton Microscopic photosynthetic bacteria and algae that are suspended in the sunlit portions of water. (pp. 253, 304)

piedmont glacier A lobe of ice that forms as a valley glacier flows onto a flat plain. (p. 558)

plane of the ecliptic The flat plane traced by the orbital paths of the planets in the solar system. (p. 74)

plate boundary The margin of a lithospheric plate. (p. 398)

plate tectonics A theory that addresses the origin, movement, and recycling of lithospheric plates and the landforms that result. (pp. 386, 396)

playa A flat, dry lake bed in a desert valley in an internal drainage basin. (p. 595)

playa lake A shallow temporary lake that forms in an internal drainage basin. (p. 595)

plucking The process by which a glacier pulls up and breaks off pieces of bedrock as it moves downslope. (p. 560)

plunge pool A bowl at the base of a waterfall created by abrasion from circulating rocks. (p. 525)

pluton A dome-shaped igneous rock mass that is a few tens of kilometers in diameter or smaller. (p. 431)

point bar An accumulation of silt, sand, and gravel that forms at the inside edge of a stream meander, where deposition exceeds erosion. (p. 535)

polar easterlies The cold, dry winds originating near both poles and flowing south and east. (p. 141)

polar high An area of cold, dense air at each pole that forms a zone of thermal high pressure. (p. 139)

polar jet stream The discontinuous narrow band of fast-flowing air found at high altitudes between 30 and 60 degrees latitude in the Northern Hemisphere. (p. 142)

population A group of organisms that interact and interbreed in the same geographic area. (p. 237)

porosity The available air space within sediments or rocks. (p. 350)

positive feedback A process in which interacting parts in a system destabilize the system. (p. 9)

potentiometric surface The elevation to which hydraulic pressure will push water in pipes or wells. (p. 354)

precipitation Solid or liquid water that falls from the atmosphere to the ground. (p. 99)

predation Consumption of one organism by another. (p. 243)

pressure-gradient force The force resulting from changes in barometric pressure across Earth's surface. (p. 133)

prevailing wind A wind that blows from the direction that is most common during a specified window of time. (p. 136)

primary forest A forest that has never been significantly modified by people. (p. 285)

primary pollutant A pollutant that enters the air or water directly from its source. (p. 50)

primary producer An organism that can convert sunlight into chemical energy through photosynthesis. (p. 253)

prime meridian Zero degrees longitude; the line of longitude that passes through Greenwich, England, and serves as the starting point from which all other lines of longitude are determined. (p. 14)

propagule Any unit, such as a seed, a mating pair, or a rooting branch, that is able to establish a new reproducing population. (p. 246)

protolith The parent or original rock from which metamorphic rock was formed. (p. 439)

pumice (pronounced PUM-iss) Lightweight, porous rock with at least 50% air content, formed from felsic lava. (p. 458)

pyroclast (or *pyroclastic material*) A fragment of solid material that is ejected from a volcano, ranging in size from ash to large boulders. (p. 453)

pyroclastic flow (or *nuée ardente*) A rapidly moving avalanche of searing hot gas and ash. (p. 464)

Q

Quaternary period The most recent 2.6 million years of Earth history; an ice age. (p. 203)

R

radar (radio detection and ranging) An active remote sensing technology used to study Earth's surface; employs rapid pulses of radio waves to create reconstructions of surface topography. (p. 24)

radiant energy (or *electromagnetic energy*) Energy propagated in the form of electromagnetic waves, including visible light and heat. (pp. 7, 67)

radiation The process by which wave energy travels through the vacuum of space or through a physical medium such as air or water. (p. 66)

radiation fog (or *valley fog*) Fog that results when the ground radiates its heat away at night, cooling the air above it to the dew point. (p. 117)

radiative equilibrium temperature The temperature of an object resulting from the balance between incoming and outgoing energy. (p. 71)

radiometric dating A method of assigning absolute ages to Earth materials based on the radioactive decay of unstable elements in those materials. (p. 379)

radiosonde An unstaffed weather balloon that automatically transmits recorded meteorological data, such as air temperature, pressure, and humidity. (p. 44)

rain shadow The dry, leeward side of a mountain range. (p. 112)

recessional moraine A moraine that forms where the toe of a glacier pauses as it is gradually retreating upslope. (p. 568)

reflection The process of returning a portion of the radiation striking a surface back in the general direction from which the radiation came. (p. 69)

reg A flat, stony plain. (p. 596)

regolith Any loose, fragmented Earth material, including soil, that covers bedrock. (p. 337)

relative age The age of one object or event in relation to the age of another object or event, without regard to the numeric age of either object or event. (p. 379)

relative humidity (RH) The ratio of water vapor content to water vapor capacity, expressed as a percentage. (p. 107)

relief The relative difference in elevation between two or more points on Earth's surface. (p. 23)

remote sensing Data collection from a distance. (p. 24)

renewable energy Energy that comes from sources that are not depleted when used, such as sunlight or wind. (p. 87)

reservoir An artificial lake. (p. 535)

reverse fault A fault that is the result of compressional force as two fault blocks are pushed together, causing one block to move upward in relationship to another block. (p. 467)

ridge push The process by which magma rising along a mid-ocean ridge forces apart oceanic crust of two separate plates. (p. 399)

rift A region where continental crust is stretching and splitting. (p. 403)

rift valley A linear valley with volcanoes formed by rifting of continental crust, sometimes filled with freshwater to form a deep lake. (p. 403)

rip current A short-lived jet of backwash formed as water from breakers flows back into the sea. (p. 618)

roche moutonnée (pronounced ROSH moo-ta-NAY) An elongated and asymmetrical ridge of glacially carved bedrock. (p. 566)

rock A solid mass composed of one or more minerals. (p. 425)

rock cycle A model of the processes by which rocks form, are transformed from one type to another, and are recycled into the mantle. (p. 427)

rock slide A landslide that consists of rocks and broken rock fragments. (p. 502)

rock varnish A thin, weathered, and darkened surface on exposed rock surfaces formed by the activities of bacteria and the accumulation of windblown clay particles. (p. 588)

rockfall A type of mass movement in which rocks tumble off a vertical or nearly vertical cliff face. (p. 504)

root wedging A physical weathering process in which plant roots break apart rocks. (p. 488)

Rossby wave (or *longwave*) A large undulation in the midlatitude geostrophic winds. (p. 143)

r-selected species Species that are general short-lived and produce many offspring during their lifetime. (p. 252)

S

Saffir-Simpson scale A hurricane ranking system based on measured wind speeds. (p. 178)

saline lake A permanent body of salty water occupying a playa. (p. 595)

salinity The concentration of dissolved minerals in seawater. (p. 308)

salt weathering A physical weathering process in which salt crystals grow in pore spaces on a rock's surface and dislodge individual mineral grains within the rock. (p. 488)

saltation The bouncing or hopping motion of sediment in moving water (or air). (p. 523)

saltwater intrusion The contamination of a well by salt water as a result of groundwater overdraft. (p. 356)

sand dune A hill or ridge formed by the accumulation of windblown sand. (p. 593)

sandspit An elongated dry bar of sand that extends from a beach into the water; usually parallel with the shore. (p. 620)

sandstone Clastic sedimentary rock composed chiefly of quartz sand grains and comprising 32% of all sedimentary rocks. (p. 434)

Santa Ana winds Winds that originate in the Great Basin and are heated adiabatically as they descend to sea level on the southern California coast and northern Baja California; often associated with major wildfires. (p. 149)

saturation The point at which an air parcel's water vapor content is equal to its water vapor capacity. (p. 104)

saturation vapor pressure The vapor pressure at which saturation occurs. (p. 105)

scattering The process by which solar radiation is redirected in random directions as it strikes physical matter. (p. 69)

sclerophyllous Having stiff, leathery, and waxy leaves adapted to reduce water loss and herbivory. (p. 282)

sea breeze A local onshore breeze created by heating and cooling differences between water and land. (p. 148)

sea stack A steep or vertical tower of rock separated from but near a rocky coastal cliff face. (p. 626)

seagrass meadow A shallow coastal ecosystem dominated by flowering plants that resemble grasses. (p. 317)

sea-level pressure Air pressure that has been adjusted to sea level. (p. 131)

seamount A mountain, often a flat-topped inactive volcano, rising from the seafloor. (p. 307)

seasonal flood A predictable period of flooding that occurs with seasonally heavy rain or snowmelt. (p. 538)

seawall An artificial hard structure designed to protect backshore environments from wave erosion during large storms. (p. 626)

secondary forest A forest that has regrown after being disturbed or cleared by people. (p. 286)

secondary pollutant A pollutant that is not directly emitted from a source but forms through chemical reactions among primary pollutants in air or water. (p. 50)

sediment An accumulation of small fragments of rock and organic material that is not cemented together. (p. 427)

sediment transport The movement of rock fragments (sediments) that have been weathered and eroded; sediment transport occurs by wind and flowing streams and glaciers. (p. 487)

sedimentary rock Rock formed from compacted and cemented sediments. (p. 427)

seining An industrial fishing method that involves using a large net to surround and catch fish. (p. 326)

seismic wave Energy released by an earthquake that travels through Earth's interior as a wave. (p. 382)

seismograph (or *seismometer*) An instrument used to detect, measure, and record ground shaking. (p. 471)

sensible heat A type of heat energy that is detectable as a change in temperature and can be measured with a thermometer. (p. 68)

September equinox The seasonal marker that occurs when the subsolar point is over the equator about September 22. (p. 78)

sere The stage of ecological succession that follows ecological disturbance. (p. 251)

severe thunderstorm A thunderstorm that produces either hail 2.54 cm (1 in) in diameter, a tornado, or wind gusts of 93 km/h (58 mph) or greater. (p. 167)

shale Clastic sedimentary rock formed from clay; accounts for 45% of all sedimentary rocks. (p. 434)

shield volcano A broad, domed volcano formed from many layers of basaltic lava. (p. 453)

shingle beach A beach composed of sediment particles larger than sand, such as gravel or cobbles. (p. 620)

shortwave radiation (SWR) Solar radiation with wavelengths shorter than 4 μm; includes visible sunlight. (p. 68)

sill A horizontal sheet of igneous rock that was injected between layers of preexisting rock. (p. 431)

single-cell thunderstorm A short-lived and rarely severe thunderstorm. (p. 165)

sinkhole A depression in Earth's surface resulting from the weathering of carbonate rock underground. (p. 492)

sinkhole lake A sinkhole that has filled with water. (p. 492)

slab pull The process by which the weight of subducting oceanic lithosphere accelerates plate movement by pulling the plate deeper into the mantle. (p. 400)

slip face The leeward side of a sand dune, where sand settles at its angle of repose. (p. 593)

slump A type of mass movement in which regolith detaches and slides downslope along a spoon-shaped failure surface and comes to rest more or less as a unit. (p. 500)

small-scale perspective The geographic scale that makes geographic features small to cover a large area of Earth's surface. (p. 5)

soil The layer of sediment closest to Earth's surface, into which plant roots extend, that has been modified by organisms and water. (p. 337)

soil creep Imperceptible downslope movement of soil and regolith as their volume changes in seasonal expansion–contraction cycles. (p. 500)

soil horizon A horizontal zone within the soil identified by its chemical and physical properties. (p. 337)

soil taxonomy A soil classification system that groups soils into 12 categories, or orders. (p. 344)

solar altitude The altitude of the Sun above the horizon, in degrees. (p. 77)

solar irradiance The amount of solar energy that reaches Earth. (p. 204)

solifluction A type of soil creep in which freeze–thaw cycles cause the soil to flow slowly downslope in overlapping sheets. (p. 500)

spatial scale The physical size, length, distance, or area of an object or the physical space occupied by a process. (p. 5)

speciation The creation of new species through evolution. (p. 237)

species A group of individuals that naturally interact and can breed and produce fertile offspring. (p. 230)

species-area curve The mathematical description of biodiversity and geographic area: larger areas tend to have more species than smaller areas. (p. 234)

specific heat (or *specific heat capacity*) The heat required to raise the temperature of 1 gram of any material by 1°C. (p. 84)

specific humidity The water vapor content of the atmosphere, expressed in grams of water per kilogram of air (g/kg). (p. 105)

speleothem A cavern formation that forms through precipitation of calcium carbonate. (p. 497)

spheroidal weathering A weathering process in which the edges of angular rocks become rounded, resulting in sphere-like boulders and outcrops. (p. 489)

spring A naturally occurring discharge of groundwater that is pushed to the ground surface by hydraulic pressure. (p. 352)

squall line A line of multicell thunderstorm cells that typically forms along a cold front. (p. 167)

stable atmosphere A condition in which air parcels are cooler and denser than the surrounding air and will not rise unless forced to do so. (p. 112)

stalactite A speleothem that grows from the ceiling of a cavern downward. (p. 497)

stalagmite A speleothem that grows from the floor of a cavern upward. (p. 497)

star dune A star-shaped dune that forms as the prevailing wind changes direction through the year. (p. 593)

stepping-stone An island in an island chain that aids in the dispersal of organisms. (p. 247)

storm surge A rise in sea level caused by the strong winds and low atmospheric pressure of a hurricane. (p. 180)

stratosphere The atmospheric layer above the troposphere, which extends between about 12 and 50 km (7.5 and 30 mi) above Earth's surface and has a permanent temperature inversion. (p. 44)

stratovolcano (or *composite volcano*) A large, potentially explosive cone-shaped volcano composed of alternating layers of lava and pyroclast. (p. 453)

stratus A cloud type characterized by low, flat sheets of clouds. (p. 116)

stream A channel in which water flows downhill by the force of gravity. (p. 515)

stream capture (or *stream piracy*) The diversion of one stream into another as headward erosion merges the two streams. (p. 529)

stream discharge The volume of water flowing past a fixed point in a stream channel; expressed in cubic meters per second or cubic feet per second. (p. 520)

stream order The numeric system used to rank stream size based on the number of tributaries flowing into a stream. (p. 518)

stream rejuvenation The process by which a stream gains downcutting energy as its base level is lowered relative to its drainage basin. (p. 525)

strike-slip fault A fault that forms as a result of shearing force as one block moves horizontally in relationship to another block. (p. 467)

subduction The process in which oceanic lithosphere bends and dives into the mantle beneath another lithospheric plate. (p. 398)

submergent coast A coast where sea level is rising or the land is sinking. (p. 619)

subpolar low A belt of low pressure roughly centered on 60 degrees north and south and made up of cyclonic systems that bring frequent precipitation. (p. 139)

subsidence Lowering of land elevation through compaction of sediments. (p. 536)

subsolar point The single point at which the Sun's rays are perpendicular to Earth's surface at or near noon; restricted to between 23.5 degrees north and south latitude. (p. 75)

subtropical high The discontinuous belt of aridity and high pressure made up of anticyclones roughly centered on 30 degrees north and south latitude. (p. 139)

sulfur dioxide (SO_2) A pungent gas, produced by volcanic eruptions and by the burning of fossil fuels, that causes human health problems and acid rain. (p. 51)

supercell thunderstorm A thunderstorm with a rotating cylindrical updraft that usually produces severe weather. (p. 168)

superposition principle A principle which states that in a sequence of rock layers, the oldest rocks are always at the bottom, and the youngest are always at the top. (p. 379)

suspended load Small particles such as clay and silt that remain suspended in flowing water or wind. (p. 521)

swash The rush of water up a beach following the collapse of a breaker. (p. 616)

sweepstakes dispersal Dispersal across an extensive region of inhospitable space. (p. 247)

syncline A fold in the crust with a U-shaped dip. (p. 413)

system A set of interacting parts or processes that function as a unit. (p. 4)

T

tafoni Pits or cavities on the surface of a rock that form through salt wedging. (p. 488)

talus (or *scree*) Pieces of angular broken rock that accumulate at the angle of repose on the base of a steep slope or cliff. (p. 504)

talus apron A continuous slope of talus formed where two or more talus cones merge. (p. 504)

talus cone A cone-shaped accumulation of talus at the foot of a cliff that settles at the angle of repose. (p. 504)

tarn A mountain lake that forms within or just below a cirque as meltwater from a glacier or snow fills a depression created by a glacier. (p. 563)

taxonomy Classification and naming of organisms based on their genetic similarities. (p. 255)

temperate deciduous forest The biome dominated by trees that shed their leaves in winter in response to low temperatures. (p. 285)

temperate grassland The grass-dominated biome characterized by significant moisture deficits, natural fires, and grazing herbivores. (p. 281)

temperate rainforest The forest biome found mostly on the west coasts of continents where annual precipitation is high; typically has large trees forming a dense canopy. (p. 286)

temperature The average kinetic movement of atoms and molecules of a substance. (p. 65)

temperature inversion The process by which air temperature increases with increased height. (p. 44)

temporal scale The window of time used to examine phenomena and processes or the length of time over which they develop or change. (p. 5)

terminal moraine A moraine that marks the farthest advance of a glacier's toe. (p. 568)

thalweg The area in a stream channel where the water flows fastest. (p. 535)

thermal air pressure Air pressure caused by heating or cooling of air. Warm air is associated with low pressure, and cold air is associated with high pressure. (p. 132)

thermal infrared (or *thermal-IR*) Solar radiation longer than near infrared and sensible as heat, such as the warmth from a fire, a warm sidewalk, or a heat lamp. (p. 68)

thermocline The transitional zone between warm surface water and cold water at depth. (p. 308)

thermosphere The atmospheric layer located from 80 to 600 km (50 to 370 mi) above Earth's surface. (p. 44)

thunder An acoustic shock wave produced when lightning rapidly heats and expands the air around it. (p. 169)

thunderstorm A cumulonimbus cloud that produces lightning and thunder. (p. 164)

tide The rise and fall of sea level caused by the gravitational effects of the Moon and the Sun. (p. 613)

tide pool A still pool that forms at low tide. (p. 320)

tidewater glacier A glacier that flows into the ocean. (p. 559)

till Any debris deposited by a glacier without the influence of running water. (p. 551)

tombolo A landform consisting of an island and a sandspit that connects it to the mainland. (p. 620)

topographic map A type of map that provides information about surface relief using contour lines. (pp. 23, A-2)

topography The shape and physical character of Earth's surface in a region. (p. 23)

tornado A violently rotating column of air that descends from a cumulonimbus cloud and touches ground. (p. 172)

tornado warning A warning issued by the National Weather Service after a tornado has been seen and called in to local authorities or when a tornado is suggested by a Doppler radar hook echo signature. (p. 175)

tornado watch An alert issued by the National Weather service when conditions are favorable for tornadic thunderstorms. (p. 175)

traction The dragging and tumbling of large rocks in a stream channel. (p. 523)

trade winds Easterly surface winds found from the ITCZ to the subtropical high, between 0 and 30 degrees north and south latitude. (p. 139)

transform plate boundary A plate boundary where one lithospheric plate slips laterally past another. (p. 402)

transmission The unimpeded movement of electromagnetic energy through a medium such as air, water, or glass. (p. 69)

transpiration The loss of water to the atmosphere by plants. (p. 99)

transverse dune A long, narrow sand dune with the ridge crest perpendicular (or transverse) to the prevailing wind. (p. 593)

trawling An industrial fishing method in which nets are dragged through the water column or along the seafloor. (p. 326)

tributary A stream that joins with other streams to form a larger stream. (p. 515)

trophic level One of the levels of an ecosystem through which energy and matter flow. (p. 254)

Tropic of Cancer The 23.5-degree north parallel; the maximum latitude of the subsolar point in the Northern Hemisphere. (p. 75)

Tropic of Capricorn The 23.5-degree south parallel; the maximum latitude of the subsolar point in the Southern Hemisphere. (p. 75)

tropical cyclone A cyclonic low-pressure system occurring in the tropics with sustained winds of 119 km/h (74 mph) or greater. (p. 176)

tropical rainforest The forest biome in the humid lowland tropics; characterized by a multilayered forest structure and high biodiversity. (p. 270)

tropical savanna The woodland biome with a wet summer and dry winter climate pattern, characterized by widely spaced trees and a grass understory. (p. 275)

tropical seasonal forest The biome in the warm lowland tropics, characterized by high biodiversity and trees that are deciduous in response to winter drought. (p. 275)

tropical storm A tropical cyclonic storm with sustained winds between 63 and 118 km/h (39 and 73 mph). (p. 178)

tropics The geographic region located between 23.5 degrees north and south latitude. (p. 14)

tropopause The boundary between Earth's troposphere and the stratosphere. (p. 44)

troposphere The lowest layer of the atmosphere, extending from Earth's surface up to about 12 km (7.5 mi), where all weather occurs. (p. 44)

trunk stream A single large stream into which smaller tributaries merge. (p. 515)

tsunami A large ocean wave triggered by an earthquake or other natural disturbance. (p. 452)

tundra The biome that occurs at any latitude where it is too cold for trees to grow. (p. 290)

typhoon A name used for tropical cyclones in Southeast Asia. (p. 176)

U

ultraviolet (UV) radiation Solar radiation that is shorter than visible wavelengths. (p. 47)

uniformitarianism The principle which states that the same imperceptible gradual processes are operating now that have operated in the past. (p. 377)

unstable atmosphere A condition in which air parcels rise on their own because they are warmer and less dense than the surrounding air. (p. 112)

upwelling The circulation of nutrient-rich water from the seafloor to the ocean surface. (p. 320)

urban heat island An urbanized region that is significantly warmer than surrounding rural areas. (p. 64)

V

valley breeze A local upslope breeze produced by heating and cooling differences in mountainous areas. (p. 147)

valley glacier A glacier that occupies a mountain valley. (p. 558)

vapor pressure The portion of air pressure exerted exclusively by molecules of water vapor. (p. 105)

ventifact A wind-sculpted desert rock shaped by abrasion by saltating sand. (p. 591)

virtual water The unseen water required to produce a manufactured item or food. (p. 363)

visible light The portion of the electromagnetic spectrum that we can see, with wavelengths between 0.40 and 0.75 μm. (p. 68)

volatile organic compound (VOC) A toxic compound of hydrogen and carbon; also called a *hydrocarbon*. (p. 53)

volcanic block A fragment of rock from a volcano's cone that is ejected during an explosive eruption. (p. 458)

volcanic crater A bowl-shaped depression that forms at the summit of a volcano. (p. 453)

volcanic explosivity index (VEI) The index used to rank volcanic eruptions based on the amount of material a volcano ejects during an eruption. (p. 463)

volcanic vent The conduit through which magma moves and reaches the surface in a volcano. (p. 453)

volcano A mountain or hill formed by eruptions of lava and rock fragments. (p. 375)

W

Wadati-Benioff zone The sloping pattern of increasingly deep earthquake foci found in a subduction zone. (p. 404)

wall cloud A cylindrical cloud that protrudes from the base of a supercell thunderstorm. (p. 172)

warm front A region where warm air advances on and flows over cooler, heavier air; not associated with severe weather. (p. 183)

water footprint The amount of water required to produce a specific item, food, or service. (p. 362)

water table The top surface of an aquifer's zone of saturation. (p. 351)

water vapor Water in a gaseous state. (p. 9)

wave of oscillation Wave in which water moves in a circular path. (p. 615)

wave of translation A wave in which water moves forward in the direction of wave movement. (p. 616)

wave refraction The process by which waves approaching the shoreline bend and maintain a nearly parallel orientation to the shore. (p. 616)

wave-cut bench A flat coastal platform resulting from wave erosion and exposed at low tide. (p. 626)

wave-cut notch A notch formed at the base of a coastal cliff as a result of weathering and erosion. (p. 626)

weather The state of the atmosphere at any given moment, comprising ever-changing events on time scales ranging from minutes to weeks. (pp. 11, 199)

weathering The process by which solid rock is dissolved and broken apart into smaller fragments. (pp. 337, 388, 487)

well A hole dug or drilled by people to gain access to groundwater. (p. 354)

westerlies Surface winds that come from the west and are found in both hemispheres between the subpolar low and the subtropical high. (p. 141)

wind shear Changes in wind speed and direction with altitude. (p. 165)

wind vane (or *weather vane*) An instrument used to measure wind direction. (p. 135)

Y

yardang An elongated ridge formed by deflation and sand abrasion in a desert; generally, two to three times longer than it is wide. (p. 591)

Younger Dryas The cold period that occurred 12,900 to 11,600 years ago. (p. 206)

Z

zone of aeration The layer of the ground that is not permanently saturated with water. (p. 351)

zone of saturation The layer of the ground that is usually saturated with water. (p. 351)

Index

Note: Page numbers followed by the letter f indicate figures; page numbers followed by the letter t indicate tables; and page numbers in *italics* indicate photographs.